BRANCH CONTROL GROUP

Jump

JMP adr	C3
JNZ adr	C2
JZ adr	CA
JNC adr	D2
JC adr	DA
JPO adr	E2
JPE adr	EA
JP adr	F2
JM adr	FA
PCHL	E9

Call

CALL adr	CD
CNZ adr	C4
CZ adr	CC
CNC adr	D4
CC adr	DC
CPO adr	E4
CPE adr	EC
CP adr	F4
CM adr	FC

Return

RET	C9
RNZ	C0
RZ	C8
RNC	D0
RC	D8
RPO	E0
RPE	E8
RP	F0
RM	F8

Restart

RST	0	C7
	1	CF
	2	D7
	3	DF
	4	E7
	5	EF
	6	F7
	7	FF

I/O AND MACHINE CONTROL

Stack Ops

PUSH	B	C5
	D	D5
	H	E5
	PSW	F5
POP	B	C1
	D	D1
	H	E1
	PSW*	F1
XTHL		E3
SPHL		F9

Input/Output

OUT byte	D3
IN byte	DB

Control

DI	F3
EI	FB
NOP	00
HLT	76

New Instructions (8085 Only)

RIM	20
SIM	30

RESTART TABLE

Name	Code	Restart Address
RST 0	C7	0000_{16}
RST 1	CF	0008_{16}
RST 2	D7	0010_{16}
RST 3	DF	0018_{16}
RST 4	E7	0020_{16}
TRAP	Hardware* Function	0024_{16}
RST 5	EF	0028_{16}
RST 5.5	Hardware* Function	$002C_{16}$
RST 6	F7	0030_{16}
RST 6.5	Hardware* Function	0034_{16}
RST 7	FF	0038_{16}
RST 7.5	Hardware* Function	$003C_{16}$

*NOTE: The hardware functions refer to the on-chip Interrupt feature of the 8085 only

USE OF THE A REGISTER BY RIM AND SIM INSTRUCTIONS (8085 ONLY)

A REGISTER AFTER EXECUTING RIM

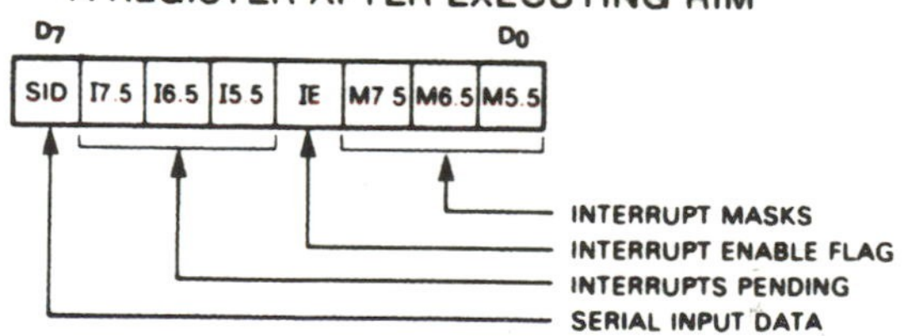

A REGISTER BEFORE EXECUTING SIM

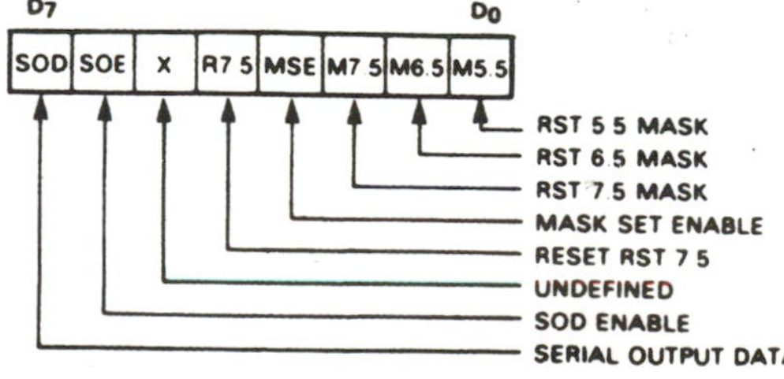

Digital and Microprocessor Fundamentals

Theory and Applications

Fourth Edition

WILLIAM KLEITZ
Tompkins Cortland Community College

Upper Saddle River, New Jersey
Columbus, Ohio

Library of Congress Cataloging-in-Publication Data

Kleitz, William.
Digital and microprocessor fundamentals : theory and applications / William Kleitz.—
4th ed.
p. cm.
Includes bibliographical references and index.
ISBN 0-13-093217-5 (alk. paper)
1. Digital electronics. 2. Logic circuits. 3. Microprocessors. I. Title.

TK7868.D5 K54 2003
621.39'5—dc21

2002022423

Editor in Chief: Stephen Helba
Assistant Vice President and Publisher: Charles E. Stewart, Jr.
Production Editor: Stephen C. Robb
Production Coordination: Holly Henjum, Clarinda Publication Services
Design Coordinator: Diane Y. Ernsberger
Cover Designer: Mark Shumaker
Cover art: Photonica
Production Manager: Matthew Ottenweller
Marketing Manager: Dave Gesell

This book was set in Times Roman and Univers by The Clarinda Company. It was printed and bound by Courier Kendallville, Inc. The cover was printed by The Lehigh Press, Inc.

Pearson Education Ltd.
Pearson Education of Australia Pty. Limited
Pearson Education Singapore Pte. Ltd.
Pearson Education North Asia Ltd.
Pearson Education Canada Ltd.
Pearson Educación de Mexico, S.A. de C.V.
Pearson Education—Japan
Pearson Education Malaysia Pte. Ltd.
Pearson Education, *Upper Saddle River, New Jersey*

10 9 8 7 6 5 4 3 2 1

ISBN: 0-13-093217-5

To Mom and Dad,
for showing me the value of hard work
and pride in a job well done.

Preface

Some college curriculums have the luxury of time for a four-semester sequence to teach Digital Electronics I and II, followed by Microprocessors I and II. The subject matter spans as many as three separate textbooks and numerous data manuals and specification sheets. Providing a working knowledge of both subjects in a single book, to be covered in one or two semesters, is quite a challenge. Before writing such a book, one might ask, "If I were a complete digital novice, what would it take to build up a working knowledge of digital and microprocessor systems in the shortest possible length of time?" To succeed, brief exposures to "nice-to-know" topics have to be omitted. Instead, the book must stick to the fundamentals that are absolutely necessary to build a solid foundation for the growth of knowledge. Advances in technology have made it easier to put together microprocessor-based designs without knowing all of the innermost details previously required to make a design work.

There are several advantages to covering both digital and microprocessor topics in a single book. The digital chapters (1–7), refer to practical applications of the theory as it will be used in the later microprocessor chapters. Then, the microprocessor chapters (8–14) revisit earlier explanations of circuits that now function as building blocks or interface devices for the microprocessor systems.

This text is intended for students of vocational two- or four-year technology or engineering technology schools. It can be used for a one-semester digital-and-microprocessor course, or for a one-semester digital course followed by a one-semester microprocessor course. Upon completion of this text, the student will be prepared to go on to advanced microprocessor topics such as 16/32-bit processors, system bus standards, and high-level-language program development. In several sections throughout the book, it is assumed that the student has an understanding of, or is concurrently enrolled in, a basic electricity course. Thus the examples and system design applications can give a *complete* explanation of circuit operation.

The microprocessor chapters use the 8085A microprocessor and 8051 microcontroller to explain the fundamentals of microprocessor architecture, programming, and hardware. The microprocessor coverage is approximately 50% software and 50% hardware. Most of the examples and applications involve some kind of Input/Output (I/O) with electronic devices such as switches, sensors, analog converters, and displays. This helps tie together the digital electronic theory learned in the earlier chapters with the hardware/software requirements needed to interface with the outside world via the most commonly used microprocessor support ICs. Solutions to practical design applications are given to illustrate I/O protocol requirements and timing analysis. The software instruction set is not given all at once—instructions are introduced as needed, to solve a particular design application. Only practical, workable designs are used so that the reader can develop a *complete* understanding of the application with no frustrating gaps.

TO THE STUDENT

As a student of Digital and Microprocessor Electronics, you are in the unique position of being able to become proficient at both the hardware *and* software aspects of microprocessor-based systems. This text is intended to provide you with the tools required to understand basic microprocessor-based systems and to provide the foundation for more advanced topics.

You'll see that the teaching style of this book is first to provide all theory required to understand a particular IC or circuit, and then to give an example of its use. By studying and reworking the examples with the solutions hidden, you can prepare yourself to solve all of the problems at the end of the chapters. You'll find the answers to half of the problems in Appendix G.

Another index that you'll find very useful is the *Supplementary Index of Integrated Circuits.* You can use this index to locate the pages in the book that give the pin configuration and function of a particular IC. Another feature is the glossary at the end of each chapter, which you can use to review the key terms presented in the chapter.

If you've written computer software in a high-level language such as BASIC, you already know how exciting it is to write a successful program. Microprocessor-level software is even more exciting because of its ability to interact directly with electronic devices. However, it can be very frustrating because each operation requires you to provide detailed instructions. As a beginner, one of the best ways to get started is to copy a program exactly as presented in this text and then modify it to suit your needs. Spend some time skimming through the *Instruction Set Reference Encyclopedia* in Appendix D. Even though you won't know what each instruction does, you'll learn what instructions are available and how to look them up.

Another important supplement is the SIM8085 software provided on the enclosed CD (see Appendix N). This will allow you to simulate most of the progams in the examples and homework on your Windows™-based PC. Many of the textbook examples are also provided on the CD. This allows you to go through the program step by step as you monitor the microprocessor registers and memory.

TO THE INSTRUCTOR

This book covers sufficient material for a two-semester sequence: first Digital Fundamentals and then Microprocessor Fundamentals. The digital chapters (1–7) cover number systems, digital signals, logic gates, combinational logic, medium-scale ICs, sequential logic circuits, and analog converters. The microprocessor chapters (8–14) cover memories, PLDs, fundamental 8085A architecture and software, intermediate hardware and software, interface applications, and the 8051 microcontroller.

If this book is used for a single-semester course in Digital and Microprocessor Fundamentals, then the following sections could be omitted without affecting the coherence of the book:

Sections 1–13 through 15
2–11 through 15
3–5 and 6

4–1 through 8
5–6, 8, 9, and 10
6–2 through 5
7–1 through 12
8–6
9–3 and 4
11–4 and 6
12–5
13–1 through 7
14–1 through 5

If the course (or courses) is intended for nonelectrical technology students, then the following sections could be omitted to eliminate the basic electricity prerequisite:

Sections 1–13 and 14
2–13 through 15
5–8 through 10
7–2, 3, 4, and 11
13–4 through 7

All of the programming and hardware interfacing were done on the SDK-85 trainer (see Appendix L) and the Primer trainer (see Appendix M), but the explanations and addressing schemes were developed so that any 8080/8085A trainer or the SIM8085 simulator (Appendix N) could be used. All of the programs and system design applications are presented as a complete package; they are completely explained, with all required hardware and software, so that they can be duplicated in the lab. The enclosed SIM8085 software is a great learning tool for the students, allowing them to simulate their programs on a Windows-based PC as they monitor registers and memory.

A worthy goal for a student of this text is to develop a solution to a complete application of a microprocessor-based system performing a task that is both practical and visually stimulating. That is the object of the last two chapters, *Interfacing and Applications* (Chapter 13) and *The 8051 Microcontroller* (Chapter 14). Chapter 13 draws on the knowledge gained in the previous twelve chapters to solve some interesting applications, including:

- Generating waveforms with a DAC
- Measuring temperature with an ADC
- Driving a multiplexed display
- Reading a keyboard
- Driving a stepper motor

Chapter 14 introduces the student to the real workhorse in modern automotive and consumer electronics, the *microcontroller.* The beauty of the 8051 microcontroller family is that it was specifically designed to provide a simple solution to the typical data acquisition and control applications faced by today's technology graduates. Assuming that the student has a good understanding of 8085A hardware and software, this chapter provides the additional information necessary to understand the basic features of the 8051. The chapter concludes with the solution to applications including time delays, switch and LED I/O, reading a keyboard, and ADC interfacing.

NEW IN THE FOURTH EDITION

I have used the first, second, and third editions of this text in my classes for the past eleven years and have incorporated many suggestions for modifying and expanding the discussion of several key points. Also, new sections were added on the following topics:

- Computer Magnetic and Optical Memory Devices (Chapter 8)
- Important WWW Sites (Appendix A)
- Expanded Review of Basic Electricity Principles (Appendix J)
- Instructions for Implementing Digital Logic with CPLDs (Appendix K)
- Circuit Design Applications Using CPLDs (Appendix K)
- Tutorial for Using Altera MAX+PLUS II CPLD Software (enclosed CD)
- Tutorial for Using Xilinx Foundation CPLD Software (enclosed CD)
- Solutions to Altera and Xilinx CPLD Applications (enclosed CD)
- Using the EMAC Primer Microprocessor Trainer (Appendix M)
- Using the SIM8085 Microprocessor Simulator on a PC (Appendix N)
- SIM8085 Microprocessor Simulator Program (enclosed CD)
- SIM8085 Simulator Solutions to Textbook Examples (enclosed CD)
- Texas Instruments Data Sheets (enclosed CD)

Acknowledgments

Thanks are due to the following professors for reviewing the manuscript and providing numerous valuable suggestions that have contributed to the realization of this fourth edition: David Hata, Portland Community College; Tseng-Yuan Tim Woo, Durham Technical Community College; Gerald Cockrell, Indiana University; Jim Morgenthien, Precision Filters Corporation; Peter Drexel, Plymouth State College; Roy Siegel, DeVry Institute of Technology; Charles Casazza, Nassau Community College; Dr. Lee Rosenthal, Fairleigh Dickenson University; Costas Vassiliadis, Ohio University; and Murray Haims, Rockland Community College.

A special thank you is extended to Aman Bhargava of ABCreations for providing the SIM8085 software that allows students to perform simulations of the 8085 microprocessor on a PC.

I am grateful to Russell Hunt from Simco Company, Kevin White and Scott Heffron from Bob Dean Corporation, Dick Quaif from DQ Systems, Alan Szary and Paul Constantini from Precision Filters, Inc., and Jim Delsignore from AT&T Corporation for their technical assistance, and to Signetics (Philips) Corporation, Intel Corporation, Texas Instruments, Inc., Hewlett-Packard Company, and Advanced Micro Devices, Inc., for providing the data sheets used in this book. Special thanks go to Texas Instruments, Inc. for permission to use their extensive data sheet library on the CD that accompanies this book. Also, thanks to my students of the past 22 years who have helped me to develop better teaching strategies and have provided suggestions for clarifying several of the explanations contained in this book.

Contents

CHAPTER 1 **DIGITAL NUMBER SYSTEMS AND REPRESENTATIONS** **1**

Objectives 1
Introduction 1
1–1 Digital Representations of Analog Quantities 2
1–2 Decimal Numbering System (Base 10) 3
1–3 Binary Numbering System (Base 2) 4
1–4 Decimal-to-Binary Conversion 5
1–5 Hexadecimal Numbering System (Base 16) 7
1–6 Hexadecimal Conversions 8
1–7 Binary-Coded-Decimal System 9
1–8 Comparison of Numbering Systems 10
1–9 The ASCII Code 10
1–10 Applications of the Numbering Systems 12
1–11 Digital Signals 14
1–12 Clock Waveform Timing 15
1–13 Switches in Electronic Circuits 16
1–14 A Transistor as a Switch 16
1–15 The TTL Integrated Circuit 18
Summary 19
Glossary 20
Problems 21

CHAPTER 2 **LOGIC GATE OPERATION AND SPECIFICATIONS** **23**

Objectives 23
Introduction 24
2–1 The AND Gate 24
2–2 The OR Gate 25
2–3 Timing Analysis 26

2–4 Using Integrated-Circuit Logic Gates 28
2–5 Introduction to Troubleshooting Techniques 30
2–6 The Inverter 33
2–7 The NAND Gate 33
2–8 The NOR Gate 35
2–9 The Exclusive-OR Gate 36
2–10 The Exclusive-NOR Gate 36
2–11 Logic Gate Waveform Generation 37
2–12 Summary of Logic Gate Operation 40
2–13 The TTL Family Specifications 43
2–14 The CMOS Family 55
2–15 Interfacing Logic Families 59
Summary 64
Glossary 65
Problems 67

CHAPTER 3 **COMBINATIONAL LOGIC CIRCUITS AND REDUCTION TECHNIQUES** **77**

Objectives 77
Introduction 77
3–1 Combinational Logic 77
3–2 Boolean Algebra Laws and Rules 80
3–3 Simplification of Combinational Logic Circuits Using Boolean Algebra 82
3–4 DeMorgan's Theorem 85
3–5 Karnaugh Mapping 92
3–6 System Design Applications 98
3–7 Arithmetic Circuits 100
3–8 Four-Bit Full-Adder ICs 103
Summary 106
Glossary 106
Problems 107

CHAPTER 4 **DATA CONTROL DEVICES** **113**

Objectives 113
Introduction 113
4–1 Comparators 113
4–2 Decoders 115
4–3 Encoders 121
4–4 Multiplexers 123
4–5 Demultiplexers 127
4–6 Multiplexer Design Applications 130
4–7 Schmitt Trigger ICs 132
4–8 System Design Applications 136
Summary 139
Glossary 140
Problems 140

CHAPTER 5 **FLIP-FLOPS AND SEQUENTIAL LOGIC** **145**

Objectives 145
Introduction 145
5–1 The *S-R* Flip-Flop 146
5–2 The Integrated-Circuit *D* Latch (7475) 150
5–3 The Integrated-Circuit *D* Flip-Flop (7474) 151
5–4 The *J-K* Flip-Flop 154
5–5 The Integrated-Circuit *J-K* Flip-Flop 156

5–6 Flip-Flop Time Parameters 160
5–7 Three-State Buffers, Latches, and Transceivers 164
5–8 Switch Debouncing 168
5–9 Oscillator Circuits and the One-Shot Multivibrator 170
5–10 Practical Input and Output Considerations 183
Summary 189
Glossary 190
Problems 192

CHAPTER 6 ***COUNTER CIRCUITS AND SHIFT REGISTERS*** **199**

Objectives 199
Introduction 199
6–1 Ripple Counters 201
6–2 Design of Divide-by-*N* Counters 203
6–3 Ripple Counter Integrated Circuits 207
6–4 System Design Applications for Counter ICs 213
6–5 Seven-Segment LED Display Decoders 218
6–6 Synchronous Counters 220
6–7 Synchronous Up/Down-Counter ICs 223
6–8 Shift Register Basics 226
6–9 Ring Shift Counter and Johnson Shift Counter 232
6–10 Shift Register ICs 234
6–11 System Design Applications for Shift Registers 238
Summary 240
Glossary 241
Problems 242

CHAPTER 7 ***INTERFACING TO THE ANALOG WORLD*** **251**

Objectives 251
Introduction 251
7–1 Digital and Analog Representations 252
7–2 Operational Amplifier Basics 252
7–3 Binary-Weighted Digital-to-Analog Converters 253
7–4 *R*/2*R* Ladder Digital-to-Analog Converters 254
7–5 Integrated-Circuit Digital-to-Analog Converters 255
7–6 IC Data Converter Specifications 259
7–7 Parallel-Encoded Analog-to-Digital Converters 260
7–8 Counter-Ramp Analog-to-Digital Converters 261
7–9 Successive-Approximation Analog-to-Digital Conversion 262
7–10 Integrated-Circuit Analog-to-Digital Converters 263
7–11 Transducers and Signal Conditioning 267
7–12 Data Acquisition Systems 271
Summary 273
Glossary 274
Problems 275

CHAPTER 8 ***MICROPROCESSOR AND COMPUTER MEMORY*** **279**

Objectives 279
Introduction 279
8–1 Memory Concepts 280
8–2 Static RAMs 283
8–3 Dynamic RAMS 287
8–4 Read-Only Memories 291
8–5 Memory Expansion and Address Decoding 294

8–6 Magnetic and Optical Storage 298
Summary 302
Glossary 302
Problems 303

CHAPTER 9 MICROPROCESSOR FUNDAMENTALS 305

Objectives 305
Introduction 305
9–1 Introduction to System Components and Buses 306
9–2 Software Control of Microprocessor Systems 308
9–3 Internal Architecture of the 8085A Microprocessor 309
9–4 Instruction Execution within the 8085A 311
Summary 313
Glossary 314
Problems 314

CHAPTER 10 INTRODUCTION TO 8085A SOFTWARE 317

Objectives 317
Introduction 317
10–1 Hardware Requirements for Basic I/O Programming 318
10–2 Writing Assembly Language and Machine Language Programs 320
10–3 Compares and Conditional Branching 322
10–4 Using the Internal Data Registers 323
10–5 Writing Time-Delay Routines 326
10–6 Using a Time-Delay Subroutine with I/O Operations 329
Summary of Instructions 332
Summary 333
Glossary 333
Problems 334

CHAPTER 11 INTRODUCTION TO 8085A SYSTEM HARDWARE 337

Objectives 337
Introduction 337
11–1 8085A Pin Definitions 338
11–2 The Multiplexed Bus and Read/Write Timing 338
11–3 Microprocessor System Design Using Memory-Mapped I/O and Standard Memories 343
11–4 CPU Instruction Timing 347
11–5 A Minimum Component 8085A-Based System Using I/O Mapped I/O 349
11–6 The 8355/8755A and 8155/8156 Programmable Support ICs 351
Summary 358
Glossary 359
Problems 360

CHAPTER 12 THE 8085A SOFTWARE INSTRUCTION SET 363

Objectives 363
Introduction 363
12–1 The Data Transfer Instruction Group 364
12–2 The Arithmetic Instruction Group 369
12–3 The Logical Instruction Group 374
12–4 Subroutines and the Stack 378
12–5 Interrupts 383
Summary of Instructions 388
Summary 389

Glossary 390
Problems 391

CHAPTER 13 ***INTERFACING AND APPLICATIONS*** **395**

Objectives 395
Introduction 395
13–1 Interfacing to a Digital-to-Analog Converter 396
13–2 Using a DAC for Waveform Generation 397
13–3 Interfacing to an Analog-to-Digital Converter 399
13–4 Designing a Digital Thermometer Using an ADC 400
13–5 Driving a Multiplexed Display 403
13–6 Scanning a Keyboard 407
13–7 Driving a Stepper Motor 408
Summary 413
Glossary 414
Problems 414

CHAPTER 14 ***THE 8051 MICROCONTROLLER*** **417**

Objectives 417
Introduction 417
14–1 The 8051 Family of Microcontrollers 418
14–2 8051 Architecture 419
14–3 Interfacing to External Memory 423
14–4 The 8051 Instruction Set 424
14–5 8051 Applications 431
Summary 435
Glossary 436
Problems 436

APPENDIX A ***WWW SITES*** **439**

APPENDIX B ***MANUFACTURERS' DATA SHEETS*** **441**

APPENDIX C ***8085A ASSEMBLY LANGUAGE REFERENCE CHART AND ALPHABETIZED MNEMONICS*** **461**

APPENDIX D ***8085A INSTRUCTION SET REFERENCE ENCYCLOPEDIA*** **465**

APPENDIX E ***8085A INSTRUCTION SET TIMING INDEX*** **481**

APPENDIX F ***8051 INSTRUCTION SET SUMMARY*** **483**

APPENDIX G ***ANSWERS TO SELECTED PROBLEMS*** **489**

APPENDIX H ***SCHEMATIC DIAGRAMS*** **505**

APPENDIX I ***8051 APPLICATION NOTES*** **515**

APPENDIX J ***REVIEW OF BASIC ELECTRICITY PRINCIPLES*** **527**

APPENDIX K ***PROGRAMMABLE LOGIC DEVICES: ALTERA AND XILINX CPLDs AND FPGAs*** **535**

APPENDIX L	***THE SDK-85 MICROPROCESSOR TRAINER***	***571***
APPENDIX M	***EMAC PRIMER 8085 TRAINER***	***575***
APPENDIX N	***SIM8085 MICROPROCESSOR SIMULATOR FOR THE PC***	***579***
	INDEX	***581***
	SUPPLEMENTARY INDEX OF INTEGRATED CIRCUITS	***587***

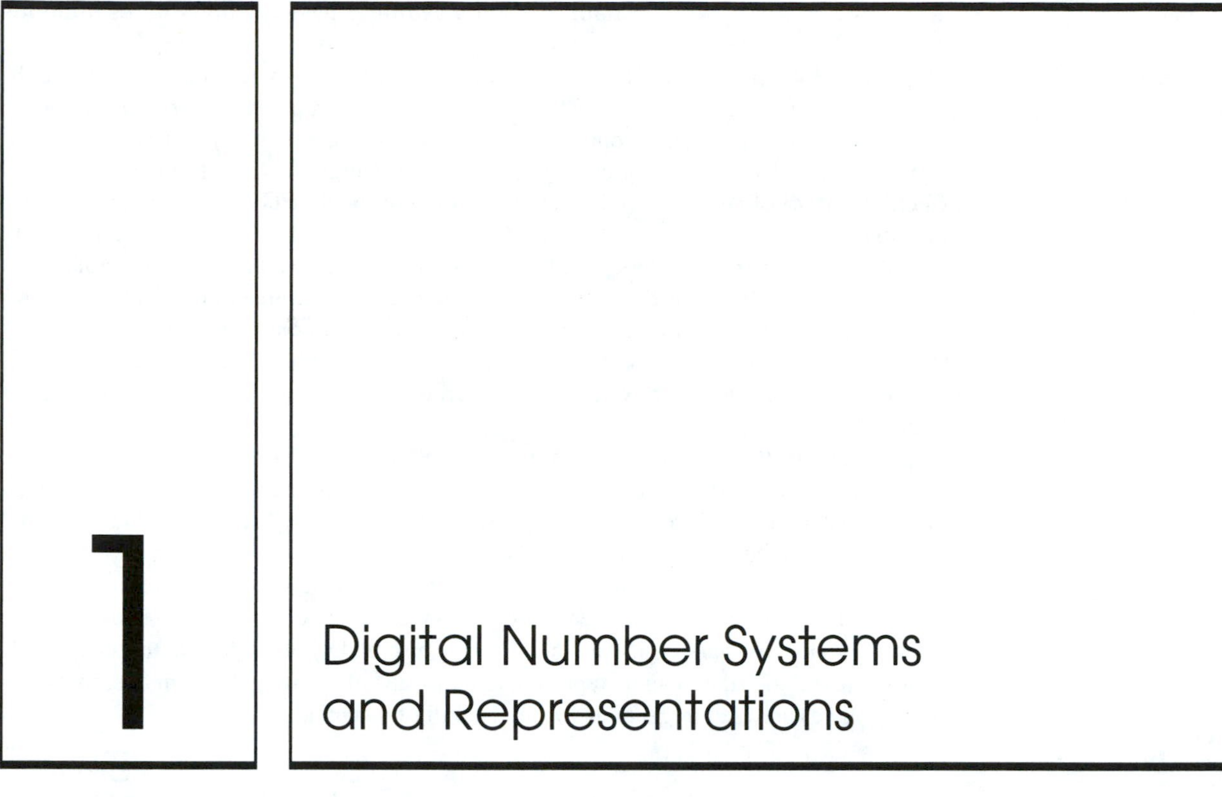

1 Digital Number Systems and Representations

OBJECTIVES

Upon completion of this chapter, you should be able to:

- Determine the weighting factor for each digit position in the decimal, binary, and hexadecimal numbering systems.
- Convert any number in one of the three numbering systems (decimal, binary, or hexadecimal) to its equivalent value in any of the remaining two numbering systems.
- Describe the format, and use, of binary-coded-decimal (BCD) numbers.
- Describe the parameters associated with digital voltage-versus-time waveforms.
- Convert between frequency and period for a periodic clock waveform.
- Explain the basic characteristics of transistors when they are forward biased and reverse biased.
- Calculate the output voltage in an electric circuit containing transistors operating as digital switches.
- Perform input/output timing analysis in electric circuits containing transistors.

INTRODUCTION

Digital circuitry is the foundation of digital computers and many automated control systems. In a modern home, digital circuitry controls the appliances, alarm systems, and heating systems. Newer automobiles, under control of digital circuitry and microprocessors, have added safety features, are more energy efficient, and are easier for a technician to diagnose and to correct malfunctions.

Other uses for digital circuitry are found in the area of automated machine control, such as numerically controlled (NC) milling machines. These machines can be programmed by a production engineer to mill a piece of stock material to prespecified dimensions with very accurate repeatability, within 0.01% accuracy. Another use is in the field of energy monitoring and control. With the high cost of energy it is very important for large industrial and commercial users to monitor the energy flows within their buildings. Effective control of heating, ventilating, and air-conditioning can reduce energy bills significantly. More and more grocery stores are using the universal product code (UPC) to take care of checking out and totaling the sale of groceries as well as taking care of inventory control and stockroom replenishment automatically. In the area of medical electronics we see more use of digital thermometers, life-support systems, and monitors. We have also seen more use of digital electronics in the reproduction of music. Digital reproduction is less susceptible to electrostatic noise and therefore can reproduce music with greater fidelity.

Digital electronics evolves from the principle that transistor circuitry can easily be fabricated and designed to output one of two voltage levels based on the levels placed at its inputs. The two distinct levels (usually +5 V and 0 V) are called "HIGH" and "LOW" and can be represented by 1 and 0, respectively.

The binary numbering system is made up of only 1s and 0s and is therefore used extensively in digital electronics. Other numbering systems and codes covered in this chapter represent groups of binary digits and therefore are also widely used. Also in this chapter we see how these logic states can be represented by means of a timing diagram and how electronic switches are used to generate meaningful digital signals.

1–1 DIGITAL REPRESENTATIONS OF ANALOG QUANTITIES

Most naturally occurring physical quantities in our world are *analog* in nature. An analog signal is a continuously variable electrical or physical quantity. Consider a mercury-filled tube thermometer. As the temperature rises, the mercury expands in analog fashion and makes a smooth, continuous motion relative to the degree scale. A baseball player swings a bat in an analog motion. The velocity and force with which a musician strikes a piano key is analog in nature. Even the resulting vibration of the piano string is an analog, sinusoidal vibration.

So why do we need to use digital representations in a world that is naturally analog? The answer is that when we want an electronic machine to interpret, communicate, and store that analog information, it is much easier to handle if we first convert it to digital. A digital value is represented by a combination of ON and OFF voltage levels that are written as a string of 1s and 0s.

For example, if an analog thermometer is registering 72°, that number could be represented in a digital circuit as a series of ON and OFF voltage levels. (We'll learn later that the number 72 converted to digital levels will be 0100 1000.) The nice thing about using ON/OFF voltage levels is that the circuitry used to generate, manipulate, and store them is very simple. Instead of having to deal with the infinite span and intervals of analog voltage levels, all we need to use is ON or OFF voltages (usually +5 V = ON and 0 V = OFF).

A good example of the use of digital representation of analog quantities is in the audio recording of music. Compact disks (CDs) and digital audio tapes (DATs) are becoming commonplace and are proving to be a superior way to record and play back music. Musical instruments and the human voice produce analog signals, and the human ear naturally responds to analog signals; so where does digital fit in? Well, it's a lot of work, but the recording industries convert the analog signals to a digital format and then store the information on a CD or DAT. The CD or DAT player then converts the digital levels back to their corresponding analog signal before playing it back for the human ear. To accurately represent a complex musical signal as a digital string, several samples of the analog signal must be taken, as shown in Figure 1–1.

The first conversion illustrated is at a point on a rising portion of the analog signal. At that point, the analog voltage is 2 V. Two volts is represented as the digital string 0000

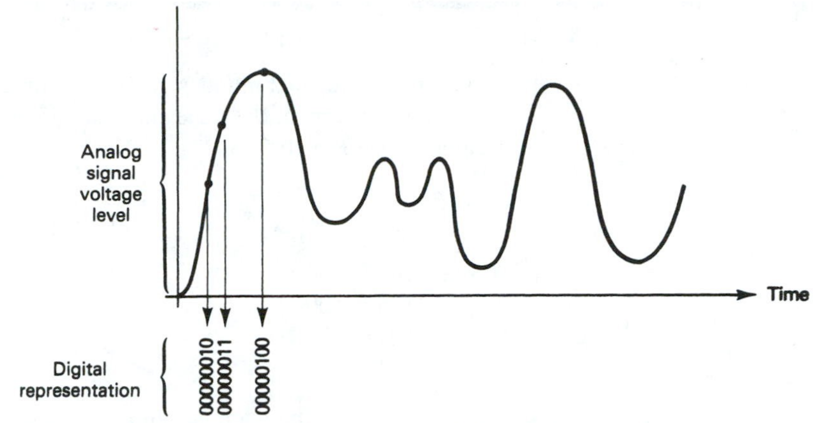

Figure 1–1 Digital representations of three data points on an analog waveform.

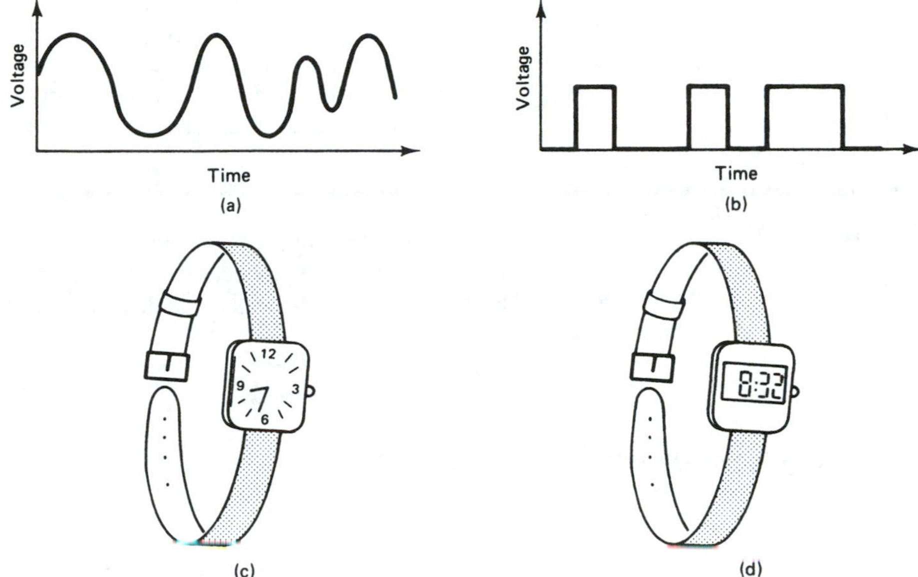

Figure 1–2 Analog versus digital: (a) analog waveform; (b) digital waveform; (c) analog watch; (d) digital watch.

0010. The next conversion is taken as the signal is still rising, and the third is taken at the highest level. This process will continue throughout the entire length of the music to be recorded. To play the music back, the process is reversed: Digital-to-analog conversions are made to re-create the original analog signal. If a high enough number of samples are taken of the original analog signal, an almost exact reproduction of the original music can be made.

It certainly is extra work, but digital recordings have virtually eliminated problems such as record wear and magnetic tape hiss associated with earlier methods of audio recording. If imperfections are introduced to a digital signal, that slight variation in the digital level will not change an ON level to an OFF level. Whereas, a slight change in an analog level is easily picked up by the human ear.

Other examples of analog versus digital are shown in Figure 1–2.

1–2 *DECIMAL NUMBERING SYSTEM (BASE 10)*

In the decimal numbering system, each position will contain 10 different possible digits: 0, 1, 2, 3, 4, 5, 6, 7, 8, and 9. Each position in a multidigit number will have a weighting factor based on a power of 10.

EXAMPLE 1–1

In a four-digit decimal number the least significant position (rightmost) will have a weighting factor of 10^0; the most significant position (leftmost) will have a weighting factor of 10^3:

10^3 10^2 10^1 10^0

where $10^3 = 1000$
$10^2 = 100$
$10^1 = 10$
$10^0 = 1$

To evaluate the decimal number 4623, the digit in each position is multiplied by the appropriate weighting factor:

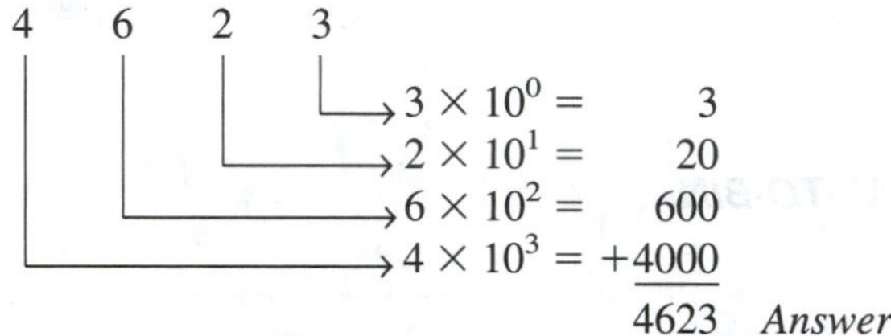

$$
\begin{aligned}
3 \times 10^0 &= 3 \\
2 \times 10^1 &= 20 \\
6 \times 10^2 &= 600 \\
4 \times 10^3 &= +\underline{4000} \\
& 4623 \quad \textit{Answer}
\end{aligned}
$$

Example 1–1 illustrates the procedure used to convert from some numbering system to its decimal (base 10) equivalent. (In that example we converted a base 10 number to a base 10 answer.) Now let's look at base 2 (binary) and base 16 (hexadecimal).

1–3 BINARY NUMBERING SYSTEM (BASE 2)

Digital electronics uses the binary numbering system because it uses only the digits 0 and 1, which can be represented simply in a digital system by two distinct voltage levels, such as +5 V = 1 and 0 V = 0.

The weighting factors for binary positions will be the powers of 2 shown in Table 1–1.

TABLE 1–1
Powers-of-2 Binary Weighting Factors

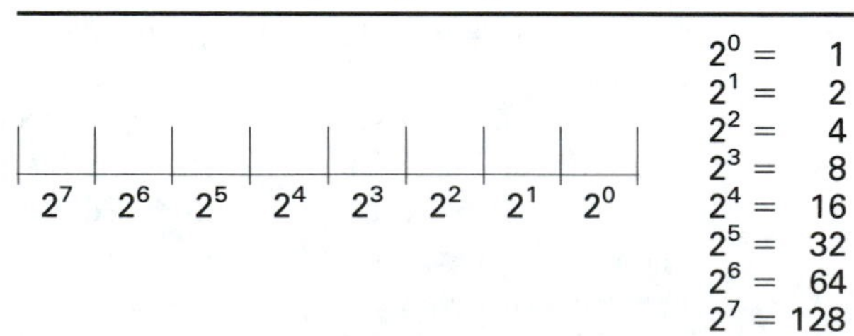

2^7 2^6 2^5 2^4 2^3 2^2 2^1 2^0

$2^0 = 1$
$2^1 = 2$
$2^2 = 4$
$2^3 = 8$
$2^4 = 16$
$2^5 = 32$
$2^6 = 64$
$2^7 = 128$

EXAMPLE 1–2

Convert the binary number 01010110_2 to a decimal. (Note the subscript 2 used to indicate that 01010110 is a base 2 number.)

Solution:

Multiply each binary digit by the appropriate weighting factor and total the results.

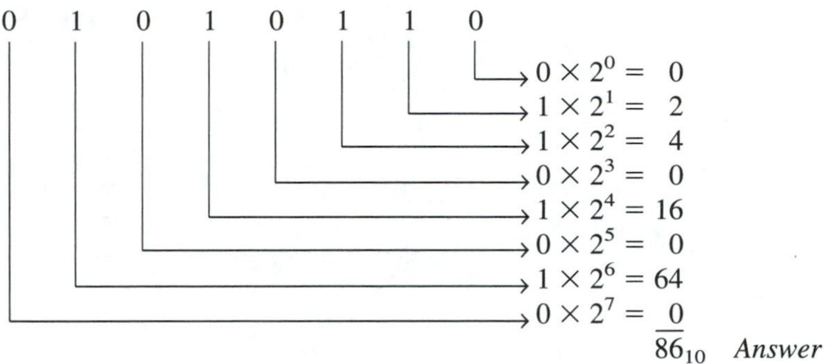

1–4 DECIMAL-TO-BINARY CONVERSION

The conversion from binary to decimal is usually performed by the digital computer for ease of interpretation by the person reading the number. On the other hand, when a person enters a decimal number into a digital computer, that number must be converted to binary before it can be operated on. Let's look at decimal-to-binary conversion.

EXAMPLE 1–3

Convert 133_{10} to binary.

Solution:

Referring to Table 1–1, we can see that the largest power of 2 that will fit into 133 is 2^7 ($2^7 = 128$). But that will still leave the value 5 ($133 - 128 = 5$) to be accounted for. Five can be taken care of by 2^2 and 2^0 ($2^2 = 4$, $2^0 = 1$). So the process looks like this:

$$
\begin{array}{r l}
133 & \\
-\underline{128} & \rightarrow 2^7 \\
5 & \\
-\underline{4} & \rightarrow 2^2 \\
1 & \\
-\underline{1} & \rightarrow 2^0 \\
0 &
\end{array}
$$

1	0	0	0	0	1	0	1
2^7	2^6	2^5	2^4	2^3	2^2	2^1	2^0

Answer: 10000101_2.

Note: The powers of 2 that fit into the number 133 were first determined. Then all other positions were filled with zeros.

EXAMPLE 1–4

Convert 122_{10} to binary.

Solution:

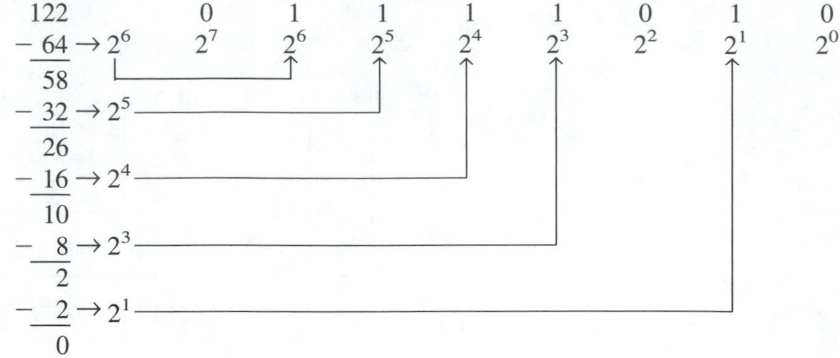

Answer: 01111010_2.

Another method of converting decimal to binary is by *successive division.* This is done by repeated division by the number of the base to which you are converting. For example, to convert 122_{10} to base 2, use the following procedure:

$122 \div 2 = 61$	with a remainder of	0	(LSB)
$61 \div 2 = 30$	with a remainder of	1	
$30 \div 2 = 15$	with a remainder of	0	
$15 \div 2 = 7$	with a remainder of	1	
$7 \div 2 = 3$	with a remainder of	1	
$3 \div 2 = 1$	with a remainder of	1	
$1 \div 2 = 0$	with a remainder of	1	(MSB)

The first remainder, 0, is the *least significant bit* (LSB) of the answer; the last remainder, 1, is the *most significant bit* (MSB) of the answer; therefore, the answer is

$$1\ 1\ 1\ 1\ 0\ 1\ 0_2$$

However, since most computers or digital systems deal with groups of 4, 8, 16, or 32 *bits* (binary digits), we should keep all our answers in that form. Adding a leading zero to the number $1\ 1\ 1\ 1\ 0\ 1\ 0_2$ will not change its numeric value; therefore, the 8-bit answer is

$$1\,1\,1\,1\,0\,1\,0_2 = 0\,1\,1\,1\,1\,0\,1\,0_2$$

EXAMPLE 1–5

Convert 152_{10} to binary using successive division.

Solution:

$152 \div 2 = 76$	remainder	0	(LSB)
$76 \div 2 = 38$	remainder	0	
$38 \div 2 = 19$	remainder	0	
$19 \div 2 = 9$	remainder	1	
$9 \div 2 = 4$	remainder	1	
$4 \div 2 = 2$	remainder	0	
$2 \div 2 = 1$	remainder	0	
$1 \div 2 = 0$	remainder	1	(MSB)

Answer: 10011000_2.

1–5 HEXADECIMAL NUMBERING SYSTEM (BASE 16)

The hexadecimal numbering system is a method of grouping bits to simplify entering and reading instructions or data present in digital computer systems. Hexadecimal uses 4-bit groupings; therefore, instructions or data used in 8-, 16-, or 32-bit computer systems can be represented as a 2-, 4-, or 8-digit hexadecimal code instead of as a long string of binary digits (see Table 1–2).

Hexadecimal (hex) uses 16 different digits and is a method of grouping binary numbers in groups of four. The 16 allowable hex digits are 0, 1, 2, 3, 4, 5, 6, 7, 8, 9, A, B, C, D, E, and F.

TABLE 1–2
Hexadecimal Numbering System

Decimal	*Binary*	*Hexadecimal*
0	0000 0000	0 0
1	0000 0001	0 1
2	0000 0010	0 2
3	0000 0011	0 3
4	0000 0100	0 4
5	0000 0101	0 5
6	0000 0110	0 6
7	0000 0111	0 7
8	0000 1000	0 8
9	0000 1001	0 9
10	0000 1010	0 A
11	0000 1011	0 B
12	0000 1100	0 C
13	0000 1101	0 D
14	0000 1110	0 E
15	0000 1111	0 F
16	0001 0000	1 0
17	0001 0001	1 1
18	0001 0010	1 2
19	0001 0011	1 3
20	0001 0100	1 4

1–6 HEXADECIMAL CONVERSIONS

To convert from binary to hexadecimal, group the binary number in groups of four (starting in the least significant position) and write down the equivalent hex digit.

EXAMPLE 1–6

Convert $0\,1\,1\,0\,1\,1\,0\,1_2$ to hex.

Solution:

$$\underbrace{0\,1\,1\,0}_{6}\quad\underbrace{1\,1\,0\,1_2}_{D}\quad = 6D_{16}\quad \textit{Answer}$$

To convert *hexadecimal to binary,* use the reverse process.

EXAMPLE 1–7

Convert $A9_{16}$ to binary.

Solution:

$$\begin{array}{cc} A & 9 \\ 1\,0\,1\,0 & 1\,0\,0\,1 = 1\,0\,1\,0\,1\,0\,0\,1_2 \quad \textit{Answer} \end{array}$$

To convert *hexadecimal to decimal,* multiply each hex digit by its appropriate weighting factor (powers of 16).

EXAMPLE 1–8

Convert $2A6_{16}$ to decimal.

Solution:

2 A 6

$$6 \times 16^0 = 6 \times 1 = 6$$
$$A \times 16^1 = 10 \times 16 = 160$$
$$2 \times 16^2 = 2 \times 256 = \underline{512}$$
$$678_{10} \quad \textit{Answer}$$

To convert from *decimal to hexadecimal,* use successive division.

EXAMPLE 1–9

Convert 151_{10} to hex.

Solution:

$$151 \div 16 = 9 \quad \text{remainder } 7 \quad \text{(LSD)}$$
$$9 \div 16 = 0 \quad \text{remainder } 9 \quad \text{(MSD)}$$
$$151_{10} = 97_{16} \quad \textit{Answer}$$

Check:

$$97_{16}: \quad 7 \times 16^0 = 7, \quad 9 \times 16^1 = 144, \quad 7 + 144 = 151 \ \checkmark$$

EXAMPLE 1–10

Convert 498_{10} to hex.

Solution:

$$\begin{aligned} 498 \div 16 &= 31 \quad \text{remainder} \quad 2 \quad \text{(LSD)} \\ 31 \div 16 &= 1 \quad \text{remainder} \quad 15 \quad (=\text{F}) \\ 1 \div 16 &= 0 \quad \text{remainder} \quad 1 \quad \text{(MSD)} \end{aligned}$$

$$498_{10} = 1\text{F}2_{16}$$

Check:

$$\begin{aligned} 1\text{F}2_{16} \qquad 2 \times 16^0 &= 2 \times 1 = 2 \\ \text{F} \times 16^1 &= 15 \times 16 = 240 \\ 1 \times 16^2 &= 1 \times 256 = \underline{256} \\ & \qquad 498 \ \checkmark \end{aligned}$$

1–7 BINARY-CODED-DECIMAL SYSTEM

The binary-coded-decimal system is used to represent each of the 10 decimal digits as a 4-bit binary code. The code is useful for outputting to displays that are always numeric (0 to 9), such as those found in digital clocks or digital voltmeters.

To form a BCD number, simply convert each decimal digit to its 4-bit binary code.

EXAMPLE 1–11

Convert 496_{10} to BCD.

Solution:

$$\underbrace{0100}_{4} \ \underbrace{1001}_{9} \ \underbrace{0110}_{6} = 0100 \ 1001 \ 0110_{\text{BCD}} \quad \textit{Answer}$$

To convert BCD to decimal, simply reverse the process.

EXAMPLE 1–12

Convert $0111\ 0101\ 1000_{\text{BCD}}$ to decimal.

Solution:

$$\underbrace{0111}_{7} \ \underbrace{0101}_{5} \ \underbrace{1000}_{8} = 758_{10} \quad \textit{Answer}$$

EXAMPLE 1–13

Convert 0110 0100 1011_{BCD} to decimal.

Solution:

0110	0100	1011
6	4	*

*This conversion is impossible because 1011 is not a valid binary-coded decimal. It is not in the range 0 to 9.

1–8 COMPARISON OF NUMBERING SYSTEMS

Table 1–3 shows a comparison of the four number systems commonly used in digital electronics and computer systems.

TABLE 1–3
Comparison of Numbering Systems

Decimal	*Binary*	*Hexadecimal*	*BCD*
0	0000 0000	0 0	0000 0000
1	0000 0001	0 1	0000 0001
2	0000 0010	0 2	0000 0010
3	0000 0011	0 3	0000 0011
4	0000 0100	0 4	0000 0100
5	0000 0101	0 5	0000 0101
6	0000 0110	0 6	0000 0110
7	0000 0111	0 7	0000 0111
8	0000 1000	0 8	0000 1000
9	0000 1001	0 9	0000 1001
10	0000 1010	0 A	0001 0000
11	0000 1011	0 B	0001 0001
12	0000 1100	0 C	0001 0010
13	0000 1101	0 D	0001 0011
14	0000 1110	0 E	0001 0100
15	0000 1111	0 F	0001 0101
16	0001 0000	1 0	0001 0110
17	0001 0001	1 1	0001 0111
18	0001 0010	1 2	0001 1000
19	0001 0010	1 3	0001 1001
20	0001 0100	1 4	0010 0000

1–9 THE ASCII CODE

To get information into and out of a computer, we need more than just numeric representations; we also have to take care of all the letters and symbols used in day-to-day processing. Information such as names, addresses, and item descriptions must be input and output in a readable format. But remember that a digital system can deal only with 1s and 0s. Therefore, we need a special code to represent all *alphanumeric* data (letters, symbols, and numbers).

Most industry has settled on an input/output (I/O) code called the American Standard Code for Information Interchange (ASCII). The *ASCII code* uses 7 bits to represent all the alphanumeric data used in computer I/O. Seven bits will yield 128 different code combinations, as listed in Table 1–4.

TABLE 1–4

American Standard Code for Information Interchange

LSB \ MSB	000	001	010	011	100	101	110	111
0000	NUL	DLE	SP	0	@	P		p
0001	SOH	DC_1	!	1	A	Q	a	q
0010	STX	DC_2	″	2	B	R	b	r
0011	ETX	DC_3	#	3	C	S	c	s
0100	EOT	DC_4	$	4	D	T	d	t
0101	ENQ	NAK	%	5	E	U	e	u
0110	ACK	SYN	&	6	F	V	f	v
0111	BEL	ETB	'	7	G	W	g	w
1000	BS	CAN	(	8	H	X	h	x
1001	HT	EM	)	9	I	Y	i	y
1010	LF	SUB	*	:	J	Z	j	z
1011	VT	ESC	+	;	K	[	k	{
1100	FF	FS	,	<	L	\	l	|
1101	CR	GS	–	=	M	]	m	}
1110	SO	RS	.	>	N	↑	n	~
1111	SI	US	/	?	O	—	o	DEL

Definitions of control abbreviations:

ACK	Acknowledge	FF	Form feed
BEL	Bell	FS	Form separator
BS	Backspace	GS	Group separator
CAN	Cancel	HT	Horizontal tab
CR	Carriage return	LF	Line feed
DC_1-DC_4	Direct control	NAK	Negative acknowledge
DEL	Delete idle	NUL	Null
DLE	Data link escape	RS	Record separator
EM	End of medium	SI	Shift in
ENQ	Enquiry	SO	Shift out
EOT	End of transmission	SOH	Start of heading
ESC	Escape	SP	Space
ETB	End of transmission block	STX	Start text
ETX	End text	SUB	Substitute
		SYN	Synchronous idle
		US	Unit separator
		VT	Vertical tab

Each time a key is depressed on an ASCII keyboard, that key is converted into its ASCII code and processed by the computer. Then, before outputting the computer contents to a display terminal or printer, all information is converted from ASCII into standard English.

To use the table, place the 4-bit group in the least significant positions and the 3-bit group in the most significant positions.

EXAMPLE 1–14

100 0111 is the code for G.

↗ (100) 3-bit group ↖ (0111) 4-bit group

EXAMPLE 1–15

Using Table 1–4, determine the ASCII code for the lowercase letter *p*.

Solution:

1110000 (*Note:* Often, a leading zero is added to form an 8-bit result, making p = 0111 0000.)

1–10 APPLICATIONS OF THE NUMBERING SYSTEMS

Because digital systems work mainly with 1s and 0s, we have spent considerable time working with the various number systems. Which system is used depends on how the data were developed and how they are to be used. In this section we will work with several applications that depend on the translation and interpretation of these digital representations.

APPLICATION 1–1

The ABC Corporation chemical processing plant uses a computer to monitor the temperature and pressure of four chemical tanks, as shown in Figure 1–3(a). Whenever a temperature or a pressure exceeds the danger limit, an internal tank sensor applies a 1 to its corresponding output to the computer. If all conditions are OK, then all outputs are 0.

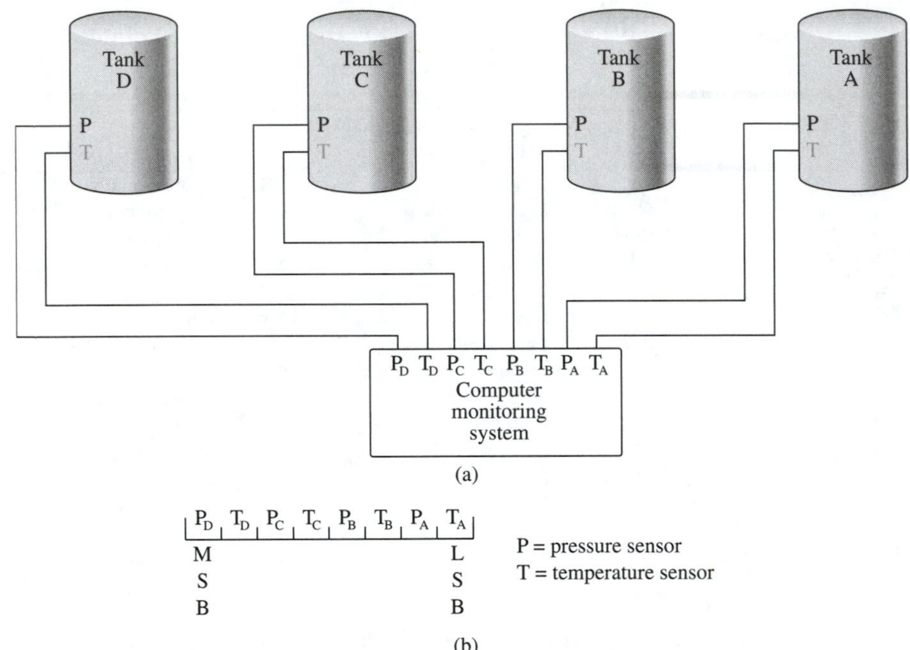

Figure 1–3 (a) Circuit connections for chemical temperature and pressure monitors at the ABC Corporation chemical processing plant; (b) layout of binary data read by the computer monitoring system.

(a) If the computer reads the binary string 0010 1000, what problems exist?

(b) What problems exist if the computer is reading 55H (55 hex)?

(c) What hexadecimal number is read by the computer if the temperature and pressure in both tanks D and B are high?

(d) Tanks A and B are taken out of use and their sensor outputs are connected to 1s. A computer programmer must write a program to ignore these new circuit conditions. The computer program must check that the value read is always less than what decimal equivalent when no problem exists?

(e) In another area of the plant, only three tanks (A, B, and C) have to be monitored. What octal number is read if tank B has a high temperature and pressure?

Solutions:

(a) Entering that binary string into the chart of Figure 1-3(b) shows us that the pressure in tanks C and B is dangerously high.

(b) 55H = 0101 0101, meaning that all temperatures are too high.

(c) CCH (1100 1100 = CCH)

(d) $<31_{10}$, because, with the 4 low-order bits HIGH, if TC goes HIGH, then the binary string will be 0001 1111, which is equal to 31_{10}.

(e) 14_8 ($001\ 100_2 = 14_8$)

APPLICATION 1–2

A particular brand of compact disk (CD) player has the capability of converting 12-bit signals from a CD into their equivalent analog values.

(a) What are the largest and smallest hex values that can be used in this CD system?

(b) How many different analog values can be represented by this system?

Solutions:

(a) Largest: FFF_{16}; smallest: 000_{16}

(b) FFF_{16} is equivalent to 4095 in decimal. Including 0, this is a total of 4096 unique representations.

APPLICATION 1–3

Typically, digital thermometers use BCD to drive their digit displays.

(a) How many BCD bits are required to drive a 3-digit thermometer display?

(b) What 12 bits are sent to the display for a temperature of 147°?

Solutions:

(a) 12; 4 bits for each digit

(b) 0001 0100 0111

APPLICATION 1–4

Most PC-compatible computer systems use a 20-bit address code to identify each of over 1 million memory locations.

(a) How many hex characters are required to identify the address of each memory location?

(b) What is the 5-digit hex address of the 200th memory location?

(c) If 50 memory locations are used for data storage starting at location 000C8H, what is the location of the last data item?

Solutions:

(a) Five (Each hex digit represents 4 bits.)

(b) 000C7H (200_{10} = C8H; but the first memory location is 00000H, so we have to subtract 1.)

(c) 000F9H (000C8H = 200_{10}, $200 + 50 = 250_{10}$, $250 - 1 = 249_{10}$, 249_{10} = F9H [We had to subtract 1 because location C8H (200_{10}) received the first data item, so we needed only 49 more memory spaces.]

APPLICATION 1–5

If the part number 651-M is stored in ASCII in a computer memory, list the binary contents of its memory locations.

Solution:

6 = 011 0110

5 = 011 0101

1 = 011 0001

- = 010 1101

M = 100 1101

Because most computer memory locations are formed by groups of 8 bits, let's add a zero to the leftmost position to fill each 8-bit memory location. (The leftmost position is sometimes filled by a parity bit, which is discussed in Chapter 6.)

Therefore, the serial number, if strung out in five memory locations, would look like

0011 0110 0011 0101 0011 0001 0010 1101 0100 1101

If you look at these memory locations in hexadecimal, they will read

36 35 31 2D 4D

APPLICATION 1–6

To look for an error in a BASIC program, a computer programmer uses a debugging utility to display the ASCII codes of a particular part of her program. The codes are displayed in hex as 474F5430203930. Assume that the leftmost bit of each ASCII string is padded with a 0.

(a) Translate the program segment that is displayed.

(b) If you know anything about programming in BASIC, try to determine what the error is.

Solution:

(a) GOT0 90.

(b) Apparently a number 0 was typed in the GOTO statement instead of the letter O. Change it and the error should go away.

1–11 DIGITAL SIGNALS

A digital signal is made up of a series of 1s and 0s that represent numbers, letters, symbols, or control signals. Figure 1–4 shows the timing diagram of a typical digital signal. Timing diagrams are used to show the HIGH and LOW (1 and 0) levels of a digital signal as it changes relative to time. In other words, it is a plot of *voltage versus time.* Figure 1–4 is a timing diagram showing the bit configuration 1 0 1 0 as it would appear on an oscilloscope. Note in Figure 1–4 that the LSB comes first in time. In this case, the LSB is transmitted first. The MSB could have been transmitted first as long as the system on the receiving end knows which method is used.

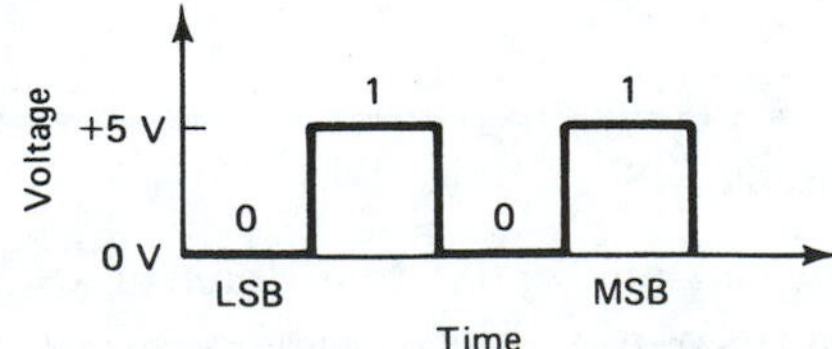

Figure 1–4 Typical digital signal.

1–12 CLOCK WAVEFORM TIMING

Most digital signals require precise timing. Special clock and timing circuits are used to produce clock waveforms to trigger the digital signals at precise intervals (timing circuit design is covered in a later chapter).

Figure 1–5 shows a typical *periodic clock waveform* as it would appear on an oscilloscope displaying voltage versus time. The term *periodic* means that the waveform is repetitive, at a specific time interval, with each successive pulse identical to the previous one.

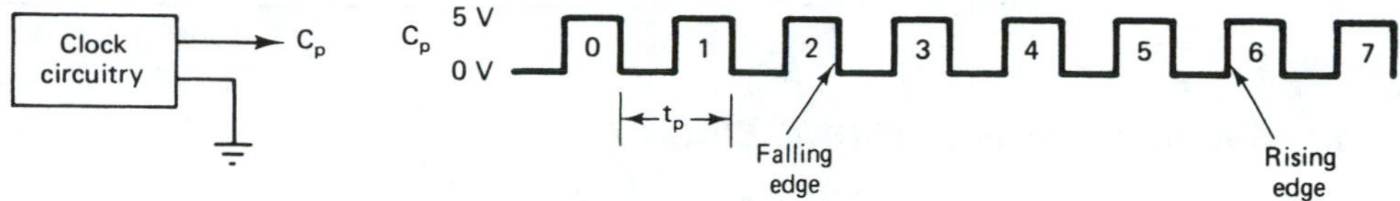

Figure 1–5 Periodic clock waveform as seen on an oscilloscope displaying voltage versus time.

Figure 1–5 shows eight clock pulses, which we will label 0, 1, 2, 3, 4, 5, 6, and 7. The *period* of the clock waveform is defined as the length of time from the falling edge of one pulse to the falling edge of the next pulse (or rising edge to rising edge) and is abbreviated t_p in Figure 1–5. The *frequency* of the clock waveform is defined as the reciprocal of the clock period. Written as a formula,

$$f = \frac{1}{t_p} \quad \text{and} \quad t_p = \frac{1}{f}$$

The basic unit for frequency is hertz and the basic unit for period is seconds.

EXAMPLE 1–16

What is the frequency of a clock waveform whose period is 2 microseconds (μs)?

Solution:

$$f = \frac{1}{t_p} = \frac{1}{2\ \mu\text{s}} = 0.5 \text{ megahertz (0.5 MHz)}$$

Hint: To review scientific notation, see Table 1–5.

EXAMPLE 1–17

If the frequency of a waveform is 4.17 MHz, what is its period?

Solution:

$$t_p = \frac{1}{f} = \frac{1}{4.17\text{ MHz}} = 0.240\ \mu\text{s}$$

TABLE 1–5
Common Scientific Prefixes

Prefix	*Abbreviation*	*Power of 10*
giga	G	10^9
mega	M	10^6
kilo	k	10^3
milli	m	10^{-3}
micro	μ	10^{-6}
nano	n	10^{-9}
pico	p	10^{-12}

1–13 SWITCHES IN ELECTRONIC CIRCUITS

The transitions between 0 and 1 digital levels are caused by switching from one voltage level to another (usually 0 V to +5 V). One way that switching is accomplished is to make and break a connection between two electrical conductors by way of a manual switch or an electromechanical relay. Another way to switch digital levels is by use of semiconductor devices such as diodes and transistors.

Manual switches and relays have almost *ideal* ON and OFF resistances in that when their contacts are closed (ON) the resistance (measured by an ohmmeter) is 0 ohms (Ω) and when their contacts are open (OFF), the resistance is infinite. Figure 1–6 shows the manual switch. When used in a digital circuit, a single-pole, double-throw manual switch can produce 0 and 1 states at some output terminal, as shown in Figures 1–7 and 1–8 by moving the switch (SW) to the up or down position.

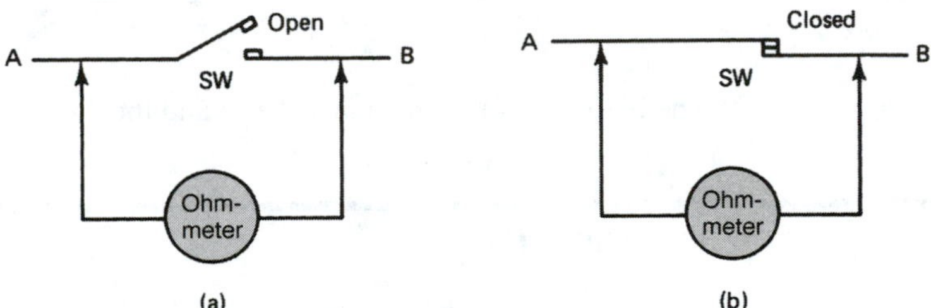

Figure 1–6 Manual switch: (a) switch open, $R = \infty$ ohms; (b) switch closed, $R = 0$ ohms.

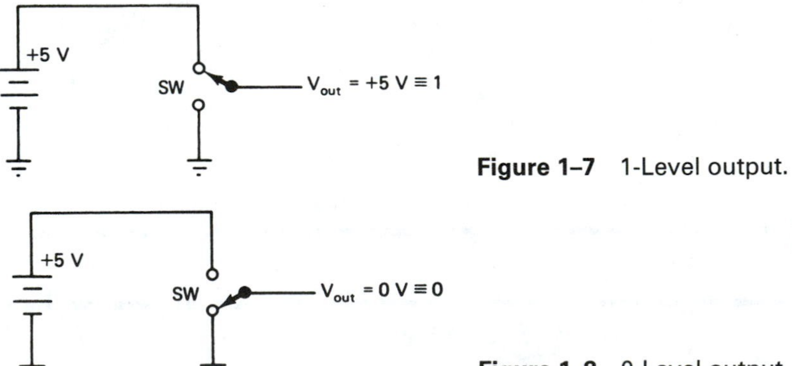

Figure 1–7 1-Level output.

Figure 1–8 0-Level output.

1–14 A TRANSISTOR AS A SWITCH

The transistor is a very commonly used switch in digital electronic circuits. It is a three-terminal semiconductor component that allows an input signal at one of its terminals to cause the other two terminals to become a short or an open circuit. The transistor is most commonly made of silicon that has been altered into *N*-type material and *P*-type material.

Three distinct regions make up a transistor: *emitter, base,* and *collector.* They can be a combination of *NPN*-type material or *PNP*-type material bonded together as a three-terminal device. Figure 1–9 shows the physical layout and symbol for an *NPN* transistor. (In a *PNP* transistor, the emitter arrow points the other way.)

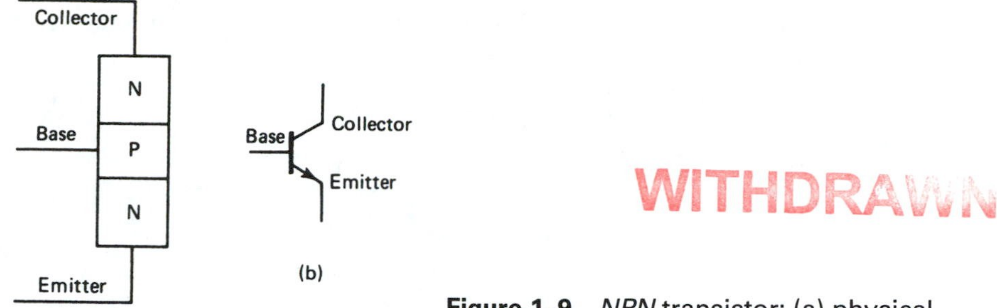

Figure 1–9 *NPN* transistor: (a) physical layout; (b) symbol.

In an electronic circuit, the input signal (1 or 0) is usually applied to the base of the transistor, which causes the collector–emitter junction to become a short or an open circuit. The rules of transistor switching are as follows:

1. In an *NPN* transistor, applying a positive voltage from base to emitter will cause the collector-to-emitter junction to short (this is called "turning the transistor ON"). Applying a negative voltage or 0 V from base to emitter will cause the collector-to-emitter junction to open (this is called "turning the transistor OFF").
2. In a *PNP*[1] transistor, applying a negative voltage from base to emitter will turn it ON. Applying a positive voltage or 0 V from base to emitter turns it OFF.

Figure 1–10 shows how an *NPN* transistor functions as a switch in an electronic circuit.

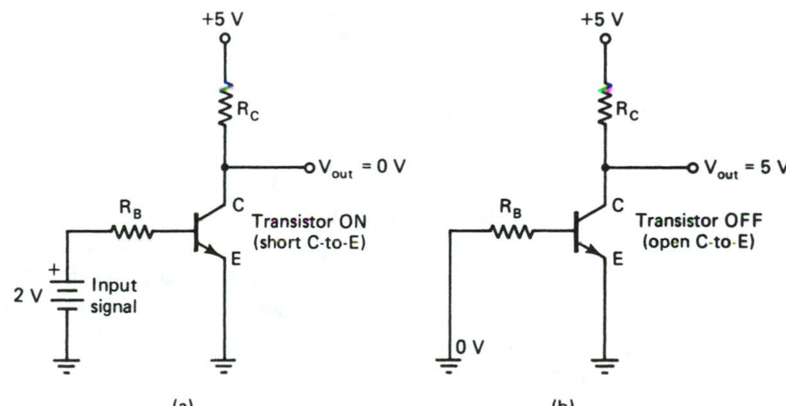

Figure 1–10 *NPN* transistor switch: (a) transistor ON; (b) transistor OFF.

In Figure 1–10 resistors R_B and R_C are used to limit the base current and the collector current. In Figure 1–10a the transistor is turned ON because the base is more positive than the emitter (input signal = +2 V). This causes the collector-to-emitter junction to short, placing ground potential at V_{out} ($V_{out} = 0$ V).

In Figure 1–10b the input signal is removed, making the base-to-emitter junction 0 V, turning the transistor OFF. With the transistor OFF there is no current through R_C, so $V_{out} = 5\text{ V} - (0\text{ A} \times R_C) = 5$ V.

[1]*PNP* transistor circuits are analyzed in the same way as *NPN* circuits except that all voltage and current polarities are reversed. *NPN* circuits are much more common in industry and will be used most often in this book.

Digital input signals are usually brought in at the base of the transistor and the output is taken off the collector or emitter. The following example uses timing analysis to compare the input and output waveforms.

EXAMPLE 1–18

An input clock waveform (C_p) is applied to the base of the transistor in Figure 1–11. Sketch the output waveform that will be produced at the transistor's collector terminal (V_c).

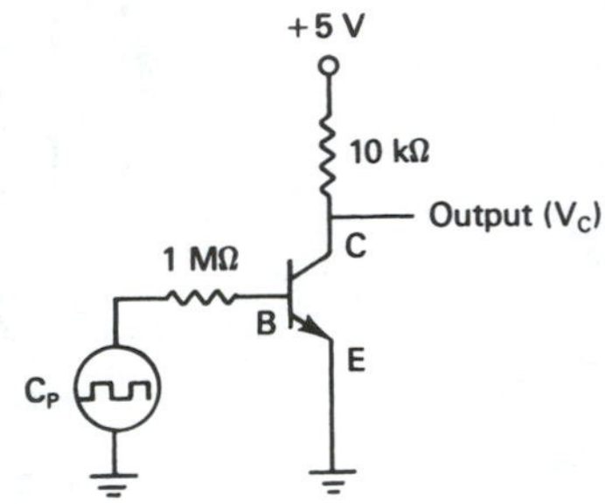

Figure 1–11

Solution:

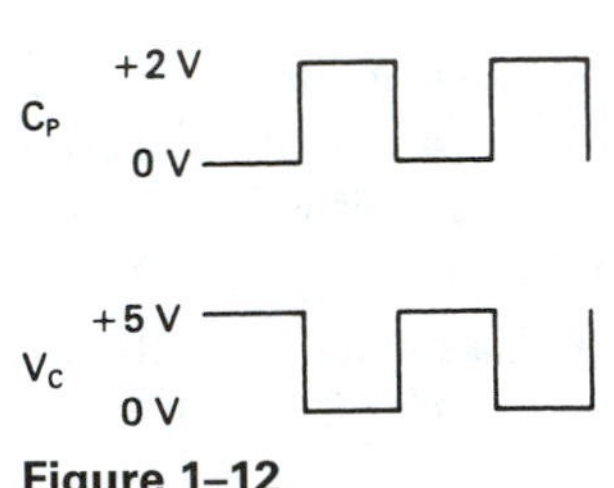

Figure 1–12

Figure 1–12 shows the input waveform, C_p, and the output waveform, V_c.

When C_p is 0 V, the transistor is OFF, and the collector-to-emitter junction is an open circuit. With the transistor OFF, there will be no current flow through, and no voltage loss across, the 10-kΩ resistor. This allows the 5 V to reach the V_c output.

When C_p is +2 V, the transistor is ON, and the collector-to-emitter junction is a short circuit. This makes a direct connection from V_c to ground, which is at 0 V, making V_c equal to 0 V. The waveform at V_c can now be drawn, showing that when C_p equals 0 V, V_c equals +5 V, and when C_p equals +2 V, V_c equals 0 V (see Figure 1–12).

1–15 THE TTL INTEGRATED CIRCUIT

Transistor–transistor logic (TTL) is one of the most widely used integrated-circuit (IC) technologies. TTL integrated circuits use a combination of several transistors, diodes, and resistors integrated together in a single package.

One basic function of a TTL integrated circuit is as a complementing switch or *inverter.* The inverter is used to take a digital level at its input and complement it to the opposite state at its output (1 becomes 0, 0 becomes 1). You may have noticed that the transistor circuit in Example 1–18 produced an inverted output at V_c relative to the input at C_p. That was a simplified inverter circuit. An actual TTL inverter circuit is more complex, using several transistors and diodes to perform the inverting action. Figure 1–13 shows the schematic diagram of a TTL inverter circuit.

TTL is a very popular family of integrated circuits. It is much more widely used than RTL (resistor–transistor logic) or DTL (diode–transistor logic) circuits, which were the forerunners of TTL. Details on the operation and specifications of TTL ICs are given in Chapter 2.

A single TTL integrated-circuit package such as the 7404 has six complete logic circuits fabricated into a single silicon chip, each logic circuit being the equivalent of Figure 1–13. The 7404 has 14 metallic pins connected to the outside of a plastic case containing the silicon chip. The 14 pins, arranged seven on each side, are aligned on 14 holes of a printed-circuit board, where they will then be soldered. The 7404 is called a 14-pin DIP (dual-in-line package) and costs less than 25 cents. Figure 1–14 shows a sketch of a 14-pin DIP IC. In subsequent chapters we will see how to use ICs in actual digital circuitry.

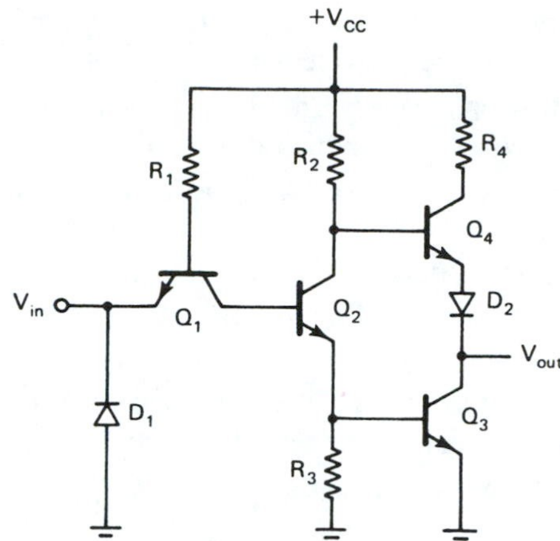

Figure 1–13 Schematic of a TTL inverter circuit.

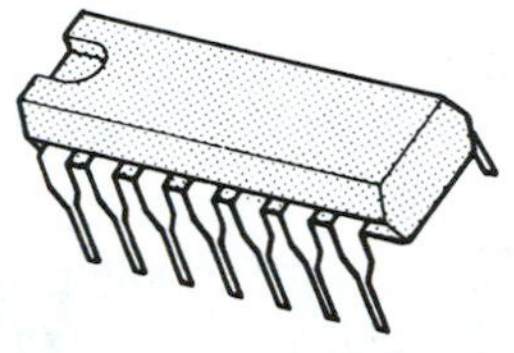

Figure 1–14 A 7404 TTL IC chip.

The pin configuration of the 7404 is shown in Figure 1–15. The power supply connections to the IC are made to pin 14 (+5 V) and pin 7 (ground), which supplies power to all six logic circuits. In the case of the 7404, the logic circuits are called *inverters.* The symbol for each inverter is a triangle with a circle at the output. The circle is used to indicate the inversion function.

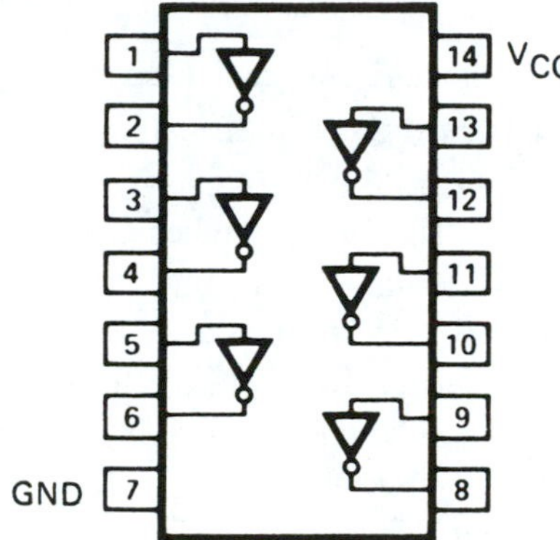

Figure 1–15 A 7404 hex inverter pin configuration.

Another common integrated-circuit technology used in digital logic is the CMOS (complementary metal-oxide semiconductor). CMOS uses a complementary pair of metal-oxide semiconductor field-effect transistors (MOSFETs) instead of the bipolar transistors used in TTL chips. (Complete coverage of TTL and CMOS is given in Chapter 2.)

SUMMARY

In this chapter we have learned that

1. Numerical quantities occur naturally in analog form but must be converted to digital form to be used by computers or digital circuitry.
2. The binary numbering system is used in digital systems because the 1s and 0s are easily represented by ON or OFF transistors which output 0 V for 0 and 5 V for 1.
3. Any number system can be converted to decimal by multiplying each digit by its weighting factor.
4. The weighting factor of the least significant digit in any numbering system is always 1.
5. The successive-division procedure can be used to convert from decimal to binary, octal or hexadecimal.

6. The binary-coded-decimal system uses groups of four bits to drive decimal displays such as those in a calculator.
7. An oscilloscope can be used to observe the rapidly changing voltage-versus-time waveform in digital systems.
8. The frequency of a clock waveform is equal to the reciprocal of the waveform's period.
9. The transistor is the basic building block of the modern digital integrated circuit. It can be switched ON or OFF by applying the appropriate voltage at its base connection.
10. TTL and CMOS integrated circuits are formed by integrating thousands of transistors in a single package. They are the most popular ICs used in digital circuitry today.

GLOSSARY

Analog: A system that deals with continuous varying physical quantities such as voltage, temperature, pressure, or velocity. Most quantities in nature occur in analog, yielding an infinite number of different levels.

BCD: Binary-coded decimal. A 4-bit code used to represent the 10 decimal digits, 0 to 9.

Binary: The base 2 numbering system. Binary numbers are made up of 1s and 0s, each position being equal to a different power of 2 (2^3, 2^2, 2^1, 2^0, etc.).

Bit: A single binary digit. The binary number 1101 is a 4-bit number.

Chip: The term given to an integrated circuit. It comes from the fact that each integrated circuit comes from a single "chip" of silicon crystal.

CMOS: Complementary metal-oxide semiconductor. A family of integrated circuits used to perform logic functions in digital circuits. The CMOS is noted for its low power consumption but sometimes slow speed.

Decimal: The base 10 numbering system. The 10 decimal digits are 0, 1, 2, 3, 4, 5, 6, 7, 8, and 9. Each decimal position is a different power of 10 (10^3, 10^2, 10^1, 10^0, etc.).

Digital: A system that deals with discrete digits or quantities. Digital electronics deals exclusively with 1s and 0s, or ONs and OFFs. Digital codes (such as ASCII) are then used to convert the 1s and 0s to a meaningful number, letter, or symbol for some output display.

Diode: A semiconductor device used to allow current flow in one direction but not the other. As an electronic switch, it acts like a short in the forward-biased condition and like an open in the reverse-biased condition.

DIP: Dual-in-line package. The most common pin layout for integrated circuits. The pins are aligned in two straight lines, one on each side of the IC.

Frequency: A measure of the number of cycles or pulses occurring each second. Its unit is the hertz and it is the reciprocal of the period.

Hexadecimal: The base 16 numbering system. The 16 hexadecimal digits are 0, 1, 2, 3, 4, 5, 6, 7, 8, 9, A, B, C, D, E, and F. Each hexadecimal position is worth a different power of 16 (16^3, 16^2, 16^1, 16^0, etc.).

Hex inverter: An integrated circuit containing six inverters on a single DIP package.

Integrated circuit: The fabrication of several semiconductor and electronic devices (transistors, diodes, and resistors) onto a single piece of silicon crystal. Integrated circuits are increasingly being used to perform the functions that used to require several hundred discrete semiconductors.

Inverter: A logic function that changes its input into the opposite logic state at its output (0 to 1 and 1 to 0).

Least significant bit (LSB): The bit having the least significance in a binary string. The LSB will be in the position of the lowest power of 2 within the binary number.

Logic state: A 1 or 0 digital level.

Most significant bit (MSB): The bit having the most significance in a binary string. The MSB will be in the position of the highest power of 2 within the binary number.

Oscilloscope: An electronic measuring device used in design and troubleshooting to display a picture of waveform magnitude (y axis) versus time (x axis).

Period: The measurement of time from the beginning of one periodic cycle or clock pulse to the beginning of the next. Its unit is the second, and it is the reciprocal of frequency.

Timing diagram: A diagram used to display the precise relationship between two or more digital waveforms as they vary relative to time.

Transistor: A semiconductor device that can be used as an electronic switch in digital circuitry. By applying an appropriate voltage at the base, the collector-to-emitter junction will act like an open or a shorted switch.

TTL: Transistor–transistor logic. The most common integrated circuit used in digital electronics today. A large family of different TTL ICs is used to perform all the logic functions necessary in a complete digital system.

PROBLEMS

1–1. Convert the following binary numbers to decimal.
(a) 0110 **(b)** 1011 **(c)** 1001 **(d)** 0111
(e) 1100 **(f)** 01001011 **(g)** 00110111 **(h)** 10110101
(i) 10100111 **(j)** 01110110

1–2. Convert the following decimal numbers to 8-bit binary.
(a) 186_{10} **(b)** 214_{10} **(c)** 27_{10} **(d)** 251_{10} **(e)** 146_{10}

1–3. Convert the following binary numbers to hexadecimal.
(a) 1011 1001 **(b)** 1101 1100 **(c)** 0111 0100
(d) 1111 1011 **(e)** 1100 0110

1–4. Convert the following hexadecimal numbers to binary.
(a) $C5_{16}$ **(b)** FA_{16} **(c)** $D6_{16}$ **(d)** $A94_{16}$ **(e)** 62_{16}

1–5. Convert the following hexadecimal numbers to decimal.
(a) 86_{16} **(b)** $F4_{16}$ **(c)** 92_{16} **(d)** AB_{16} **(e)** $3C5_{16}$

1–6. Convert the following decimal numbers to hexadecimal.
(a) 127_{10} **(b)** 68_{10} **(c)** 107_{10} **(d)** 61_{10} **(e)** 29_{10}

1–7. Convert the following BCD numbers to decimal.
(a) 10011000_{BCD} **(b)** 01101001_{BCD} **(c)** 01110100_{BCD}
(d) 00110110_{BCD} **(e)** 10000001_{BCD}

1–8. Convert the following decimal numbers to BCD.
(a) 87_{10} **(b)** 142_{10} **(c)** 94_{10} **(d)** 61_{10} **(e)** 44_{10}

1–9. Determine the period of a clock waveform whose frequency is:
(a) 2 MHz **(b)** 500 kHz **(c)** 4.27 MHz **(d)** 17 MHz

1–10. Determine the frequency of a clock waveform whose period is:
(a) 2 μs **(b)** 100 μs **(c)** 0.75 ms **(d)** 1.5 μs

1–11. What is the resistance of an open switch? A closed switch?

1–12. To use a common-emitter transistor circuit as an inverter, the input signal is connected to the (base, collector, or emitter) and the output signal is taken from the (base, collector, or emitter).

1–13. Find V_{out1} and V_{out2} for the circuits of Figure P1–13.

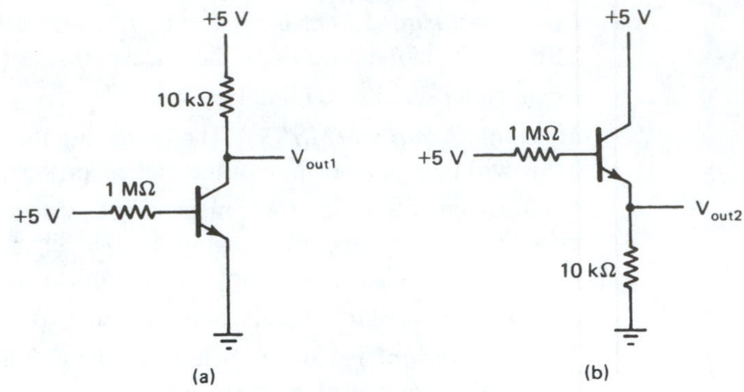

Figure P1–13

1–14. Sketch the output waveform at V_c for the circuit of Figure P1–14.

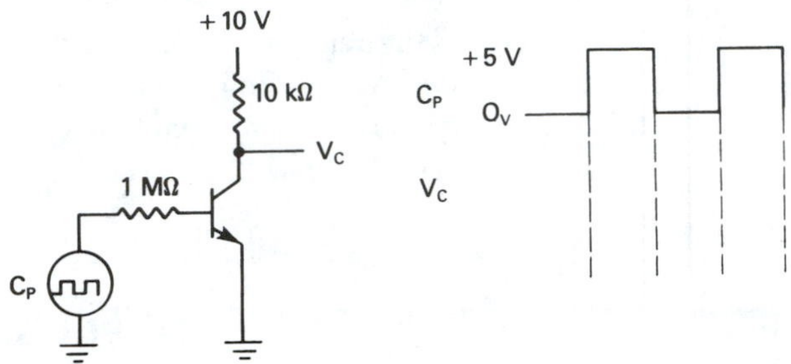

Figure P1–14

1–15. Which pins of the 7404 IC are connected to the power supply?

1–16. How many inverters are contained within a 7404 IC?

SCHEMATIC INTERPRETATION PROBLEMS

Note: Appendix H contains four schematic diagrams of actual digital systems. At the end of each chapter you will have the opportunity to work with these diagrams to gain experience with real-world circuitry and observe the application of digital logic that was presented in the chapter.

1–17. Locate the HC11D0 Master Board schematic in Appendix H. Determine the component name and grid coordinates of the following components (Example: Q3 is a 2N2907 located at A3):
(a) U1 **(b)** U16 **(c)** Q1 **(d)** P2

1–18. Find the Date and Revision number for the HC11D0 Master Board Schematic.

1–19. Find the quantity of the following components that are used on the Watchdog Timer schematic:
(a) 74HC85 **(b)** 74HC08 **(c)** 74HC74 **(d)** 74HC32

1–20. Y1 in the 4096/4196 Control Card schematic sheet 1 is a crystal used to generate a very specific frequency.
(a) What is its rated frequency?
(b) What time period does that create?

1–21. Repeat Problem 1–20 for the crystal X1 in the HC11D0 Master Board schematic.

1–22. The transistor Q1 in the HC11D0 schematic is turned ON and OFF by the level of pin 2 on U3 : A. What level must pin 2 be to turn Q1 ON and what will happen to the level on the line labeled RESET_B when that happens?

2 Logic Gate Operation and Specifications

OBJECTIVES

Upon completion of this chapter, you should be able to:

- Describe the operation and use of AND, OR, NAND, NOR, and INVERT gates.
- Construct truth tables for two-, three-, and four-input gates.
- Draw timing diagrams for AND, OR, NAND, NOR, and INVERT gates.
- Sketch the external connections to integrated-circuit chips to implement logic functions.
- Explain how to use a logic pulser and a logic probe to troubleshoot digital integrated circuits.
- Use the outputs of a Johnson shift counter to generate specialized waveforms utilizing various combinations of the five basic gates.
- Develop a comparison of the Boolean equations and truth tables for the five basic gates.
- Describe the operation and use of exclusive-OR and exclusive-NOR gates.
- Construct truth tables and draw timing diagrams for exclusive-OR and exclusive-NOR gates.
- Determine IC input and output voltage and current ratings from the manufacturer's data manual.
- Explain gate loading, fan-out, noise margin, and time parameters.
- Discuss the differences and proper use of the various subfamilies within both the TTL and CMOS lines of ICs.
- Describe the reasoning and various techniques for interfacing between the TTL and CMOS families of ICs.

INTRODUCTION

Logic gates are the basic building blocks for forming digital electronic circuitry. A logic gate has one output terminal and one or more input terminals. Its output will be HIGH (1) or LOW (0) depending on the digital level(s) at the input terminal(s). Through the use of logic gates we can design digital systems that will evaluate digital input levels and produce a specific output response based on that particular logic circuit design. The seven logic gates are AND, OR, NAND, NOR, INVERT, exclusive-OR, and exclusive-NOR.

Integrated-circuit logic gates (small-scale integration, SSI), combinational logic circuits (medium-scale integration, MSI), and microprocessor systems (large-scale integration and very-large-scale integration, LSI and VLSI) are readily available from several manufacturers through distributors and electronic parts suppliers. Basically, there are three commonly used families of digital IC logic: TTL (transistor–transistor logic), CMOS (complementary metal-oxide semiconductor), and ECL (emitter-coupled logic). Within each family there are several subfamilies (or series) of logic types available with different ratings for speed, power consumption, temperature range, voltage levels, and current levels. To use integrated circuits correctly, we must learn to understand the specifications presented in the manufacturers' data manuals. In this chapter we will learn to interpret these specifications and become aware of IC limitations and interfacing considerations.

2–1 THE AND GATE

Let's start by looking at the two-input AND gate shown in Figure 2–1. The operation of the AND gate is simple and is defined as follows: *The output, X, will be HIGH if input A AND input B are* both *HIGH.* In other words, if $A = 1$ *AND* $B = 1$, then $X = 1$. If either *A* or *B* or both are LOW, the output will be LOW.

Figure 2–1 Two-input AND gate.

TABLE 2–1
Truth Table for a Two-Input AND Gate

Inputs		Output
A	*B*	*X*
0	0	0
0	1	0
1	0	0
1	1	1

The best way to illustrate how the output level of a gate responds to all the possible input-level combinations is with a *truth table.* Table 2–1 is a truth table for a two-input AND gate. On the left side of the truth table, all possible input-level combinations are listed, and on the right side the resultant output is listed.

From the truth table we can see that the output at *X* is HIGH *only* when *both A* AND *B* are HIGH. If this AND gate is a TTL integrated circuit, HIGH means +5 V and LOW means 0 V (i.e., 1 is defined as +5 V and 0 is defined as 0 V).

One example of how an AND gate might be used is in a bank burglary alarm system. The output of the AND gate would go HIGH to turn on the alarm if the alarm activation key is in the ON position *AND* the front door is opened. This setup is illustrated in Figure 2–2.

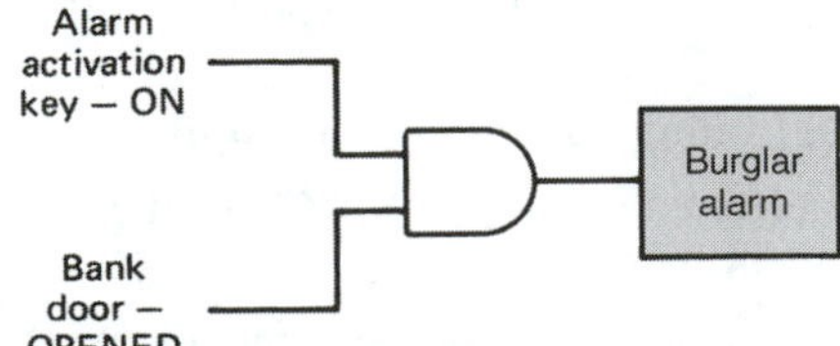

Figure 2–2 AND gate used to activate a burglar alarm.

Another way to illustrate the operation of an AND gate is by use of a series electric circuit. In Figure 2–3, using manual and transistor switches, the output at *X* will be HIGH if *both* switches *A AND B* are HIGH (1).

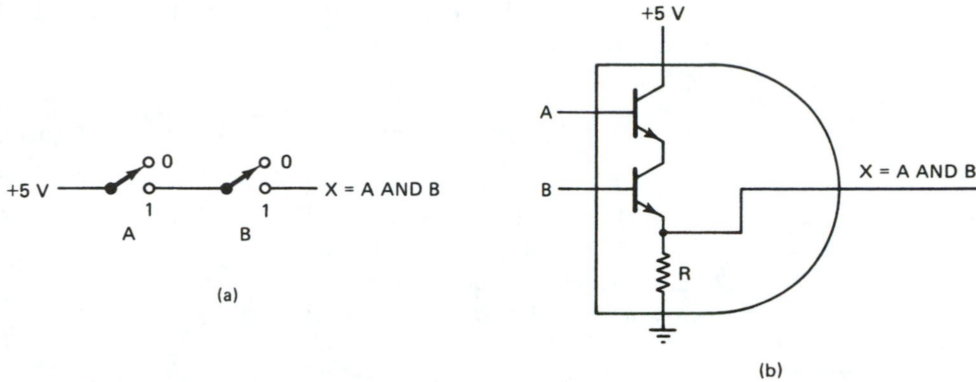

Figure 2–3 Electrical analogy of an AND gate: (a) using manual switches; (b) using transistor switches.

TABLE 2–2
Truth Table for a Four-Input AND Gate

A	B	C	D	X
0	0	0	0	0
0	0	0	1	0
0	0	1	0	0
0	0	1	1	0
0	1	0	0	0
0	1	0	1	0
0	1	1	0	0
0	1	1	1	0
1	0	0	0	0
1	0	0	1	0
1	0	1	0	0
1	0	1	1	0
1	1	0	0	0
1	1	0	1	0
1	1	1	0	0
1	1	1	1	1

Figure 2–3 also shows what is known as the *Boolean equation* for the AND function, $X = A$ and B, which can be thought of as *X equals 1 if A AND B both equal 1.* The Boolean equation for the AND function can more simply be written as $X = A \cdot B$ or just $X = AB$. *Boolean equations* will be used throughout the rest of the book to depict algebraically the operation of a logic gate or combination of logic gates.

AND gates can have more than two inputs. Figure 2–4 shows a four-input and an eight-input AND gate. The truth table for an AND gate with four inputs is shown in Table 2–2. To determine the total number of different combinations to be listed in the truth table, use the equation

$$\text{Number of combinations} = 2^N \qquad \text{where } N = \text{number of inputs} \qquad (2\text{–}1)$$

Therefore, in the case of a four-input AND gate, the number of possible input combinations is $2^4 = 16$.

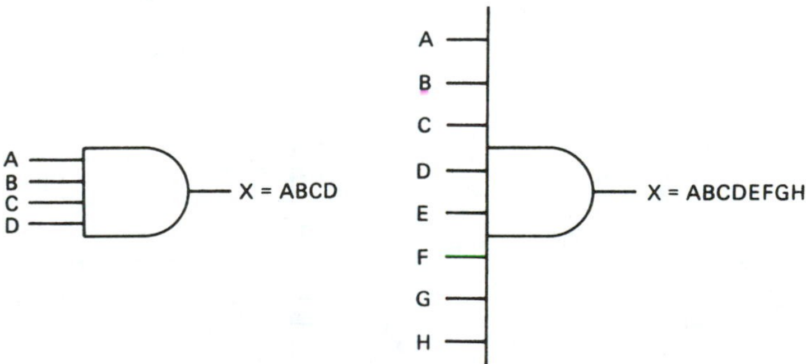

Figure 2–4 Multiple-input AND gate symbols.

When building the truth table, be sure to list all 16 *different* combinations of input levels. One easy way to ensure that you do not miss a combination or duplicate a combination is to list the inputs in the order of a binary counter (0000, 0001, 0010, ... , 1111). Also note in Table 2–2 that the A column lists eight 0s, then eight 1s; the B column lists four 0s, four 1s, four 0s, four 1s; the C column lists two 0s, two 1s, two 0s, two 1s, etc.; and the D column lists one 0, one 1, one 0, one 1, etc.

2–2 THE OR GATE

The OR gate also has two or more inputs and a single output. The symbol for a two-input OR gate is shown in Figure 2–5. The operation of the two-input OR gate is defined as follows: *The output at X will be HIGH whenever input A OR input B is HIGH or both are*

HIGH. As a Boolean equation this can be written $X = A + B$. Note the use of the + symbol to represent the OR function.

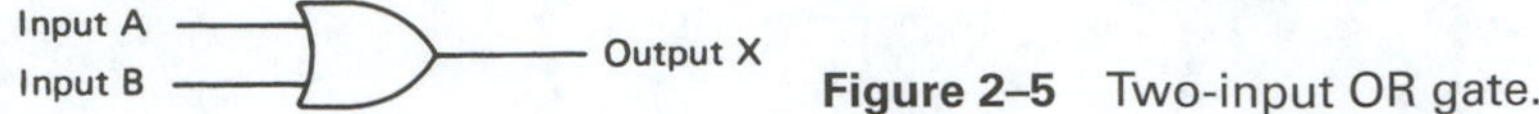

Figure 2–5 Two-input OR gate.

TABLE 2–3
Truth Table for a Two-Input OR Gate

Inputs		Output
A	*B*	*X*
0	0	0
0	1	1
1	0	1
1	1	1

The truth table for a two-input OR gate is shown in Table 2–3.

From the truth table you can see that *X* is 1 whenever *A OR B* is 1 or if *both A* and *B* are 1. Using manual or transistor switches in an electric circuit as shown in Figure 2–6, we can observe the electrical analogy to an OR gate. From the figure we see that the output at *X* will be 1 if *A or B*, or *both*, are HIGH (1).

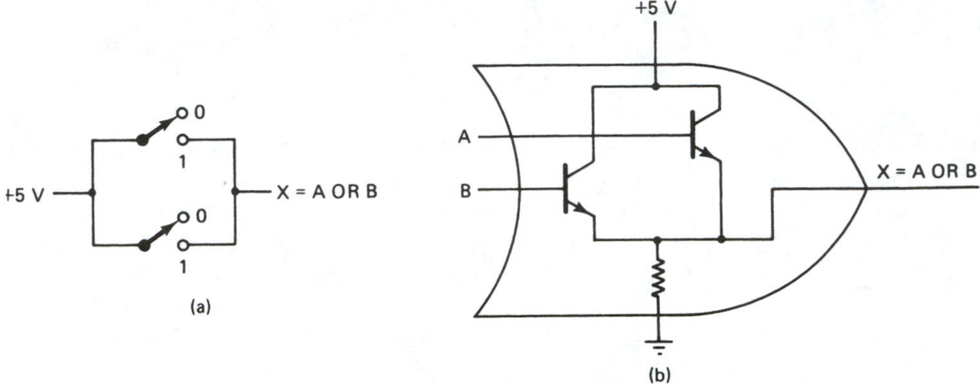

Figure 2–6 Electrical analogy of an OR gate: (a) using manual switches; (b) using transistor switches.

OR gates can also have more than two inputs. Figure 2–7 shows a three-input OR gate and Figure 2–8 shows an eight-input OR gate. The truth table for the three-input OR gate will have eight entries ($2^3 = 8$) and the eight-input OR gate will have 256 entries ($2^8 = 256$).

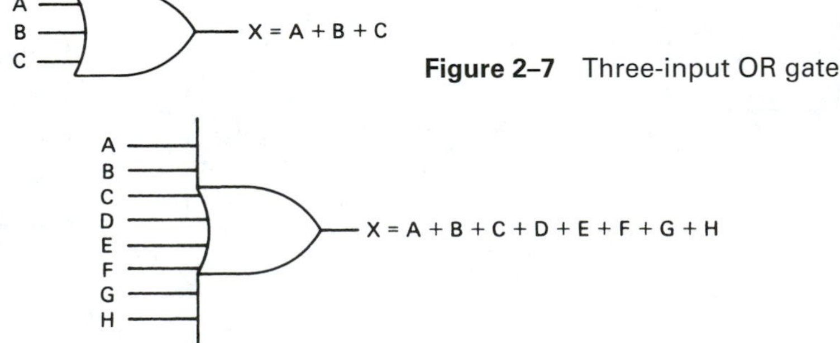

Figure 2–7 Three-input OR gate symbol.

Figure 2–8 Eight-input OR gate symbol.

TABLE 2–4
Truth Table for a Three-Input OR Gate

A	*B*	*C*	*X*
0	0	0	0
0	0	1	1
0	1	0	1
0	1	1	1
1	0	0	1
1	0	1	1
1	1	0	1
1	1	1	1

Let's build a truth table for the three-input OR gate.

The truth table of Table 2–4 is built by first using Equation 2–1 to determine that there will be eight entries, then listing the eight combinations of inputs in the order of a binary counter (000 to 111), then filling in the output column (*X*) by realizing that *X* will always be HIGH as long as at least one of the inputs is HIGH. When you look at the completed truth table, you can see that the only time the output is LOW is when *all* the inputs are LOW.

2–3 TIMING ANALYSIS

Another useful means of analyzing the output response of a gate to varying input-level changes is by means of a *timing diagram.* A timing diagram, as described in Chapter 1, is used to illustrate graphically how the output levels change in response to input-level changes.

The timing diagram in Figure 2–9 shows the two input waveforms (*A* and *B*) that are applied to a two-input AND gate, and the *X* output that will result from the AND operation. (For TTL and most CMOS logic gates, 1 = +5 V, 0 = 0 V.) As you can see, timing analysis is very useful for visually illustrating the level at the output for varying input-level changes.

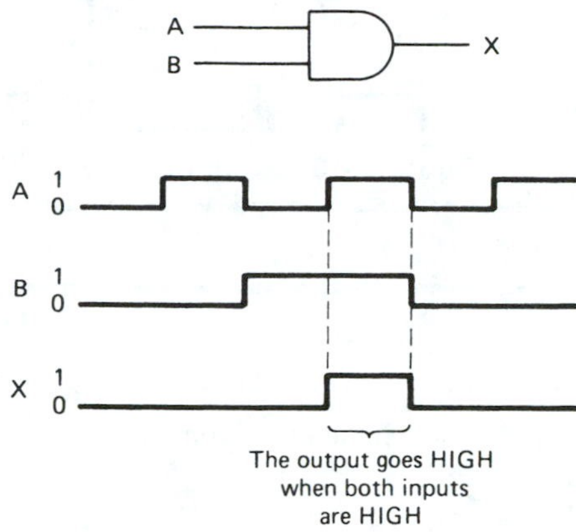

Figure 2–9 Timing analysis of an AND gate.

Timing waveforms are observed on an *oscilloscope* or a *logic analyzer.* A dual-trace oscilloscope is capable of displaying *two* voltage-versus-time waveforms on the same *x* axis. That is ideal for comparing the relationship of one waveform relative to another. The other timing analysis tool is the logic analyzer. It can, among other things, display 8 or 16 voltage-versus-time waveforms on the same *x* axis. It can also display the levels of the digital signals in a *state table,* which lists the binary levels of all the waveforms, at predefined intervals, in binary, hexadecimal, or octal. Timing analysis of 8 or 16 channels concurrently is very important when analyzing advanced digital and microprocessor systems, where the interrelationship of several digital signals is critical for proper circuit operation.

EXAMPLE 2–1

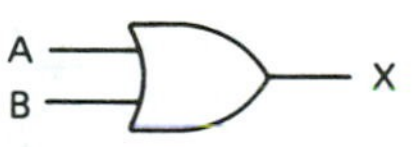

Figure 2–10

Sketch the output waveform at *X* for the two-input OR gate shown in Figure 2–10, with the given *A* and *B* input waveforms.

Solution:

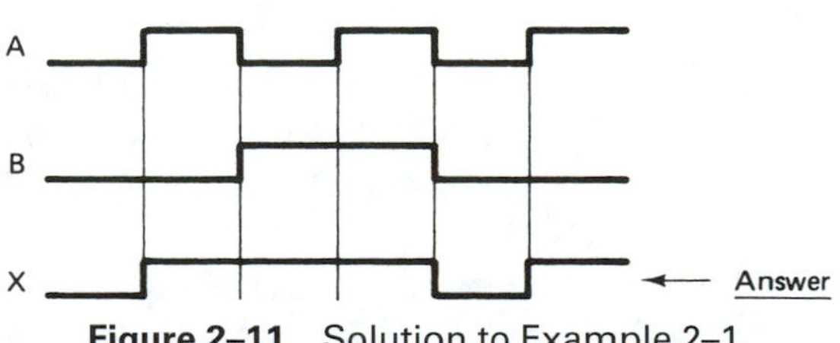

Figure 2–11 Solution to Example 2–1.

EXAMPLE 2–2

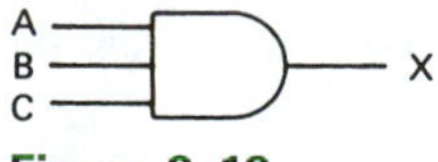

Figure 2–12

Sketch the output waveform at *X* for the three-input AND gate shown in Figure 2–12, with the given *A*, *B*, and *C* input waveforms.

Solution:

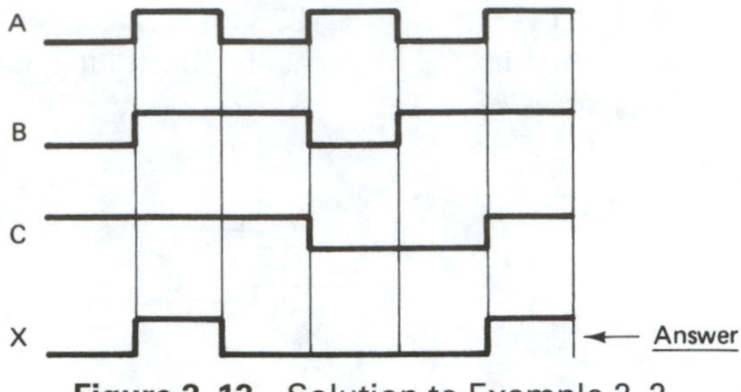

Figure 2–13 Solution to Example 2–2.

EXAMPLE 2–3

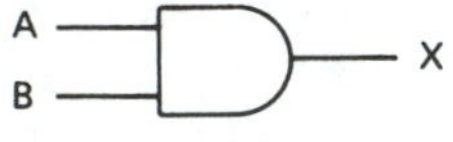

Figure 2–14

The input waveform at *A* and the output waveform at *X* are given for the AND gate in Figure 2–14. Sketch the input waveform that is required at *B* to produce the output at *X*.

Solution:

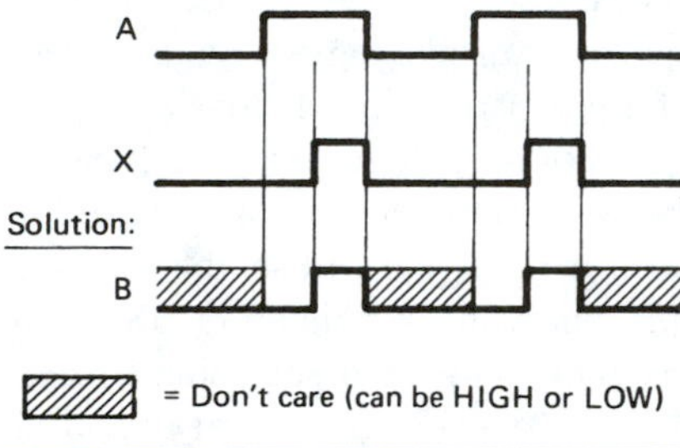

Figure 2–15 Solution to Example 2–3.

2–4 USING INTEGRATED-CIRCUIT LOGIC GATES

AND and OR gates are available as integrated circuits (ICs). The IC pin layout, logic gate type, and technical specifications are all contained in the logic data manual supplied by the manufacturer of the IC. For example, referring to a TTL or a CMOS logic data manual, we can see that there are several AND and OR gate ICs. To list just a few:

1. The 7408 (74HC08) is a quad two-input AND gate.
2. The 7411 (74HC11) is a triple three-input AND gate.
3. The 7421 (74HC21) is a dual four-input AND gate.
4. The 7432 (74HC32) is a quad two-input OR gate.

In each case, the HC stands for "high-speed CMOS." For example, the 7408 is a TTL AND gate and the 74HC08 is the equivalent CMOS AND gate. The terms *quad* (four), *triple* (three), and *dual* (two) refer to the number of separate gates on a single IC.

Let's look in more detail at one of these ICs, the 7408 (Figure 2–16). The 7408 is a 14-pin dual-in-line package (DIP) IC. The power supply connections are made to pins 7 and 14. This supplies the operating voltage for all four AND gates on the IC. Let's make the external connections to the IC to form a clock oscillator enable circuit (Figure 2–17).

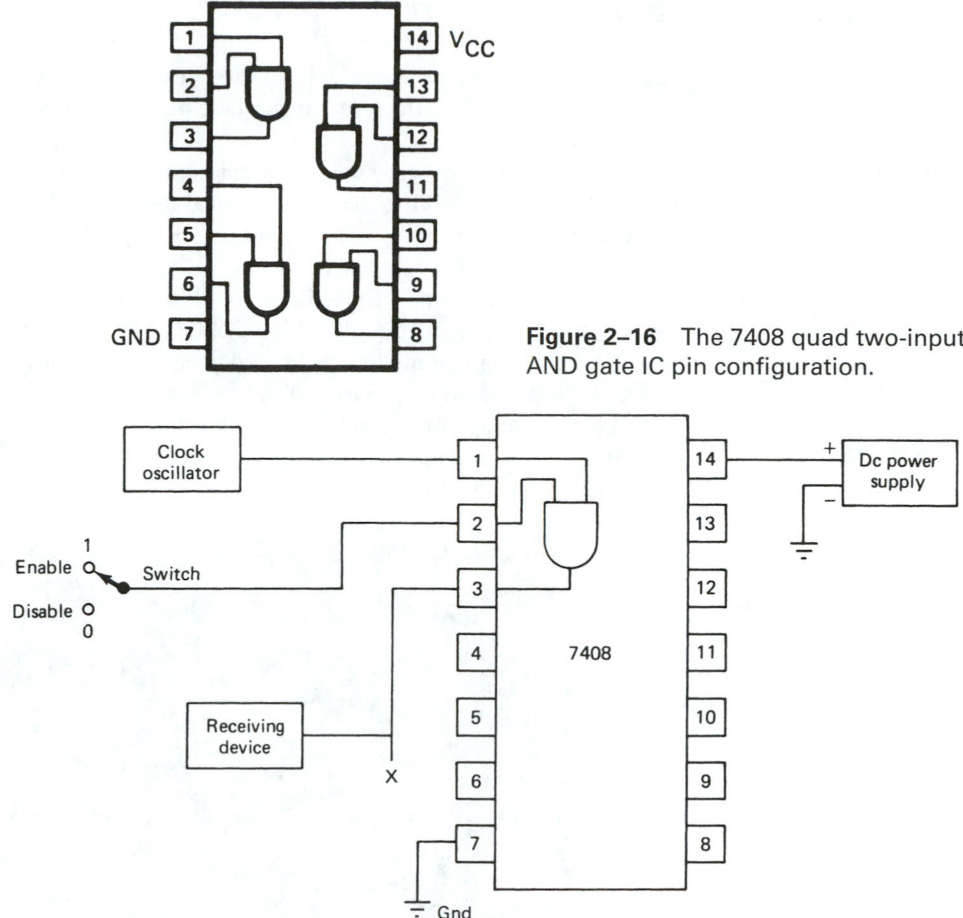

Figure 2–16 The 7408 quad two-input AND gate IC pin configuration.

Figure 2–17 Using the 7408 TTL IC in a clock enable circuit.

In Figure 2–17 the first AND gate in the IC was used and the other three are ignored. The IC is powered by connecting pin 14 to the positive power supply and pin 7 to ground. The other connections are made to form a clock enable circuit. The clock oscillator signal passes on to the receiving device when the switch is in the *enable* (1) position, and it stops when in the *disable* (0) position.

The pin configurations for some other logic gates are shown in Figure 2–18.

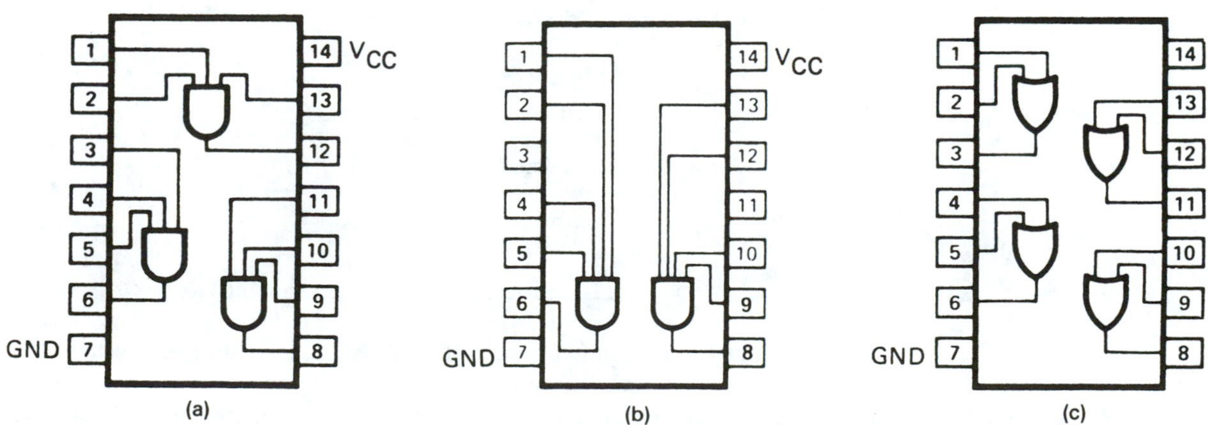

Figure 2–18 Pin configurations for other popular TTL and CMOS AND and OR gate ICs: (a) 7411(74HC11); (b) 7421(74HC21); (c) 7432(74HC32).

2–5 INTRODUCTION TO TROUBLESHOOTING TECHNIQUES

Like any other electronic device, integrated circuits and digital electronic circuits can go bad. *Troubleshooting* is the term given to the procedure used to find the *fault* or *trouble* in the circuits.

To be a good troubleshooter you must first *understand the theory and operation* of the circuit, devices, and ICs that are suspected to be bad. If you understand how a particular IC is *supposed* to operate, it is a simple task to put the IC through a test or to exercise its functions to see if it operates as you expect.

There are two simple tools that we will start with to test the ICs and digital circuits: the logic pulser and logic probe (Figure 2–19). The *logic probe* has a metal tip that is placed on the IC pin, printed-circuit-board trace, or device lead that you want to test. It also has an indicator lamp that glows, telling you the digital level at that point. If the level is HIGH (1), the lamp glows brightly. If the level is LOW (0), the lamp goes out. If the level is floating (open circuit, neither HIGH nor LOW), the lamp is dimly lit. Table 2–5 summarizes the states of the logic probe.

Figure 2–19 Logic pulser and logic probe. (Photo courtesy of Hewlett-Packard Company)

TABLE 2–5
Logic Probe States

Logic level	*Indicator lamp*
HIGH (1)	On
LOW (0)	Off
Float	Dim

The *logic pulser* is used to provide digital pulses to a circuit being tested. By applying a pulse to a circuit and simultaneously observing a logic probe, you can tell if the pulse signal is getting through the IC or device as you would expect. As you become more and more experienced at troubleshooting, you will find that most IC and device faults are due to an open or short at the input or output terminals. The following troubleshooting examples will illustrate some basic troubleshooting techniques using the logic probe and pulser.

EXAMPLE 2–4

The integrated-circuit AND gate in Figure 2–20 is suspected of having a fault and you want to test it. What procedure should you follow?

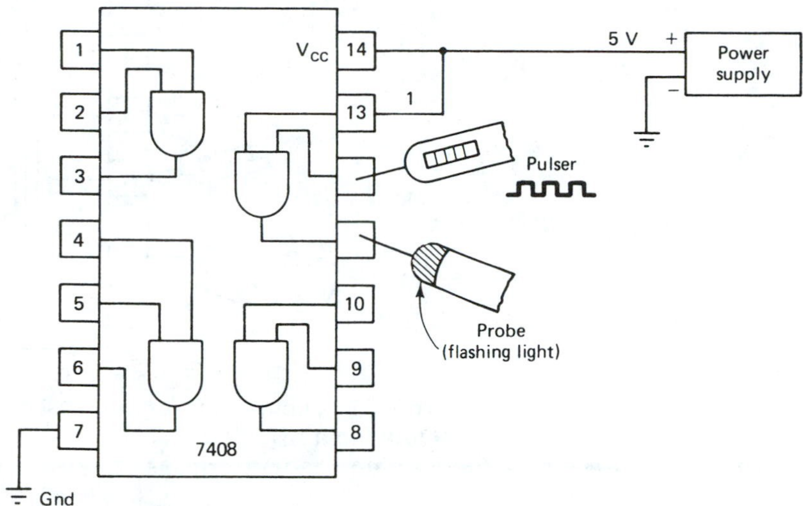

Figure 2–20 Connections for troubleshooting one gate of a quad AND IC.

Solution:

First you apply power to V_{CC} (pin 14) and Gnd (pin 7). Next you want to check each AND gate with the pulser/probe. Since it takes a HIGH (1) on *both* inputs to an AND gate to make the output go HIGH, if we put a HIGH (+5 V) on one input and pulse the other, we would expect to get pulses at the output of the gate. Figure 2–20 shows the connections to test one of the gates of a quad AND IC. When the pulser is put on pin 12, the light in the end of the probe flashes at the same speed as the pulser, indicating that the AND gate is passing the pulses through the gate (similar in operation to the clock enable circuit of Figure 2–17).

The next check is to reverse the connections to pins 12 and 13 and check the probe. If the probe still flashes, that gate is okay. Proceed to the other three gates and follow the same procedure. When one of the gate outputs does not flash, you have found the fault.

As mentioned earlier, *the key to troubleshooting an IC is understanding how the IC works.*

EXAMPLE 2–5

Sketch the connections for troubleshooting the first gate of a 7421 dual AND gate.

Solution:

The connections are shown in Figure 2–21. The probe should be flashing if the gate is good. Check each of the four inputs with the pulser by keeping three inputs high and pulsing the fourth while you look at the probe. In any case, if the probe does not flash, you have found a bad gate.

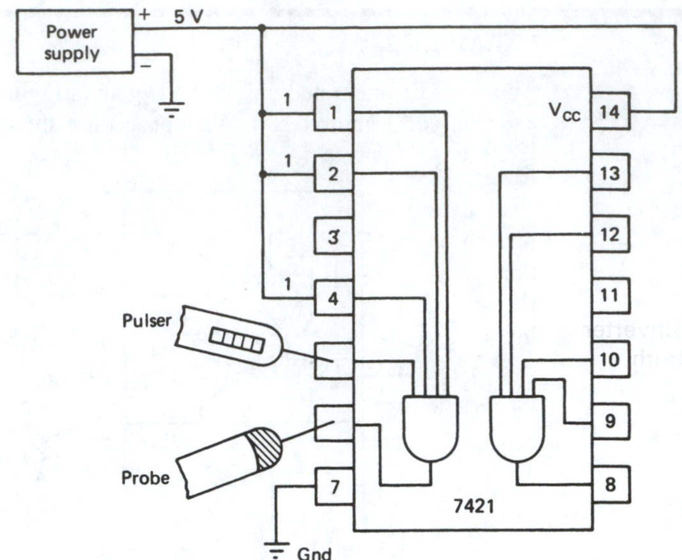

Figure 2–21 Connections for troubleshooting one gate of a 7421 dual four-input AND gate.

EXAMPLE 2–6

Sketch the connections for troubleshooting the first gate of a 7432 quad OR gate.

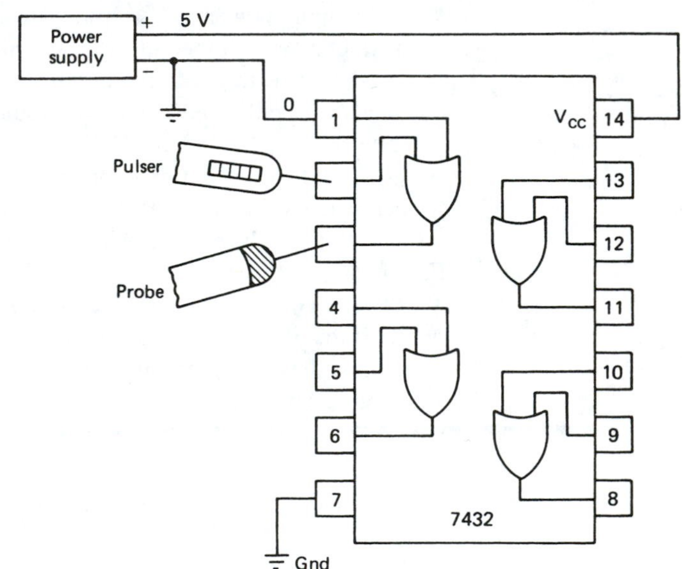

Figure 2–22 Connections for troubleshooting one OR gate of a 7432 IC.

Solution:

The connections are shown in Figure 2–22. The probe should be flashing if the gate is good. Note that the second input to the OR gate being checked is connected to a LOW (0) instead of a HIGH. The reason for this is that the output would *always* be HIGH if one input was connected HIGH. Since one input is connected LOW instead, the output will flash together with the pulses from the logic pulser if the gate is good.

2–6 THE INVERTER

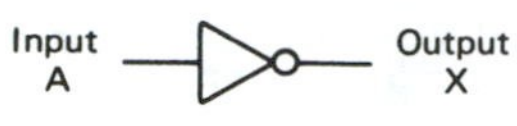

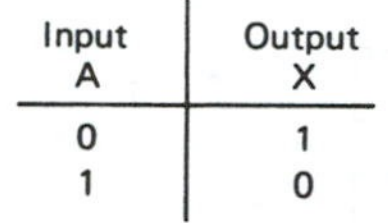

Input A	Output X
0	1
1	0

Figure 2–23 Inverter symbol and truth table.

The inverter is used to complement or invert a digital signal. It has a single input and a single output. If a HIGH level (1) comes in, it produces a LOW level (0) output. If a LOW level (0) comes in, it produces a HIGH level (1) output. The symbol and truth table for the inverter gate are shown in Figure 2–23.

The operation of the inverter is very simple and can be illustrated further by studying the timing diagram of Figure 2–24. The timing diagram graphically shows the operation of the inverter. When the input is HIGH, the output is LOW, and when the input is LOW, the output is HIGH. The output waveform is therefore the exact complement of the input.

The Boolean equation for an inverter is written $X = \overline{A}$ (X = NOT A). The *bar* over the A is an inversion bar, used to signify the *complement.*

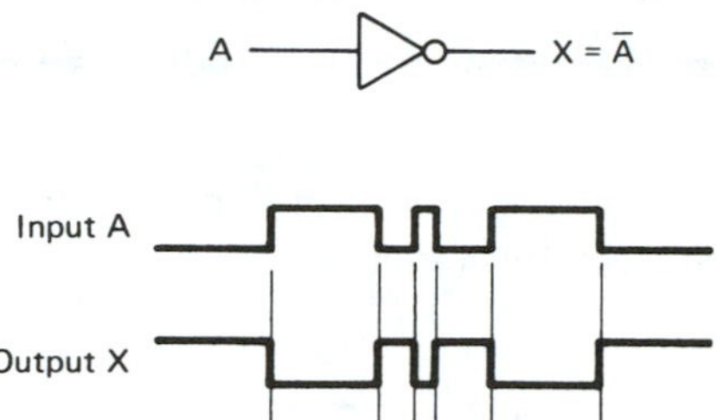

Figure 2–24 Timing analysis of an inverter gate.

2–7 THE NAND GATE

The operation of the NAND gate is the same as the AND gate except that its output is inverted. You can think of a NAND gate as an AND gate with an inverter at its output. The symbol for a NAND gate is an AND gate with a small circle (bubble) or triangle at its output, as shown in Figure 2–25.

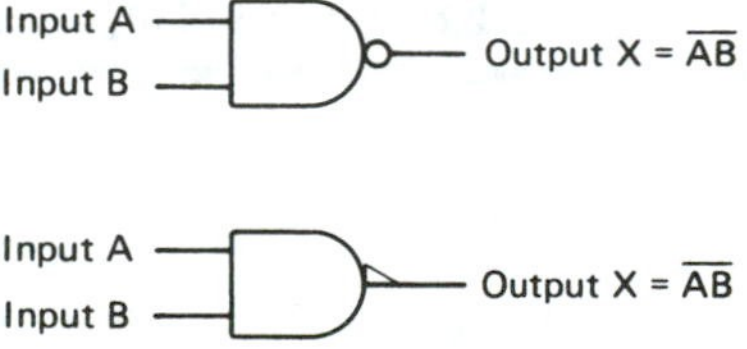

Figure 2–25 Symbols for a NAND gate.

In digital circuit diagrams you will find the small circle or triangle used whenever complementary action (inversion) is to be indicated. (The circle is more commonly used, however, and will be used most often in this text.) The circle or triangle at the output acts just like an inverter, so a NAND gate can be drawn symbolically as an AND gate with an inverter connected to its output, as shown in Figure 2–26.

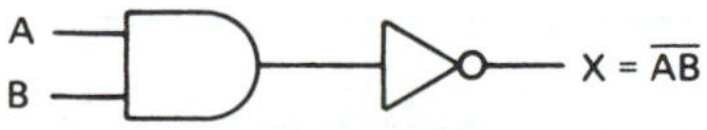

Figure 2–26 AND–INVERT equivalent of a NAND gate.

TABLE 2–6
Two-Input NAND Gate Truth Table

A	*B*	*X*
0	0	1
0	1	1
1	0	1
1	1	0

The Boolean equation for the NAND gate is written $X = \overline{AB}$. The inversion bar is drawn over the (A and B) meaning that the output of the NAND is the complement of (A and B) [NOT (A and B)]. Since we are inverting the output, the truth table outputs in Table 2–6 will be the complement of the AND gate truth table outputs. The easy way to construct the truth table is to think of how an AND gate would respond to the inputs, then invert your answer. From Table 2–6 we can see that the output is LOW when *both* inputs *A and B* are HIGH (just the opposite of an AND gate). Also, the output is HIGH whenever either input is LOW.

TABLE 2–7
Truth Table for a Three-Input NAND Gate

A	*B*	*C*	*X*
0	0	0	1
0	0	1	1
0	1	0	1
0	1	1	1
1	0	0	1
1	0	1	1
1	1	0	1
1	1	1	0

NAND gates can also have more than two inputs. Figure 2–27 shows three-input and eight-input NAND gate symbols. The truth table for a three-input NAND gate (Table 2–7) shows that the output is always HIGH unless *all* inputs go HIGH.

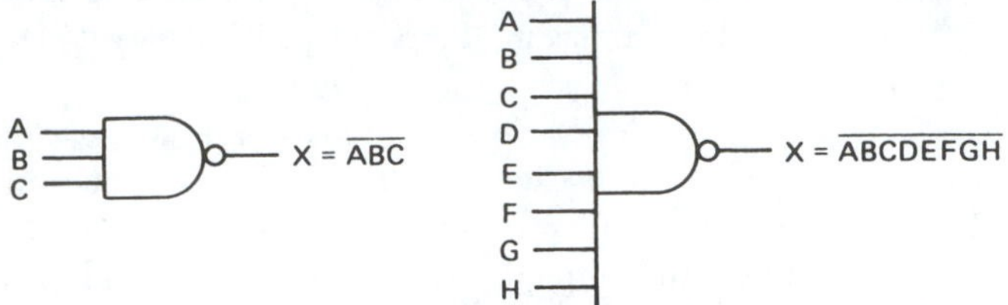

Figure 2–27 Symbols for three-input and eight-input NAND gates.

Timing analysis can also be used to illustrate the operation of NAND gates. The following examples will contribute to your understanding.

EXAMPLE 2–7

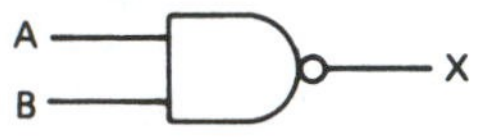

Figure 2–28

Sketch the output waveform at *X* for the NAND gate shown in Figure 2–28, with the given input waveforms.

Solution:

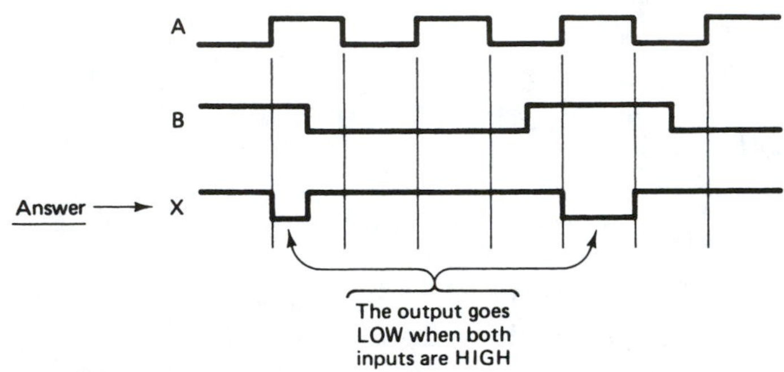

Figure 2–29 Timing analysis of a NAND gate.

EXAMPLE 2–8

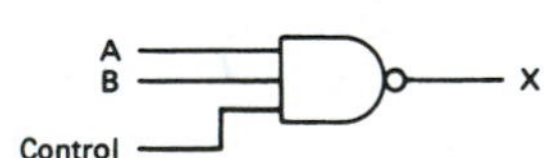

Figure 2–30

Sketch the output waveform at *X* for the NAND gate shown in Figure 2–30, with the given input waveforms at *A*, *B*, and Control.

Solution:

In Figure 2–31 the Control input waveform is used to *enable/disable* the NAND gate. When it is LOW, the output is stuck HIGH. When it goes HIGH, the output will respond LOW when *A* and *B* go HIGH.

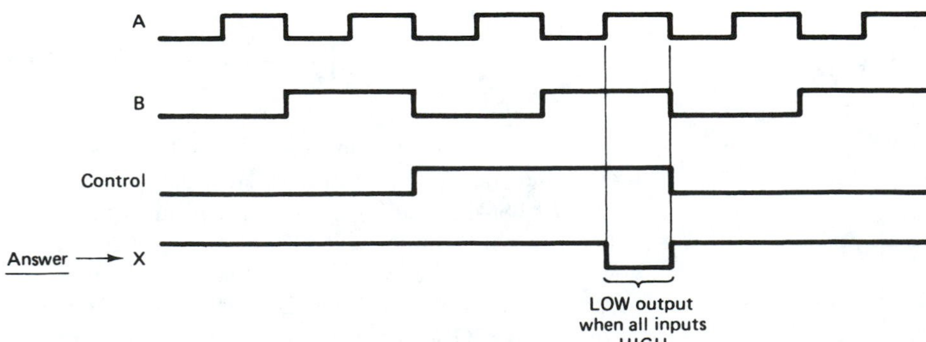

Figure 2–31 Timing analysis of a NAND gate with a "Control" input.

2–8 THE NOR GATE

The operation of the NOR gate is the same as that of the OR gate except that its output is inverted. You can think of a NOR gate as an OR gate with an inverter at its output. The symbols for a NOR gate and its equivalent OR–INVERT symbol are shown in Figure 2–32.

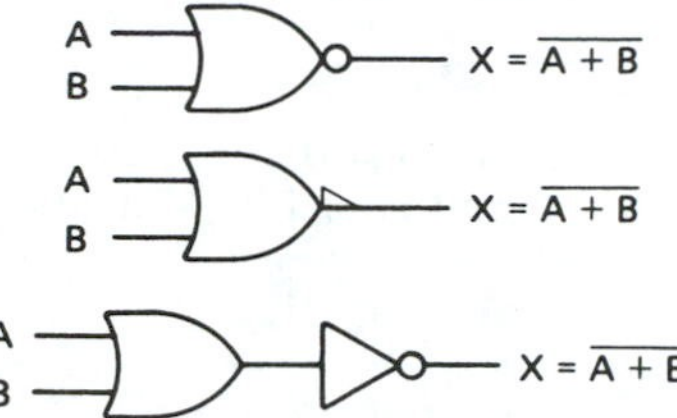

Figure 2–32 NOR gate symbols and its OR–INVERT equivalent.

TABLE 2–8
Truth Table for a NOR Gate

A	B	$X = \overline{A + B}$
0	0	1
0	1	0
1	0	0
1	1	0

The Boolean equation for the NOR function is $X = \overline{A + B}$. The equation is stated "X equals *not* (A or B)." In other words, X is LOW if A or B is HIGH. The truth table for a NOR gate is given in Table 2–8. Note that the output column is the complement of the OR gate truth table output column.

Now let's study some timing analysis examples to get a better grasp of NOR gate operation.

EXAMPLE 2–9

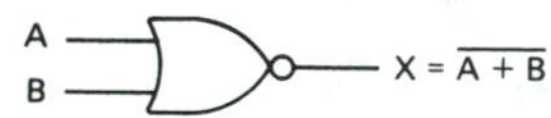

Figure 2–33

Sketch the output waveform at X for the NOR gate shown in Figure 2–33, with the given input waveforms.

Solution:

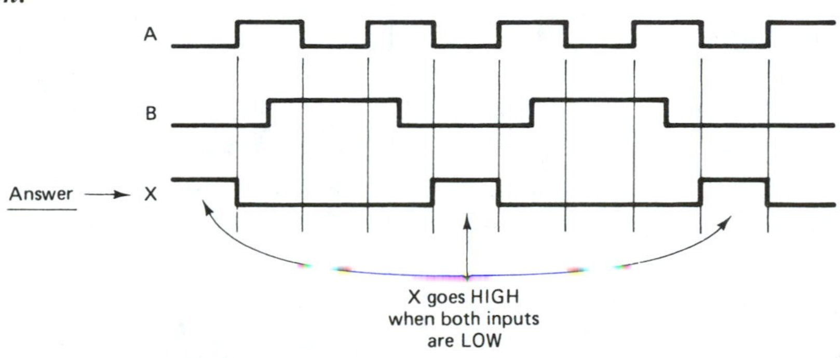

Figure 2–34 NOR gate timing analysis.

EXAMPLE 2–10

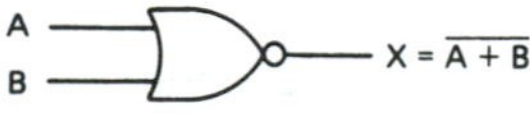

Figure 2–35

Sketch the waveform at the B input of the gate shown in Figure 2–35 that will produce the output waveform shown for X.

Solution:

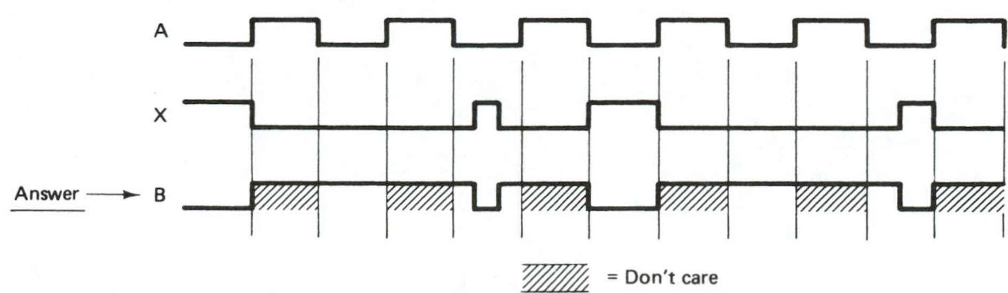

Figure 2–36 Input waveform requirement to produce a specific output.

2–9 THE EXCLUSIVE-OR GATE

Remember, the OR gate provides a HIGH output if one input or the other input is HIGH, *or if both inputs are HIGH.* The *exclusive-OR,* on the other hand, provides a HIGH output if one input or the other input is HIGH *but not both.* This point is made clearer by comparing the truth tables for an OR gate versus an exclusive-OR gate, as shown in Table 2–9.

TABLE 2–9
Truth Tables for an OR Gate versus an Exclusive-OR Gate

A	B	X
0	0	0
0	1	1
1	0	1
1	1	1

(OR)

A	B	X
0	0	0
0	1	1
1	0	1
1	1	0

(Exclusive-OR)

The Boolean equation for the Ex-OR function is written $X = \overline{A}B + A\overline{B}$ and can be constructed using the combinational logic shown in Figure 2–37. By experimenting and using Boolean reduction, we can find several other combinations of the basic gates that provide the Ex-OR function. For example, the combination of AND, OR, and NAND gates shown in Figure 2–38 will reduce to the "one-or-the-other-but-not-both" (Ex-OR) function.

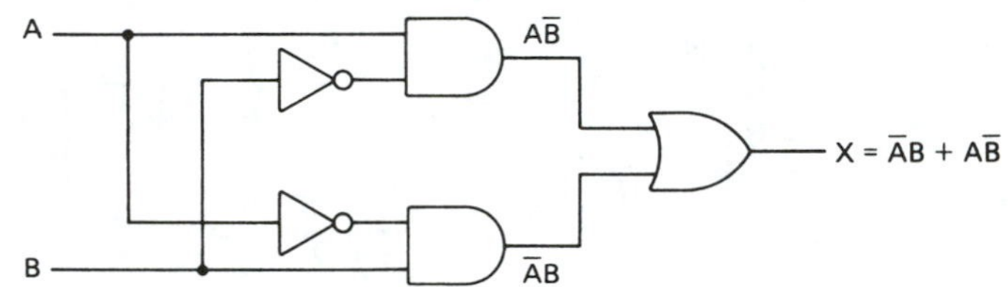

Figure 2–37 Logic circuit for providing the exclusive-OR function.

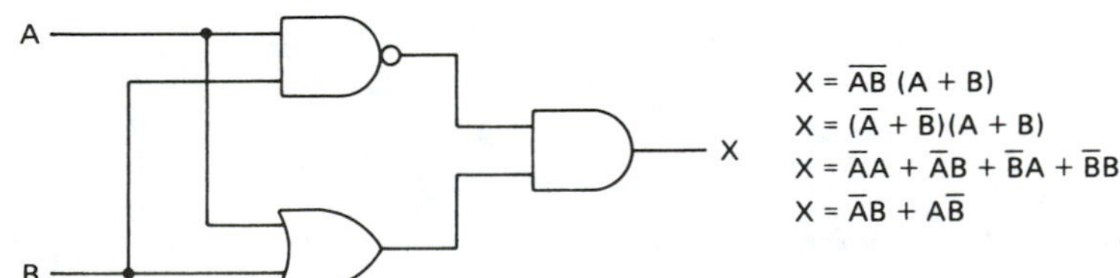

Figure 2–38 Exclusive-OR built with an AND–OR–NAND combination.

The exclusive-OR gate is common enough to deserve its own logic symbol and equation, shown in Figure 2–39. (Note the shorthand method of writing the Boolean equation is to use a plus sign with a circle around it.)

A, B — $X = A \oplus B = \overline{A}B + A\overline{B}$

Figure 2–39 Logic symbol and equation for the exclusive-OR.

2–10 THE EXCLUSIVE-NOR GATE

The exclusive-NOR is the complement of the exclusive-OR. A comparison of the truth tables in Table 2–10 illustrates that point.

The truth table for the Ex-NOR shows a HIGH output for both inputs LOW or both inputs HIGH. The Ex-NOR is sometimes called the "equality gate" because both inputs

TABLE 2–10
Truth Tables of the Exclusive-NOR versus an Exclusive-OR

$X = AB + \overline{A}\overline{B}$			$X = \overline{A}B + A\overline{B}$		
A	B	X	A	B	X
0	0	1	0	0	0
0	1	0	0	1	1
1	0	0	1	0	1
1	1	1	1	1	0
Exclusive-NOR			(Exclusive-OR)		

must be equal to get a HIGH output. The basic logic circuit and symbol for the Ex-NOR is shown in Figure 2–40.

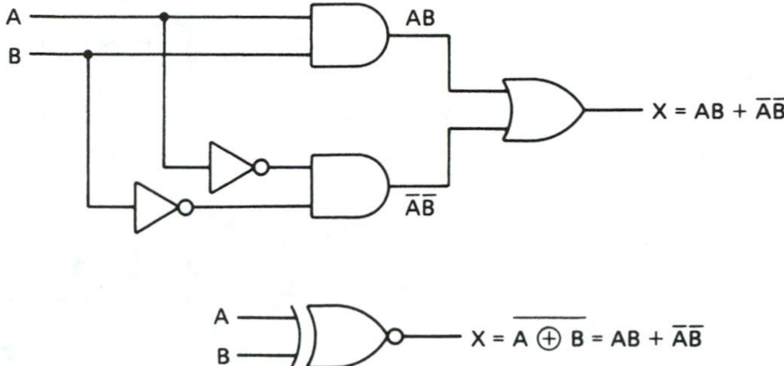

Figure 2–40 Exclusive-NOR logic circuit and logic symbol.

Summary

The exclusive-OR and exclusive-NOR gates are two-input logic gates that provide a very important commonly used function that we will see in upcoming chapters. Basically, the operation of the gates is as follows:

1. The exclusive-OR gate provides a HIGH output for one or the other inputs HIGH, but not both ($X = \overline{A}B + A\overline{B}$).
2. The exclusive-NOR gate provides a HIGH output for both inputs HIGH or both inputs LOW ($X = AB + \overline{A}\overline{B}$).

Also, the Ex-OR and Ex-NOR gates are available in both TTL and CMOS integrated-circuit packages. For example, the 7486 is a TTL quad Ex-OR and the 4077 is a CMOS quad Ex-NOR.

2–11 LOGIC GATE WAVEFORM GENERATION

Using the basic gates, a clock oscillator, and a repetitive waveform generator circuit, we can create specialized waveforms to be used in digital control and sequencing circuits. A popular general-purpose repetitive waveform generator is the Johnson shift counter. We will use the output waveforms from it to create our own, specialized waveforms.

The Johnson shift counter we will use outputs eight separate repetitive waveforms: A, B, C, D and their complements, $\overline{A}$, $\overline{B}$, $\overline{C}$, $\overline{D}$. The input to the Johnson shift counter is a clock oscillator (C_p). Figure 2–41 shows a Johnson shift counter with its input and output waveforms.

The clock oscillator produces the C_p waveform, which is input to the Johnson shift counter. The shift counter uses C_p and internal circuitry to generate the eight repetitive output waveforms shown.

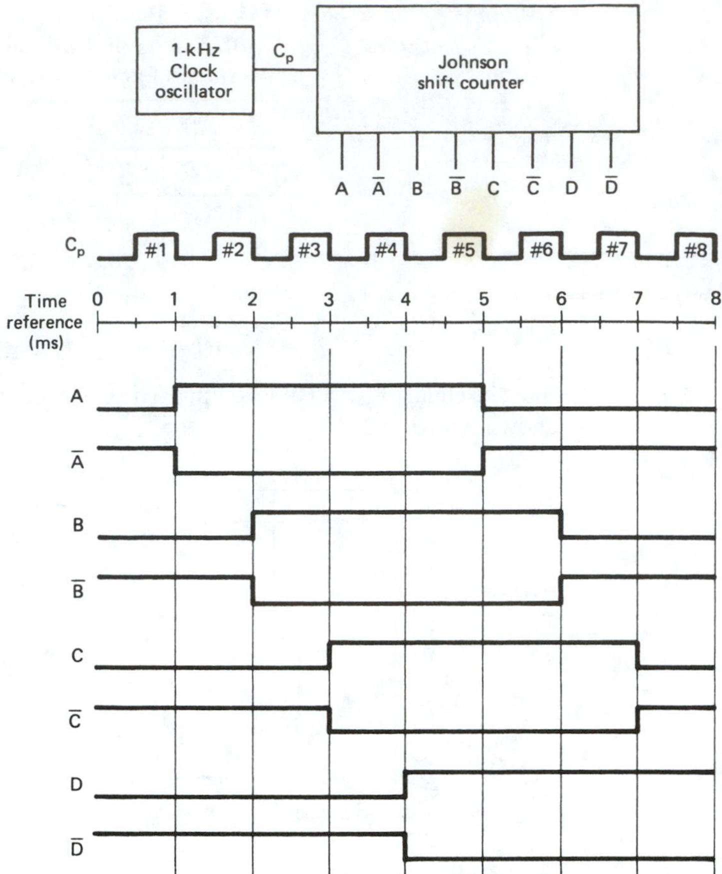

Figure 2–41 Johnson shift counter waveform generator.

Now, if one of those waveforms is exactly what you want, you are all set. But let's say we need a waveform that is HIGH for 3 ms, from 2 until 5 on the millisecond time reference scale. Looking at Figure 2–41, we can see that that waveform is not available.

Using some logic gates, however, will enable us to get any waveform we desire. In this case, if we feed the *A* and *B* waveforms into an AND gate, we will get our HIGH level from 2 to 5, as shown in Figure 2–42.

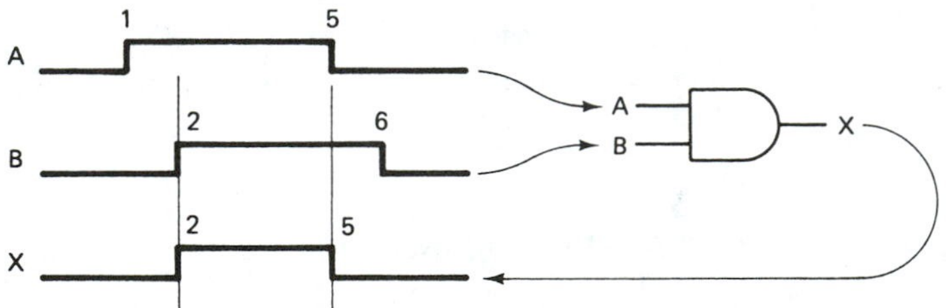

Figure 2–42 Generating a 3-ms HIGH pulse using an AND gate and a Johnson shift counter.

Working through the following examples will help you understand logic gate operation and waveform generation.

EXAMPLE 2–11

Which Johnson counter outputs will you connect to an AND gate to get a 1-ms HIGH-level output from 4 to 5 ms?

Solution:

Referring to Figure 2–41, we see that the two waveforms that are *both* HIGH from 4 to 5 ms are A and D; therefore, the circuit of Figure 2–43 will give us the required output.

Figure 2–43 Solution to Example 2–11.

EXAMPLE 2–12

Which Johnson counter outputs must be connected to a three-input AND gate to enable just the C_p 4 pulse to be output?

Solution:

Referring to Figure 2–41, we see that the C and $\overline{D}$ waveforms are both HIGH only during the C_p 4 *period.* To get just the C_p 4 *pulse,* you must provide C_p as the third input. Now when you look at all three input waveforms, you will see that they are all HIGH only during the C_p 4 *pulse* (see Figure 2–44).

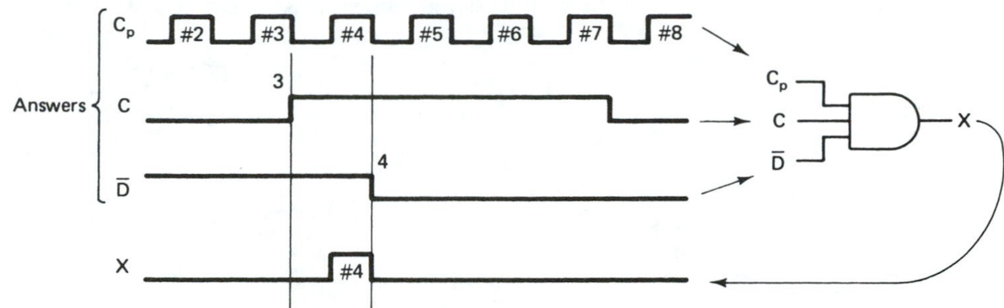

Figure 2–44 Solution to Example 2–12.

EXAMPLE 2–13

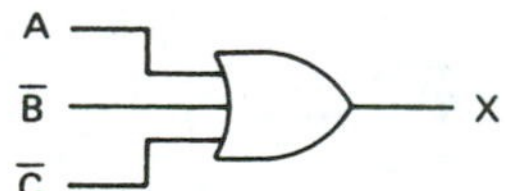

Figure 2–45

Sketch the output waveform that will result from inputting A, $\overline{B}$, and $\overline{C}$ into the three-input OR gate shown in Figure 2–45.

Solution:

The output of an OR gate is always HIGH unless *all* inputs are LOW. Therefore, the output is always HIGH except between 5 and 6, as shown in Figure 2–46.

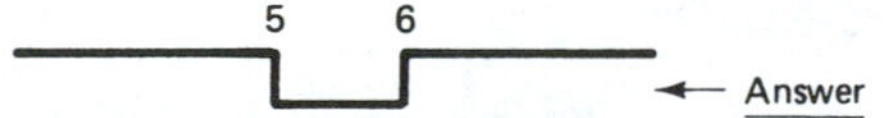

Figure 2–46 Solution to Example 2–13.

EXAMPLE 2–14

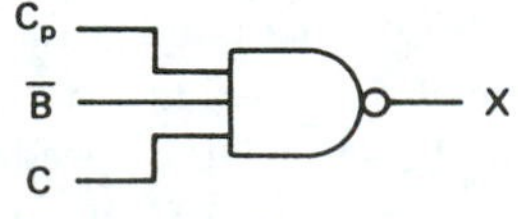

Figure 2–47

Sketch the output waveform that will result from inputting C_p, $\overline{B}$, and C into the NAND gate shown in Figure 2–47.

Solution:

From reviewing the truth table of a NAND gate we determine that the output is always HIGH unless *all* inputs are HIGH. Therefore, the output will always be HIGH except during pulse 7, as shown in Figure 2–48 on the next page.

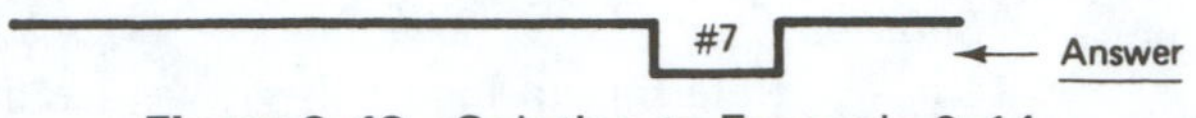

Figure 2–48 Solution to Example 2–14.

EXAMPLE 2–15

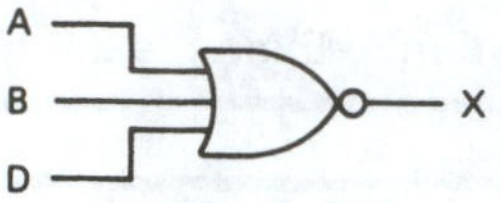

Figure 2–49

Sketch the output waveforms that will result from inputting *A*, *B*, and *D* into the NOR gate shown in Figure 2–49.

Solution:

Reviewing the truth table for a NOR gate, we determine that the output is always LOW except when *all* inputs are LOW. Therefore, the output will always be LOW except from 0 to 1, as shown in Figure 2–50.

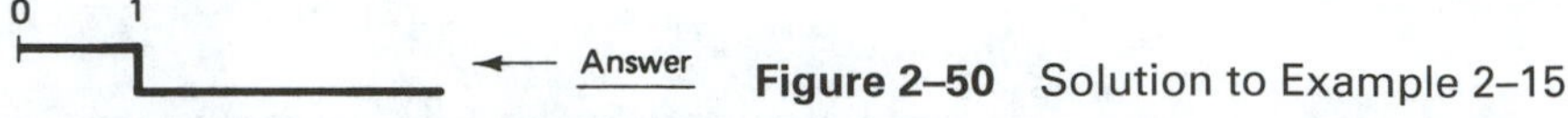

Figure 2–50 Solution to Example 2–15.

EXAMPLE 2–16

By using combinations of gates, we can obtain more specialized waveforms. Sketch the output waveforms for the circuit shown in Figure 2–51.

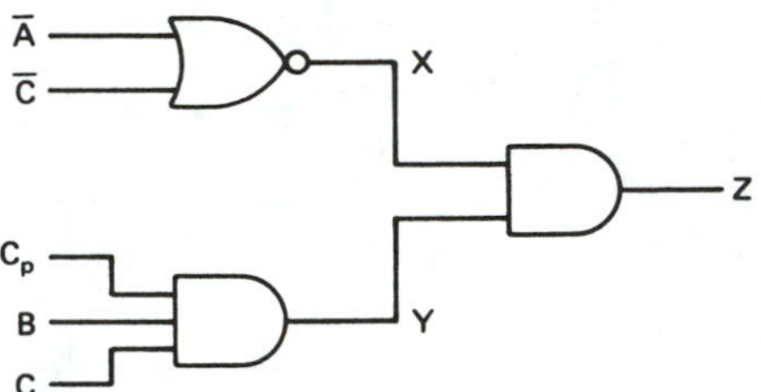

Figure 2–51

Solution:

The output waveforms are shown in Figure 2–52.

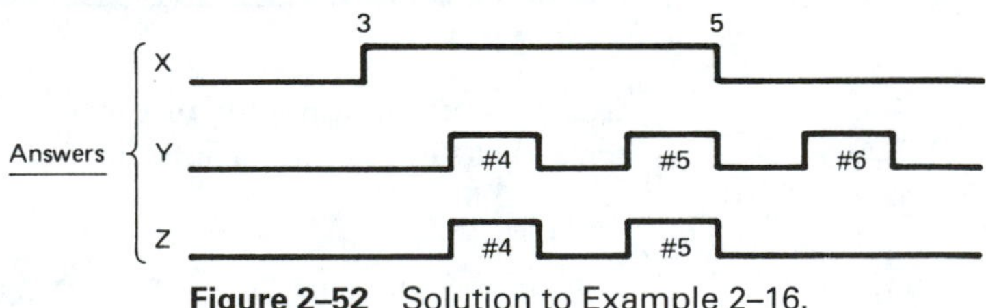

Figure 2–52 Solution to Example 2–16.

2–12 SUMMARY OF LOGIC GATE OPERATION

All the logic gates are available in various configurations in the TTL and CMOS families. To list just a few: the 7404 TTL and the 4049 CMOS are hex (six) inverter ICs, the 7400 TTL and the 4011 CMOS are quad (four) two-input NAND ICs, and the 7402 TTL and the 4001 CMOS are quad two-input NOR ICs. Other popular NANDs and NORs are available in three-, four-, and eight-input configurations. Consult a TTL or CMOS data manual for availability and pin configuration of those ICs. The pin configurations for the hex inverter, the quad NOR, and the quad NAND are given in Figures 2–53 and 2–54. (High-speed CMOS 74HC04, 74HC00, and 74HC02 have the same pin configuration as the TTL ICs.) The pin configurations for popular AND and OR gates were given earlier in this chapter.

By now you should have a thorough understanding of the basic logic gates: inverter, AND, OR, NAND, NOR, Ex-OR, and Ex-NOR. In Chapter 3 we will be combining

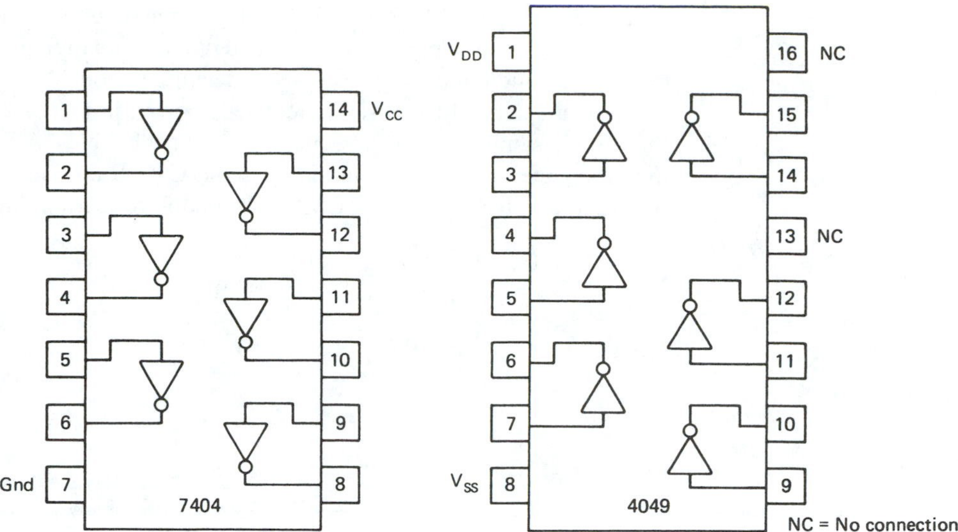

Figure 2–53 7404 TTL and 4049 CMOS inverter pin configurations.

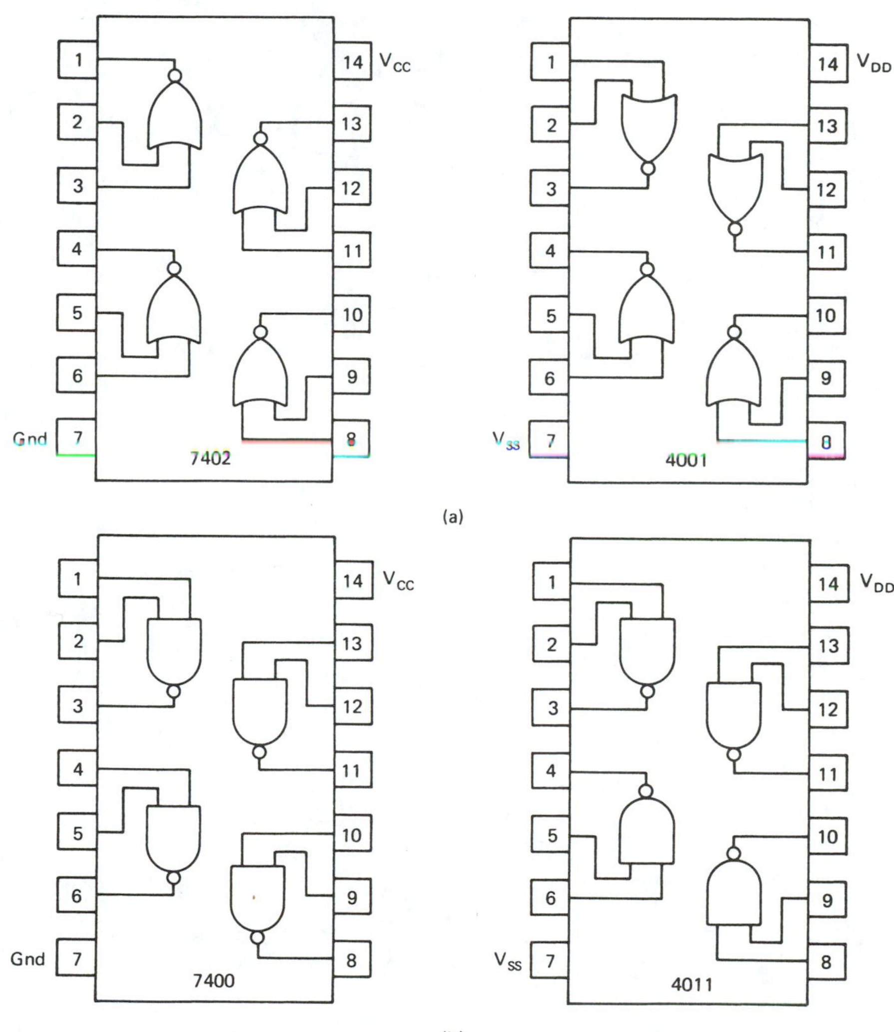

Figure 2–54 (a) 7402 TTL NOR and 4001 CMOS NOR pin configurations; (b) 7400 TTL NAND and 4011 CMOS NAND pin configurations.

several gates together to form complex logic functions. Since the basic logic gates are the building blocks for larger-scale integrated circuits and digital systems, it is very important that the operation of these gates is second nature to you.

A summary of the basic logic gates is given in Figure 2–55. You should memorize those logic symbols, Boolean equations, and truth tables. Also, a table of the most common integrated-circuit gates in the TTL and CMOS families is given in Table 2–11. You will need to refer to a TTL or CMOS data book for the pin layout and specifications.

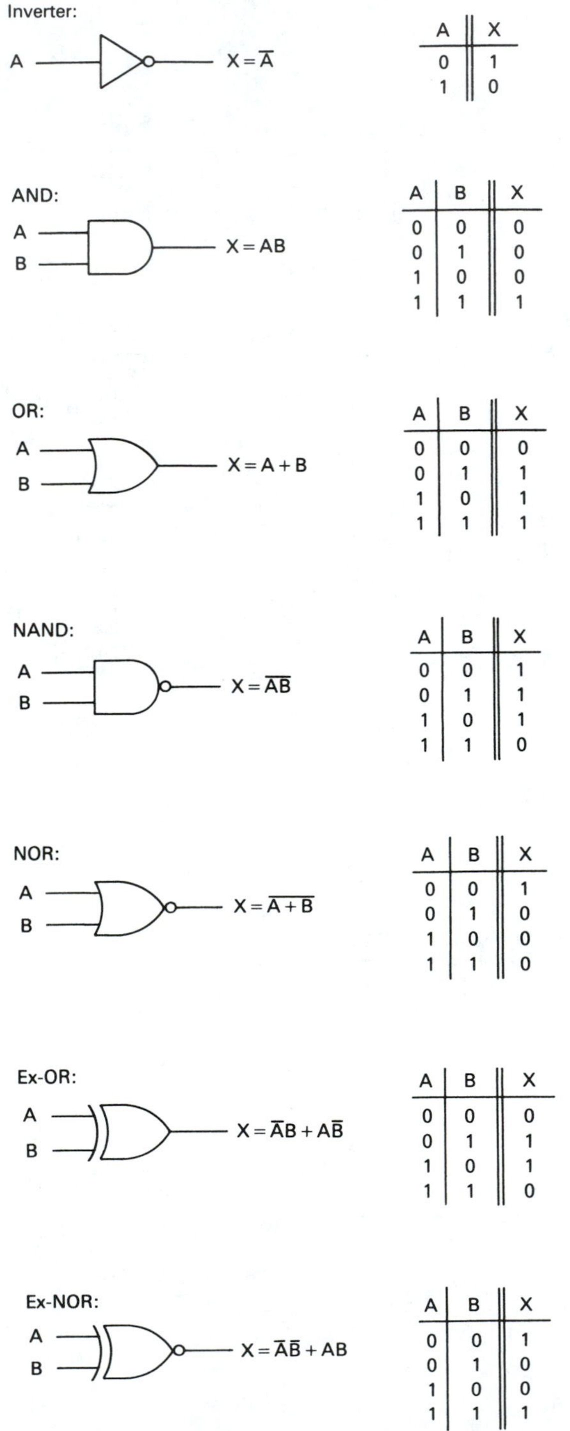

A	X
0	1
1	0

A	B	X
0	0	0
0	1	0
1	0	0
1	1	1

A	B	X
0	0	0
0	1	1
1	0	1
1	1	1

A	B	X
0	0	1
0	1	1
1	0	1
1	1	0

A	B	X
0	0	1
0	1	0
1	0	0
1	1	0

A	B	X
0	0	0
0	1	1
1	0	1
1	1	0

A	B	X
0	0	1
0	1	0
1	0	0
1	1	1

Figure 2–55 Summary of logic gates, Boolean equations, and truth tables.

TABLE 2–11
Common IC Gates in the TTL and CMOS Families

			Part number		
Gate name	*Number of inputs per gate*	*Number of gates per chip*	*Standard TTL*	*Standard CMOS*	*High-speed CMOS*
Inverter	1	6	7404	4069	74HC04
AND	2	4	7408	4081	74HC08
	3	3	7411	4073	74HC11
	4	2	7421	4082	—
OR	2	4	7432	4071	74HC32
	3	3	—	4075	74HC4075
	4	2	—	4072	—
NAND	2	4	7400	4011	74HC00
	3	3	7410	4013	74HC10
	4	2	7420	4012	74HC20
	8	1	7430	4068	—
	12	1	74134	—	—
	13	1	74133	—	—
NOR	2	4	7402	4001	74HC02
	3	3	7427	4025	74HC27
	4	2	7425	4002	74HC4002
	5	2	74260	—	—
	8	1	—	4078	—
Ex-OR	2	4	7486	4030	74HC86
Ex-NOR	2	4	74266	4077	74HC266

2–13 THE TTL FAMILY SPECIFICATIONS

Fortunately, the different manufacturers of digital logic ICs have standardized a numbering scheme so that basic part numbers will be the same regardless of the manufacturer. The prefix of the part number, however, will differ because it is the manufacturer's abbreviation. For example, a typical TTL part number might be S74F08N. The 7408 is the basic number used by all manufacturers for a quad *AND* gate. The F stands for the *FAST* TTL subfamily and the S prefix is the manufacturer's code for *Signetics.* National Semiconductor uses the prefix DM, and Texas Instruments uses the prefix SN. The N suffix at the end of the part number is used to specify the package type. N is used for the plastic dual-in-line package (DIP), W is used for the ceramic flatpack, and D is used for the surface-mounted SO plastic package. The best source of information on the available package styles and their dimensions is given in the manufacturers' data manuals. Most data manuals will list the 7408 as 5408/7408. The 54XX series is the military version, which has less stringent power supply requirements and an extended temperature range of −55° to +125°C, whereas the 74XX is the commercial version with a temperature range of 0 to +70°C and strict power supply requirements.

For the purposes of this text, reference is usually made to the 74XX commercial version and the manufacturer's prefix code and package style suffix code are ignored. The XX is used in this book to fill the space normally occupied by the actual part number. For example, the part number for an inverter in the 74XX series is 7404.

The standard 74XX TTL IC family has evolved through several stages since the late 1960s. Along the way, improvements have been made to reduce the internal time delays and power consumption. At the same time, each manufacturer has been introducing chips with new functions and applications.

The fundamental operation of a TTL chip can be explained by studying the internal circuitry of the basic two-input 7400 NAND gate shown in Figure 2–56. The diodes D_1 and D_2 are negative clamping diodes used to protect the inputs from any short-term negative input voltages. The input transistor, Q_1, acts like an AND gate and is usually fabricated

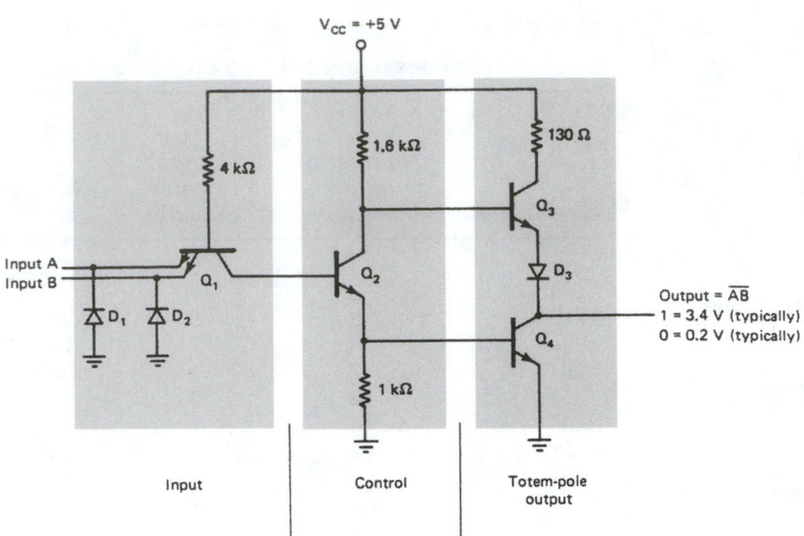

Figure 2–56 Internal circuitry of a 7400 two-input NAND gate.

with a *multiemitter* transistor that characterizes TTL technology. (To produce two-, three-, four-, and eight-input NAND gates, the manufacturer uses two-, three-, four-, and eight-emitter transistors.) Q_2 provides control and current boosting to the totem pole output stage.

The reasoning for the totem pole setup is to improve the output characteristics. Basically, when the output is HIGH (1), Q_4 is OFF (open) and Q_3 is ON (short). When the output is LOW (0), Q_4 is ON and Q_3 is OFF. Since one, or the other, transistor is always OFF, the current flow from V_{CC} to ground in that section of the circuit is minimized.

We like to think of TTL circuits as operating at 0-V and 5-V levels, but that is not true. As we draw more and more current out of the HIGH-level output, the output voltage drops lower and lower until finally it will not be recognized as a HIGH level anymore by the other TTL gates that it is feeding.

Input/Output Current and Fan-Out

The *fan-out* of a subfamily is defined as the number of gates of the same subfamily that can be connected to a single output without exceeding the current ratings of the gate. (A typical fan-out for most TTL subfamilies is 10.) Figure 2–57 shows an example of fan-out with 10 gates driven from a single gate.

To determine fan-out, you must know how much input current a gate load draws (I_I) and how much output current the driving gate can supply (I_O). In Figure 2–57 the single 7400 is the driving gate, supplying current to 10 other gate loads. The output current capability for the HIGH condition is abbreviated I_{OH} and is called a *source* current. I_{OH} for the 7400 is $-400\ \mu$A maximum. (The minus sign signifies current *leaving* the gate.)

The input current requirement for the HIGH condition is abbreviated I_{IH} and for the 74XX subfamily is equal to 40 μA maximum. To find the fan-out, divide the source current ($-400\ \mu$A) by the input requirements for a gate (40 μA). The fan-out is 400 μA/40 μA = 10.

For the LOW condition the maximum output current for the 74XX subfamily is 16 mA, and the input requirements for each 74XX gate is -1.6 mA maximum, also for a fan-out of 10. The fan-out is usually the same for both the HIGH and LOW conditions for the 74XX gates; if not, we use the lower of the two.

Since a LOW output level is close to 0 V, the current will actually flow into the output terminal and sink down to ground. This is called a *sink* current and is illustrated in Figure 2–58. In the figure two gates are connected to the output of gate 1. The total current that gate 1 must sink in this case is 2×1.6 mA = 3.2 mA. Since the maximum current a gate can sink in the LOW condition (I_{OL}) is 16 mA, gate 1 is well within its maximum rating of I_{OL}. (Gate 1 could sink the current from as many as *10* gate inputs.)

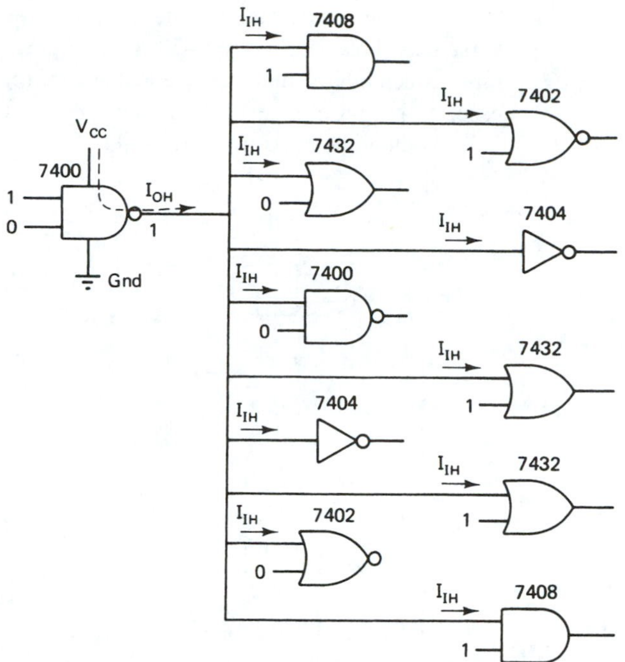

Figure 2–57 Ten gates driven from a single source.

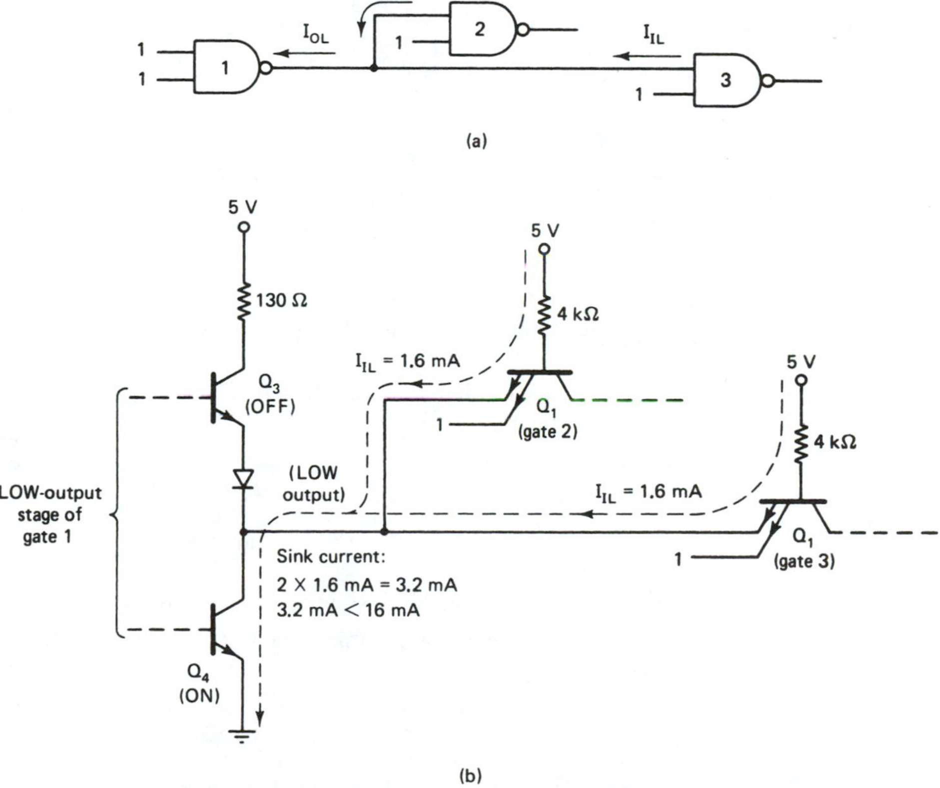

Figure 2–58 Totem pole LOW output of a TTL gate sinking the input currents from two gate inputs: (a) logic gate symbols; (b) logic gate internal circuitry.

For the HIGH-output condition, the circuitry is the same but the current flow is reversed, as shown in Figure 2–59. In the figure you can see that the 40 μA going into each input is actually a small reverse leakage current flowing against the emitter arrow. In this case, the output of gate 1 is sourcing $-80\ \mu$A to the inputs of gates 2 and 3. This $-80\ \mu$A is well below the maximum allowed HIGH-output current rating of $-400\ \mu$A.

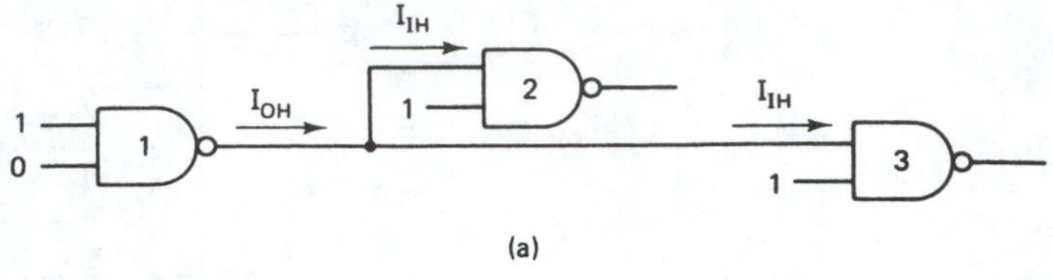

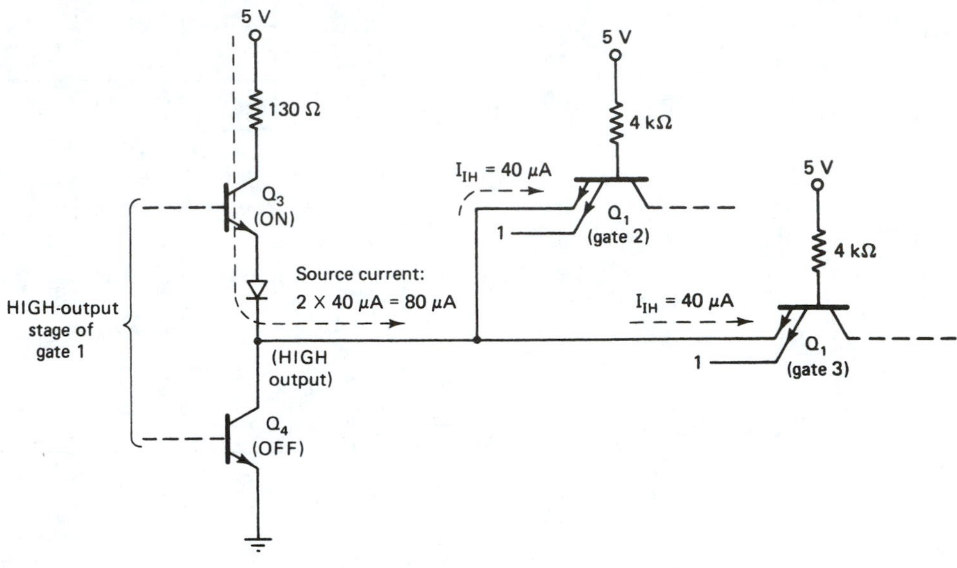

Figure 2–59 Totem pole HIGH output of a TTL gate sourcing current to two gate inputs.

To summarize input/output current and fan-out:

1. The maximum current that an input to a *standard* (i.e., 74XX) TTL gate will sink or source is

$$I_{IL} = -1.6 \text{ mA}$$
$$I_{IH} = 40\ \mu\text{A}$$

   (The minus sign signifies current *leaving* the gate.)

2. The maximum current that the output of a *standard* TTL gate can sink or source is

$$I_{OL} = 16 \text{ mA}$$
$$I_{OH} = -400\ \mu\text{A or } -800\ \mu\text{A}$$

   (The *actual* output current depends on the number and type of loads connected.)

3. The maximum number of gate inputs that can be connected to a *standard* TTL gate output is

$$\text{Fan-out} = 10$$

Input/Output Voltages and Noise Margin

We must also concern ourselves with the specifications for the acceptable input and output voltage levels. For the *LOW output condition,* the lower transistor (Q_4) in the totem pole output stage is saturated (ON) and the upper one (Q_3) is cut off (OFF). V_{out} for the LOW condition (V_{OL}) is the voltage across the saturated Q_4, which has a typical value of 0.2 V and a maximum value of 0.4 V, as specified in the manufacturer's data manual.

For the *HIGH output condition,* the upper transistor (Q_3) is saturated and the lower transistor (Q_4) is cut off. The voltage that reaches the output (V_{OH}) is V_{CC} minus the drop across the 130-Ω resistor, minus the collector-emitter drop, minus the diode drop. Manufacturers' data sheets specify that the HIGH-level output will typically be 3.4 V, and they will guarantee that the worst-case minimum value will be 2.4 V. This means that the next gate input must interpret any voltage from 2.4 V up to 5.0 V as a HIGH level. Therefore, we must also consider the *input* voltage level specifications (V_{IH}, V_{IL}).

Manufacturers will guarantee that any voltage between a minimum of 2.0 V up to 5.0 V will be interpreted as a HIGH (V_{IH}). Also, any voltage from a maximum of 0.8 V down to 0 V will be interpreted as a LOW (V_{IL}).

These values leave us a little margin for error, what is called the *noise margin.* For example, V_{OL} is guaranteed not to exceed 0.4 V, and V_{IL} can be as high as 0.8 V to still be interpreted as a LOW. Therefore, we have 0.4 V (0.8 V − 0.4 V) of leeway (noise margin), as illustrated in Figure 2–60.

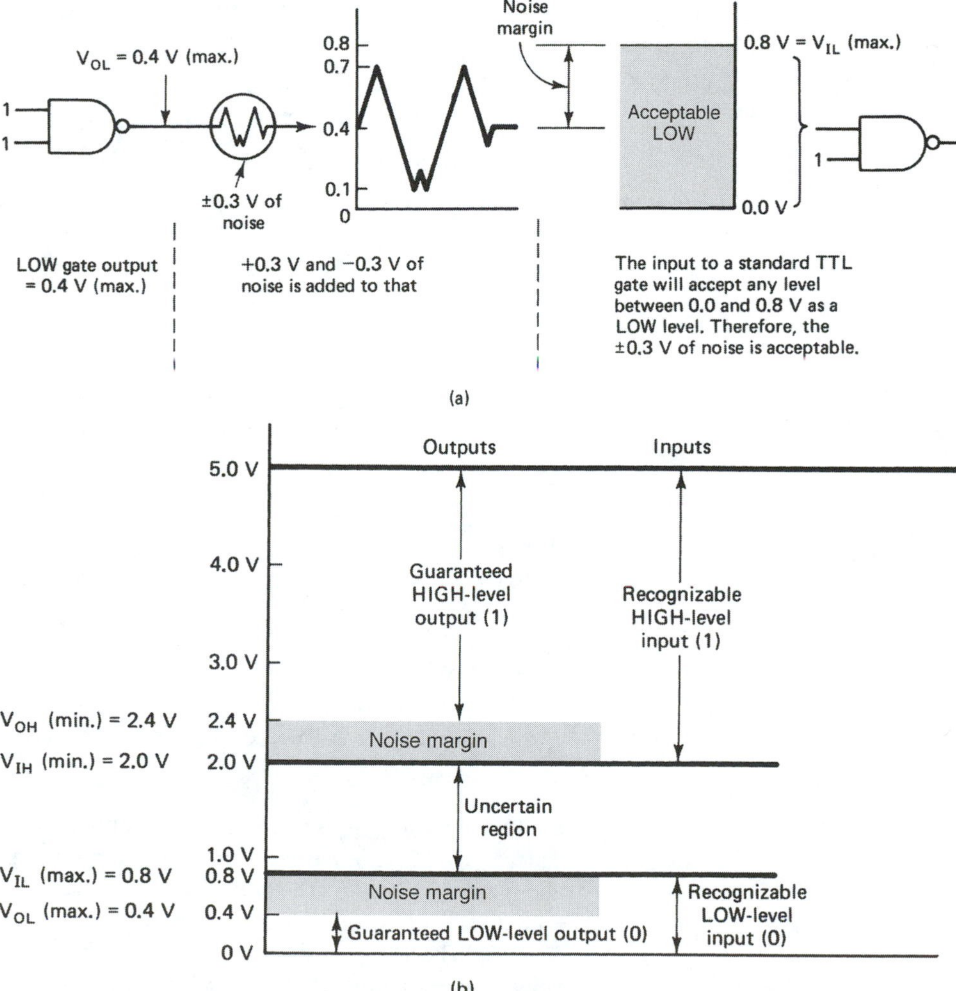

Figure 2–60 (a) Adding noise to a LOW-level output; (b) graphical illustration of the input/output voltage levels for the standard 74XX TTL series.

Table 2–12 is a summary of input/output voltage levels and noise margin for the standard family of TTL ICs.

TABLE 2–12
Standard 74XX Series Voltage Levels

Parameter	*Minimum*	*Typical*	*Maximum*	
V_{OL}		0.2 V	0.4 V	Noise margin = 0.4 V
V_{IL}			0.8 V	
V_{OH}	2.4 V	3.4 V		Noise margin = 0.4 V
V_{IH}	2.0 V			

The following examples illustrate the use of the current and voltage ratings for establishing acceptable operating conditions for TTL logic gates.

EXAMPLE 2–17

Find the voltages and currents that are asked for in Figure 2–61 if the gates are all standard (74XX) TTL.

(a) Find V_a and I_a for Figure 2–61a.
(b) Find V_a, V_b, and I_b for Figure 2–61b.
(c) Find V_a, V_b, and I_b for Figure 2–61c.

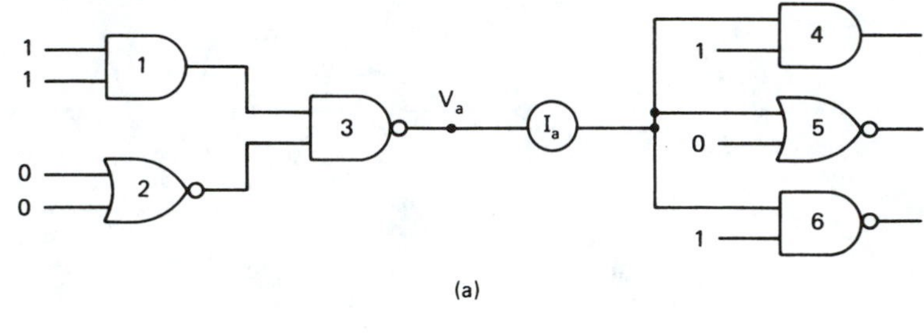

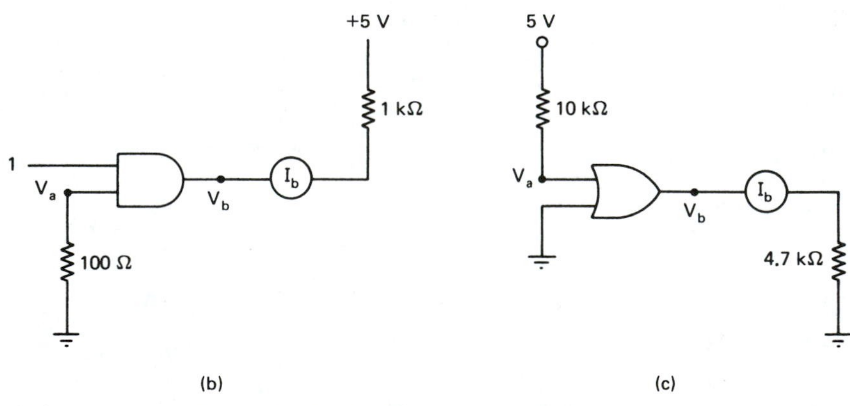

Figure 2–61 Voltage and current ratings.

Solution:

(a) The input to gate 3 is a 1–1, so the output will be LOW. Using the *typical* value, $V_a = 0.2$ V. Since gate 3 is LOW, it will be sinking current from the three other gates: 4, 5, and 6. The typical value for each I_{IL} is -1.6 mA; therefore, $I_a = -4.8$ mA (-1.6 mA $-$ 1.6 mA $-$ 1.6 mA).

(b) The 100-Ω resistor to ground will place a LOW level at that input. I_{IL} typically is -1.6 mA, which flows down through the 100-Ω resistor, making $V_a =$ 0.16 V (1.6 mA × 100 Ω). The 0.16 V at V_a will be recognized as a LOW level ($V_{IL} = 0.8$ V maximum), so the AND gate will output a LOW level; $V_b = 0.2$ V Ω (typical). The AND gate will sink current from the 1-kΩ resistor; I_b = 4.8 mA

[(5 V − 0.2 V)/1 kΩ]. The 4.8 mA is well below the maximum allowed current of 16 mA (I_{OL}), so the AND gate will not burn out.

(c) I_{IH} into the OR gate is 40 μA; therefore, the voltage at V_a = 4.6 V [5 V − (10 kΩ × 40 μA)]. The output level of the OR gate will be HIGH (V_{OH}), making V_b = 3.4 V and I_b = 3.4 V/4.7 kΩ = 723 μA. Since 723 μA is below the maximum rating of the OR gate (I_{OH} = −800 μA maximum), *the OR gate will not burn out.*

Pulse-Time Parameters: (Rise Time, Fall Time, and Propagation Delay)

We have been using ideal pulses for the input and output waveforms up until now. Actually, however, the pulse is not perfectly square; it takes time for the digital level to rise from 0 up to 1 and to fall from 1 down to 0.

As shown in Figure 2–62(a), the *rise time* (t_r) is the length of time it takes for a pulse to rise from its 10% point up to its 90% point. For a 5-V pulse, the 10% point is 0.5 V (10% × 5 V) and the 90% point is 4.5 V (90% × 5 V). The *fall time* (t_f) is the length of time it takes to fall from its 90% point to its 10% point.

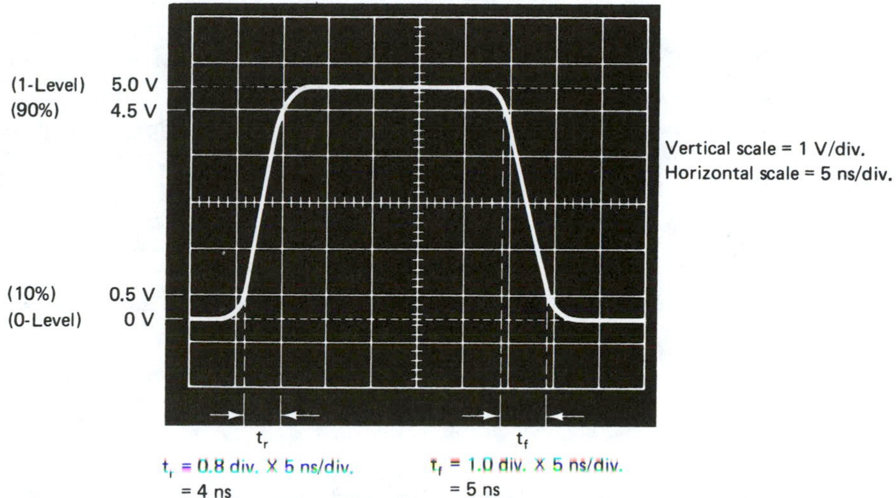

Figure 2–62 (a) Oscilloscope display of pulse rise and fall times.

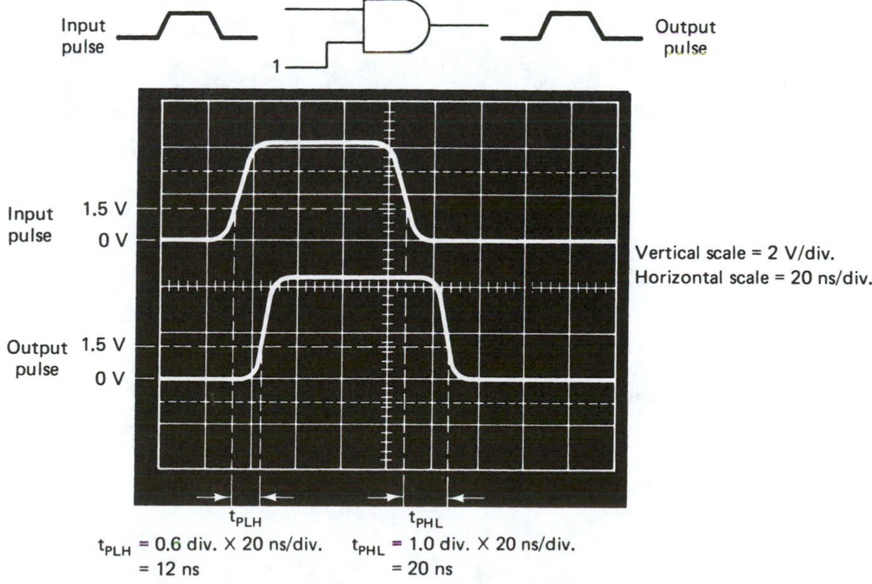

Figure 2–62 (b) Oscilloscope display of propagation delay times.

Signetics

7400, LS00, S00
Gates

Quad Two-Input NAND Gate
Product Specification

Logic Products

Typical switching speed

Typical power supply requirements

TYPE	TYPICAL PROPAGATION DELAY	TYPICAL SUPPLY CURRENT (TOTAL)
7400	9ns	8mA
74LS00	9.5ns	1.6mA
74S00	3ns	15mA

ORDERING CODE ← Gives part numbers for various package styles.

PACKAGES	COMMERCIAL RANGE $V_{CC} = 5V \pm 5\%$; $T_A = 0°C$ to $+70°C$
Plastic DIP	N7400N, N74LS00N, N74S00N
Plastic SO	N74LS00D, N74S00D

NOTE:
For information regarding devices processed to Military Specifications, see the Signetics Military Products Data Manual.

Shows every Input/Output combination → **FUNCTION TABLE**

INPUTS		OUTPUT
A	B	Y
L	L	H
L	H	H
H	L	H
H	H	L

H = HIGH voltage level
L = LOW voltage level

INPUT AND OUTPUT LOADING AND FAN-OUT TABLE

PINS	DESCRIPTION	74	74S	74LS
A, B	Inputs	1ul	1Sul	1LSul
Y	Output	10ul	10Sul	10LSul

NOTE:
Where a 74 unit load (ul) is understood to be 40μA I_{IH} and −1.6mA I_{IL}, a 74S unit load (Sul) is 50μA I_{IH} and −2.0mA I_{IL}, and 74LS unit load (LSul) is 20μA I_{IH} and −0.4mA I_{IL}.

Means that the output can drive 10 unit load inputs of the same family (fan-out = 10)

Dependency Notation symbol

Traditional Logic symbol

Gives IC wiring information → **PIN CONFIGURATION**

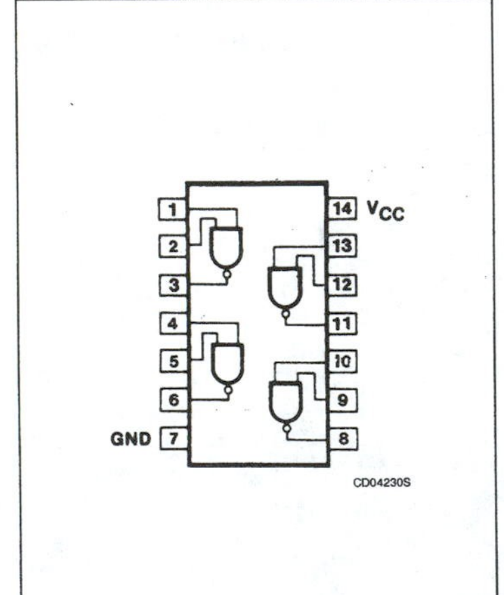

LOGIC SYMBOL

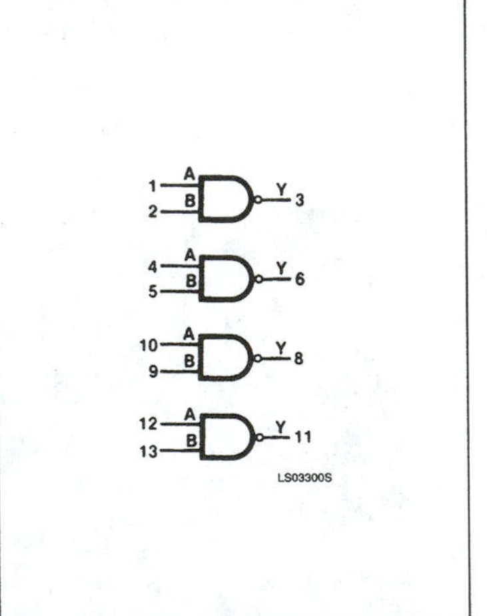

LOGIC SYMBOL (IEEE/IEC)

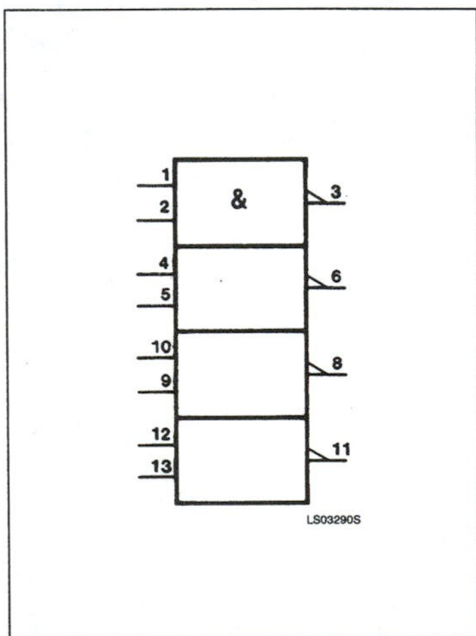

Figure 2–63 The 7400 data sheet.

Range not to be exceeded

ABSOLUTE MAXIMUM RATINGS (Over operating free-air temperature range unless otherwise noted.)

PARAMETER		74	74LS	74S	UNIT
V_{CC}	Supply voltage	7.0	7.0	7.0	V
V_{IN}	Input voltage	−0.5 to +5.5	−0.5 to +7.0	−0.5 to +5.5	V
I_{IN}	Input current	−30 to +5	−30 to +1	−30 to +5	mA
V_{OUT}	Voltage applied to output in HIGH output state	−0.5 to $+V_{CC}$	−0.5 to $+V_{CC}$	−0.5 to $+V_{CC}$	V
T_A	Operating free-air temperature range	0 to 70			°C

mal range
e used

Specs for each
7400 series

RECOMMENDED OPERATING CONDITIONS

PARAMETER		74			74LS			74S			UNIT
		Min	Nom	Max	Min	Nom	Max	Min	Nom	Max	
V_{CC}	Supply voltage	4.75	5.0	5.25	4.75	5.0	5.25	4.75	5.0	5.25	V
V_{IH}	HIGH-level input voltage	2.0			2.0			2.0			V
V_{IL}	LOW-level input voltage			+0.8			+0.8			+0.8	V
I_{IK}	Input clamp current			−12			−18			−18	mA
I_{OH}	HIGH-level output current			−400			−400			−1000	μA
I_{OL}	LOW-level output current			16			8			20	mA
T_A	Operating free-air temperature	0		70	0		70	0		70	°C

Stay within this range but use 5.0 V nominal

ut voltage
ecs

ut current
s

Shows the
esults of
n input
ulse
pplied to a
400 Device
Jnder Test
DUT)

TEST CIRCUITS AND WAVEFORMS

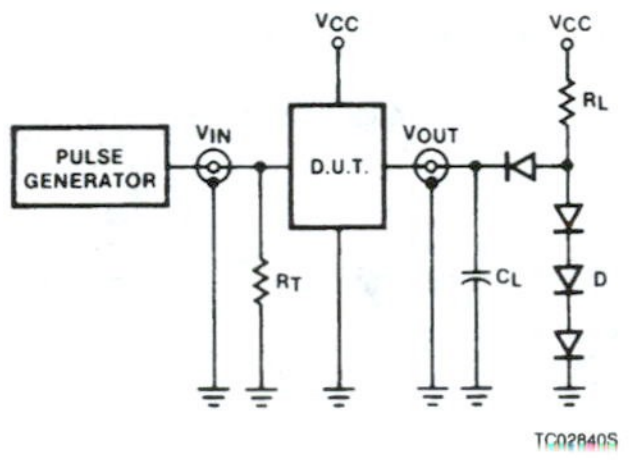

Test Circuit For 74 Totem-Pole Outputs

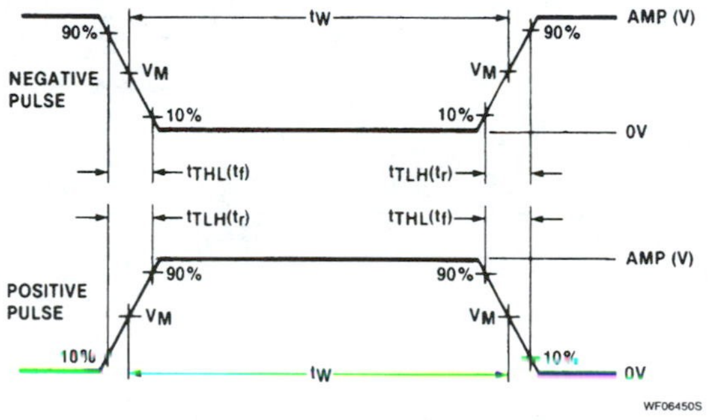

V_M = 1.3V for 74LS; V_M = 1.5V for all other TTL families.

Input Pulse Definition

DEFINITIONS
R_L = Load resistor to V_{CC}; see AC CHARACTERISTICS for value.
C_L = Load capacitance includes jig and probe capacitance; see AC CHARACTERISTICS for value.
R_T = Termination resistance should be equal to Z_{OUT} of Pulse Generators.
D = Diodes are 1N916, 1N3064, or equivalent.
t_{TLH}, t_{THL} Values should be less than or equal to the table entries.

FAMILY	INPUT PULSE REQUIREMENTS				
	Amplitude	Rep. Rate	Pulse Width	t_{TLH}	t_{THL}
74	3.0V	1MHz	500ns	7ns	7ns
74LS	3.0V	1MHz	500ns	15ns	6ns
74S	3.0V	1MHz	500ns	2.5ns	2.5ns

Figure 2–63 *(Continued)*

Gates 7400, LS00, S00

The Min and Max are guaranteed limits but you can expect the typ (typical)

DC ELECTRICAL CHARACTERISTICS (Over recommended operating free-air temperature range unless otherwise noted.)

PARAMETER		TEST CONDITIONS[1]		7400 Min	7400 Typ[2]	7400 Max	74LS00 Min	74LS00 Typ[2]	74LS00 Max	74S00 Min	74S00 Typ[2]	74S00 Max	UNIT
V_{OH}	HIGH-level output voltage	V_{CC} = MIN, V_{IH} = MIN, V_{IL} = MAX, I_{OH} = MAX		2.4	3.4		2.7	3.4		2.7	3.4		V
V_{OL}	LOW-level output voltage	V_{CC} = MIN, V_{IH} = MIN	I_{OL} = MAX		0.2	0.4		0.35	0.5			0.5	V
			I_{OL} = 4mA (74LS)					0.25	0.4				V
V_{IK}	Input clamp voltage	V_{CC} = MIN, I_I = I_{IK}				−1.5			−1.5			−1.2	V
I_I	Input current at maximum input voltage	V_{CC} = MAX	V_I = 5.5V			1.0						1.0	mA
			V_I = 7.0V						0.1				mA
I_{IH}	HIGH-level input current	V_{CC} = MAX	V_I = 2.4V			40							μA
			V_I = 2.7V						20			50	μA
I_{IL}	LOW-level input current	V_{CC} = MAX	V_I = 0.4V			−1.6			−0.4				mA
			V_I = 0.5V									−2.0	mA
I_{OS}	Short-circuit output current[3]	V_{CC} = MAX		−18		−55	−20		−100	−40		−100	mA
I_{CC}	Supply current (total)	V_{CC} = MAX	I_{CCH} Outputs HIGH		4	8		0.8	1.6		10	16	mA
			I_{CCL} Outputs LOW		12	22		2.4	4.4		20	36	mA

Output voltage specs (V_OH, V_OL); Input current specs (I_IH, I_IL)

NOTES:
1. For conditions shown as MIN or MAX, use the appropriate value specified under recommended operating conditions for the applicable type.
2. All typical values are at V_{CC} = 5V, T_A = 25°C.
3. I_{OS} is tested with V_{OUT} = +0.5V and V_{CC} = V_{CC} MAX + 0.5V. Not more than one output should be shorted at a time and duration of the short circuit should not exceed one second.

Waveform shows definitions for propagation time specs

AC WAVEFORM

V_{IN} V_M V_M t_{PHL} t_{PLH} V_{OUT} V_M V_M

WF07570S

V_M = 1.3V for 74LS; V_M = 1.5V for all other TTL families.

Waveform 1. Waveform For Inverting Outputs

AC ELECTRICAL CHARACTERISTICS T_A = 25°C, V_{CC} = 5.0V

PARAMETER		TEST CONDITIONS	74 (C_L = 15pF, R_L = 400Ω) Min	74 Max	74LS (C_L = 15pF, R_L = 2kΩ) Min	74LS Max	74S (C_L = 15pF, R_L = 280Ω) Min	74S Max	UNIT
t_{PLH} t_{PHL}	Propagation delay	Waveform 1		22 15		15 15		4.5 5.0	ns

Propagation delay specs

Figure 2–63 *(Continued)*

Not only are input and output waveforms sloped on their rising and falling edges, but there is also a delay time for an input wave to propagate through an IC to the out-put, called the *propagation delay* (t_{PLH} and t_{PHL}). The propagation delay is due to limitations in transistor switching speeds caused by undesirable internal capacitive stored charges.

Figure 2–62(b) shows that it takes a certain length of time for an input pulse to reach the output of an IC gate. A specific measurement point (1.5 V for the standard TTL series) is used as a reference. The propagation delay time for the *output* to respond in the LOW-to-HIGH direction is labeled t_{PLH} and in the HIGH-to-LOW direction is labeled t_{PHL}. Figure 2–63 on pp. 50-52 shows a data sheet giving the complete specifications for a 7400 IC.

EXAMPLE 2–18

The propagation delay times for the 7402 NOR gate shown in Figure 2–64 are listed in a TTL data manual as t_{PLH} = 22 ns and t_{PHL} = 15 ns. Sketch and label the input and output pulses to a 7402.

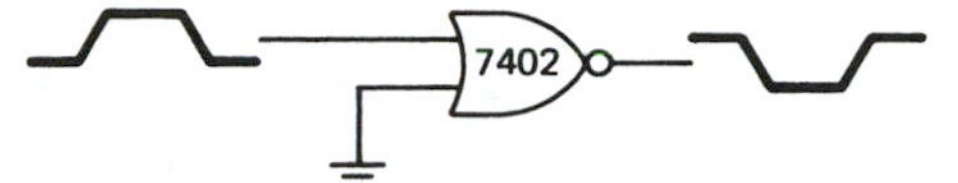

Figure 2–64

Solution:

The input and output pulses are shown in Figure 2–65.

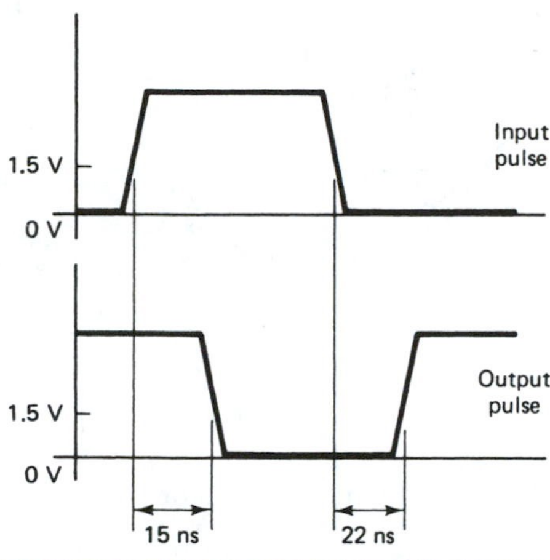

Figure 2–65 Solution to Example 2–18.

Open-Collector Outputs

Instead of using a totem pole arrangement in the output stage of a TTL gate, another arrangement, called the *open-collector* output, is available. Remember that with the totem pole output stage, for a LOW output the lower transistor is ON and the upper transistor is OFF, and vice versa for a HIGH output; whereas with the open-collector output, the upper transistor is *removed,* as shown in Figure 2–66. Now the output will be *LOW* when Q_4 is ON and the output will *float* (not HIGH or LOW) when Q_4 is OFF. This means that an open-collector (OC) output can sink current *but it cannot source current.*

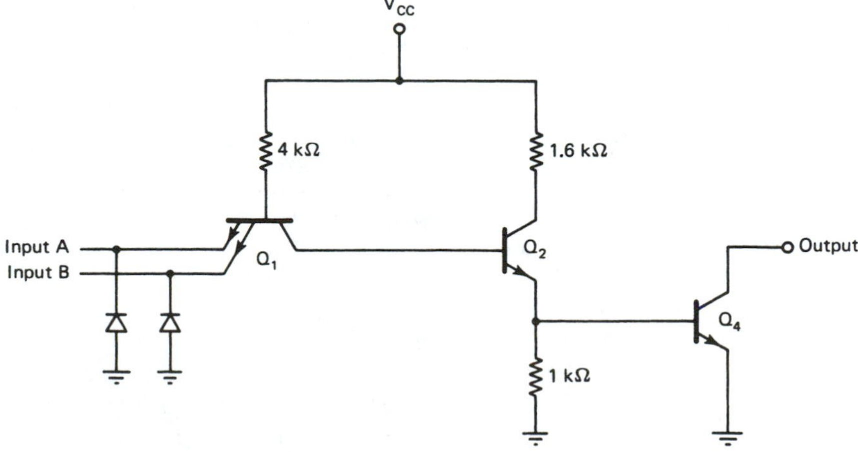

Figure 2–66 TTL NAND with an open-collector output.

To get an OC output to produce a HIGH, an external resistor (called a *pull-up* resistor) must be used, as shown in Figure 2–67. Now when Q_4 is OFF (open), the output is approximately 5 V (HIGH) and when Q_4 is ON (short) the output is approximately 0 V (LOW). The optimum size for a pull-up resistor depends on the size of the gate load. Usually, a good size for a pull-up resistor is 10 kΩ; 10 kΩ is not too small to allow excessive current flow when Q_4 is ON, and it is not too large to cause an excessive voltage drop across itself when Q_4 is OFF.

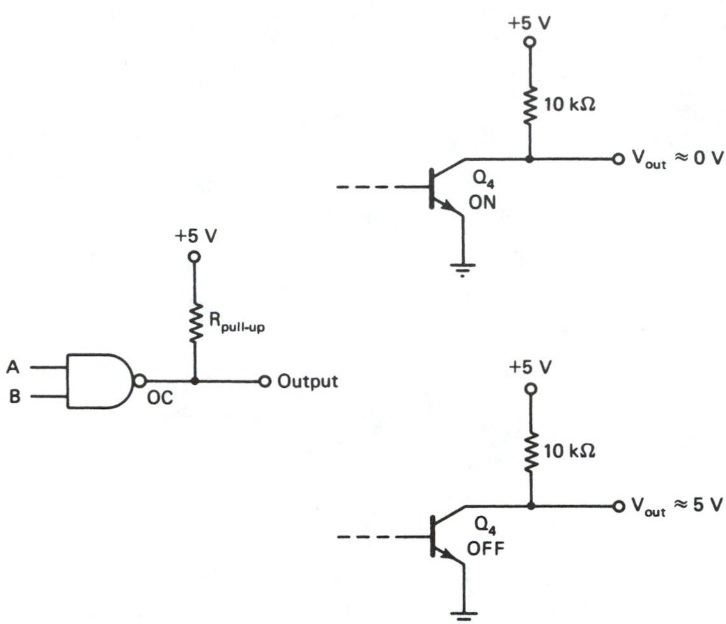

Figure 2–67 Using a pull-up resistor with an open-collector output.

Improved TTL Series

Integrated-circuit design engineers have constantly been working on the improvement of the standard TTL series. A simple improvement that was made early on was simply reducing all the internal resistor values of the standard TTL series. That increased the power consumption (or dissipation), which was bad, but it reduced the internal $R \times C$ time constants that cause propagation delays. The result was the *74HXX* series, which has almost half the propagation delay time but almost double the power consumption of the standard TTL series. The product of delay time × power (the speed–power product), which is a figure of merit for IC families, remained approximately the same, however.

Another series, the *74LXX*, was developed using just the opposite approach. The internal resistors were increased, thus reducing the power consumption, but the propagation delay increased, keeping the speed–power product about the same. The 74HXX and 74LXX series have, for the most part, been replaced now by the Schottky TTL and CMOS series of ICs.

Schottky TTL

The major speed limitation of the standard TTL series is due to the capacitive charge in the base region of the transistors. The transistors basically operate at either cutoff or saturation. When the transistor is saturated, charges build up at the base region, and when it comes time to switch to cutoff, the stored charges must be dissipated, which takes time, causing propagation delay.

Schottky logic overcomes the saturation and stored charges problem by placing a Schottky diode across the base-to-collector junction, as shown in Figure 2–68. With the Schottky diode in place, any excess charge on the base is passed on to the collector, and the transistor is held just below saturation. The Schottky diode has a special metal junction that

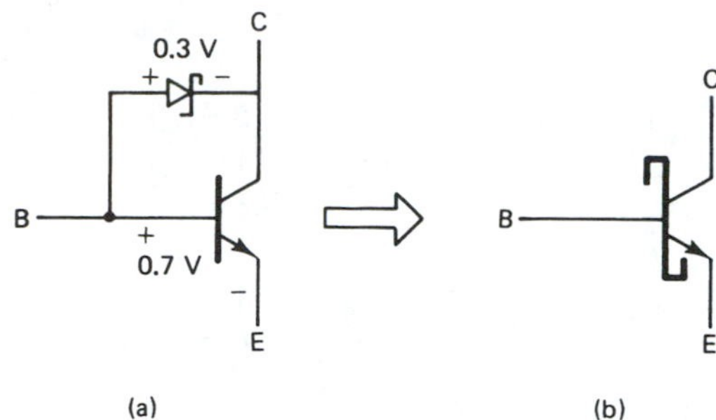

Figure 2–68 Schottky-clamped transistor: (a) Schottky diode reduces stored charges; (b) symbol.

minimizes its own capacitive charge and increases its switching speed. Using Schottky-clamped transistors and decreased resistor values, the propagation delay is reduced by a factor of 4 and the power consumption is only doubled. Therefore, the speed–power product of the 74SXX TTL series is improved to about half that of the 74XX TTL series (the lower, the better).

Low-Power Schottky (LS). By using different integration techniques and increasing the values of the internal resistors, the power dissipation of the Schottky TTL is reduced significantly. The speed–power product of the 74LSXX TTL series is about one-third that of the 74SXX series and about one-fifth that of the 74XX series.

Advanced Low-Power Schottky (ALS). Further improvement on the 74LSXX series reduced the propagation delay time from 9 to 4 ns and the power dissipation from 2 to 1 mW per gate. The 74ALSXX and 74LS series are rapidly replacing the standard 74XX and 74SXX series because of the speed and power improvements. As with any new technology, they are slightly more expensive and do not yet provide all the functions available from the standard 74XX series.

Fast (F)

It was long clear to TTL IC design engineers that new processing technology was needed to improve the speed of the LS series. A new process of integration, called *oxide isolation* (also used by the ALS series), has reduced the propagation delay in the 74FXX series to below 3 ns. In this process, transistors are isolated from each other, not by a reverse-biased junction, but by an actual channel of oxide. This dramatically reduces the size of the devices, which in turn reduces their associated capacitances and thus reduces propagation delay.

2–14 THE CMOS FAMILY

The CMOS family of integrated circuits uses an entirely different type of transistor as its basic building block. The TTL family uses bipolar transistors (*NPN* and *PNP*). CMOS (complementary metal-oxide semiconductor) uses complementary pairs of transistors (*N* type and *P* type) called MOSFETs (metal-oxide semiconductor field-effect transistors). MOSFETs are also used in other families of MOS ICs, including PMOS, NMOS, and VMOS, which are most commonly used for large-scale memories and microprocessors in the LSI and VLSI (large-scale and very-large-scale integration) category. One advantage that MOSFETs have over bipolar transistors is that the input to a MOSFET is electrically isolated from the rest of the MOSFET (see Figure 2–69b), giving it a high input impedance.

The *N*-channel MOSFET is similar to the *NPN* bipolar transistor in that it is two back-to-back *NP* junctions, and current will not flow down through it until a positive voltage is applied to the base (or gate in the case of the MOSFET). The silicon dioxide (SiO_2)

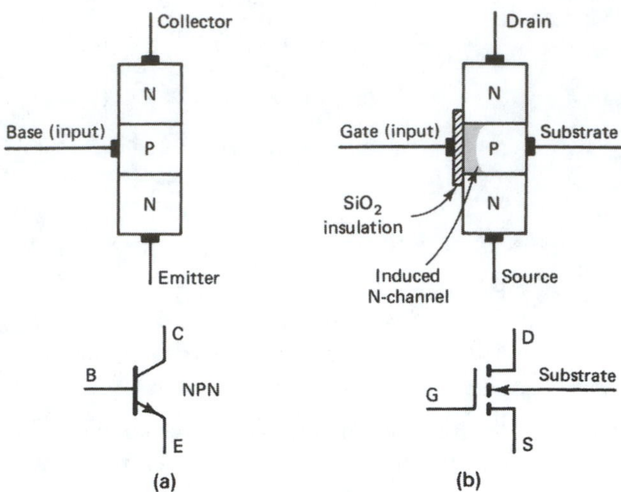

Figure 2–69 Simplified diagrams of bipolar and field-effect transistors: (a) *NPN* bipolar transistor used in TTL ICs; (b) *N*-channel MOSFET used in CMOS ICs.

layer between the gate material and the *P* substrate (base) of the MOSFET prevents any gate current from flowing, which provides a high input impedance and low power consumption.

The MOSFET shown in Figure 2–69b is a normally OFF device because there are no negative carries in the *P* material for current flow to occur. However, conventional current will flow down from drain to source if a positive voltage is applied to the gate with respect to the substrate. That voltage induces an electric field across the SiO_2 layer, which repels enough of the positive charges in the *P* material to form a channel of negative charges on the left side of the *P* material. This allows electrons to flow from source to drain (conventional current flows from drain to source). The channel that is formed is called an *N* channel because it contains negative carriers.

P-channel MOSFETS are just the opposite, constructed from *PNP* materials. The channel is formed by placing a *negative* voltage at the gate with respect to the substrate.

Using an *N*-channel MOSFET with its complement, the *P*-channel MOSFET, a simple complementary-MOS (CMOS) inverter can be formed as shown in Figure 2–70.

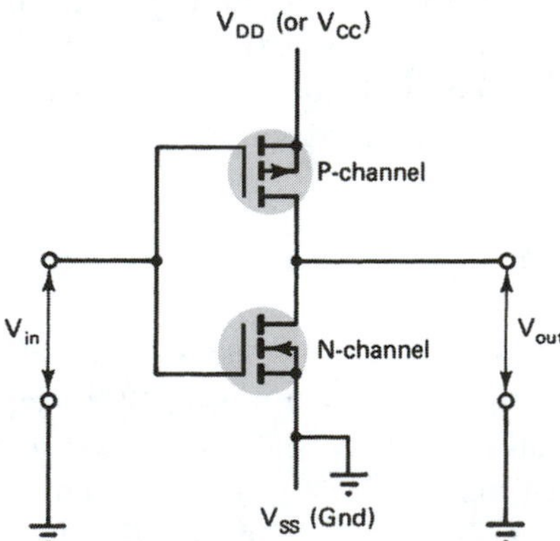

Figure 2–70 CMOS inverter formed from complementary *N*-channel/*P*-channel transistors.

We can think of MOSFETs as ON/OFF switches just as we did for the bipolar transistors. Table 2–13 summarizes the ON/OFF operation of *N*- and *P*-channel MOSFETS.

We can use Table 2–13 to prove that the circuit of Figure 2–70 operates as an inverter. With $V_{in} = 1$, the *N*-channel transistor is ON and the *P*-channel transistor is OFF, so $V_{out} = 0$. With $V_{in} = 0$, the *N*-channel transistor is OFF and the *P* channel is ON, so $V_{out} = 1$. Therefore, $V_{out} = \overline{V_{in}}$. Note that this complementary action is very similar to the TTL totem pole output stage, but much simpler to understand.

TABLE 2–13
Basic MOSFET Switching Characteristics

Gate level[a]	*N channel*	*P channel*
1	ON	OFF
0	OFF	ON

[a]1 = V_{DD} (or V_{CC}); 0 = V_{SS} (Gnd).

Handling MOS Devices

The silicon dioxide layer that isolates the gate from the substrate is so thin that it is very susceptible to burn-through from electrostatic charges. You must be very careful and use the following guidelines when handling MOS devices:

1. Store the integrated circuits in a conductive foam or leave in their original container.
2. Work on a conductive surface (e.g., metal tabletop) that is properly grounded.
3. Ground all test equipment and soldering irons.
4. Connect your wrist to ground with a length of wire and a 1-MΩ series resistor.
5. Do not connect signals to the inputs while the device power supply is off.
6. Connect all unused inputs to V_{DD} or Gnd.
7. Don't wear electrostatic-prone clothing such as wool, silk, or synthetic fibers.
8. Don't remove or insert an IC with the power on.

CMOS Availability

The CMOS family of integrated circuits provides almost all the same functions that are available in the TTL family, plus CMOS has available several "special-purpose" functions not provided by TTL. Like TTL, the CMOS family has evolved into several different subfamilies, or series, each having better performance specifications than the previous one.

4000 Series. The 4000 series (or the improved 4000B) is the original CMOS line. It became popular because it offered very low power consumption and could be used in battery-powered devices. It is much slower than any of the TTL series and has a low level of electrostatic discharge protection. The power supply voltage to the IC can range anywhere from +3 to +15 V with the minimum one-level input equal to $\frac{2}{3}V_{CC}$ and the maximum 0-level input equal to $\frac{1}{3}V_{CC}$.

40H00 Series. This series was designed to be faster than the 4000 series. It did overcome some of the speed limitations but is still much slower than LSTTL.

74C00 Series. This series was developed to be pin compatible with the TTL family, making interchangeability easier. It uses the same numbering scheme as TTL except that it begins with 74C. It has a low-power advantage over the TTL family but is still much slower.

74HC00 and 74HCT00 Series. The 74HC00 (high-speed CMOS) and 74HCT00 (high-speed CMOS, TTL compatible) offer a vast improvement over the original 74C00 series. The HC/HCT series are as speedy as the LSTTL series and still consume less power, depending on the operating frequency. They are pin compatible (the HCT is also input/output voltage level compatible) with the TTL family, yet offer greater noise immunity and greater voltage and temperature operating ranges. Further improvements to the HC/HCT series have led to the Advanced CMOS Logic (ACL) and Fairchild advanced CMOS Technology (FACT) series, which have even better operating characteristics.

74-BiCMOS Series. Several IC manufacturers have developed technology that combines the best features of bipolar transistors and CMOS transistors, forming *BiCMOS* logic. The high-speed characteristics of bipolar *P–N* junctions are integrated with the low-power characteristics of CMOS to form an extremely low-power, high-speed family of digital logic. Each manufacturer uses different suffixes to identify their BiCMOS line. For example, Texas Instruments uses 74BCTXXX, Harris uses 74FCTXXX, and Signetics (Philips) uses 74ABTXXX.

The product line is especially well suited for and is mostly limited to microprocessor bus interface logic. This logic is mainly available in octal (8-bit) configurations used to interface 8-, 16-, and 32-bit microprocessors with high-speed peripheral devices such as memories and displays. An example is the 74ABT244 octal buffer from Signetics. Its logic is equivalent to the 74244 of other families, but it has several advanced characteristics. It has TTL-compatible input and output voltages, gate input currents less than 0.01 μA, and output sink and source current capability of 64 and -32 mA, respectively. It is extremely fast, having a typical propagation delay of 2.9 ns.

One of the most desirable features of these bus-interface ICs is the fact that, when their outputs are inactive ($\overline{OE} = 1$) or HIGH, the current draw from the power supply (I_{CCZ} or I_{CCH}) is only 0.5 μA. Since interface logic spends a great deal of its time in an inactive (idle) state, this can translate into a power dissipation as low as 2.5 μW! The actual power dissipation depends on how often the IC is inactive and on the HIGH/LOW duty cycle of its outputs when it is active.

74-Low Voltage Series. A new series of logic using a nominal supply voltage of 3.3 V has been developed to meet the extremely low power design requirements of battery-powered and hand-held devices. These ICs are being designed into the circuits of notebook computers, mobile radios, hand-held video games, telecom equipment, and high-performance workstation computers. Some of the more common low-voltage families are identified by the following suffixes:

LV: low-voltage HCMOS
LVC: low-voltage CMOS
LVT: low-voltage technology
ALVC: advanced low-voltage CMOS
HLL: high-speed low-power low-voltage

The power consumption of CMOS logic ICs decreases approximately with the square of power supply voltage. The propagation delay increases slightly at this reduced voltage but the speed is restored, and even increased, by using finer geometry and submicron CMOS technology that is tailored for low-power and low-voltage applications.

The supply voltage of LV logic can range from 1.2 to 3.6 V, which makes it well suited for battery-powered applications. When operated between 3.0 and 3.6 V, it can be interfaced directly with TTL levels. The switching speed of LV logic is extremely fast, ranging from about 9 ns for the LV series, down to 2.1 ns for the ALVC. Like BiCMOS logic, the power dissipation of LV logic is negligible in the idle state or at low frequencies. At higher frequencies the power dissipation is down to half as much as BiCMOS, depending on the power supply voltage used on the LV logic. Another key benefit of LV logic is its high output drive capability. The highest capability is provided by the LVT series, which can sink up to 64 mA and source up to 32 mA.

74AHC and 74AHCT Series. The advanced, high-speed CMOS is an enhanced version of the 74HC and 74HCT series. Designers who previously upgraded to the 74HC/HCT series can take the next step and migrate to this advanced version. It provides superior speed and low power consumption, and has a broad product selection. 74AHC has half the static power consumption, one-third the propagation delay, high-output drive current, and can operate at a V_{CC} of 3.3 or 5 V.

Two new forms of packaging have emerged with this series: *Single-gate logic* and *Widebus*™. Single-gate logic has a lower pin count and takes up less area on a printed circuit board by having only a single gate on the IC instead of the two-, four-, or six-gate versions. For example, the 74AHC1G00 is the single-gate version of the 74AHC00 quad NAND. It is a five-pin IC containing a single NAND gate.

The *Widebus*™ version* of logic is an extension of the octal ICs commonly found in microprocessor applications. They provide 16-bit I/O capability in a single IC package. For example, the 74AHC16244 is the *Widebus*™ version of the 74AHC244 octal buffer. It provides sixteen buffers in a 48-pin IC package.

Emitter-Coupled Logic

Another family designed for extremely high speed applications is emitter-coupled logic (ECL). ECL comes in two series, ECL 10K and ECL 100K. ECL is extremely fast, with propagation delay times as low as 0.8 ns. That speed makes it well suited for large mainframe computer systems that require a high number of operations per second, but are not as concerned about an increase in power dissipation.

The high speed of ECL is achieved by never letting the transistors saturate; in fact, the whole basis for HIGH and LOW levels is determined by which transistor in a differential amplifier is conducting more.

Developing New Digital Logic Technologies

The quest for logic devices that can operate at even higher frequencies and can be packed more densely in an IC package is a continuing process. Designers have high hopes for other new technologies, such as integrated injection logic (I^2L), silicon-on-sapphire (SOS), gallium arsenide (GaAs), and Josephen junction circuits. Eventually, propagation delays will be measured in picoseconds, and circuit densities will enable the supercomputer of today to become the desktop computer of tomorrow.

2–15 *INTERFACING LOGIC FAMILIES*

Throughout the years, system designers have been given a wide variety of digital logic to choose from. The main parameters to consider include speed, power dissipation, availability, types of functions, noise immunity, operating frequency, output-drive capability, and interfacing. First and foremost, however, are the basic speed and power concerns. Table 2–14 shows the propagation delay, power dissipation, and speed–power product for the most popular families.

TABLE 2–14

Typical Single-Gate Performance Specification

Family	*Propagation delay (ns)*	*Power dissipation (mW)*	*Speed-power product pWs (picowatt-seconds)*
74	10	10	100
74S	3	20	60
74LS	9	2	18
74ALS	4	1	4
74F	2.7	4	11
4000B (CMOS)	105	1 at 1 MHz	105
74HC (CMOS)	10	1.5 at 1 MHz	15
100K (ECL)	0.8	40	32

Courtesy of Philips Components

**Widebus* is a registered trademark of Texas Instruments, Inc.

Often, the need arises to interface (connect) between the various TTL and CMOS families. You have to make sure that a HIGH out of a TTL gate looks like a HIGH to the input of a CMOS gate, and vice versa. The same holds true for the LOW logic levels. You also have to make sure that the driving gate can sink or source enough current to meet the input current requirements of the gate being driven.

TTL to CMOS

Let's start by looking at the problems that might arise when interfacing a standard 7400 series TTL to 4000B series CMOS. Figure 2–71 shows the input and output voltage specifications for both, assuming that the 4000B is powered by a 5-V supply.

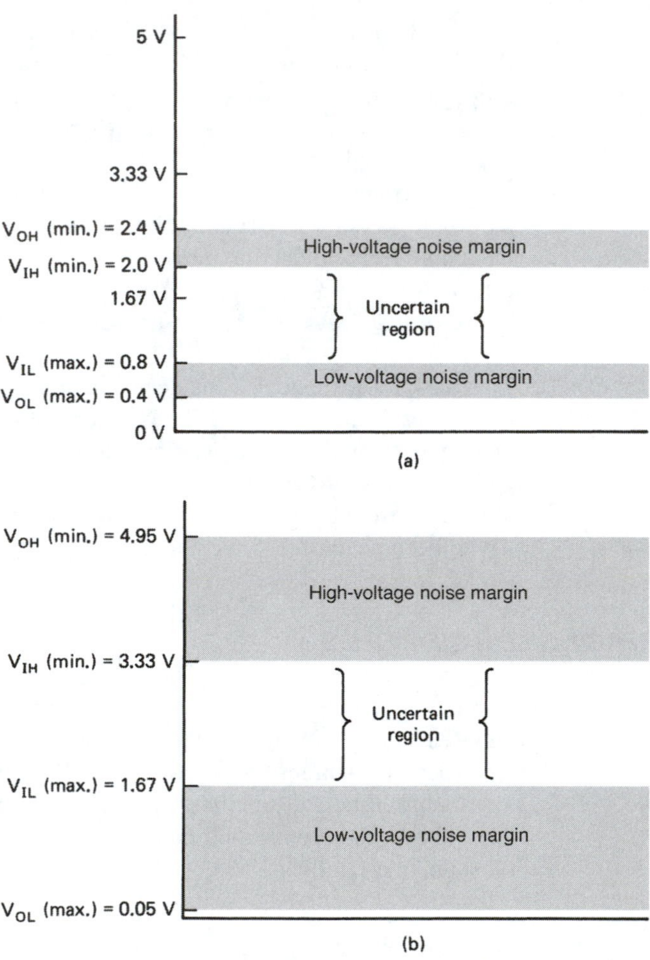

Figure 2–71 Input and output voltage specifications: (a) 7400 series TTL; (b) 4000B series CMOS (5-V supply).

When the TTL gate is used to drive the CMOS gate, there is no problem for the LOW-level output because the TTL guarantees a maximum LOW-level output of 0.4 V, and the CMOS will accept any voltage up to 1.67 V ($\frac{1}{3}V_{CC}$) as a LOW-level input.

But for the HIGH level, the TTL may output as little as 2.4 V as a HIGH. The CMOS expects at least 3.33 V as a HIGH-level input. Therefore, 2.4 V is unacceptable because it falls within the uncertain region. However, a resistor can be connected between the CMOS input to V_{CC} as shown in Figure 2–72 to solve the HIGH-level input problem.

In Figure 2–72, with V_{out1} *LOW*, the 7404 will sink current from the 10-kΩ resistor and the I_{IL} from the 4069B making V_{out2} HIGH. With V_{out1} *HIGH* the 10-kΩ resistor will "pull" the voltage at V_{in2} up to 5 V, causing V_{out2} to go LOW. The 10-kΩ resistor is called a *pull-up resistor* and is used to raise the output of the TTL gate closer to 5 V when it is in a HIGH output state. With V_{out1} HIGH the voltage at V_{in2} will be almost 5 V because

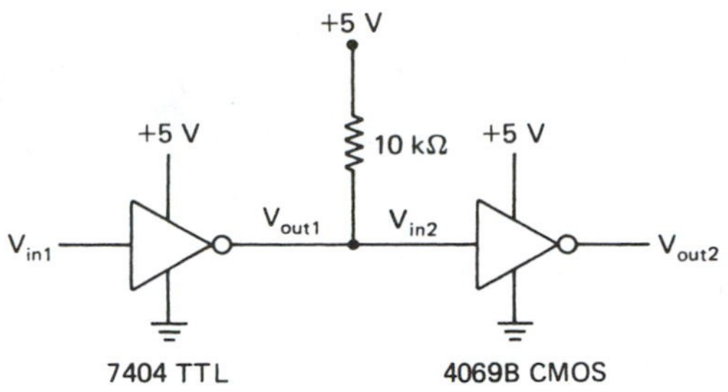

Figure 2–72 Using a pull-up resistor to interface TTL to CMOS.

current into the 4069B is so LOW ($\approx 1\ \mu$A) that the voltage drop across the 10 kΩ is insignificant, leaving almost 5 V at V_{in2} ($V_{\text{in2}} = 5\text{ V} - 1\ \mu\text{A} \times 10\text{ k}\Omega = 4.99\text{ V}$).

The other thing to look at when interfacing is the current levels of all gates that are involved. In this case, the 7404 can sink (I_{OL}) 16 mA, which is easy enough for the I_{IL} of the 4069B (1 μA) plus the current from the 10-kΩ resistor (5 V/10 kΩ = 0.5 mA). I_{OH} of the 7404 ($-400\ \mu$A) is no problem either because with the pull-up resistor the 7404 will not have to source current.

CMOS to TTL

When driving TTL from CMOS, the voltage levels are no problem because the CMOS will output about 4.95 V for a HIGH and 0.05 V for a LOW, which is easily interpreted by the TTL gate.

But the current levels can be a real concern because 4000B CMOS has severe output current limitations. (The 74C and 74HC series have much better output current capabilities, however.) Figure 2–73 shows the input/output currents that flow when interfacing CMOS to TTL.

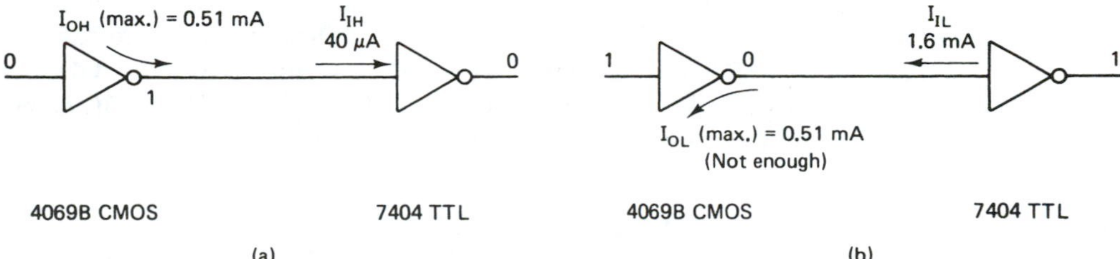

Figure 2–73 Current levels when interfacing CMOS to TTL: (a) CMOS I_{OH}; (b) CMOS I_{OL}.

For the HIGH output condition (Figure 2–73a), the 4069B CMOS can source a maximum current of 0.51 mA, which is enough to supply the HIGH-level input current (I_{IH}) to one 7404 inverter. But for the LOW output condition, the 4069B can also sink only 0.51 mA, which is not enough for the 7404 LOW-level input current (I_{IL}).

Most of the 4000B series has the same problem of low-output-drive current capability. To alleviate the problem, two special gates, the 4050 buffer and the 4049 inverting buffer, are specifically designed to provide high output current to solve many interfacing problems. They have drive capabilities of $I_{\text{OL}} = 4.0$ mA and $I_{\text{OH}} = -0.9$ mA, which is enough to drive two 74XXTTL loads, as shown in Figure 2–74.

If the CMOS buffer was used to drive another TTL series, let's say the 74LS series, we would have to refer to a TTL data book to determine how many loads could be connected without exceeding the output current limits. (The 4050B can actually drive *ten* 74LS loads.) Table 2–15 summarizes the input/output voltage and current specifications of some popular TTL and CMOS series, which enables us to determine interface parameters and family characteristics easily.

By reviewing Table 2–15 we can see that the 74HCMOS has relatively low input current requirements compared to the bipolar TTL series. Its HIGH output can source

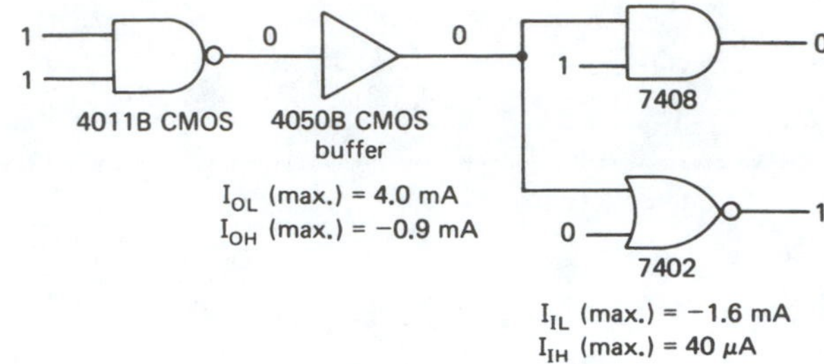

Figure 2–74 Using the 4050B CMOS buffer to supply sink and source current to two standard TTL loads.

TABLE 2–15
Worst-Case Values for Interfacing Considerations[a]

Parameter	*4000B CMOS*	*74HCMOS*	*74HCTMOS*	*74TTL*	*74LSTTL*	*74ALSTTL*
V_{IH} (min.) (V)	3.33	3.5	2.0	2.0	2.0	2.0
V_{IL} (max.) (V)	1.67	1.0	0.8	0.8	0.8	0.8
V_{OH} (min.) (V)	4.95	4.9	4.9	2.4	2.7	2.7
V_{OL} (max.) (V)	0.05	0.1	0.1	0.4	0.4	0.4
I_{IH} (max.) (μA)	1	1	1	40	20	20
I_{IL} (max.) (μA)	−1	−1	−1	−1600	−400	−100
I_{OH} (max.) (mA)	−0.51	−4	−4	−0.4	−0.4	−0.4
I_{OL} (max.) (mA)	0.51	4	4	16	8	4

[a]All values are for V_{supply} = 5.0 V.

4 mA, which is 10 times the capability of the TTL series. Also, the noise margin for the 74HCMOS is much wider than any of the TTL series (1.4 V HIGH, 0.9 V LOW).

Because of the low input current requirements, any of the TTL series can drive several of the 74HCMOS loads. An interfacing problem occurs in the voltage level, however. The 74HCMOS logic expects 3.5 V at a minimum for a HIGH-level input. The worst case (which we must always assume could happen) for the HIGH output level of a 74LSTTL is 2.7 V, so we will need to use a pull-up resistor at the 74LSTTL output to ensure an adequate HIGH level for the 74HCMOS input as shown in Figure 2–75.

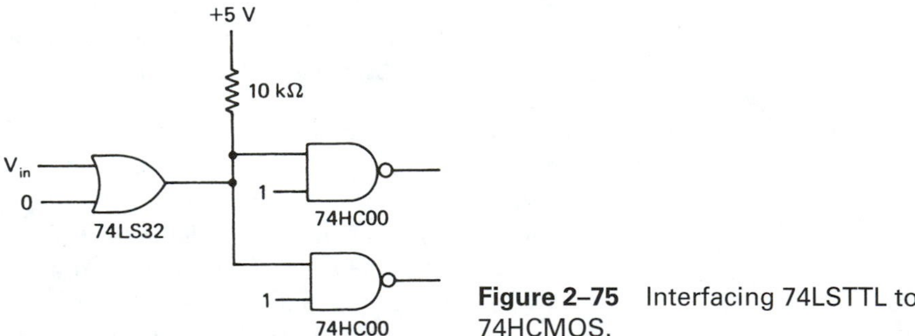

Figure 2–75 Interfacing 74LSTTL to 74HCMOS.

The combinations of interfacing situations are quite extensive (74HCMOS to 74ALSTTL, 74TTL to 74LSTTL, etc.). In each case, reference to a data book must be made to check the worst-case voltage and current parameters, as we will see in upcoming examples. In general, a pull-up resistor is required when interfacing TTL to CMOS to bring the HIGH-level TTL output up to a suitable level for the CMOS input. (The exception to the rule is when using *74HCTMOS,* which is designed for TTL voltage levels.) The disadvantage of using a pull-up resistor is that it takes up valuable room on a printed-circuit (PC) board and it dissipates power in the form of heat.

Different series within the TTL family and the TTL-compatible 74HCTMOS series can be interfaced directly. The main concern here is determining how many gate loads can be connected to a single output.

EXAMPLE 2–19

Determine from Table 2–15 how many 74LSTTL logic gates can be driven by a single 74TTL logic gate.

Solution:

The output voltage levels (V_{OL}, V_{OH}) of the 74TTL series are compatible with the input voltage levels (V_{IL}, V_{IH}) of the 74LSTTL series. The voltage noise margin for the LOW level is 0.4 V (2.4 − 2.0) and for the HIGH level is 0.4 V (0.8 − 0.4).

The HIGH-level output current (I_{OH}) for the 74TTL series is −400 μA. Each 74LSTTL gate draws 20 μA of input current for the HIGH level (I_{IH}), so one 74TTL gate can drive 20 74LSTTl loads in the HIGH state (400 μA/20 μA = 20).

For the LOW state, the 74TTL I_{OL} is 16 mA and the 74LSTTL I_{IL} is −400 μA, meaning that for the LOW condition, one 74TTL can drive 40 74LSTTL loads (16 mA/400 μA = 40). Therefore, considering both the LOW and HIGH conditions, a single 74TTL can drive twenty 74LSTTL gates.

EXAMPLE 2–20

One 74HCT04 inverter is to be used to drive one input to each of the following gates: 7400 (NAND), 7402 (NOR), 74LS08 (AND), 74ALS32 (OR). Draw the circuit and label input and output worst-case voltages and currents. Will there be total voltage and current compatibility?

Solution:

The circuit is shown in Figure 2–76. Figure 2–76a shows the worst-case HIGH-level values. If you sum all the input currents, the total that the 74HCT04 must supply is 120 μA (40 + 40 + 20 + 20 μA), which is well below the −4-mA maximum source capability of the 74HCT04. Also, the 4.9-V output voltage of the 74HCT04 *is* compatible with the 2.0-V minimum requirement of the TTL inputs, leaving a noise margin of 2.9 V (4.9 − 2.0 V).

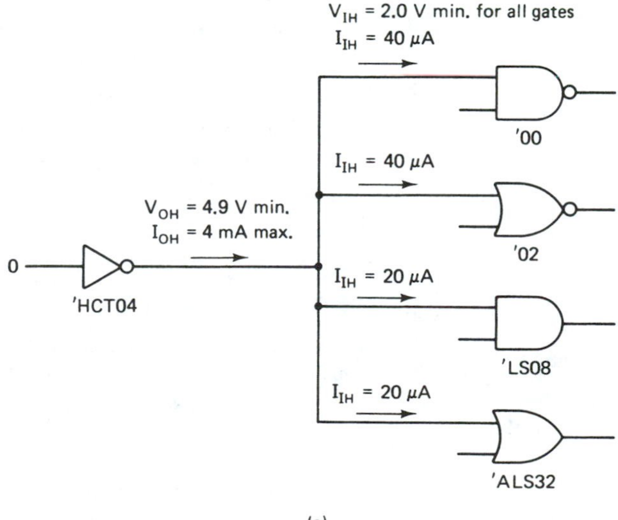

Figure 2–76 Interfacing 74HCTMOS to several different TTL series: (a) HIGH-level values.

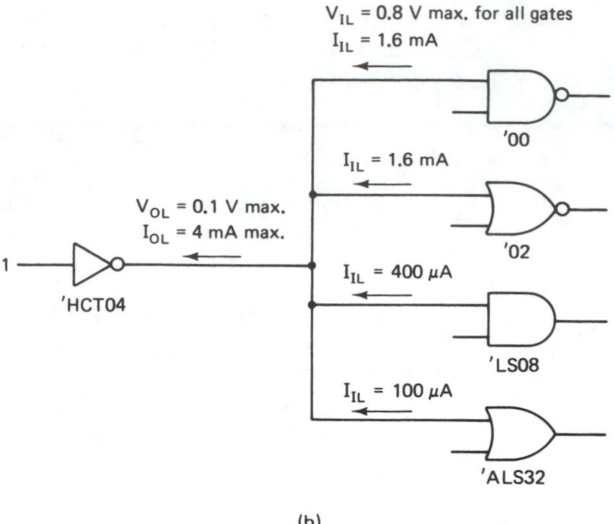

Figure 2–76 *(Continued)* (b) LOW-level values.

Figure 2–76b shows the worst-case LOW-level values. The sum of all the TTL input currents is 3.7 mA (1.6 mA + 1.6 mA + 400 μA + 100 μA), which is less than the 4-mA maximum sink capability of the 74HCT04. Also, the 0.1-V output of the 74HCT04 *is* compatible with the 0.8-V maximum requirement of the TTL inputs, leaving a noise margin of 0.7 V (0.8V − 0.1 V).

SUMMARY

In this chapter we have learned that

1. The AND gate requires that all inputs be HIGH in order to get a HIGH output.
2. The OR gate will output a HIGH if *any* of its inputs are HIGH.
3. An effective way to measure the precise timing relationships of digital waveforms is with an oscilloscope or a logic analyzer.
4. There are several integrated circuits available in both TTL and CMOS that provide the basic logic functions.
5. Two important troubleshooting tools are the logic pulser and the logic probe. The pulser is used to inject pulses into a circuit under test. The probe reads the level at a point in a circuit to determine if it is HIGH, LOW, or floating.
6. An inverter provides an output that is the complement of its input.
7. A NAND gate outputs a LOW when all of its inputs are HIGH.
8. A NOR gate outputs a HIGH when all of its inputs are LOW.
9. Manufacturers' data manuals are used by the technician to find the pin configuration and operating characteristics for the ICs used in modern circuitry.
10. The Exclusive-OR gate outputs a HIGH if one or the other of the inputs is HIGH, but *not both* HIGH.
11. The Exclusive-NOR gate outputs a HIGH if both inputs are HIGH or if both inputs are LOW.
12. There are basically three stages of internal circuitry in a TTL (transistor–transistor logic) IC: input, control, and output.

13. The input current (I_{IL} or I_{IH}) to an IC gate is a constant value specified by the IC manufacturer.
14. The output current of an IC gate depends on the size of the load connected to it. Its value cannot exceed the maximum rating of the chip, I_{OL} or I_{OH}.
15. The HIGH- and LOW-level output voltages of the standard TTL family are *not* 5 V and 0 V, but are typically 3.4 V and 0.2 V.
16. The propagation delay is the length of time that it takes for the output of a gate to respond to a stimulus at its input.
17. The rise and fall times of a pulse describe how long it takes for the voltage to travel between its 10 and 90% levels.
18. Open-collector outputs are required whenever logic outputs are connected to a common point.
19. Several improved TTL and CMOS families providing decreased power consumption and decreased propagation delay are available and continue to be introduced each year.
20. The CMOS family uses Complementary Metal-Oxide Semiconductor transistors instead of the bipolar transistors used in TTL ICs. Traditionally, CMOS was a lower power consuming family but slower than TTL. However, recent advances in both technologies have narrowed the differences.
21. Emitter-coupled logic (ECL) provides the highest speed ICs. Its drawback is its very high power consumption.
22. A figure of merit of IC families is the product of their propagation delay and power consumption, called the speed-power product (the lower the better).
23. When interfacing logic families several considerations must be made. The output voltage level of one family must be high enough, or low enough, to meet the input requirements of the receiving family. Also, the output current capability of the driving gate must be high enough for the input draw of the receiving gate, or gates.

GLOSSARY

Binary string: Two or more binary bits used collectively to form a meaningful binary representation.

Bipolar transistor: Three-layer *NPN* or *PNP* junction transistor.

Boolean equation: An algebraic expression that illustrates the functional operation of a logic gate or combination of logic gates.

Buffer: A device placed between two other devices that provides isolation and current amplification. The input logic level is equal to the output logic level.

CMOS: Complementary metal-oxide semiconductor.

Complement: A change to the opposite digital state. A 1 becomes a 0, a 0 becomes a 1.

Disable: To disallow or deactivate a function or circuit.

ECL: Emitter-coupled logic.

Enable: To allow or activate a function or circuit.

Exclusive-NOR: A gate that produces a HIGH output for both inputs HIGH or both inputs LOW.

Exclusive-OR: A gate that produces a HIGH output for one or the other of the inputs HIGH, but not both.

Fall time: The time required for a digital pulse to fall from 90% down to 10% of its maximum voltage level.

Fan-out: The number of logic gate inputs that can be driven from a single gate output of the same subfamily.

Fault: The problem in a nonfunctioning electrical circuit. It is usually due to an open circuit, short circuit, or defective component.

Float: A logic level in a digital circuit that is neither HIGH nor LOW. It acts like an open circuit to anything connected to it.

Gate: The basic building block of digital electronics. The basic logic gate has one or more inputs and one output and is used to perform one of the following logic functions: AND, OR, NOR, NAND, INVERT, exclusive-OR, or exclusive-NOR.

Hex: When dealing with integrated circuits, this term specifies that there are *six* gates on a single IC package.

Inversion: A change to the opposite digital state.

Inversion bar: A line over variables in a Boolean equation signifying that the digital state of the variables is to be complemented. For example, the output of a two-input NAND gate is written $X = \overline{AB}$.

Johnson shift counter: A digital circuit that produces several repetitive digital waveforms useful for specialized waveform generation.

Logic probe: An electronic tool used in the troubleshooting procedure to indicate a HIGH, LOW, or float level at a particular point in a circuit.

Logic pulser: An electronic tool used in the troubleshooting procedure to inject a pulse or pulses into a particular point in a circuit.

MOSFET: Metal-oxide semiconductor field-effect transistor.

Noise margin: The voltage difference between the guaranteed output voltage level and the required voltage level of a logic gate.

NOT: When reading a Boolean equation, the word "NOT" is used to signify an inversion bar. For example, the equation $X = \overline{AB}$ would read "*X* equals NOT *AB*."

Open-collector output: A special output stage of the TTL family that has the upper transistor of a totem pole configuration removed.

Power dissipation: The electrical power (watts) that is consumed by a device and given off (dissipated) in the form of heat.

Propagation delay: The time required for a change in logic level to travel from the input to the output of a logic gate.

Pull-up resistor: A resistor with one end connected to V_{CC} and the other end connected to a point in a logic circuit that needs to be raised to a voltage level closer to V_{CC}.

Quad: When dealing with integrated circuits, this term specifies that there are *four* gates on a single IC package.

Repetitive waveform: A waveform that repeats itself after each cycle.

Rise time: The time required for a digital pulse to rise from 10% up to 90% of its maximum voltage level.

Sink current: Current entering the output or input of a logic gate.

Source current: Current leaving the output or input of a logic gate.

Substrate: The silicon supporting structure or framework of an integrated circuit.

Totem pole: The term used to describe the output stage of most TTL integrated circuits. The totem pole stage consists of one transistor in series with another, configured in such a way that when one transistor is saturated, the other is cut off.

Troubleshooting: The work done to find the problem in a faulty electrical circuit.

Truth table: A tabular listing used to illustrate all the possible combinations of digital input levels to a gate and the output that will result.

TTL: Transistor–transistor logic.

Waveform generation: The production of specialized digital waveforms.

PROBLEMS

2–1. Build the truth table for a three-input AND gate.

2–2. Build the truth table for a four-input AND gate.

2–3. If we were to build a truth table for an eight-input AND gate, how many different combinations of inputs would we have?

2–4. Describe, in words, the operation of an AND gate.

2–5. Describe, in words, the operation of an OR gate.

2–6. Write the Boolean equation for:

(a) A three-input AND gate

(b) A four-input AND gate

(c) A three-input OR gate

2–7. Sketch the output waveform at *X* for the two-input OR gates shown in Figure P2–7.

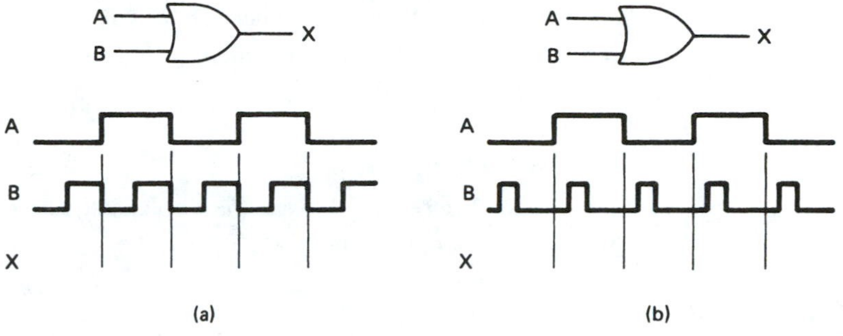

Figure P2–7

2–8. Sketch the output waveform at *X* for the three-input AND gates shown in Figure P2–8.

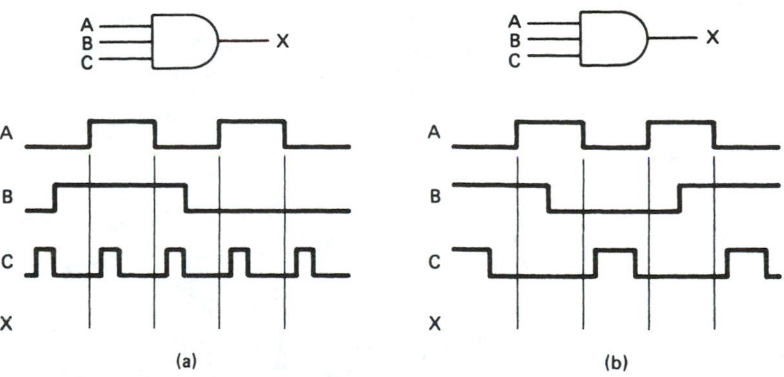

Figure P2–8

2–9. The input waveform at *A* is given for the two-input AND gates shown in Figure P2–9. Sketch the input waveform at *B* that will produce the output at *X*.

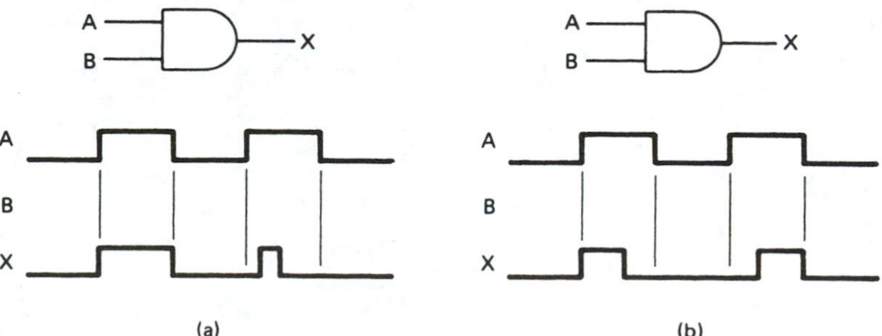

Figure P2–9

2–10. Repeat Problem 2–9 for the two input OR gates shown in Figure P2–10.

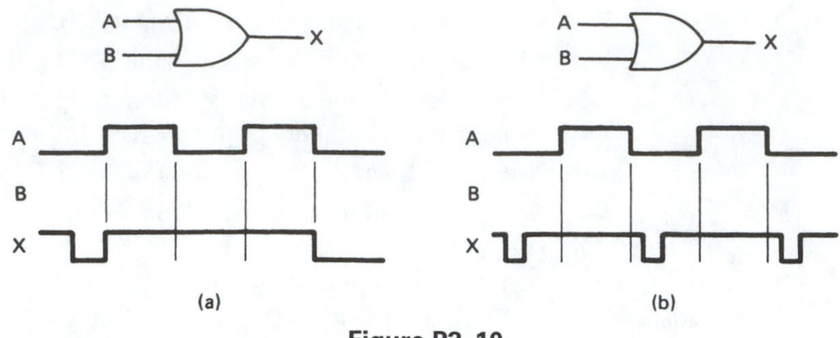

Figure P2–10

2–11. How many separate OR gates are contained within the 7432 TTL IC?

2–12. How many inputs are there on each AND gate of a 7421 TTL IC?

2–13. The 7421 IC is a 14-pin dual-in-line package (DIP). How many of the pins are *not* used for anything?

Inverting Gates

2–14. For Figure P2–14, write the Boolean equation at *X*. If $A = 1$, what is *X*?

2–15. For Figure P2–15, write the Boolean equation at *X* and *Z*. If $A = 0$, what is *X*? What is *Z*?

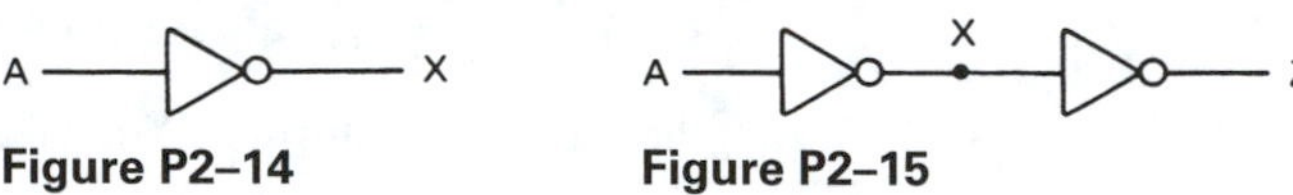

Figure P2–14 Figure P2–15

2–16. Using Figure P2–15, sketch the output waveform at *X* and *Z* if the timing waveform shown in Figure P2–16 is input at *A*.

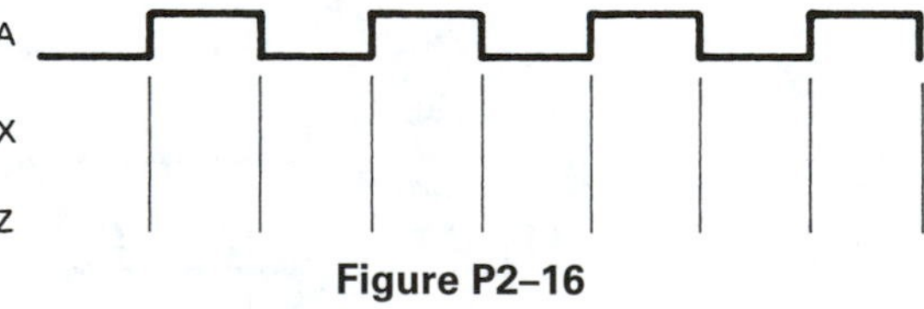

Figure P2–16

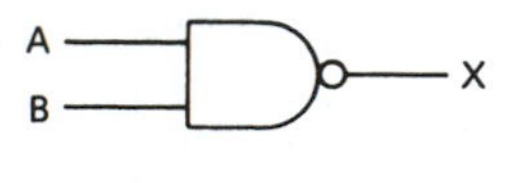

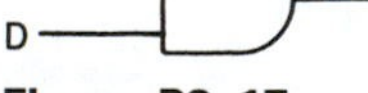

Figure P2–17

2–17. For Figure P2–17, write the Boolean equation at *X* and *Y*.

2–18. Build a truth table for each gate in Figure P2–17.

2–19. Using Figure P2–17, sketch the output waveforms for *X* and *Y* given the input waveforms shown in Figure P2–19.

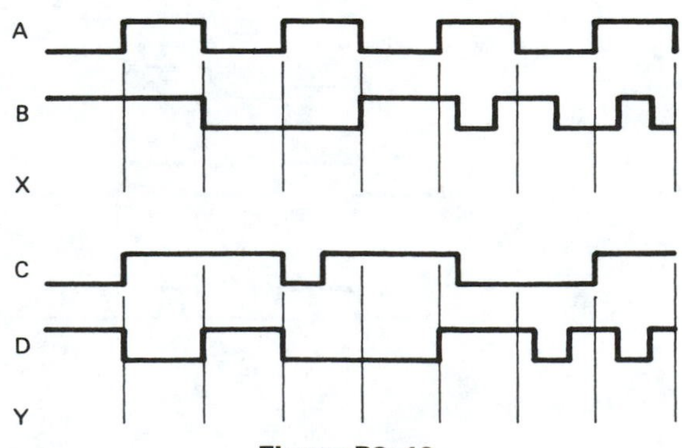

Figure P2–19

2–20. Using Figure P2–20, sketch the waveforms at X and Y with the switches in the down (0) position. Repeat with the switches in the up (1) position.

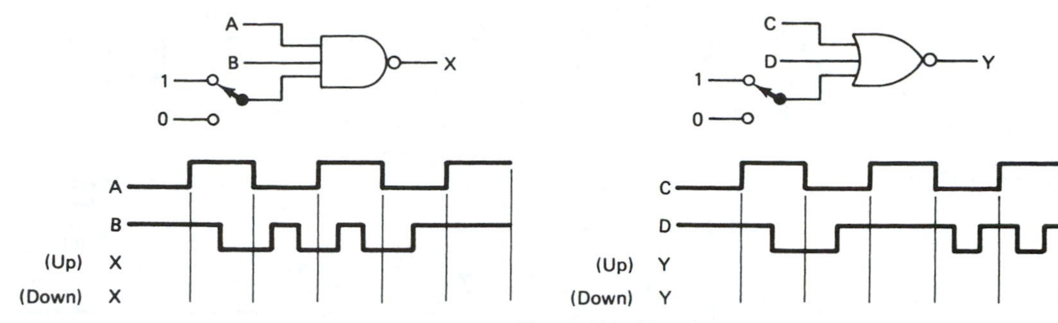

Figure P2–20

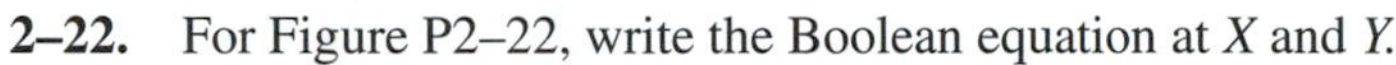

2–21. In words, what effect does the switch have on each circuit in Figure P2–20?

2–22. For Figure P2–22, write the Boolean equation at X and Y.

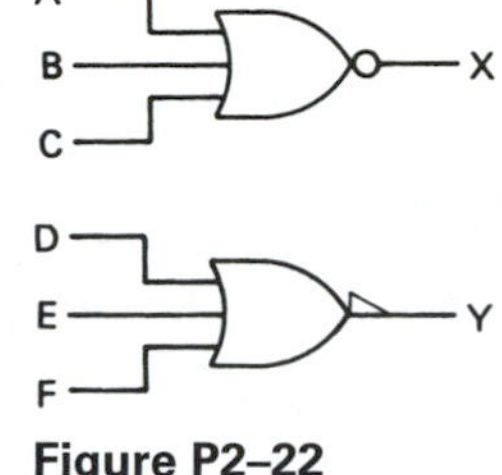

Figure P2–22

2–23. Make a truth table for the first NOR gate in Figure P2–22.

2–24. Referring to Figure P2–22, sketch the output at X and Y given the input waveforms in Figure P2–24.

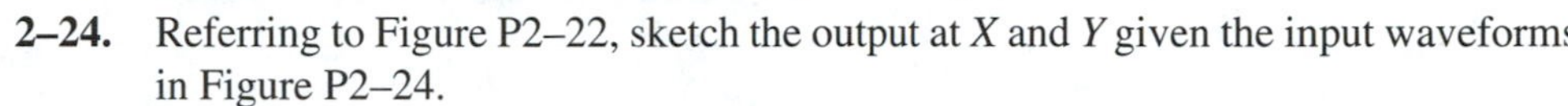

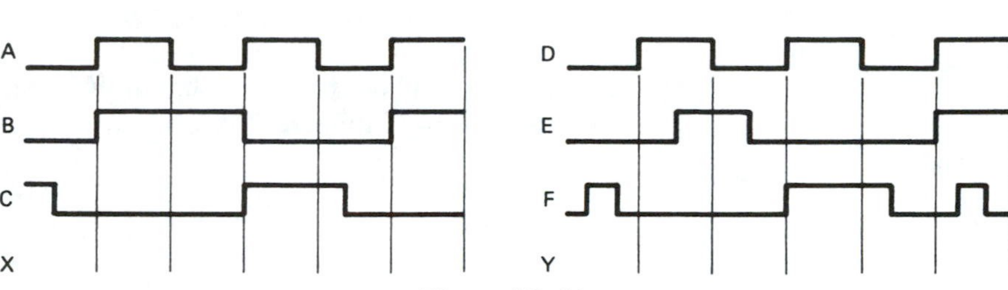

Figure P2–24

2–25. Describe, in words, the operation of an exclusive-OR gate; an exclusive-NOR gate.

2–26. Describe, in words, the difference between:

(a) An exclusive-OR and an OR gate

(b) An exclusive-NOR and an AND gate

2–27. Complete the timing diagram in Figure P2–27 for the exclusive-OR and the exclusive-NOR.

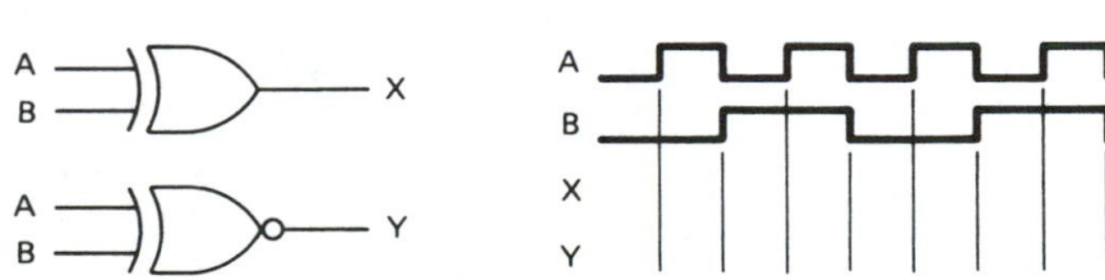

Figure P2–27

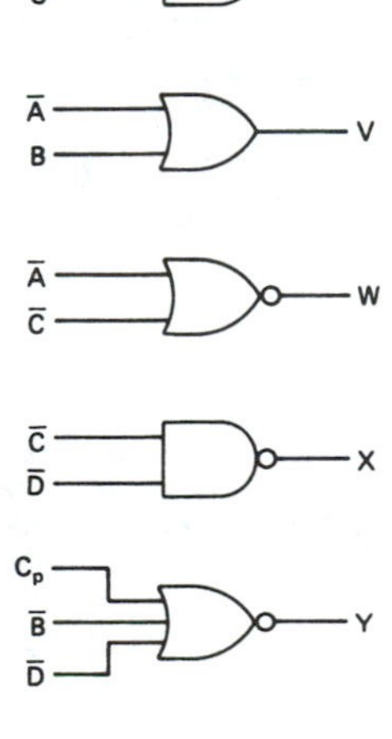

Figure P2–28

Waveform Generation

2–28. The Johnson shift counter outputs shown in Figure 2–41 are connected to the inputs of the logic gates shown in Figure P2–28. Sketch and label the output waveform at U, V, W, X, Y, and Z.

2–29. Repeat Problem 2–28 for the gates shown in Figure P2–29.

2–30. Using the Johnson shift counter outputs from Figure 2–41, label the inputs to the logic gates shown in Figure P2–30, so that they will produce the indicated output.

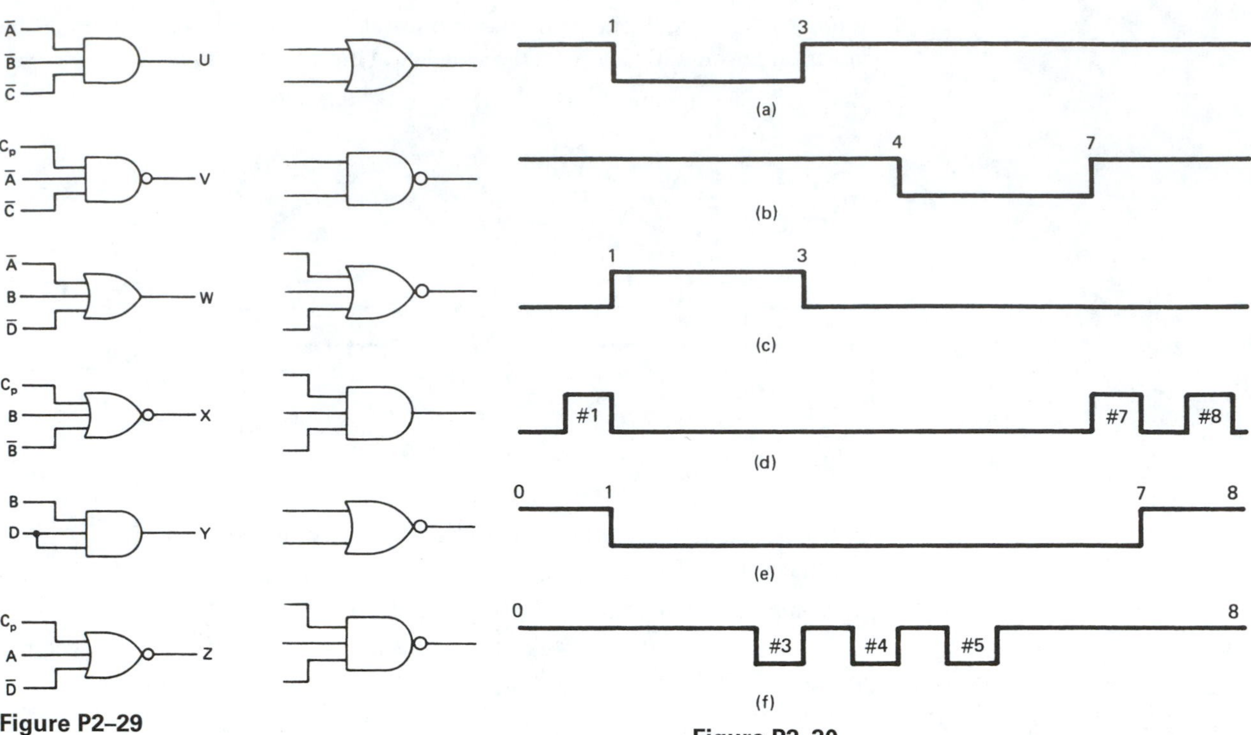

Figure P2–29

Figure P2–30

2–31. Determine which lines from the Johnson shift counter are required at the inputs of the circuits shown in Figure P2–31 to produce the waveforms at *U, V, W,* and *X*.

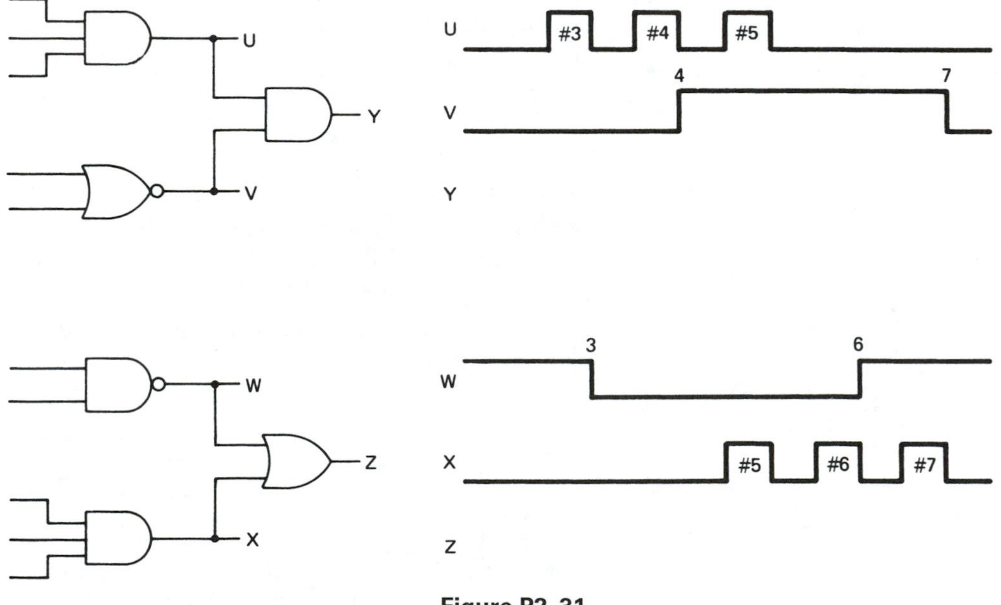

Figure P2–31

2–32. The waveforms at *U, V, W,* and *X* are given in Figure P2–31. Sketch the waveforms at *Y* and *Z*.

2–33. Make the external connections to a 7404 inverter IC and a 7402 NOR IC to implement the function $X = \overline{\overline{A} + B}$.

Troubleshooting

2–34. What are the three logic levels that can be indicated by a logic probe?

2–35. What is the function of the logic pulser?

2–36. When troubleshooting an OR gate such as the 7432, when the pulser is applied to one input, should the other input be connected HIGH or LOW? Why?

2–37. When troubleshooting an AND gate such as the 7408, when the pulser is connected to one input, should the other input be connected HIGH or LOW? Why?

2–38. The clock enable circuit shown in Figure P2–38 is not working. The enable switch is up in the "enable" position. A logic probe is placed on the following pins and gets the following results. Find the cause of the problem.

Probe on pin	*Indicator lamp*
1	Flashing
2	On
3	Off
7	Off
14	On

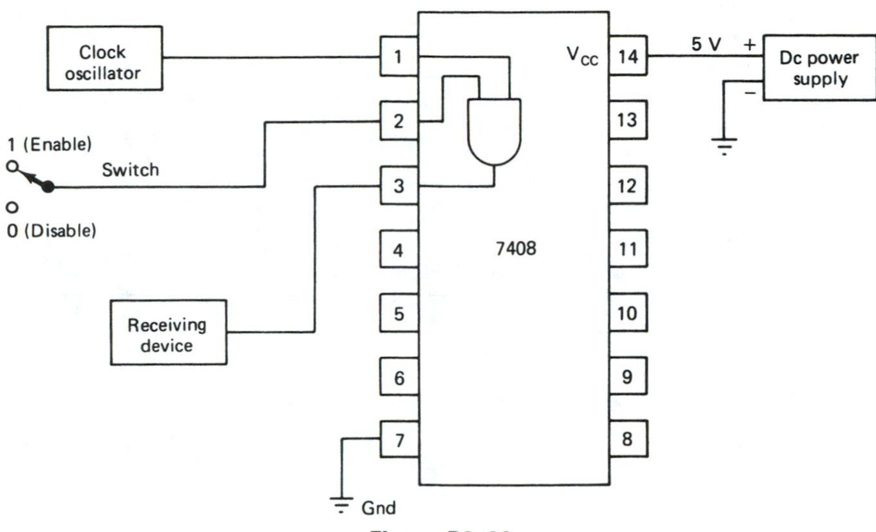

Figure P2–38

2–39. Repeat Problem 2–38 for the following troubleshooting results.

Probe on pin	*Indicator lamp*
1	Flashing
2	Off
3	Off
7	Off
14	On

2–40. Repeat Problem 2–38 for the following troubleshooting results.

Probe on pin	*Indicator lamp*
1	Flashing
2	On
3	Off
7	Dim
14	On

2–41. When troubleshooting a NOR gate like the 7402, with the logic pulser applied to one input, should the other input be held HIGH or LOW? Why?

2–42. When troubleshooting a NAND gate like the 7400, with the logic pulser applied to one input, should the other input be held HIGH or LOW? Why?

2–43. The following data table was built by putting a logic probe on every pin of the hex inverter shown in Figure P2–43. Are there any problems with the chip? If so, which gate(s) are bad?

Pin	*Logic level*
1	HIGH
2	LOW
3	LOW
4	LOW
5	LOW
6	HIGH
7	LOW
8	HIGH
9	LOW
10	LOW
11	LOW
12	LOW
13	HIGH
14	HIGH

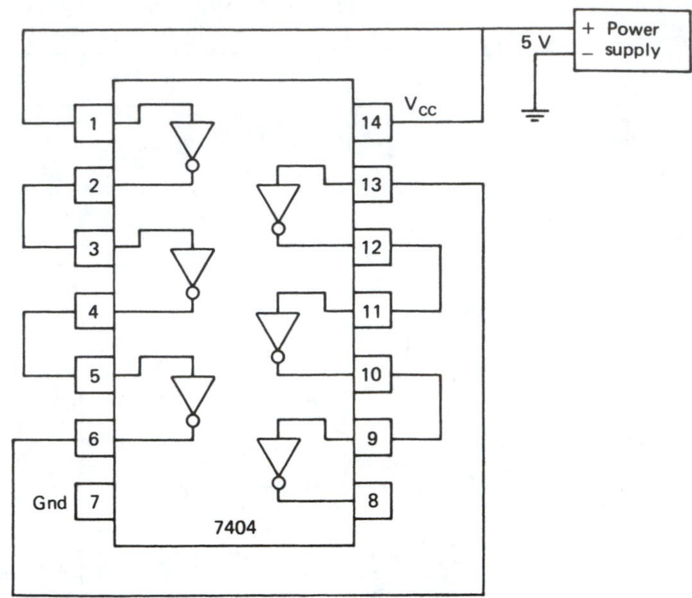

Figure P2–43

2–44. The logic probe in Figure P2–44 is always OFF (0) whether the switch is in the up position or the down position. Is the problem with the inverter, the NOR, or is there no problem?

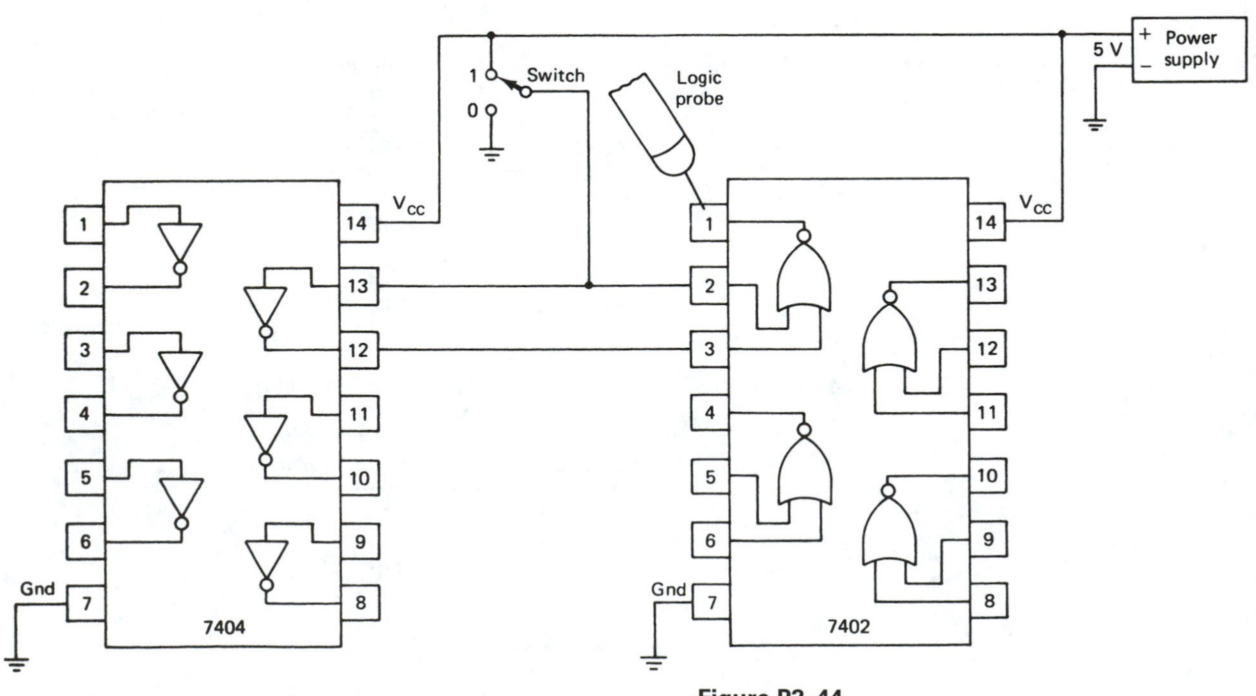

Figure P2–44

2–45. Another circuit constructed the same way as Figure P2–44 causes the logic probe to come on when the switch is in the down (0) position. Further testing with the probe shows that pins 2 and 3 of the NOR IC are both LOW. Is anything wrong? If so, where is the fault?

2–46. Your company has purchased several of the 7430 eight-input NANDS shown in Figure P2–46. List the steps that you would follow to determine if they are all good ICs.

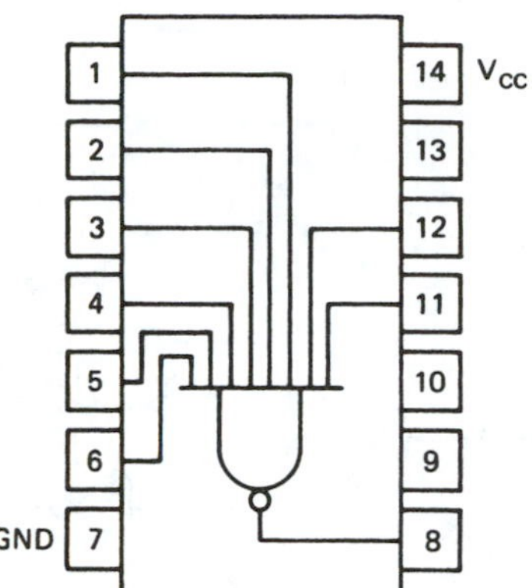

Figure P2–46

2–47. The following data table was built by putting a logic probe on every pin of the 7427 NOR IC shown in Figure P2–47 while it was connected in a digital circuit. Which gates, if any, are bad, and why?

Pin	*Logic level*
1	LOW
2	LOW
3	LOW
4	LOW
5	LOW
6	HIGH
7	LOW
8	Flashing
9	HIGH
10	LOW
11	Flashing
12	HIGH
13	HIGH
14	HIGH

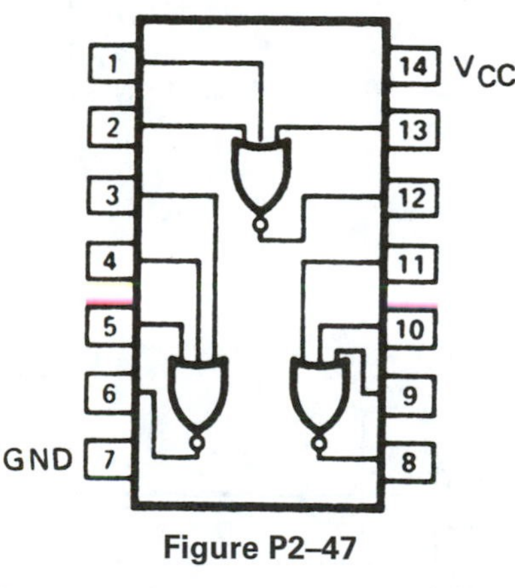

Figure P2–47

IC Specifications

2–48. What does the negative sign in the rating of source current (e.g., $I_{OH} = -400\ \mu A$) signify?

2–49. For TTL outputs, which is higher, the source current or the sink current?

2–50. **(a)** Find V_a and I_a in the circuits of Figure P2–50 using the following specifications:

$I_{IL} = -1.6$ mA $\quad$ $I_{IH} = 40\ \mu A$ $\quad$ $I_{OL} = 16$ mA
$I_{OH} = -400\ \mu A$ $\quad$ $V_{IL} = 0.8$ V max. $\quad$ $V_{IH} = 2.0$ V min.
$V_{OL} = 0.2$ V typ. $\quad$ $V_{OH} = 3.4$ V typ.

(b) Repeat part (a) using input/output specifications that you gather from a TTL data book assuming that all gates are 74LSXX series.

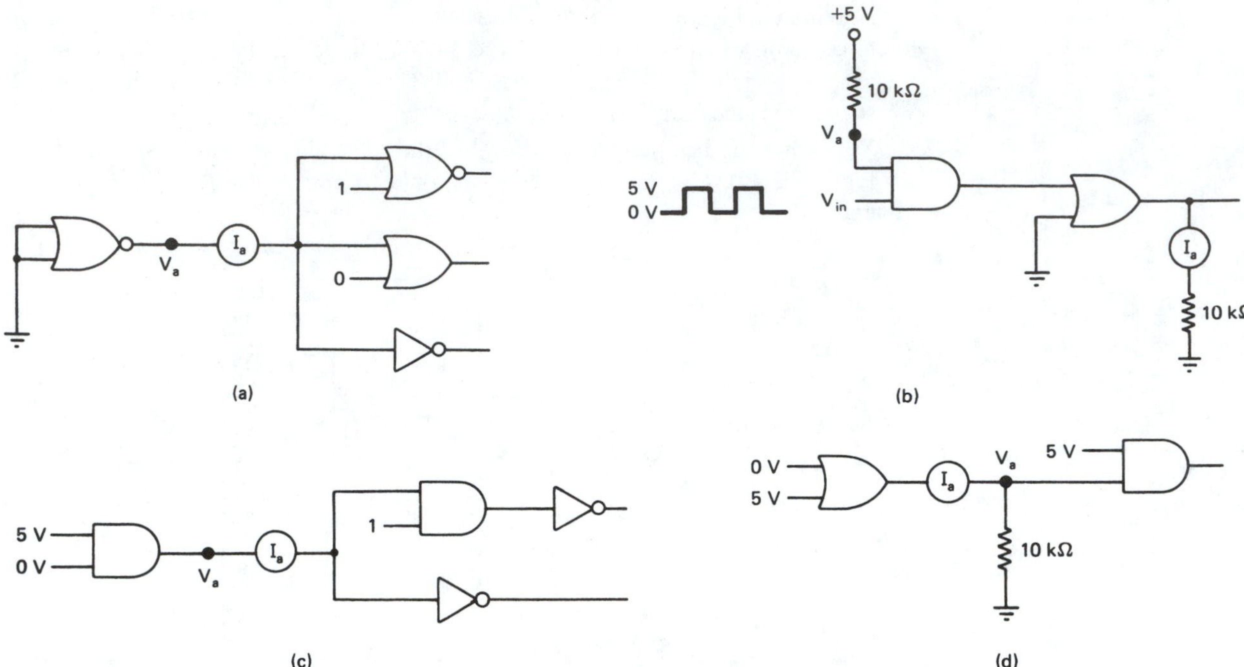

Figure P2–50

2–51. The input and output waveforms to an OR gate are given in Figure P2–51. Determine:

(a) The period and frequency of V_{in}

(b) The rise and fall times (t_r, t_f) of V_{in}

(c) The propagation delay times of (t_{PLH}, t_{PHL}) of the OR gate

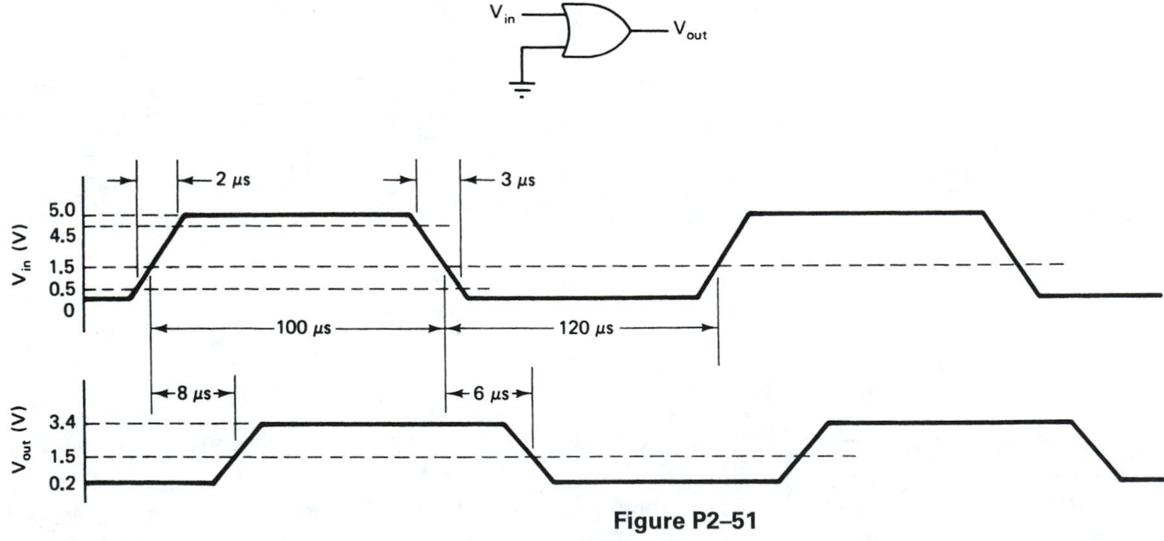

Figure P2–51

2–52. The propagation delay times for a 74LS08 AND gate are t_{PLH} = 15 ns, t_{PHL} = 20 ns and for a 7402 NOR gate are t_{PLH} = 22 ns, t_{PHL} = 15 ns. Sketch V_{out1} and V_{out2} showing the effects of propagation delay. (Assume 0 ns for the rise and fall times.)

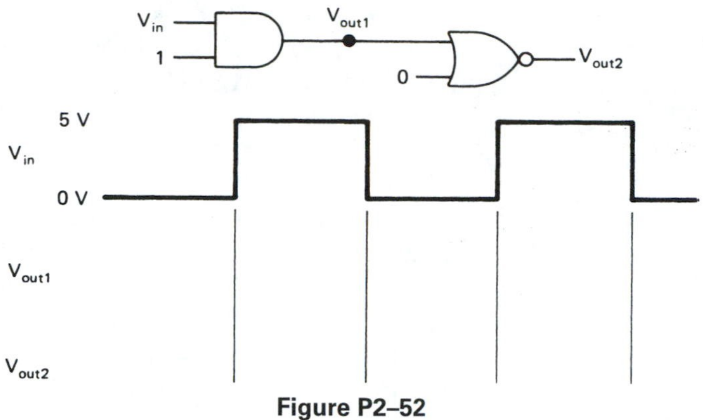

Figure P2–52

2–53. Repeat Problem 2–52 for the circuit of Figure P2–53.

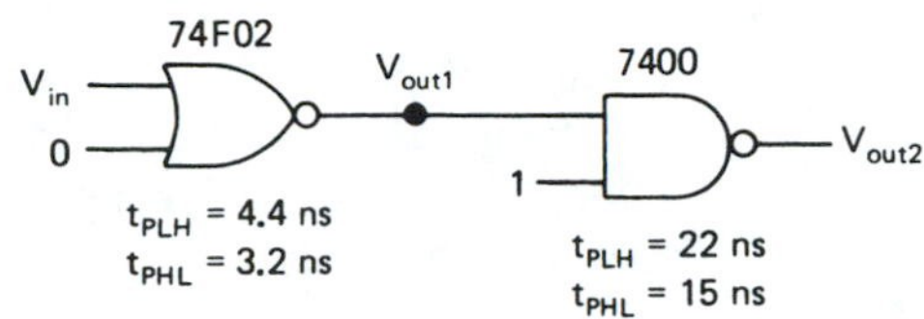

Figure P2–53

2–54. Refer to a TTL data sheet to compare the typical LOW-level output voltage (V_{OL}) at maximum output current for a 7400 versus a 74LS00.

2–55. **(a)** Refer to a TTL data sheet to determine the noise margins for the HIGH state and LOW state of both the 7400 and 74LS00.
(b) Which has better noise margins, the 7400 or 74LS00?

2–56. **(a)** Refer to a TTL data sheet to determine which can sink more current at its output, the commercial 74LS00 or the military 54LS00.
(b) Which has a wider range of recommended V_{CC} supply voltage, the 7400 or the 5400?

2–57. Why is a pull-up resistor required at the output of an open-collector gate to achieve a HIGH-level output?

2–58. Make a general comparison of both the switching speed and power dissipation of the 7400 TTL series versus the 4000B CMOS series.

2–59. Which type of transistor, bipolar or field-effect, is used in:
(a) TTL ICs?
(b) CMOS ICs?

2–60. Why is it important to store MOS ICs in antistatic conductive foam?

2–61. **(a)** Using the data in Table 2–15, draw a graph of input and output specifications, similar to Figure 2–71 for the 74HCMOS and the 74ALSTTL IC series.
(b) From your graphs of the two IC series, compare the HIGH-level and LOW-level noise margins.
(c) From your graphs, can you see a problem in *directly* interfacing:
1 The 74HCMOS to the 74ALSTTL?
2 The 74ALSTTL to the 74HCMOS?

2–62. Refer to Table 2–15 to determine which of the following interfacing situations (driving gate-to-gate load) will require a pull-up resistor, and why?

(a) 74TTL to 74ALSTTL
(b) 74HCMOS to 74TTL
(c) 74TTL to 74HCMOS
(d) 74LSTTL to 74HCTMOS
(e) 74LSTTL to 4000B CMOS

2–63. Of the interfacing situations given in Problem 2–62, will any of the driving gates have trouble sinking or sourcing current to a single connected gate load?

2–64. From Table 2–15 determine:

(a) How many 74LSTTL loads can be driven by a single 74HCTMOS gate
(b) How many 74HCTMOS loads can be driven by a single 74LSTTL gate

SCHEMATIC INTERPRETATION PROBLEMS

2–65. What is the component name and grid location of the AND gate and the OR gate in the Watchdog Timer schematic?

2–66. A logic probe is used to check the operation of the AND and OR gates in the Watchdog Timer circuit. If the probe indicator is ON for pin 2 of both gates and flashing on pin 1, what will pin 3 be for:

(a) the AND gate
(b) the OR gate

2–67. On the 4096/4196 schematic there are several gates labeled U1. Why are they all labeled the same?

2–68. Describe a method that you could use to check the operation of the inverter labeled U4:A of the Watchdog Timer. Assume that you have a dual-trace oscilloscope available for troubleshooting.

Combinational Logic Circuits and Reduction Techniques

OBJECTIVES

Upon completion of this chapter, you should be able to:

- Write Boolean equations for combinational logic applications.
- Utilize Boolean algebra laws and rules for simplifying combinational logic circuits.
- Apply DeMorgan's theorem to complex Boolean equations to arrive at simplified equivalent equations.
- Troubleshoot combinational logic circuits.
- Implement sum-of-products expressions.
- Utilize the Karnaugh mapping procedure to reduce systematically complex Boolean equations to their simplest form.
- Describe the steps involved in solving a complete system design application.

INTRODUCTION

Generally you will find that the simple gate functions AND, OR, NAND, NOR, and INVERT by themselves are not enough to implement the complex requirements of digital systems. The basic gates will be used as the building blocks for the more complex logic that is implemented by using combinations of gates called *combinational logic.*

3–1 COMBINATIONAL LOGIC

Combinational logic employs the use of two or more of the basic logic gates to form a more useful, complex function. For example, let's design the logic for an automobile warning buzzer using combinational logic. The criterion for the activation of the warning buzzer is as follows: The buzzer will activate if the headlights are on *and* and driver's door is opened, *or* if the key is in the ignition *and* the door is opened.

The logic function for the automobile warning buzzer is illustrated symbolically in Figure 3–1. The figure illustrates a "combination" of logic functions that can be written as a Boolean equation in the form

$$B = K \text{ and } D \qquad \text{or} \qquad H \text{ and } D$$

also written as

$$B = KD + HD$$

That equation can be stated: B is HIGH if K *and* D are HIGH, *or* if H *and* D are HIGH.

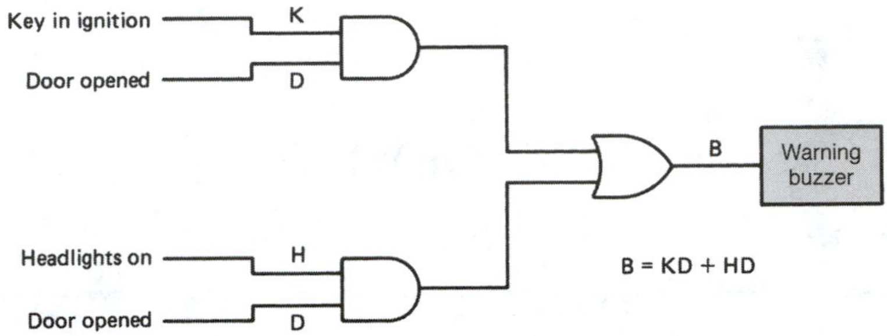

Figure 3–1 Combinational logic requirements for an automobile warning buzzer.

When you think about the operation of the warning buzzer, you may realize that it is activated whenever the door is opened *and* either the key is in the ignition *or* the headlights are on. If you can realize that, you have just performed your first *Boolean reduction* using Boolean algebra. (The systematic reduction of logic circuits is performed using Boolean algebra, named after the nineteenth-century mathematician George Boole.)

The new Boolean equation becomes $B = D$ and (K or H), also written as $B = D(K + H)$. (Note the use of parentheses. Without them, the equation would imply that the buzzer activates if the door is opened with the key in the ignition or any time the headlights are on, which is invalid. $B \neq DK + H$.) The new equation represents the same logic operation but is a simplified implementation because it requires only two logic gates, as shown in Figure 3–2.

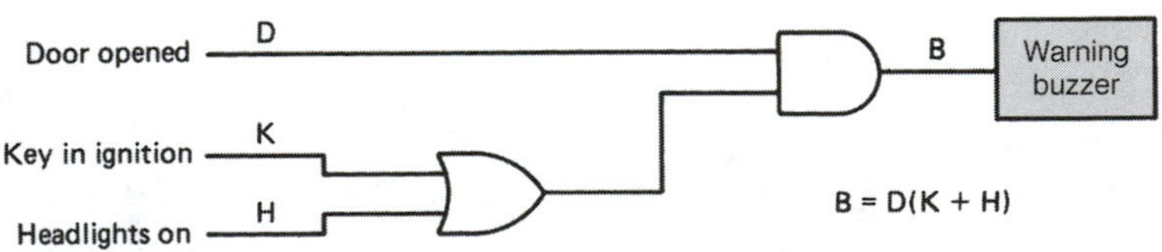

Figure 3–2 Reduced logic circuit for the automobile buzzer.

EXAMPLE 3–1

Write the Boolean logic equation and draw the logic circuit that represent the following function: A bank burglar alarm (A) is to activate if it is after banking hours (H) *and* the front door (F) is opened, *or* if it is after banking hours (H) and the vault door is opened (V).

Solution:

$A = HF + HV$. The logic circuit is shown in Figure 3–3.

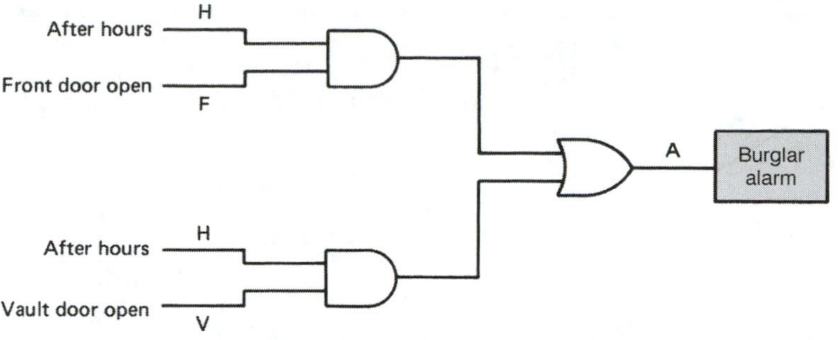

Figure 3–3 Solution to Example 3–1.

EXAMPLE 3–2

Using common reasoning, reduce the logic function described in Example 3–1 to a simpler form.

Solution:

The alarm is activated if it is after banking hours *and* the front door is opened *or* the vault door is opened (see Figure 3–4). The simplified equation is written as

$$A = H(F + V) \qquad \text{(note the use of parentheses)}$$

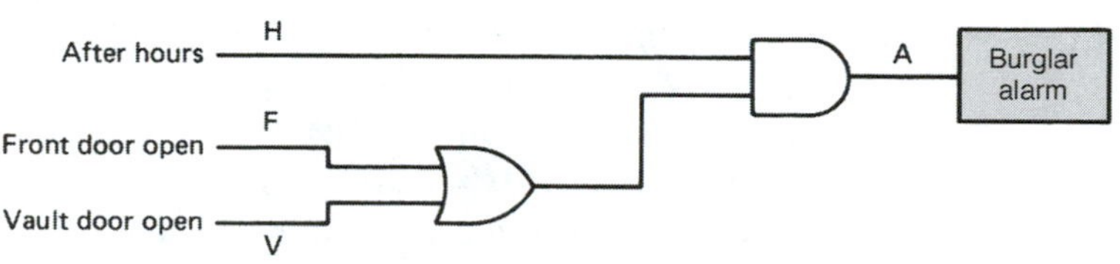

Figure 3–4 Solution to Example 3–2.

EXAMPLE 3–3

Draw the logic circuit that could be used to implement the following Boolean equation:

$$X = AB + C(M + N)$$

Solution:

The logic circuit is shown in Figure 3–5.

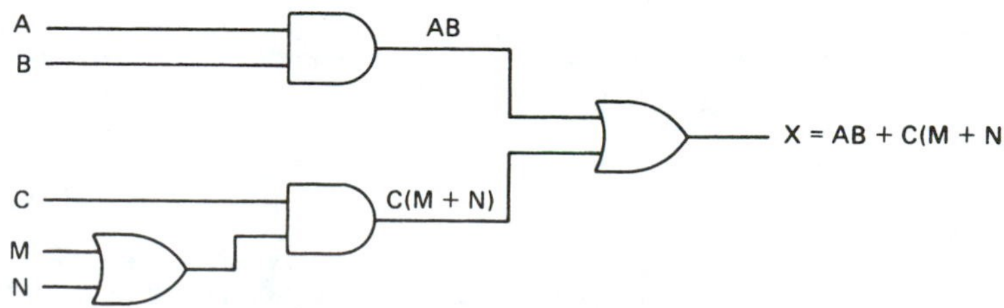

Figure 3–5 Solution to Example 3–3.

EXAMPLE 3–4

Write the Boolean equation for the logic circuit shown in Figure 3–6.

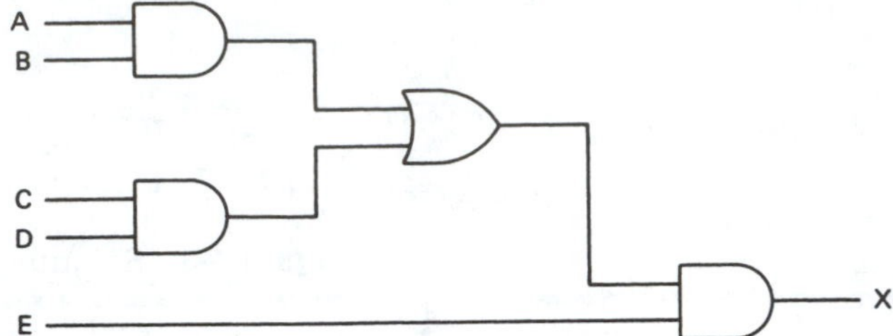

Figure 3–6 Combinational logic circuit for Example 3–4.

Solution:

$X = (AB + CD)E.$

3–2 BOOLEAN ALGEBRA LAWS AND RULES

Boolean algebra uses many of the same laws as those of ordinary algebra. The OR function ($X = A + B$) is *Boolean addition* and the AND function ($X = AB$) is *Boolean multiplication.* The following three laws are the same for Boolean algebra as they are for ordinary algebra:

1. *The commutative law of addition:* $A + B = B + A$, *and multiplication:* $AB = BA$. This means that the order of ORing or ANDing does not matter.
2. *The associative law of addition:* $A + (B + C) = (A + B) + C$, *and multiplication:* $A(BC) = (AB)C$. This means that the grouping of several variables ORed or ANDed together does not matter.
3. *The distributive law:* $A(B + C) = AB + AC$, *and,* $(A + B)(C + D) = AC + AD + BC + BD$. This shows the method for expanding an equation containing ORs and ANDs.

Those three laws hold true for any number of variables. For example, the associative law can be applied to $X = A + BC + D$ to form the equivalent equation, $X = BC + A + D$.

You may wonder when you will need to use one of the laws. Later in this chapter you will see that by using these laws to rearrange Boolean equations, you will be able to change some combinational logic circuits to simpler equivalent circuits using fewer gates.

Besides the three basic laws, there are several rules concerning Boolean algebra. The rules of Boolean algebra allow us to combine or eliminate certain variables in the equation to form simpler equivalent circuits.

The following example illustrates the use of the first Boolean rule, which states: Anything ANDed with a 0 will always output a 0.

EXAMPLE 3–5

A bank burglar alarm (B) will activate if it is after banking hours (A) and someone opens the front door (D). The logic level of the variable A is 1 after banking hours and 0 during banking hours. Also, the logic level of the variable D is 1 if the door sensing switch is opened and 0 if the door sensing switch is closed. The Boolean equation is therefore $B = AD$. The logic circuit to implement this function is shown in Figure 3–7a.

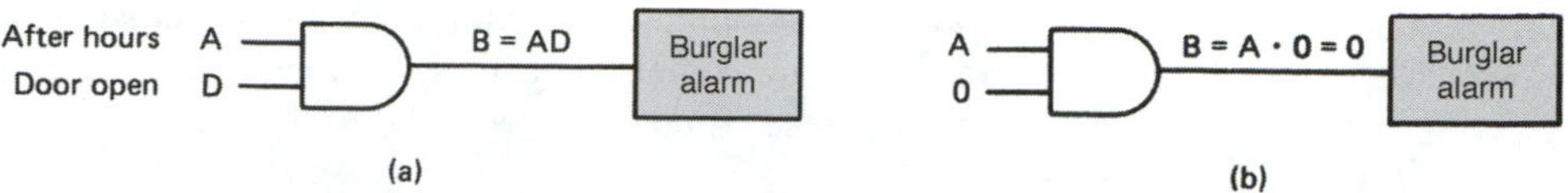

Figure 3–7 (a) Logic circuit for a simple burglar alarm; (b) disabling the burglar alarm by making $D = 0$.

Later, a burglar comes along and puts tape on the door sensing switch, holding it closed so that it always puts out a 0 logic level. Now the Boolean equation ($B = AD$) becomes $B = A \cdot 0$ because the door sensing switch is always 0. The alarm will never sound in this condition because one input to the AND gate is always 0. The burglar must have studied the Boolean rules and realized that anything ANDed with a 0 will output a 0, as shown in Figure 3–7b.

Example 3–5 helped illustrate the reasoning for Boolean rule 1. The other nine rules can be derived using common sense and knowing basic gate operation.

Rule 1: Anything ANDed with a 0 is equal to 0 ($A \cdot 0 = 0$).

Rule 2: Anything ANDed with a 1 is equal to itself ($A \cdot 1 = A$). From Figure 3–8 we can see that with one input tied to a 1, if the A input is 0, the X output is 0; if A is 1, X is 1; therefore, X is equal to whatever the logic level of A is ($X = A$).

A, 1 — $X = A \cdot 1 = A$

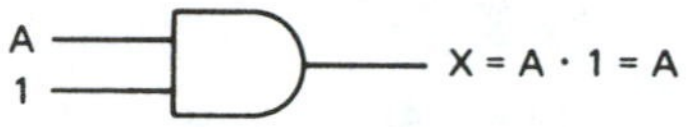

Figure 3–8 Logic circuit illustrating Rule 2.

Rule 3: Anything ORed with a 0 is equal to itself ($A + 0 = A$). In Figure 3–9 since one input is always 0, if $A = 1$, $X = 1$, and if $A = 0$, $X = 0$; therefore, X is equal to whatever the logic level of A is ($X = A$).

A, 0 — $X = A + 0 = A$

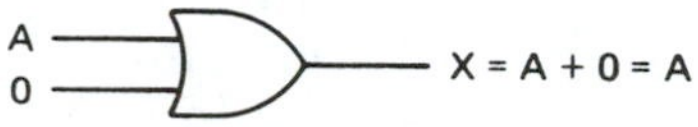

Figure 3–9 Logic circuit illustrating Rule 3.

Rule 4: Anything ORed with a 1 is equal to 1 ($A + 1 = 1$). In Figure 3–10 since one input to the OR gate is always 1, the output will always be 1, no matter what A is ($X = 1$).

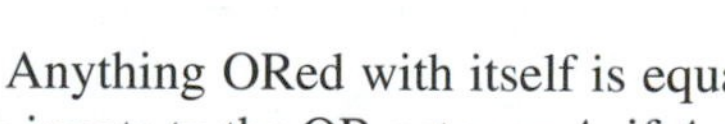

A, 1 — $X = A + 1 = 1$

Figure 3–10 Logic circuit illustrating Rule 4.

Rule 5: Anything ANDed with itself is equal to itself ($A \cdot A = A$). In Figure 3–11 since both inputs to the AND gate are A, if $A = 1$, 1 and 1 equals 1, and if $A = 0$, 0 and 0 equals 0. Therefore, X will be equal to whatever the logic level of A is ($X = A$).

A, A — $X = A \cdot A = A$

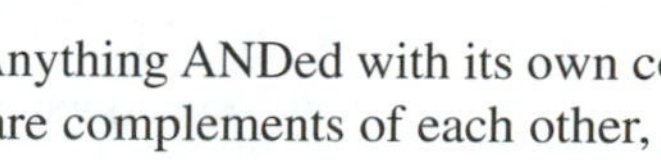

Figure 3–11 Logic circuit illustrating Rule 5.

Rule 6: Anything ORed with itself is equal to itself ($A + A = A$). In Figure 3–12 since both inputs to the OR gate are A, if $A = 1$, 1 or 1 equals 1, and if $A = 0$, 0 or 0 equals 0. Therefore, X will be equal to whatever the logic level of A is ($X = A$).

A, A — $X = A + A = A$

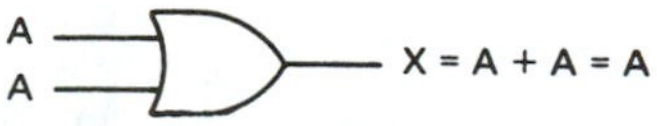

Figure 3–12 Logic circuit illustrating Rule 6.

Rule 7: Anything ANDed with its own complement equals 0. In Figure 3–13 since the inputs are complements of each other, one of them will always be 0. With a 0 at the input, the output will always be 0 ($X = 0$).

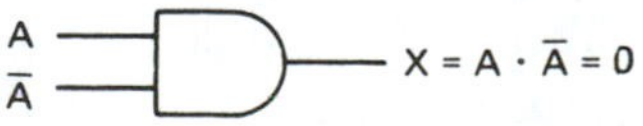

A, $\overline{A}$ — $X = A \cdot \overline{A} = 0$

Figure 3–13 Logic circuit illustrating Rule 7.

Rule 8: Anything ORed with its own complement equals 1. In Figure 3–14 since the inputs are complements of each other, one of them will always be 1. With a 1 at the input, the output will always be 1 ($X = 1$).

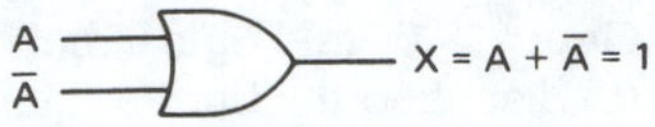

A, $\overline{A}$ — $X = A + \overline{A} = 1$

Figure 3–14 Logic circuit illustrating Rule 8.

Rule 9: A variable that is complemented twice will return to its original logic level. As shown in Figure 3–15, when a variable is complemented once, it changes to the opposite logic level. When it is complemented a second time, it changes back to its original logic level ($\overline{\overline{A}} = A$).

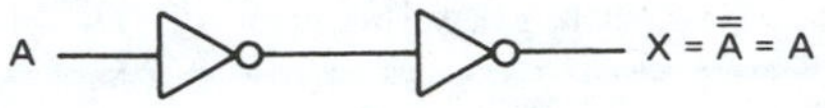

A — $X = \overline{\overline{A}} = A$

Figure 3–15 Logic circuit illustrating Rule 9.

Rule 10: $A + \overline{A}B = A + B$ and $\overline{A} + AB = \overline{A} + B$. This rule differs from the others because it involves two variables. It is useful because when an equation is in this form, one variable in the second term can be eliminated. Proof of this rule is performed very simply by using a Karnaugh map, which will be done later in this chapter.

Table 3–1 summarizes the laws and rules that relate to Boolean algebra. By using them, we can reduce complicated combinational logic circuits to their simplest form, as we will see in the next sections.

TABLE 3–1

Boolean Laws and Rules for the Reduction of Combinational Logic Circuits

Laws	
1	$A + B = B + A$
	$AB = BA$
2	$A + (B + C) = (A + B) + C$
	$A(BC) = (AB)C$
3	$A(B + C) = AB + AC$
	$(A + B)(C + D) = AC + AD + BC + BD$
Rules	
1	$A \cdot 0 = 0$
2	$A \cdot 1 = A$
3	$A + 0 = A$
4	$A + 1 = 1$
5	$A \cdot A = A$
6	$A + A = A$
7	$A \cdot \overline{A} = 0$
8	$A + \overline{A} = 1$
9	$\overline{\overline{A}} = A$
10	$A + \overline{A}B = A + B$
	$\overline{A} + AB = \overline{A} + B$

3–3 *SIMPLIFICATION OF COMBINATIONAL LOGIC CIRCUITS USING BOOLEAN ALGEBRA*

Quite often in the design and development of digital systems, a designer will start with simple logic gate requirements but add more and more complex gating, making the final design a complex combination of several gates, some having the same inputs. At that point the designer must step back and review the combinational logic circuit that has been developed and see if there are ways of reducing the number of gates without changing the function of the circuit. If an equivalent circuit can be formed with fewer gates, the cost of the circuit is reduced and its reliability is improved. This process is called the *reduction* or *simplification of combinational logic circuits* and is performed by using the laws and rules of Boolean algebra presented in the preceding section.

The following examples illustrate the use of Boolean algebra and present some techniques for the simplification of logic circuits.

EXAMPLE 3–6

The logic circuit shown in Figure 3–16 is used to turn on a warning buzzer at X based on the input conditions at A, B, and C. A simplified equivalent circuit that will perform the same function can be formed by using Boolean algebra. Write the equation of the circuit in Figure 3–16, simplify the equation, and draw the logic circuit of the simplified equation.

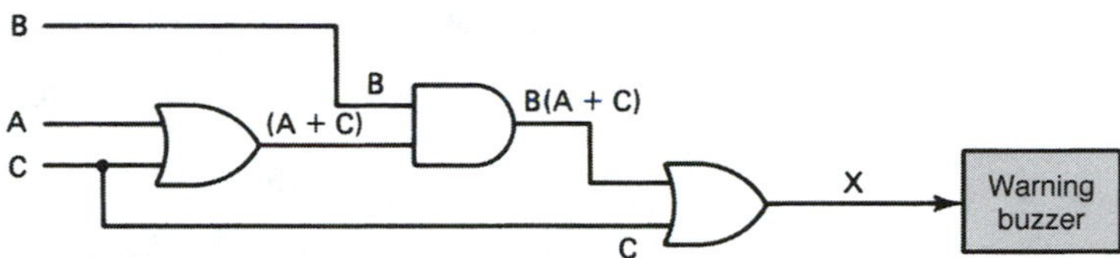

Figure 3–16 Logic circuit for Example 3–6.

Solution:

The Boolean equation for X is

$$X = B(A + C) + C$$

To simplify, first apply Law 3 [$B(A + C) = BA + BC$]:

$$X = BA + BC + C$$

Next, factor a C from terms 2 and 3:

$$X = BA + C(B + 1)$$

Apply Rule 4 ($B + 1 = 1$):

$$X = BA + C \cdot 1$$

Apply Rule 2 ($C \cdot 1 = C$):

$$X = BA + C$$

Apply Law 1 ($BA + AB$):

$$X = AB + C \leftarrow \text{simplified equation}$$

The logic circuit of the simplified equation is shown in Figure 3–17.

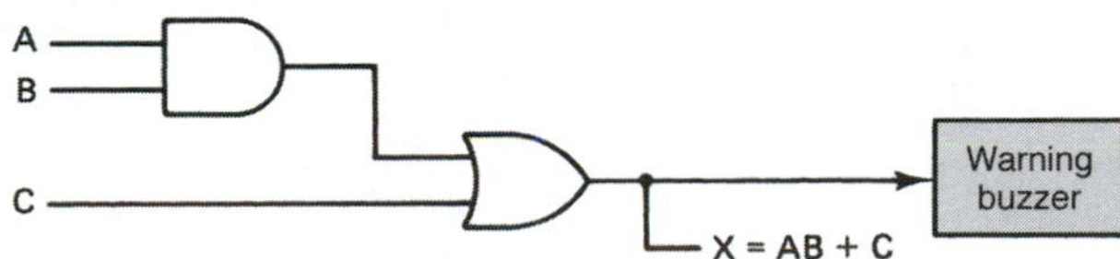

Figure 3–17 Simplified logic circuit for Example 3–6.

EXAMPLE 3–7

Repeat Example 3–6 for the logic circuit shown in Figure 3–18.

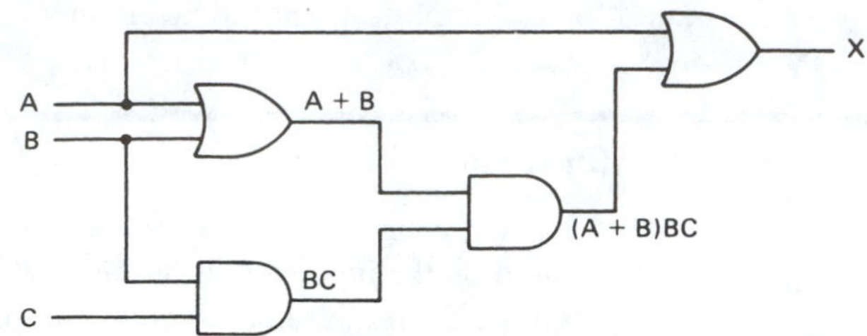

Figure 3–18 Logic circuit for Example 3–7.

Solution:

The Boolean equation for X is

$$X = (A + B)BC + A$$

To simplify, first apply Law 3 [$(A + B)BC = ABC + BBC$]:

$$X = ABC + BBC + A$$

Apply Rule 5 ($B \cdot B = B$):

$$X = ABC + BC + A$$

Factor a BC from terms 1 and 2:

$$X = BC(A + 1) + A$$

Apply Rule 4 ($A + 1 = 1$):

$$X = BC \cdot 1 + A$$

Apply Rule 2 ($BC \cdot 1 = BC$):

$$X = BC + A \leftarrow \text{simplified equation}$$

The logic circuit for the simplified equation is shown in Figure 3–19.

Figure 3–19 Simplified logic circuit for Example 3–7.

EXAMPLE 3–8

Repeat Example 3–6 for the logic circuit shown in Figure 3–20.

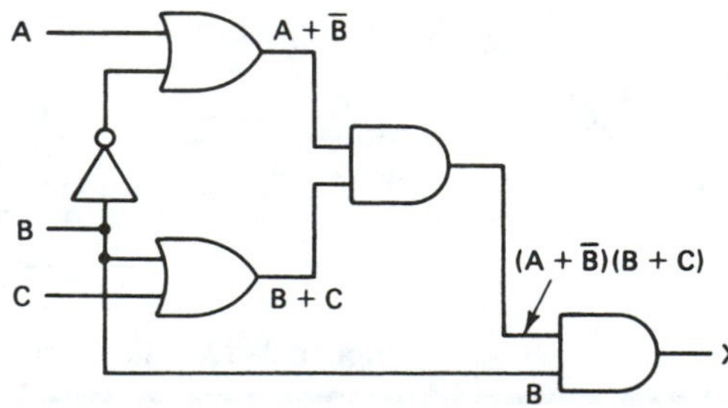

Figure 3–20 Logic circuit for Example 3–8.

Solution:

The Boolean equation for X is

$$X = [(A + \overline{B})(B + C)]B$$

To simplify, first apply Law 3:

$$X = (AB + AC + \overline{B}B + \overline{B}C)B$$

The $\overline{B}B$ term can be eliminated using Rule 7, then Rule 3:

$$X = (AB + AC + \overline{B}C)B$$

Apply Law 3 again:

$$X = ABB + ACB + \overline{B}CB$$

Apply Law 1:

$$X = ABB + ABC + \overline{B}BC$$

Apply Rule 5 and Rule 7:

$$X = AB + ABC + 0 \cdot C$$

Apply Rule 1:

$$X = AB + ABC$$

Factor an AB from both terms:

$$X = AB(1 + C)$$

Apply Rule 4, then Rule 2:

$$X = AB \leftarrow \text{simplified equation}$$

The logic circuit of the simplified equation is shown in Figure 3–21.

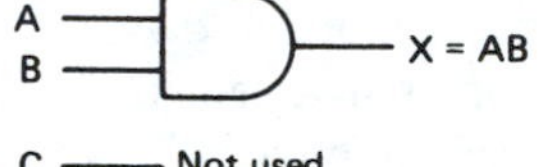

Figure 3–21 Simplified logic circuit for Example 3–8.

3–4 DeMORGAN'S THEOREM

You may have noticed that we did not use NANDs or NORs in any of the logic circuits in Section 3–3. In order to simplify circuits containing NANDs and NORs, we need to use a theorem developed by the mathematician DeMorgan. This theorem allows us to convert an expression having an inversion bar over two or more variables into an expression having inversion bars over single variables only. That allows us to use the rules presented in the preceding section for the simplification of the equation.

In the form of an equation, DeMorgan's theorem is stated as follows:

$$\overline{A \cdot B} = \overline{A} + \overline{B}$$

$$\overline{A + B} = \overline{A} \cdot \overline{B}$$

Also, for three or more variables,

$$\overline{A \cdot B \cdot C} = \overline{A} + \overline{B} + \overline{C}$$

$$\overline{A + B + C} = \overline{A} \cdot \overline{B} \cdot \overline{C}$$

Basically, to use the theorem, you break the bar over the variables and either change the AND to an OR or change the OR to an AND.

To prove to ourselves that this works, let's apply the theorem to a NAND gate and then compare the truth table of the equivalent circuit to that of the original NAND gate. As you can see in Figure 3–22, to use DeMorgan's theorem on a NAND gate, first break the bar over the $A \cdot B$, then change the AND symbol to an OR. The new equation becomes $X = \overline{A} + \overline{B}$. Note that inversion bubbles are used on the OR gate instead of using inverters. By observing the truth tables of the two equations, we can see that the result in the X column is the same for both, which proves that they provide an equivalent output result.

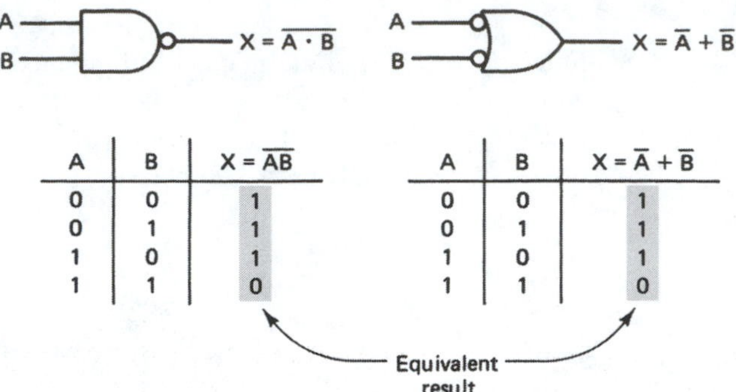

***Figure 3–22** DeMorgan's theorem applied to a NAND gate produces two identical truth tables.

Also, by looking at the two circuits, we can say that "an AND gate with its output inverted is equivalent to an OR gate with its inputs inverted." Therefore, the OR gate with inverted inputs is sometimes used as an alternative symbol for a NAND gate.

By applying DeMorgan's theorem to a NOR gate, we will also produce two identical truth tables, as shown in Figure 3–23. Therefore, we can also think of an OR gate with its output inverted as being equivalent to an AND gate with its inputs inverted. The inverted input AND gate symbol is also sometimes used as an alternative to the NOR gate symbol.

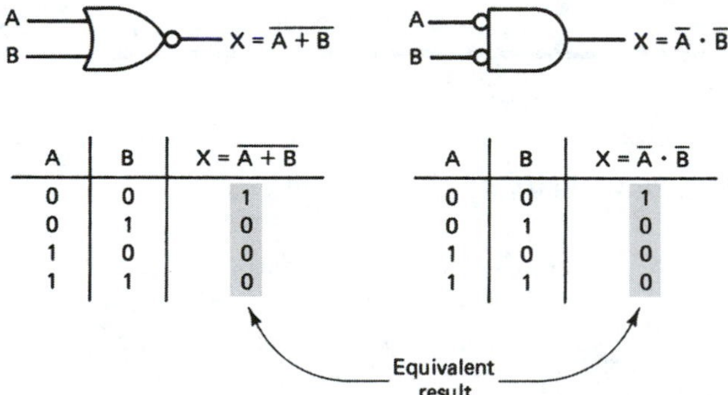

Figure 3–23 DeMorgan's theorem applied to a NOR gate produces two identical truth tables.

When you write the equation for an AND gate with its inputs inverted, be careful to keep the inversion bar over each individual variable (not both) because $\overline{A \cdot B}$ is not equal to $\overline{A} \cdot \overline{B}$. (Prove that to yourself by building a truth table for both. Also, $\overline{A + B}$ is not equal to $\overline{A} + \overline{B}$.)

The question always arises: Why would a designer ever use an *inverted-input-OR-gate* symbol instead of a NAND? Or, why use an *inverted-input-AND-gate* symbol instead of a NOR? In complex logic diagrams you will see both the inverted-input and the inverted-output symbols being used. The designer will use whichever symbol makes more sense for the particular application.

For example, referring to Figure 3–22, let's say you need a HIGH output level whenever either A or B is LOW. It makes sense to draw that circuit as an OR gate with inverted

*See Application K-1 in Appendix K for a CPLD implementation of this circuit.

A and *B* inputs, but you will save two inverters by using a NAND gate to implement the function.

Also, referring to Figure 3–23, let's say you need a HIGH output whenever both *A* and *B* are LOW. You would probably use the *inverted-input-AND gate* for your logic diagram because it makes sense logically, but you would use a NOR gate actually to implement the circuit because you could eliminate the inverters.

The alternative methods of drawing NANDs and NORs are also useful for the simplification of logic circuits. Take, for example, the circuit of Figure 3–24. By changing the NOR gate to an *inverted-input-AND gate,* the inversion bubbles cancel and the equation becomes simply $X = ABCD$. Figure 3–25 summarizes the alternative representations for the inverter, NAND, and NOR gates.

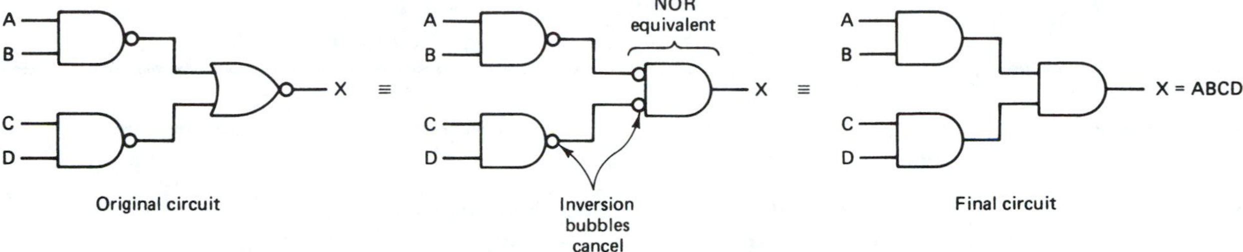

Figure 3–24 Using the alternative NOR symbol eases circuit simplification.

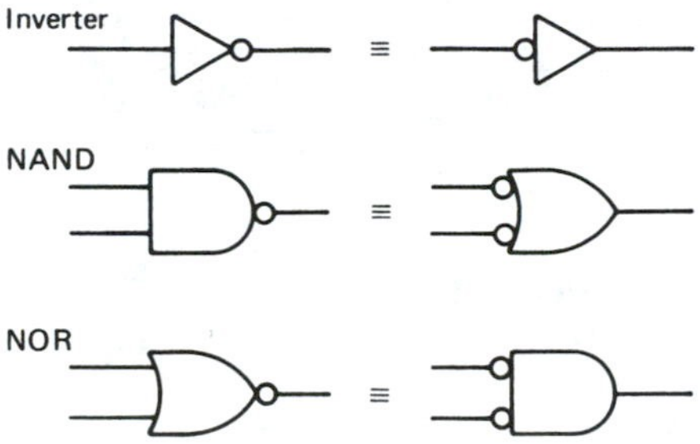

Inverter

NAND

NOR

Figure 3–25 Summary of alternative gate symbols.

The following examples illustrate the application of DeMorgan's theorem for the simplification of logic circuits.

EXAMPLE 3–9

Write the Boolean equation for the circuit shown in Figure 3–26. Use DeMorgan's theorem, then Boolean algebra rules to simplify the equation. Draw the simplified circuit.

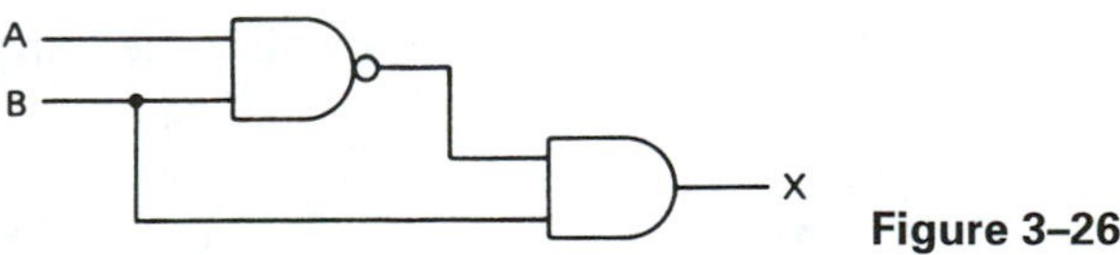

Figure 3–26

Solution:

The Boolean equation at *X* is

$$X = \overline{AB} \cdot B$$

Applying DeMorgan's theorem produces

$$X = (\overline{A} + \overline{B}) \cdot B$$

(Note the use of parentheses to maintain proper grouping.) Using Boolean algebra rules produces

$$X = \overline{A}B + \overline{B}B$$
$$= \overline{A}B + 0$$
$$= \overline{A}B \leftarrow \text{simplified equation}$$

The simplified circuit is shown in Figure 3–27.

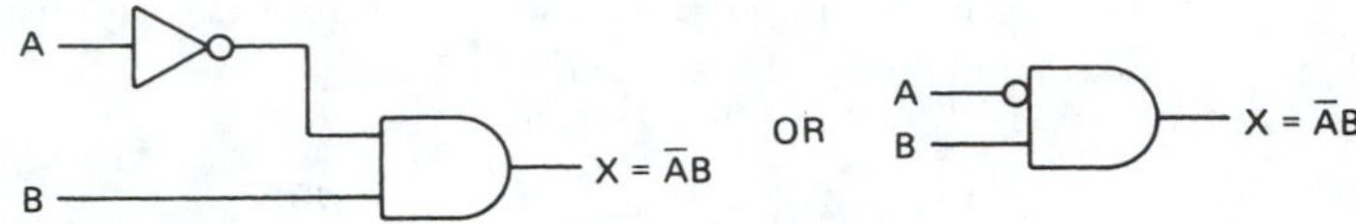

Figure 3–27 Simplified logic circuit for Example 3–9.

EXAMPLE 3–10

Repeat Example 3–9 for the circuit shown in Figure 3–28.

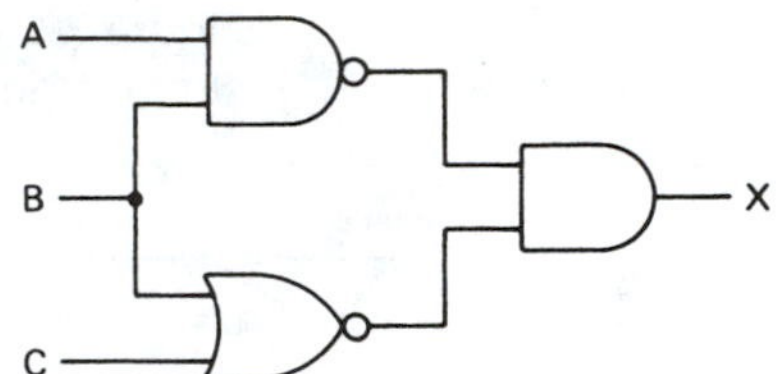

Figure 3–28

Solution:

The Boolean equation at X is

$$X = \overline{AB} \cdot \overline{B + C}$$

Applying DeMorgan's theorem produces

$$X = (\overline{A} + \overline{B}) \cdot \overline{B}\,\overline{C}$$

(Note the use of parentheses to maintain proper grouping.) Using Boolean algebra rules produces

$$X = \overline{A}\,\overline{B}\,\overline{C} + \overline{B}\,\overline{B}\,\overline{C}$$
$$= \overline{A}\,\overline{B}\,\overline{C} + \overline{B}\,\overline{C}$$
$$= \overline{B}\,\overline{C}(\overline{A} + 1)$$
$$= \overline{B}\,\overline{C} \leftarrow \text{simplified equation}$$

The simplified circuit is shown in Figure 3–29.

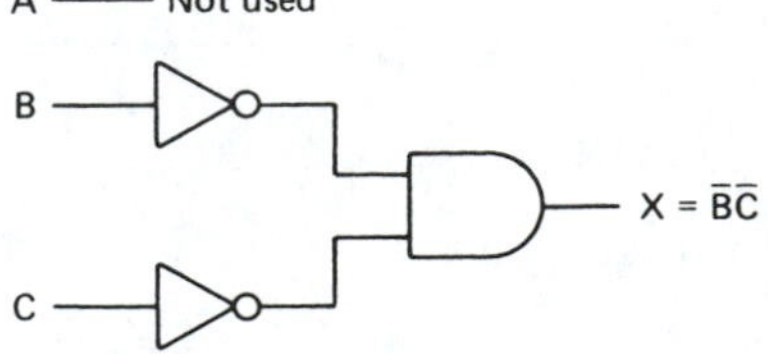

Figure 3–29 Simplified logic circuit for Example 3–10.

Also remember from Figure 3–23 that an AND gate with inverted inputs is equivalent to a NOR gate. Therefore, an equivalent solution to Example 3–10 would be a NOR gate with B and C as inputs, as shown in Figure 3–30.

B C X = $\overline{B} \cdot \overline{C}$ Is equivalent to B C X = $\overline{B + C} = \overline{B} \cdot \overline{C}$

Figure 3–30 Equivalent solution to Example 3–10.

EXAMPLE 3–11

Use DeMorgan's theorem and Boolean algebra on the circuit shown in Figure 3–31 to develop an equivalent circuit that has inversion bars covering only single variables.

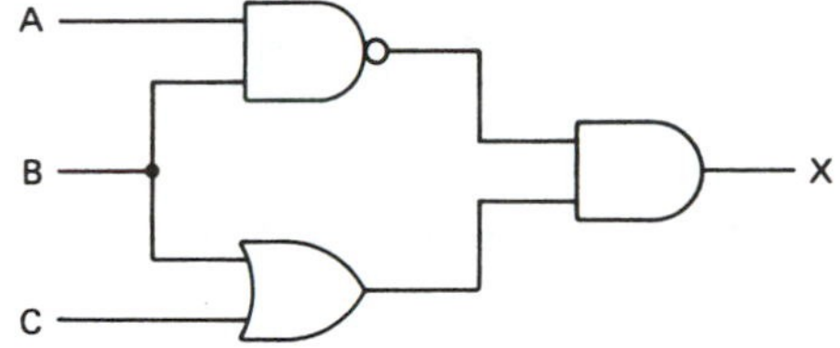

Figure 3–31

Solution:

The Boolean equation at X is

$$X = \overline{AB} \cdot (B + C)$$

Applying DeMorgan's theorem produces

$$X = (\overline{A} + \overline{B}) \cdot (B + C)$$

(Note the use of parentheses to maintain proper grouping.) Using Boolean algebra rules produces

$$X = \overline{A}B + \overline{A}C + \overline{B}B + \overline{B}C$$

$$= \overline{A}B + \overline{A}C + \overline{B}C \leftarrow \text{final equation (sum-of-products form)}$$

The equivalent circuit is shown in Figure 3–32.

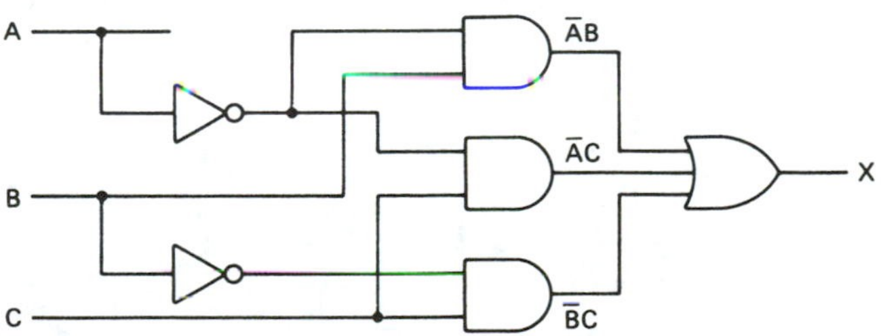

Figure 3–32 Logic circuit equivalent for Example 3–11.

Note that the final equation actually produces a circuit that is more complicated than the original. In fact, if a technician were to build a circuit, he or she would choose the original because it is simpler and has fewer gates. However, the final equation is in a form called the *sum of products*. That form of the equation was achieved by using Boolean algebra and is very useful for building truth tables and Karnaugh maps, which are covered later in this chapter.

***EXAMPLE 3–12**

Draw the logic circuit for the following equation, simplify the equation, and construct a truth table for the simplified equation.

$$X = \overline{A \cdot \overline{B}} + \overline{A \cdot (\overline{A} + C)}$$

*Application K-2 in Appendix K shows a CPLD implementation that is similar to this example.

Solution:

To draw the circuit, we have to reverse our thinking from the previous examples. When we study the equation we see that we need two NANDS feeding into an OR gate, as shown in Figure 3–33a. Then we have to provide the inputs to the NAND gates as shown in Figure 3–33b. Next, we will use DeMorgan's theorem and Boolean algebra to simplify the equation:

$$
\begin{aligned}
X &= \overline{A \cdot \overline{B}} + \overline{A \cdot (\overline{A} + C)} \\
&= (\overline{A} + \overline{\overline{B}} + (\overline{A} + \overline{\overline{A} + C}) \\
&= \overline{A} + B + \overline{A} + \overline{\overline{A}} \cdot \overline{C} \\
&= \overline{A} + \overline{A} + A\overline{C} + B \\
&= \overline{A} + A\overline{C} + B
\end{aligned}
$$

Apply Rule 10:

$$X = \overline{A} + \overline{C} + B \leftarrow \text{simplified equation}$$

Now, to construct a truth table (Table 3–2), we need three input columns (A, B, C), eight entries ($2^3 = 8$), and fill in a 1 for X when $A = 0$ or $C = 0$ or $B = 1$.

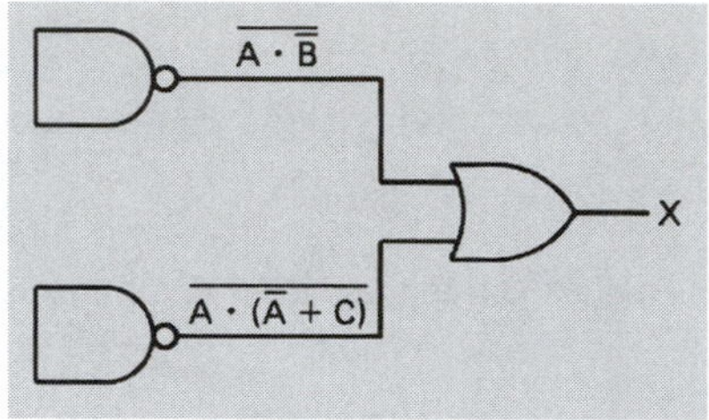

(a)

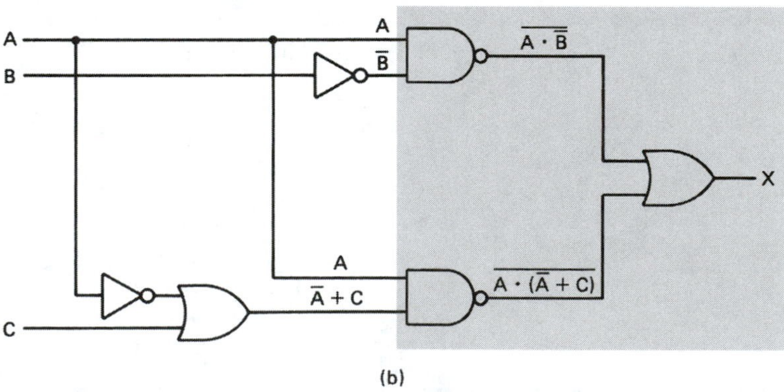

(b)

Figure 3–33 (a) Partial solution to Example 3–12; (b) logic circuit of the equation for Example 3–12.

TABLE 3–2

Truth Table for Example 3–12

A	B	C	$X = \overline{A} + \overline{C} + B$
0	0	0	1
0	0	1	1
0	1	0	1
0	1	1	1
1	0	0	1
1	0	1	0
1	1	0	1
1	1	1	1

*EXAMPLE 3–13

Complete the truth table and timing diagram for the following simplified Boolean equation:

$$X = AB + B\overline{C} + \overline{A}\,\overline{B}C$$

Solution:

The required truth table and timing diagram are shown in Figure 3–34. To fill in the truth for X, we first put a 1 for X when $A = 1$, $B = 1$. Then $X = 1$ for $B = 1$, $C = 0$; then $X = 1$ for $A = 0$, $B = 0$, $C = 1$. All other entries for X are 0.

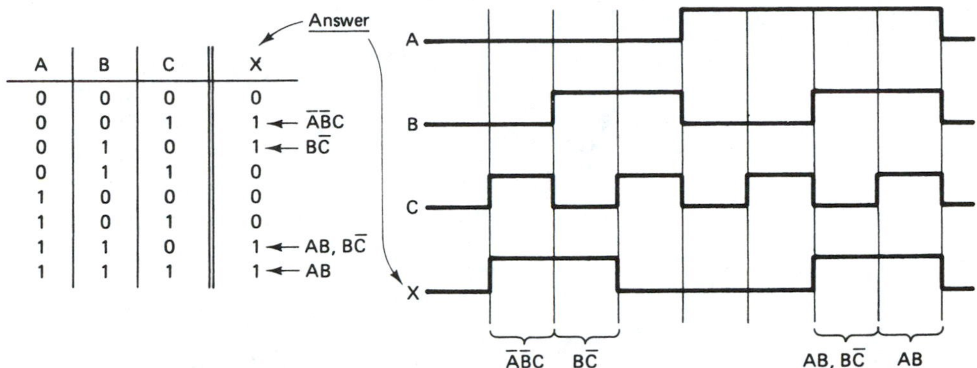

A	B	C	X
0	0	0	0
0	0	1	1 ← $\overline{A}\,\overline{B}C$
0	1	0	1 ← $B\overline{C}$
0	1	1	0
1	0	0	0
1	0	1	0
1	1	0	1 ← AB, $B\overline{C}$
1	1	1	1 ← AB

Figure 3–34 Truth table and timing diagram depicting the logic levels at X for all combinations of inputs.

The timing diagram performs the same functions as the truth table except that it is a more graphic illustration of the HIGH and LOW logic levels of X as the A, B, C inputs change over time. The logic levels at X are filled in the same way as they were for the truth table.

Bubble Pushing

Another trick that can be used, based on DeMorgan's theorem, is called *bubble pushing* and is illustrated in Figure 3-35. As you can see, to form the equivalent logic circuit, you must:

1. Change the logic gate (AND to OR or OR to AND).
2. Add bubbles to the inputs and outputs where there were none, and remove the original bubbles.

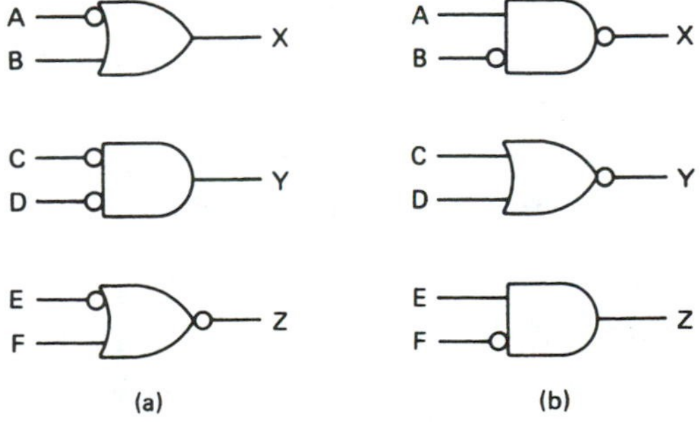

Figure 3–35 (a) Original logic circuits; (b) equivalent logic circuits.

Prove to yourself that this method works by comparing the truth table of each equivalent circuit to the truth table of each original circuit.

*See Application K-3 in Appendix K for a CPLD implementation of this example.

Sum-of-Products (SOP) Expressions

Most Boolean reductions result in an equation in one of two forms:

1. *Product-of-sums* (POS) expression
2. *Sum-of-products* (SOP) expression

The POS expression usually takes the form of two or more ORed variables within parentheses ANDed with two or more other variables within parentheses. Examples of POS expressions are:

$$X = (A + \overline{B}) \cdot (B + C)$$

$$X = (B + \overline{C} + \overline{D}) \cdot (BC + \overline{E})$$

$$X = (A + \overline{C}) \cdot (\overline{B} + E) \cdot (C + B)$$

The SOP expression usually takes the form of two or more variables ANDed together ORed with two or more other variables ANDed together. Examples of SOP expressions are:

$$X = A\overline{B} + AC + \overline{A}BC$$

$$X = AC\overline{D} + \overline{C}D + B$$

$$X = B\overline{C}\,\overline{D} + A\overline{B}DE + CD$$

The SOP expression is used most often because it lends itself nicely to the development of truth tables, timing diagrams, and Karnaugh maps.

For example, let's work with the equation

$$X = \overline{A\overline{B} + \overline{C}D}$$

Using DeMorgan's theorem yields

$$X = \overline{A\overline{B}} \cdot \overline{\overline{C}D}$$

Using DeMorgan's theorem again puts it into a POS format:

$$X = (\overline{A} + B) \cdot (C + \overline{D}) \leftarrow \text{POS}$$

Using the distributive law produces an equation in the SOP format:

$$X = \overline{A}C + \overline{A}\,\overline{D} + BC + B\overline{D} \leftarrow \text{SOP}$$

Now, to fill in a truth table for X using the *SOP* expression, we would put a 1 at X for $A = 0, C = 1$; and for $A = 0, D = 0$; and for $B = 1, C = 1$; and for $B = 1, D = 0$.

3–5 KARNAUGH MAPPING

We learned in previous sections that by using Boolean algebra and DeMorgan's theorem, we can minimize the number of gates that are required to implement a particular logic function. This is very important for the reduction of circuit cost, physical size, and gate failures. You may have found some of the steps in the Boolean reduction process to require ingenuity on your part, and a lot of practice.

Karnaugh mapping, named for its originator, is another method of simplifying logic circuits. It still requires that you reduce the equation to an SOP form, but from there you follow a *systematic approach* that will always produce the simplest configuration possible for the logic circuit.

A Karnaugh map (K-map) is similar to a truth table in that it graphically shows the output level of a Boolean equation for each of the possible input variable combinations. Each output level is placed in a separate *cell* of the K-map. K-maps can be used to simplify equations having two, three, four, five, or six different input variables. Solving five- and six-variable K-maps is extremely cumbersome and can be more practically solved using advanced computer techniques. In this book we will solve two-, three-, and four-variable K-maps.

Determining the number of cells in a K-map is the same as finding the number of combinations or entries in a truth table. A two-variable map will require $2^2 = 4$ cells. A three-variable map will require $2^3 = 8$ cells. A four-variable map will require $2^4 = 16$ cells. The three different K-maps are shown in Figure 3–36.

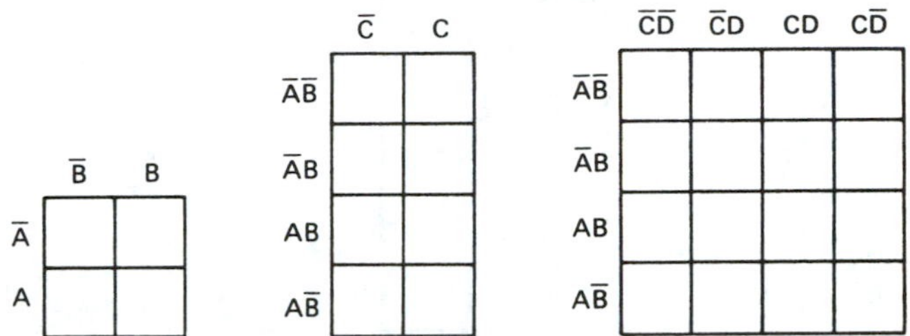

Figure 3–36 Two-variable, three-variable, and four-variable Karnaugh maps.

Each cell within the K-map corresponds to a particular combination of the input variables. For example, in the two-variable K-map, the upper-left cell corresponds to $\overline{A}\,\overline{B}$, the lower-left cell is $A\overline{B}$, the upper-right cell is $\overline{A}B$, and the lower-right cell is AB.

Also notice that when moving from one cell to an adjacent cell, only one variable changes. For example, look at the three-variable K-map. The upper-left cell is $\overline{A}\,\overline{B}\,\overline{C}$, the adjacent cell just below it is $\overline{A}B\overline{C}$. In that case the $\overline{A}\,\overline{C}$ remained the same and only the $\overline{B}$ changed to B. The same holds true for each adjacent cell.

To use the K-map reduction procedure, you must perform the following steps:

1. Transform the Boolean equation to be reduced into an SOP expression.
2. Fill in the appropriate cells of the K-map.
3. Encircle adjacent cells in groups of two, four, or eight. (The more adjacent cells encircled, the simpler the final equation.)
4. Find each term of the final SOP equation by determining which variables remain constant within each circle.

Now, let's consider the equation

$$X = \overline{A}(\overline{B}C + \overline{B}\,\overline{C}) + \overline{A}B\overline{C}$$

First, transform the equation to an SOP expression:

$$X = \overline{A}\overline{B}C + \overline{A}\,\overline{B}\,\overline{C} + \overline{A}B\overline{C}$$

The terms of that SOP expression can be put into a truth table, then transferred to a K-map, as shown in Figure 3–37. Working with the K-map, we will now encircle adjacent 1s in groups of two, four, or eight. We end up with two circles of two cells each, as shown in Figure 3–38. The first circle surrounds the two 1s at the top of the K-map and the second circle surrounds the two 1s in the left column of the K-map.

Once the circles have been drawn encompassing all the 1s in the map, the final simplified equation is obtained by determining *which variables remain the same within each*

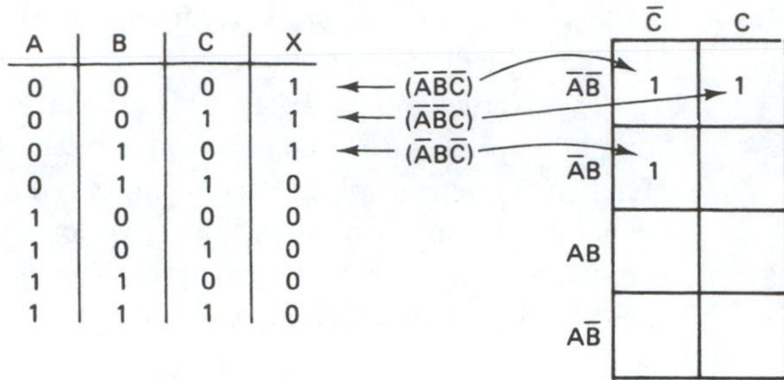

A	B	C	X
0	0	0	1
0	0	1	1
0	1	0	1
0	1	1	0
1	0	0	0
1	0	1	0
1	1	0	0
1	1	1	0

Figure 3–37 Truth table and Karnaugh map of $X = \overline{A}\,\overline{B}\,\overline{C} + \overline{A}\,\overline{B}\,C + \overline{A}B\overline{C}$.

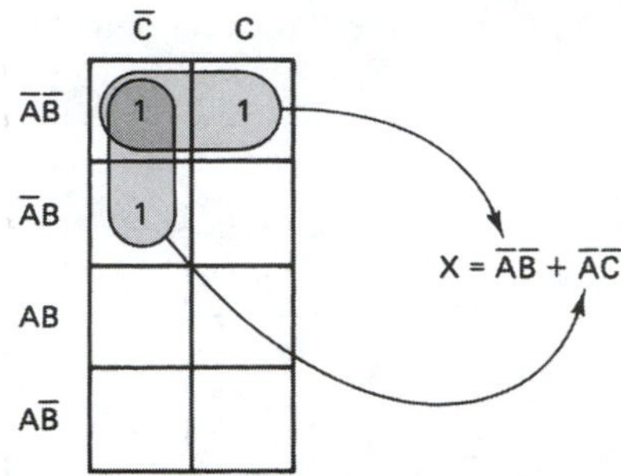

Figure 3–38 Encircling adjacent cells in a Karnaugh map.

circle. Well, the first circle (across the top) encompasses $\overline{A}\,\overline{B}\,\overline{C}$ and $\overline{A}\,\overline{B}C$. The variables that remain the same within the circle are $\overline{A}\,\overline{B}$. Therefore, $\overline{A}\,\overline{B}$ becomes one of the terms in the final SOP equation. The second circle (left column) encompasses $\overline{A}\,\overline{B}\,\overline{C}$ and $\overline{A}B\overline{C}$. The variables that remain the same within that circle are $\overline{A}\,\overline{C}$. Therefore, the second term in the final equation is $\overline{A}\,\overline{C}$.

Since the final equation is always written in the SOP format, the answer is $X = \overline{A}\,\overline{B} + \overline{A}\,\overline{C}$. Actually, the original equation was simple enough that we could have reduced it using standard Boolean algebra. Let's do it just to check our answer:

$$
\begin{aligned}
X &= \overline{A}\,\overline{B}C + \overline{A}\,\overline{B}\,\overline{C} + \overline{A}B\overline{C} \\
&= \overline{A}\,\overline{B}(C + \overline{C}) + \overline{A}B\overline{C} \\
&= \overline{A}\,\overline{B} + \overline{A}B\overline{C} \\
&= \overline{A}(\overline{B} + B\overline{C}) \\
&= \overline{A}(\overline{B} + \overline{C}) \\
&= \overline{A}\,\overline{B} + \overline{A}\,\overline{C} \ \checkmark
\end{aligned}
$$

There are several other points to watch out for when applying the Karnaugh mapping technique. The following examples will be used to illustrate several important points in filling in the map, determining adjacencies, and obtaining the final equation. Work through these examples carefully so that you do not miss any special techniques.

EXAMPLE 3–14

Simplify the following SOP equation using the Karnaugh mapping technique:

$$X = \overline{A}B + \overline{A}\,\overline{B}\,\overline{C} + AB\overline{C} + A\overline{B}\,\overline{C}$$

Solution:

1. Construct an eight-cell K-map (Figure 3–39) and fill in a 1 in each cell that corresponds to a term in the original equation. (Note that $\overline{A}B$ has no C variable in it. Therefore, $\overline{A}B$ is satisfied whether C is HIGH or LOW, so $\overline{A}B$ will fill in two cells: $\overline{A}BC + \overline{A}B\overline{C}$.)

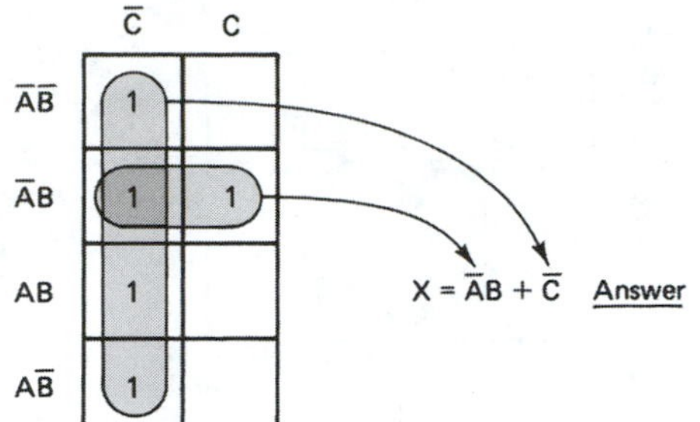

Figure 3–39 Karnaugh map and final equation for Example 3–14.

2. Encircle adjacent cells in the largest group of two or four or eight.
3. Identify the variables that remain the same within each circle and write the final simplified SOP equation by ORing them together.

EXAMPLE 3–15

Simplify the following equation using the Karnaugh mapping procedure:

$$X = \overline{A}B\overline{C}D + A\overline{B}\,\overline{C}D + \overline{A}\,\overline{B}\,\overline{C}D + AB\overline{C}D + AB\overline{C}\,\overline{D} + ABCD$$

Solution:

Since there are four different variables in the equation, we need a 16-cell map ($2^4 = 16$), as shown in Figure 3–40.

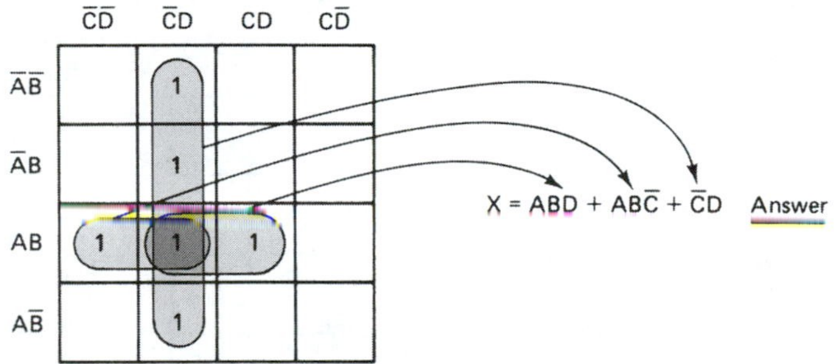

Figure 3–40 Solution to Example 3–15.

EXAMPLE 3–16

Simplify the following equation using the Karnaugh mapping procedure:

$$X = B\overline{C}\,\overline{D} + \overline{A}B\overline{C}D + AB\overline{C}D + \overline{A}BCD + ABCD$$

Solution:

Note in Figure 3–41 that the $B\overline{C}\,\overline{D}$ term in the original equation fills in *two* cells: $AB\overline{C}\,\overline{D} + \overline{A}B\overline{C}\,\overline{D}$. Also note in Figure 3–41 that we could have encircled four cells, then two cells, but that would not have given us the simplest final equation. By encircling four cells, then four cells, we will be sure to get the simplest final equation. (Always encircle the largest number of cells possible, even if some of the cells have already been encircled in another group.)

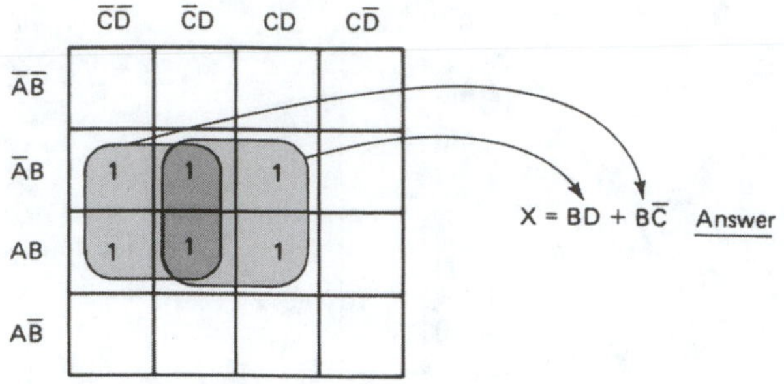

Figure 3–41 Solution to Example 3–16.

EXAMPLE 3–17

Simplify the following equation using the Karnaugh mapping procedure:

$$X = \overline{A}\,\overline{B}\,\overline{C} + A\overline{C}\,\overline{D} + A\overline{B} + ABC\overline{D} + \overline{A}\,\overline{B}C$$

Solution:

Note in Figure 3–42 that a new technique called *wraparound* is introduced. You have to think of the K-map as a continuous cylinder in the horizontal direction, like the label on a soup can. This makes the left row of cells adjacent to the right row of cells. Also, in the vertical direction, a continuous cylinder like a soup can lying on its side makes the top row of cells adjacent to the bottom row of cells. In Figure 3–42, for example, the four top cells are adjacent to the four bottom cells, to combine as eight cells having the variable $\overline{B}$ in common.

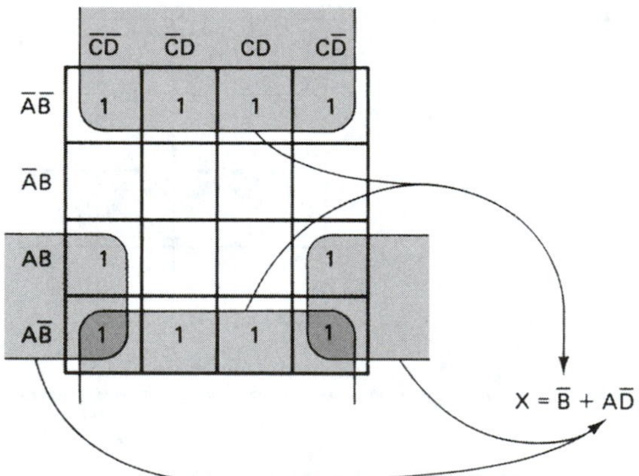

Figure 3–42 Solution to Example 3–17 illustrating the wraparound feature.

Another circle of four is formed by the wraparound adjacencies of the lower-left and lower-right pairs combining to have $A\overline{D}$ in common. The final equation becomes $X = \overline{B} + A\overline{D}$. Compare that simple equation with the original equation that had five terms in it.

EXAMPLE 3–18

Simplify the following equation using the Karnaugh mapping procedure:

$$X = \overline{B}(CD + \overline{C}) + C\overline{D}(\overline{A + B} + AB)$$

Solution:

Before filling in the K-map, an SOP expression must be formed:

$$X = \overline{B}CD + \overline{B}\,\overline{C} + C\overline{D}(\overline{A}\,\overline{B} + AB)$$
$$= \overline{B}CD + \overline{B}\,\overline{C} + \overline{A}\,\overline{B}C\overline{D} + ABC\overline{D}$$

The group of four 1s can be encircled to form $\overline{A}\,\overline{B}$, as shown in Figure 3–43. Another group of four can be encircled using wraparound to form $\overline{B}\,\overline{C}$. That leaves two 1s that are not combined with any others. The unattached 1 in the bottom row can be combined within a group of four, as shown, to form $\overline{B}D$.

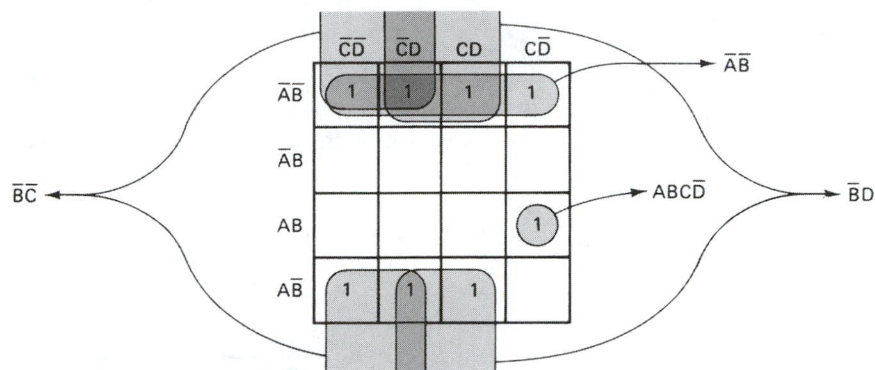

Figure 3–43 Solution to Example 3–18.

The last 1 is not adjacent to any other, so it must be encircled by itself to form $ABC\overline{D}$. The final simplified equation is

$$X = \overline{A}\,\overline{B} + \overline{B}\,\overline{C} + \overline{B}D + ABC\overline{D}$$

EXAMPLE 3–19

Simplify the following equation using the Karnaugh mapping procedure:

$$X = \overline{A}\,\overline{D} + A\overline{B}\,\overline{D} + \overline{A}\,\overline{C}D + \overline{A}CD$$

Solution:

First, the group of eight cells can be encircled, as shown in Figure 3–44. $\overline{A}$ is the only variable present in each cell within the circle, so that the circle of eight simply reduces to $\overline{A}$. (Note that larger circles will reduce to fewer variables in the final equation.)

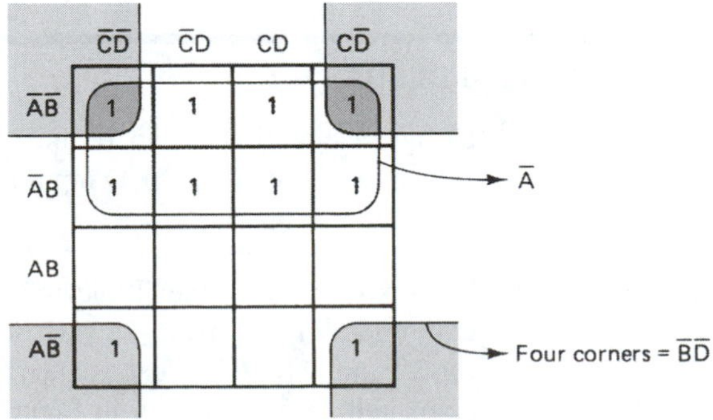

Figure 3–44 Solution to Example 3–19.

Also, all four corners are adjacent to each other because the K-map can be wrapped around in both the vertical *and* horizontal directions. Encircling the four corners results in $\overline{B}\,\overline{D}$. The final equation is

$$X = \overline{A} + \overline{B}\,\overline{D}$$

EXAMPLE 3–20

Simplify the following equation using the Karnaugh mapping procedure:

$$X = \overline{A}\,\overline{B}\,\overline{D} + A\overline{C}\,\overline{D} + \overline{A}B\overline{C} + AB\overline{C}D + A\overline{B}C\overline{D}$$

Solution:

Encircling the four corners forms $\overline{B}\,\overline{D}$, as shown in Figure 3–45. The other group of four forms $B\overline{C}$. You may be tempted to encircle the $\overline{C}\,\overline{D}$ group of four as shown by the dotted line, but that would be *redundant* because each of those 1s is already contained within an existing circle. Therefore, the final equation is

$$X = \overline{B}\,\overline{D} + B\overline{C}$$

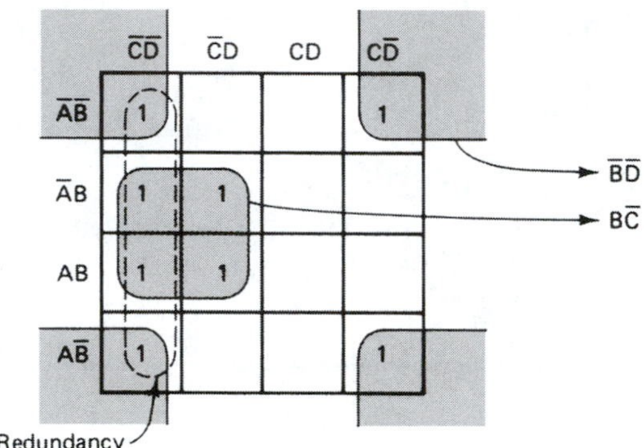

Figure 3–45 Solution to Example 3–20.

3–6 SYSTEM DESIGN APPLICATIONS

Let's summarize the entire chapter now by working through two complete design problems. The following examples illustrate practical applications of a K-map to ensure that when we implement the circuit we will have the simplest possible solution.

Note: The construction of digital circuits with higher complexity than those of these examples will be more practically suited for implementation using what is called *programmable logic devices,* which are discussed in Chapter 8 and Appendix K.

SYSTEM DESIGN 3–1

Design a circuit that can be built using logic gates that will output a HIGH (1) whenever the 4-bit BCD input is an odd number from 0 to 9.

Solution:

First, build a truth table (Table 3–3) to identify which BCD codes from 0 to 9 produce odd numbers. (Use the variable A to represent the 2^0 BCD input, B for 2^1, C for 2^2, and D for 2^3.) Next, reduce that equation into its simplest form by using a Karnaugh map, as shown in Figure 3–46a. Finally, using basic logic gates, the circuit can be constructed, as shown in Figure 3–46b.

TABLE 3–3

Truth Table Used to Determine the Equation for Odd Numbers[a] from 0 to 9

D	C	B	A	DEC	
0	0	0	0	0	
0	0	0	1	1	$\leftarrow A\overline{B}\,\overline{C}\,\overline{D}$
0	0	1	0	2	
0	0	1	1	3	$\leftarrow AB\overline{C}\,\overline{D}$
0	1	0	0	4	
0	1	0	1	5	$\leftarrow A\overline{B}C\overline{D}$
0	1	1	0	6	
0	1	1	1	7	$\leftarrow ABC\overline{D}$
1	0	0	0	8	
1	0	0	1	9	$\leftarrow A\overline{B}\,\overline{C}D$

[a]Odd number = $A\overline{B}\,\overline{C}\,\overline{D} + AB\overline{C}\,\overline{D} + A\overline{B}C\overline{D} + ABC\overline{D} + A\overline{B}\,\overline{C}D$.

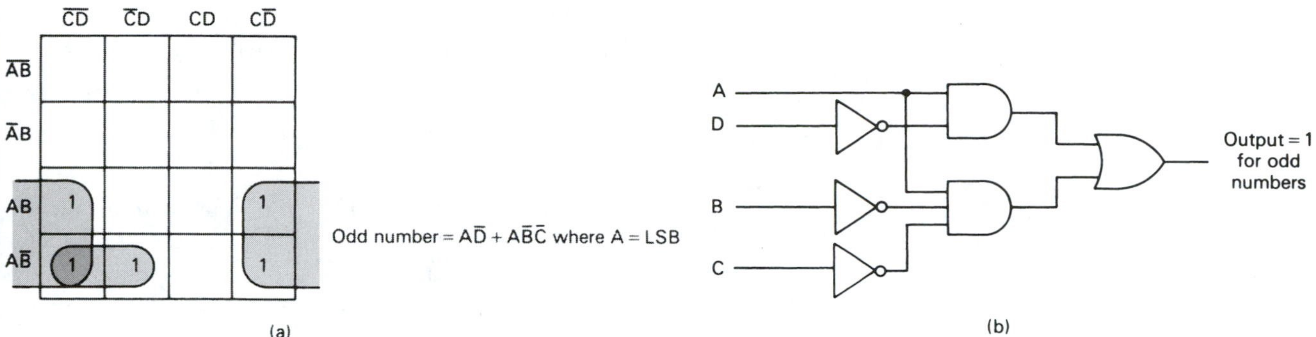

Figure 3–46 (a) Simplified equation derived from a Karnaugh map; (b) logic circuit for the "odd-number decoder."

SYSTEM DESIGN 3–2

A chemical plant needs an alarm system developed to warn of critical conditions in one of its chemical tanks. The tank has four HIGH/LOW (1/0) switches, monitoring temperature (T), pressure (P), fluid level (L), and weight (W). Design a system that will activate an alarm when any of the following conditions arise:

1. A high fluid level with a high temperature and a high pressure
2. A low fluid level with a high temperature and a high weight
3. A low fluid level with a low temperature and a high pressure
4. A low fluid level with a low weight and a high temperature

Solution:

First, write in Boolean equation form, the conditions that will activate the alarm:

$$\text{Alarm} = LTP + \overline{L}TW + \overline{L}\,\overline{T}P + \overline{L}\,\overline{W}T$$

Next, factor the equation into its simplest form by using a Karnaugh map, as shown in Figure 3–47a. Finally, the logic circuit can be constructed, as shown in Figure 3–47b.

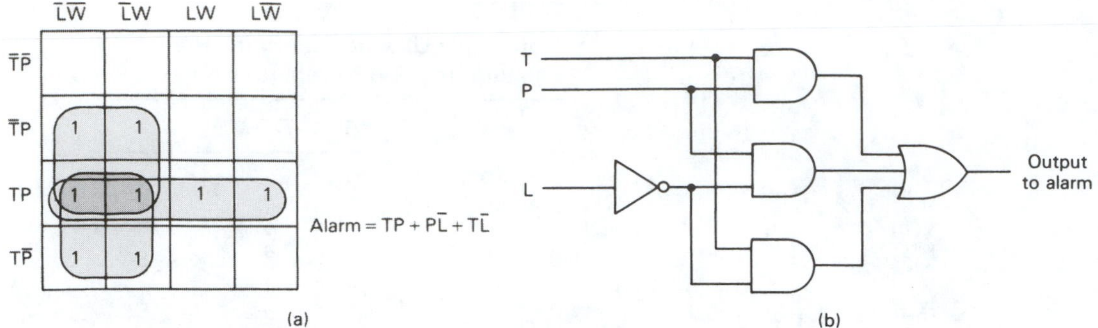

Figure 3–47 (a) Simplified equation derived from a Karnaugh map; (b) logic circuit for the "chemical tank alarm."

3–7 ARITHMETIC CIRCUITS

Binary adder circuits are the building blocks of the modern computer. The basic gates covered in previous sections can be used to implement hardware that will input binary data and provide a binary output of their sum.

Basic Adder Circuit

By reviewing the truth table in Figure 3–48, we can determine the input conditions that produce each combination of sum and carry output bits. Figure 3–48 shows the addition of two 2-bit numbers. This could easily be expanded to cover 4-, 8-, or 16-bit addition. Note that addition in the least significant bit column requires analyzing only two inputs (A_0 plus

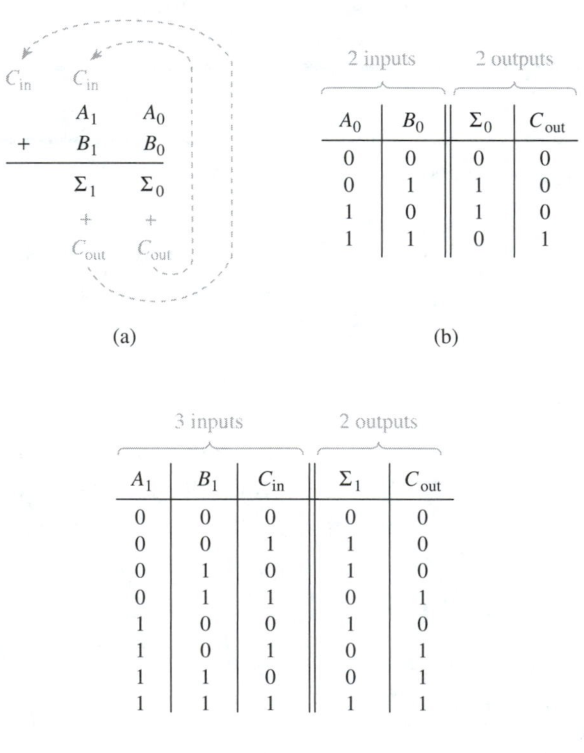

2 inputs		2 outputs	
A_0	B_0	Σ_0	C_{out}
0	0	0	0
0	1	1	0
1	0	1	0
1	1	0	1

(b)

3 inputs			2 outputs	
A_1	B_1	C_{in}	Σ_1	C_{out}
0	0	0	0	0
0	0	1	1	0
0	1	0	1	0
0	1	1	0	1
1	0	0	1	0
1	0	1	0	1
1	1	0	0	1
1	1	1	1	1

(c)

Figure 3–48 (a) Addition of two 2-bit binary numbers; (b) truth table for the LSB addition; (c) truth table for the more significant column.

B_0) to determine the output sum (Σ_0) and carry (C_{out}). But any more significant columns (2^1 column and up) require the inclusion of a third input, which is the carry (C_{in}) from the column to its right. For example, the carry-out (C_{out}) of the 2^0 column becomes the carry-in (C_{in}) to the 2^1 column. Figure 3–48(c) shows the inclusion of a third input for the truth table for the more significant column additions.

Half-Adder

Designing logic circuits to automatically implement the desired outputs for these truth tables is simple. Look at the LSB truth table; for what input conditions is the Σ_0 bit HIGH? The answer is *A or B* HIGH but *not both* (exclusive-OR function). For what input condition is the C_{out} bit HIGH? The answer is *A and B* HIGH (AND function). Therefore, the circuit design to perform addition in the LSB column can be implemented using an exclusive-OR and an AND gate. That circuit is called a *half-adder* and is shown in Figure 3–49. If the exclusive-OR function in Figure 3–49 is implemented using an AND–NOR–NOR configuration, we can tap off the AND gate for the carry, as shown in Figure 3–50.

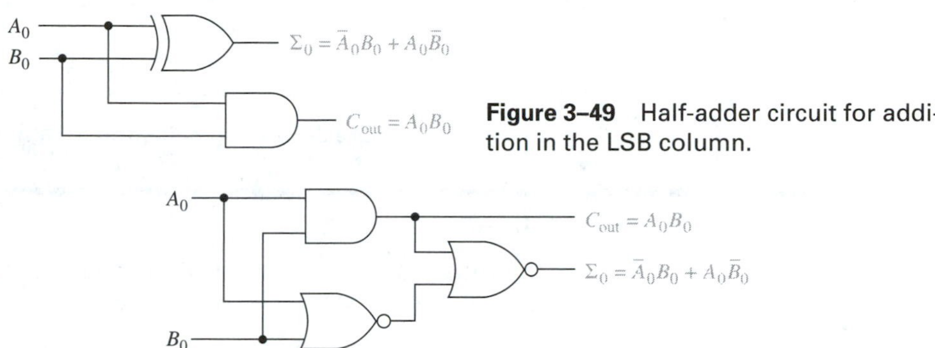

Figure 3–49 Half-adder circuit for addition in the LSB column.

Figure 3–50 Alternative half-adder circuit built from an AND–NOR–NOR configuration.

Full-Adder

As you can see in Figure 3–48, addition in the 2^1 (or higher) column requires three inputs to produce the sum (Σ_1) and carry (C_{out}) outputs. Look at the truth table [Figure 3–48(c)]; for what input conditions is the sum output (Σ_1) HIGH? The answer is that the Σ_1 bit is HIGH whenever the three inputs (A_1, B_1, C_{in}) are *odd.* The circuit in Figure 3–51 produces a HIGH output whenever the sum of the inputs is odd. We will use this circuit to generate our Σ_1 output bit.

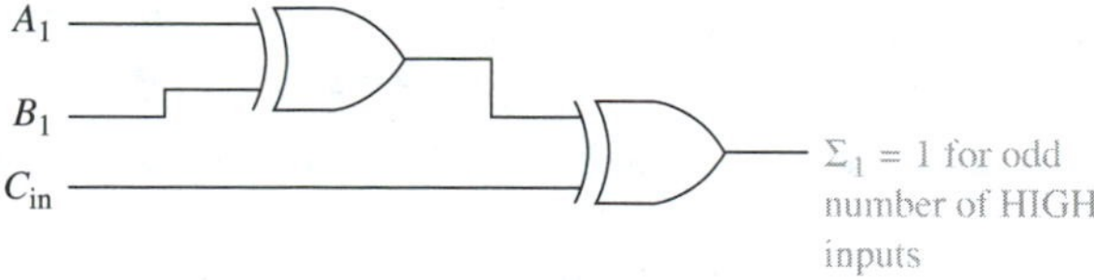

Figure 3–51 The sum (Σ_1) function of the full-adder.

How about the carry-out (C_{out}) bit? What input conditions produce a HIGH at C_{out}? The answer is that C_{out} is HIGH whenever any two of the inputs are HIGH. Therefore, we can take care of C_{out} with three ANDs and an OR, as shown in Figure 3–52.

The two parts of the full-adder circuit shown in Figures 3–51 and 3–52 can be combined to form the complete *full-adder* circuit shown in Figure 3–53. In the figure the Σ_1 function is produced using the same logic as that in Figure 3–51 (an Ex-OR feeding an Ex-OR). The C_{out} function comes from A_1B_1 or C_{in} $(A_1\overline{B}_1 + \overline{A}_1B_1)$. Prove to yourself that the Boolean equation at C_{out} will produce the necessary result. [*Hint:* Write the equation for C_{out} from the truth table in Figure 3–48(c).] Also, Example 3–21 will help you better understand the operation of the full-adder.

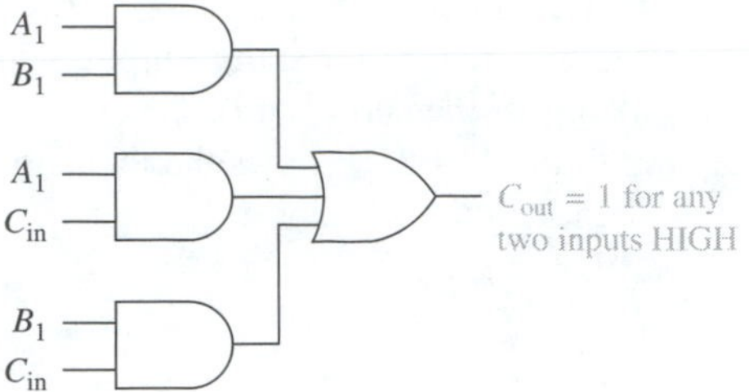

Figure 3–52 Carry-out (C_{out}) function of the full-adder.

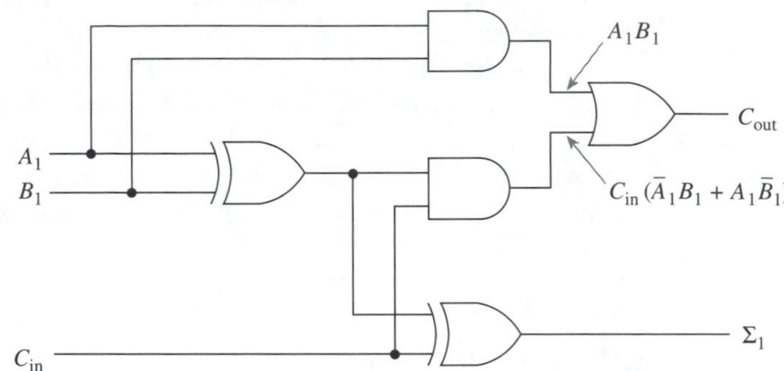

***Figure 3–53** Logic diagram of a full-adder.

EXAMPLE 3–21

Apply the following input bits to the full-adder of Figure 3–53 to verify its operation ($A_1 = 0, B_1 = 1, C_{in} = 1$).

Solution:

The full-adder operation is shown in Figure 3–54.

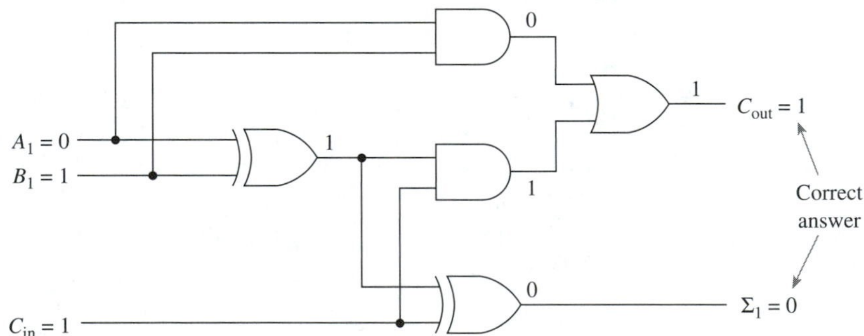

Figure 3–54 Full-adder operation for Example 3–21.

Block Diagrams

Now that we know the construction of half-adder and full-adder circuits, we can simplify their representation by just drawing a box with the input and output lines, as shown in Figure 3–55. When drawing multibit adders, a *block diagram* is used to represent the addition in each column. For example, in the case of a 4-bit adder, the 2^0 column needs only a half-adder, because there will be no carry-in. Each of the more significant columns requires a full-adder, as shown in Figure 3–56.

Note in Figure 3–56 that the LSB half-adder has no carry-in. The carry-out (C_{out}) of the LSB becomes the carry-in (C_{in}) to the next full-adder to its left. The carry-out (C_{out}) of the MSB full-adder is actually the highest-order sum output (Σ_4).

*See Application K-5 in Appendix K for a CPLD implementation of this circuit.

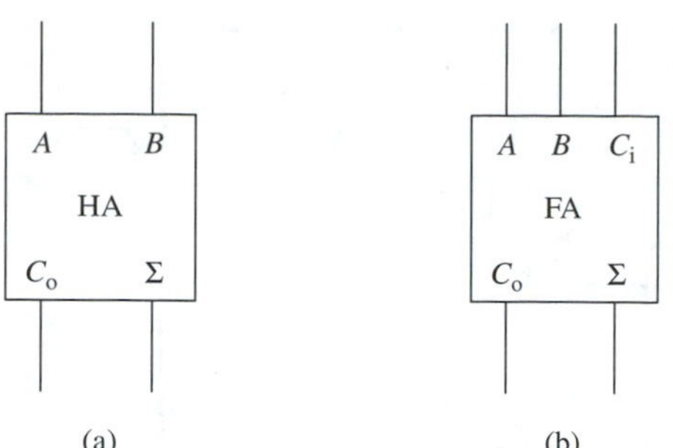

Figure 3–55 Block diagrams of (a) half-adder; (b) full-adder.

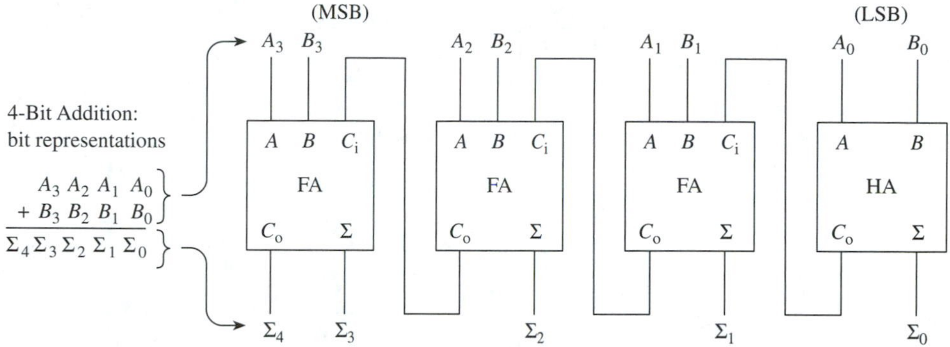

Figure 3–56 Block diagram of a 4-bit binary adder.

3–8 FOUR-BIT FULL-ADDER ICs

Medium-scale-integration (MSI) ICs are available with four full-adders in a single package. Table 3–4 lists the most popular adder ICs. Each adder in the table contains four full-adders, and all are functionally equivalent; however, their pin layouts differ (refer to your data manual for the pin layouts). They each will add two 4-bit *binary words* plus one incoming carry. The binary sum appears on the sum outputs (Σ_1 to Σ_4) and the outgoing carry.

TABLE 3–4
MSI Adder ICs

Device	*Family*	*Description*
7483	TTL	4-Bit binary full-adder, fast carry
74HC283	CMOS	4-Bit binary full-adder, fast carry
4008	CMOS	4-Bit binary full-adder, fast carry

Figure 3–57 shows the functional diagram, the logic diagram, and the logic symbol for the 7483. In the figure the least significant binary inputs (2^0) come into the A_1B_1 terminals, and the most significant (2^3) come into the A_4B_4 terminals. (Be careful; depending on which manufacturer's data manual you are using, the inputs may be labeled A_1B_1 to A_4B_4 or A_0B_0 to A_3B_3). The carry-out (C_{out}) from each full-adder is *internally connected* to the carry-in of the next full-adder. The carry-out of the last full-adder is brought out to a terminal to be used as the sum$_5$ (Σ_5) output or to be used as a carry-in (C_{in}) to the next full-adder IC if more than 4 bits are to be added (as in Example 3–22).

Something else that we have not seen before is the *fast-look-ahead carry* [see Figure 3–57(a)]. This is very important for speeding up the arithmetic process. For example, if we were adding two 8-bit numbers using two 7483s, the fast-look-ahead carry evaluates the four *low-order* inputs (A_1B_1 to A_4B_4) to determine if they are going to produce a carry-out of the fourth full-adder to be passed on to the next-higher-order adder IC (see Example 3–22). In this way the addition of the *high-order* bits (2^4 to 2^7) can take place concurrently

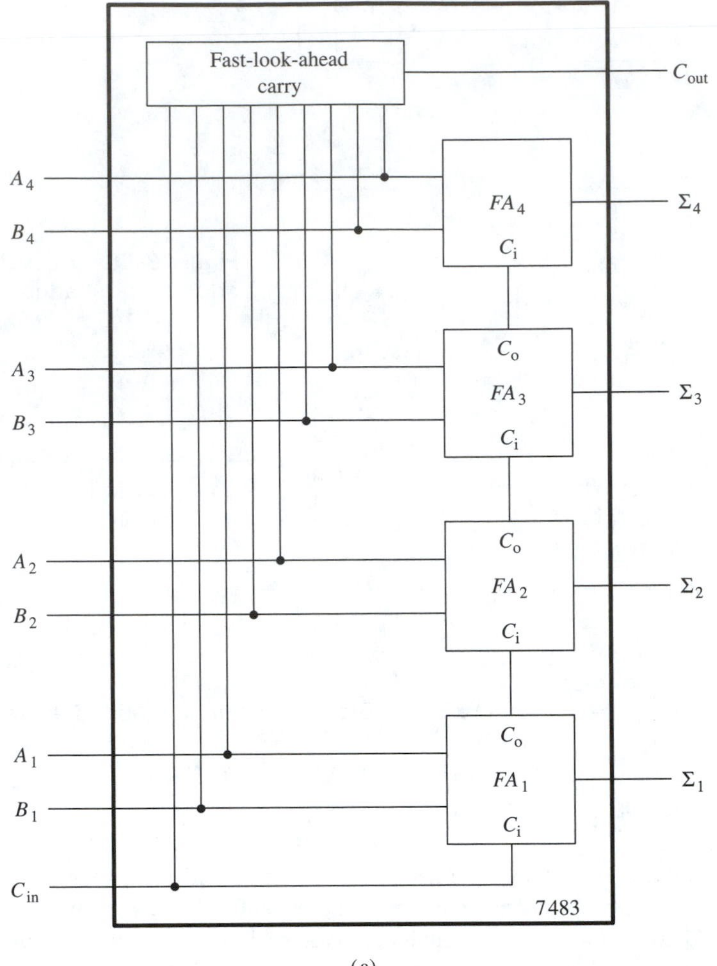

(a)

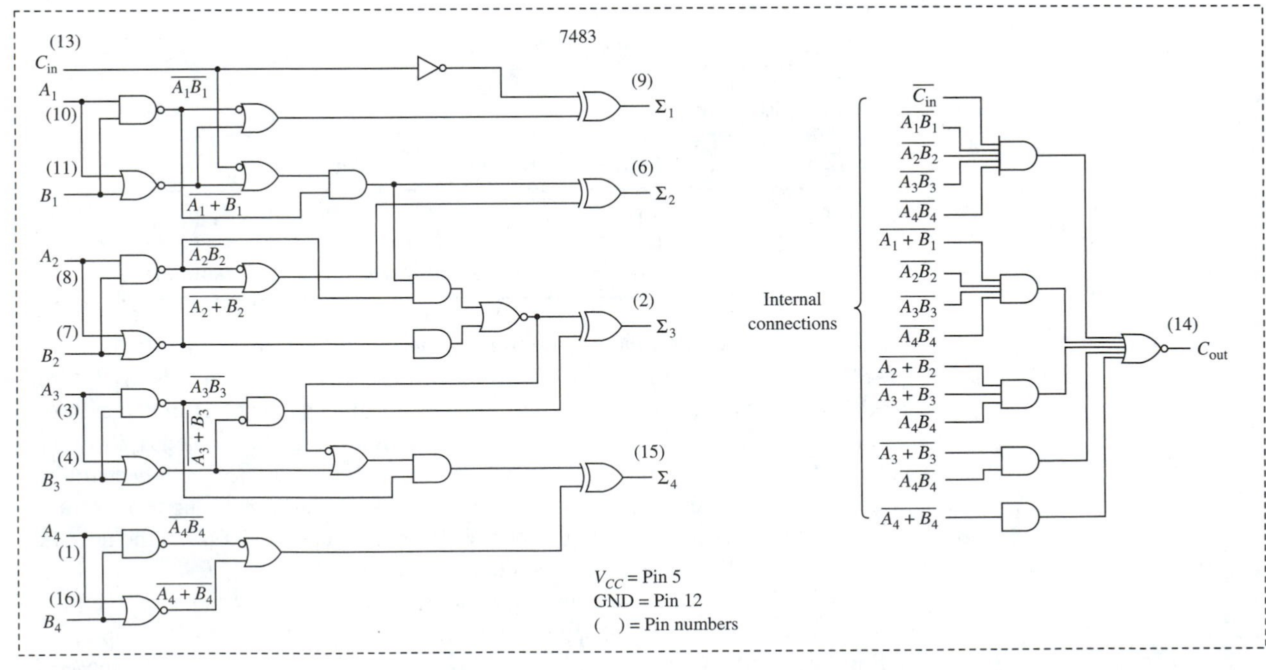

(b)

Figure 3–57 The 7483 4-bit full-adder: (a) functional diagram; (b) logic diagram. [(b) Courtesy of Philips Semiconductors.]

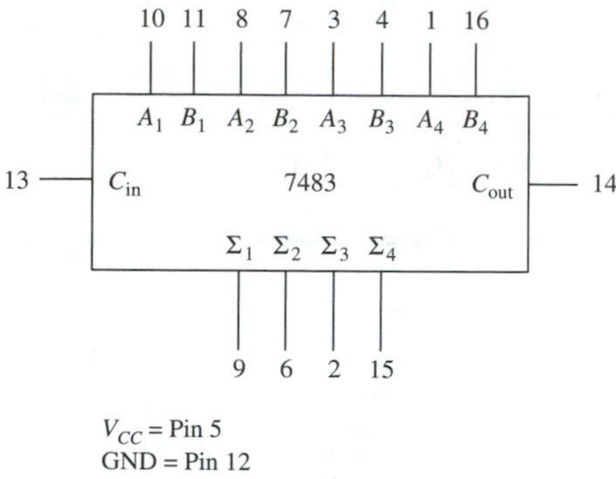

Figure 3–57 *(Continued)* (c) logic symbol.

with the low-order (2^0 to 2^3) addition *without having to wait* for the carries to propagate, or *ripple* through FA_1 to FA_2 to FA_3 to FA_4 to become available to the high-order addition. A discussion of the connections for the addition of two 8-bit numbers using two 7483s is presented in the following example.

*EXAMPLE 3–22

Show the external connections to two 4-bit adder ICs to form an 8-bit adder capable of performing the following addition:

$$\begin{array}{r} A_7A_6A_5A_4A_3A_2A_1A_0 \\ +\ \underline{B_7B_6B_5B_4B_3B_2B_1B_0} \\ \Sigma_8\Sigma_7\Sigma_6\Sigma_5\Sigma_4\Sigma_3\Sigma_2\Sigma_1\Sigma_0 \end{array}$$

Solution:

We can choose any of the IC adders listed in Table 3–4 for our design. Let's choose the 74HC283, which is the high-speed CMOS version of the 4-bit adder (it has the same logic symbol as the 7483). As you can see in Figure 3–58, the two 8-bit numbers are brought into the A_1B_1-to-A_4B_4 inputs of each chip, and the sum output comes out of the Σ_4-to-Σ_1 outputs of each chip.

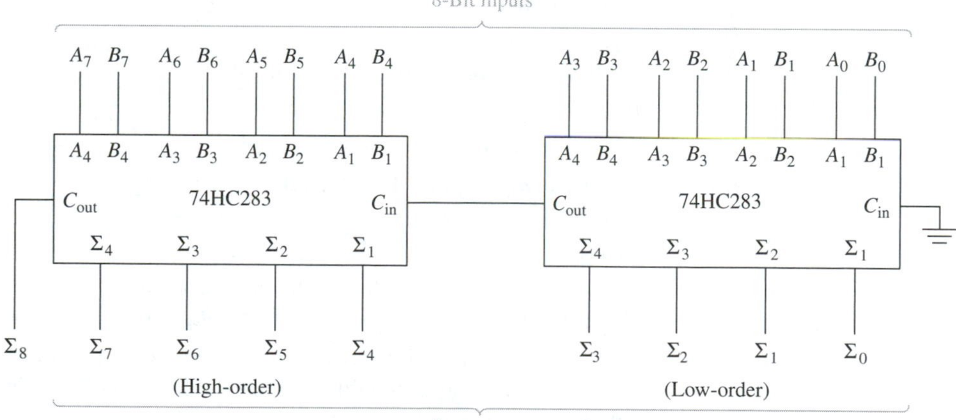

Figure 3–58 Eight-bit binary adder using two 74HC283 ICs.

The C_{in} of the least significant addition ($A_0 + B_0$) is grounded (0) because there is no carry-in (it acts like a half-adder), and if it were left floating, the IC would not know whether to assume a 1 state or 0 state.

The carry-out (C_{out}) from the addition of $A_3 + B_3$ must be connected to the carry-in (C_{in}) of the $A_4 + B_4$ addition, as shown. The fast-look-ahead carry

*See Application K-6 in Appendix K for a CPLD implementation of the 74283.

circuit ensures that the carry-out (C_{out}) signal from the low-order addition is provided in the carry-in (C_{in}) of the high-order addition within a very short period of time so that the $A_4 + B_4$ addition can take place without having to wait for all the internal carries to propagate through all four of the low-order additions first.

SUMMARY

In this chapter we have learned that

1. Several logic gates can be connected together to form combinational logic.
2. There are several Boolean laws and rules that provide the means to form equivalent circuits.
3. Boolean algebra is used to reduce logic circuits to simpler equivalents that function exactly like the original circuit.
4. DeMorgan's theorem is required in the reduction process whenever inversion bars cover more than one variable in the original Boolean equation.
5. NAND and NOR gates are sometimes referred to as *universal gates* because they can be used to form any of the other gates.
6. AND-OR-INVERT gates are often used to implement sum-of-products equations.
7. Karnaugh mapping provides a systematic method of reducing logic circuits.

GLOSSARY

Adjacent Cell: Cells within a Karnaugh map are considered adjacent if they border each other on one side or the top or bottom of the cell.

Boolean Reduction: An algebraic technique that follows specific rules in order to convert a Boolean equation into a simpler form.

Cell: Each box within a Karnaugh map is a cell. Each cell corresponds to a particular combination of input variable logic levels.

Combinational Logic: Logic circuits formed by combining several of the basic logic gates together to form a more complex function.

DeMorgan's Theorem: A Boolean law used for equation reduction that allows the user to convert an equation having an inversion bar over several variables into an equivalent equation having inversion bars over single variables only.

Don't Care: A variable appearing in a truth table or timing waveform that will have no effect on the final output regardless of the logic level of the variable. Therefore, don't-care variables can be ignored.

Equivalent Circuit: A simplified version of a logic circuit that can be used to perform the exact logic function of the original complex circuit.

Inversion Bubbles: An alternative to drawing the triangular inversion symbol. The bubble (or circle) can appear at the input or output of a logic gate.

Karnaugh Map: A two-dimensional table of Boolean output levels used as a tool to perform a systematic reduction of complex logic circuits into simplified equivalent circuits.

Product-of-Sums (POS) Form: A Boolean equation in the form of a group of ORed variables ANDed with another group of ORed variables, for example,

$$X = (A + \overline{B} + C)(B + D)(\overline{A} + \overline{C})$$

Redundancy: Once all filled-in cells in a Karnaugh map are contained within a circle, the final simplified equation can be written. Drawing another circle around a different group of cells is redundant.

Sum-of-Products (SOP) Form: A Boolean equation in the form of a group of ANDed variables ORed with another group of ANDed variables, for example,

$$X = ABC + \overline{B}DE + \overline{A}\,\overline{D}$$

Wraparound: The left and right cells and the top and bottom cells of a Karnaugh map are actually adjacent to each other by means of the wraparound feature.

PROBLEMS

3–1. Write the Boolean equation for each of the logic circuits shown in Figure P3–1.

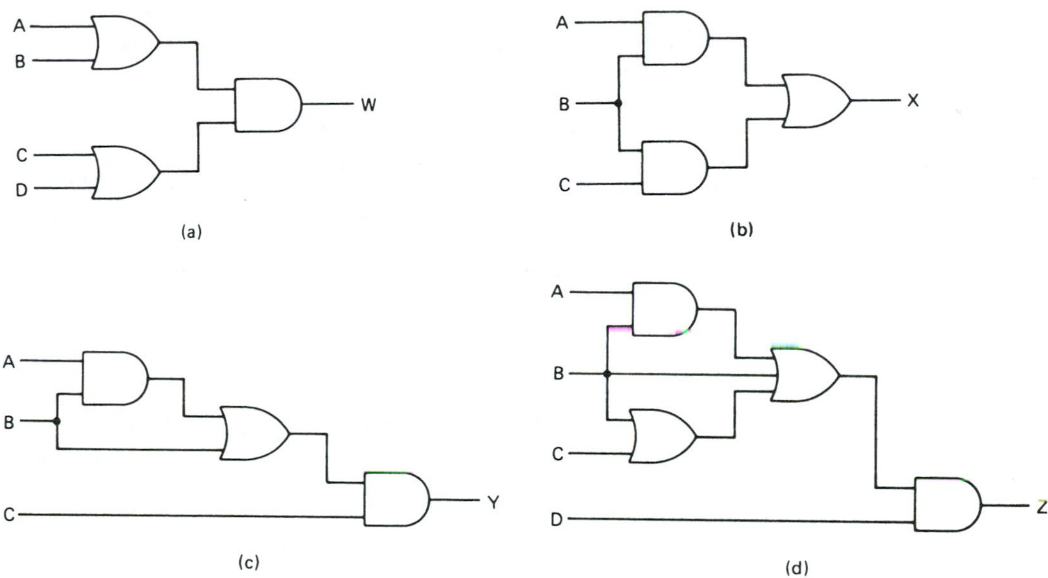

Figure P3–1

3–2. Draw the logic circuit that would be used to implement the following Boolean equations.

(a) $M = (AB) + (C + D)$
(b) $N = (A + B + C)D$
(c) $P = (AC + BC)(A + C)$
(d) $Q = (A + B)BCD$
(e) $R = BC + D + AD$
(f) $S = B(A + C) + AC + D$

3–3. Construct a truth table for each of the equations given in Problem 3–2.

3–4. Write the Boolean equation, then complete the timing diagram at *W*, *X*, *Y*, and *Z* for the logic circuits shown in Figure P3–4.

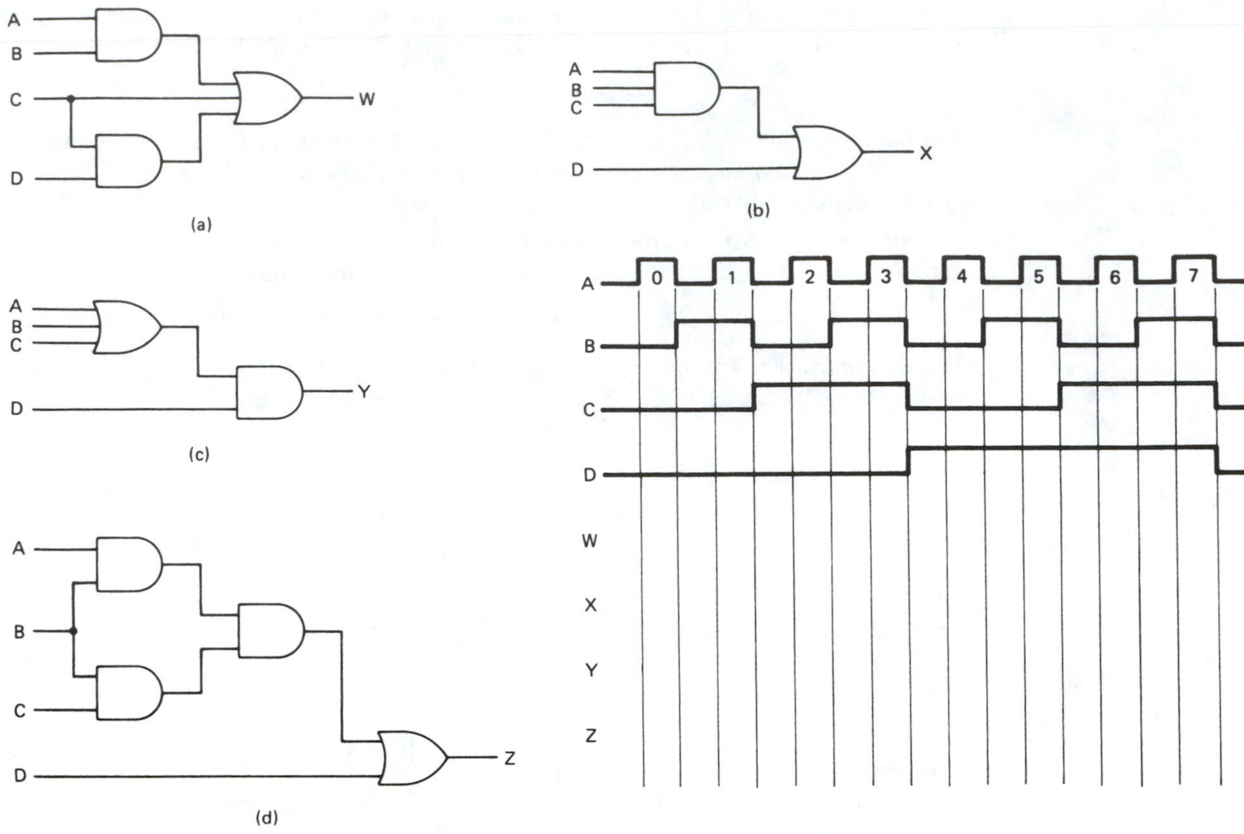

Figure P3–4

3–5. Using the 10 Boolean rules presented in Table 3–1, determine the outputs of the logic circuits shown in Figure P3–5.

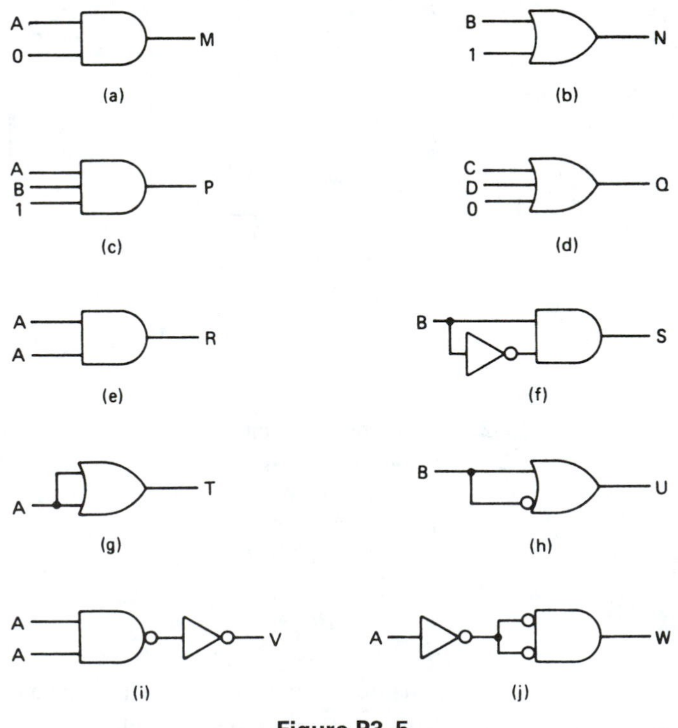

Figure P3–5

3–6. Draw the logic circuit for the following equations. Simplify the equations and draw the simplified logic circuit.

(a) $V = AC + ACD + CD$

(b) $W = (BCD + C)CD$

(c) $X = (B + D)(A + C) + ABD$

(d) $Y = AB + BC + ABC$

(e) $Z = ABC + CD + CDE$

3–7. Construct a truth table for each of the simplified equations of Problem 3–6.

3–8. The pin layouts for a 74HCT08 CMOS AND gate and a 74HCT32 CMOS OR gate are given in Figure P3–8. Make the external connections to the chips to implement the following logic equation. (Simplify the logic equation first.)

$$X = (A + B)(D + C) + ABD$$

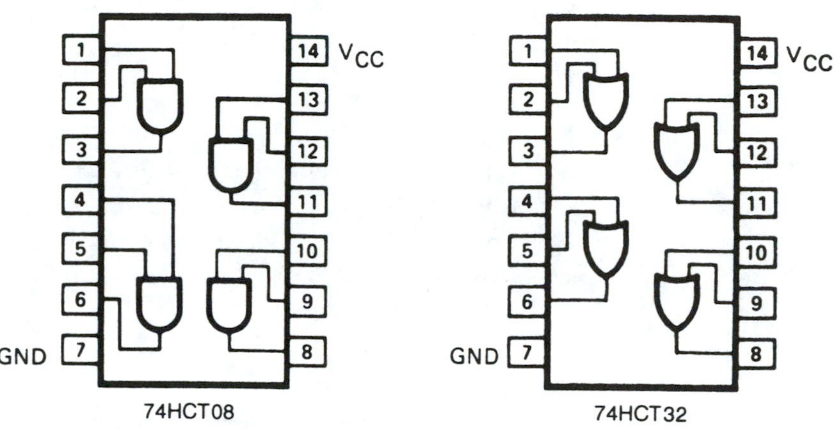

Figure P3–8

3–9. Repeat Problem 3–8 for the following equation:

$$Y = AB(C + BD) + BD$$

3–10. Write a sentence describing how DeMorgan's theorem is applied in the simplification of a logic equation.

3–11. **(a)** DeMorgan's theorem can be used to prove that an OR gate with inverted inputs is equivalent to what type of gate?

(b) An AND gate with inverted inputs is equivalent to what type of gate?

3–12. Use DeMorgan's theorem to prove that a NOR gate with inverted inputs is equivalent to an AND gate.

3–13. Draw the logic circuit for the following equations. Apply DeMorgan's theorem and Boolean algebra rules to reduce them to equations having inversion bars over single variables only. Draw the simplified circuit.

(a) $W = \overline{AB} + \overline{A + C}$

(b) $X = A\overline{B + C} + \overline{BC}$

(c) $Y = \overline{(AB) + C} + B\overline{C}$

(d) $Z = \overline{AB + (\overline{A} + C)}$

3–14. Write the Boolean equation for the circuits of Figure P3-14. Use DeMorgan's theorem and Boolean algebra rules to simplify the equation. Draw the simplified circuit.

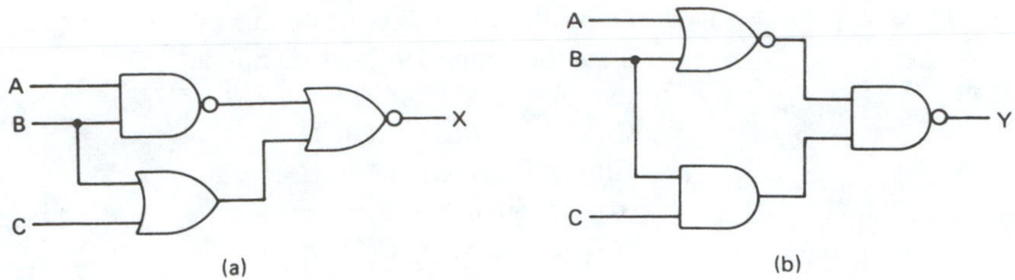

Figure P3–14

3–15. Draw a logic circuit that will put out a 1 (HIGH) if A and B are both 1, while either C or D is 1.

3–16. Draw a logic circuit that will put out a 0 if A or B is 0.

3–17. Draw a logic circuit that will put out a LOW if A or B is HIGH while C or D is LOW.

3–18. Draw a logic circuit that will put out a HIGH if only one of the inputs A, B, or C is LOW.

3–19. Complete a truth table for the following simplified Boolean equations.

(a) $W = A\overline{B}\,\overline{C} + \overline{B}C + \overline{A}B$

(b) $X = \overline{A}\,\overline{B} + A\overline{B}C + B\overline{C}$

(c) $Y = \overline{C}D + \overline{A}\,\overline{B}\,\overline{C}\,\overline{D} + BCD + \overline{A}C\overline{D}$

(d) $Z = \overline{A}C + C\overline{D} + \overline{B}\,\overline{C} + \overline{A}BC\overline{D}$

3–20. Complete the timing diagram in Figure P3–20 for the following simplified Boolean equations.

(a) $X = \overline{A}\,\overline{B}\,\overline{C} + ABC + A\overline{C}$

(b) $Y = \overline{B} + \overline{A}B\overline{C} + AC$

(c) $Z = B\overline{C} + A\overline{B} + \overline{A}BC$

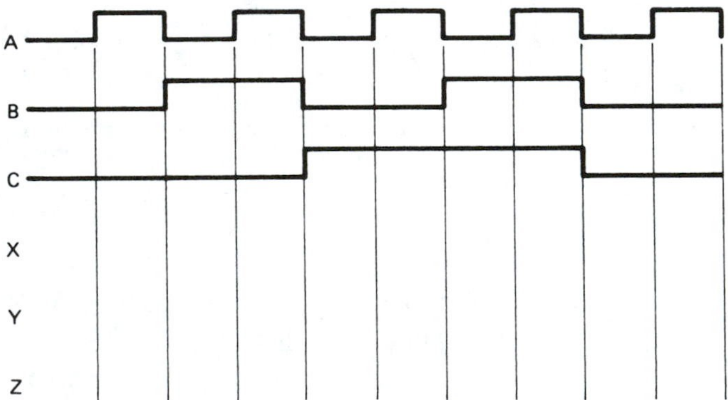

Figure P3–20

3–21. Draw the connections required to convert:

(a) A NAND gate into an inverter

(b) A NOR gate into an inverter

3–22. Identify each of the following Boolean equations as a product-of-sums (POS) expression or sum-of-products (SOP) expression, or both.

(a) $U = A\overline{B}C + BC + \overline{A}C$

(b) $V = (A + C)(\overline{B} + \overline{C})$

(c) $W = A\overline{C}(\overline{B} + C)$

(d) $X = AB + \overline{C} + BD$

(e) $Y = (A\overline{B} + D)(A + \overline{C}D)$

(f) $Z = (A + \overline{B})(BC + A) + \overline{A}B + CD$

3–23. Simplify the circuit of Figure P3–23 down to its SOP form, then draw the logic circuit of the simplified form.

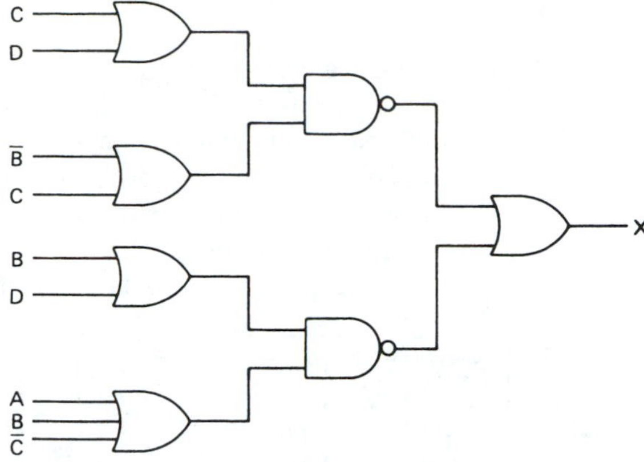

Figure P3–23

3–24. Using a Karnaugh map, reduce the following equations to a minimum sum-of-products form.

(a) $X = AB\overline{C} + \overline{A}B + \overline{A}\,\overline{B}$

(b) $Y = BC + \overline{A}\,\overline{B}C + B\overline{C}$

(c) $Z = ABC + A\overline{B}\,\overline{C} + \overline{A}\,\overline{B}C + AB\overline{C}$

3–25. Using a Karnaugh map, reduce the following equations to a minimum sum-of-products form.

(a) $W = \overline{B}(C\overline{D} + \overline{A}D) + \overline{B}\,\overline{C}(A + \overline{A}\,\overline{D})$

(b) $X = \overline{A}\,\overline{B}\,\overline{D} + B(\overline{C}\,\overline{D} + ACD) + A\overline{B}\,\overline{D}$

(c) $Y = A(C\overline{D} + \overline{C}\,\overline{D}) + A\overline{B}D + \overline{A}\,\overline{B}C\overline{D}$

(d) $Z = \overline{B}\,\overline{C}D + B\overline{C}D + \overline{C}\,\overline{D} + C\overline{D}(B + \overline{A}\,\overline{B})$

3–26. Use a Karnaugh map to simplify the circuits in Figure P3-26.

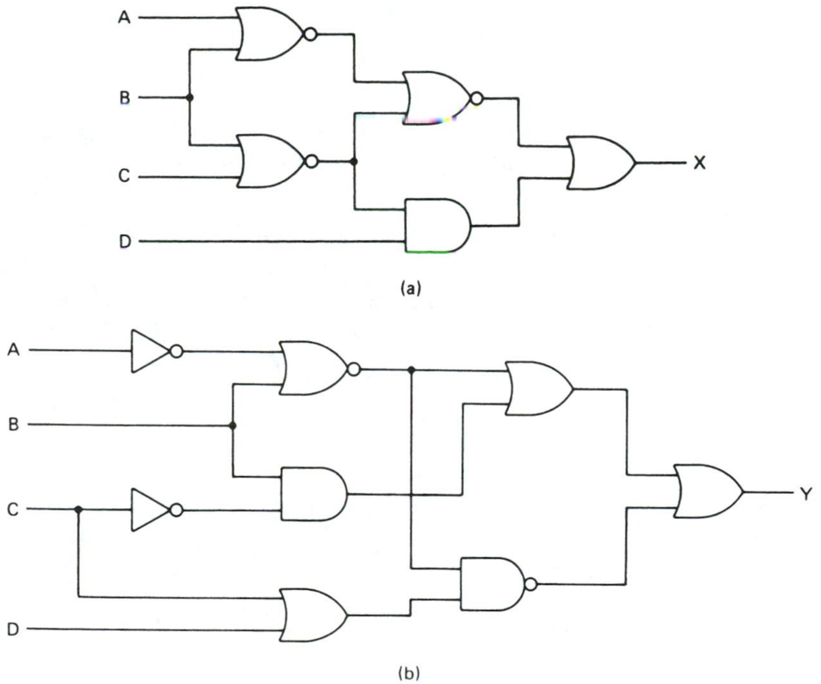

Figure P3–26

Design

3–27. Seven-segment displays are commonly used in calculators to display each of the decimal digits. Each segment of a digit is controlled separately and when all seven of the segments are on, the number 8 is displayed. The upper-right segment of the display comes on when displaying the numbers 0, 1, 2, 3, 4, 7, 8, and 9. (The numerical designation for each of the digits 0 to 9 is shown in Figure P3–27.) Design a circuit that puts out a HIGH (1) whenever a 4-bit BCD code translates to a number that uses the upper-right segment.

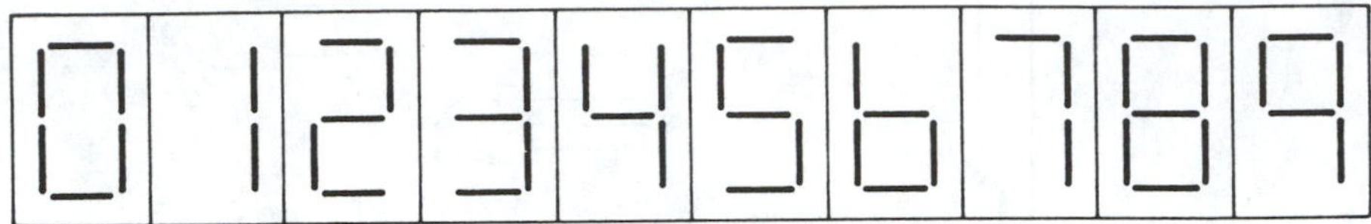

Figure P3–27

3–28. Repeat Problem 3–27 for the lower-left segment of a seven-segment display (0, 2, 6, 8).

SCHEMATIC INTERPRETATION PROBLEMS

3–29. Find U8 in the HC11D0 schematic. Pins 11 and 12 are unused so they are connected to V_{CC}. What if they were connected to ground instead?

3–30. Find U1 : A in the Watchdog Timer schematic. This device is called a flip-flop and will be explained in Chapter 5. It has two inputs: D and CLK, and two outputs: Q and CD. Write the Boolean equation at the output (pin 3) of U2 : A.

3–31. Write the Boolean equation at the output (pin 3) of U12 : A in the Watchdog Timer schematic. (*Hint:* Use the information given in Problem 3–30.)

3–32. Locate the U14 gates in the 4096/4196 schematic.
- **(a)** Write the Boolean equation of the output at pin 6 of U14.
- **(b)** What kind of gate does it turn into if you use the bubble-pushing technique?
- **(c)** This is a 74HC08. What kind of logic gate is that?
- **(d)** Complete the following sentence: Pin 3 of U14 : A goes LOW if ______ OR if ______ .

3–33. U10 of the 4096/4196 schematic is a RAM memory IC. Its operation will be discussed in Chapter 8. To enable the chip to work, the Chip Enable input at pin 20 must be made LOW. Write a sentence describing the logic operation that makes that line go LOW. (*Hint:* Pin 20 of U10 goes LOW if ______ .)

Data Control Devices

OBJECTIVES

Upon completion of this chapter, you should be able to:

- Utilize an integrated-circuit magnitude comparator to perform binary comparisons.
- Describe the function of a decoder and an encoder.
- Understand the internal circuitry for encoding and decoding.
- Utilize manufacturers' data sheets to determine the operation of IC decoder and encoder chips.
- Describe the function and uses of multiplexers and demultiplexers.
- Design circuits that employ multiplexer and demultiplexer ICs.
- Explain the wave-shaping capability and operating characteristics of Schmitt trigger ICs.

INTRODUCTION

Information, or data, that is used by digital devices comes in many formats. The mechanisms for conversion, transfer, and selection of data are handled by combinational logic ICs.

In this chapter we first take a general approach to the understanding of data-handling circuits, then deal with the specific operation and application of practical data-handling MSI chips. The MSI chips covered include comparators, decoders, encoders, multiplexers, demultiplexers, and Schmitt triggers.

4–1 COMPARATORS

Quite often in the evaluation of digital information it is important to compare two binary strings (or binary words) to determine if they are exactly equal. This comparison process is performed by a digital *comparator.*

The basic comparator will evaluate two binary strings bit by bit and output a 1 if they are exactly equal. An exclusive-NOR gate is the easiest way to compare the equality of 2 bits. If both bits are equal (0–0 or 1–1), the Ex-NOR puts out a 1.

To compare more than just 2 bits, we need additional Ex-NORs, and the output of all of them must be 1. For example, to design a comparator to evaluate two 4-bit numbers, we need four Ex-NORs. To determine total equality, connect all four outputs into an AND gate. That way, if all four outputs are 1s, the AND gate puts out a 1. Figure 4–1 shows a comparator circuit built from exclusive-NORs and an AND gate.

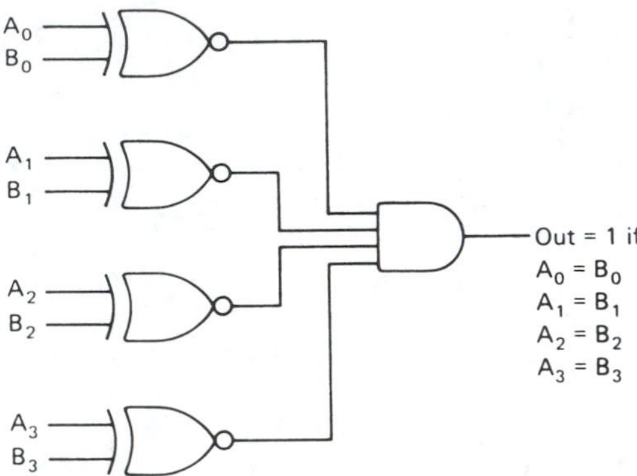

***Figure 4–1** Binary comparator for comparing two 4-bit binary strings.

Studying Figure 4–1, you should realize that if A_0–B_0 equals 1–1 or 0–0, the top Ex-NOR will output a 1. The same holds true for the second, third, and fourth Ex-NOR gates. If all of them output a 1, the AND gate outputs a 1, indicating equality.

Integrated-circuit *magnitude comparators* are available in both the TTL and CMOS families. A magnitude comparator not only determines if *A* equals *B*, but also if *A is greater than B* or *A is less than B.*

The 7485 is a TTL 4-bit magnitude comparator. The pin configuration and logic symbol for the 7485 are given in Figure 4–2. The 7485 can be used just like the basic

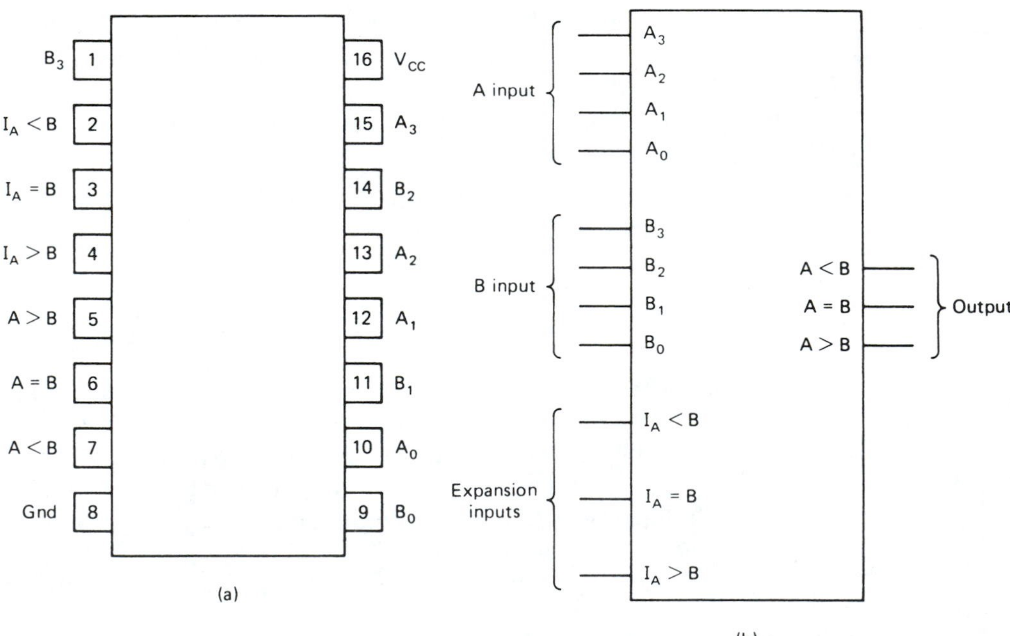

Figure 4–2 The 7485 4-bit magnitude comparator: (a) pin configuration; (b) logic symbol.

*See Application K-4 in Appendix K for a CPLD implementation of this circuit.

comparator of Figure 4–1 by using the A inputs, B inputs, and the equality output ($A = B$). The 7485 has the additional feature of telling you which number is larger if the equality is not met. The $A > B$ output is 1 if A is larger than B, and the $A < B$ output is 1 if B is larger than A.

The expansion inputs $I_A < B$, $I_A = B$, and $I_A > B$ are used for expansion to a system capable of comparisons greater than 4 bits. For example, to set up a circuit capable of comparing two *8-bit words,* two 7485s are required. The $A > B$, $A = B$, $A < B$ outputs of the low-order (least significant) comparator are connected to the expansion inputs of the high-order comparator. That way the comparators act together, comparing two entire 8-bit words, outputting the result from the high-order comparator outputs. For proper operation, the expansion inputs to the low-order comparator should be tied as follows: $I_A > B$ = LOW, $I_A = B$ = HIGH, and $I_A = B$ = LOW. Figure 4–3 shows the connections for magnitude comparison of two 8-bit binary strings. In that figure, if the high-order A inputs are equal to the high-order B inputs, then the expansion inputs are used as the "tie-breaker." Expansion to greater than 8 bits using multiple 7485s is also possible.

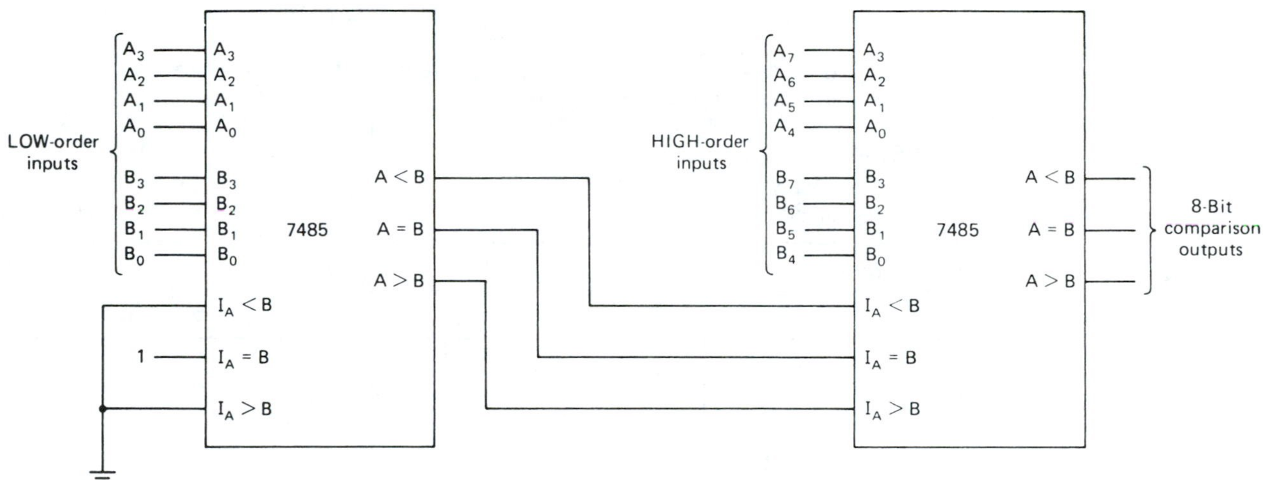

Figure 4–3 Magnitude comparison of two 8-bit binary strings (or binary words).

4–2 DECODERS

Decoding is the process of converting some code (such as binary, BCD, or hex) into some recognizable number or character. Take, for example, a system that reads a 4-bit BCD code and converts it to its appropriate decimal number by turning on a decimal indicating lamp. Figure 4–4 illustrates such a system. This decoder is made up of a combination of logic gates that produce a HIGH at one of the 10 outputs, based on the levels at the four inputs.

In this section we learn how to use decoder ICs by studying the internal logic, the function tables, and the input/output waveforms of several popular 7400-series decoder ICs.

3-Bit Binary-to-Octal Decoding

To design a decoder, it is useful first to make a truth table of all possible input/output combinations. An octal (base 8) decoder must provide eight outputs, one for each of the eight different combinations of inputs, as shown in Table 4–1.

Before the design is made, we must decide if we want an *active-HIGH-level* output or an *active-LOW-level* output to indicate the value selected. For example, the

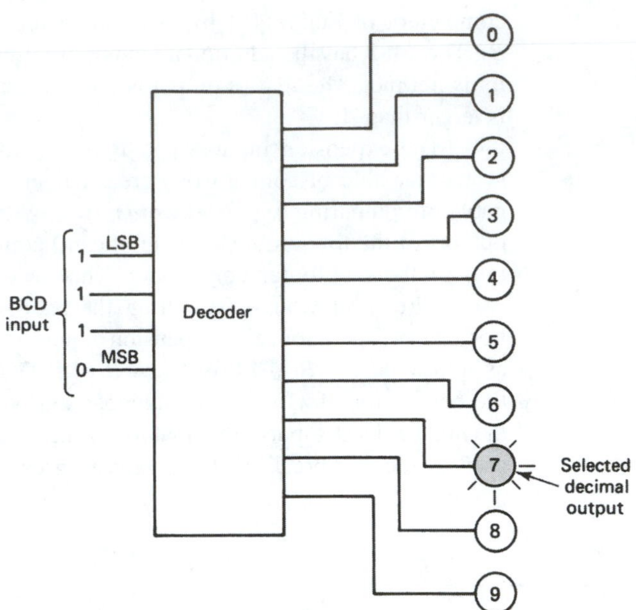

Figure 4–4 A BCD decoder selects the correct decimal indicating lamp based on the BCD input.

TABLE 4–1

Truth Tables for an Octal Decoder

(a) Active-HIGH outputs

Input			*Output*							
2^2	2^1	2^0	*0*	*1*	*2*	*3*	*4*	*5*	*6*	*7*
0	0	0	1	0	0	0	0	0	0	0
0	0	1	0	1	0	0	0	0	0	0
0	1	0	0	0	1	0	0	0	0	0
0	1	1	0	0	0	1	0	0	0	0
1	0	0	0	0	0	0	1	0	0	0
1	0	1	0	0	0	0	0	1	0	0
1	1	0	0	0	0	0	0	0	1	0
1	1	1	0	0	0	0	0	0	0	1

(b) Active-LOW outputs

Input			*Output*							
2^2	2^1	2^0	*0*	*1*	*2*	*3*	*4*	*5*	*6*	*7*
0	0	0	0	1	1	1	1	1	1	1
0	0	1	1	0	1	1	1	1	1	1
0	1	0	1	1	0	1	1	1	1	1
0	1	1	1	1	1	0	1	1	1	1
1	0	0	1	1	1	1	0	1	1	1
1	0	1	1	1	1	1	1	0	1	1
1	1	0	1	1	1	1	1	1	0	1
1	1	1	1	1	1	1	1	1	1	0

active-HIGH truth table in Table 4–1a shows us that for an input of 011 (3), output 3 will be HIGH while all other outputs are LOW. The *active-LOW* truth table is just the opposite (output 3 is LOW, all other outputs are HIGH).

Therefore, we have to know whether the indicating lamp (or other receiving device) requires a HIGH level to activate or a LOW level. Most devices used in digital electronics are designed to activate from a LOW-level signal, so most decoder designs use *active-LOW* outputs, as shown in Table 4–1b.

The octal decoder is sometimes referred to as a *1-of-8 decoder* because, based on the input code, one of the eight outputs will be active. It is also known as a *3-line-to-8-line decoder* because it has three input lines and eight output lines.

Integrated-circuit decoder chips provide basic decoding as well as several other useful functions. Manufacturers' data books list several decoders and give function tables illustrating the input/output operation and special functions. Rather than designing decoders using combinational logic, it is much more important to be able to use a data book to find the decoder that you need and to determine the proper pin connections and operating procedure to perform a specific decoding task. Table 4–2 lists some of the more popular TTL decoder ICs. (Equivalent CMOS ICs are also available.)

TABLE 4–2
Decoder ICs

Device number	*Function*
74138	1-of-8 octal decoder (3 line-to-8 line)
7442	1-of-10 BCD decoder (4 line-to-10 line)
74154	1-of-16 hex decoder (4 line-to-16 line)
7447	BCD-to-seven segment decoder

Octal Decoder IC*

The 74138 is an octal decoder capable of decoding the eight possible octal codes into eight separate active-LOW outputs. It also has three enable inputs for additional flexibility. Figure 4–5 shows information presented in a data book for the 74138.

Just by looking at the logic symbol (Figure 4–5b) and function table (Figure 4–5d) we can figure out the complete operation of the chip. First of all, the inversion bubbles on the decoded outputs indicate active-LOW operation. The three inputs, $\overline{E_1}$, $\overline{E_2}$, and E_3, are used to *enable* the chip. The function table shows that the chip is disabled (all outputs HIGH) *unless* $\overline{E_1}$ = *LOW and* $\overline{E_2}$ = LOW *and* E_3 = HIGH. The enables are useful for go/no-go operation of the chip based on some external control signal.

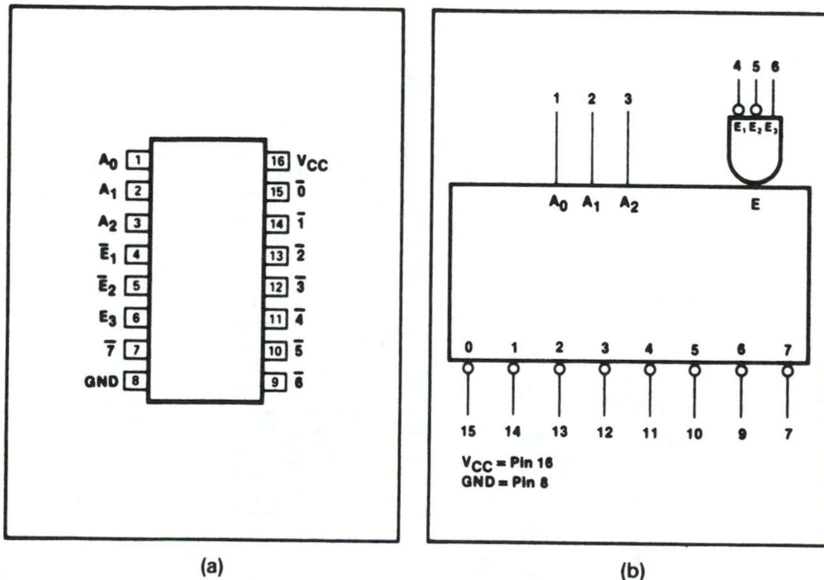

Figure 4–5 The 74138 octal decoder: (a) pin configuration; (b) logic symbol. (Courtesy of Philips Components–Signetics)

*See Application K-7 in Appendix K for a CPLD implementation of an octal decoder.

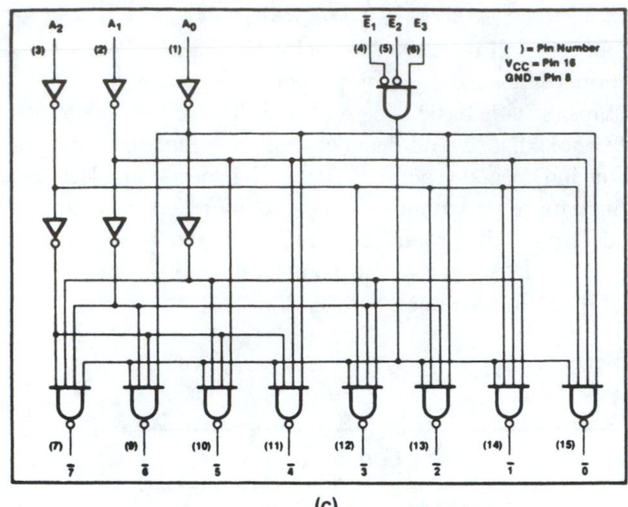

(c)

INPUTS						OUTPUTS							
$\overline{E}_1$	$\overline{E}_2$	E_3	A_0	A_1	A_2	$\overline{0}$	$\overline{1}$	$\overline{2}$	$\overline{3}$	$\overline{4}$	$\overline{5}$	$\overline{6}$	$\overline{7}$
H	X	X	X	X	X	H	H	H	H	H	H	H	H
X	H	X	X	X	X	H	H	H	H	H	H	H	H
X	X	L	X	X	X	H	H	H	H	H	H	H	H
L	L	H	L	L	L	L	H	H	H	H	H	H	H
L	L	H	H	L	L	H	L	H	H	H	H	H	H
L	L	H	L	H	L	H	H	L	H	H	H	H	H
L	L	H	H	H	L	H	H	H	L	H	H	H	H
L	L	H	L	L	H	H	H	H	H	L	H	H	H
L	L	H	H	L	H	H	H	H	H	H	L	H	H
L	L	H	L	H	H	H	H	H	H	H	H	L	H
L	L	H	H	H	H	H	H	H	H	H	H	H	L

NOTES
H = HIGH voltage level
L = LOW voltage level
X = Don't care

(d)

Figure 4–5 *(Continued)* (c) logic diagram; (d) function table. (Courtesy of Philips Components–Signetics)

When the chip is *disabled,* the Xs in the binary input columns A_0, A_1, and A_2 indicate *don't-care* levels, meaning the outputs will all be HIGH no matter what level A_0, A_1, and A_2 are. When the chip is *enabled,* the binary inputs A_0, A_1, and A_2 are used to select which output goes LOW. In this case, A_0 is the least significant bit (LSB) input. Be aware that some manufacturers label the inputs A, B, C instead of A_0, A_1, A_2 and assume that A is the LSB.

The logic diagram in Figure 4–5c shows the actual internal combinational logic required to perform the decoding. The extra inverters on the inputs are required to prevent excessive loading of the driving source(s). Those internal inverters supply the driving current to the eight NAND gates instead of the driving source(s) having to do it. The three enable inputs ($\overline{E_1}$, $\overline{E_2}$, E_3) are connected to an AND gate, which can disable all the output NANDs by sending them a LOW input level if $\overline{E_1}$, $\overline{E_2}$, E_3 is not 001.

EXAMPLE 4–1

Sketch the output waveforms of the 74138 in Figure 4–6. Figure 4–7 shows the input waveforms to the 74138.

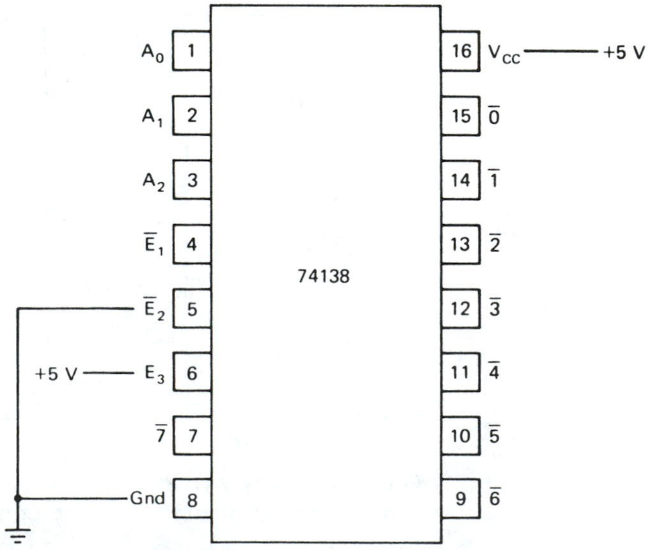

Figure 4–6 The 74138 connections for Example 4–1.

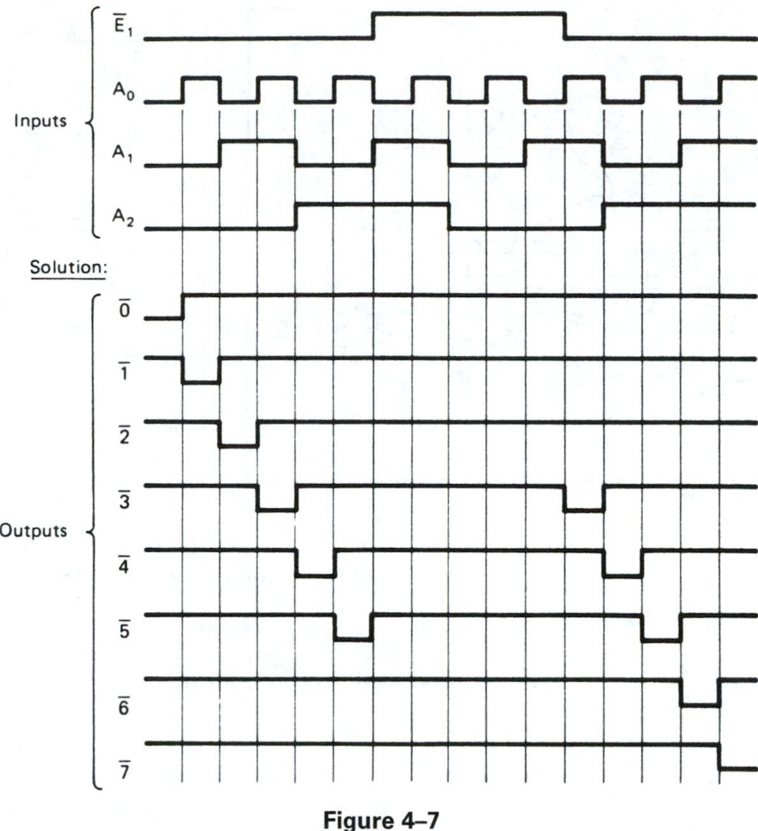

Figure 4–7

BCD Decoder IC

The 7442 is a BCD-to-decimal (1-of-10) decoder. It has four pins for the BCD input bits (0000 to 1001) and has 10 active-LOW outputs for the decoder decimal numbers. Figure 4–8 gives the operational information for the 7442 from a manufacturer's data book.

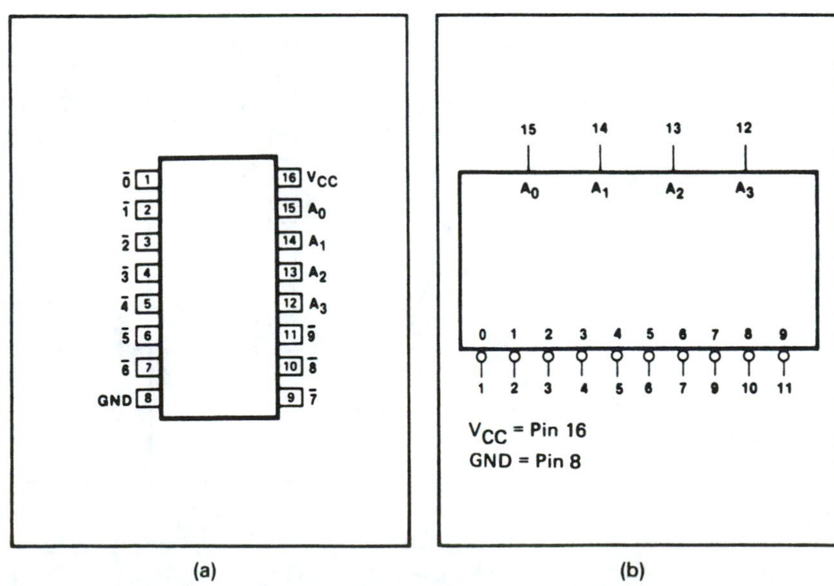

Figure 4–8 The 7442 BCD-to-DEC decoder: (a) pin configuration; (b) logic symbol. (Courtesy of Philips Components–Signetics)

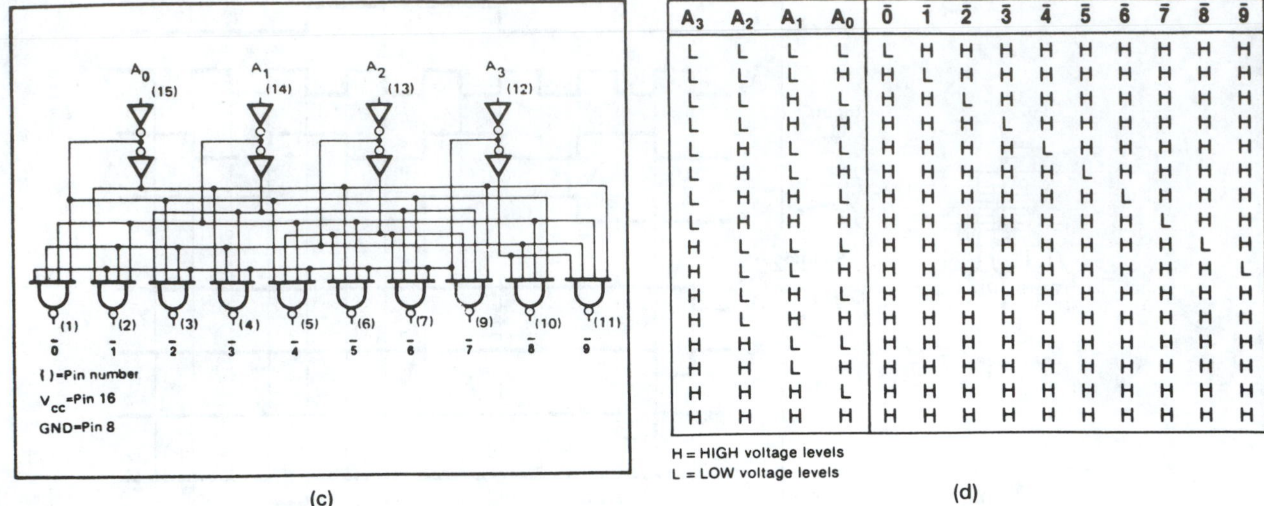

A_3	A_2	A_1	A_0	$\bar{0}$	$\bar{1}$	$\bar{2}$	$\bar{3}$	$\bar{4}$	$\bar{5}$	$\bar{6}$	$\bar{7}$	$\bar{8}$	$\bar{9}$
L	L	L	L	L	H	H	H	H	H	H	H	H	H
L	L	L	H	H	L	H	H	H	H	H	H	H	H
L	L	H	L	H	H	L	H	H	H	H	H	H	H
L	L	H	H	H	H	H	L	H	H	H	H	H	H
L	H	L	L	H	H	H	H	L	H	H	H	H	H
L	H	L	H	H	H	H	H	H	L	H	H	H	H
L	H	H	L	H	H	H	H	H	H	L	H	H	H
L	H	H	H	H	H	H	H	H	H	H	L	H	H
H	L	L	L	H	H	H	H	H	H	H	H	L	H
H	L	L	H	H	H	H	H	H	H	H	H	H	L
H	L	H	L	H	H	H	H	H	H	H	H	H	H
H	L	H	H	H	H	H	H	H	H	H	H	H	H
H	H	L	L	H	H	H	H	H	H	H	H	H	H
H	H	L	H	H	H	H	H	H	H	H	H	H	H
H	H	H	L	H	H	H	H	H	H	H	H	H	H
H	H	H	H	H	H	H	H	H	H	H	H	H	H

H = HIGH voltage levels
L = LOW voltage levels

(d)

Figure 4–8 *(Continued)* (c) logic diagram; (d) function table. (Courtesy of Philips Components–Signetics)

Hexadecimal 1-of-16 Decoder IC

The 74154 is a 1-of-16 decoder. It accepts a 4-bit binary input (0000 to 1111), decodes it, and provides an active-LOW output to one of the 16 output pins. It also has a two-input active-LOW enable gate for disabling the outputs. If either enable input ($\overline{E_0}$ *or* $\overline{E_1}$) is made HIGH, the outputs are forced HIGH regardless of the A_0 to A_3 inputs. The operational information for the 74154 is given in Figure 4–9.

The logic diagram in Figure 4–9c shows the actual combinational logic circuit used to provide the decoding. The inverted-input AND gate is used in the circuit to disable all output NAND gates if either $\overline{E_0}$ or $\overline{E_1}$ is made HIGH. Follow the logic levels through the circuit for several combinations of inputs to A_0 though A_3 to prove its operation.

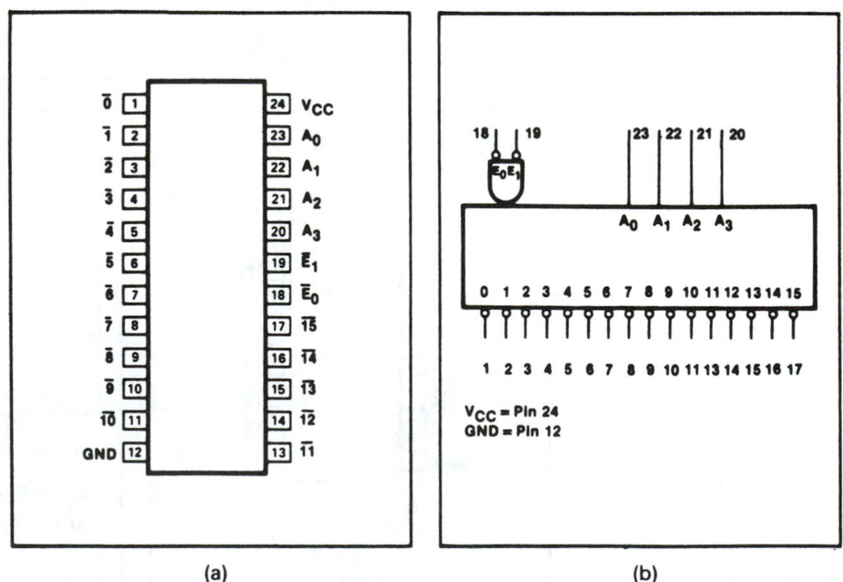

(a) (b)

Figure 4–9 The 74154 1-of-16 decoder: (a) pin configuration; (b) logic symbol; (c) logic diagram; (d) function table. (Courtesy of Philips Components–Signetics)

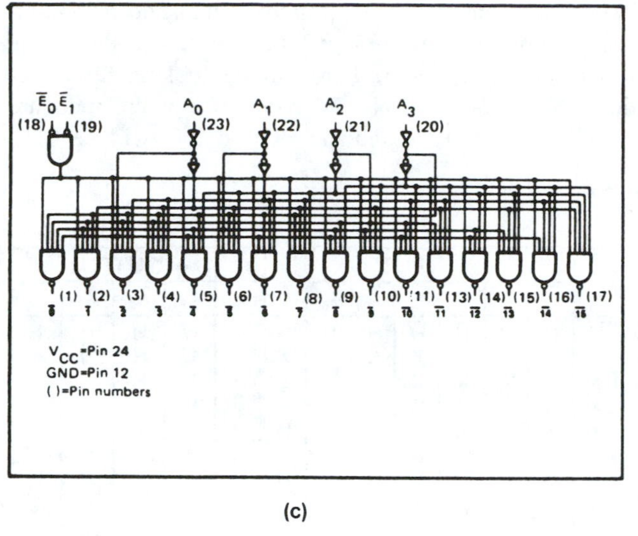

(c)

INPUTS						OUTPUTS															
$\overline{E}_0$	$\overline{E}_1$	A_3	A_2	A_1	A_0	$\overline{0}$	$\overline{1}$	$\overline{2}$	$\overline{3}$	$\overline{4}$	$\overline{5}$	$\overline{6}$	$\overline{7}$	$\overline{8}$	$\overline{9}$	$\overline{10}$	$\overline{11}$	$\overline{12}$	$\overline{13}$	$\overline{14}$	$\overline{15}$
L	H	X	X	X	X	H	H	H	H	H	H	H	H	H	H	H	H	H	H	H	H
H	L	X	X	X	X	H	H	H	H	H	H	H	H	H	H	H	H	H	H	H	H
H	H	X	X	X	X	H	H	H	H	H	H	H	H	H	H	H	H	H	H	H	H
L	L	L	L	L	L	L	H	H	H	H	H	H	H	H	H	H	H	H	H	H	H
L	L	L	L	L	H	H	L	H	H	H	H	H	H	H	H	H	H	H	H	H	H
L	L	L	L	H	L	H	H	L	H	H	H	H	H	H	H	H	H	H	H	H	H
L	L	L	L	H	H	H	H	H	L	H	H	H	H	H	H	H	H	H	H	H	H
L	L	L	H	L	L	H	H	H	H	L	H	H	H	H	H	H	H	H	H	H	H
L	L	L	H	L	H	H	H	H	H	H	L	H	H	H	H	H	H	H	H	H	H
L	L	L	H	H	L	H	H	H	H	H	H	L	H	H	H	H	H	H	H	H	H
L	L	L	H	H	H	H	H	H	H	H	H	H	L	H	H	H	H	H	H	H	H
L	L	H	L	L	L	H	H	H	H	H	H	H	H	L	H	H	H	H	H	H	H
L	L	H	L	L	H	H	H	H	H	H	H	H	H	H	L	H	H	H	H	H	H
L	L	H	L	H	L	H	H	H	H	H	H	H	H	H	H	L	H	H	H	H	H
L	L	H	L	H	H	H	H	H	H	H	H	H	H	H	H	H	L	H	H	H	H
L	L	H	H	L	L	H	H	H	H	H	H	H	H	H	H	H	H	L	H	H	H
L	L	H	H	L	H	H	H	H	H	H	H	H	H	H	H	H	H	H	L	H	H
L	L	H	H	H	L	H	H	H	H	H	H	H	H	H	H	H	H	H	H	L	H
L	L	H	H	H	H	H	H	H	H	H	H	H	H	H	H	H	H	H	H	H	L

H = HIGH voltage level
L = LOW voltage level
X = Don't care

(d)

Figure 4–9 *(Continued)* (c) logic diagram; (d) function table. (Courtesy of Philips Components–Signetics)

4–3 *ENCODERS*

Encoding is the opposite process from decoding. Encoding is used to generate a coded output (such as BCD or binary) from a numeric input such as decimal or octal. For example, Figure 4–10 shows a typical block diagram for a decimal-to-BCD encoder and an octal-to-binary encoder.

The 74147 Decimal-to-BCD Encoder

The 74147 logic symbol looks like the typical block diagram for an encoder except for two major differences.

1. The inputs *and* outputs are all active-LOW (see the bubbles on the logic symbol, Figure 4–11a).

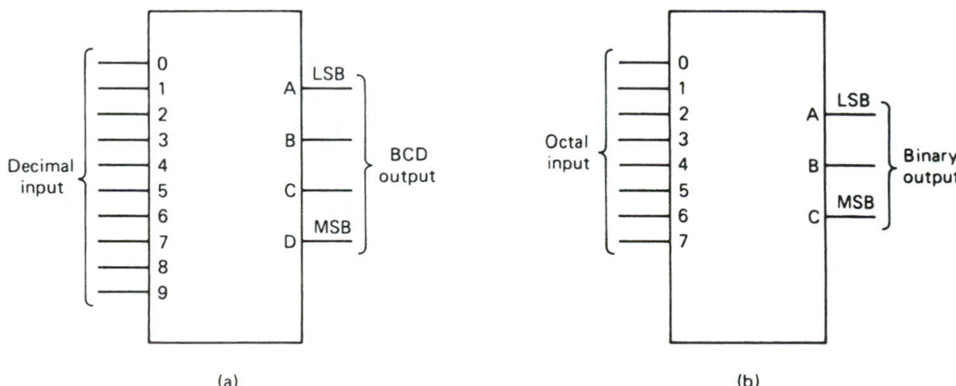

Figure 4–10 Typical block diagrams for encoders: (a) decimal-to-BCD encoder; (b) octal-to-binary encoder.

2. The 74147 is a *priority* encoder. That means that if more than one decimal number is input, the highest numeric input has *priority* and will be encoded to the output (see the function table, Figure 4–11b). For example, looking at the second line in the function table, if $\overline{I_9}$ is LOW (decimal 9), all other inputs are "don't care" (could be HIGH *or* LOW), and the BCD output will be 0110 (active-LOW BCD 9).

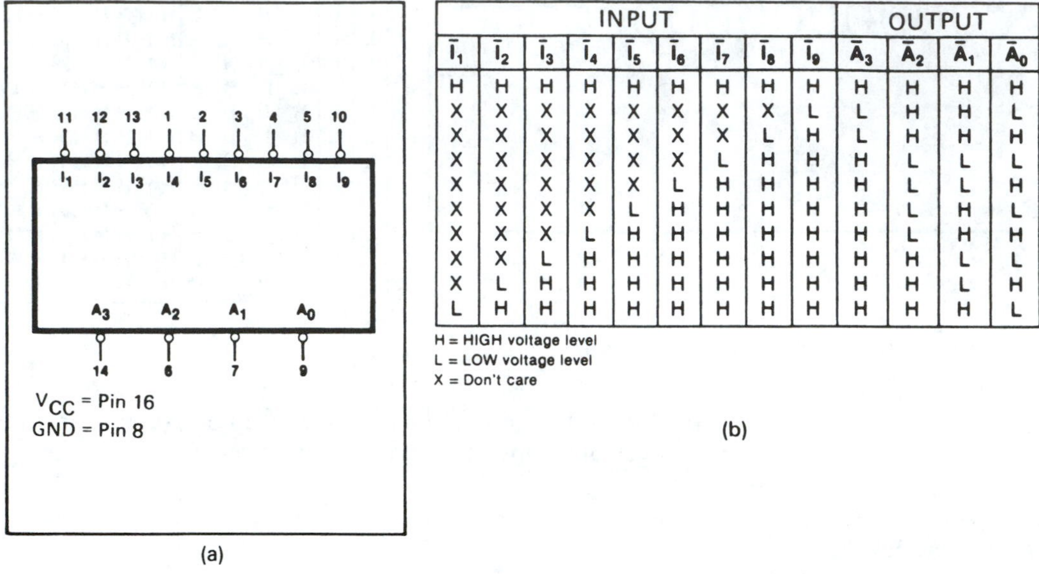

INPUT									OUTPUT			
$\overline{I_1}$	$\overline{I_2}$	$\overline{I_3}$	$\overline{I_4}$	$\overline{I_5}$	$\overline{I_6}$	$\overline{I_7}$	$\overline{I_8}$	$\overline{I_9}$	$\overline{A_3}$	$\overline{A_2}$	$\overline{A_1}$	$\overline{A_0}$
H	H	H	H	H	H	H	H	H	H	H	H	H
X	X	X	X	X	X	X	X	L	L	H	H	L
X	X	X	X	X	X	X	L	H	L	H	H	H
X	X	X	X	X	X	L	H	H	H	L	L	L
X	X	X	X	X	L	H	H	H	H	L	L	H
X	X	X	X	L	H	H	H	H	H	L	H	L
X	X	X	L	H	H	H	H	H	H	L	H	H
X	X	L	H	H	H	H	H	H	H	H	L	L
X	L	H	H	H	H	H	H	H	H	H	L	H
L	H	H	H	H	H	H	H	H	H	H	H	L

H = HIGH voltage level
L = LOW voltage level
X = Don't care

(b)

Figure 4–11 The 74147 decimal-to-BCD (10-line-to-4-line) encoder: (a) logic symbol; (b) function table.

EXAMPLE 4–2

For simplicity, the 74147 IC shown in Figure 4–12 is set up for encoding just three of its inputs (7, 8, and 9). Using the function table from Figure 4–11b, sketch the outputs at $\overline{A_0}$, $\overline{A_1}$, $\overline{A_2}$, and $\overline{A_3}$ as the $\overline{I_7}$, $\overline{I_8}$, and $\overline{I_9}$ inputs are switching as shown in Figure 4–13.

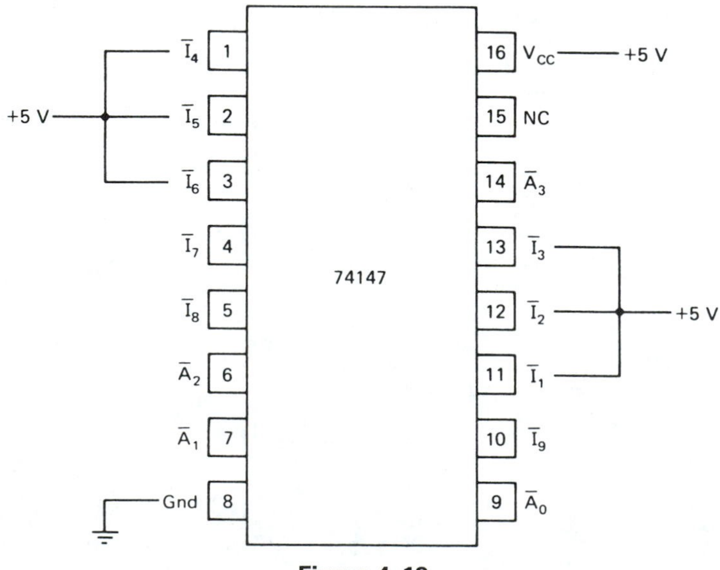

Figure 4–12

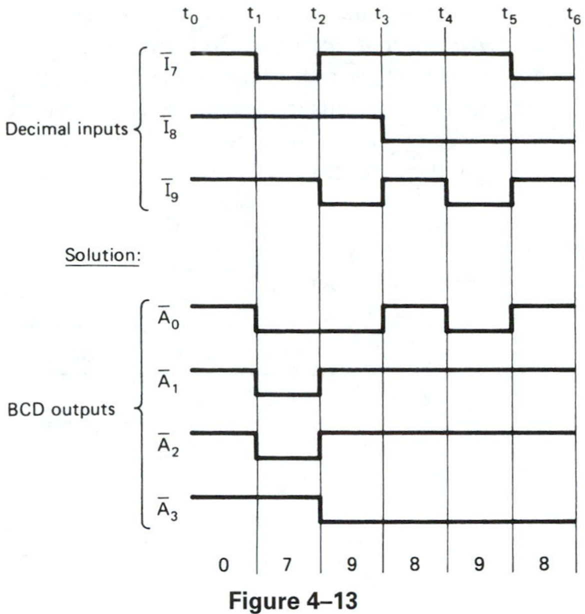

Figure 4–13

The $\overline{I_1}$ to $\overline{I_6}$ inputs are all tied HIGH and will have no effect on the output.

t_0–t_1: Decimal inputs are all HIGH; BCD outputs represent a 0.
t_1–t_2: $\overline{I_7}$ is LOW; BCD outputs represent a 7.
t_2–t_3: $\overline{I_9}$ is LOW; BCD outputs represent a 9.
t_3–t_4: $\overline{I_8}$ is LOW; BCD outputs represent an 8.
t_4–t_5: $\overline{I_8}$ and $\overline{I_9}$ are LOW; $\overline{I_9}$ has priority; BCD outputs represent a 9.
t_5–t_6: $\overline{I_7}$ and $\overline{I_8}$ are LOW; $\overline{I_8}$ has priority; BCD outputs represent an 8.

4–4 MULTIPLEXERS

A multiplexer is a device capable of funneling several data lines into a single line for transmission to another point. The multiplexer will have two or more digital input signals connected to its input. Control signals will also be input to tell which data input line to select for transmission (data selection). Figure 4–14 illustrates the function of a multiplexer.

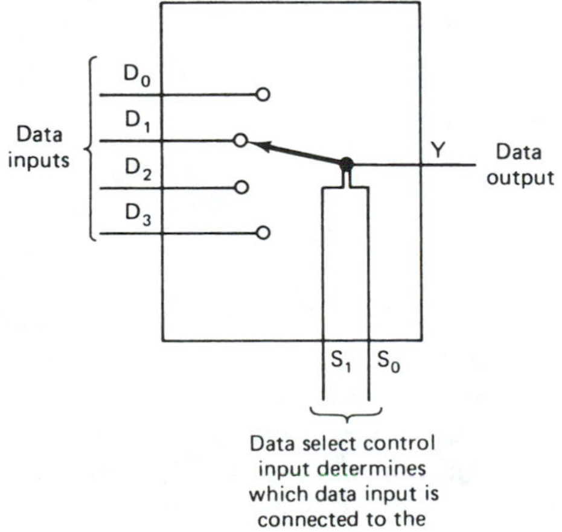

Figure 4–14 Functional diagram of a four-line multiplexer.

The multiplexer is also known as a *data selector.* Figure 4–14 shows that the *data select control inputs* (S_1, S_0) are responsible for determining which data input (D_0 to D_3) is selected to be transmitted to the data output line (Y). The S_1, S_0 inputs will be a binary code that corresponds to the data input line that you want to select. If $S_1 = 0$, $S_0 = 0$, then D_0 is selected; if $S_1 = 0$, $S_0 = 1$, then D_1 is selected; and so on. Table 4–3 lists the codes for input data selection.

TABLE 4–3
Data Select Input Codes for Figure 4–14

Data select control inputs		
S_1	S_0	*Data input selected*
0	0	D_0
0	1	D_1
1	0	D_2
1	1	D_3

A sample four-line multiplexer built from SSI logic gates is shown in Figure 4–15. The control inputs (S_1, S_0) take care of enabling the correct AND gate to pass just one of the data inputs through to the output. In Figure 4–15, 1s and 0s were placed on the diagram to show the levels that occur when selecting data input, D_1. Note that AND gate 1 will be enabled, passing D_1 to the output, while all other AND gates are disabled.

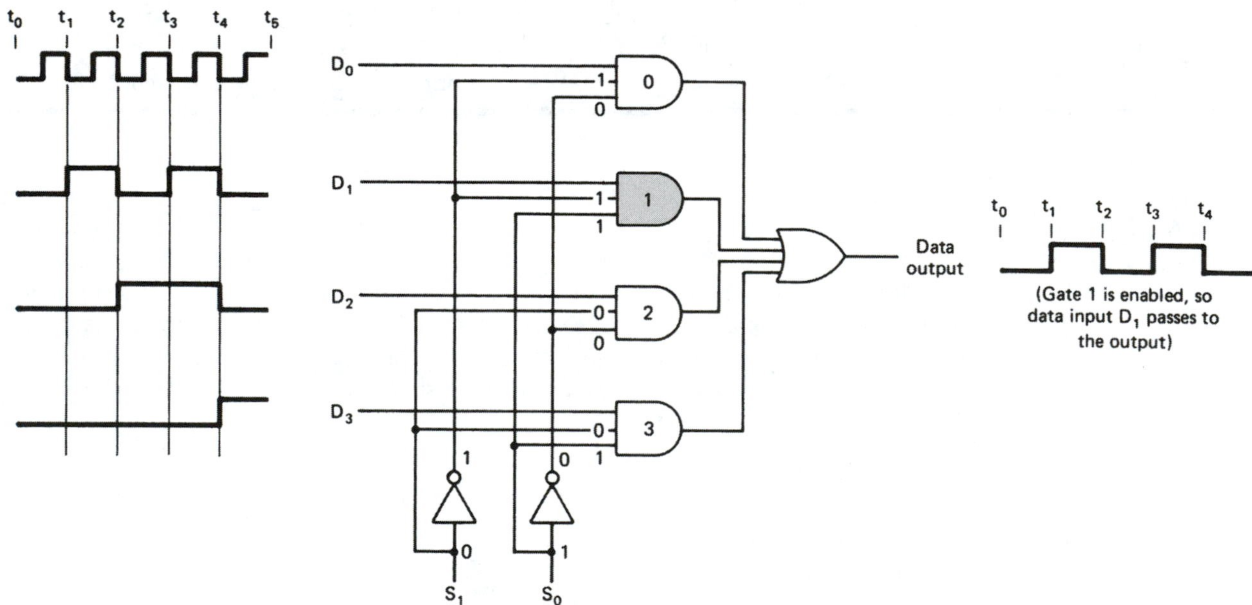

Figure 4–15 Logic diagram for a four-line multiplexer.

Two-, 4-, 8-, and 16-input multiplexers are readily available in MSI packages. Table 4–4 lists some popular TTL and CMOS multiplexers.

The logic symbol and logic diagram for the 74151 are given in Figure 4–16. Since the 74151 has eight lines to select from (I_0 to I_7), it requires three data select inputs—S_2, S_1, S_0—to determine which input to choose ($2^3 = 8$). True (Y) and complemented ($\overline{Y}$) outputs are provided. The active-LOW enable input ($\overline{E}$) disables all inputs when it is HIGH and forces Y LOW regardless of all other inputs.

TABLE 4–4
TTL and CMOS Multiplexers

Function	*Device*	*Logic family*
Quad 2-input	74157	TTL
	74HC157	H-CMOS
	4019	CMOS
Dual 8-input	74153	TTL
	74HC153	H-CMOS
	4539	CMOS
8-input	74151	TTL
	74HC151	H-CMOS
	4512	CMOS
16-input	74150	TTL

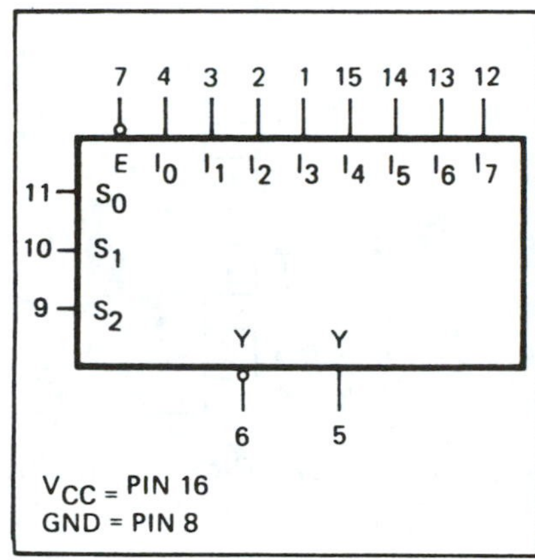

(a)

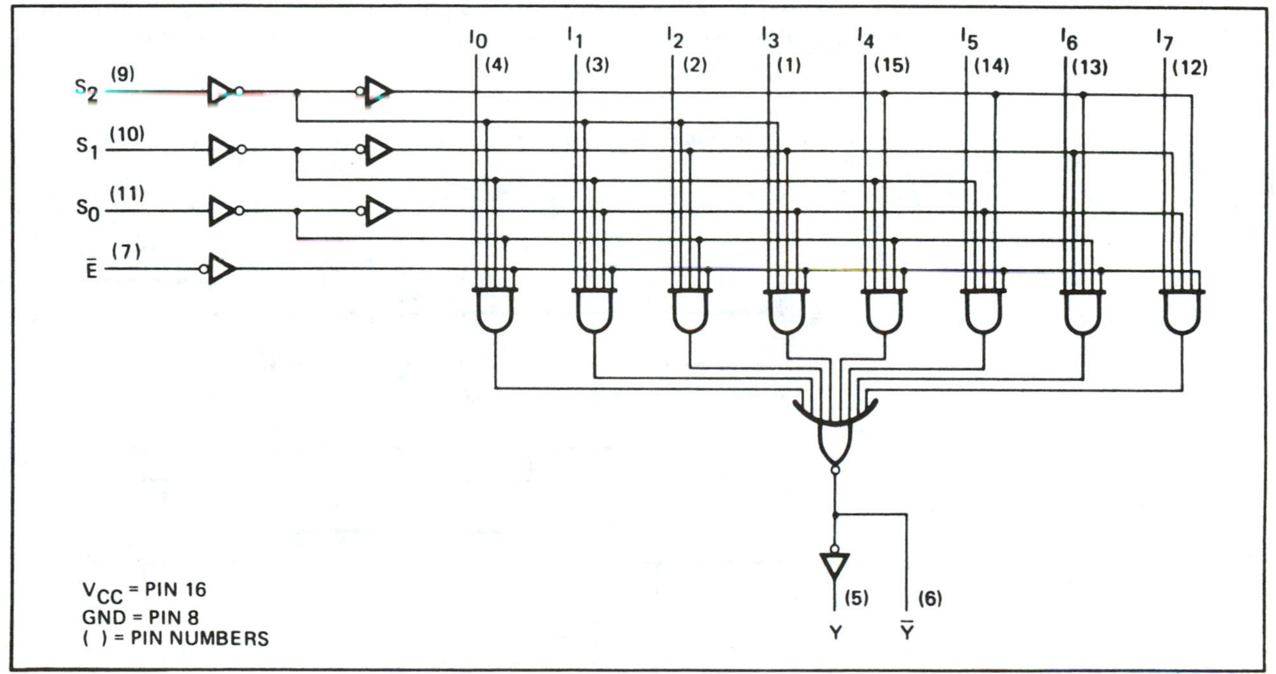

(b)

***Figure 4–16** The 74151 eight-line multiplexer: (a) logic symbol; (b) logic diagram. (Courtesy of Philips Components–Signetics)

*See Application K-8 in Appendix K for a CPLD implementation of this multiplexer.

EXAMPLE 4–3

Sketch the output waveforms at Y for the 74151 shown in Figure 4–17. For this example, the eight input lines (I_0 to I_7) are each connected to a constant level, and the data select lines (S_0 to S_2) and input enable ($\overline{E}$) are given as input waveforms.

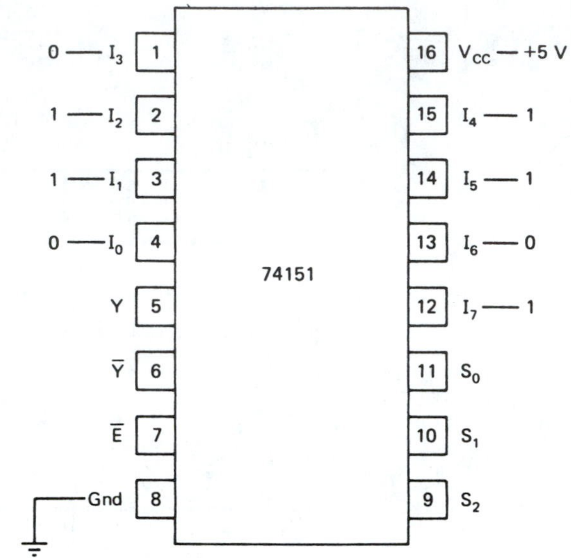

Figure 4–17 The 74151 multiplexer pin connections for Example 4–3.

See Figure 4–18. From t_0 to t_8 the waveforms at S_0, S_1, S_2 form a binary counter from 000 to 111. Therefore, the output at Y will be selected from I_0, then I_1, then I_2, and so on, up to I_7. From t_8 to t_9 the S_0, S_1, S_2 inputs are back to 000, so I_0 will be selected for output. From t_9 to t_{11} the $\overline{E}$ enable line goes HIGH, disabling all inputs and forcing Y LOW.

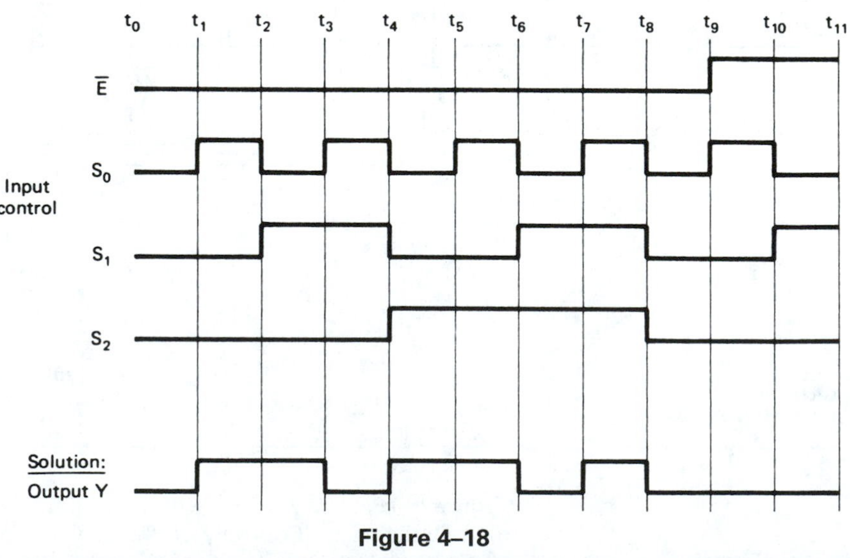

Figure 4–18

EXAMPLE 4–4

Using two 74151s, design a 16-line multiplexer controlled by four data select control inputs.

Solution:

The multiplexer is shown in Figure 4–19. Since there are 16 data input lines, we must use four data select inputs ($2^4 = 16$). (A is the LSB data select line and D is the MSB.)

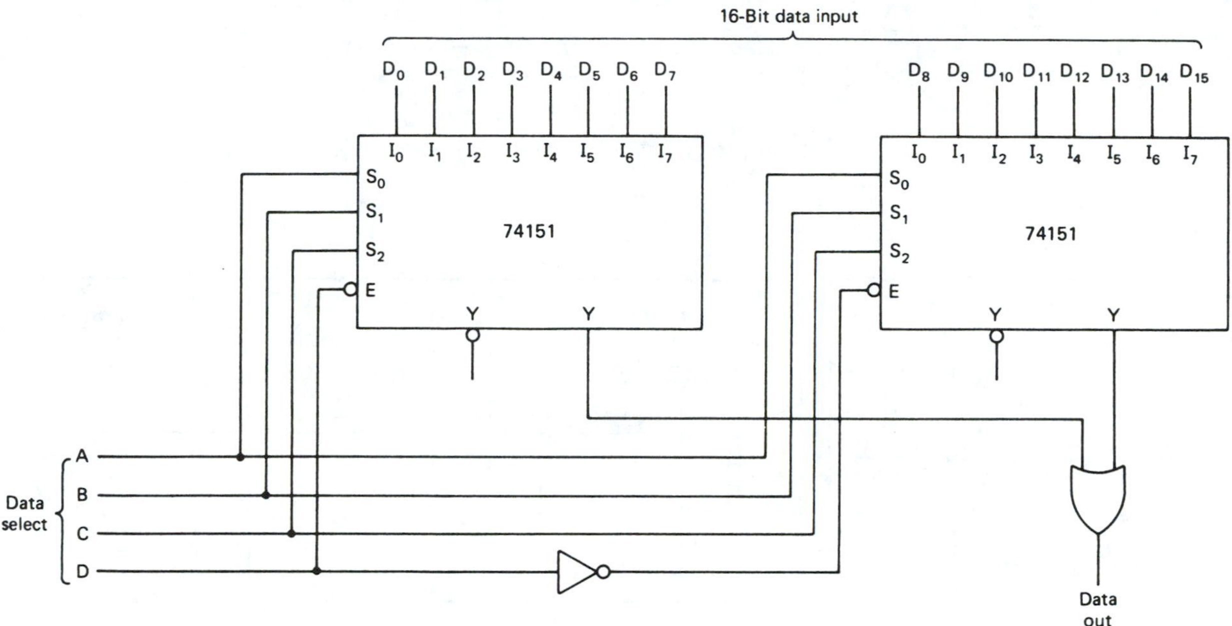

Figure 4–19 Design solution for Example 4–4.

When the data select is in the range 0000 to 0111, the D line is 0, which enables the low-order (left) multiplexer selecting the D_0 to D_7 inputs and disables the high-order (right) multiplexer.

When the data select inputs are in the range 1000 to 1111, the D line is 1, which disables the low-order multiplexer and enables the high-order multiplexer, allowing D_8 to D_{15} to be selected. Since the Y output of a disabled multiplexer is 0, an OR gate is used to combine the two outputs, allowing the output from the enabled multiplexer to pass through.

4–5 DEMULTIPLEXERS

Demultiplexing is the opposite procedure from multiplexing. We can think of a demultiplexer as a *data distributor.* It takes a single input data value and routes it to one of several outputs, as illustrated in Figure 4–20.

Integrated-circuit demultiplexers come in several configurations of inputs/outputs. The two that we discuss in this section are the 74139 dual four-line demultiplexer and the 74154 16-line demultiplexer.

The logic diagram and logic symbol for the 74139 are given in Figure 4–21. Note that the 74139 is divided into two equal sections. By looking at the logic diagram, you will see that the schematic is the same as that of a 2-line-to-4-line decoder. Decoders and

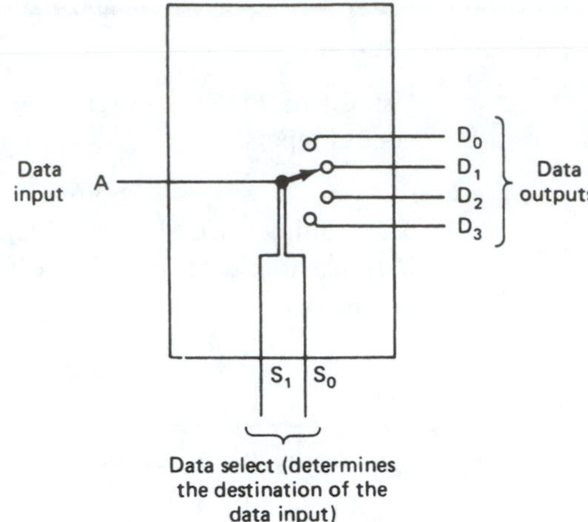

Figure 4–20 Functional diagram of a four-line demultiplexer.

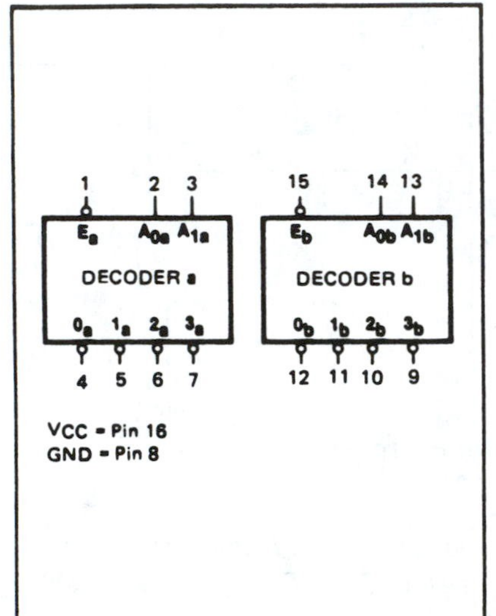

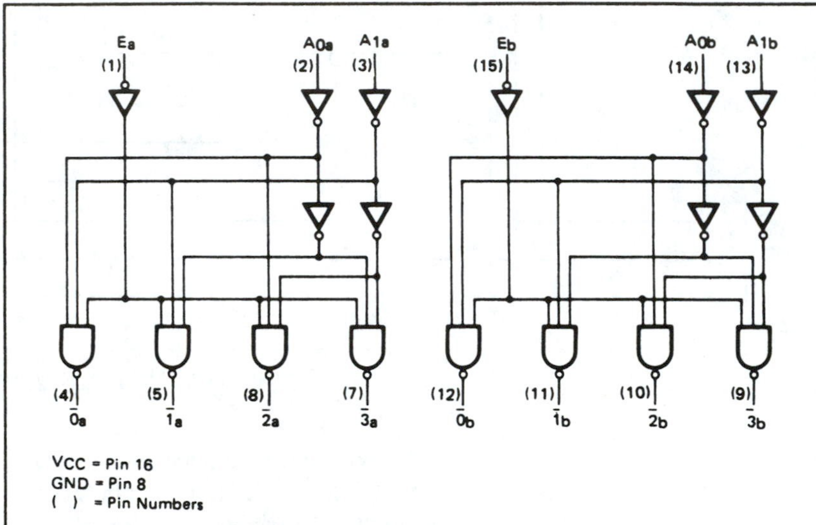

Figure 4–21 The 74139 dual four-line demultiplexer: (a) logic symbol; (b) logic diagram. (Courtesy of Philips Components–Signetics)

demultiplexers are the same, except with a decoder you hold the $\overline{E}$ enable line LOW and enter a code at the A_0A_1 inputs. As a demultiplexer, the A_0A_1 inputs are used to select the destination of input data. The input data are brought in via the $\overline{E}$ line. The 74138 3-line-to-8-line decoder that we covered earlier in this chapter can also function as an 8-line demultiplexer.

To use the 74139 as a demultiplexer to route some input data signal to, let's say, the $\overline{2a}$ output, the connections shown in Figure 4–22 would be made. In the figure the destination $\overline{2a}$ is selected by making $A_{1a} = 1, A_{0a} = 0$. The input signal is brought into the enable line ($\overline{E_a}$). When $\overline{E_a}$ goes LOW, the selected output line goes LOW; when $\overline{E_a}$ goes HIGH, the selected output line goes HIGH. (All nonselected lines remain HIGH continuously.)

The 74154 was used earlier in the chapter as a 4-line-to-16-line hexadecimal decoder. It can also be used as a 16-line demultiplexer. Figure 4–23 shows how it can be connected to route an input data signal to the $\overline{5}$ output.

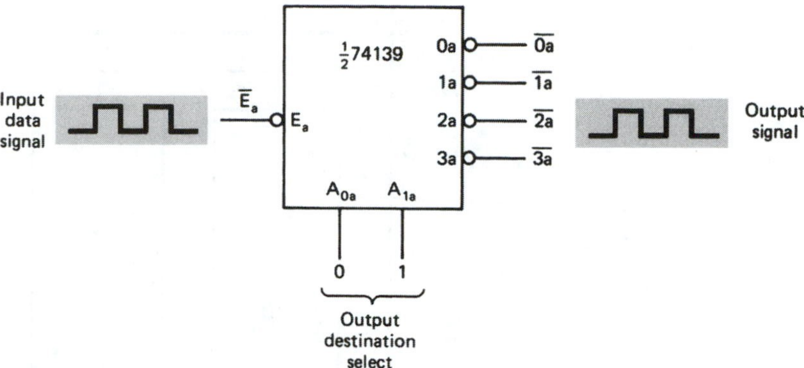

Figure 4–22 Connections to route an input data signal to the $\overline{2a}$ output of a 74139 demultiplexer.

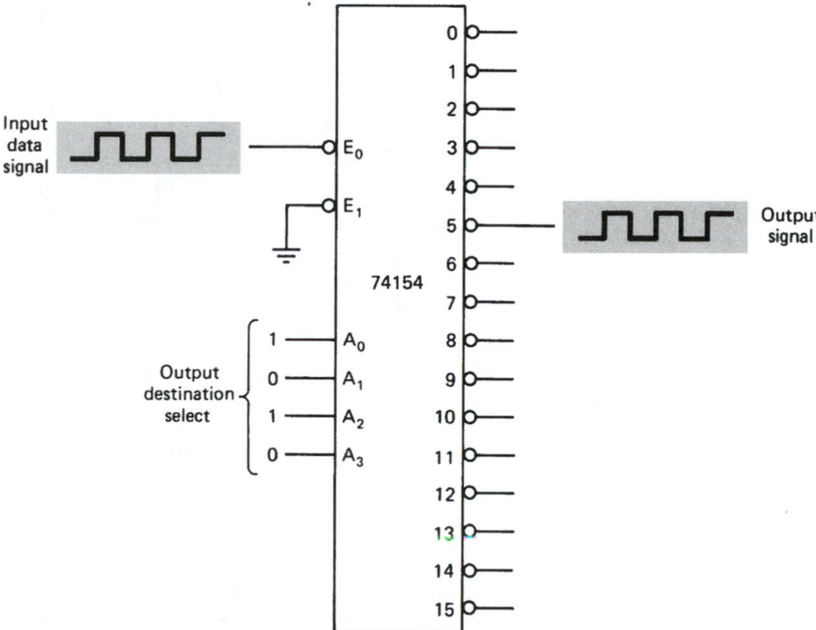

Figure 4–23 The 74154 demultiplexer connections to route an input signal to the $\overline{5}$ output.

Analog Multiplexer/Demultiplexer

Several analog multiplexers/demultiplexers are available in the CMOS family. The 4051, 4052, and 4053 are combination multiplexer *and* demultiplexer CMOS ICs. They can function in either configuration because their inputs and outputs are *bidirectional,* meaning that the flow can go in either direction. Also, they are *analog,* meaning that they can input and output levels other than just 1 and 0. The input/output levels can be any analog voltage between the positive and negative supply levels.

The functional diagram for the 4051 eight-channel multiplexer/demultiplexer is given in Figure 4–24. The eight square boxes in the functional diagram represent the bidirectional I/O lines. Used as a multiplexer, the analog levels will come in on the Y_0 to Y_7 lines, and the decoder will select which of these inputs are outputted to the Z line. As a demultiplexer, the connections are reversed, with the input coming into the Z line and the output going out on one of the Y_0 to Y_7 lines.

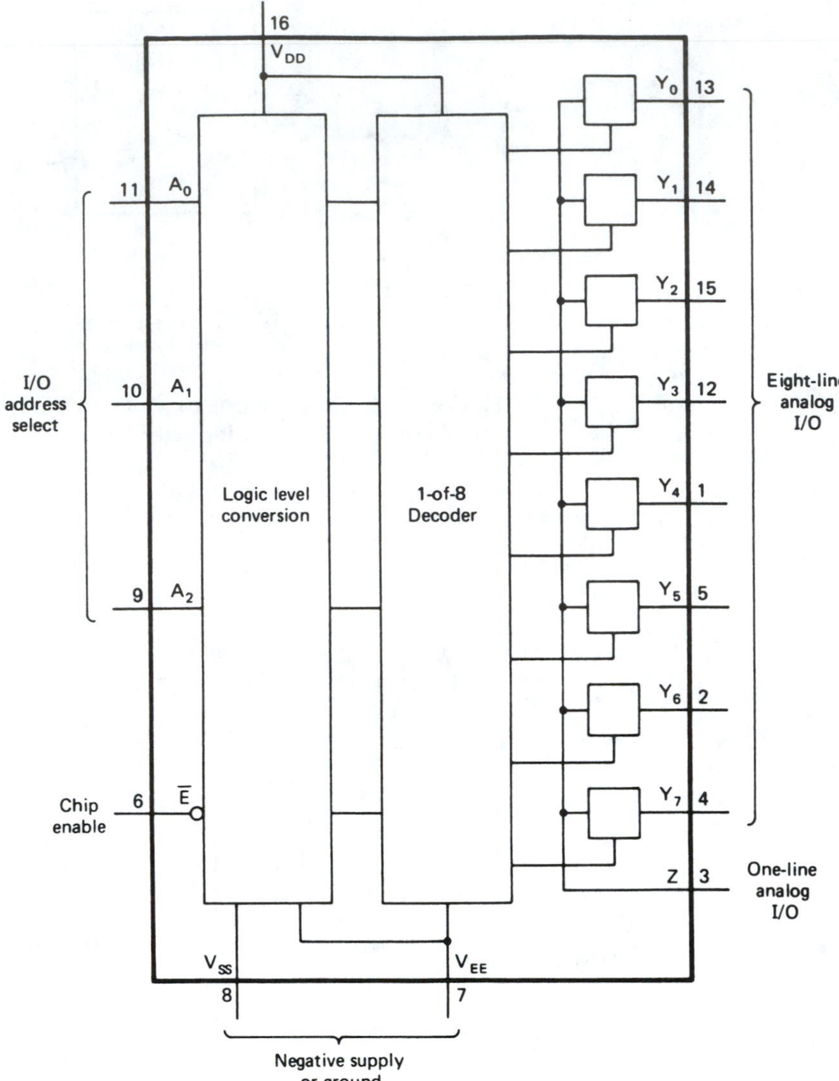

Figure 4–24 The 4051 CMOS analog multiplexer/demultiplexer. (Courtesy of Philips Components–Signetics)

4–6 MULTIPLEXER DESIGN APPLICATIONS

Analog Multiplexer Application

The 4051 is very versatile. One use is in the design of multitrace oscilloscope displays for displaying as many as eight traces on the same display screen. To do that, each input signal to be displayed must be superimposed on (added to) a different voltage level so that each trace will be at a different level on the display screen.

The 4051 can be set up to sequentially output eight different voltage levels repeatedly if connected as shown in Figure 4–25. The resistor voltage-divider network in Figure 4–25 is set up to drop 0.5 V across each 100-Ω resistor. That will put 0.5 V at Y_0, 1.0 V at Y_1, and so on. The binary counter outputs a binary progression from 000 up to 111, which causes each of the Y_0 to Y_7 inputs to be selected for Z out, one at a time, in order. The result is the staircase waveform shown in Figure 4–25, which can superimpose a different voltage level on each of eight separate digital input signals that are brought in via the 74151 eight-line *digital* multiplexer (not shown) driven by the same binary counter.

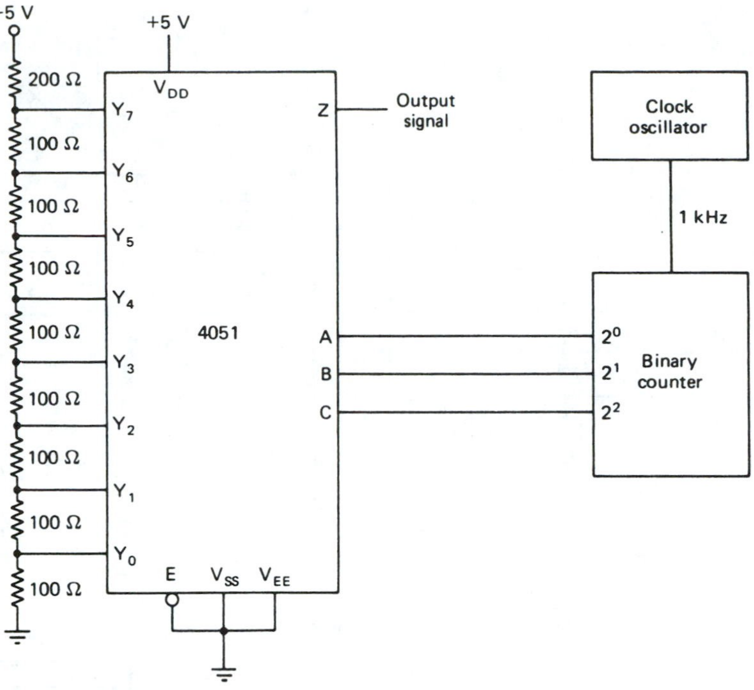

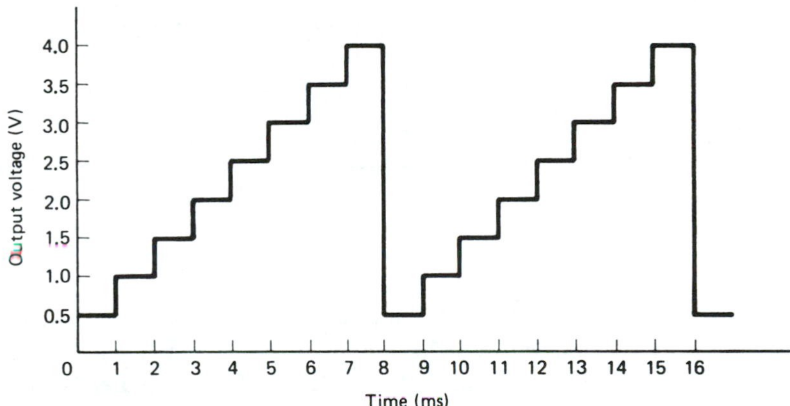

Figure 4–25 The 4051 analog multiplexer used as a staircase generator.

Multiplexed Display Application

Figure 4–26 shows a common method of using multiplexing to reduce the cost of producing a multidigit display in a digital system or computer.

Theory of Operation

Multiplexing multidigit displays reduces circuit cost and failure rate by *sharing* common ICs, components, and conductors. The seven-segment digit displays, decoders, and drivers will be covered in detail in Chapter 6. For now, we need to know that a decoding process must take place to convert the BCD digit information to a recognizable digit display.

The digit bus and display bus are each just a common set of conductors *shared by* the digit storage registers and display segments. The four-digit registers are therefore

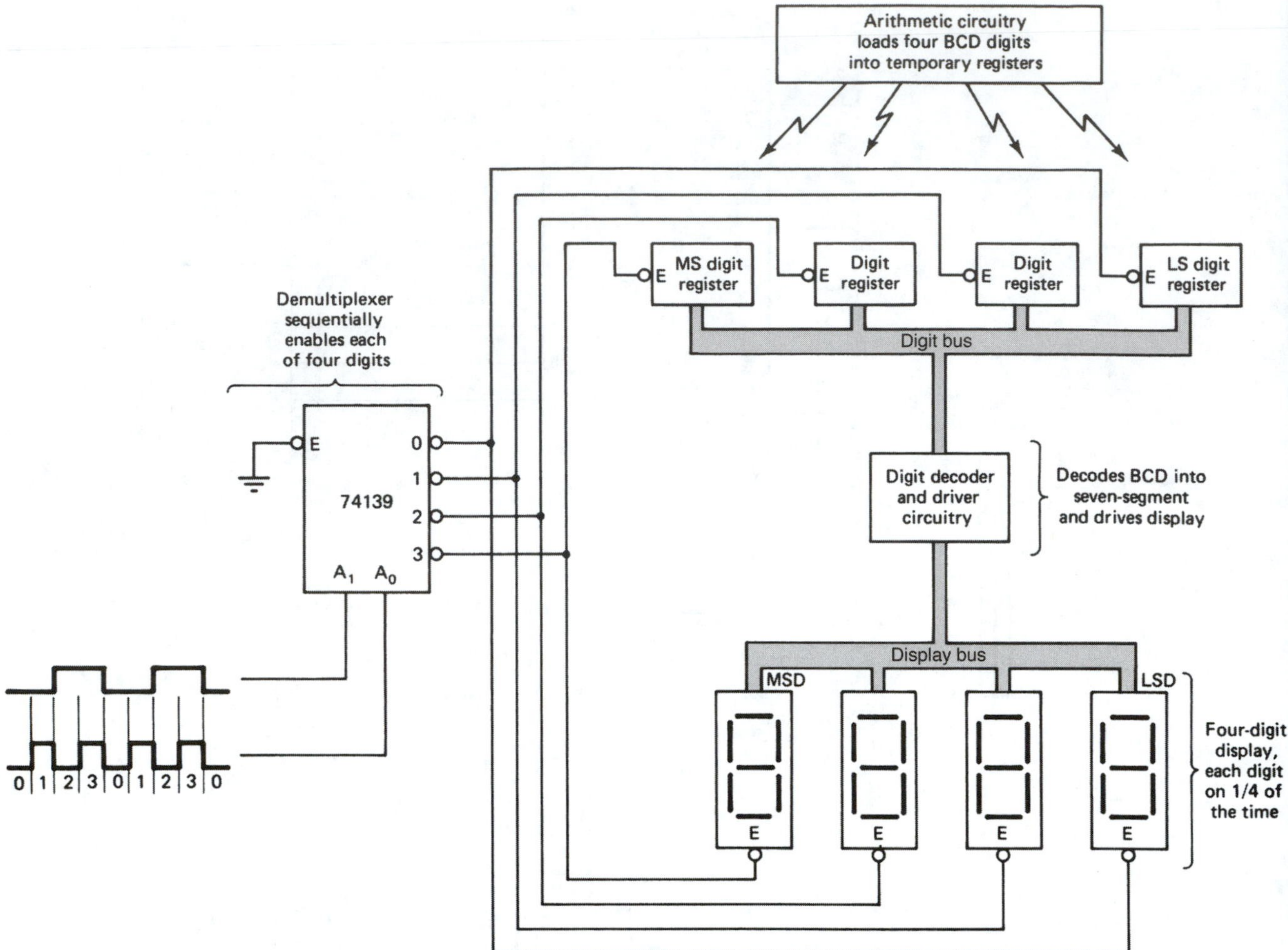

Figure 4–26 Multiplexed four-digit-display block diagram.

multiplexed into a single-digit bus, and the display bus is demultiplexed into the four-digit displays.

The 74139 four-line demultiplexer takes care of sequentially accessing each of the four digits. It first outputs a LOW on the $\overline{0}$ line. This enables the LS digit register *and* the LS digit display. The LS BCD information travels down the digit bus to the decoder/driver, which *decodes* the BCD into the special seven-segment code used by the LS digit display, and *drives* the LS digit display.

Next, the second digit register and display are enabled, then the third, then the fourth. This process continues repeatedly, each digit being ON one-fourth of the time. The circulation is set up fast enough (1 kHz or more) that it appears that all four digits are on at the same time. The external arithmetic circuitry is free to change the display at any time simply by reloading the temporary digit registers.

4–7 SCHMITT TRIGGER ICs

A Schmitt trigger is a special type of integrated circuit that is used to transform slowly changing waveforms into sharply defined, jitter-free output signals. They are useful for changing clock edges that may have slow rise and fall times into straight vertical edges.

The Schmitt trigger employs a technique called *positive feedback* internally to speed up the level transitions, and also to introduce an effect called *hysteresis.* "Hysteresis" means that the switching threshold on a positive-going input signal is at a higher level than the switching threshold on a negative-going input signal (see Figure 4–27). This is useful

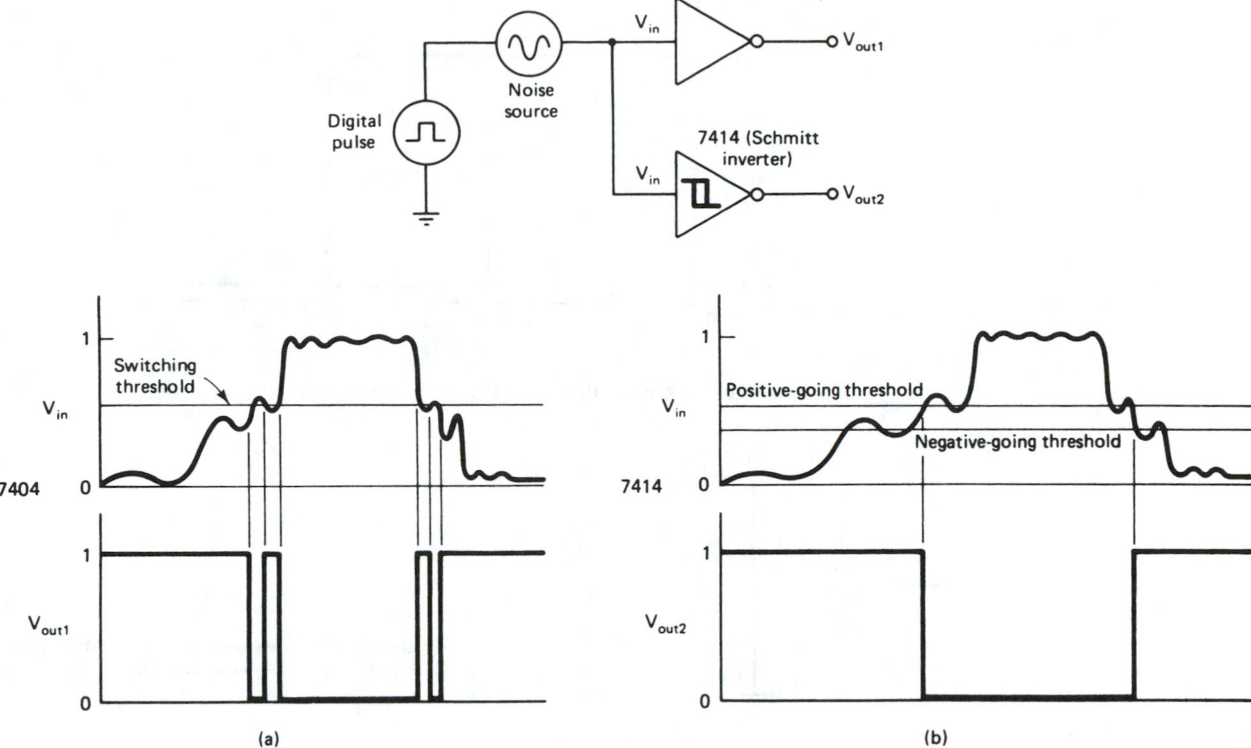

Figure 4–27 Edge-sharpening, jitter-free operation of a Schmitt trigger: (a) regular inverter; (b) Schmitt inverter.

for devices that have to ignore small amounts of jitter, or electrical noise on input signals. Note in Figure 4–27 that when the positive- and negative-going thresholds are exactly the same, as with standard gates, and a small amount of noise causes the input to jitter slightly, the output will switch back and forth several times until the input level is far above the threshold voltage.

Figure 4–27 illustrates the difference in the output waveforms for a standard 7404 inverter and a 7414 Schmitt trigger inverter. As you can see in Figure 4–27b, the output (V_{out2}) is an inverted, jitter-free pulse. On the other hand, just think if V_{out1} were fed into a counting device; it would count three times (three negative edges) instead of once as was intended.

The difference between the positive-going threshold and the negative-going threshold is defined as the hysteresis voltage. For the 7414, the positive-going threshold (V_{T+}) is typically 1.7 V, and the negative-going threshold (V_{T-}) is typically 0.9 V, yielding a hysteresis voltage (ΔV_T) of 0.8 V. The small box symbol (⌶) inside the 7414 symbol is used to indicate that it is a Schmitt trigger inverter instead of a regular inverter.

The most important specification for Schmitt trigger devices is illustrated by use of a "transfer function" graph, which is a plot of V_{out} versus V_{in}. From the transfer function, we can determine the HIGH- and LOW-level output voltages (typically 3.4 and 0.2 V, respectively, the same as most TTL gates), as well as V_{T+}, V_{T-}, and ΔV_T.

Figure 4–28 shows the transfer function for the 7414. The transfer function graph is produced experimentally by using a variable voltage source at the input to the Schmitt and a voltmeter (VOM) at V_{in} and V_{out}, as shown in Figure 4–29.

As the V_{in} of Figure 4–29 is increased from 0 V up toward 5 V, V_{out} will start out at approximately 3.4 V (1) and switch to 0.2 V (0) when V_{in} exceeds the positive-going threshold ($\approx$1.7 V). The output transition from HIGH to LOW is indicated in Figure 4–28 by the downward arrow. As V_{in} is increased up to 5 V, V_{out} remains at 0.2 V (0).

As the input voltage is then decreased down toward 0 V, V_{out} will remain LOW until the negative-going threshold is passed ($\approx$0.9 V). At that point the output will switch up to

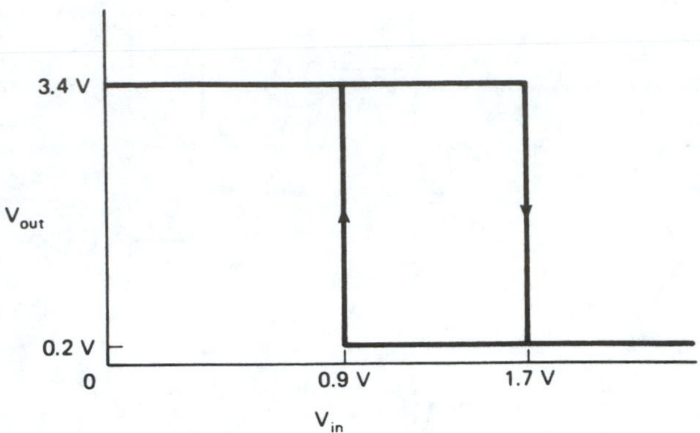

Figure 4–28 Transfer function for a 7414 Schmitt trigger inverter.

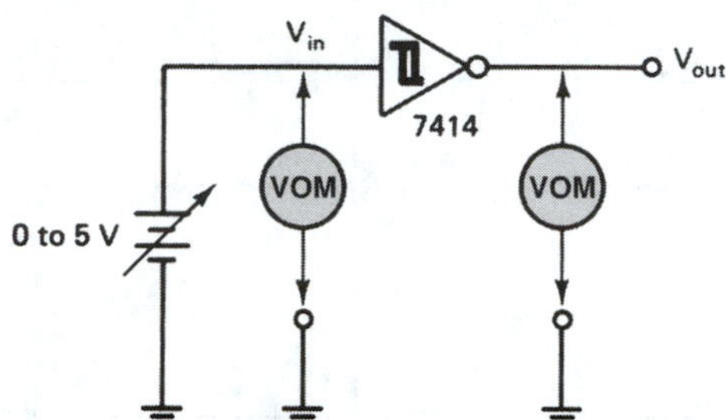

Figure 4–29 Circuit used to produce a Schmitt trigger transfer function experimentally.

3.4 V (1), as indicated by the upward arrow in Figure 4–28. As V_{in} continues to 0 V, V_{out} remains HIGH at 3.4 V. The hysteresis in this case is $1.7 - 0.9 = 0.8$ V.

The following three examples illustrate the operation of Schmitt triggers.

EXAMPLE 4–5

Let's use the Schmitt trigger to convert a small signal sine wave (E_s) into a square wave (V_{out}).

Solution:

The diode is used to short the negative 4 V from E_s to ground to protect the Schmitt input, as shown in Figure 4–30a. The 1-kΩ resistor will limit the current through the diode when it is conducting. [$I_{diode} = (4 - 0.7 \text{ V})/1 \text{ k}\Omega = 3.3$ mA, which is well within the rating of most silicon diodes.]

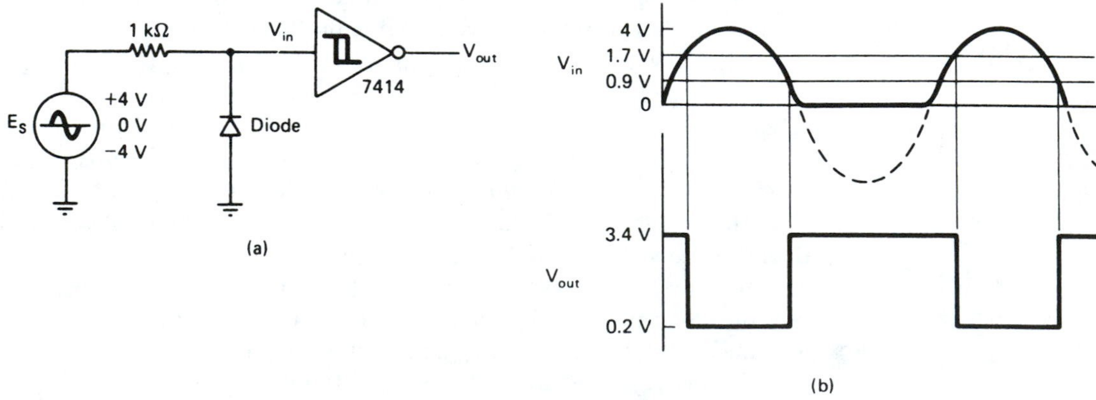

Figure 4–30

Also, the HIGH-level input current to the Schmitt (I_{IH}) is only 40 μA, causing a voltage drop of $40\ \mu\text{A} \times 1\ \text{k}\Omega = 0.04$ V when V_{in} is HIGH. (We can assume that 0.04 V is negligible compared to +4 V.)

The input to the Schmitt will therefore be a half-wave signal with a 4-V peak. The output will be a square wave, as shown in Figure 4–30b.

EXAMPLE 4–6

Figure 4–31

Sketch V_{out} of the 7414 in Figure 4–31 given the V_{in} waveform shown in Figure 4–32.

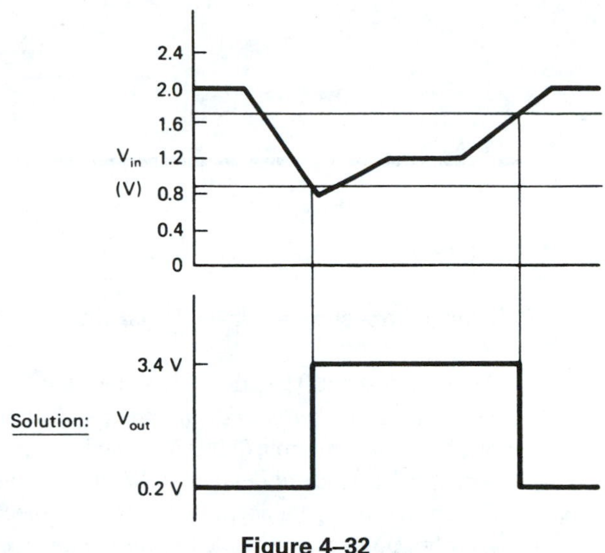

Figure 4–32

EXAMPLE 4–7

Draw and completely label the V_{out} versus V_{in} transfer function for the Schmitt trigger device whose V_{in} and V_{out} waveforms are given in Figure 4–33.

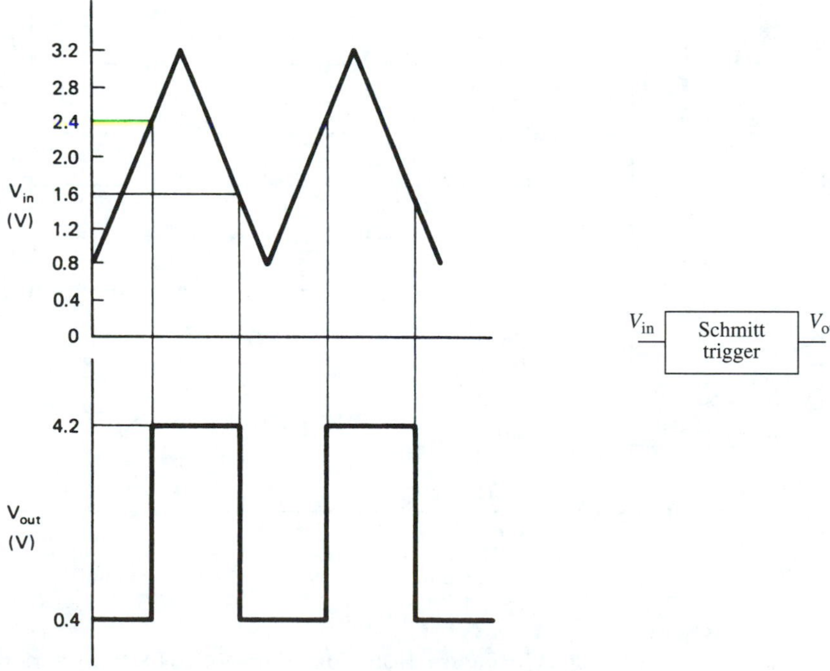

Figure 4–33

Solution:

The transfer function is shown in Figure 4–34.

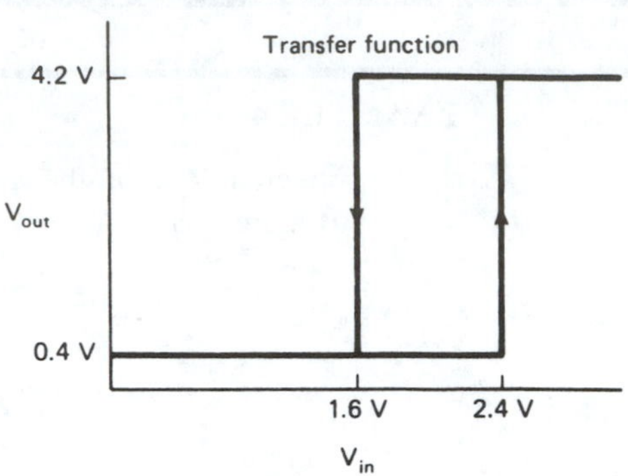

Figure 4–34

4–8 SYSTEM DESIGN APPLICATIONS

Microprocessor Address Decoding

The 74138 and its CMOS version, the 74HCT138, are popular choices for decoding the address lines in *microprocessor* circuits. A typical 8-bit microprocessor such as the Intel 8085A or the Motorola 6809 has 16 address lines (A_0–A_{15}) for designating unique *addresses* for all the peripheral devices and memory connected to it. When a microprocessor-based system has a large amount of memory connected to it, a designer often chooses to set the memory up in groups, called *memory banks.* For example, Figure 4–35 shows a decoding

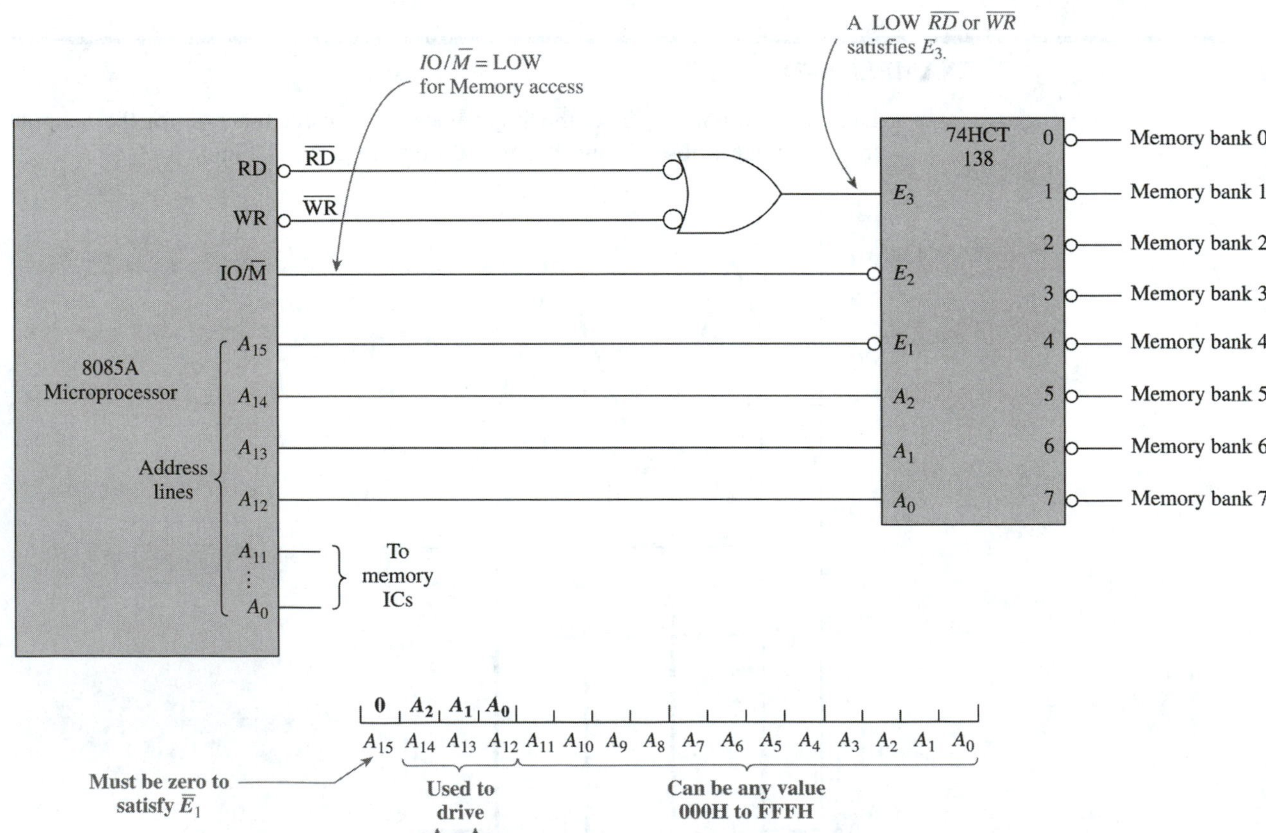

Figure 4–35 Using the 74HCT138 for a memory address decoder in an 8085A microprocessor system.

scheme that can be used to select one of eight separate memory banks within a microprocessor-based system. The high-order bits of the address (A_{12}–A_{15}) are output by the microprocessor to designate which memory bank is to be accessed. In this design, A_{15} must be LOW for the decoder IC to be enabled. The three other high-order bits, A_{12}–A_{14}, are then used to select the designated memory bank.

The 8085A also outputs control signals that are used to enable/disable memory operations. First, if we are performing a memory operation, we must be doing a read ($\overline{RD}$) or a write ($\overline{WR}$). The inverted-input OR gate (NAND) provides the HIGH to the E_3 enable if it receives a LOW $\overline{RD}$ *or* a LOW $\overline{WR}$. The other control signal, IO/$\overline{M}$, is used by the 8085A to distinguish between input/output (IO) to peripheral devices versus memory operations. If IO/$\overline{M}$ is HIGH, an IO operation is to take place, and if it is LOW, a memory operation is to occur. Therefore, one of the memory banks will be selected if IO/$\overline{M}$ is LOW and address line A_{15} is LOW while either $\overline{RD}$ is LOW or $\overline{WR}$ is LOW.

EXAMPLE 4–8

What is the range of addresses that can be specified by the 8085A in Figure 4–35 to access memory within bank 2?

Solution:

Referring to the chart in Figure 4–35, address bit A_{13} must be HIGH in order to make A_1 in the decoder HIGH to select bank 2. The other address bits, A_0–A_{11}, are not used by the decoder IC and can be any value. Therefore, any address within the range of 2000H through 2FFFH will select memory bank 2.

Alarm Encoder for a Microcontroller

This design application uses a 74148 encoder to monitor the fluid level of eight chemical tanks. If any level exceeds a predetermined height, a sensor outputs a LOW level to the input of the encoder. The encoder encodes the active input into a 3-bit binary code to be read by a *microcontroller.* This way, the microcontroller needs to use only three input lines to monitor eight separate points. Figure 4–36 shows the circuit connections.

A microcontroller differs from a microprocessor in that it has several input/output ports and memory built into its architecture, making it better suited for monitoring and control applications. The microcontroller used here is the Intel 8051.* We will use one of its 8-bit ports to read the encoded alarm code, and we will use its interrupt input, $\overline{INT0}$, to receive notification that an alarm has occurred. The 8051 will be programmed to be in a HALT mode (or, in some versions, a low-power SLEEP mode) until it receives an interrupt signal to take a specific action. In this case it will perform the desired response to the alarm when it receives a LOW at $\overline{INT0}$. This LOW interrupt signal is provided by $\overline{GS}$, which goes LOW whenever any of the 74148 inputs becomes active.

Serial Data Multiplexing for a Microcontroller

Multiplexing and demultiplexing are very useful for data communication between a computer system and serial data terminals. The advantage is that only one serial receive line and one serial transmit line are required by the computer to communicate with several data terminals. A typical configuration is shown in Figure 4–37.

Again the 8051 is used because of its built-in control and communication capability. Its RXD and TXD pins are designed to receive (RXD) and transmit (TXD) serial data at a speed and in a format dictated by a computer program written by the user.

The selected data terminal to be read from is routed through the 74HCT151 multiplexer to the microcontroller's serial input terminal, RXD. First, the computer program writes the appropriate hex code to Port 1 to enable the 151 to route the serial data stream

*For an in-depth study of the 8085A microprocessor and the 8051 microcontroller refer to Chapters 9–14.

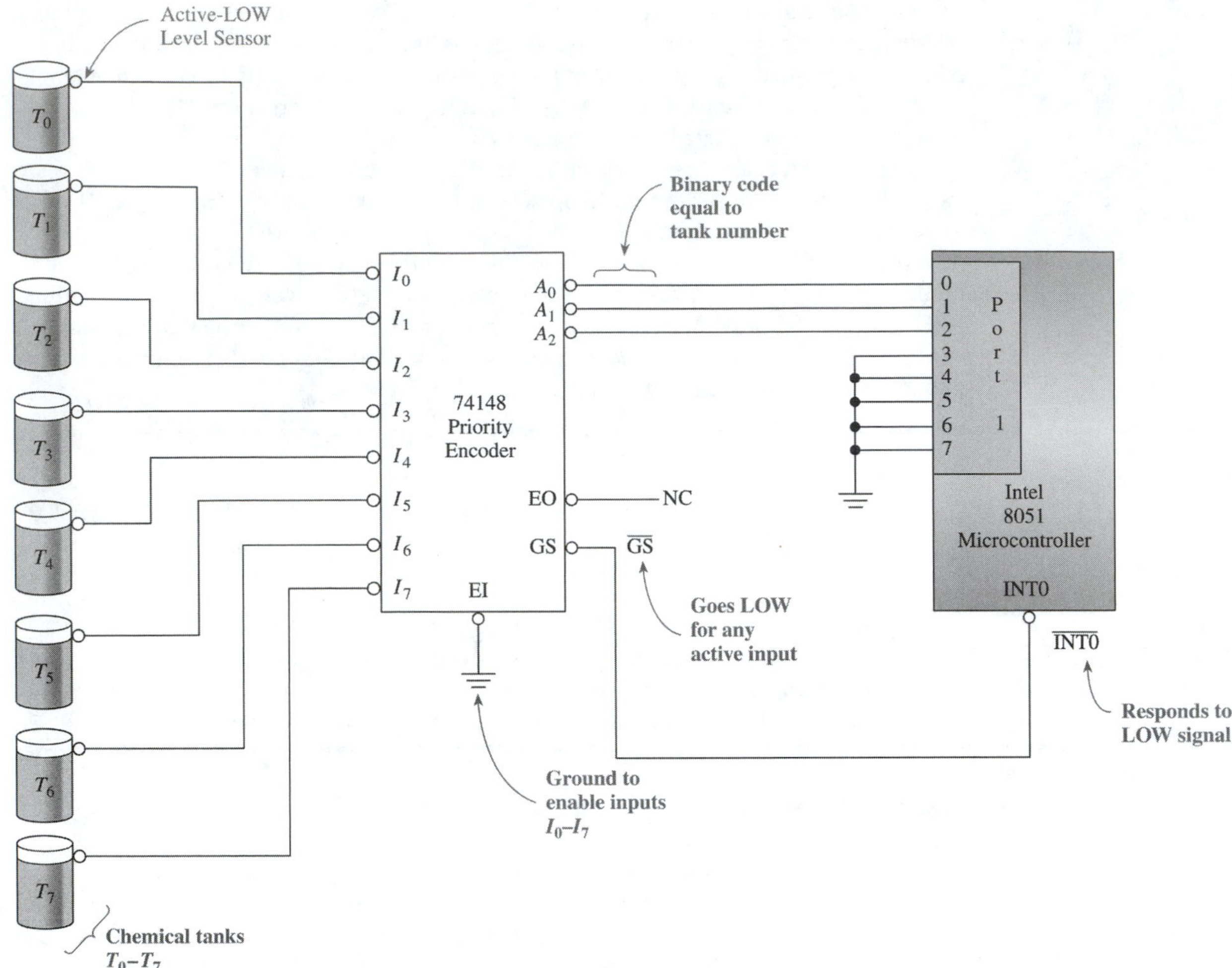

Figure 4–36 Using a 74148 to encode an active alarm to be monitored by a microcontroller.

from the selected data terminal through to its *Y* output. The 8051 then reads the serial data at its RXD input and performs the desired action.

To output to one of the data terminals, the 8051 must first output the appropriate hex code to Port 1 to select the correct data terminal; then it outputs serial data on its TXD pin. The 74HCT238 is used as a demultiplexer (data distributor) in this application. The 74HCT238 is identical to the 74138 decoder/demultiplexer except the '238 has noninverting outputs. The HCT versions are used to match the high speed of the 8051 and keep the power requirements to a minimum (see Section 2–14). The hex code output at Port 1 must provide a LOW to $\overline{E_2}$ and the proper data-routing select code to A_2–A_0. The selected *Y* output then duplicates the HIGH/LOW levels presented at the E_3 pin.

EXAMPLE 4–9

Determine the correct hex codes that must be output to Port 1 in Figure 4–37 to accomplish the following action: **(a)** read from data terminal 3 (DT3); **(b)** write to data terminal 6 (DT6).

Solution:

The 0, 1, and 2 outputs of Port 1 are used to control the S_0, S_1, and S_2 input-select pins of the '151 and the A_0, A_1, and A_2 output-select pins of the '238. Output 7 of Port 1 is used to determine which IC is enabled; 0 enables the '151 multiplexer, and 1 enables the '238 demultiplexer. **(a)** Assuming that the NC (no connection) lines are 0s, the hex code to read from data terminal 3 is 03H (0000 0011). **(b)** The hex code to write to data terminal 6 is 86H (1000 0110).

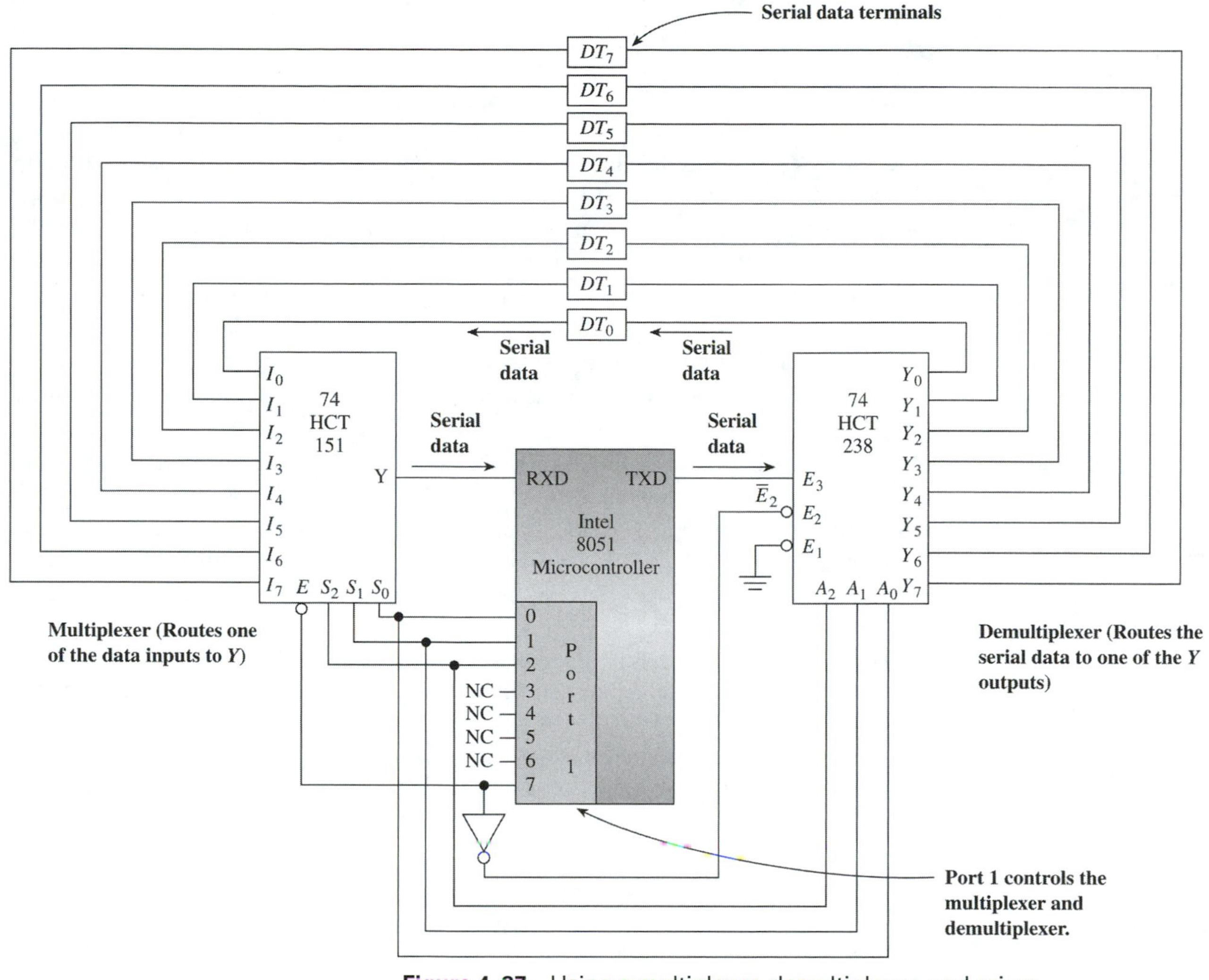

Figure 4–37 Using a multiplexer, demultiplexer, and microcontroller to provide communication capability to several serial data terminals.

SUMMARY

In this chapter we have learned that

1. Comparators can be used to determine equality or which of two binary strings is larger.
2. Decoders can be used to convert a binary code into a singular active output representing its numeric value.
3. Encoders can be used to generate a coded output from a singular active numeric input line.
4. Multiplexers are used to select one of several data inputs and pass it to a single data output.

5. Demultiplexers are used to take a single data value or waveform and route it to one of several outputs.
6. The two key features of Schmitt trigger ICs are that they output extremely sharp edges and they have two distinct input threshold voltages. The difference between the threshold voltages is called the hysteresis voltage.

GLOSSARY

Active-HIGH: Means that the input to, or the output from, a terminal must be HIGH to be enabled or "active."

Active-LOW: Means that the input to, or the output from, a terminal must be LOW to be enabled or "active."

Bidirectional: A device capable of functioning in either of two directions, thus being able to reverse its input/output functions.

Bus: A common set of conductors shared by several devices or ICs.

Comparator: A device used to compare the magnitude or size of two binary bit strings or words.

Decoder: A device that converts a digital code such as hex or octal into a single output representing its numeric value.

Don't Care (×): A variable that is signified in a function table as a don't care, or ×, can take on either value, HIGH *or* LOW, without having any effect on the output.

Encoder: A device that converts a weighted numeric input line to an equivalent digital code, such as hex or octal.

Hysteresis: In digital Schmitt trigger ICs, hysteresis is the difference in voltage between the positive-going switching threshold and the negative-going switching threshold at the input.

Jitter: A term used in digital electronics to describe a waveform that has some degree of electrical noise on it, causing it to rise and fall slightly between and during level transitions.

Positive Feedback: A technique employed by Schmitt triggers that involves taking a small sample of the output of a circuit and feeding it back into the input of the same circuit to increase its switching speed and introduce hysteresis.

Priority: When more than one input to a device is active and only one can be acted on, the one with the highest "priority" will be acted on.

Schmitt Trigger: A circuit used in digital electronics to provide ultrafast level transitions and introduce hysteresis for improving jittery or slowly rising waveforms.

Superimpose: Combining two waveforms together such that the result is the sum of their levels at each point in time.

Threshold: The exact voltage level at the input to a digital IC that causes it to switch states. In Schmitt trigger ICs, there are two different threshold levels: the positive-going threshold (LOW to HIGH) and the negative-going threshold (HIGH to LOW).

Transfer Function: A plot of V_{out} versus V_{in} that is used to determine graphically the operating specifications of a Schmitt trigger.

PROBLEMS

4–1. Design a binary comparator circuit using exclusive-ORs and a NOR gate that will compare two 8-bit binary strings.

4–2. Label all the lines in your design for Problem 4–1 with digital levels that will occur when comparing $A = 1101\ 1001$ and $B = 1101\ 1001$.

4–3. Label the digital levels on all the lines in Figure 4–3 that would occur when comparing the two 8-bit strings $A = 1011\ 0101$ and $B = 1100\ 0011$.

4–4. Write a two-sentence description of the function of a decoder.

4–5. Construct a truth table similar to Table 4–1 for an active-LOW output BCD (1-of-10) decoder.

4–6. What state must the inputs $\overline{E_1}$, $\overline{E_2}$, E_3 be in in order to *enable* the 74138 decoder?

4–7. What does the × signify in the function table for the 74138?

4–8. Describe the difference between active-LOW outputs and active-HIGH outputs.

4–9. Sketch the output waveforms ($\overline{0}$ to $\overline{7}$) given the inputs shown in Figure P4–9b to the 74138 of Figure P4–9a.

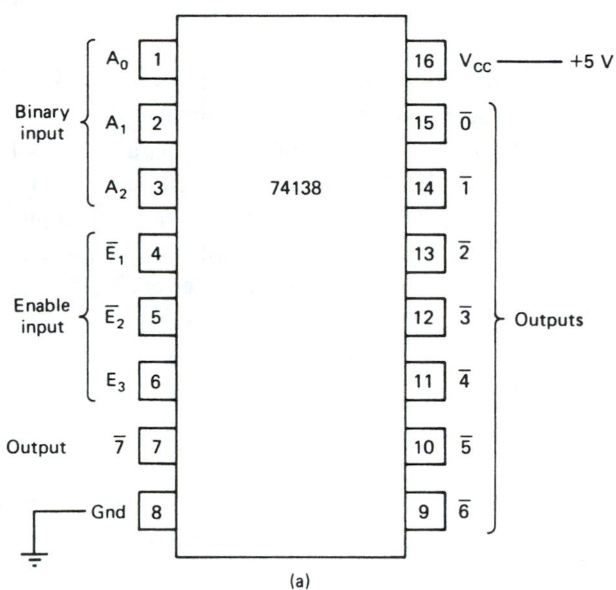

(a)

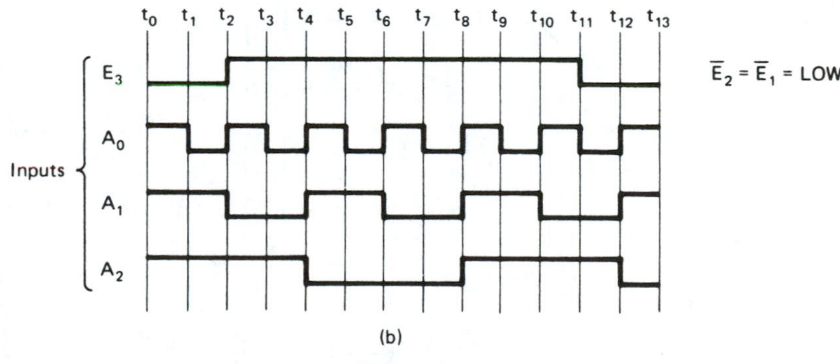

(b)

Figure P4–9

4–10. Repeat Problem 4–9 for the input waveforms shown in Figure P4–10.

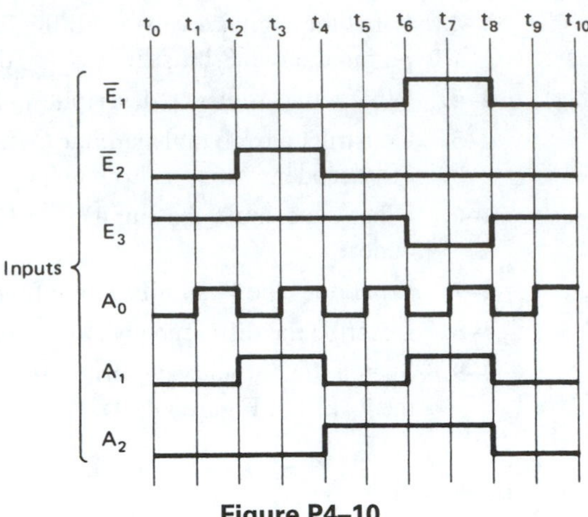

Figure P4–10

4–11. What state do the outputs of a 7442 BCD decoder go to when an invalid BCD number (10 to 15) is input to A_0 to A_3?

4–12. Design a circuit, based on a 74154 4-line-to-16-line decoder, that will output a HIGH whenever the 4-bit binary input is greater than 12. (When the binary input is less than or equal to 12, it will output a LOW.)

4–13. With the 74147 priority encoder, if two different decimal numbers are input at the same time, which will be encoded?

4–14. A 74147 is connected with $\overline{I_1} = \overline{I_2} = \overline{I_3} =$ LOW and $\overline{I_4} = \overline{I_5} = \overline{I_6} = \overline{I_7} = \overline{I_8} = \overline{I_9} =$ HIGH. Determine $\overline{A_0}$, $\overline{A_1}$, $\overline{A_2}$, and $\overline{A_3}$.

4–15. The connectors shown in Figure P4–15 are made to the 74151 eight-line multiplexer. Determine Y and $\overline{Y}$.

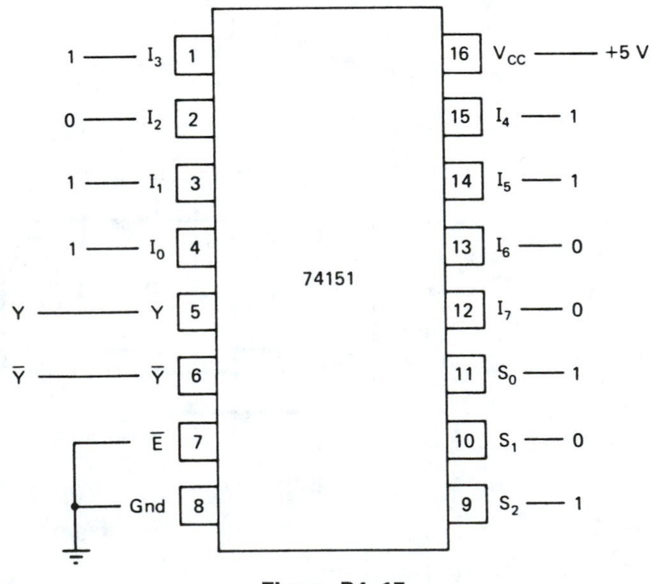

Figure P4–15

Design

4–16. Using a technique similar to that presented in Figure 4–19, design a 32-bit multiplexer using four 74151s.

4–17. Design an 8-bit demultiplexer using one 74139.

4–18. Design a 16-bit demultiplexer using two 74138s.

Troubleshooting

4–19. There is a malfunction in a digital system that contains several multiplexer and demultiplexer ICs. A reading was taken at each pin with a logic probe and the results were recorded in Table 4–5. Which IC, or ICs, are not working correctly?

TABLE 4–5
IC Logic States for Troubleshooting Problem 4–19

74138		*74151*		*74139*		*74154*	
Pin	*Level*	*Pin*	*Level*	*Pin*	*Level*	*Pin*	*Level*
1	1	1	1	1	0	1	1
2	1	2	0	2	1	2	1
3	0	3	0	3	0	3	1
4	0	4	1	4	0	4	1
5	1	5	1	5	1	5	1
6	1	6	0	6	1	6	1
7	1	7	0	7	1	7	1
8	0	8	0	8	0	8	1
9	1	9	0	9	0	9	1
10	1	10	0	10	0	10	1
11	1	11	0	11	1	11	1
12	0	12	0	12	0	12	0
13	1	13	1	13	1	13	1
14	1	14	0	14	0	14	1
15	1	15	0	15	1	15	1
16	1	16	1	16	1	16	1
						17	1
						18	1
						19	0
						20	0
						21	0
						22	1
						23	1
						24	1

Schmitt Trigger

4–20. One particular Schmitt trigger inverter has a positive-going threshold of 1.9 V and a negative-going threshold of 0.7 V. Its V_{OH} (typical) is 3.6 V and V_{OL} (typical) is 0.2 V. Sketch the transfer function (V_{out} versus V_{in}) for that Schmitt trigger.

4–21. If the input waveform (V_{in}) shown in Figure P4–21 is fed into the Schmitt trigger described in Problem 4–20, sketch its output waveform (V_{out}).

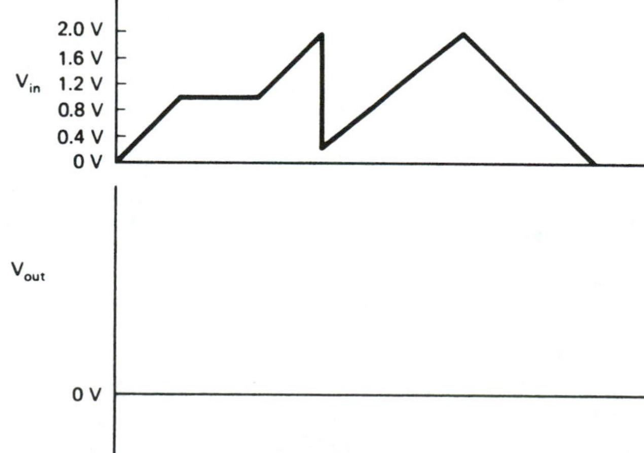

Figure P4–21

4–22. If the V_{in} and V_{out} waveforms shown in Figure P4–22 are observed on a Schmitt trigger device, determine its characteristics and sketch the transfer function (V_{out} versus V_{in}).

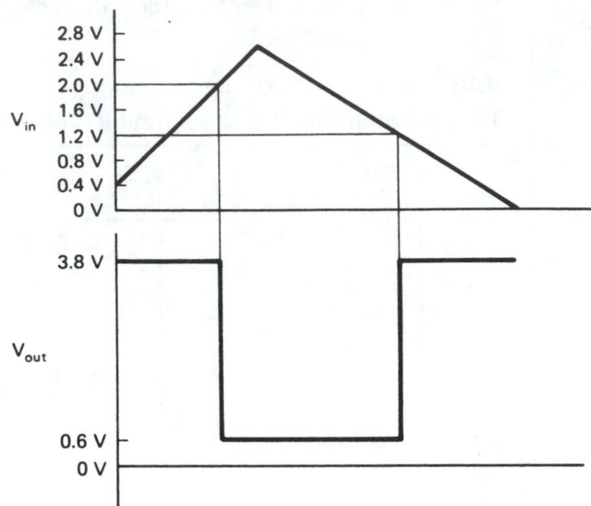

Figure P4–22

SCHEMATIC INTERPRETATION PROBLEMS

4–23. If you wanted to check the power supply connections for the 8031 IC (U8) on the 4096/4196 circuit, which pins would you check, and what level should they be?

4–24. Find the decoders U28 and U29 on sheet 2 of the 4096/4196 schematic. They are cascaded together to form a 1-of-18 decoder for the lines labeled ICS1-ICS18.

(a) Determine the levels on pins 2, 5, 6, 9, and 12 of U31 to provide an active-LOW output at ICS5.

(b) Repeat for ICS18.

Flip-Flops and Sequential Logic

OBJECTIVES

Upon completion of this chapter, you should be able to:

- Explain the operation of *S-R* and gated *S-R* flip-flops.
- Compare the operation of *D* latches and *D* flip-flops by using timing diagrams.
- Describe the characteristics of edge-triggered flip-flops.
- Connect integrated-circuit *J-K* flip-flops as toggle and *D* flip-flops.
- Use timing diagrams to illustrate the synchronous and asynchronous operation of *J-K* flip-flops.
- Use manufacturers' data sheets to determine IC operating specifications such as setup time, hold time, and propagation delay.
- Describe the problems caused by switch bounce and how to eliminate its effects.
- Sketch the waveforms and calculate voltage and time values for astable and monostable multivibrators.
- Connect integrated-circuit monostable multivibrators to output a waveform with a specific pulse width.
- Explain the operation of the internal components of the 555 IC timer.
- Connect a 555 IC timer as an astable multivibrator and as a monostable multivibrator.
- Discuss the operation and application of crystal oscillator circuits.

INTRODUCTION

The logic circuits studied in the previous chapters have consisted mainly of logic gates (AND, OR, NAND, NOR, INVERT) and combinational logic. Starting in this chapter, we will be dealing with data storage circuitry that will "latch" on to (remember) a digital state (1 or 0).

This new type of digital circuitry is called *sequential logic* because it is controlled by, and is used for, controlling other circuitry in a specific sequence dictated by a control clock or enable/disable control signals.

The simplest form of data storage is the Set–Reset (*S-R*) flip-flop. These circuits are called *transparent latches* because the outputs respond immediately to changes at the input and the input state will be remembered, or "latched" on to. The latch will sometimes have an "enable input," which is used to control the latch to accept or ignore the *S-R* input states.

More sophisticated flip-flops use a clock as the control input and are used wherever the input and output signals must occur within a particular sequence. Later in the chapter we will examine the oscillation circuits used to generate the clock waveforms.

5–1 THE S-R FLIP-FLOP

The *S-R* flip-flop is a data storage circuit that can be constructed using the basic gates covered in previous chapters. Using a cross-coupling scheme with two NOR gates, we can form the flip-flop shown in Figure 5–1.

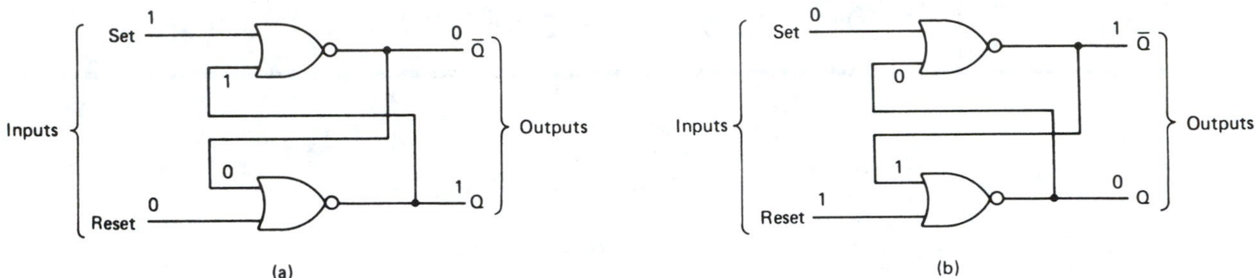

Figure 5–1 Cross-NOR *S-R* flip-flop: (a) Set condition; (b) Reset condition.

Let's start our analysis by placing a 1 (HIGH) on the Set and a 0 (LOW) on the Reset (Figure 5–1a). This is defined as the Set condition and should make the Q output 1 and $\overline{Q}$ output 0. Well, a HIGH on the Set will make the output of the upper NOR equal 0 ($\overline{Q} = 0$) and that 0 is fed down to the lower NOR, which together with a LOW on the Reset input will cause the lower NOR's output to equal a 1 ($Q = 1$). [Remember, a NOR gate is always 0 output except when *both* inputs are 0 (Chapter 2).]

Now, when the 1 is removed from the Set input, the flip-flop should "remember" that it is Set (i.e., $Q = 1, \overline{Q} = 0$). So with Set = 0, Reset = 0, and $Q = 1$ from previously being Set, let's continue our analysis. The upper NOR has a 0–1 at its inputs, making $\overline{Q} = 0$; while the lower NOR has a 0–0 at its inputs, keeping $Q = 1$. Great—the flip-flop remained Set even after the Set input was returned to 0.

Now we should be able to Reset the flip-flop by making $S = 0, R = 1$ (Figure 5–1b). Well, with $R = 1$, the lower NOR will output a 0 ($Q = 0$), placing a 0–0 on the upper NOR, making its output 1 ($\overline{Q} = 1$); thus the flip-flop "flipped" to its Reset state.

The only other input condition is when both S and R inputs are HIGH. In this case both NORs will put out a LOW, making Q *and* $\overline{Q}$ equal 0, which is a condition that is not used. (Why would anyone want to Set *and* Reset at the same time, anyway!)

From the previous analysis we can construct the *S-R* flip-flop function table shown in Table 5–1, which will list all input and output conditions.

An *S-R* flip-flop can also be made from cross-NAND gates, as shown in Figure 5–2. Prove to yourself that Figure 5–2 will produce the function table shown in Table 5–2. (Start with $S = 1$, $R = 0$ and remember that a NAND is LOW out only when *both* inputs are HIGH.) The symbols used for an *S-R* flip-flop are shown in Figure 5–3. The symbols show

TABLE 5–1
Function Table for Figure 5–1

S	*R*	*Q*	$\overline{Q}$	*Comments*
0	0	*Q*	$\overline{Q}$	Hold condition (no change)
1	0	1	0	Flip-flop Set
0	1	0	1	Flip-flop Reset
1	1	0	0	Not used

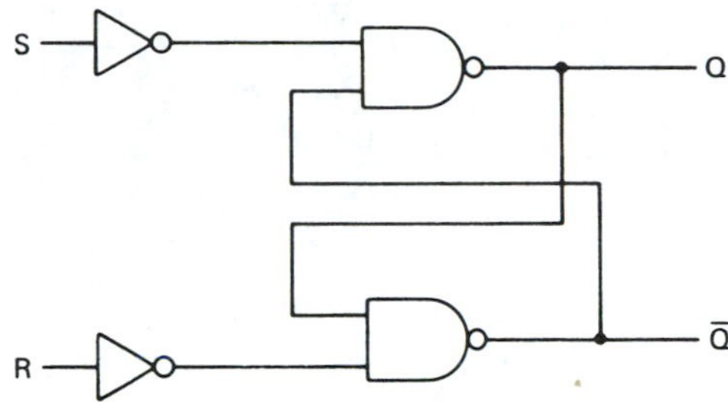

Figure 5–2 Cross-NAND *S-R* flip-flop.

TABLE 5–2
Function Table for Figure 5–2

S	*R*	*Q*	$\overline{Q}$	*Comments*
0	0	*Q*	$\overline{Q}$	Hold condition
1	0	1	0	Flip-flop Set
0	1	0	1	Flip-flop Reset
1	1	1	1	Not used

Figure 5–3 Symbol for an *S-R* flip-flop.

that both *true* and *complemented* *Q* outputs are available. The second symbol is technically more accurate, but the first symbol is found most often in manufacturers' data manuals and throughout this book.

Now let's get practical and find an integrated-circuit TTL NOR gate and draw the actual wiring connections to form a cross-NOR like Figure 5–1 so that we may check it in the lab.

The TTL data manual shows a quad NOR gate 7402. Looking at its pin layout in conjunction with Figure 5–1, we can draw the circuit of Figure 5–4. To check out the operation of Figure 5–4 in the lab, apply 5 V to pin 14 and ground pin 7. Set the flip-flop by placing a HIGH (5 V) to the Set input and a LOW (0 V, ground) to the Reset input. A logic probe attached to the *Q* output should register a HIGH. When the *S-R* inputs are returned to the 0–0 state, the *Q* output should remain "latched" in the 1 state. The Reset function can be checked using the same procedure.

S-R Timing Analysis

By performing a timing analysis on the *S-R* flip-flop, we can see why it is called "transparent," and also observe the "latching" phenomenon.

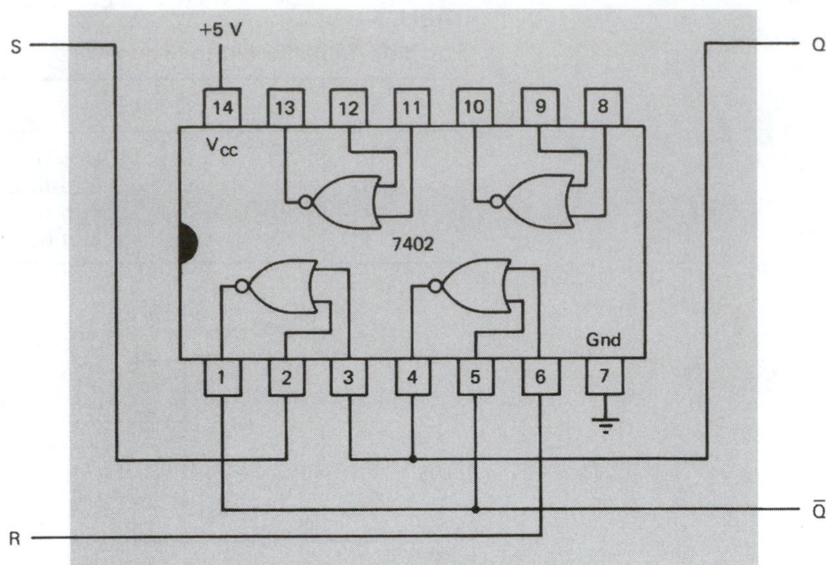

Figure 5–4 *S-R* flip-flop connections using a 7402 TTL IC.

EXAMPLE 5–1

The *S* and *R* waveforms given in Figure 5–6 are connected to an *S-R* flip-flop shown in Figure 5-5. Sketch the *Q* output waveform that will result.

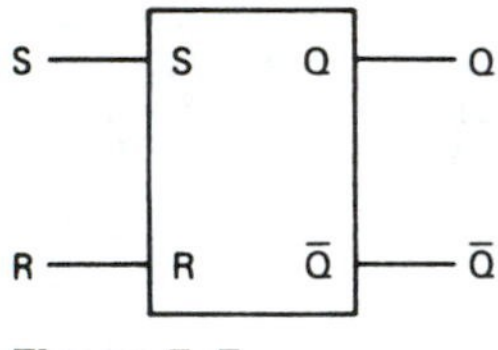

Figure 5–5

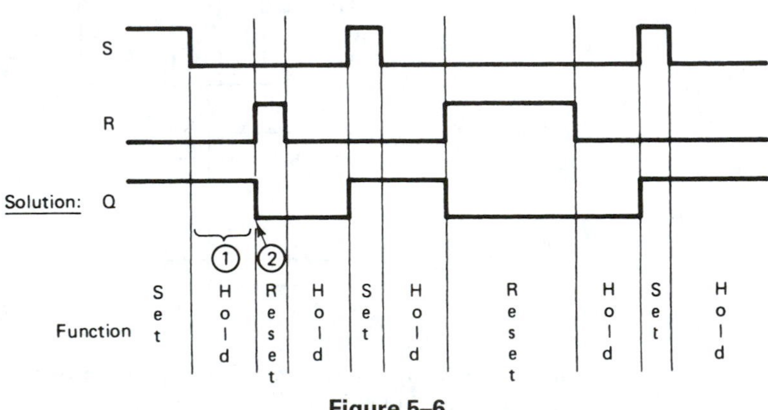

Figure 5–6

Note:

1. The flip-flop is "latched" in the Set condition even after the HIGH is removed from the *S* input.
2. The flip-flop is considered "transparent" because the *Q* output responds immediately to input changes.

Gated *S-R* Flip-Flop

Simple gate circuits, combinational logic, and transparent *S-R* flip-flops are called *asynchronous* (not synchronous) because the output responds immediately to input changes. *Synchronous* circuits operate sequentially, in step, with a control input. To make an *S-R* flip-flop synchronous, we add a gated input to enable and disable the *S* and *R* inputs. Figure 5–7 shows the connections that make the cross-NOR *S-R* flip-flop into a gated *S-R* flip-flop.

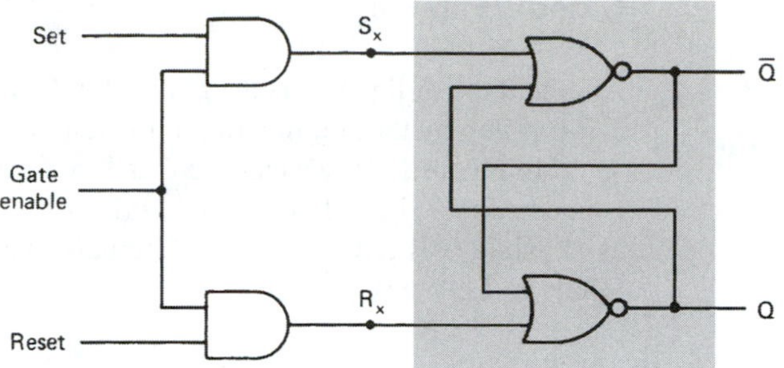

Figure 5–7 Gated *S-R* flip-flop.

The S_x and R_x lines in Figure 5–7 are the original Set and Reset inputs. With the addition of the AND gates, however, the S_x and R_x lines will be kept LOW-LOW (Hold condition) as long as the Gate Enable is LOW. The flip-flop will operate normally while the Gate Enable is HIGH. The function chart (Figure 5–8b) and Example 5–2 illustrate the operation of the gated *S-R* flip-flop.

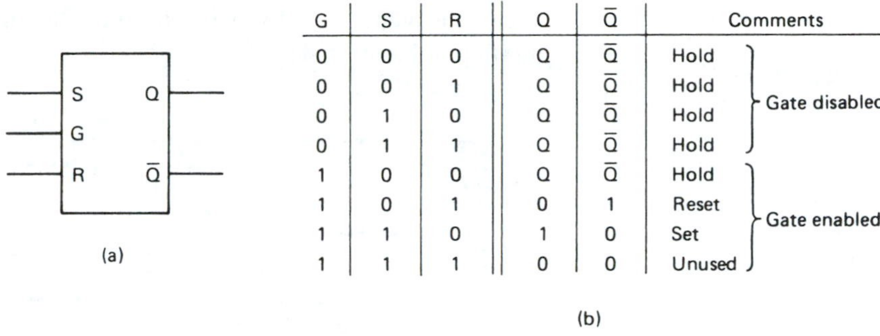

G	S	R	Q	$\bar{Q}$	Comments	
0	0	0	Q	$\bar{Q}$	Hold	Gate disabled
0	0	1	Q	$\bar{Q}$	Hold	
0	1	0	Q	$\bar{Q}$	Hold	
0	1	1	Q	$\bar{Q}$	Hold	
1	0	0	Q	$\bar{Q}$	Hold	Gate enabled
1	0	1	0	1	Reset	
1	1	0	1	0	Set	
1	1	1	0	0	Unused	

(b)

Figure 5–8 Gated *S-R* flip-flop: (a) symbol; (b) function table of Figure 5–7.

EXAMPLE 5–2

Feed the following *G*, *S*, and *R* inputs in Figure 5–9 into the gated *S-R* flip-flop, sketch the output wave at *Q*, and list the flip-flop functions.

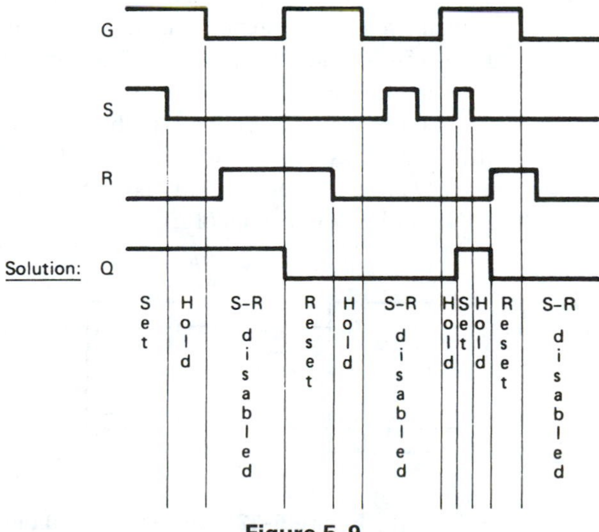

Figure 5–9

Gated *D* Flip-Flop

Another type of flip-flop is the *D* flip-flop (*Data* flip-flop). It can be formed from the gated *S-R* flip-flop by the addition of an inverter.

In Figure 5–10 we can see that *S* and *R* will be complements of each other, and *S* is connected to a single line labeled *D* (Data). The operation is such that *Q* will be the same as *D*, while *G* is HIGH and *Q* will remain "latched" in whatever state it was in before the HIGH-to-LOW transition on *G*.

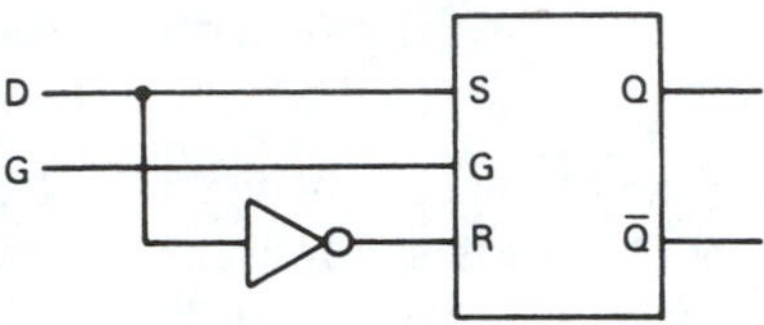

Figure 5–10 Gated *D* flip-flop.

EXAMPLE 5–3

Sketch the output waveform at *Q* for the following inputs at *D* and *G* of a gated *D* flip-flop in Figure 5–11.

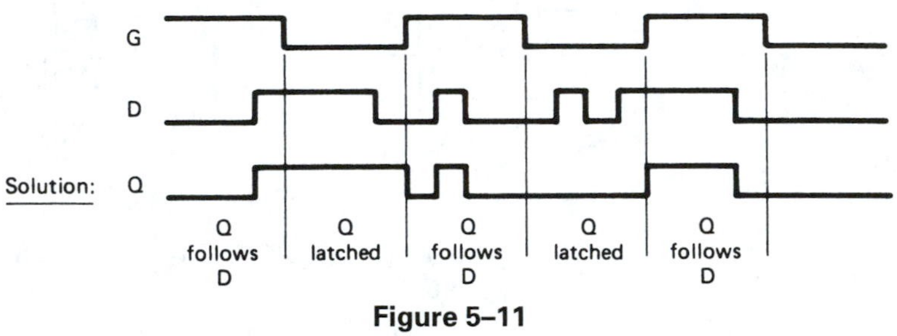

Figure 5–11

5–2 THE INTEGRATED-CIRCUIT D LATCH (7475)

The 7475 is an example of an integrated-circuit *D* latch (also called a *bistable latch*). It contains *four* transparent *D* latches. Its logic symbol and pin configuration are given in Figure 5–12. Latches 0 and 1 share a common Enable (E_{0-1}), and latches 2 and 3 share a common Enable (E_{2-3}).

From the function table (Table 5–3) we can see that the *Q* output will follow *D* (transparent) as long as the enable line (*E*) is HIGH (called active-HIGH enable). When *E* goes LOW, the *Q* output will become "latched" to the value that *D* was just before the HIGH-to-LOW transition of *E*.

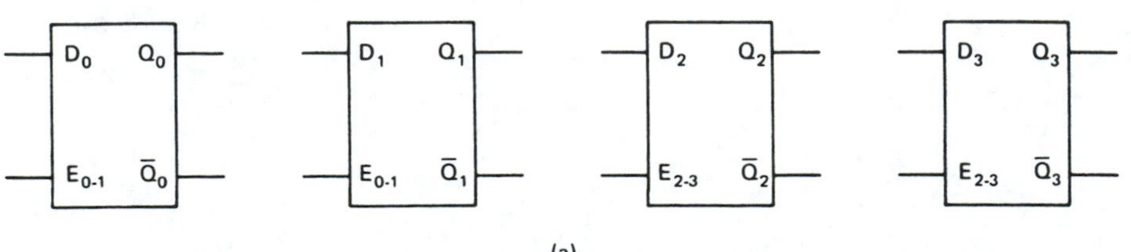

(a)

Figure 5–12 The 7475 quad bistable *D* latch: (a) logic symbol.

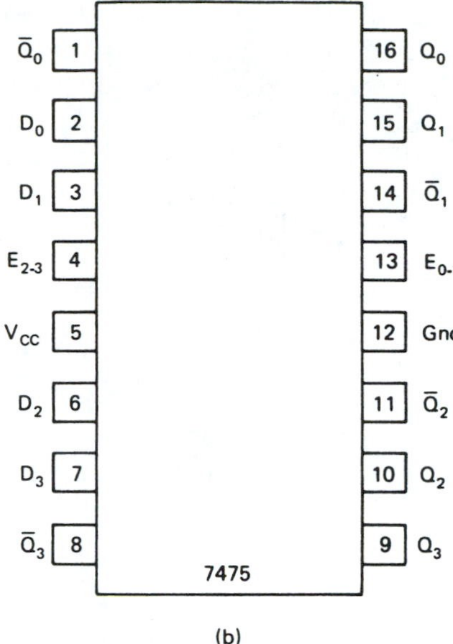

(b)

Figure 5–12 *Continued* (b) pin configuration.

TABLE 5–3
Function Table for a 7475[a]

	Inputs		Outputs	
Operating mode	*E*	*D*	*Q*	$\overline{Q}$
Data enabled	H	L	L	H
	H	H	H	L
Data latched	L	×	*q*	$\overline{q}$

[a]*q* = State of *Q* before the HIGH-to-LOW edge of *E*; × = don't care.

EXAMPLE 5–4

Sketch the output waveform at Q_0 for the following inputs at D_0 and E_{0-1} for the 7475 *D* latch shown in Figures 5–13 and 5–14.

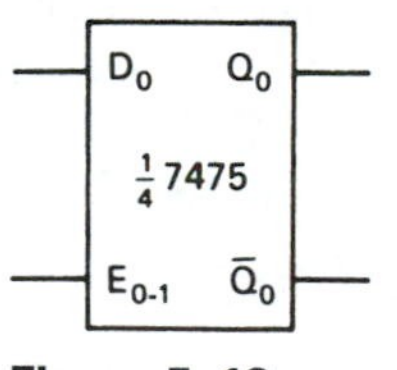

Figure 5–13

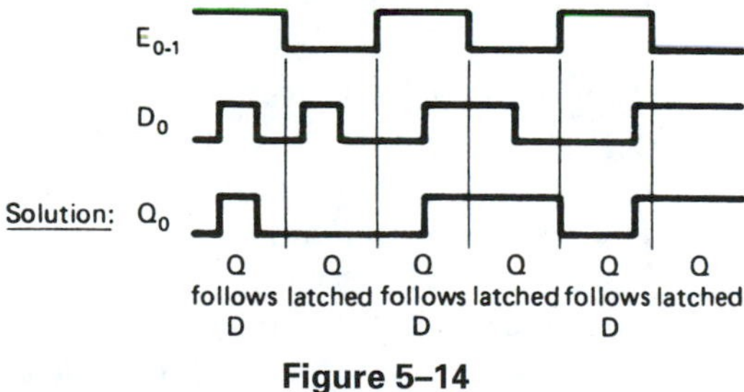

Figure 5–14

5–3 THE INTEGRATED-CIRCUIT D FLIP-FLOP (7474)

The 7474 *D* flip-flop differs from the 7475 *D* latch in several ways. Most important, the 7474 is an edge-triggered device. That means that transitions in *Q* will occur only at the edge of the input trigger pulse. The trigger pulse is usually a clock or timing signal instead of an enable line. In the case of the 7474, the trigger point is at the "positive" edge of C_p

(LOW-to-HIGH transition). The small triangle on the *D* flip-flop symbol (Figure 5–15a) is used to indicate that it is edge-triggered.

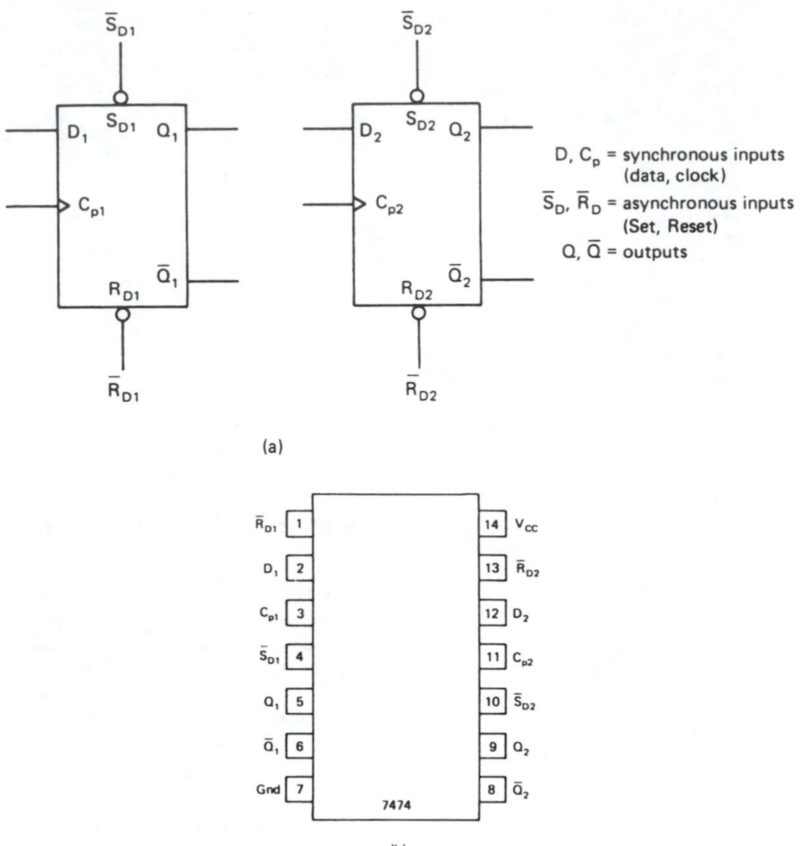

Figure 5–15 The 7474 dual *D* flip-flop: (a) logic symbol; (b) pin configuration.

The 7474 has two distinct types of inputs: synchronous and asynchronous. The *synchronous inputs* are the *D* (Data) and C_p (Clock) inputs. The state at the *D* input will be transferred to *Q* at the positive edge of the input trigger (LOW-to-HIGH edge of C_p). The *asynchronous inputs,* $\overline{S_D}$ (Set) and $\overline{R_D}$ (Reset), operate independent of *D* and C_p. Being asynchronous means that they are *not* in sync with the clock pulse, and the *Q* outputs will respond *immediately* to input changes at $\overline{S_D}$ and $\overline{R_D}$. The little circle at S_D and R_D means that they are active-LOW inputs, and since the circles act like inverters, the external pin on the IC is labeled as the complement of the internal label.

All of that sounds complicated, but it really is not. Just realize that *a LOW on* $\overline{S_D}$ *will immediately Set the flip-flop,* and *a LOW on* $\overline{R_D}$ *will immediately Reset the flip-flop regardless of the states at the synchronous* (*D*, C_p) *inputs.*

The function table (Table 5–4) and following examples will help illustrate the operation of the 7474 *D* flip-flop.

The lowercase h in the *D* column indicates that in order to do a synchronous Set, the *D* must be in a HIGH state at least one setup time prior to the positive edge of the clock. The same rules apply for the lowercase l (Reset).

The setup time for this flip-flop is 20 ns, which means that if *D* is changing while C_p is LOW, that's okay, but *D* must be held stable (HIGH or LOW) at least 20 ns *before* the LOW-to-HIGH transition of C_p. (We discuss setup time in greater detail in Section 5–6.) Also realize that the only digital level on the *D* input that is used is the level that is present at the positive edge of C_p.

We have learned a lot of new terms in regard to the 7474 (active-LOW, edge triggered, asynchronous, etc.). Those terms are important because they apply to almost all of the ICs that are used in the building of sequential circuits.

TABLE 5–4
Function Table for a 7474 *D* Flip-Flop[a]

Operating mode	*Inputs* $\overline{S_D}$	$\overline{R_D}$	C_p	D	*Outputs* Q	$\overline{Q}$
Asynchronous Set	L	H	×	×	H	L
Asynchronous Reset	H	L	×	×	L	H
Not used	L	L	×	×	H	H
Synchronous Set	H	H	↑	h	H	L
Synchronous Reset	H	H	↑	l	L	H

[a]↑ = Positive edge of clock; H = HIGH; h = HIGH level one setup time prior to positive clock edge; L = LOW; l = LOW level one setup time prior to positive clock edge; × = don't care.

EXAMPLE 5–5

Sketch the output waveform at *Q* for a 7474 *D* flip-flop shown in Figures 5–16 and 17 whose input waveforms are as given.

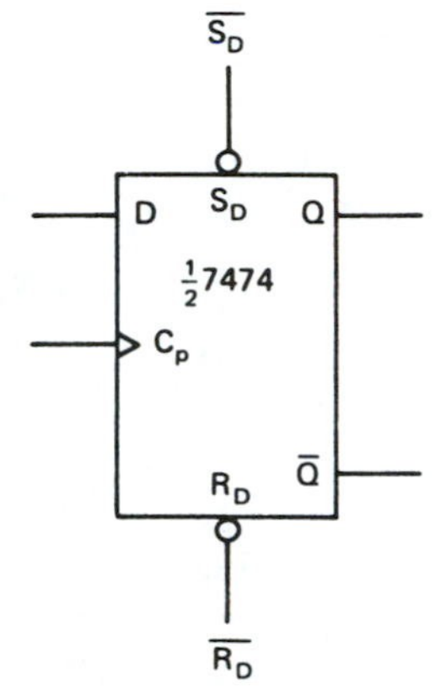

Figure 5–16

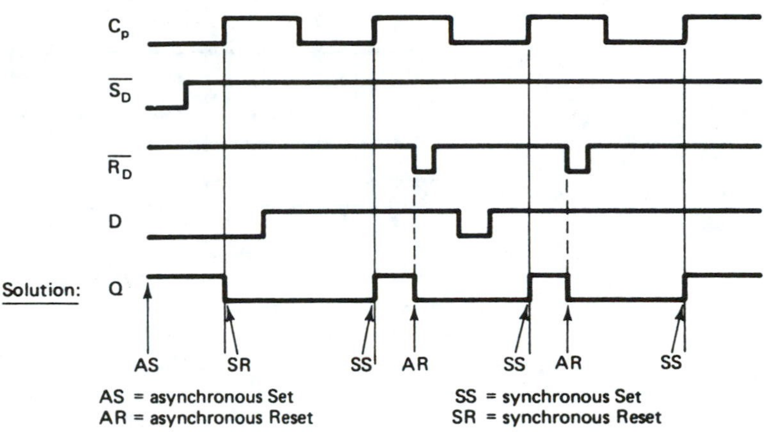

Figure 5–17

EXAMPLE 5–6

Sketch the output waveforms at *Q* for the 7474 *D* flip-flops shown in Figures 5–18 and 5–19 whose input waveforms are as given.

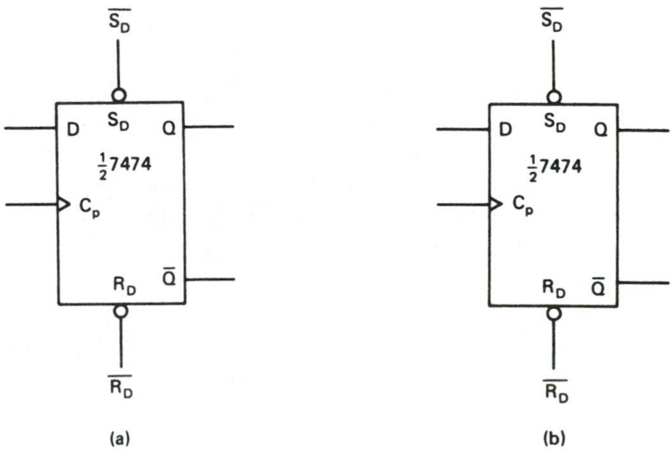

Figure 5–18

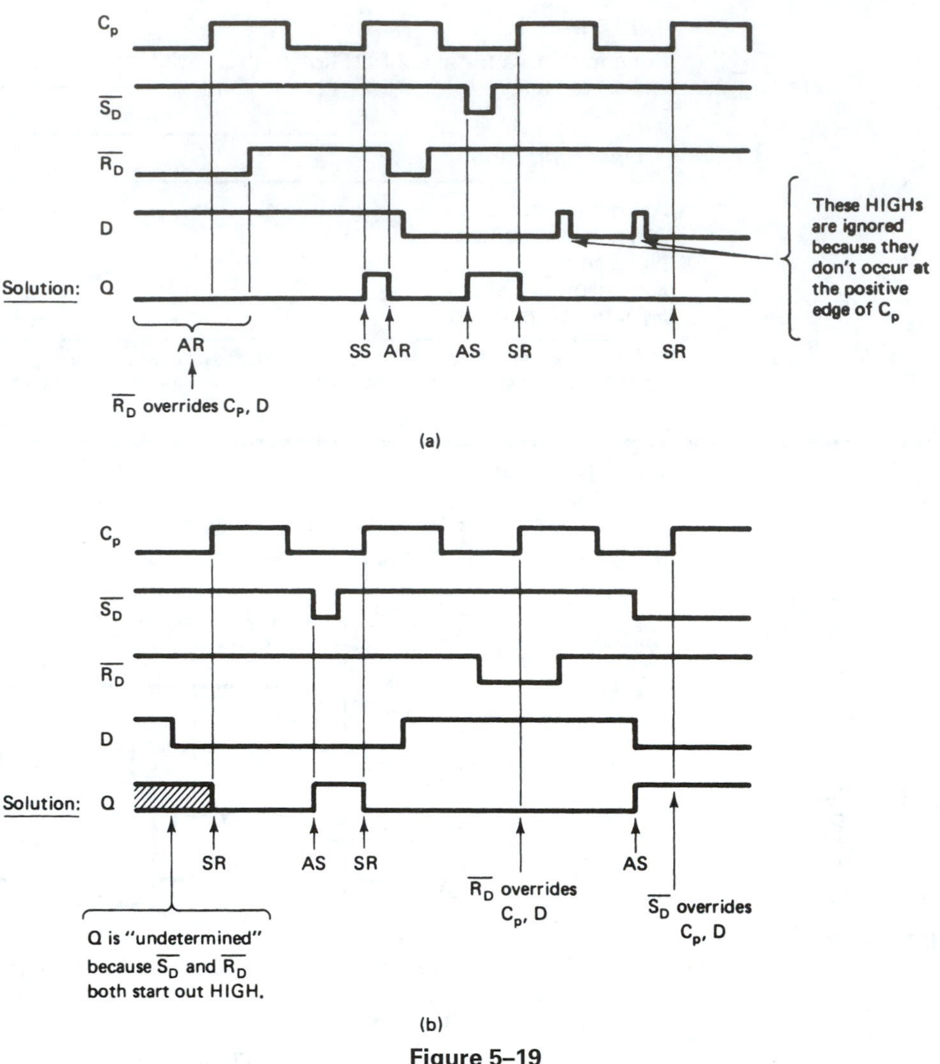

Figure 5–19

5–4 THE J-K FLIP-FLOP

Another type of flip-flop is the *J-K* flip-flop. It differs from the *S-R* flip-flop in that it has one new mode of operation, called *toggle.* Toggle means that Q and $\overline{Q}$ will switch to their *opposite* state at the active clock edge. The synchronous inputs to the *J-K* flip-flop are labeled *J*, *K*, and C_p. *J* acts like the *S* input to an *S-R* flip-flop and *K* acts like the *R* input in an *S-R* flip-flop. The toggle mode is achieved by making *both J* and *K* HIGH before the active clock edge. Table 5–5 shows the four synchronous operating modes of *J-K* flip-flops.

Most *J-K* flip-flops are operated using edge triggering. With edge triggering, the flip-flop only accepts data on the *J* and *K* inputs that are present at the active clock edge (either the HIGH-to-LOW edge of C_p or the LOW-to-HIGH edge of C_p). This gives the design engineer the ability to accept input data on *J* and *K* at a precise instant in time. Transitions of the level on *J* and *K* before or after the active clock trigger edge are ignored. The logic symbols for edge-triggered flip-flops use a small triangle at the clock input to signify that it is an edge-triggered device (see Figure 5–20).

Transitions of the *Q* output for the positive edge-triggered flip-flop shown in Figure 5–20a will occur when the C_p input goes from LOW-to-HIGH (positive edge). Figure

TABLE 5–5
Synchronous Operating Modes of a *J-K* Flip-Flop

Operating Mode	*J*	*K*
Hold	0	0
Set	1	0
Reset	0	1
Toggle	1	1

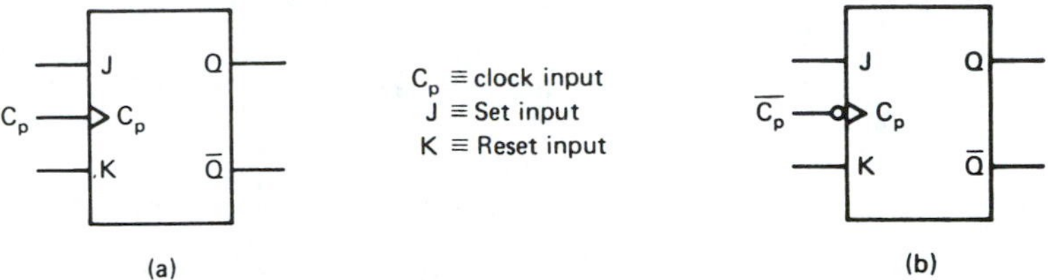

Figure 5–20 Symbols for edge-triggered *J-K* flip-flops: (a) positive edge triggered; (b) negative edge triggered.

5–20b shows a negative edge-triggered flip-flop. The input clock signal will connect to the IC pin labeled $\overline{C_p}$. The small circle indicates that transitions in the output will occur at the HIGH-to-LOW edge (negative edge) of the $\overline{C_p}$ input.

The function table for a negative edge-triggered *J-K* flip-flop is shown in Figure 5–21.

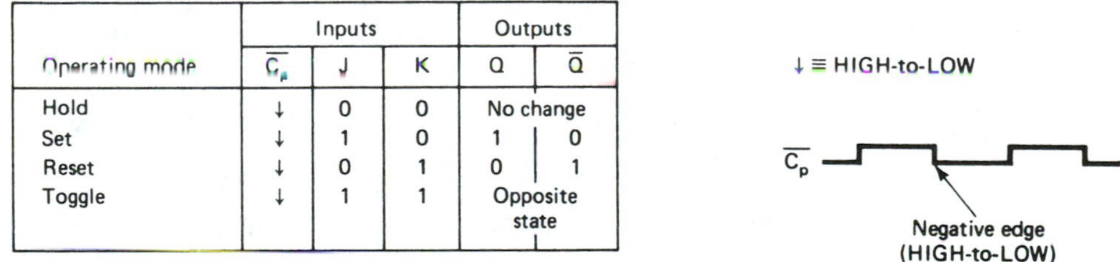

Operating mode	Inputs			Outputs	
	$\overline{C_p}$	J	K	Q	$\overline{Q}$
Hold	↓	0	0	No change	
Set	↓	1	0	1	0
Reset	↓	0	1	0	1
Toggle	↓	1	1	Opposite state	

Figure 5–21 Function table for a negative edge-triggered *J-K* flip-flop.

The downward arrow in the $\overline{C_p}$ column indicates that the flip-flop is triggered by the HIGH-to-LOW transition (negative edge) of the clock.

EXAMPLE 5–7

To illustrate edge triggering, let's draw the *Q* output for the negative edge-triggered *J-K* flip-flop shown in Figures 5–22 and 5–23. (Assume that *Q* is initially 0.)

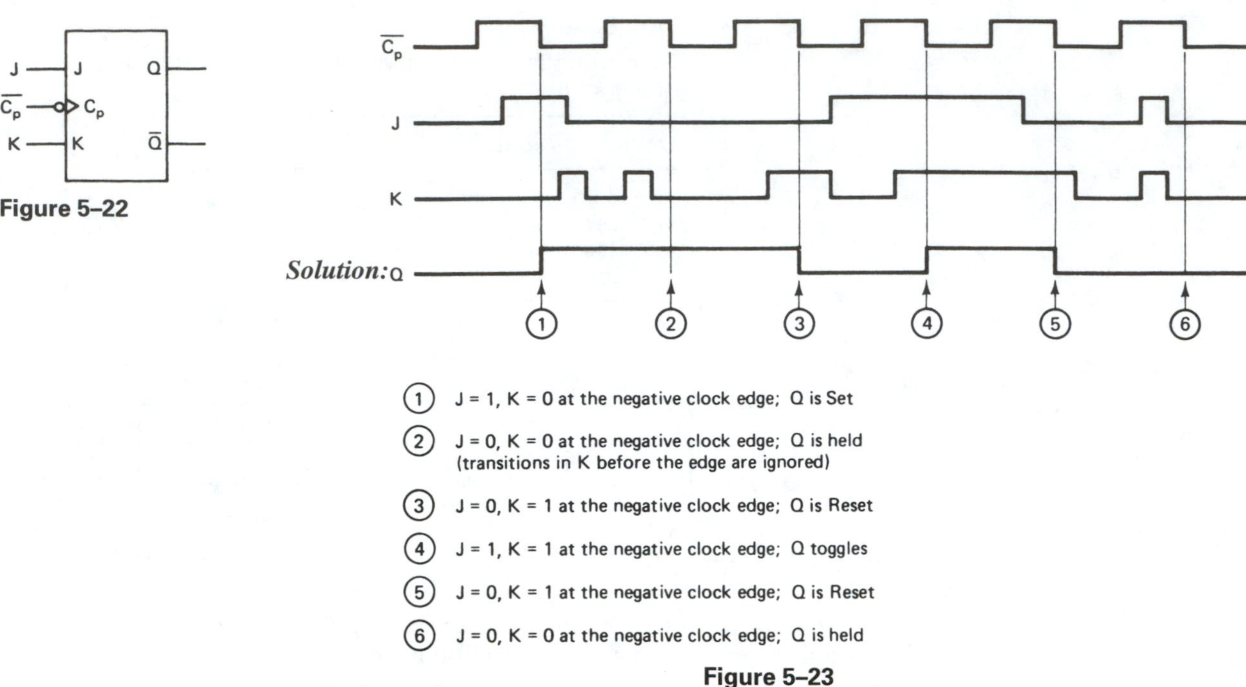

Figure 5–22

(1) J = 1, K = 0 at the negative clock edge; Q is Set

(2) J = 0, K = 0 at the negative clock edge; Q is held (transitions in K before the edge are ignored)

(3) J = 0, K = 1 at the negative clock edge; Q is Reset

(4) J = 1, K = 1 at the negative clock edge; Q toggles

(5) J = 0, K = 1 at the negative clock edge; Q is Reset

(6) J = 0, K = 0 at the negative clock edge; Q is held

Figure 5–23

5–5 THE INTEGRATED-CIRCUIT J-K FLIP-FLOP

Now let's take a look at actual *J-K* flip-flop ICs. The 7476 and 74LS76 are popular *J-K* flip-flops because they are both dual flip-flops (two flip-flops in each IC package), and they have asynchronous inputs ($\overline{R_D}$ and $\overline{S_D}$) as well as synchronous inputs ($\overline{C_p}$, *J*, *K*). The 7476 is a positive pulse-triggered (master–slave) flip-flop, and the 74LS76 is a negative edge-triggered flip-flop, a situation that can trap the unwary technician who attempts to replace the 7476 with the 74LS76! We will limit our discussion to the more popular 74LS76.

From Figure 5–24a and Table 5–6 we can see that the asynchronous inputs $\overline{S_D}$ and $\overline{R_D}$ are *active-LOW.* That is, a LOW on $\overline{S_D}$ (Set) will Set the flip-flop ($Q = 1$) and a LOW on $\overline{R_D}$ will Reset the flip-flop ($Q = 0$). Remember, the asynchronous inputs will cause the flip-flop to respond immediately *without* regard to the clock trigger input.

For synchronous operations using *J*, *K*, and $\overline{C_p}$, the asynchronous inputs must be disabled by putting a HIGH level on both $\overline{S_D}$ and $\overline{R_D}$. The *J* and *K* inputs are read one setup

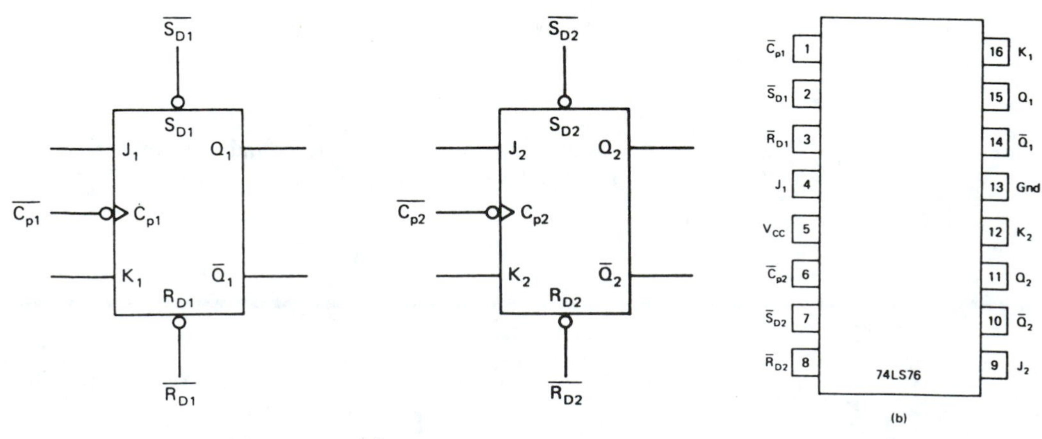

Figure 5–24 The 74LS76 negative edge-triggered flip-flop: (a) logic symbol; (b) pin configuration.

TABLE 5–6
Function Table for the 74LS76[a]

Operating mode	*Inputs* $\overline{S_D}$	$\overline{R_D}$	$\overline{C_p}$	J	K	*Outputs* Q	$\overline{Q}$
Asynchronous Set	L	H	×	×	×	H	L
Asynchronous Reset	H	L	×	×	×	L	H
Synchronous Hold	H	H	↓	l	l	q	$\overline{q}$
Synchronous Set	H	H	↓	h	l	H	L
Synchronous Reset	H	H	↓	l	h	L	H
Synchronous Toggle	H	H	↓	h	h	$\overline{q}$	q

[a]H = HIGH-voltage steady state; L = LOW-voltage steady state; h = HIGH voltage one setup time prior to negative clock edge; l = LOW voltage one setup time prior to negative clock edge; × = don't care; q = state of Q prior to negative clock edge; ↓ = HIGH-to-LOW (negative) clock edge.

time prior to the HIGH-to-LOW edge of the clock ($\overline{C_p}$). One setup time for the 74LS76 is 20 ns. That means that the state of J and K, 20 ns *before* the negative edge of the clock, is used to determine the synchronous operation to be performed. (The 7476 is different from the 74LS76 in that it will read the state of J and K during the *entire* positive pulse.)

Also note that in the toggle mode ($J = K = 1$), after a negative clock edge, Q becomes whatever $\overline{Q}$ was before the clock edge, and vice versa (i.e., if $Q = 1$ before the negative clock edge, then $Q = 0$ after the negative clock edge).

Now let's work through several timing analysis examples to be sure that we fully understand the operation of *J-K* flip-flops.

EXAMPLE 5–8

Sketch the Q waveform for the 74LS76 negative edge-triggered *J-K* flip-flop shown in Figures 5–25 and 5–26 with the given input waveforms.

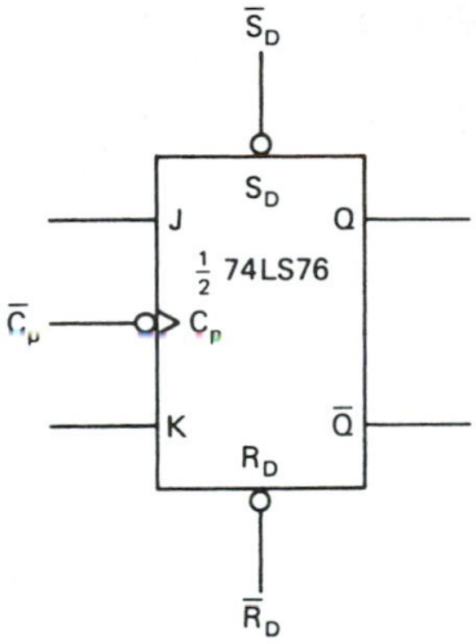

Figure 5–25

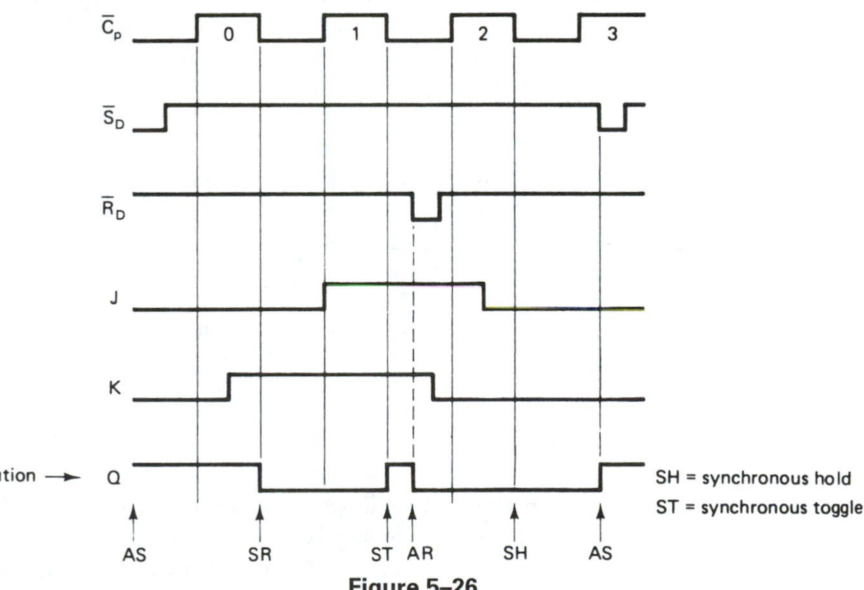

Figure 5–26

Note: Q changes only on the negative edge of $\overline{C_p}$ except when asynchronous operations ($\overline{S_D}$, $\overline{R_D}$) are taking place.

EXAMPLE 5–9

The 74109 is a positive edge-triggered J-$\overline{K}$ flip-flop. The logic symbol (Figure 5–27) and input waveforms (Figure 5–28) are given; sketch Q.

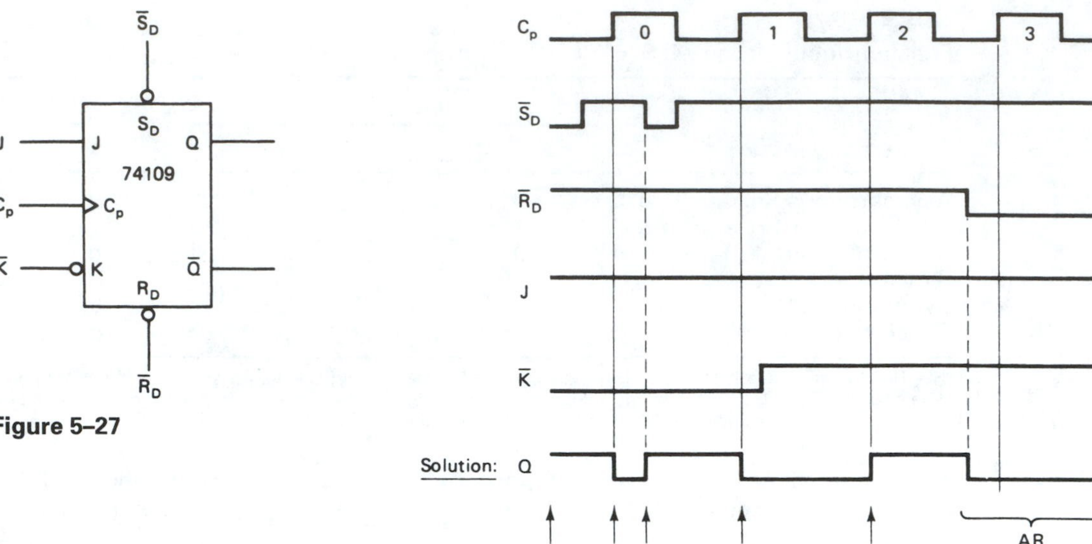

Figure 5–27

Figure 5–28

Note:

1. *Positive* edge triggering.
2. $\overline{K}$ instead of K; therefore, for a toggle, $J = 1$, $K = 0$.

The *J-K* flip-flop can be used to form other flip-flops by making the appropriate external connections. For example, to form a *D* flip-flop, add an inverter between the *J* and *K* inputs and bring the data into the *J* input, as shown in Figure 5–29.

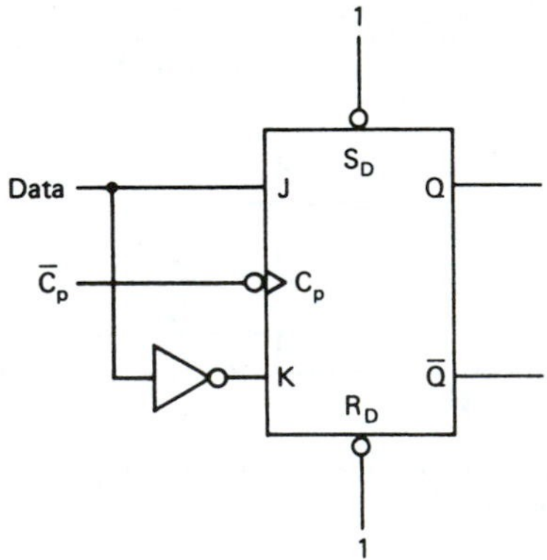

Figure 5–29 *D* flip-flop made from a *J-K* flip-flop.

Figure 5–29 will operate as a *D* flip-flop because the data are brought in on the *J* terminal and its complement is at the *K*; so if Data = 1, the flip-flop will be Set after the clock edge; if Data = 0, the flip-flop will be Reset after the clock edge. (*Note:* You lose the toggle mode and hold mode using this configuration.)

Also, quite often it is important for a flip-flop to operate in the toggle mode. This can be done simply by connecting both *J* and *K* to 1. This will cause the flip-flop to change states at each active clock edge, as shown in Figure 5–30. Note that the frequency of the output waveform at Q will be one-half the frequency of the input waveform at $\overline{C_p}$.

As we have seen, there is a variety of flip-flops, each with their own operating characteristics. In Chapter 6, we learn how to use these ICs to perform sequential operations such as counting, data shifting, and sequencing.

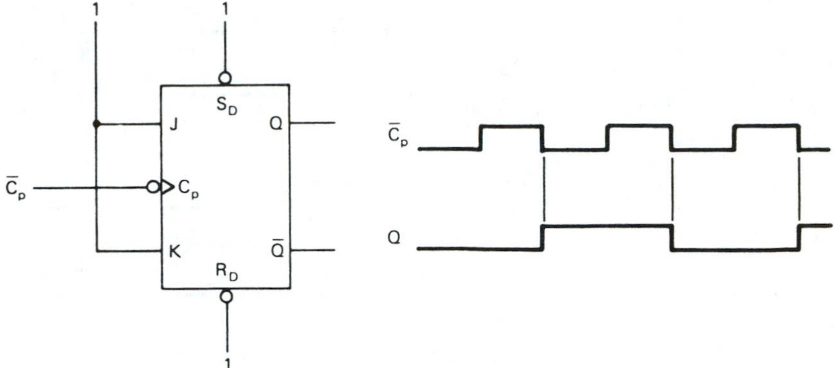

Figure 5–30 *J-K* connected as a toggle flip-flop.

First, let's summarize what we have learned about flip-flops by utilizing three common flip-flops in the same circuit, and supplying input signals and sketching the Q outputs of each (Example 5–10).

EXAMPLE 5–10

Sketch the Q outputs for each of the flip-flops shown in Figure 5–31. (Their input waveforms are given in Figure 5–32.)

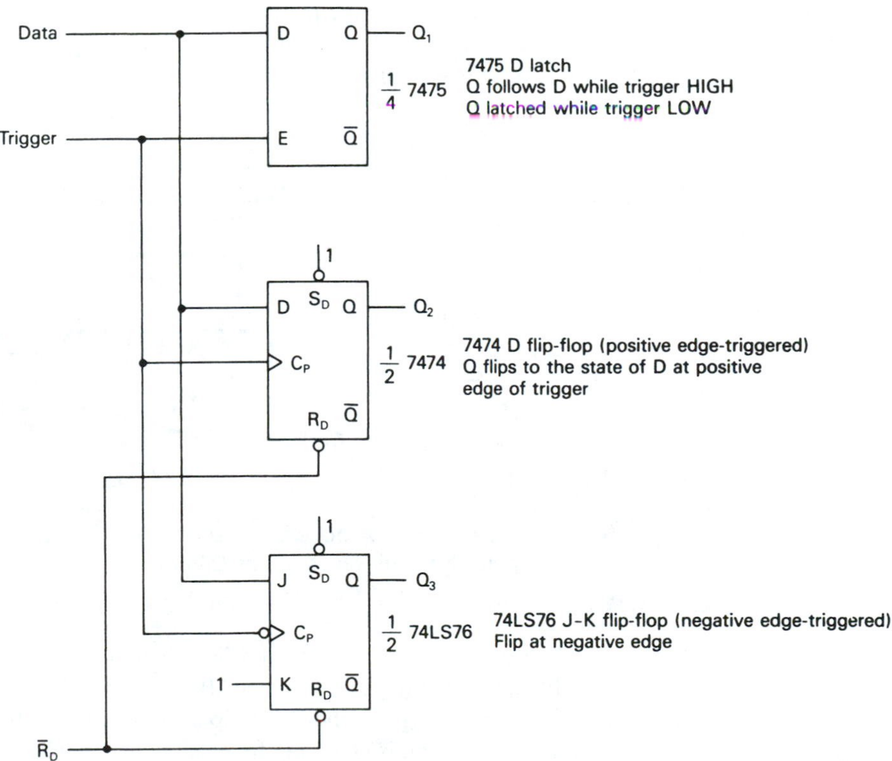

Figure 5–31

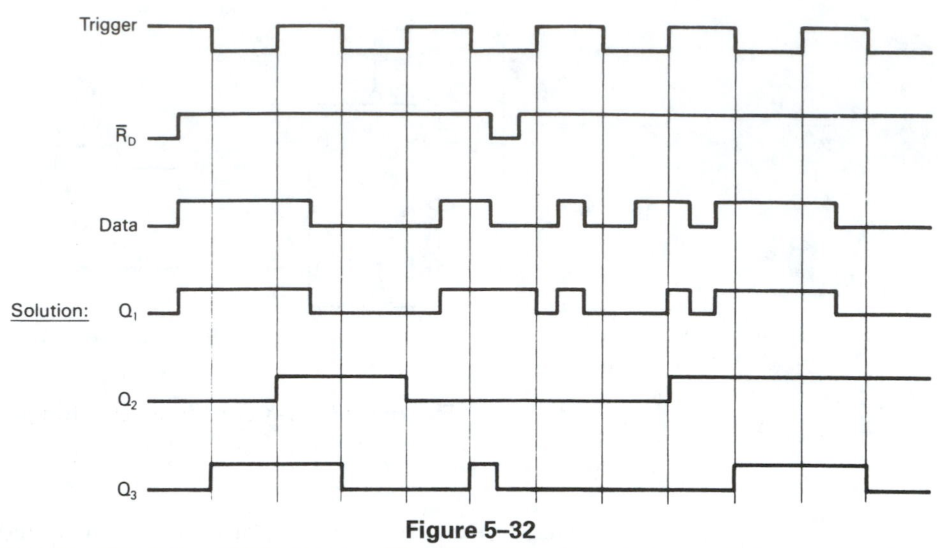

Figure 5–32

5–6 FLIP-FLOP TIME PARAMETERS

Now we have the major building blocks required to form sequential circuits. There are a few timing considerations that we have to deal with first, before we connect ICs together to form sequential logic.

For instance, ideally a 74LS76 flip-flop switches on the negative edge of the input clock, but actually it could take the output as long as 30 ns to switch. Thirty nanoseconds (30×10^{-9} s) does not sound like much, but when you cascade several flip-flops end to end, or any time you have combinational logic with flip-flops that rely on a high degree of accurate timing, the IC delay times could cause serious design problems.

There are several time parameters listed in IC manufacturers' data manuals that require careful analysis. For example, let's look at Figure 5–33, which uses a 74LS76 flip-flop with the *J* and $\overline{C_p}$ inputs brought in from some external circuit.

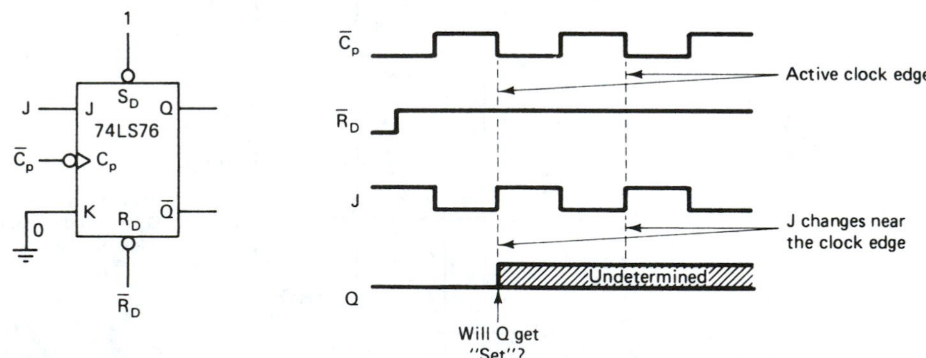

Figure 5–33 A possible "race" condition on a *J-K* flip-flop creates an undetermined result at *Q*.

The waveform shown for *J* and $\overline{C_p}$ will create a *race condition. Race* is the term used when the inputs to a triggerable device (like a flip-flop) are changing at the same time that the active trigger edge of the input clock is making its transition. In the case of Figure 5–33, the *J* waveform is changing from LOW to HIGH exactly at the negative edge of the clock; so what is *J* at the negative edge of the clock—LOW or HIGH?

Now when you look at Figure 5–33, you should ask the question, "Will Q ever get Set?" Remember from Section 5–5 that J must be HIGH at the negative edge of $\overline{C_p}$ in order to set the flip-flop. Actually, J must be HIGH "one *setup time*" prior to the negative edge of the clock.

The *setup time* is the length of time prior to the active clock edge that the flip-flop looks back to determine the levels to use at the inputs. In other words, for Figure 5–33, the flip-flop will look back one setup time prior to the negative clock edge to determine the level at J and K.

The setup time for the 74LS76 is 20 ns, so we must ask, "Were J and K HIGH or LOW 20 ns prior to the negative clock edge?" Well, K is tied to ground, so it was LOW, and depending on when J changed from LOW to HIGH, the flip-flop may have Set ($J = 1$, $K = 0$) or Held ($J = 0$, $K = 0$).

In a data manual, the manufacturer will give you "ac waveforms," which illustrate the measuring points for all the various time parameters. The illustration for setup time will look something like Figure 5–34.

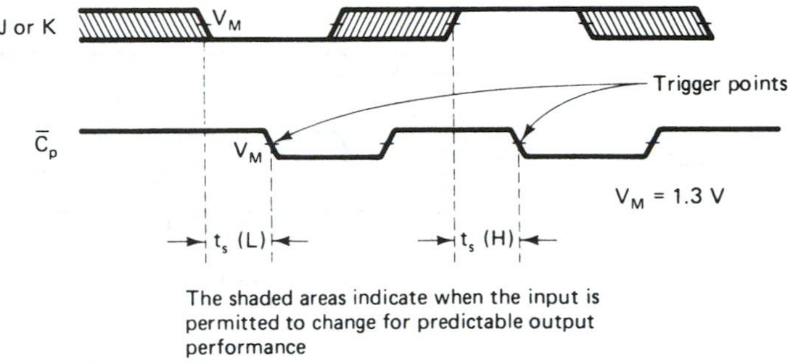

Figure 5–34 Setup time waveform specifications for a 74LS76.

The active transition (trigger point) of the $\overline{C_p}$ input (clock) occurs when $\overline{C_p}$ goes from above to below the 1.3-V level.

Setup time (LOW), t_s(L), is given as 20 ns. This means that J and K can be changing states 21 ns or more before the active transition of $\overline{C_p}$, but in order to be interpreted as a LOW, they must be 1.3 V *or less* at 20 ns *before* the active transition of $\overline{C_p}$.

Setup time (HIGH), t_s(H), is given as 20 ns also. This means that J and K can be changing states 21 ns or more before the active edge of $\overline{C_p}$, but to be interpreted as a HIGH, they must be 1.3 V *or more* at 20 ns *before* the active transition of $\overline{C_p}$.

Did you follow all of that? If not, go back and read it again! Sometimes, material like this has to be read over and over again, carefully, to be fully understood.

Not only does the input have to be set up some definite time *before* the clock edge, but it also has to be *held* for a definite time after the clock edge. This time is called the *hold time* [t_h(L) and t_h(H)].

The hold time for the 74LS76 (and most other flip-flops) is given as 0 ns. This means that the desired levels at J and K must be held 0 ns *after* the active clock edge. In other words, the levels do not have to be held beyond the active clock edge for most flip-flops. In the case of 74LS76, the desired level for J and K must be present from 20 ns before the negative clock edge to 0 ns after the clock edge.

For example, for a 74LS76 to have a LOW level on J and K, the waveforms in Figure 5–35 illustrate the *minimum* setup and hold times allowed to still have the LOW reliably interpreted as a LOW.

Figure 5–35 shows us that J and K are allowed to change states any time greater than 20 ns before the negative clock edge, and since the hold time is zero, they are permitted to change immediately after the negative clock edge.

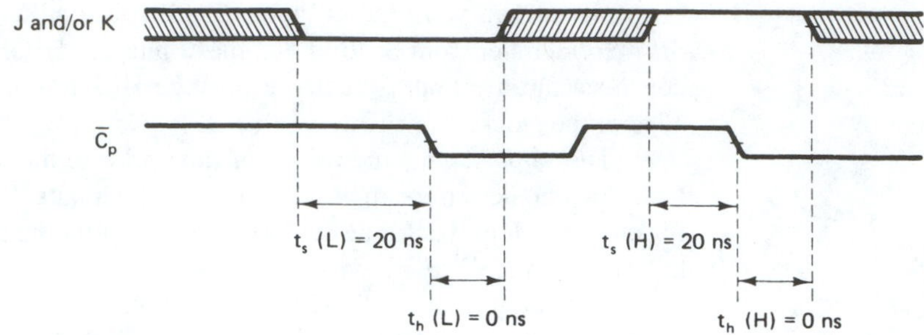

Figure 5–35 Setup and hold parameters for a 74LS76 flip-flop.

EXAMPLE 5–11

Follow the rules for setup and hold times, and sketch the waveform at Q for the 74H106 shown in Figure 5–36 [t_s(L) = 13 ns, t_s(H) = 10 ns, t_h(L) = t_h(H) = 0 ns] given the waveforms in Figure 5–37.

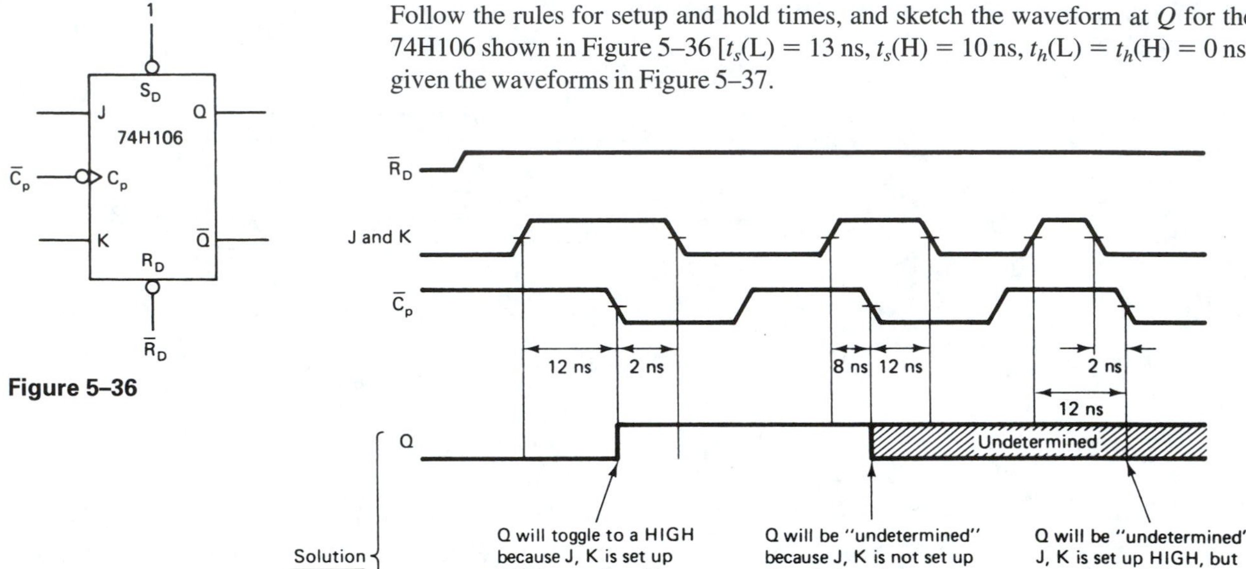

Figure 5–36

Figure 5–37

EXAMPLE 5–12

Sketch the Q output for a 74H106 shown in Figure 5–38, with the given input waveforms in Figure 5–39 [t_s(L) = 13 ns, t_s(H) = 10 ns, t_h(L) = t_h(H) = 0 ns].

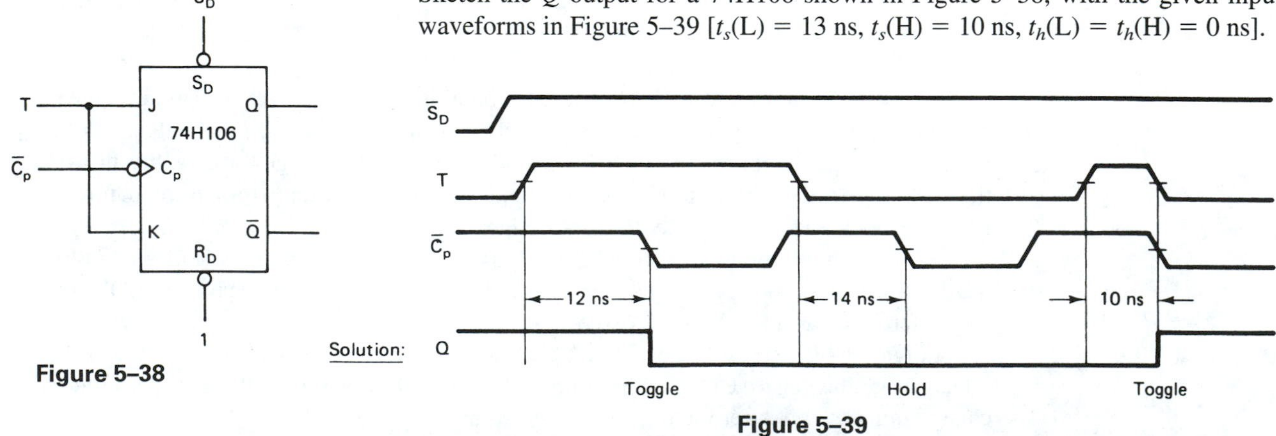

Figure 5–38

Figure 5–39

Have you noticed in Examples 5–11 and 5–12 that the Q output changes *exactly* on the negative clock edge? Do you really think that it will? It won't! There are electrical charges built up inside any digital logic circuit that won't allow it to change states instantaneously as the inputs change. This delay from input to output is called *propagation delay.* There are propagation delays from the synchronous inputs to the output and also the asynchronous inputs to the output.

For example, there is a propagation delay period from the instant the $\overline{R_D}$ or $\overline{S_D}$ goes LOW until the Q output responds accordingly. The data manual shows a *maximum* propagation delay for $\overline{S_D}$ to Q of 20 ns and $\overline{R_D}$ to Q of 30 ns. Since a LOW on $\overline{S_D}$ causes Q to go *LOW to HIGH,* the propagation delay is abbreviated t_{PLH}. A LOW on $\overline{R_D}$ causes Q to go *HIGH to LOW;* therefore, use t_{PHL} for that propagation delay, as illustrated in Figure 5–40.

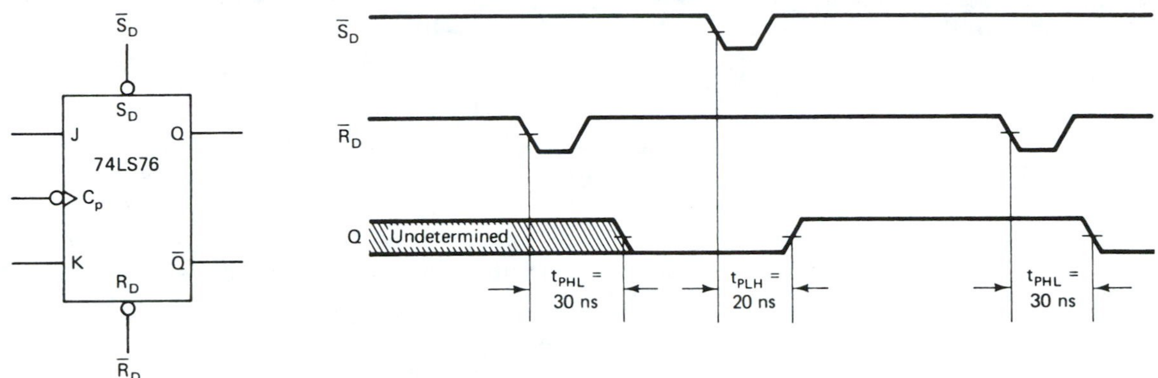

Figure 5–40 Propagation delay for the asynchronous input-to-Q output for a 74LS76.

The propagation delay from the clock trigger point to the Q output is also called t_{PLH} or t_{PHL}, depending on whether the Q output is going LOW to HIGH or HIGH to LOW. For the 74LS76, clock to output, t_{PLH} = 20 ns and t_{PHL} = 30 ns. Figure 5–41 illustrates the synchronous propagation delays.

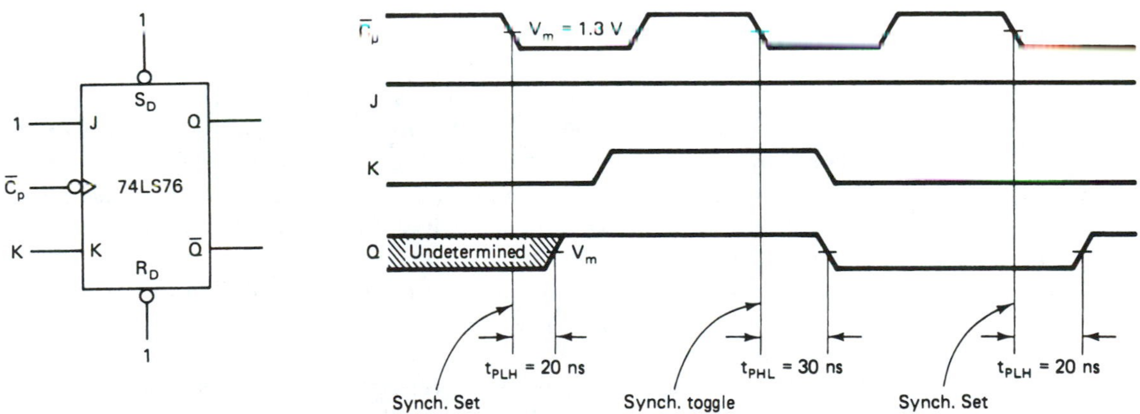

Figure 5–41 Propagation delay for the clock-to-output of the 74LS76.

Besides setup, hold, and propagation delay times, the manufacturer's data manual will also give:

1. *Maximum frequency* (f_{max}). This is the maximum frequency allowed at the clock input. Any frequency above this limit will yield unpredictable results.
2. *Clock pulse width (LOW)* [t_w(L)]. This is the minimum width (in nanoseconds) that is allowed at the clock input during the LOW level for reliable operation.

3. *Clock pulse width (HIGH)* [t_w(H)]. This is the minimum width (in nanoseconds) that is allowed at the clock input during the HIGH level for reliable operation.
4. *Set or Reset pulse width (LOW)* [t_w (L)]. This is the minimum width (in nanoseconds) of the LOW pulse at the Set ($\overline{S_D}$) or Reset ($\overline{R_D}$) inputs.

Figure 5–42 shows the measurement points for those specifications.

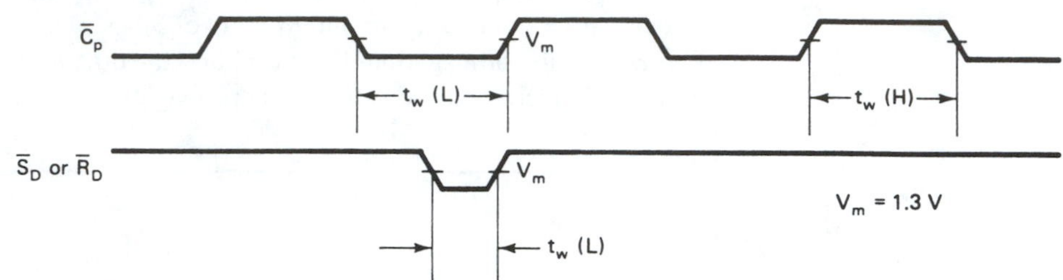

Figure 5–42 Minimum pulse-width specifications.

Complete specifications for the 74LS76 flip-flop are given in a manufacturer's data manual. If you have one, compare the specifications for the 74LS76 with those of other flip-flops.

5–7 THREE-STATE BUFFERS, LATCHES, AND TRANSCEIVERS

When we start studying microprocessor hardware, we'll see a need for transmitting a number of bits simultaneously as a group. A single flip-flop will not do. What we need is a group of flip-flops, called a *register,* to facilitate the movement and temporary storage of binary information. The most commonly used registers are 8-bits wide and function as either a buffer, a latch, or a transceiver.

Three-State Buffers

In microprocessor systems several input and output devices have to share the same data lines going into the microprocessor IC. (These "shared" data lines are called the *data bus.*) For example, if an 8-bit microprocessor has to interface with four separate 8-bit input devices, we must provide a way to enable just one of the devices to place its data on the data bus at a time while the other three are disabled. One way to accomplish this is to use three-state octal buffers.

In Figure 5–43, the second buffer is enabled, which allows the eight data bits from input device 2 to reach the data bus. The other three buffers are disabled, keeping their outputs in the "float" condition.

A buffer is simply a device that, when enabled, passes a digital level from its input to its output, unchanged. It provides isolation, or a "buffer," between the input device and the data bus. A buffer also provides the sink or source current required by any devices connected to its output without loading down the input device. An octal buffer IC has eight individual buffers within a single package.

The term *three-state* refers to the fact that the output can have one of three levels: HIGH, LOW, or float. The symbol and function table for an individual three-state buffer are shown in Figure 5–44.

From Figure 5–44 we can see that the circuit acts like a straight buffer (output = input) when $\overline{OE}$ is LOW (active-LOW output enable). When the output is disabled ($\overline{OE}$ = HIGH), the output level is placed in the "float" or "high-impedance" state. In the high-impedance state the output looks like an open circuit to anything else connected to it. In

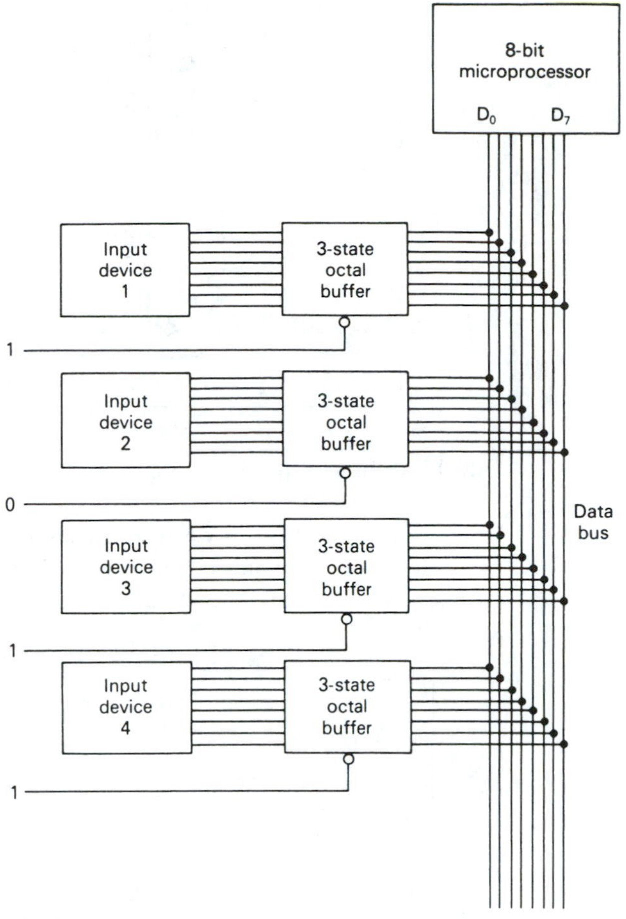

Figure 5–43 Using a three-state octal buffer (like the 74LS244) to pass eight data bits from input device 2 to the data bus.

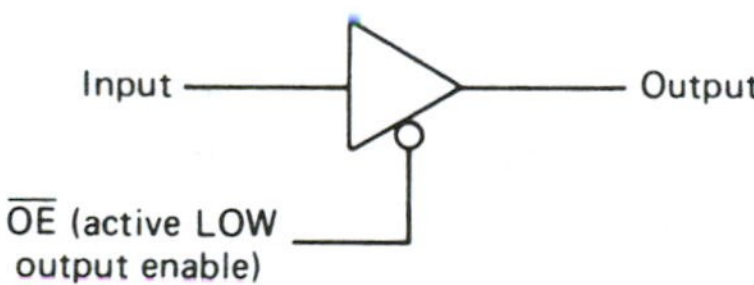

Input	$\overline{OE}$	Output
1	0	1
0	0	0
1	1	Float
0	1	Float

Figure 5–44 Three-state buffer symbol and function table.

other words, in the float state the output is neither HIGH nor LOW and cannot sink nor source current.

A popular three-state octal buffer is the 74LS244 shown in Figure 5–45.

The buffers are configured in two groups of four. The first group (group *a*) is controlled by $\overline{OE_a}$ and the second group (group *b*) is controlled by $\overline{OE_b}$. Here also $\overline{OE}$ is active-LOW, meaning that it takes a LOW to allow data to pass from the inputs (*I*) to the outputs (*Y*). Other features of the 74LS244 are that it has Schmitt trigger hysteresis and very high sink and source current capabilities (24 and 15 mA, respectively).

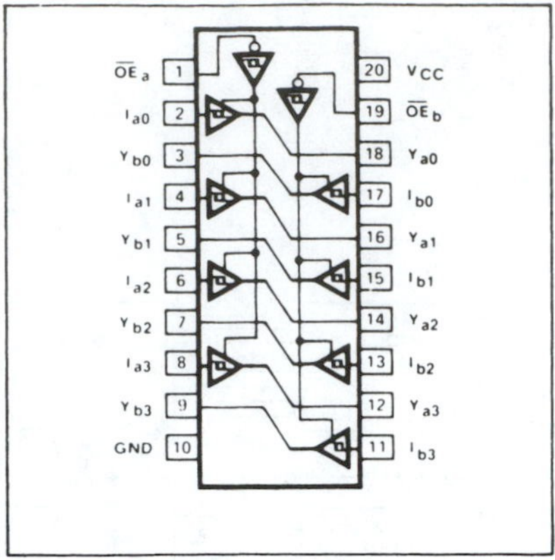

Figure 5–45 Pin configuration for the 74LS244 three-state octal buffer.

Octal Latches/Flip-Flops

In microprocessor systems we need latches and flip-flops to "remember" digital states that the microprocessor issues before it goes on to other tasks. Take, for example, a microprocessor system that is used to drive two separate 8-bit output devices, as shown in Figure 5–46.

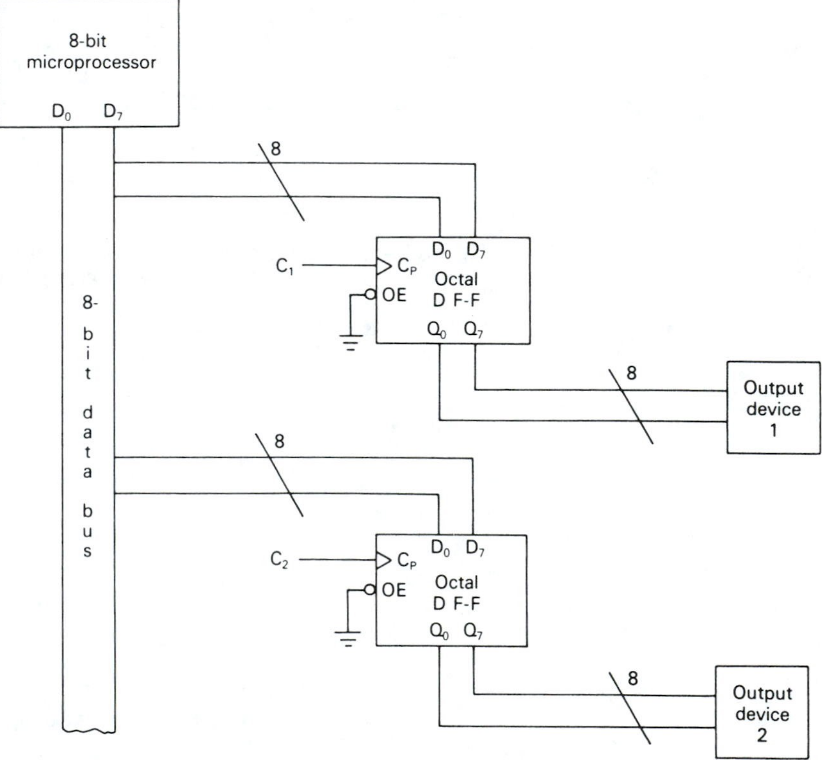

Figure 5–46 Using octal *D* flip-flops (like the 74LS374) to capture data that appear momentarily on a microprocessor data bus.

To send information to output device 1, the microprocessor first sets up the data bus (D_0–D_7) with the appropriate data, then issues a LOW-to-HIGH pulse on line C_1. The positive edge of the pulse causes the data that is at D_0–D_7 of the flip-flop to get stored at Q_0–Q_7. Since $\overline{\text{OE}}$ is tied LOW, those data are sent on to output device 1. (The diagonal line with the number 8 above it is a shorthand method used to indicate eight separate lines or conductors.)

Next, the microprocessor sets up the data bus with data for output device 2 and issues a LOW-to-HIGH pulse on C_2. Now the second octal *D* flip-flop is loaded with valid data. The outputs of the *D* flip-flops will remain at those digital levels, allowing the microprocessor to go on to perform other tasks.

Earlier in this chapter we studied the 7475 transparent latch and the 7474 *D* flip-flop. The 74LS373 and 74LS374 shown in Figure 5–47 operate the same way except they were developed to handle 8-bit data operations.

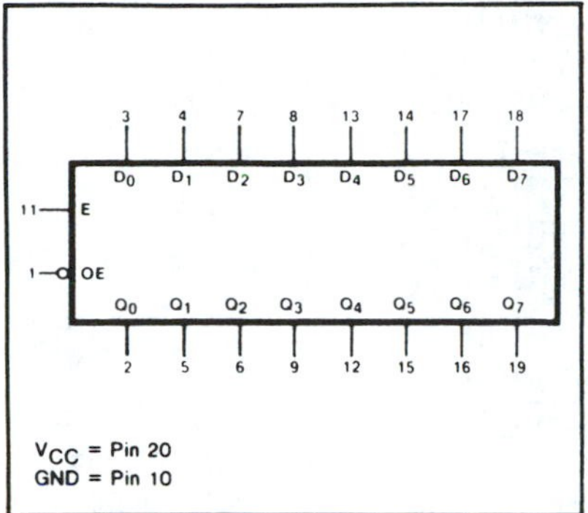

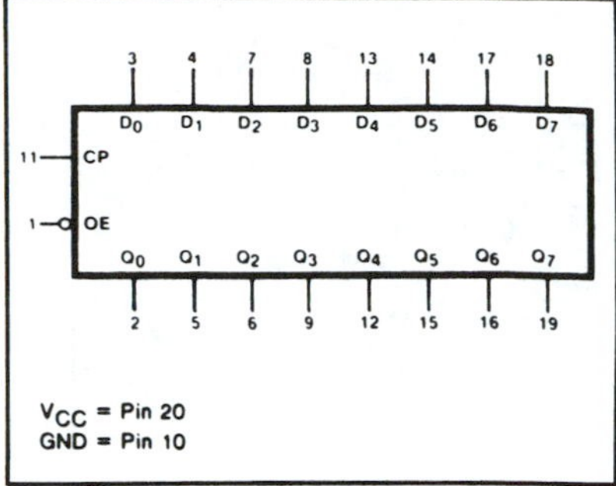

Figure 5–47 Logic symbol for the 74LS373 octal latch and the 74LS374 octal *D* flip-flop.

Transceivers

Another way to connect devices to a shared data bus is to use a transceiver (transmitter/receiver). The transceiver differs from a buffer or latch because it is *bidirectional.* This is necessary for interfacing devices that are used for *both input and output* to the microprocessor. Figure 5–48 shows a common way to connect an I/O device to a data bus via a transceiver.

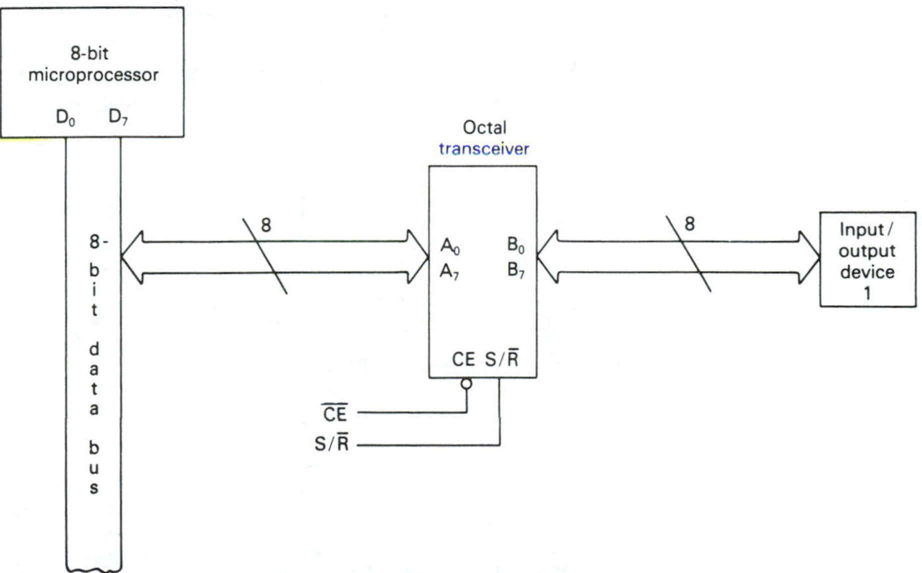

Figure 5–48 Using an octal transceiver (like the 74LS245) to interface an input/output device to an 8-bit data bus.

To make input/output device 1 the active interface, the $\overline{CE}$ (chip enable) line must first be made LOW. If $\overline{CE}$ is HIGH, the transceiver disconnects the I/O device from the bus by making the connection float.

After making $\overline{CE}$ LOW, the microprocessor then issues the appropriate level on the $S/\overline{R}$ line depending on whether it wants to *send data to* the I/O device or *receive data from* the I/O device. If $S/\overline{R}$ is made HIGH, the transceiver allows data to pass to the I/O device (from A to B). If $S/\overline{R}$ is made LOW, the transceiver allows data to pass to the microprocessor data bus (from B to A).

To see how the transceiver is able to both send and receive data, study the internal logic of the 74LS245 shown in Figure 5–49.

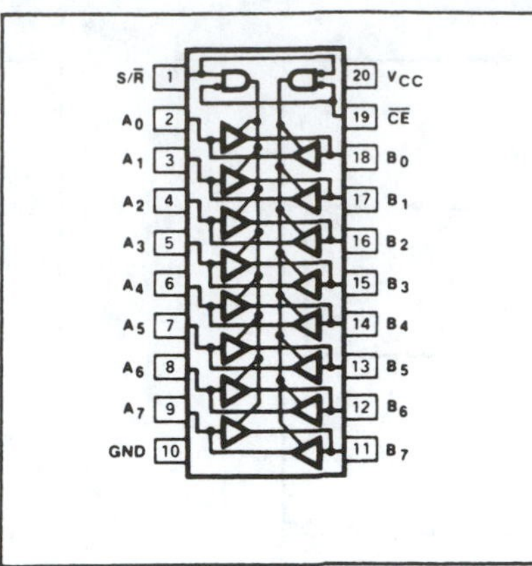

Figure 5–49 Pin configuration and internal logic of the 74LS245 octal three-state transceiver.

5–8 SWITCH DEBOUNCING

Quite often, mechanical switches are used in the design of digital circuits. Unfortunately, however, most switches exhibit a phenomenon called *switch bounce.* Switch bounce is the action that occurs when a mechanical switch is opened or closed. For example, when the contacts of a switch are closed together, the electrical and mechanical connection is first made, but due to a slight springing action of the contacts, they will bounce back open, then close, then open, then close repeatedly until they finally settle down in the closed position. This bouncing action will typically take place for as long as 50 ms.

A typical connection for a single-pole, single-throw (SPST) switch is shown in Figure 5–50. This is a poor design because if we expect the toggle to operate only once when we close the switch, we will be out of luck because of switch bounce. Why do I say that? Let's look at the waveform at $\overline{C_p}$ to see what actually happens when a switch is closed.

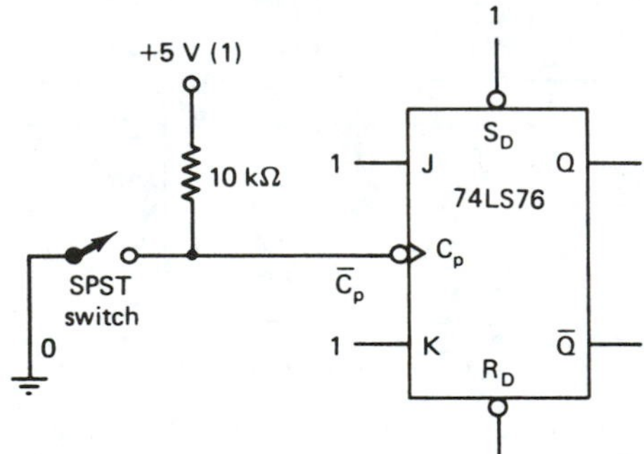

Figure 5–50 Switch used as a clock input to a toggle flip-flop.

Figure 5–51 shows that $\overline{C_p}$ will receive several LOW pulses each time the switch is closed instead of the single pulse that we expect.

(The 10-kΩ pull-up resistor in Figure 5–50 is necessary to hold the voltage level at $\overline{C_p}$ up close to +5 V while the switch is open. If the pull-up resistor were not used, the voltage at $\overline{C_p}$ with the switch open would be undetermined; but *with* the 10-kΩ, and realizing that the current into the $\overline{C_p}$ terminal is negligible, the level at the $\overline{C_p}$ terminal will be held at approximately +5 V while the switch is open.)

There are several ways to eliminate the effects of switch bounce. If you need to debounce a single-pole, single-throw switch or pushbutton, the Schmitt trigger scheme shown in Figure 5–52 can be used. With the switch open, the capacitor will be charged to

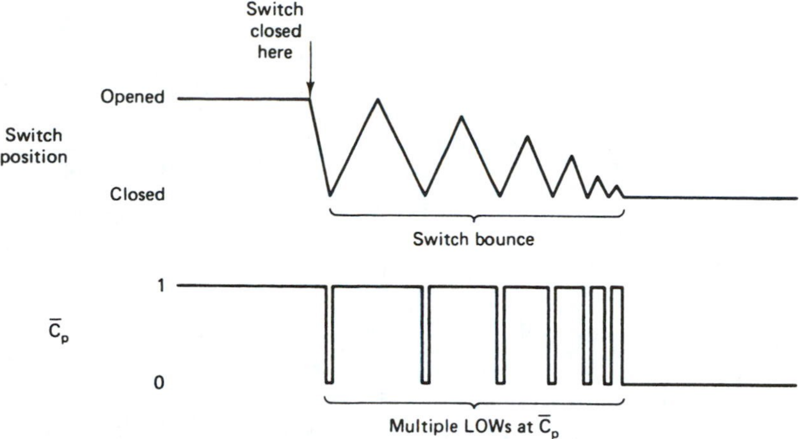

Figure 5–51 Waveform at point $\overline{C_p}$ for Figure 5–50.

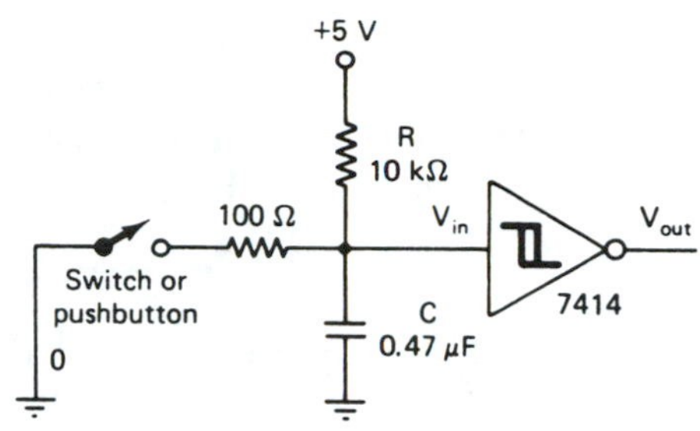

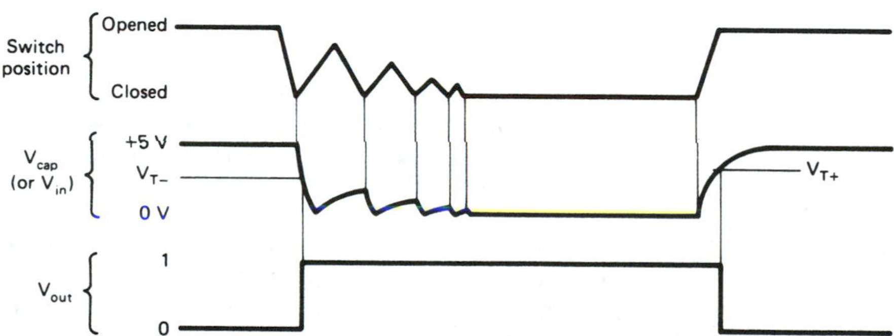

Figure 5–52 Debouncing a single-pole, single-throw switch or pushbutton.

+5 V (1), keeping $V_{out} = 0$. When the switch is closed, the capacitor will discharge rapidly to 0 via the 100-Ω current-limiting resistor, making V_{out} equal to 1. Then, as the switch bounces, the capacitor will repeatedly try to charge slowly back up to a HIGH, then discharge rapidly to 0. The *RC* charging time constant (10 kΩ × 0.47 μF) is long enough that the capacitor will not get the chance to charge up high enough (above V_{T+}) before the switch bounces back to the closed position. This keeps V_{out} equal to 1.

When the switch is reopened, the capacitor is allowed to charge all the way up to +5 V. When it crosses V_{T+}, V_{out} will switch to 0, as shown in Figure 5–52. The result is that by closing the switch or pushbutton once, *you will get only a single pulse at the output even though the switch is bouncing.*

To debounce single-pole, double-throw switches, a different method is required, as illustrated in Figure 5–53. The single-pole, double-throw switch shown in Figure 5–53 actually has three positions: (1) position A, (2) in between position A and position B while it is making its transition, and (3) position B. The cross-NAND debouncer works very similar to the cross-NAND *S-R* flip-flop presented in Figure 5–2.

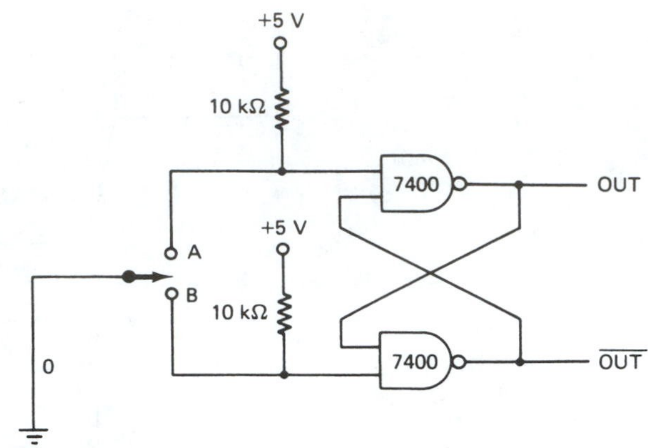

Figure 5–53 Cross-NAND method of debouncing a single-pole, double-throw switch.

When the switch is in position A, OUT will be Set (1). When the switch is moved to position B, it bounces, causing OUT to Reset, Hold, Reset, Hold, Reset, Hold repeatedly until the switch stops bouncing and settles into position B. From the time the switch first touched position B until it is returned to position A, OUT will be Reset even though the switch is bouncing. (See Figure 5–54.)

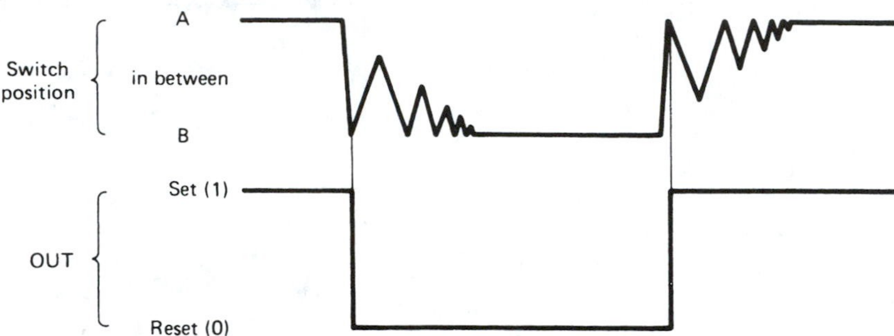

Figure 5–54 Waveforms for the switch debouncer circuit of Figure 5–53.

When the switch is returned to position A, it will bounce, causing OUT to be Set, Hold, Set, Hold, Set, Hold repeatedly until the switch stops bouncing. In this case, OUT will be Set, and remain Set, from the moment the switch first touched position A, even though the switch is bouncing.

An alternative to using the cross-NAND setup is to use the asynchronous $\overline{S_D}$, $\overline{R_D}$ inputs to a *D* or *J-K* flip-flop.

5–9 OSCILLATOR CIRCUITS AND THE ONE-SHOT MULTIVIBRATOR

We have seen that timing is very important in digital electronics. Clock oscillators, used to drive counters and shift registers, must be designed to oscillate at a specific frequency. Specially designed pulse-stretching and time-delay circuits are also required to produce specific pulse widths and delay periods.

Multivibrators

Multivibrator circuits have been around for years, designed from various technologies, to fulfill electronic circuit timing requirements. A multivibrator is a circuit that changes between the two digital levels on a continuous, "free-running" basis or on demand from some external trigger source. Basically there are three types of multivibrators: bistable, astable, and monostable.

The *bistable* multivibrator is triggered into one of the two digital states by an external source, and stays in that state until it is triggered into the opposite state. The *S-R* flip-flop is a bistable multivibrator; it is in either the Set or Reset state.

The *astable* multivibrator is a free-running oscillator that alternates between the two digital levels at a specific frequency and duty cycle.

The *monostable* multivibrator, also known as a *one-shot,* provides a single output pulse of a specific time length when it is triggered from an external source.

The bistable multivibrator (*S-R* flip-flop) was discussed earlier. The astable and monostable multivibrators discussed in this section can be built from basic logic gates, or from special ICs designed specifically for timing applications. In either case, the charging and discharging rate of a capacitor is used to provide the specific time durations required for the circuits to operate.

Astable Multivibrator Oscillator

A very simple astable multivibrator (free-running oscillator) can be built from a single Schmitt trigger inverter and an *RC* circuit as shown in Figure 5–55. The oscillator of Figure 5–55 operates as follows:

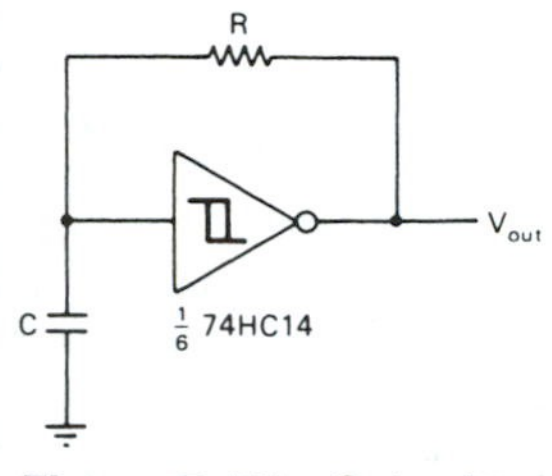

Figure 5–55 Schmitt trigger astable multivibrator.

1. When the IC supply power is first turned on, V_{cap} is 0 V, so V_{out} will be HIGH ($\approx$5.0 V for high-speed CMOS).
2. The capacitor will start charging toward the 5 V at V_{out}.
3. When V_{cap} reaches the positive-going threshold (V_{T+}) of the Schmitt trigger, the output of the Schmitt will change to a LOW ($\approx$0 V).
4. Now with $V_{out} \approx 0$ V, the capacitor will start discharging toward 0 V.
5. When V_{cap} drops below the negative-going threshold (V_{T-}), the output of the Schmitt will change back to a HIGH.
6. The cycle repeats now with the capacitor charging back up to V_{T+}, then down to V_{T-}, then up to V_{T+}, and so on.

The waveform at V_{out} will be a square wave oscillating between V_{OH} and V_{OL}, as shown in Figure 5–56. To calculate t_{HI} and t_{LO} use the following equations:

$$t_{HI} = RC \ln\left(\frac{V_{OH} - V_{T-}}{V_{OH} - V_{T+}}\right) \tag{5–1}$$

$$t_{LO} = RC \ln\left(\frac{V_{OL} - V_{T+}}{V_{OL} - V_{T-}}\right) \tag{5–2}$$

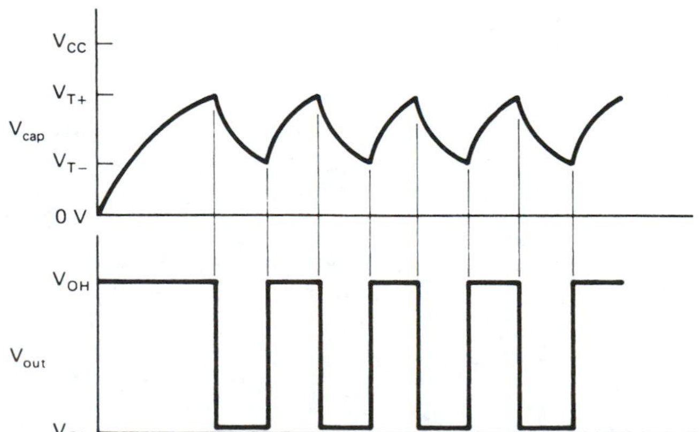

Figure 5–56 Waveforms from the oscillator circuit of Figure 5–55.

EXAMPLE 5–13

(a) Sketch and label the waveforms for the Schmitt *RC* oscillator of Figure 5–55 given the following specifications for a 74HC14 high-speed CMOS Schmitt inverter ($V_{CC} = 5.0$ V).

$$V_{OH} = 5.0 \text{ V} \qquad V_{OL} = 0.0 \text{ V}$$

$$V_{T+} = 2.75 \text{ V} \qquad V_{T-} = 1.67 \text{ V}$$

(b) Calculate the time HIGH (t_{HI}), time LOW (t_{LO}), duty cycle, and frequency if $R = 10$ kΩ and $C = 0.022$ μF.

Solution:

(a) The waveforms for the oscillator are shown in Figure 5–57.

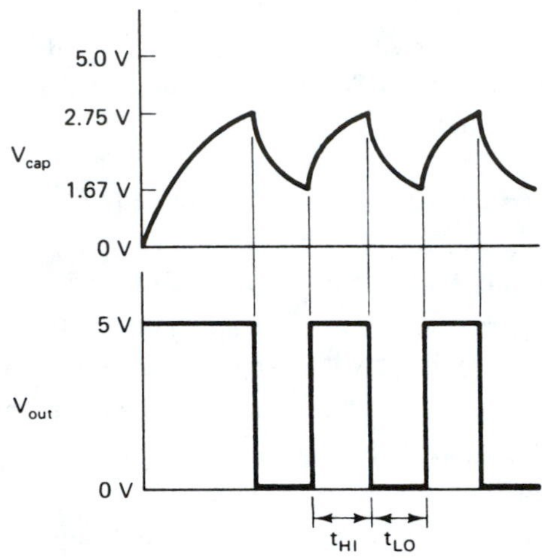

Figure 5–57 Solution to Example 5–13.

(b) To solve for t_{HI}:

$$t_{HI} = RC \ln\left(\frac{V_{OH} - V_{T-}}{V_{OH} - V_{T+}}\right)$$

$$= (10 \text{ k}\Omega)(0.022\ \mu\text{F}) \ln\left(\frac{5.0 - 1.67 \text{ V}}{5.0 - 2.75 \text{ V}}\right)$$

$$= 86.2\ \mu\text{s}$$

To solve for t_{LO}:

$$t_{LO} = RC \ln\left(\frac{V_{OL} - V_{T+}}{V_{OL} - V_{T-}}\right)$$

$$= (10 \text{ k}\Omega)(0.022\ \mu\text{F}) \ln\left(\frac{0.0 - 2.75 \text{ V}}{0.0 - 1.67 \text{ V}}\right)$$

$$= 110\ \mu\text{s}$$

To solve for duty cycle, duty cycle is a ratio of the length of time a square wave is HIGH versus the total period:

$$D = \frac{t_{HI}}{t_{HI} + t_{LO}}$$

$$= \frac{86.2\ \mu\text{s}}{86.2 + 110\ \mu\text{s}} = 0.439 = 43.9\%$$

To solve for frequency:

$$f = \frac{1}{t_{HI} + t_{LO}}$$

$$= \frac{1}{86.2 + 110\ \mu s}$$

$$= 5.10\ kHz$$

One-Shot Monostable Multivibrator

The block diagram and I/O waveforms for a monostable multivibrator (commonly called a one-shot) are shown in Figure 5–58. The one-shot has one *stable state* that is Q = LOW and $\overline{Q}$ = HIGH. The outputs switch to their opposite state for a length of time t_w only when a trigger is applied to the $\overline{A}$ input. $\overline{A}$ is a negative edge trigger in this case (other one-shots use a positive edge trigger or both). The input/output waveforms in Figure 5–58 show the effect that $\overline{A}$ has on the Q output. Q is LOW until the HIGH-to-LOW edge of $\overline{A}$ causes Q to go HIGH for the length of time t_w. The output pulse width (t_w) is determined by the discharge rate of a capacitor in an RC circuit.

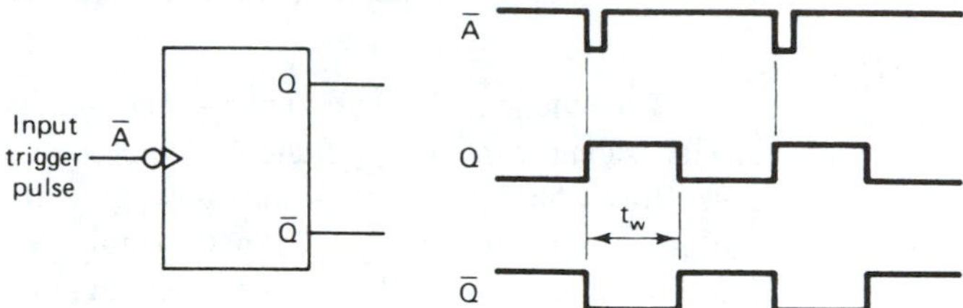

Figure 5–58 Block diagram and input/output waveforms for a monostable multivibrator.

Monostable multivibrators are available in an integrated-circuit package. Two popular ICs are the 74121 and the 74123 (retiggerable) monostable multivibrators. To use these ICs, you need to connect the RC timing components to achieve the proper pulse width. The 74121 provides for two active-LOW and one active-HIGH trigger inputs ($\overline{A}_1$, $\overline{A}_2$, B) and true and complemented outputs (Q, $\overline{Q}$). Figure 5–59 shows the 74121 block diagram and function table that we can use to figure out its operation. To trigger the multivibrator at point T in Figure 5–59, the inputs to the Schmitt AND gate must both be HIGH. To do that, you need B with ($\overline{A}_1$ or $\overline{A}_2$). Holding $\overline{A}_1$ or $\overline{A}_2$ LOW and bringing the input trigger in on B is useful if the trigger signal is slow rising or if it has noise on it because the Schmitt input will provide a definite trigger point.

The RC timing components are set up on pins 9, 10, and 11. If you can use the 2-kΩ *internal* resistor, just connect pin 9 to V_{CC} and put a timing capacitor between pins 10 and 11. An *external* timing resistor can be used instead by placing it between pin 11 and V_{CC}, and putting the timing capacitor between pins 10 and 11. If the external timing resistor is used, pin 9 must be left open. The allowable range of R_{ext} is 1.4 to 40 kΩ and C_{ext} is 0 to 1000 μF. If an electrolytic capacitor is used, its positive side must be connected to pin 11.

The formula that the IC manufacturer gives for determining the output pulse width is

$$t_w = R_{ext} C_{ext} \ln 2 \tag{5–3}$$

(Substitute 2 kΩ for R_{ext} if the internal timing resistor is used.)

For example, if the external timing RC components are 10 kΩ and 0.047 μF, then t_w will equal 10 kΩ × 0.047 μF × ln 2, which works out to be 326 μs. Using the maximum allowed values of $R_{ext}C_{ext}$, the maximum pulse width is almost 28 s (40 kΩ × 1000 μF × ln 2).

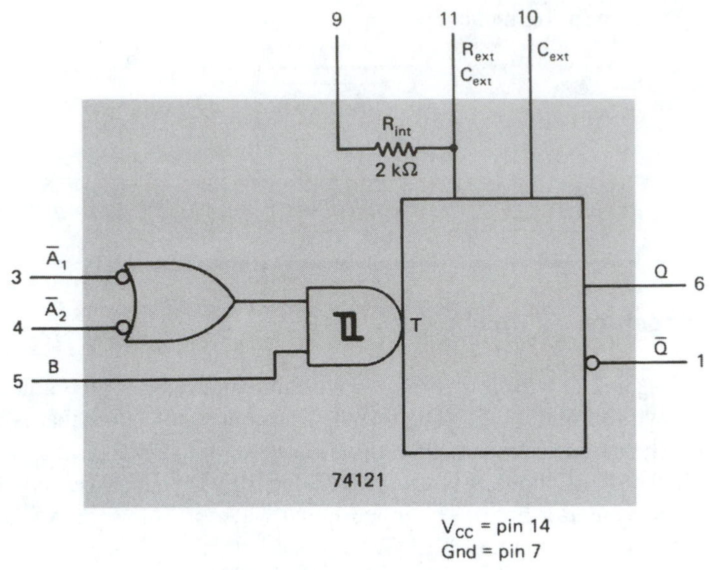

INPUTS			OUTPUTS	
$\overline{A}_1$	$\overline{A}_2$	B	Q	$\overline{Q}$
L	X	H	L	H
X	L	H	L	H
X	X	L	L	H
H	H	X	L	H
H	↓	H	⊓	⊔
↓	H	H	⊓	⊔
↓	↓	H	⊓	⊔
L	X	↑	⊓	⊔
X	L	↑	⊓	⊔

H = HIGH voltage level
L = LOW voltage level
X = Don't care
↑ = LOW-to-HIGH transition
↓ = HIGH-to-LOW transition

(b)

Figure 5–59 The 74121 monostable multivibrator one-shot: (a) block diagram; (b) function table. (Courtesy of Philips Components–Signetics)

The function table in Figure 5–59 shows that the Q output is LOW and the $\overline{Q}$ output is HIGH as long as the $\overline{A}_1, \overline{A}_2, B$ inputs do not provide a HIGH–HIGH to the Schmitt-AND inputs. But by holding B HIGH and applying a HIGH-to-LOW edge to $\overline{A}_1$ or $\overline{A}_2$, the outputs will produce a pulse. Also, the function table shows, in its last two entries, that a LOW-to-HIGH edge at input B will produce an output pulse as long as either $\overline{A}_1$ or $\overline{A}_2$ is held LOW.

The following examples illustrate the use of the 74121 for a one-shot operation.

EXAMPLE 5–14

Design a circuit using a 74121 to convert a 50-kHz, 80% duty cycle square wave to a 50-kHz, 50% duty cycle square wave. (In other words, "stretch" the negative-going pulse to cover 50% of the total period.)

Solution:

First, let's draw the original square wave (Figure 5-60a) to see what we have to work with ($t = 1/50$ kHz $= 20\ \mu$s, $t_{HI} = 80\% \times 20\ \mu$s $= 16\ \mu$s).

Now, we want to stretch the 4-μs negative pulse out to 10 μs to make the duty cycle 50%. If we use the HIGH-to-LOW edge on the negative pulse to trigger the $\overline{A}_1$ input to a 74121 and set the output pulse width (t_w) to 10 μs, we should have the solution. The output will be taken from $\overline{Q}$ because it provides a negative pulse when triggered.

Using the formula given in the IC specifications, we can calculate an appropriate $R_{ext}C_{ext}$ to yield 10 μs.

$$t_w = R_{ext}C_{ext} \ln (2)$$

$$10\ \mu\text{s} = R_{ext}C_{ext}(0.693)$$

$$R_{ext}C_{ext} = 14.4\ \mu\text{s}$$

Pick $C_{ext} = 0.001\ \mu$F (1000 pF); then

$$R_{ext} = \frac{14.4\ \mu\text{s}}{0.001\ \mu\text{F}}$$

$$= 14.4\ \text{k}\Omega \quad \text{(use a 10-k}\Omega\text{ fixed resistor with a 5-k}\Omega\text{ potentiometer)}$$

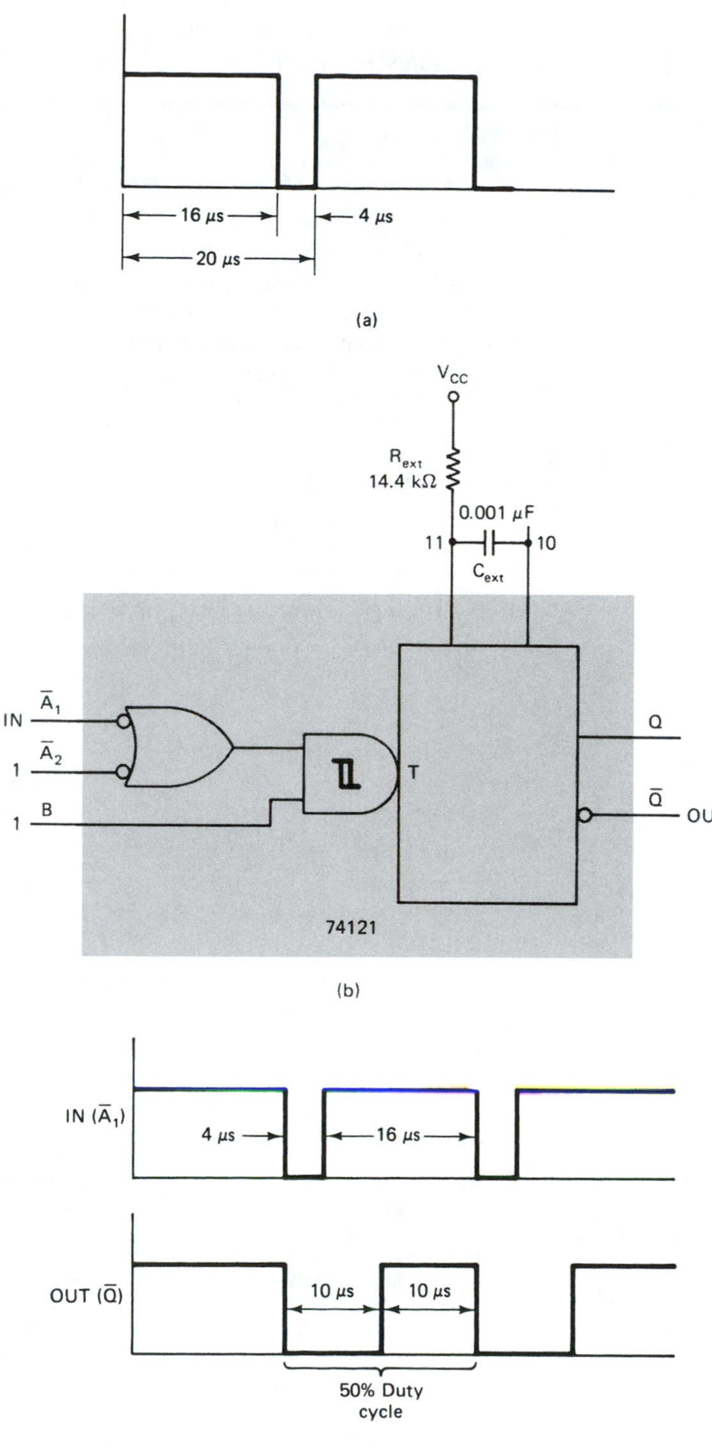

Figure 5–60 (a) Original square wave for Example 5–14; (b) monostable multivibrator circuit connections; (c) input/output waveforms.

The value 0.001 μF is a good choice for C_{ext} because it is much larger than any stray capacitance that might be encountered in a typical circuit. Values of capacitance less than 100 pF (0.0001 μF) may be unsuitable because it is not uncommon for there to be 50 pF of stray capacitance between traces in a printed-circuit board. Also, resistances in the kilohm range are a good choice because they are big enough to limit current flow but not so big to be susceptible to electrostatic noise. The final circuit design and waveforms are given in Figures 5–60b and c.

EXAMPLE 5–15

In microprocessor systems, most control signals are active-LOW, and quite often one-shots are required to introduce delays for certain devices to wait for other, slower devices to respond. For example, to read from a memory device, a line called $\overline{\text{READ}}$ goes LOW to enable the memory device. Most systems have to introduce a delay after the memory device is enabled (to allow for internal propagation delays), before the microprocessor actually reads the data. Design a system using two 74121s to output a 200-ns LOW pulse (called $\overline{\text{Data-Ready}}$) 500 ns after the $\overline{\text{READ}}$ line goes LOW.

Solution:

The first 74121 will be used to produce the 500-ns delay pulse as soon as the $\overline{\text{READ}}$ line goes LOW (see Figure 5–61). The second 74121 will be triggered by the end of the 500-ns delay pulse and will output its own 200-ns LOW pulse for the $\overline{\text{Data-Ready}}$ line. (The 74121s are edge-triggered, so they will trigger only on a HIGH-to-LOW or LOW-to-HIGH *edge.*)

For $t_w = 500$ ns (output for first 74121):

$$t_w = R_{ext}C_{ext} \ln (2)$$

$$500 \text{ ns} = R_{ext}C_{ext}(0.693)$$

$$R_{ext}C_{ext} = 0.722 \ \mu\text{s}$$

Pick $C_{ext} = 100$ pF; then

$$R_{ext} = \frac{0.722 \ \mu\text{s}}{0.0001 \ \mu\text{F}} = 7.22 \text{ k}\Omega$$

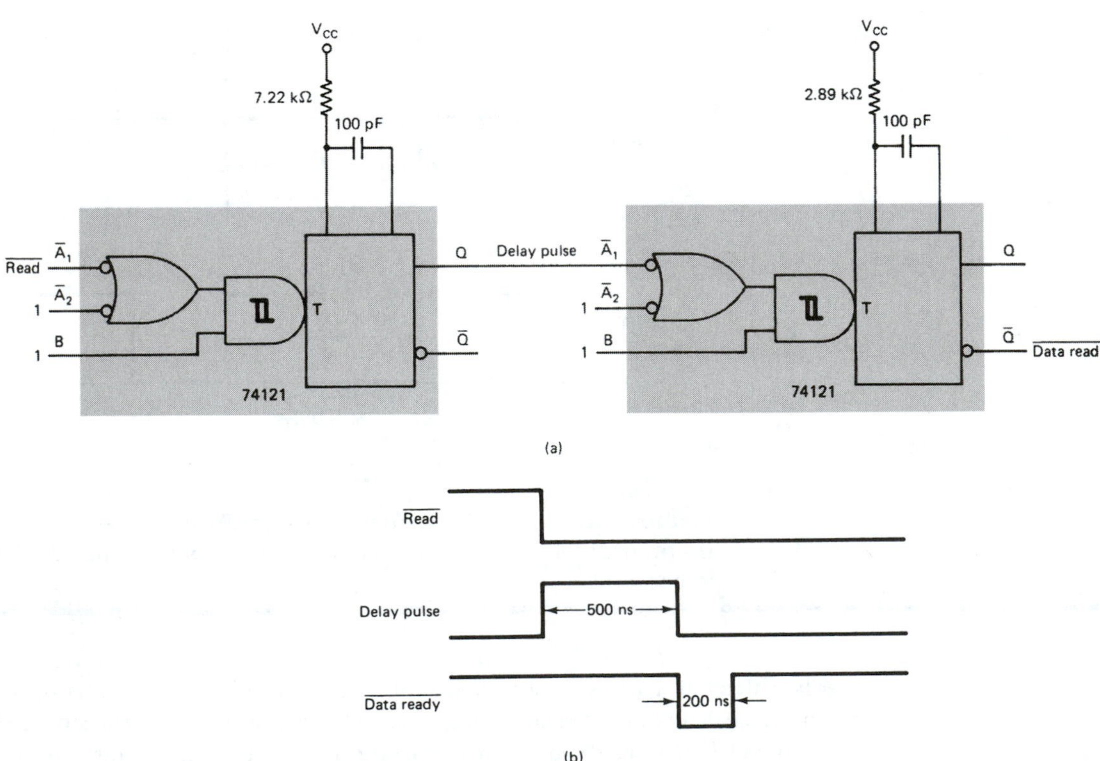

Figure 5–61 Solution to Example 5–15: (a) circuit connections for creating a delayed pulse; (b) input/output waveforms.

For $t_w = 200$ ns (output for second 74121):

$$t_w = R_{ext}C_{ext} \ln(2)$$

$$200 \text{ ns} = R_{ext}C_{ext}(0.693)$$

$$R_{ext}C_{ext} = 0.289\ \mu s$$

Pick $C_{ext} = 100$ pF; then

$$R_{ext} = 2.89\ k\Omega$$

The 555 IC Timer as an Astable Oscillator

The 555 is a very popular, general-purpose timer IC. It can be connected as a one-shot or an astable oscillator as well as being used for a multitude of custom designs. Figure 5–62 shows a block diagram of the chip with its internal components and the external components that are required to set it up as an astable oscillator.

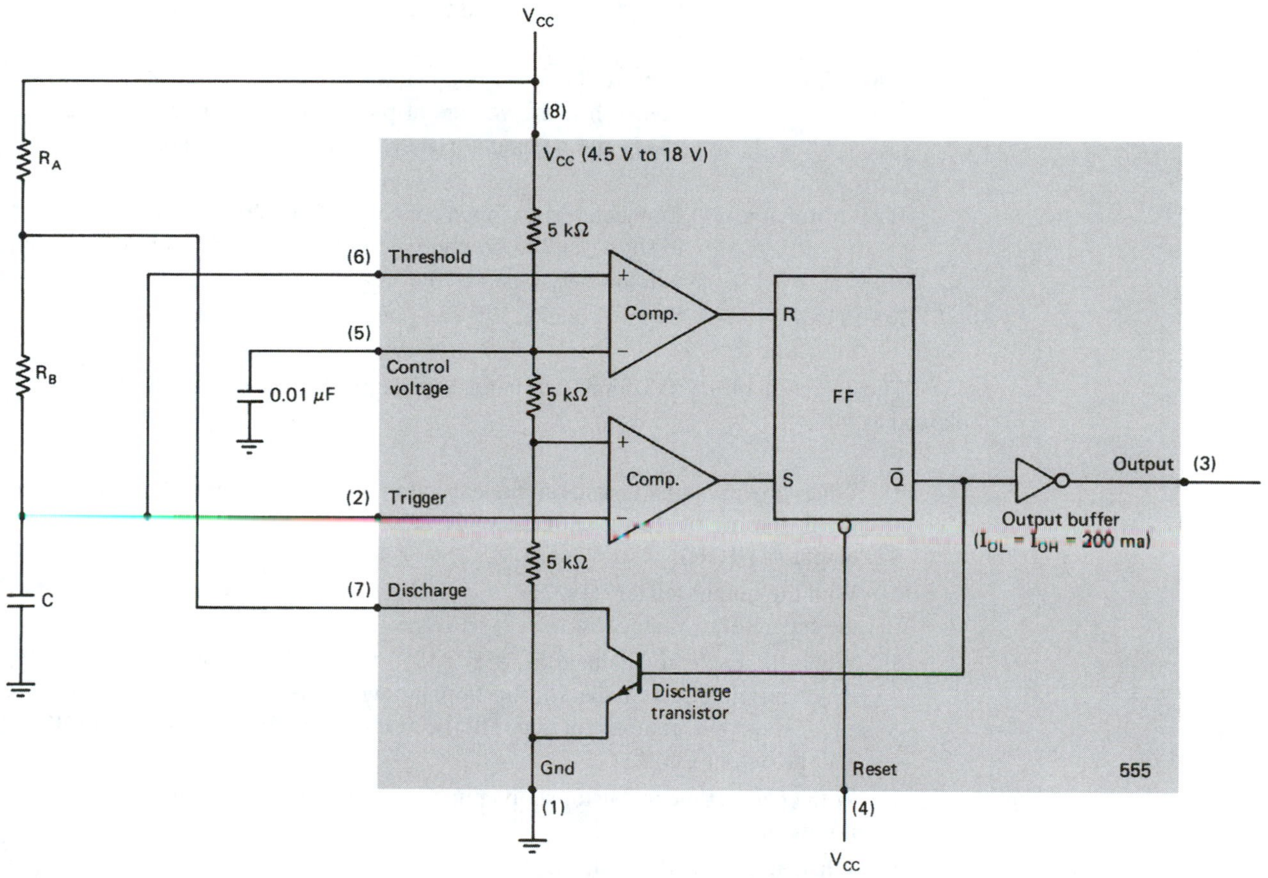

Figure 5–62 Simplified block diagram of a 555 timer with the external timing components to form an astable multivibrator.

The 555 got its name from the three 5-kΩ resistors. They are set up as a voltage divider from V_{CC} to ground. The top of the lower 5 kΩ is at $\frac{1}{3}V_{CC}$ and the top of the middle 5 kΩ is at $\frac{2}{3}V_{CC}$. For example, if V_{CC} is 6 V, each resistor will drop 2 V.

The triangle-shaped symbols represent *comparators.* A comparator simply outputs a HIGH or LOW based on a comparison of the analog voltage levels at its input. If the + input is *more positive* than the − input, it outputs a HIGH. If the + input is *less positive* than the − input, it outputs a LOW.

The *S-R flip-flop* is driven by the two comparators. It has an active-LOW Reset and its output is taken from the $\overline{Q}$.

The *discharge transistor* is an *NPN*, which is used to short pins 7 to 1 when $\overline{Q}$ is HIGH.

The operation and function of the 555 pins are as follows:

Pin 1 (ground): System ground.

Pin 2 (trigger): Input to the lower comparator, which is used to Set the flip-flop. When the voltage at pin 2 crosses from above to below $\frac{1}{3}V_{CC}$, the comparator switches to a HIGH, setting the flip-flop.

Pin 3 (output): The output of the 555 is driven by an inverting buffer capable of sinking or sourcing 200 mA. The output voltage levels are dependent on the output current but are approximately $V_{OH} = V_{CC} - 1.5$ V and $V_{OL} = 0.1$ V.

Pin 4 (Reset): Active-LOW Reset, which forces $\overline{Q}$ HIGH and pin 3 (output) LOW.

Pin 5 (control): Used to override the $\frac{2}{3}V_{CC}$ level, if required. Usually, it is connected to a grounded 0.01-μF capacitor to bypass noise on the V_{CC} line.

Pin 6 (threshold): Input to the upper comparator, which is used to Reset the flip-flop. When the voltage at pin 6 crosses from below to above $\frac{2}{3}V_{CC}$, the comparator switches to a HIGH, resetting the flip-flop.

Pin 7 (discharge): Connected to the open collector of the *NPN* transistor. It is used to short pin 7 to ground when $\overline{Q}$ is HIGH (pin 3 LOW), which will discharge the external capacitor.

Pin 8 (V_{CC}): Supply voltage. V_{CC} can range from 4.5 to 18 V.

The operation of the 555 connected in the astable mode shown in Figure 5–62 is explained as follows.

1. When power is first turned on, the capacitor is discharged, which places 0 V at pin 2, forcing the lower comparator HIGH. This sets the flip-flop ($\overline{Q}$ = LOW, output = HIGH).
2. With the output HIGH ($\overline{Q}$ LOW) the discharge transistor is open, which allows the capacitor to charge toward V_{CC} via $R_A + R_B$.
3. When the capacitor voltage exceeds $\frac{1}{3}V_{CC}$, the lower comparator goes LOW, which has no effect on the *S-R* flip-flop, but when the capacitor voltage exceeds $\frac{2}{3}V_{CC}$, the upper comparator goes HIGH, resetting the flip-flop, forcing $\overline{Q}$ HIGH and the output LOW.
4. With $\overline{Q}$ HIGH the transistor shorts pin 7 to ground, which discharges the capacitor via R_B.
5. When the capacitor voltage drops below $\frac{1}{3}V_{CC}$, the lower comparator goes back HIGH again, setting the flip-flop and making $\overline{Q}$ LOW, and the output HIGH.
6. Now, with $\overline{Q}$ LOW, the transistor opens again, allowing the capacitor to start charging up again.
7. The cycle repeats with the capacitor charging up to $\frac{2}{3}V_{CC}$, then discharging down to $\frac{1}{3}V_{CC}$ continuously. While the capacitor is charging, the output is HIGH, and when the capacitor is discharging, the output is LOW.

The waveforms, depicting the operation of the 555 as an astable oscillator, are shown in Figure 5–63.

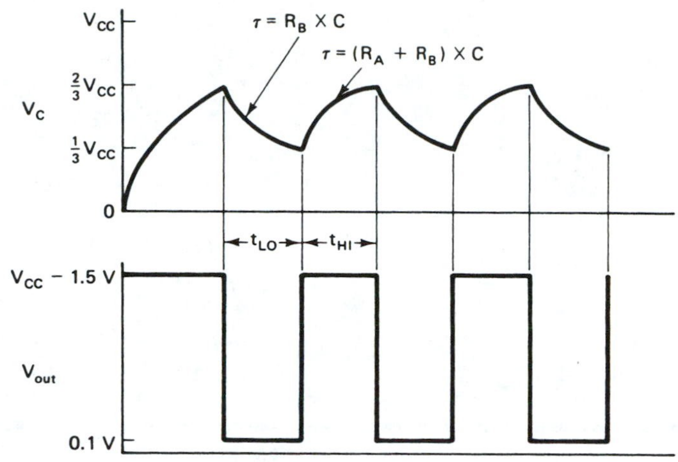

Figure 5–63 Waveforms for the 555 astable multivibrator circuit of Figure 5–62.

The formulas for the time durations t_{LO} and t_{HI} are as follows:

$$t_{LO} = 0.693 R_B C \tag{5–4}$$

$$t_{HI} = 0.693(R_A + R_B)C \tag{5–5}$$

EXAMPLE 5–16

Determine t_{HI}, t_{LO}, duty cycle, and frequency for the 555 astable multivibrator circuit of Figure 5–64.

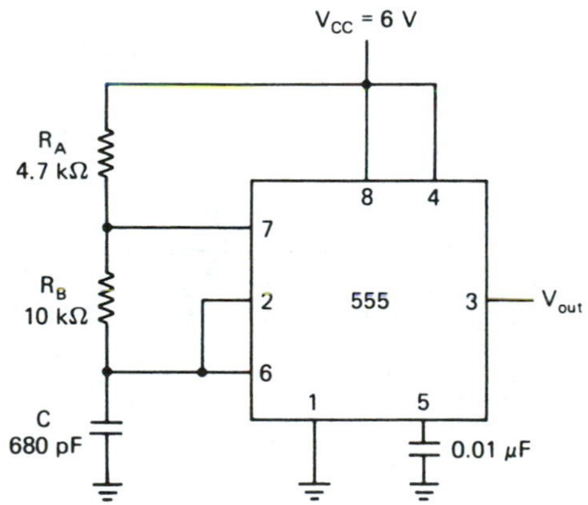

Figure 5–64 The 555 astable connections for Example 5–16.

Solution:

$$\begin{aligned} t_{LO} &= 0.693 R_B C \\ &= 0.693(10\text{ k}\Omega)680\text{ pF} \\ &= 4.71\ \mu\text{s} \\ t_{HI} &= 0.693(R_A + R_B)C \\ &= 0.693(4.7 + 10\text{ k}\Omega)680\text{ pF} \\ &= 6.93\ \mu\text{s} \end{aligned}$$

$$\text{Duty cycle} = \frac{t_{HI}}{t_{HI} + t_{LO}}$$
$$= \frac{6.93\ \mu s}{6.93 + 4.71\ \mu s}$$
$$= 59.5\%$$
$$\text{Frequency} = \frac{1}{t_{HI} + t_{LO}}$$
$$= \frac{1}{6.93 + 4.71\ \mu s}$$
$$= 85.9\ \text{kHz}$$

The 555 IC Timer as a One-Shot Multivibrator

Another common use for the 555 is as a monostable multivibrator, as shown in Figure 5–65. The one-shot of Figure 5–65 operates as follows:

1. Initially (before the trigger is applied), V_{out} is LOW, shorting pin 7 to ground and discharging C.
2. Pin 2 is normally held HIGH by the 10-kΩ pull-up resistor. To trigger the one-shot, a negative-going pulse (less than $\frac{1}{3}V_{CC}$) is applied to pin 2.
3. The trigger forces the lower comparator HIGH (see Figure 5-62), which sets the flip-flop, making V_{out} HIGH and opening the discharge transistor (pin 7).
4. Now the capacitor is free to charge from 0 V up toward V_{CC} via R_A.
5. When V_C crosses the threshold of $\frac{2}{3}V_{CC}$, the upper comparator goes HIGH, resetting the flip-flop making V_{out} LOW and shorting the discharge transistor.
6. The capacitor discharges rapidly to 0 V and the one-shot is held in its stable state (V_{out} = LOW) until another trigger is applied.

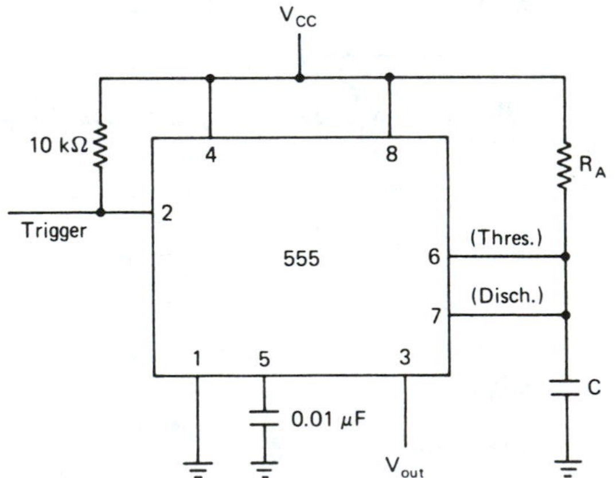

Figure 5–65 The 555 connections for one-shot operation.

The waveforms that are generated for the one-shot operation are shown in Figure 5–66.

The equation for t_w is

$$t_w = 1.10 R_A C \tag{5–6}$$

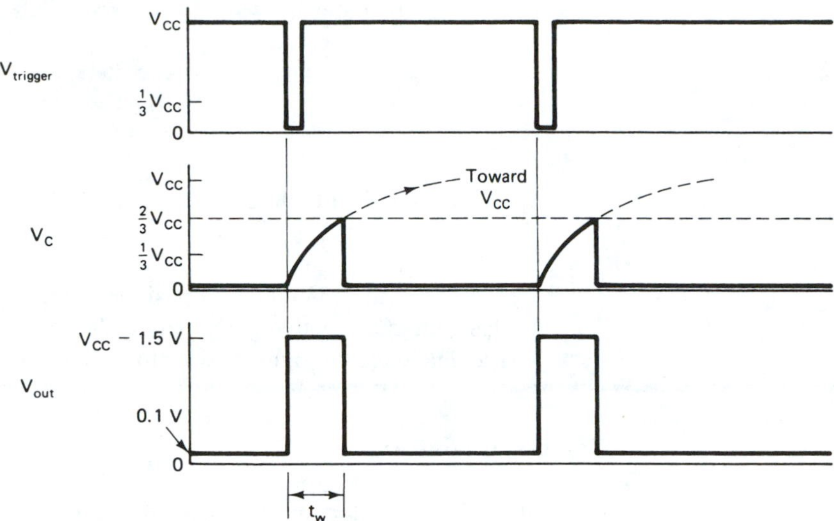

Figure 5–66 Waveforms for the one-shot circuit of Figure 5–65.

EXAMPLE 5–17

Design a circuit using a 555 one-shot that will stretch a 1-μs negative-going pulse that occurs every 60 μs into a 10-μs negative-going pulse.

Solution:

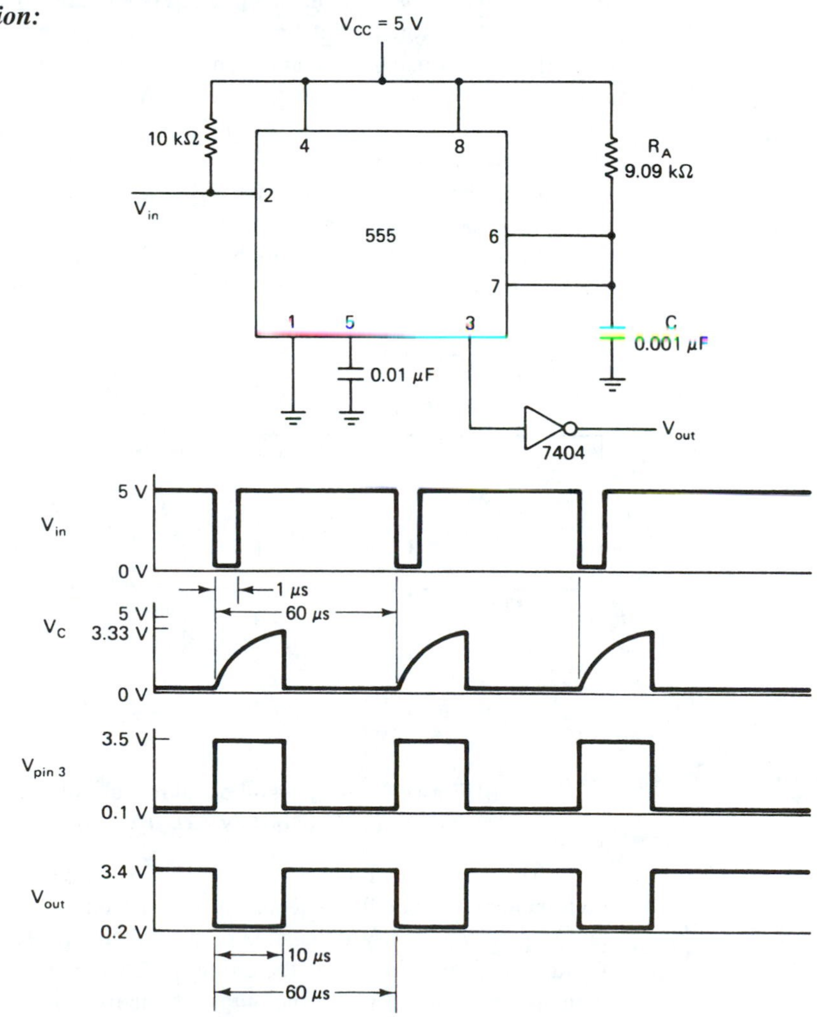

Figure 5–67 Solution to Example 5–17.

To set the output pulse width to 10 μs:

$$t_w = 1.10 R_A C$$

$$10\ \mu\text{s} = 1.10 R_A C$$

$$R_A C = 9.09\ \mu\text{s}$$

Pick $C = 0.001\ \mu$F; then

$$R_A = 9.09\ \text{k}\Omega$$

Also, since the 555 outputs a positive-going pulse, an inverter must be added to change it to a negative-going pulse. (7404 inverter: $V_{OH} = 3.4$ V, $V_{OL} = 0.2$ V.) The final circuit design and waveforms are shown in Figure 5–67.

Crystal Oscillators

None of the *RC* oscillators or one-shots presented in the previous sections are extremely stable. In fact, the standard procedure for building those timing circuits is to prototype them based on the *R* and *C* values calculated using the formulas, then "tweak" (or make adjustments to) the resistor values while observing the time periods on an oscilloscope. Normally, standard values are chosen for the capacitors and potentiometers are used for the resistors.

However, even after a careful calibration of the time period, changes in the components and IC occur as the devices age and as the ambient temperature varies. To partially overcome this problem, some manufacturers will allow their circuits to "burn in" or age for several weeks before the final calibration and shipment.

Instead of using *RC* components, another timing component is available to the design engineer when extremely critical timing is required. This highly stable and accurate timing component is the *quartz crystal.* A piece of quartz crystal is cut to a specific size and shape to vibrate at a specific frequency, similar to an *RLC* resonant circuit. Its frequency is typically in the range 10 kHz to 10 MHz. Accuracy of more than *five significant digits* can easily be achieved using this method.

Crystal oscillators are available as an integrated-circuit package or can be built using an external quartz crystal in circuits such as those shown in Figure 5–68. The circuits shown in the figure will oscillate at a frequency dictated by the crystal chosen. In Figure 5–68a the 100-kΩ pot may need adjustment to start oscillation.

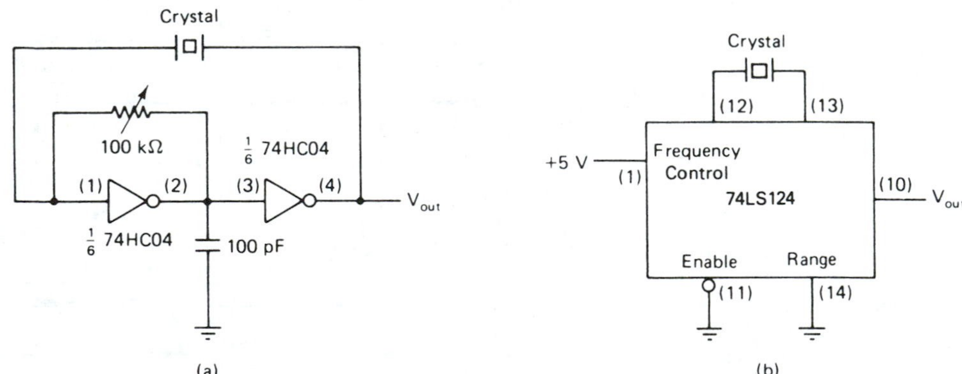

Figure 5–68 Crystal oscillator circuits: (a) high-speed CMOS oscillator; (b) Schottky-TTL oscillator.

The 74S124 TTL chip in Figure 5–68b is a *voltage-controlled oscillator* (VCO), set up to generate a specific frequency at V_{out}. By changing the crystal to a capacitor, the output frequency will vary, depending on the voltage level at the frequency control (pin 1) and frequency range (pin 14) inputs. Using specifications presented in the manufacturer's data manual, you can determine the output frequency of the VCO, based on the voltage level applied to its inputs.

5–10 PRACTICAL INPUT AND OUTPUT CONSIDERATIONS

Before designing and building the practical digital circuits in the next few chapters, let's study some circuit designs for (1) a simple 5-V power supply, (2) a clock to drive synchronous trigger inputs, and (3) circuit connections for interfacing to the outputs and inputs of integrated-circuit chips.

A 5-V Power Supply

For now, we limit our discussion to the TTL family of integrated circuits. From the data manual we can see that TTL requires a constant supply voltage of 5.0 V ± 5%. Also, the total supply current requirement into the V_{CC} terminal ranges from 20 to 100 mA for most TTL ICs.

For the power supply, the 78XX series of integrated-circuit *voltage regulators* is inexpensive and easy to use. To construct a regulated 5.0-V supply, we use the 7805 (the 05 designates 5 V; a 7808 would designate an 8.0-V supply). The 7805 is a three-terminal device (input, ground, output) capable of supplying 5.0 V ± 0.2% at 1000 mA. Figure 5–69 shows how a 7805 voltage regulator is used in conjunction with an ac-to-dc *rectifier* circuit.

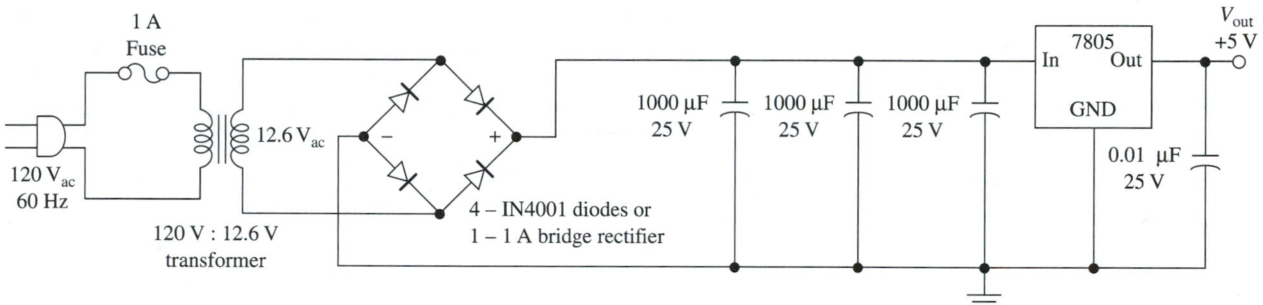

Figure 5–69 Complete 5-V, 1-A TTL power supply.

In Figure 5–69 the 12.6-V ac rms is rectified by the diodes (or a four-terminal bridge rectifier) into a full-wave dc of approximately 20 V. The 3000 μF of capacitance is required to hold the dc level into the 7805 at a high, steady level. The 7805 will automatically decrease the 20-V dc input to a solid, *ripple*-free 5.0-V dc output.

The 0.01-μF capacitor is recommended by TTL manufacturers for *decoupling* the power supply. Tantalum capacitors work best and should be mounted as close as possible to the V_{CC}-to-ground pins on every TTL IC used in your circuit. Their size should be between 0.01 and 0.1 μF with a voltage rating ≥5 V. The purpose of the capacitor is to eliminate the effects of voltage spikes created from the internal TTL switching and electrostatic noise generated on the power and ground lines.

The 7805 will get very hot when your circuit draws more than 0.5 A. In that case, it should be mounted on a heat sink to help dissipate the heat.

A 60-Hz Clock

Figure 5–70 shows a circuit design for a simple 60-Hz TTL-level (0 to 5 V) clock that can be powered from the same transformer used in Figure 5–69 and used to drive the clock inputs to our synchronous ICs. Our electric power industry supplies us with accurate 60-Hz ac voltages. It is a simple task to reduce the voltage to usable levels and still maintain a 60-Hz [60-pulse-per-second (pps)] signal.

From analog electronics courses, you may remember that a zener diode will conduct normally in the forward-biased direction, and in the reverse-biased direction it will start conducting when a voltage level equal to its *reverse zener breakdown* rating is reached (4.3 V for the IN749).

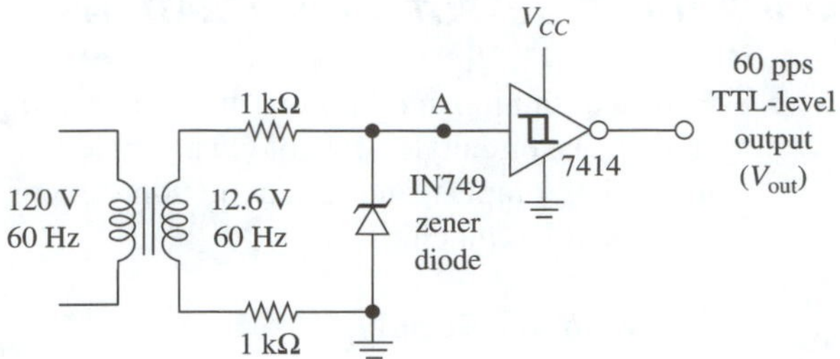

Figure 5–70 Accurate 60-Hz, TTL-level clock pulse generator.

The 1-kΩ resistors are required to limit the zener current to reasonable levels. Figure 5–71 shows the waveform that will appear at point A and V_{out} of Figure 5–70. The zener breaks down at 4.3 V, which is high enough for a one-level input to the Schmitt trigger but not too high to burn out the chip. The V_{out} waveform will be an accurate 60-pulse-per-second, approximately 50% duty cycle square wave. As we will see in Chapter 6, this frequency can easily be divided down to 1 pulse per second by using toggle flip-flops. One pulse per second is handy because it is slow enough to see on visual displays (like LEDs) and accurate enough to use as a trigger pulse on a digital clock.

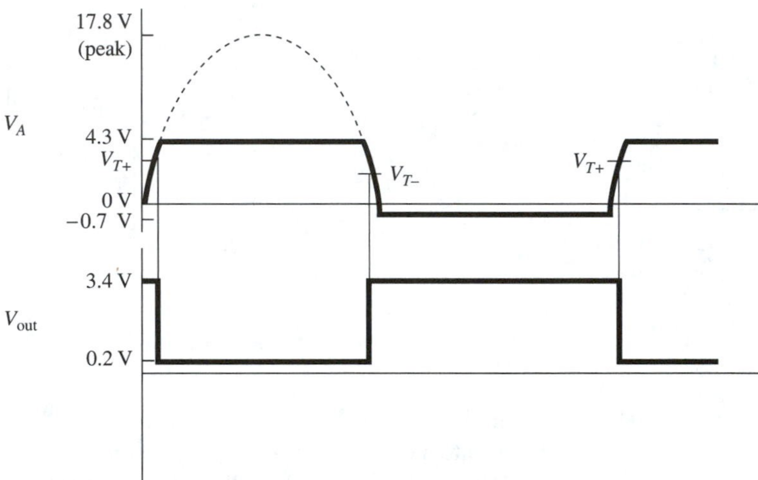

Figure 5–71 Voltage waveform at point A and V_{out} of Figure 5–70.

Driving Light-Emitting Diodes

Light-emitting diodes (LEDs) are good devices to visually display a HIGH (1) or LOW (0) digital state. A typical red LED will drop 1.7 V cathode to anode when forward biased (positive anode-to-cathode voltage) and will illuminate with 10 to 20 mA flowing through it. In the reverse-biased direction (zero or negative anode-to-cathode voltage), the LED will block current flow and not illuminate.

Because it takes 10 to 20 mA to illuminate an LED, we may have trouble driving it with a TTL output. From the TTL data manual, we can determine that most ICs can sink (0-level output) a lot more current than they can source (1-level output). Typically, the maximum sink current, I_{OL}, is 16 mA and the maximum source current, I_{OH}, is only 0.4 mA. Therefore, we had better use a LOW level (0) to turn on our LED instead of a HIGH level.

Figure 5–72 shows how we can drive an LED from the output of a TTL circuit (a *J-K* flip-flop in this case). The *J-K* flip-flop is set up in the toggle mode so that Q will flip states once each second.

When Q is LOW (0 V), the LED is forward biased and current will flow through the LED and resistor and sink into the Q output. The 330-Ω resistor is required to limit the series current to 10 mA [$I = (5\text{ V} - 1.7\text{ V})/330\ \Omega = 10\text{ mA}$], and 10 mA into the LOW-level

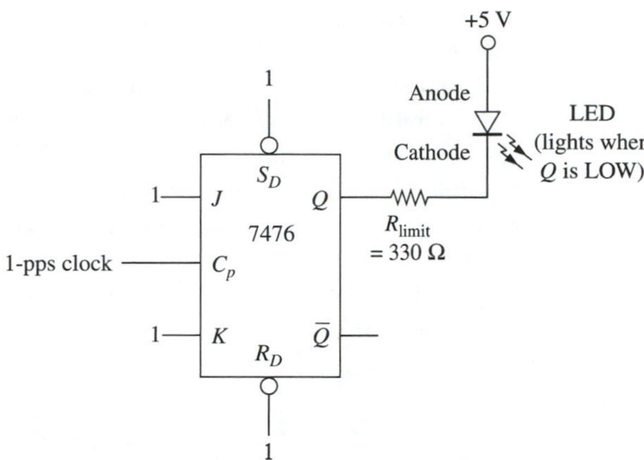

Figure 5–72 Driving an LED.

Q output will not burn out the flip-flop. If, however, we were trying to turn the LED on with a HIGH-level output, we would turn the LED around and connect the cathode to ground. But 10 mA would exceed the limit of I_{OH} on the 7476 and either burn it out or just not illuminate the LED.

Phototransistor Input to a Latching Alarm System

A *phototransistor* is made to turn off and on by shining light on its base region. It is encased in clear plastic and is turned on when light strikes its base region or if the base connection is forward biased with an external voltage. The resistance from collector to emitter for an OFF transistor is typically 1 to 10 MΩ. An ON transistor will range from 1000 Ω to as low as 10 Ω depending on the light intensity.

The circuit of Figure 5–73 uses a phototransistor in an alarm system. The phototransistor could be placed in a doorway and positioned so that light is normally striking it. This will keep its resistance low and the voltage at point A low. When a person walks through the doorway, the light is interrupted, making the voltage at point A momentarily high. The 74HCT14 Schmitt inverters will react by outputting a LOW-to-HIGH pulse. This creates the clock trigger to the D flip-flop, which will latch HIGH, turning on the alarm. The alarm will remain on until the Reset pushbutton is pressed.

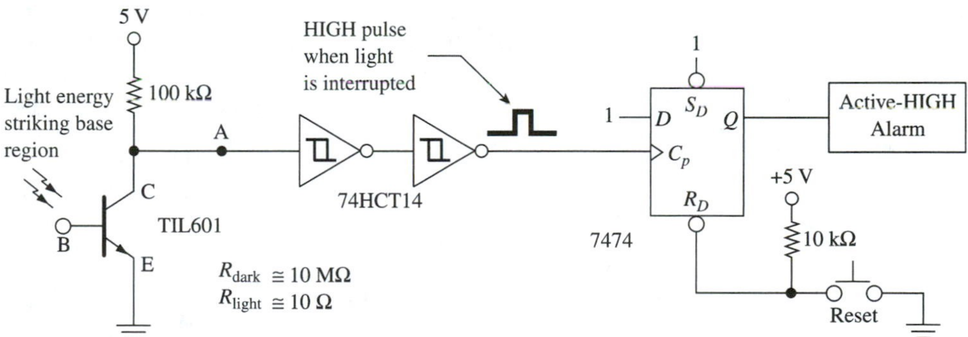

Figure 5–73 Phototransistor used as an input to a latching alarm system.

Using an Optocoupler for Level Shifting

An *optocoupler* (or optoisolator) is an integrated circuit with an LED and phototransistor encased in the same package. The phototransistor has a very high OFF resistance (dark) and a low ON resistance (light), which are controlled by the amount of light striking its base from the LED. The terms *optocoupler* and *optoisolator* come from the fact that the output side of the device is electrically *isolated* from the input side and can therefore be used to *couple* one circuit to another without being concerned about incompatible or harmful voltage levels.

Figure 5–74 shows how an optocoupler can be used to transmit TTL-level data to another circuit having 25-V logic levels. The 7408 is used to sink current through the optocoupler's LED each time the input signal goes LOW. Each time the LED illuminates, the phototransistor will exhibit a low resistance, making V_{out} LOW. Therefore, the output signal will be in phase with the input signal, but its HIGH/LOW levels will be approximately 25 V/0 V. Note that the 25-V circuit is totally separate from the 5-V TTL circuit, providing complete isolation from the potentially damaging higher voltage.

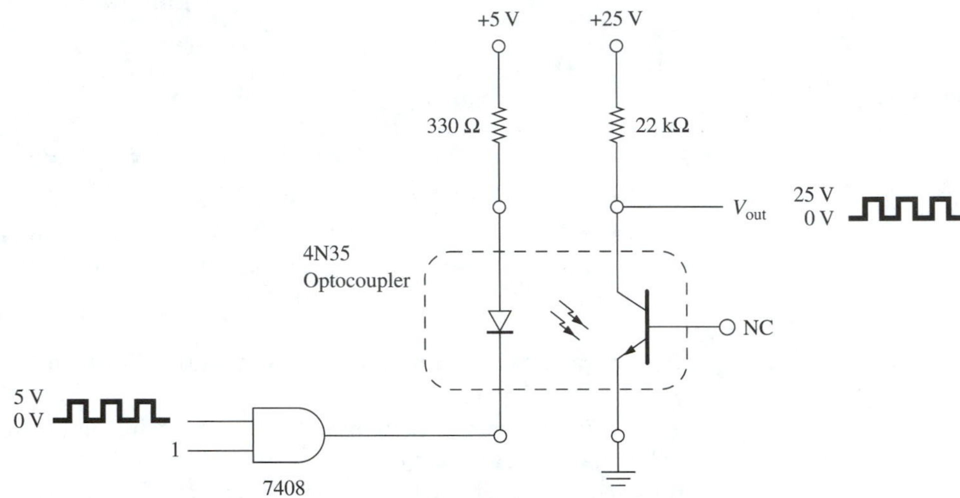

Figure 5–74 An optocoupler provides isolation in a level-shifting application.

A Power MOSFET Used to Drive a Relay and AC Motor

The output drive capability of digital logic severely limits the size of the load that can be connected. An LS-TTL buffer such as the 74LS244 can sink up to 24 mA, and the BiCMOS 74ABT244 can sink up to 64 mA, but this is still far below the current requirements of some loads. A common way to boost the current capability is to use a *power MOSFET,* which is particularly well suited for these applications. This is a transistor specifically designed to have a very high input impedance to limit current draw into its gate and also be capable of passing a high current through its drain to source. (See Section 2–14 to review MOSFET operation.)

Figure 5–75 shows a circuit that could be used to drive a $\frac{1}{3}$-hp ac motor from a digital logic circuit. Because the starting current of a motor can be extremely high, we will use

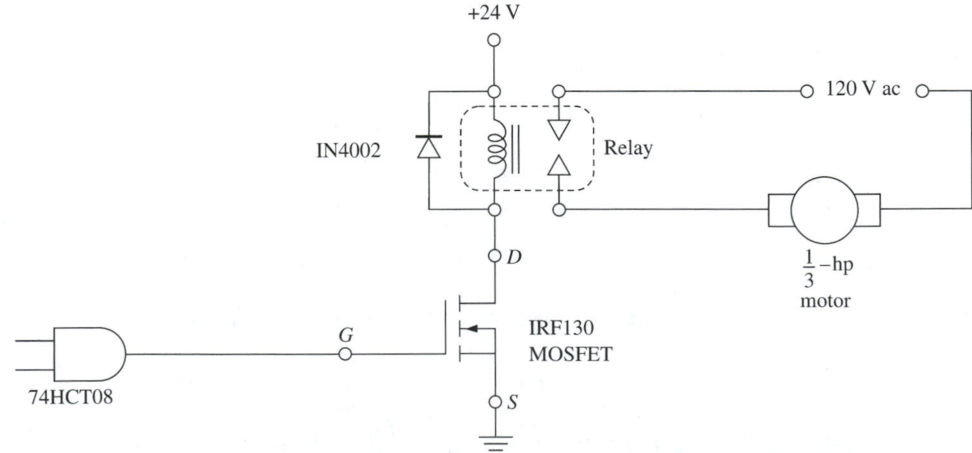

Figure 5–75 Using a power MOSFET to interface digital logic to high-power ac circuitry.

a relay with a 24-V dc coil and a 50-A contact rating. A relay of this size may require as much as 200 mA to energize the coil to pull in the contacts. A MOSFET such as the IRF130 can pass up to 12 A through its drain to source, so it can easily handle this relay coil requirement.

When the 74HCT08 outputs a HIGH (5 V) to the gate of the MOSFET, the drain to source becomes a short (approximately 0.2 Ω). This allows current to flow through the relay coil, creating the magnetic flux required to pull in the contacts. The motor will start. The 1N4002 diode provides arc protection across the coil when it is de-energized.

Level Detecting with an Analog Comparator

Analog comparators such as the LM339 are commonly used to interface to digital circuitry. A comparison of the two analog voltages at the comparator's input are used to determine the device's output logic level (1 or 0). If the analog voltage at the (+) input is greater than the voltage at the (−) input, then the output is a logic 1. Otherwise, the output is a logic 0. The output of the LM339 acts like an open-collector gate, so a pullup resistor is required to make the output HIGH.

The circuit in Figure 5–76 is used to detect when the temperature of a furnace exceeds 100°C. The LM35 is a temperature sensor used to monitor the furnace temperature. It outputs 10 mV for each degree Celcius. (For example, if it is 20°C, it will output 200 mV.)

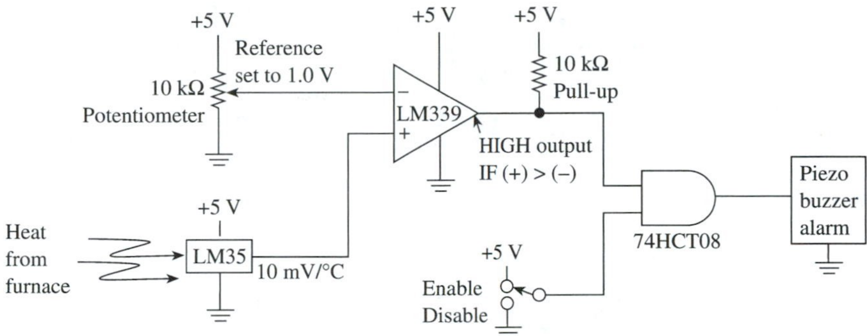

Figure 5–76 Using an LM339 analog comparator to interface to digital logic.

This circuit is set up to sound an alarm when the temperature exceeds 100°C. The 10-k potentiometer can be set at any value as a reference to compare to. In this case, we want to set it at 1.00 V. When the temperature exceeds 100°C the (+) input becomes greater than the (−) and the LM339 outputs a HIGH. This HIGH will find its way to the buzzer as long as the Enable switch is in the UP position. Piezo alarm buzzers are noted for their very small current draw and can easily be driven by the digital logic gate.

Many other types of sensors could be monitored by the comparator instead of the temperature sensor. Sensors that output levels in the 0 to 5-V range are available for monitoring such quantities as pressure, velocity, light intensity, and displacement. The reference level set by the potentiometer allows you to select exactly what level triggers the alarm.

Using an Octal *D* Flip-Flop in a Microcontroller Application

Most of the basic latches and flip-flops are also available as *octal* ICs. In this configuration, there are eight latches or flip-flops in a single IC package. If all eight latches or flip-flops are controlled by a common clock, it is called an 8-bit *register.* An example of an 8-bit *D* flip-flop register is the high-speed CMOS 74HCT273 (also available in the TTL LS and S families). The ′273 contains eight *D* flip-flops, all controlled by a common edge-triggered clock, C_p (see Figure 5–77). At the positive edge of C_p, the 8 bits of data at D_0 through D_7 are stored in the eight *D* flip-flops and output at Q_0 through Q_7. The ′273 also has an active-LOW master reset ($\overline{MR}$), which provides asynchronous Reset capability to all eight flip-flops.

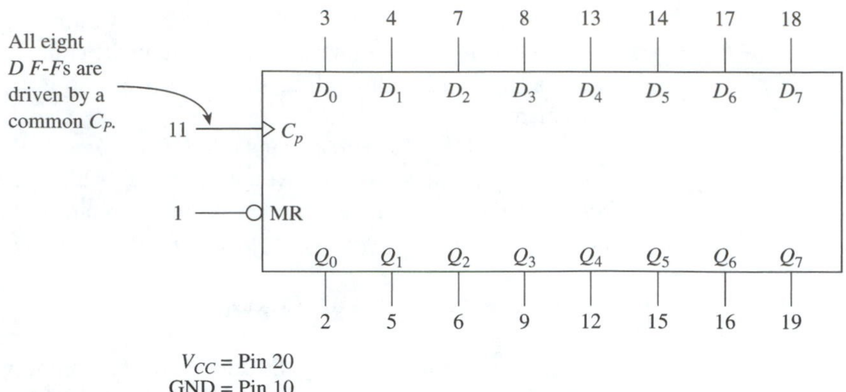

Figure 5–77 Logic diagram for a 74HCT273 octal *D* flip-flop.

An application of the '273 octal *D* flip-flop is shown in Figure 5–78. Here it is used as an *update and hold* register. Every 10 s it receives a clock pulse from the Motorola 68HC11 microcontroller. The data that are on D_0–D_7 at each positive clock edge are stored in the register and output at Q_0–Q_7.

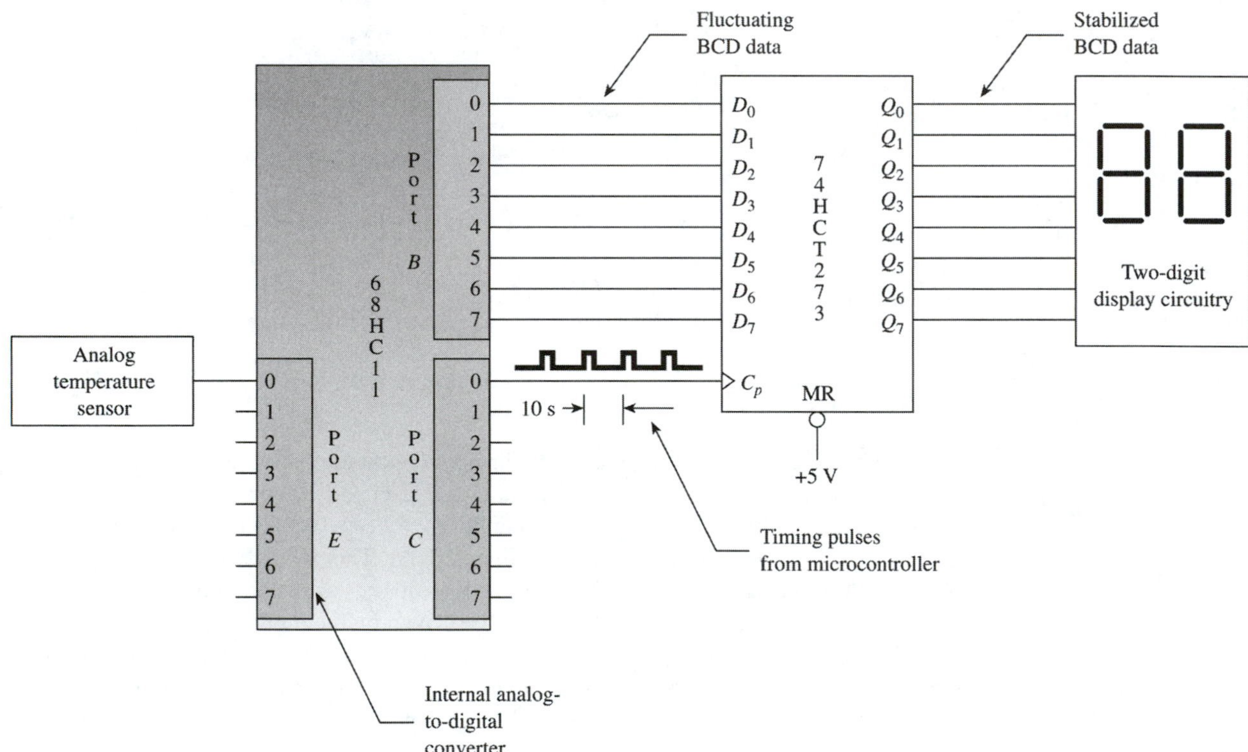

Figure 5–78 Using an octal *D* flip-flop to interface a display to a microcontroller.

The analog temperature sensor is designed to output a voltage that is proportional to degrees centigrade. (See Section 7–11 for the design of temperature sensors.) The 68HC11 microcontroller has the capability to read analog voltages and convert them into their equivalent digital value. A software program is written for the microcontroller to translate this digital string into a meaningful two-digit BCD output for the display.

The BCD output of the 68HC11 is constantly changing as the temperature fluctuates. One way to stabilize this fluctuating data is to use a *storage register* like the 74HCT273. Because the '273 only accepts the BCD data every 10 s, it will hold the two-digit display constant for that length of time, making it easier to read.

SUMMARY

In this chapter we have learned that

1. The *S-R* flip-flop is a single-bit data storage circuit that can be constructed using basic gates.
2. Adding gate enable circuitry to the *S-R* flip-flop makes it *synchronous*. This means that it will operate only under the control of a clock or enable signal.
3. The *D*-flip-flop operates similarly to the *S-R* except it has only a single data input, *D*.
4. The 7475 is an integrated-circuit *D* latch. The output (*Q*) follows the data input (*D*) while the enable (*E*) is HIGH. When *E* goes LOW, *Q* remains latched.
5. The 7474 is an integrated-circuit *D* flip-flop. It has two *synchronous* inputs, *D* and C_P, and two *asynchronous* inputs, $\overline{S_D}$ and $\overline{R_D}$. *Q* changes to the level of *D* at the *positive edge* of C_P. *Q* responds *immediately* to the asynchronous inputs regardless of the synchronous operations.
6. The *J-K* flip-flop differs from the *S-R* flip-flop because it can also perform a *toggle* operation. Toggle means that *Q* flips to its opposite state.
7. The 74LS76 is an edge-triggered *J-K* flip-flop IC. It has synchronous and asynchronous inputs.
8. Mechanical switches exhibit a phenomenon called switch bounce, which can cause problems in most kinds of logic circuits.
9. Three-state buffers, latches, and transceivers are an integral part of microprocessor interface circuitry. They allow the microprocessor system to have external control of 8-bit groups of data. The *buffer* can be used to allow multiple input devices to feed a common point or to provide high output current to a connected load. The *latch* can be used to remember momentary data from the microprocessor that needs to be held for other devices in the system. The *transceiver* provides bidirectional (input or output) control of interface circuitry.
10. Multivibrator circuits are used to produce free-running clock oscillator waveforms or to produce a timed digital level change triggered by an external source.
11. Capacitor voltage charging and discharging rates are the most common way to produce predictable time duration for oscillator and timing operations.
12. An astable multivibrator is a free-running oscillator whose output oscillates between two voltage levels at a rate determined by a capacitor in an attached *R-C* circuit.
13. A monostable multivibrator is used to produce an output pulse that starts when the circuit receives an input trigger and lasts for a length of time dictated by the attached *R-C* circuit.
14. The 74121 is an IC monostable multivibrator with two active-LOW and one active-HIGH input trigger sources and an active-HIGH and an active-LOW pulse output terminal.
15. The 555 IC is a general-purpose timer that can be used to make astable and monostable multivibrators and perform a number of other timing functions.
16. Crystal oscillators are much more accurate and stable than *R-C* timing circuits. They are used most often for micropocessor and digital communication timing.

GLOSSARY

Active Clock Edge: A clock edge is the point in time where the waveform is changing from HIGH to LOW (negative edge) or LOW to HIGH (positive edge). The *active* clock edge is the edge (either positive or negative) used to trigger a synchronous device to accept input digital states.

Asynchronous: (not synchronous). A condition in which the output of a device will switch states instantaneously as the inputs change without regard to an input clock signal.

Buffer: A logic device connected between two digital circuits, providing isolation, high sink, and source current and usually three-state control.

Burn-In: A step near the end of the production process in which a manufacturer "exercises" the functions of an electronic circuit and "ages" the components before the final calibration step.

Clock: A device used to produce a periodic digital signal that repeatedly switches from LOW to HIGH and back at a predetermined rate.

Combinational Logic: The use of several of the basic gates (AND, OR, NOR, NAND) together to form more complex logic functions.

Comparator: As used in a 555 timer, it compares the analog voltage level at its two inputs and outputs a HIGH or a LOW, depending on which input was higher. (If the voltage level on the + input is higher than the voltage level on the − input, the output is HIGH; otherwise, it is LOW.)

Complement: Opposite digital state (i.e., the complement of 0 is 1, and vice versa).

Crystal: A material, usually made from quartz, that can be cut and shaped to oscillate at a very specific frequency. It is used in highly accurate clock and timing circuits.

Data Bus: A group of eight lines or electrical conductors usually connected to a microprocessor and shared by a number of other devices connected to it.

Digital State: The logic levels within a digital circuit (HIGH level = 1 state, and LOW level = 0 state).

Disabled: The condition in which a digital circuit's inputs or outputs are not allowed to accept or transmit digital states.

Duty Cycle: A ratio of the lengths of time that a digital signal is HIGH versus its total period:

$$\text{Duty cycle} = \frac{t_{HI}}{t_{HI} + t_{LO}}$$

Edge Triggered: The term given to a digital device that can accept inputs and change outputs only on the positive or negative *edge* of some input control signal or clock.

Enabled: The condition in which a digital circuit's inputs or outputs are allowed to accept or transmit digital states normally.

Flip-Flop: A circuit capable of storing a digital 1 or 0 level based on sequential digital levels input to it.

Float: A digital output level that is neither HIGH nor LOW but instead is in a *high-impedance state.* In this state, the output acts like a high impedance with respect to ground and will float to any voltage level that happens to be connected to it.

Function Chart: A diagram that illustrates all the possible combinations of input and output states for a given digital IC or device.

High-Impedance State: *See* Float.

Hold Time: The length of time *after* the active clock edge that the input data to be recognized (usually, *J* and *K*) must be held stable to ensure its recognition.

Latch: The ability to "hold" onto a particular digital state. A latch circuit will hold the level of a digital pulse even after the input is removed.

Level Triggered: *See* Pulse triggered.

Multivibrator: An electronic circuit or IC used in digital electronics to generate HIGH and LOW logic states. The *bistable* multivibrator is an *S-R* flip-flop triggered into its HIGH or LOW state. The *astable* multivibrator is a free-running oscillator that continuously alternates between its HIGH and LOW states. The *monostable* multivibrator is a one-shot that, when triggered, outputs a single pulse of a specific time duration.

Negative Edge: The edge on a clock or trigger pulse that is making the transition from HIGH to LOW.

Noise: Any fluctuations in power supply voltages, switching surges, or electrostatic charges will cause irregularities in the HIGH- and LOW-level voltages of a digital signal. These irregularities or fluctuations in voltage levels are called electrical "noise" and can cause false readings of digital levels.

Octal: When referring to an IC, octal means that there are *eight* logic devices in a single package.

Oscillator: An electronic circuit whose output continuously alternates between HIGH and LOW states at a specific frequency.

Output Enable: An input pin on an IC that can be used to enable or disable the outputs. When disabled, the outputs are in the "float condition."

Positive Edge: The edge of a clock or trigger pulse that is making the transition from LOW to HIGH.

Propagation Delay: The length of time that it takes for an input level change to pass through an IC and appear as a level change at the output.

Pull-Up Resistor: A resistor with one end connected to a HIGH voltage level and the other end connected to an input or output line, so that when that line is in a float condition (not HIGH or LOW), the voltage level on that line will instead be "pulled up" to a HIGH state.

Pulse Stretching: Increasing the time duration of a pulse width.

Pulse Triggered: The term given to a digital device that can accept inputs during an entire positive or negative pulse of some input control signal or clock. (Also called "level triggered.")

Race Condition: The condition that occurs when a digital input level (1 or 0) is changing states at the same time as the active clock edge of a synchronous device, making the input level at that time undetermined.

RC Circuit: A simple series circuit consisting of a resistor and a capacitor used to provide time delay.

Register: Two or more flip-flops (or storage units) connected as a group and operated simultaneously.

Reset: A condition that produces a digital LOW (0) state.

Sequential Logic: Digital circuits that involve the use of a sequence of timing pulses in conjunction with storage devices such as flip-flops and latches, and functional ICs such as counters and shift registers.

Set: A condition that produces a digital HIGH (1) state.

Setup Time: The length of time *prior to* the active clock edge that the input data to be recognized (usually, *J* and *K*) must be held stable to ensure its recognition. [That is, if the setup time of a device is 20 ns, the inputs must be held stable (and will be read) 20 ns before the trigger edge.]

SPST Switch: The abbreviation for single pole, single throw. An SPST switch is used simply to make or break contact in a single electrical line.

Switch Bounce: An undesirable characteristic of most switches because they will physically make and break contact several times (bounce), each time they are opened or closed.

Synchronous: A condition in which the output of a device will operate only in synchronization (in step with) a specific HIGH or LOW timing pulse or trigger signal.

Three-State Output: A feature on some ICs that allows you to connect several outputs to a common point. When one of the outputs is HIGH or LOW, all others will be in the float condition (the three output levels are HIGH, LOW, and float).

Toggle: In a flip-flop, a toggle is when Q changes to the level of $\overline{Q}$ and $\overline{Q}$ changes to the level of Q.

Transceiver: A data transmission device that is bidirectional, allowing data to flow through it in either direction.

Transition: The instant of change in digital state from HIGH to LOW or LOW to HIGH.

Transparent Latch: An asynchronous device whose outputs will "hold" onto the most recent digital state of the inputs. The outputs immediately follow the state of the inputs (transparent) while the trigger input is active, then latch onto that information when the trigger is removed.

Trigger: The input control signal to a digital device that is used to specify the instant that the device is to accept input or change outputs.

Voltage-Controlled Oscillator (VCO): An oscillator whose output frequency is dependent on the analog voltage level at its input.

PROBLEMS

5–1. Make the necessary connections to a 7400 quad NAND gate IC (Figure P5–1) to form the cross-NAND S-R flip-flop of Figure 5–2. *Hint:* An inverter can be formed from a NAND.

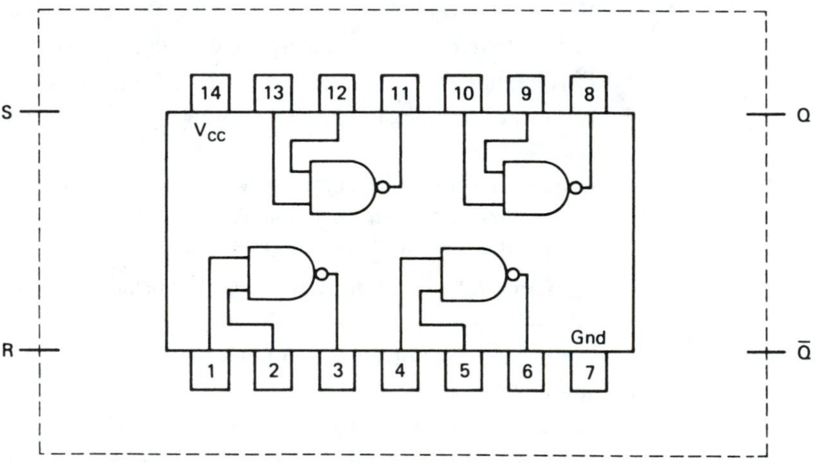

Figure P5–1

5–2. Sketch the Q output waveform for a gated S-R flip-flop (Figure 5–7) given the inputs at S, R, and G shown in Figure P5–2.

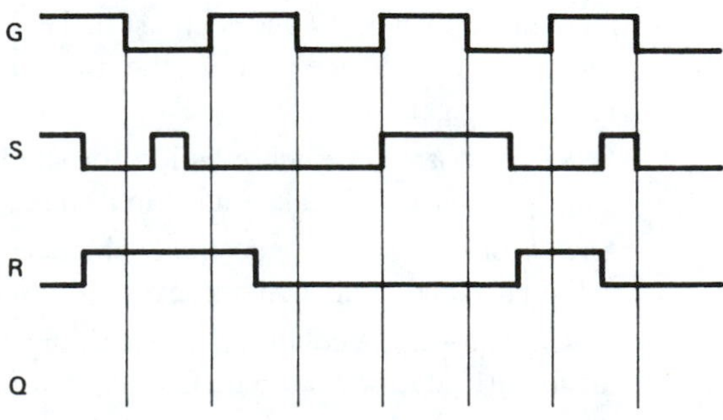

Figure P5–2

5–3. Sketch the Q output waveform for the gated D flip-flop of Figure 5–10 given the D and G inputs shown in Figure P5–3.

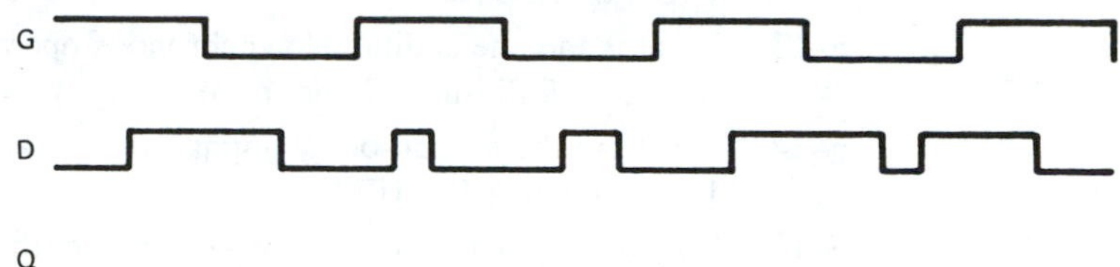

Figure P5–3

5–4. The logic symbol for one-fourth of a 7475 transparent D latch is given in Figure P5–4. Sketch the Q output waveform given the inputs at E and D.

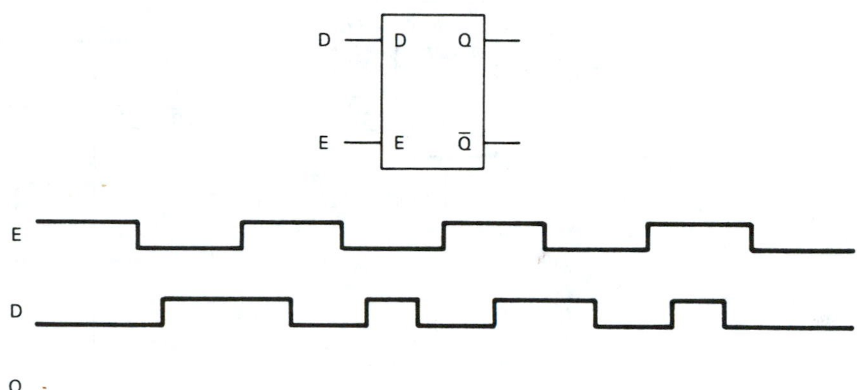

Figure P5–4

5–5. Explain why the 7475 is called "transparent" and why it is called a "latch."

5–6. The 7475 is transparent while the E input is ____________ (LOW or HIGH) and it is latched while E is ____________ (LOW or HIGH).

5–7. **(a)** What are the asynchronous inputs to the 7474 D flip-flop?

(b) What are the synchronous inputs to the 7474 D flip-flop?

5–8. The logic symbol for one-half of a 7474 dual D flip-flop is given in Figure P5–8. Sketch the Q output wave given the inputs at C_p, D, $\overline{S_D}$, and $\overline{R_D}$ shown in Figure P5–8b.

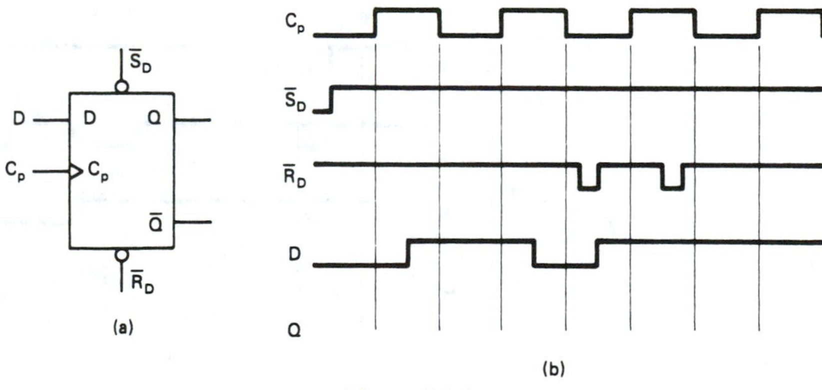

Figure P5–8

5–9. Describe several differences between the 7474 D flip-flop and the 7475 D latch.

5–10. Describe the differences between the asynchronous inputs and the synchronous inputs of the 7474.

5–11. What does the small triangle on the C_p line of the 7474 indicate?

5–12. To disable the asynchronous inputs to the 7474, should they be connected to a HIGH or a LOW?

5–13. What is the one additional synchronous operating mode that the *J-K* flip-flop has that the *S-R* flip-flop did not have?

5–14. What are the asynchronous inputs to the 74LS76 *J-K* flip-flop? Are they active-LOW or active-HIGH?

5–15. The logic symbol and input waveforms for the 74LS76 are given in Figure P5–15. Sketch the waveform at the *Q* output.

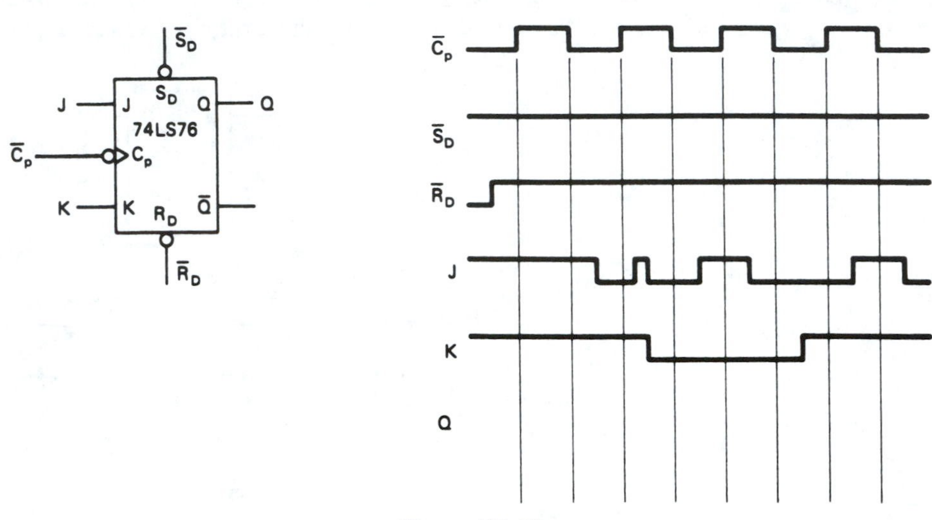

Figure P5–15

5–16. Repeat Problem 5–15 for the input waveforms shown in Figure P5–16.

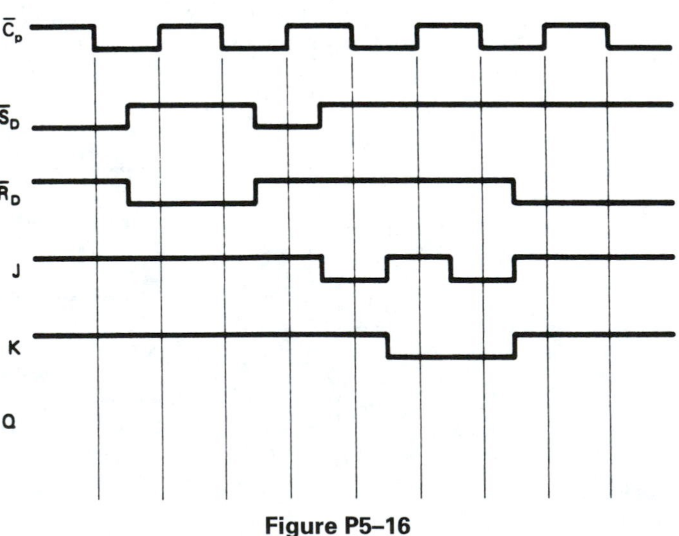

Figure P5–16

5–17. Sketch the output waveform at *Q* for Figure P5–17.

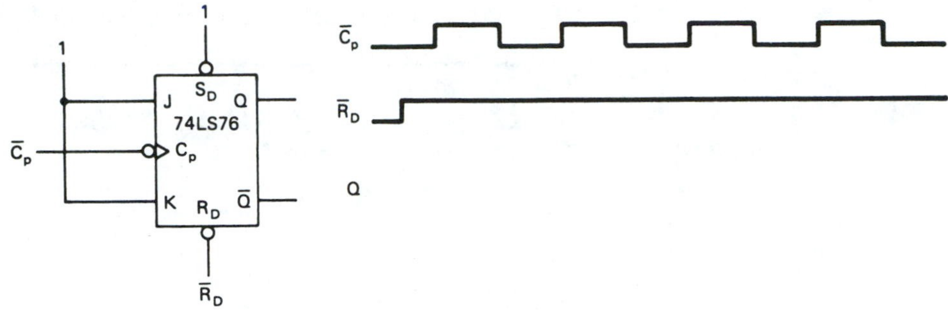

Figure P5–17

5–18. Sketch the output waveform at Q for Figure P5–18.

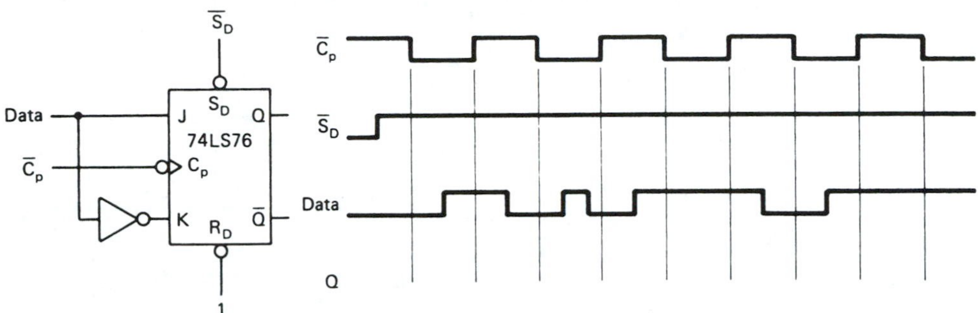

Figure P5–18

5–19. Sketch the output waveform at Q for Figure P5–19.

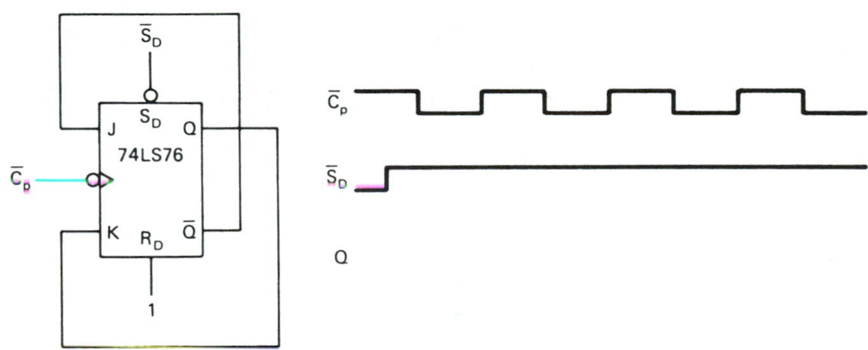

Figure P5–19

5–20. Sketch the Q output waveform for a 74LS76 given the input waveforms shown in Figure P5–20. (Ignore propagation delay times.)

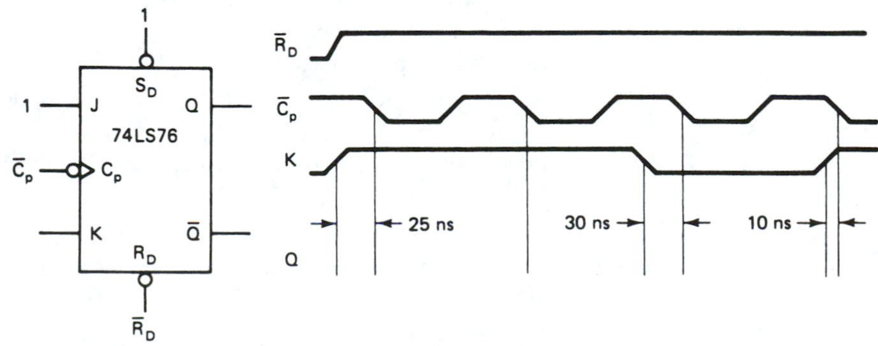

Figure P5–20

5–21. Repeat Problem 5–20 for the waveforms shown in Figure P5–21.

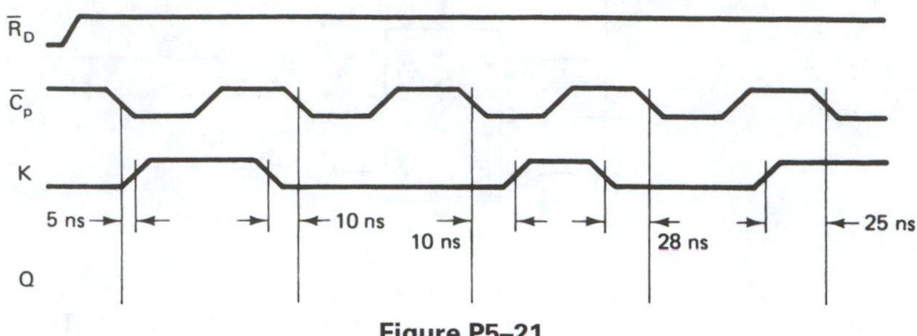

Figure P5–21

5–22. Using the specifications for the 74LS76, label the propagation delay times on the waveforms shown in Figure P5–22.

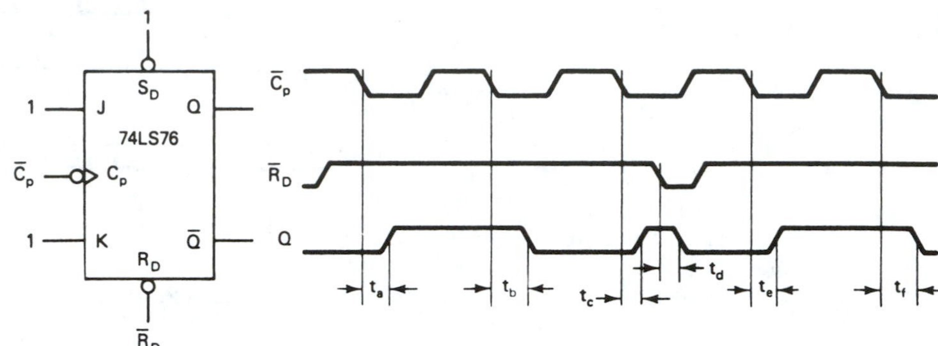

Figure P5–22

5–23. Describe the difference between the following: a buffer and a latch; a buffer and a transceiver.

5–24. Why is it important to use devices with three-state outputs when interfacing to a microprocessor data bus?

5–25. Why are designers of digital circuitry concerned about switch bounce?

Multivibrators

5–26. Which type of multivibrator is also known as a(n):
- **(a)** One-shot?
- **(b)** *S-R* flip-flop?
- **(c)** Free-running oscillator?

5–27. Why is a Schmitt trigger inverter used for the astable multivibrator circuit of Figure 5–55 instead of a regular inverter like a 74HC04?

5–28. In a Schmitt trigger astable multivibrator, if the hysteresis voltage ($V_{T+} - V_{T-}$) decreases due to a temperature change, what happens to:
- **(a)** The output frequency?
- **(b)** The output voltage?

5–29. Specifications for the 74HC14 Schmitt inverter when powered from a 6-V supply are as follows: $V_{OH} = 6.0$ V, $V_{OL} = 0.0$ V, $V_{T+} = 3.3$ V, and $V_{T-} = 2.0$ V.
- **(a)** Sketch and label the waveforms for V_{cap} and V_{out} in the astable multivibrator circuit of Figure 5–55. (Use $R = 68$ kΩ and $C = 0.0047$ μF.)
- **(b)** Calculate t_{HI}, t_{LO}, duty cycle, and frequency.

5–30. Make the external connections to a 74121 monostable multivibrator to convert a 100-kHz, 30% duty cycle square wave to a 100-kHz, 50% duty cycle square wave.

5–31. Use two 74121s as a delay line to reproduce the waveforms shown in Figure P5–31. (The output pulse will look just like the input pulse but delayed by 30 μs.)

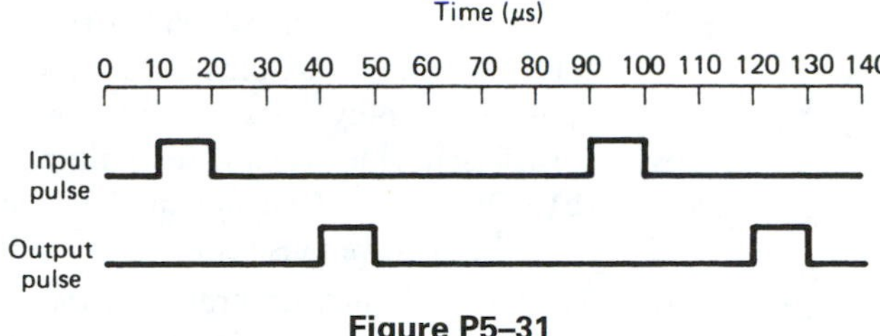

Figure P5–31

5–32. Sketch and label the waveforms at V_{out} for the 555 circuit of Figure P5–32 with the potentiometer set at 0 Ω.

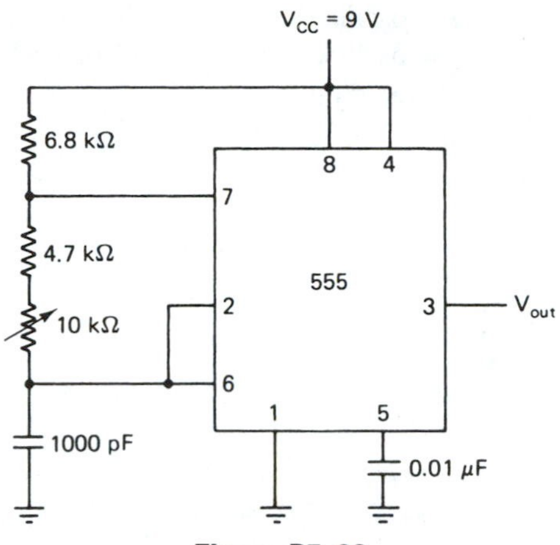

Figure P5–32

5–33. Determine the maximum and minimum frequency and the maximum and minimum duty cycle that can be achieved by adjusting the potentiometer in Figure P5–32.

5–34. Derive formulas for duty cycle and frequency in terms of R_A, R_B, and C for a 555 astable multivibrator. (Test your formulas by resolving Problem 5–33.)

5–35. Using a 555, design an astable multivibrator that will oscillate at 50 kHz, 60% duty cycle. (So that we all get the same answer, let's pick $C = 0.0022\ \mu$F.)

5–36. Sketch and label the waveforms at $V_{trigger}$, V_{cap}, and V_{out} for the 555 one-shot circuit of Figure 5–65. Assume that $V_{trigger}$ is a 5-μs negative-going pulse that occurs every 100 μs and $V_{CC} = 5$ V, $R_A = 47$ kΩ, $C = 1000$ pF.

SCHEMATIC INTERPRETATION PROBLEMS

5–37. Assume that the inverter U4:A in the Watchdog Timer schematic has the following propagation delay times: $t_{PLH} = 7.0$ ns, $t_{PLH} = 9.0$ ns. Also assume that WATCHDOG_CLK is a 10-MHz square wave. Sketch the waveforms at WATCHDOG_CLK and the input labeled CLK on U1:B on the same time axis.

5–38. Repeat Problem 5–37 with a 7404 used in place of the 74HC04. Assume that the 7404 has the following propagation delay times: $t_{PHL} = 15.0$ ns, $t_{PHL} = 22.0$ ns.

5–39. Find U1:A of the Watchdog Timer schematic. Assume that, initially, WATCHDOG_EN = LOW and /CPU_RESET is pulsed LOW.

(a) What is the output level of U2:A?
(b) When WATCHDOG_EN goes HIGH does the output of U2:A go LOW?
(c) What must happen to U1:A to make the output of U2:A go LOW?

5–40. In the Watchdog Timer schematic, both U14 flip-flops are Reset when there is a LOW /CPU_RESET (and, or) a LOW from U14:B.

5–41. After being Reset, U14:A will be set as soon as ________ .

5–42. Find the section of the Watchdog Timer schematic that shows U14 : A, U15 : A, and U14 : B. Assume that pins 2 and 13 of U15 : A are both HIGH and U14 : A is initially reset. Apply a positive pulse on the line labeled WATCHDOG_SEL.

(a) Discuss the possible setup time problems that may occur with U14 : B.

(b) Discuss how the situation changes if pin 1 of U15 : A is already HIGH and the positive pulse comes in on pin 2 instead.

5–43. The 68HC11 microcontroller in the HC11D0 master board schematic provides a clock output signal at the pin labeled E. This clock signal is used as the input to the LCD controller, M1 (grid location E-7). The frequency of this signal is 9.8304 MHz, as dictated by the crystal on the 68HC11. To experiment with different clock speeds on the LCD controller you want to design a variable frequency oscillator that can scan the frequency range of 100 kHz to 1 MHz. Design this oscillator using a 555 that will output its signal to pins 6 and 10 of the LCD controller.

Counter Circuits and Shift Registers

OBJECTIVES

Upon completion of this chapter, you should be able to:

- Use timing diagrams for the analysis of counter and shift register circuits.
- Design any modulus counter and frequency divider using counter ICs.
- Describe the difference between ripple counters and synchronous counters.
- Solve various counter design applications using 4-bit counter ICs and external gating.
- Connect seven-segment LEDs and BCD decoders to form multidigit numeric displays.
- Cascade counter ICs to provide for higher counting and frequency division.
- Connect *J-K* flip-flops as serial or parallel-in to serial or parallel-out multibit shift registers.
- Draw timing waveforms to illustrate shift register operation.
- Explain the operation and application of ring and Johnson shift counters.
- Make external connections of MSI shift register ICs to perform conversions between serial and parallel data formats.
- Discuss the operation of circuit design applications that employ shift registers.

INTRODUCTION

Now that we understand the operation of flip-flops and latches, we can apply our knowledge to the design and application of sequential logic circuits. One common application of sequential logic arrives from the need to count events and time the duration of various processes. These applications are called *sequential* because they follow a predetermined sequence of digital states and are triggered by a timing pulse or clock.

To be useful in digital circuitry and microprocessor systems, counters normally count in binary and can be made to stop or recycle to the beginning at any time. In a recycling counter, the number of different binary states defines the modulus (MOD) of the counter. For example, a counter that counts from 0 to 7 is called a MOD-8 counter. For a counter to count from 0 to 7 it must have three binary outputs and one clock trigger input, as shown in Figure 6–1.

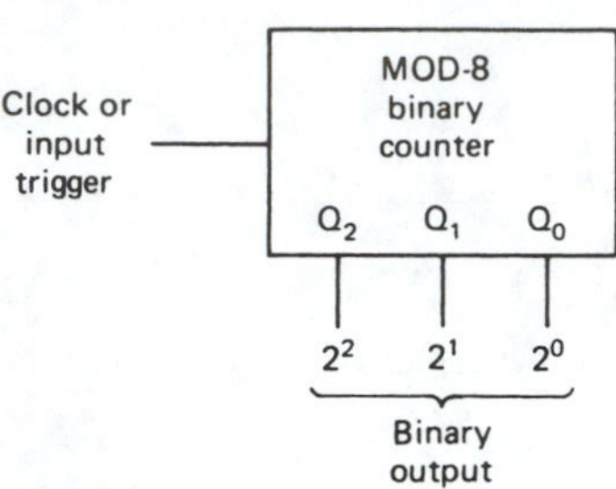

Figure 6–1 Simplified block diagram of a MOD-8 binary counter.

Normally, each binary output will come from the Q output of a flip-flop. Flip-flops are used because they can hold, or remember, a binary state until the next clock or trigger pulse comes along. The count sequence of a 1 to 7 binary counter is shown in Table 6–1 and Figure 6–2.

TABLE 6–1
Binary Count Sequence of a MOD-8 Binary Counter

Q_2	Q_1	Q_0	Count	
0	0	0	0	eight different binary states
0	0	1	1	
0	1	0	2	
0	1	1	3	
1	0	0	4	
1	0	1	5	
1	1	0	6	
1	1	1	7	
0	0	0	0	
0	0	1	1	
0	1	0	2	
0	1	1	3	
	Etc.			

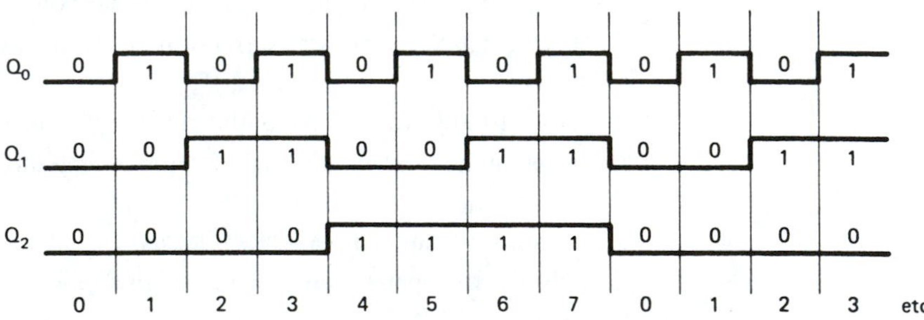

Figure 6–2 Waveforms for a MOD-8 binary counter.

The other group of sequential circuits covered in this chapter are called *shift registers*. Registers are required in digital systems for the temporary storage of a group of bits. Data bits (1s and 0s) traveling through a digital system sometimes have to be temporarily stopped, copied, moved, or even shifted to the right or left one or more positions.

A shift register facilitates this movement and storage of data bits. Most shift registers can handle parallel movement (simultaneous, as a group) of data bits, as well as serial movement (one at a time, on a single line). They can also be used to convert from parallel to serial and serial to parallel.

6–1 RIPPLE COUNTERS

Flip-flops can be used to form binary counters. The counter output waveforms discussed in the beginning of this chapter (Figure 6–2) could be generated by using three flip-flops cascaded together (cascaded means connecting the Q output of one flip-flop to the clock input of the next). Three flip-flops are needed to form a 3-bit counter (each flip-flop will represent a different power of 2: 2^2, 2^1, 2^0). With three flip-flops we can produce 2^3 different combinations of binary outputs ($2^3 = 8$). The eight different binary outputs from a 3-bit binary counter will be 000, 001, 010, 011, 100, 101, 110, 111.

If we have a 4-bit binary counter, we would count from 0000 up to 1111, which is 16 different binary outputs. As it turns out, we can determine the number of different binary output states (modulus) by using the following formula:

$$\text{Modulus} = 2^N \qquad \text{where } N = \text{number of flip-flops}$$

To form a 3-bit binary counter, we cascade three *J-K* flip-flops, each operating in the *toggle mode,* as shown in Figure 6–3. The clock input used to increment the binary count comes into the $\overline{C_p}$ input of the first flip-flop. Each flip-flop will toggle every time its clock input receives a HIGH-to-LOW edge.

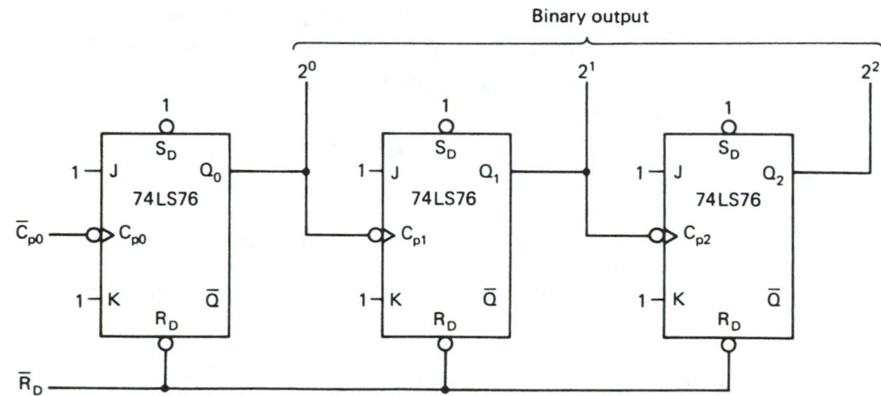

Figure 6–3 Three-bit binary ripple counter.

When we analyze the circuit and waveforms, we see that Q_0 toggles at each negative edge of $\overline{C_{p0}}$, Q_1 toggles at each negative edge of Q_0, and Q_2 toggles at each negative edge of Q_1. The result is that the outputs will "count" from 000 up to 111, then 000 to 111 repeatedly, as shown in Figure 6–4. The term *ripple* is derived from the fact that the input clock trigger is not connected to each flip-flop directly but instead has to propagate down through each flip-flop to reach the next.

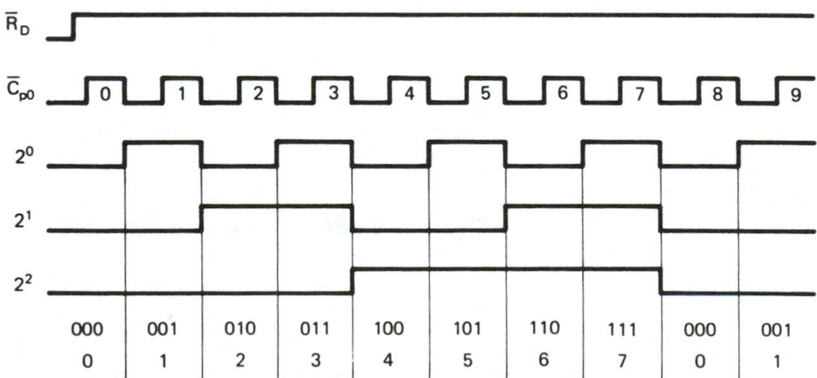

Figure 6–4 Waveforms generated from the 3-bit binary ripple counter.

For example, look at clock pulse 7. The negative edge of $\overline{C_{p0}}$ causes Q_0 to toggle LOW . . . which causes Q_1 to toggle LOW . . . which causes Q_2 to toggle LOW. There will definitely be a propagation delay between the time that $\overline{C_{p0}}$ goes LOW until Q_2 finally goes LOW. Because of this delay, ripple counters are called *asynchronous counters,* which means that each flip-flop is not triggered at exactly the same time.

Synchronous counters can be formed by driving each flip-flop's clock by the same clock input. Synchronous counters are more complicated, however, and will be covered after we have a thorough understanding of asynchronous ripple counters.

The propagation delay inherent in ripple counters places limitations on the maximum frequency allowed by the input trigger clock. The reason is that if the input clock has an active trigger edge before the previous trigger edge has propagated through all the flip-flops, you will get an erroneous binary output.

Let's look at the 3-bit counter waveforms in more detail, now taking into account the propagation delays of the 74LS76 flip-flops. In reality, the 2^0 waveform will be delayed to the right ("skewed") by the propagation of the first flip-flop. The 2^1 waveform will be skewed to the right from the 2^0 waveform, and the 2^2 waveform will be skewed to the right from the 2^1 waveform. This is an accumulative effect that causes the 2^2 waveform to be skewed to the right of the original $\overline{C_{p0}}$ waveform by three propagation delays. [Remember, however, that the propagation delay for most flip-flops is in the 20-ns range, which will not hurt us until the input clock period is very short, 100 to 200 ns (5 to 10 MHz).] Figure 6–5 illustrates the effect of propagation delay on the output waveform.

From Figure 6–5 we can see that the length of time that it takes to change from binary 011 to 100 (3 to 4) will be

$$t_{PHL1} + t_{PHL2} + t_{PLH3} = 30 + 30 + 20 = 80 \text{ ns}$$

As we cascade more and more flip-flops to form higher modulus counters, the accumulative effect of the propagation delay becomes more of a problem.

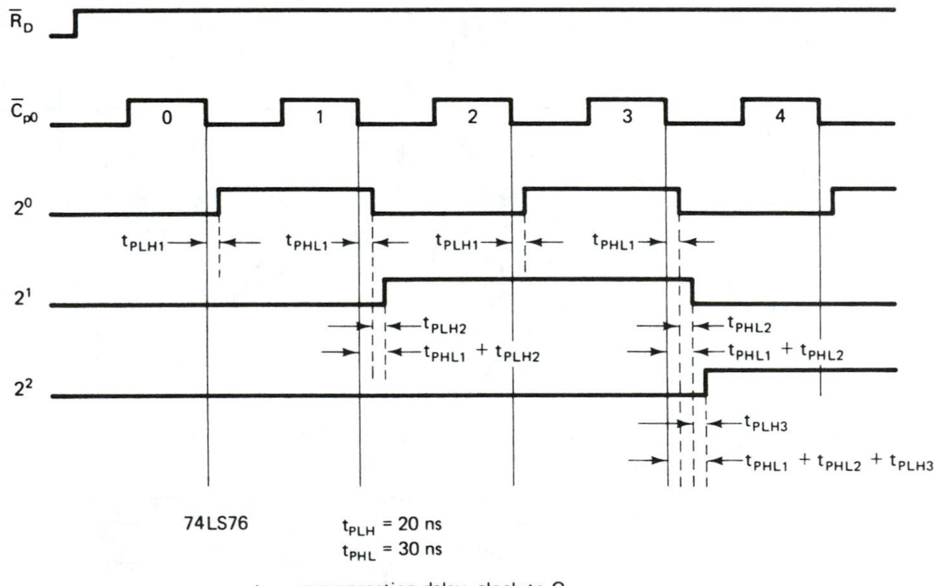

Figure 6–5 Effect of propagation delay on ripple counter outputs.

A MOD-16 ripple counter can be built using four ($2^4 = 16$) flip-flops. Figures 6–6 and 6–7 show the circuit design and waveforms for a MOD-16 ripple counter. From the waveforms we can see that the 2^1 line toggles at every negative edge of the 2^0 line, the 2^2 line toggles at every negative edge of the 2^1 line, and so on down through each successive flip-flop. When the count reaches 15 (1111), the next negative edge of $\overline{C_{p0}}$ causes all four flip-flops to toggle and changes the count to 0 (0000).

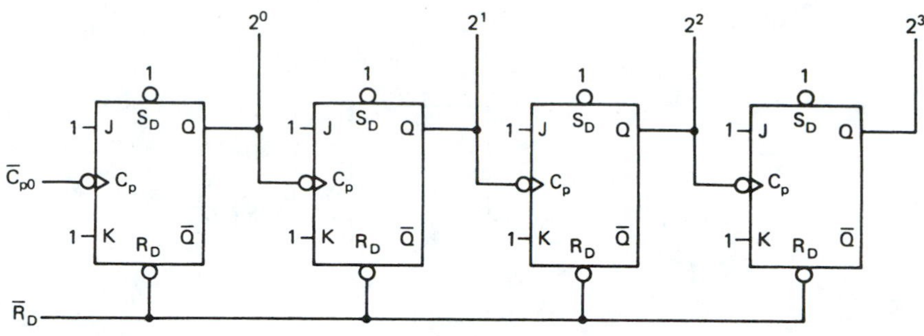

Figure 6–6 MOD-16 ripple counter.

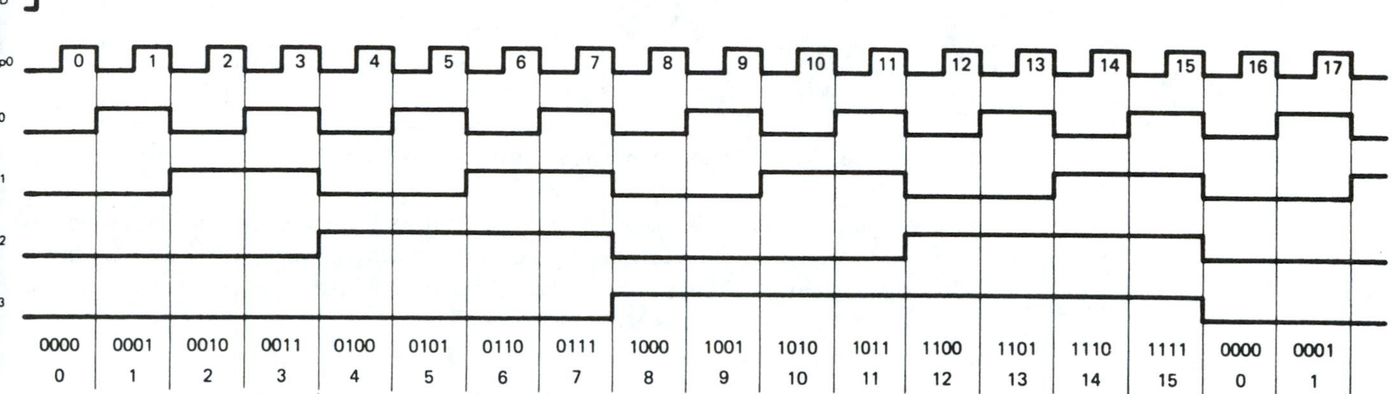

Figure 6–7 MOD-16 ripple counter waveforms.

Down-Counters

On occasion there is a need to count down in binary instead of counting up. To form a down-counter, simply take the binary outputs from the $\overline{Q}$ outputs instead of the Q outputs.

6–2 DESIGN OF DIVIDE-BY-N COUNTERS

Counter circuits are also used as frequency dividers to reduce the frequency of periodic waveforms. For example, if we study the waveforms generated by the MOD-8 counter of Figure 6–4, we can see that the frequency of the 2^2 output line is one-eighth of the frequency of the $\overline{C_{p0}}$ input clock line. This concept is illustrated in the block diagram of Figure 6–8, assuming that the input frequency is 24 kHz. So as it turns out, a MOD-8 counter can be used as a divide-by-8 frequency divider and a MOD-16 can be used as a divide-by-16 frequency divider.

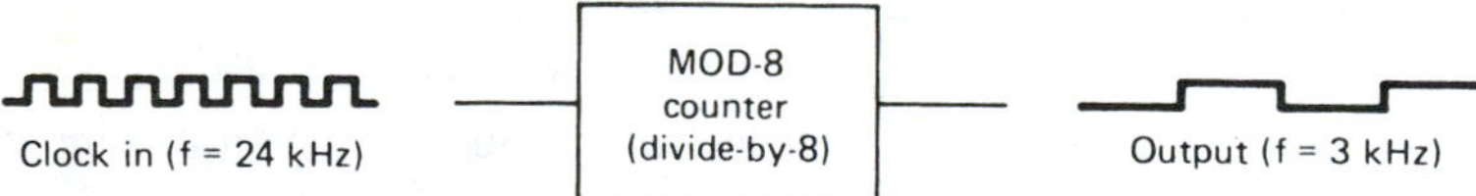

Figure 6–8 Block diagram of a divide-by-8 counter.

What if we need a divide-by-5 (MOD-5) counter? We can modify the MOD-8 counter so that when it reaches the number 5 (101) all flip-flops will be Reset. The new count sequence will be 0–1–2–3–4–0–1–2–3–4–0–etc. To get the counter to Reset at number 5 (binary 101), you will have to monitor the 2^0 and 2^2 lines and when they are both HIGH, put out a LOW Reset pulse to all flip-flops. Figure 6–9 shows a circuit that can do that for us.

As you can see, the inputs to the NAND gate are connected to the 2^0 and 2^2 lines, so that when the number 5 (101) comes up, the NAND puts out a LOW level to Reset all flip-flops. The waveforms in Figure 6–10 illustrate the operation of the MOD-5 counter of Figure 6–9.

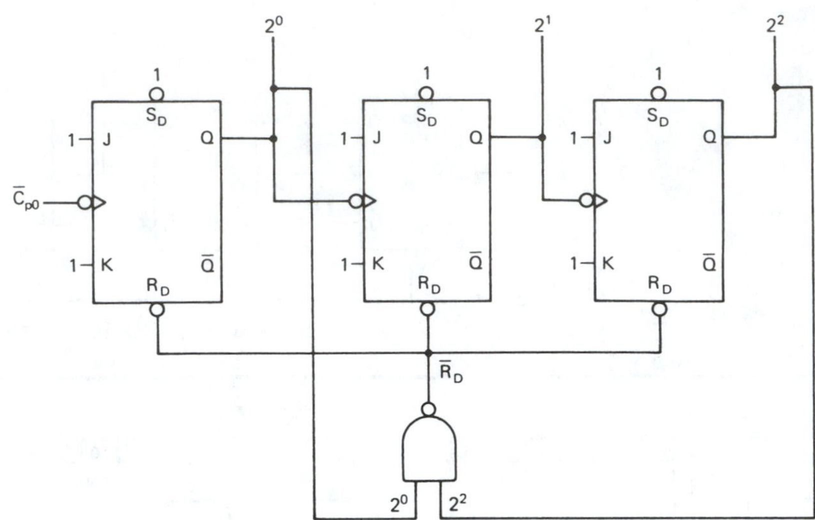

Figure 6–9 Connections to form a divide-by-5 (MOD-5) binary counter.

As we can see in Figure 6–10, the number 5 will appear at the outputs for a short duration, just long enough to Reset the flip-flops. The resulting short pulse on the 2^0 line is called a *glitch*. Do you think you could determine how long the glitch is? (Assume that the flip-flop is a 74LS76 and the NAND gate is a 7400.)

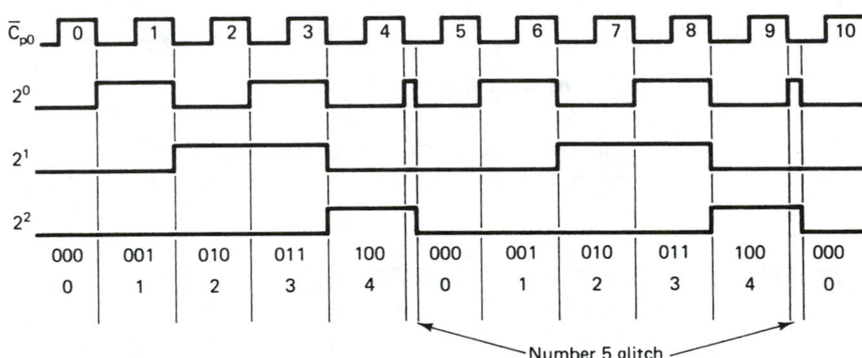

Figure 6–10 Waveforms for the MOD-5 counter.

Since t_{PHL} of the NAND gate is 15 ns, it takes that long just to drive the $\overline{R_D}$ inputs LOW. But then it also takes 30 ns (t_{PHL}) for the LOW on $\overline{R_D}$ to Reset the Q output to LOW. Therefore, the total length of the glitch is 45 ns. If the input clock period is in the microsecond range, then 45 ns is insignificant, but at extremely high clock frequencies that glitch could give us erroneous results.

Any modulus counter (divide-by-N counter) can be formed by using external gating to Reset at a predetermined number. The following examples illustrate the design of some other divide-by-N counters.

EXAMPLE 6–1

Design a MOD-6 ripple up-counter that can be manually Reset by an external pushbutton.

Solution:

The ripple up-counter is shown in Figure 6–11. The count sequence will be 0–1–2–3–4–5. When 6 (binary 110) is reached, the output of the AND gate will go HIGH, causing the NOR gate to put a LOW on the $\overline{R_D}$ line, resetting all flip-flops to 0.

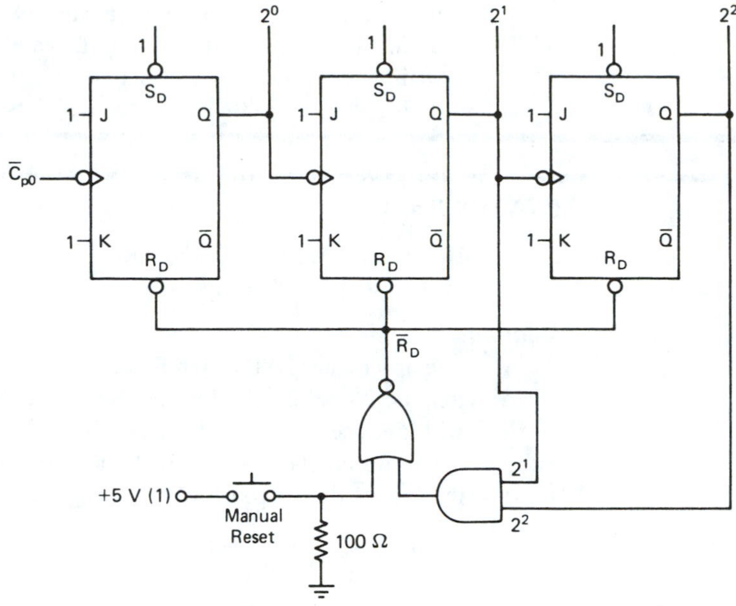

Figure 6–11

As soon as all outputs return to 0, the AND gate will go back to a LOW output, causing the NOR and $\overline{R_D}$ to return to a HIGH, allowing the counter to count again.

This cycle continues to repeat until the manual Reset pushbutton is pressed. The HIGH from the pushbutton will also cause the counter to Reset. The 100-Ω pull-down resistor will keep the input to the NOR gate LOW when the pushbutton is in the open position. [I_{IL}(NOR) $= -1.6$ mA, $V_{100\Omega} = 1.6 \text{ mA} \times 100\ \Omega = 0.160 \text{ V} \equiv \text{LOW}$.]

EXAMPLE 6–2

Design a MOD-10 ripple up-counter with a manual pushbutton Reset.

Solution:

The ripple up-counter is shown in Figure 6–12. Four flip-flops are required to give us a possibility of $2^4 = 16$ binary states ($2^3 = 8$ would not be enough). We

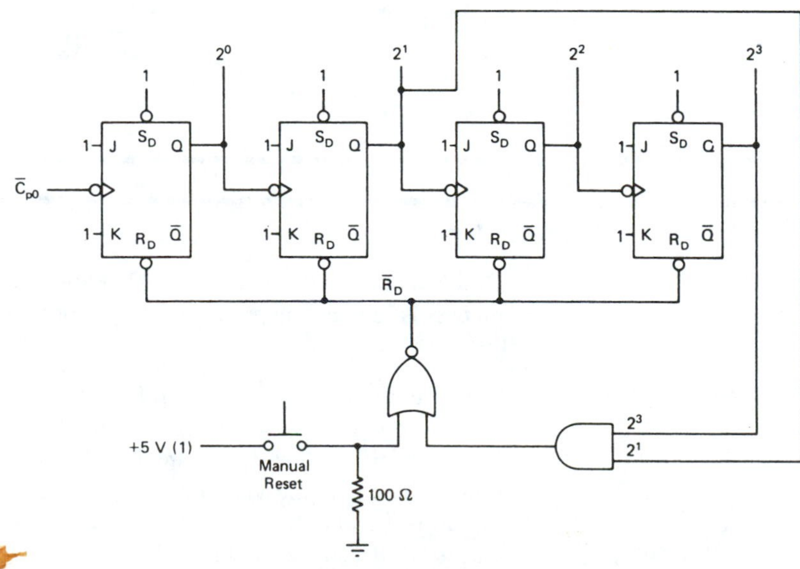

Figure 6–12

want to stop the count and automatically Reset when 10 (binary 1010) is reached. This is taken care of by the AND gate feeding into the NOR, making the $\overline{R_D}$ line go LOW when 10 is reached. The count sequence will be 0–1–2–3–4–5–6–7–8–9–0–1–etc., which is a MOD-10 up-counter.

EXAMPLE 6–3

Design a MOD-5 up-counter that counts in the sequence 6–7–8–9–10–6–7–8–9–10–6–etc.

Solution:

The up-counter is shown in Figure 6–13. By pressing the manual Preset pushbutton, the 2^1 and 2^2 flip-flops get Set while the 2^0 and 2^3 flip-flops get Reset. This will give the number 6 (binary 0110) at the output. In the count mode, when the count reaches 11 (binary 1011), the output of the AND gate goes HIGH, causing the $\overline{\text{Preset}}$ line to go LOW, recycling the count to 6 again.

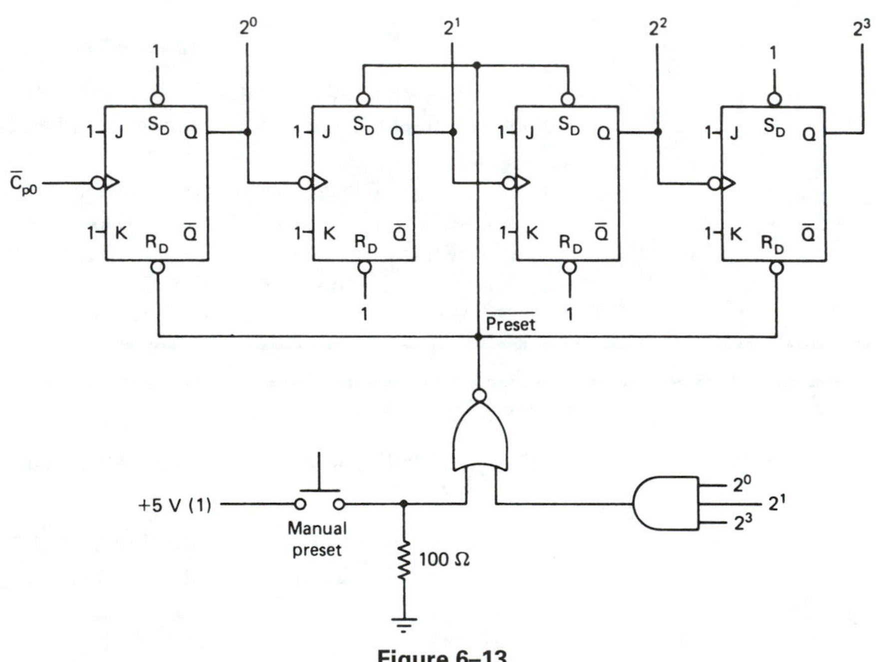

Figure 6–13

EXAMPLE 6–4

Design a counter that counts 0–1–2–3–4–5, then stops and turns on an LED. The process is initiated by pressing a "start" pushbutton.

Solution:

The required counter is shown in Figure 6–14. When power is first applied to the circuit ("power-up"), the capacitor will charge up toward 5 V. It starts out at a 0 level, however, which causes the 7474 to Reset ($Q_D = 0$). The LOW at Q_D will remain there until the start button is pressed. With a LOW at Q_D the three counter flip-flops are all held in the Reset state (binary 000). The output of the NAND gate is HIGH, so the LED is OFF.

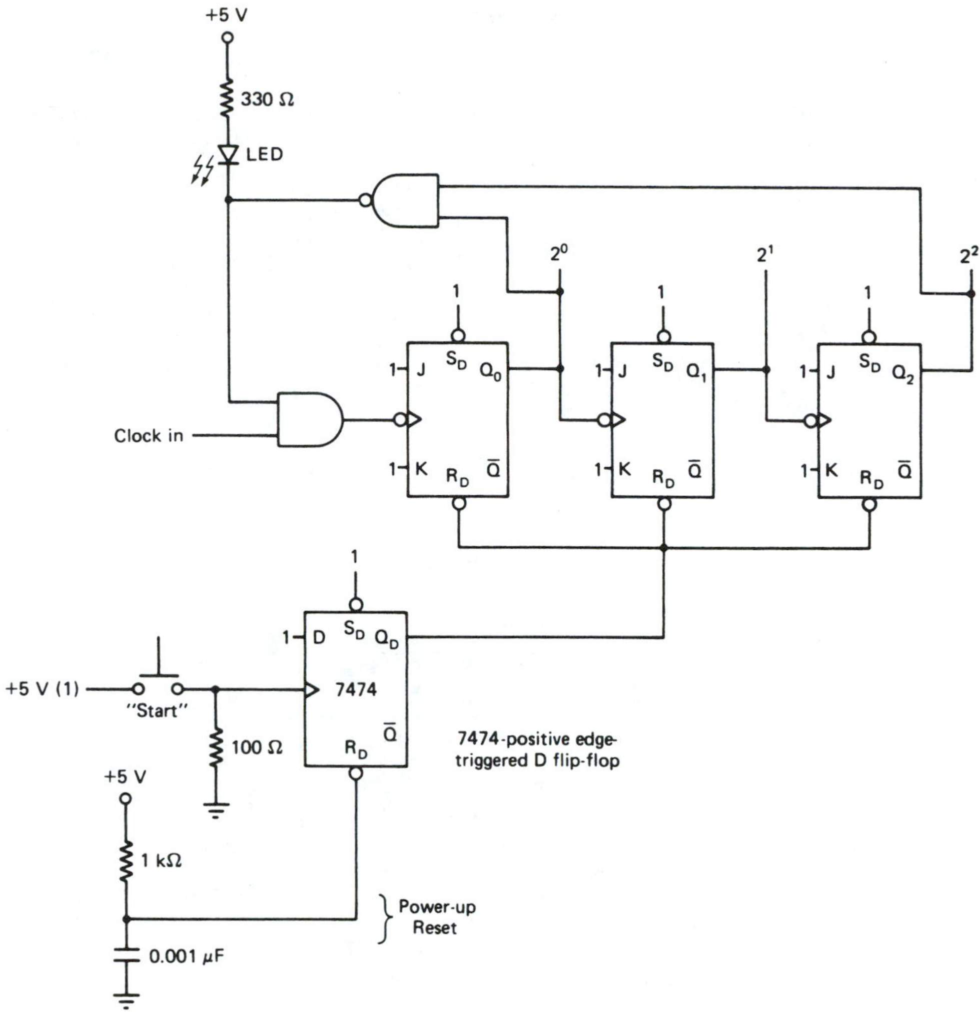

Figure 6–14

When the start button is pressed, Q_D goes HIGH and stays HIGH after the button starts bouncing and is released. With Q_D HIGH the counter begins counting: 0–1–2–3–4–5. When 5 is reached, the output of the NAND gate goes LOW, turning on the LED. The current through the LED will be (5 V − 1.7 V)/330 Ω = 10 mA. The NAND gate can sink a maximum of 16 mA (I_{OL} = 16 mA), so 10 mA will not burn it out.

The LOW output of the NAND gate is also fed to the input of the AND gate, which will disable the clock input. Since the clock cannot get through the AND gate to the first flip-flop, the count stays at 5 and the LED stays lit.

If you want to Reset the counter to 0 again, you could put a pushbutton in parallel across the capacitor so that when it was pressed, Q_D would go LOW and stay LOW until the "start" button was pressed again.

6–3 RIPPLE COUNTER INTEGRATED CIRCUITS

Four-bit binary ripple counters are available in a single integrated-circuit package. The most popular are the 7490, 7492, and 7493 TTL ICs.

Figure 6–15 shows the internal logic diagram for the 7493 4-bit binary ripple counter. The 7493 has four *J-K* flip-flops in a single package. It is divided into two

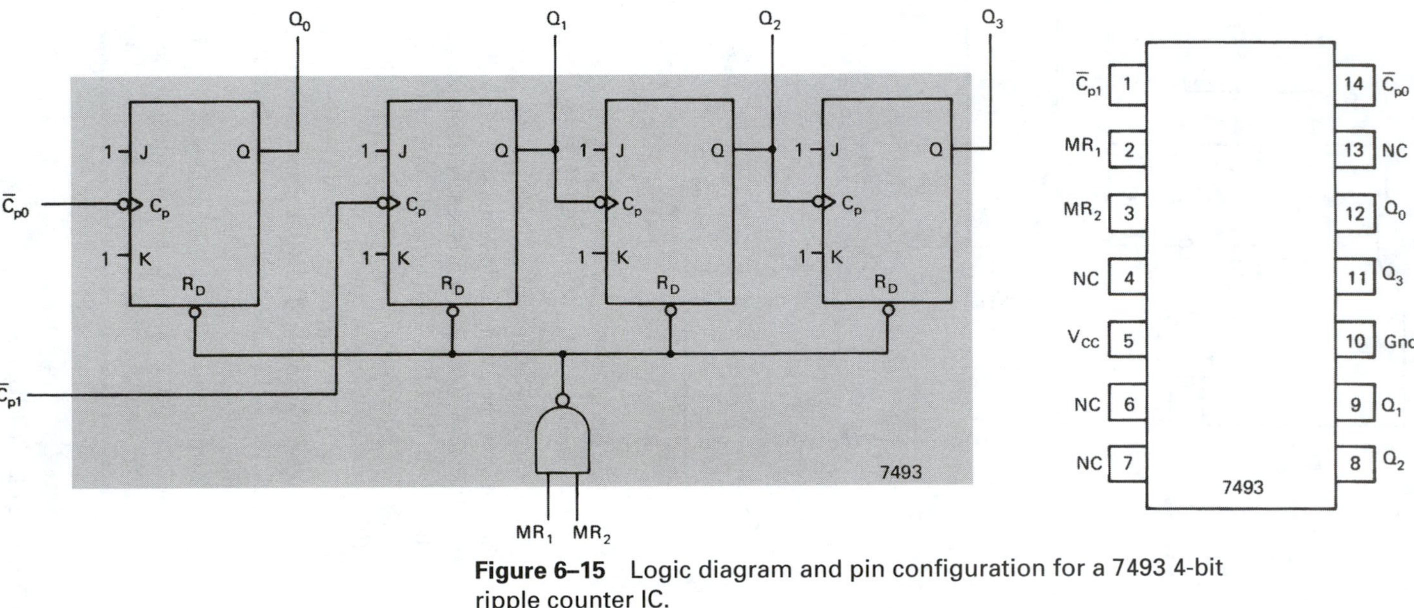

Figure 6–15 Logic diagram and pin configuration for a 7493 4-bit ripple counter IC.

sections: a divide-by-2 and a divide-by-8. The first flip-flop provides the divide-by-2 with its $\overline{C_{p0}}$ input and Q_0 output. The second group of three flip-flops are cascaded to each other and provide the divide-by-8 via the $\overline{C_{p1}}$ input and Q_1 Q_2 Q_3 outputs. To get a divide-by-16 you can *externally* connect Q_0 to $\overline{C_{p1}}$ so that all four flip-flops are cascaded end to end as shown in Figure 6–16. Note that two Master Reset inputs (MR_1, MR_2) are provided to asynchronously Reset all four flip-flops. When MR_1 and MR_2 are both HIGH, all Qs will be Reset to 0. (MR_1 to MR_2 must be held LOW to enable the count mode.)

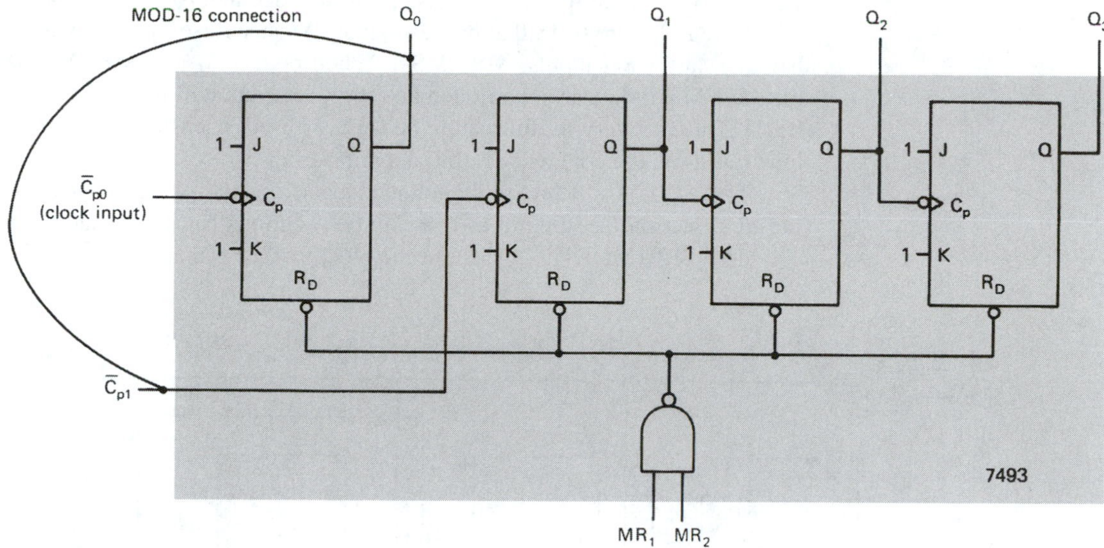

Figure 6–16 A 7493 connected as a MOD-16 ripple counter.

With the MOD-16 connection, the frequency output at Q_0 is equal to one-half the frequency input at $\overline{C_{p0}}$. Also, $f_{Q1} = \frac{1}{4}f_{\overline{Cp0}}$, $f_{Q2} = \frac{1}{8}f_{\overline{Cp0}}$, and $f_{Q3} = \frac{1}{16}f_{\overline{Cp0}}$.

The 7493 can be used to form any modulus counter less than or equal to MOD-16 by utilizing the MR_1 and MR_2 inputs. For example, to form a MOD-12 counter, simply make the external connections shown in Figure 6–17.

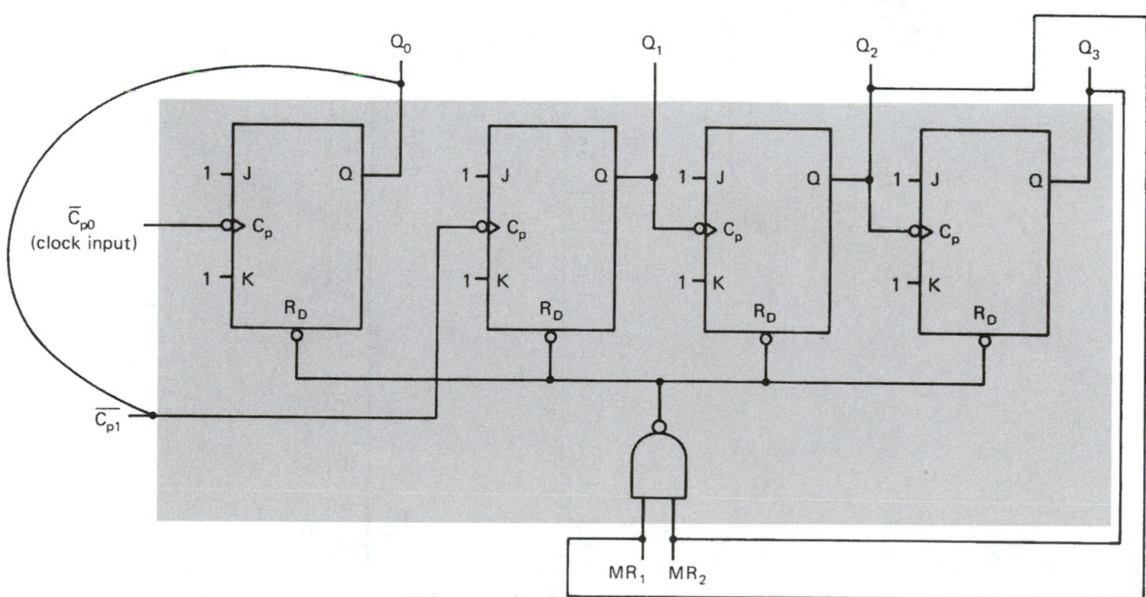

Figure 6–17 External connections to a 7493 to form a MOD-12 counter.

The count sequence of the MOD-12 counter will be 0–1–2–3–4–5–6–7–8–9–10–11–0–1–etc. Each time 12 (1100) tries to appear at the outputs, a HIGH–HIGH is placed on MR_1–MR_2 and the counter resets to 0.

Two other common ripple counter ICs are the 7490 and 7492. They both have four internal flip-flops like the 7493 but through the application of internal gating, they automatically recycle to 0 after 9 and 11, respectively.

The 7490 is a 4-bit ripple counter consisting of a divide-by-2 section and a divide-by-5 section (see Figure 6–18). The two sections can be cascaded together to form a divide-by-10 (decade or BCD) counter by connecting Q_0 to $\overline{C_{p1}}$ externally. The 7490 is most commonly used for applications requiring a decimal (0 to 9) display.

Note in Figure 6–18 that besides having Master Reset inputs (MR_1–MR_2) the 7490 also has Master Set inputs (MS_1–MS_2). When both MS_1 and MS_2 are made HIGH, the clock and MR inputs are overridden and the Q outputs will be asynchronously Set to a 9 (1001). This is a very useful feature because, if used, it ensures that *after* the first active clock transition the counter will start counting from 0.

The 7492 is a 4-bit ripple counter consisting of a divide-by-2 section and a divide-by-6 section (see Figure 6–19). The two sections can be cascaded together to form a divide-by-12 (MOD-12) by connecting Q_0 to $\overline{C_{p1}}$ and using $\overline{C_{p0}}$ as the clock input.

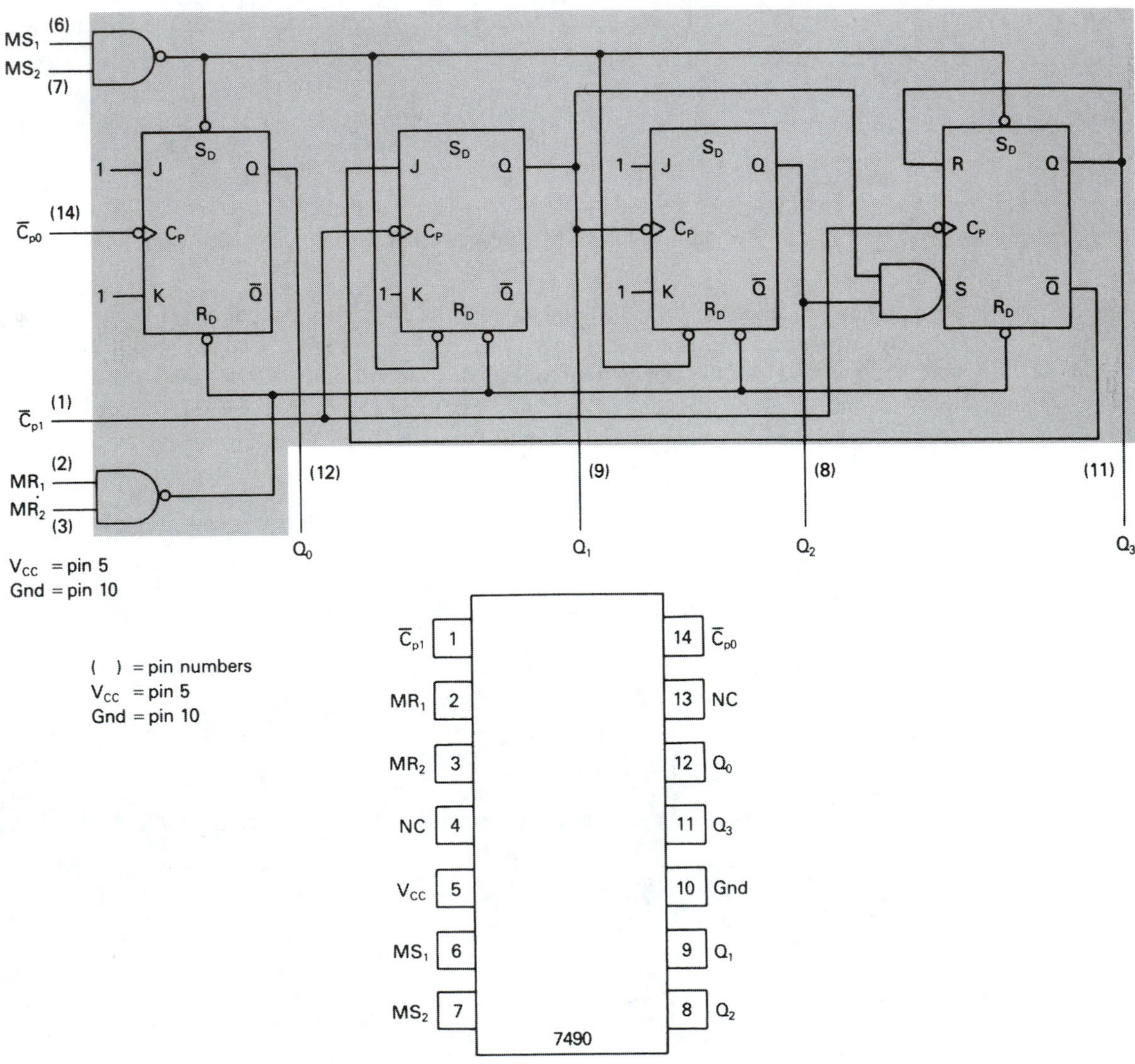

Figure 6–18 Logic diagram and pin configuration for a 7490 decade counter.

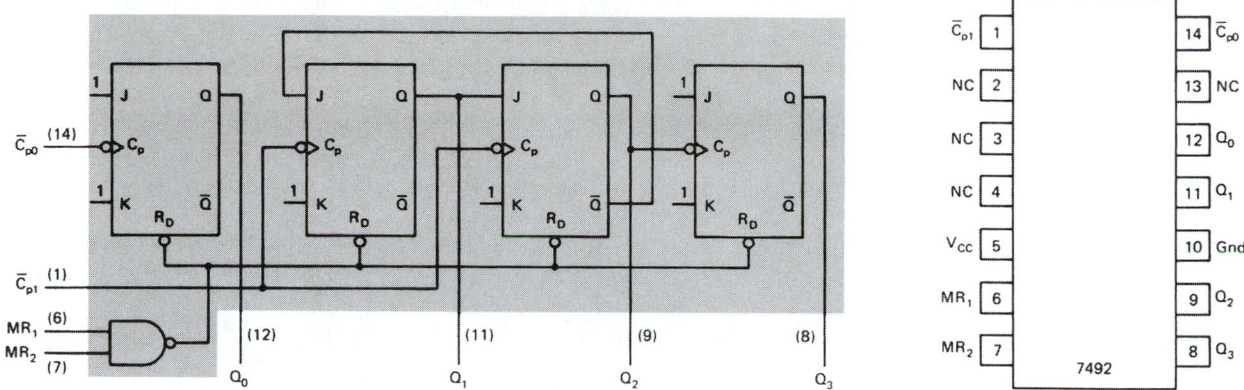

Figure 6–19 Logic diagram and pin configuration for a 7492 counter.

The 7492 is most commonly used for applications requiring MOD-12 and MOD-6 frequency dividing such as in digital clocks. You can get a divide-by-6 frequency divider simply by ignoring the $\overline{C_{p0}}$ input of the first flip-flop and, instead, bringing the clock input into $\overline{C_{p1}}$, which is the input to the divide-by-6 section. (One peculiarity of the 7492 is that when connected as a MOD-12, it does *not* count sequentially from 0 to 11. Instead, it counts from 0 to 13, skipping 6 and 7, but still functions as a divide-by-12.)

EXAMPLE 6–5

Make the necessary external connections to a 7490 to form a MOD-10 counter.

Solution:

The MOD-10 counter is shown in Figure 6–20.

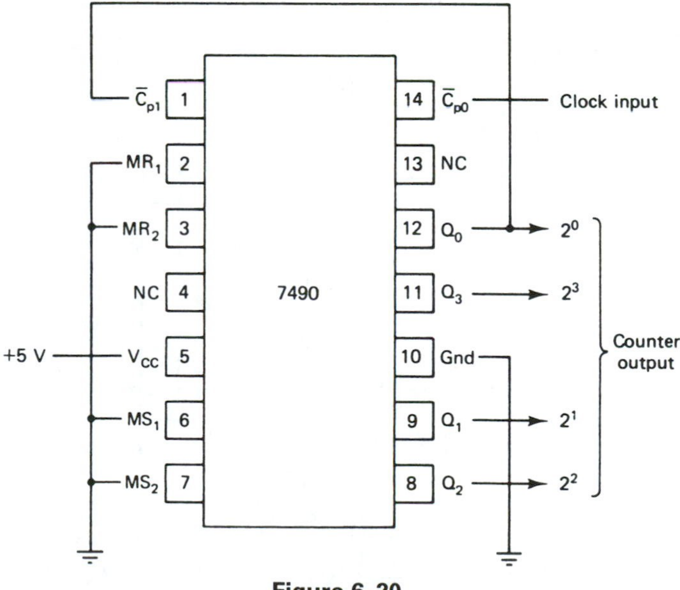

Figure 6–20

EXAMPLE 6–6

Make the necessary external connections to a 7492 to form a divide-by-6 frequency divider ($f_{out} = \frac{1}{6}f_{in}$).

Solution:

The frequency divider is shown in Figure 6–21.

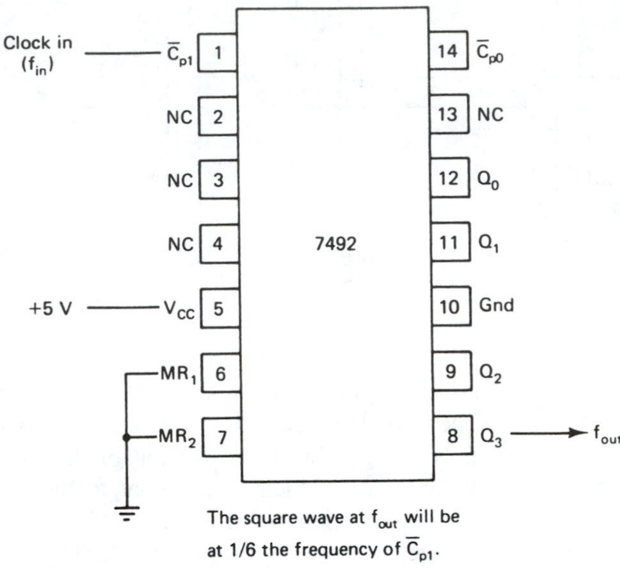

Figure 6–21

EXAMPLE 6–7

Make the necessary external connections to a 7490 to form a MOD-8 counter (0 to 7). Also, upon initial power-up, Set the counter at 9 so that after the first active input clock edge the output will be 0 and the count sequence will proceed from there.

Solution:

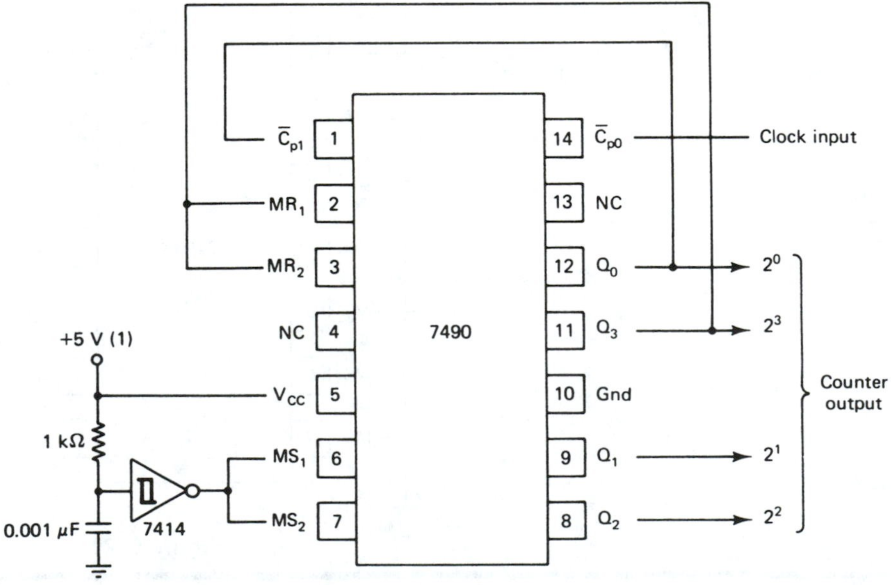

Figure 6–22

The MOD-8 counter is shown in Figure 6–22. The output of the 7414 Schmitt inverter will initially be HIGH when power is first turned on because the capacitor feeding its input is initially discharged to 0. This HIGH on MS_1 and MS_2 will Set the counter to 9. Then, as the capacitor charges up above 1.7 V, the Schmitt will switch to a LOW output, allowing the counter to start its synchronous counting sequence.

Q_0 is connected to $\overline{C_{p1}}$ so that all four flip-flops are cascaded. When the counter reaches 8 (1000), the 2^3 line, which is connected to MR_1 and MR_2, causes the counter to Reset to 0. The counter will continue to count in the sequence 0–1–2–3–4–5–6–7–0–1–2–etc. continuously.

6–4 SYSTEM DESIGN APPLICATIONS FOR COUNTER ICs

Integrated-circuit counter chips are used in a multitude of applications dealing with timing operations, counting, sequencing, and frequency division. To implement a complete system application, output devices such as light-emitting diode (LED) indicators, seven-segment LED displays, relay drivers, and alarm buzzers must be configured to operate from the counter outputs. The synchronous and asynchronous inputs can be driven by such devices as a clock oscillator, a pushbutton switch, the output from another digital IC, or control signals provided by a microprocessor.

APPLICATION 6–1

For example, let's consider an application that requires an LED indicator to illuminate for 1 s once every 13 s to signal an assembly line worker to perform some manual operation.

Solution:

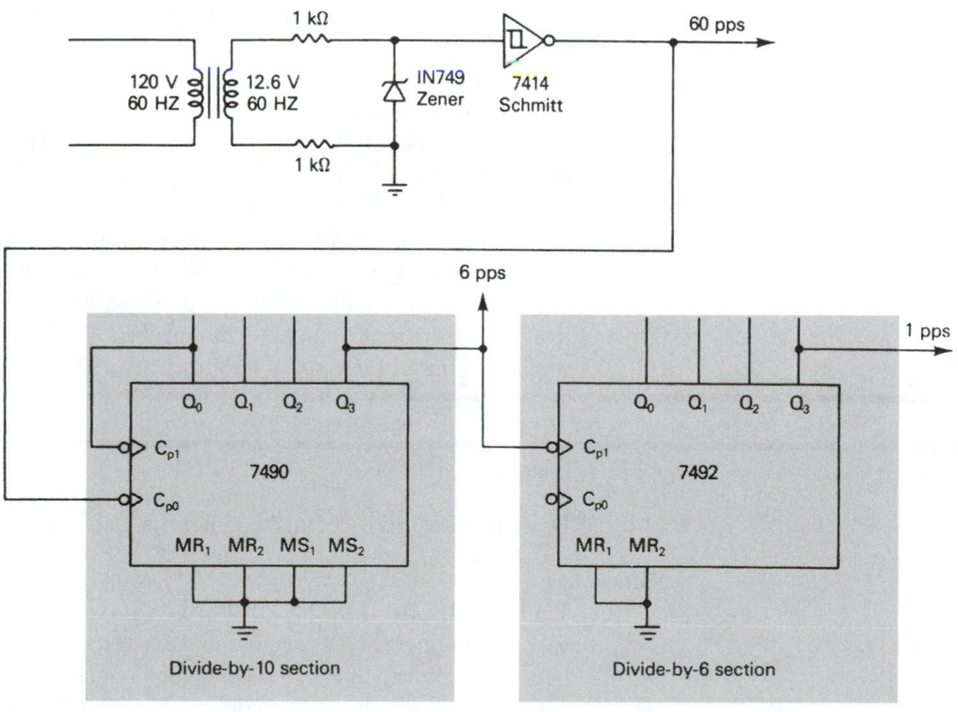

Figure 6–23 (a) Circuit used to produce 1 pulse per second

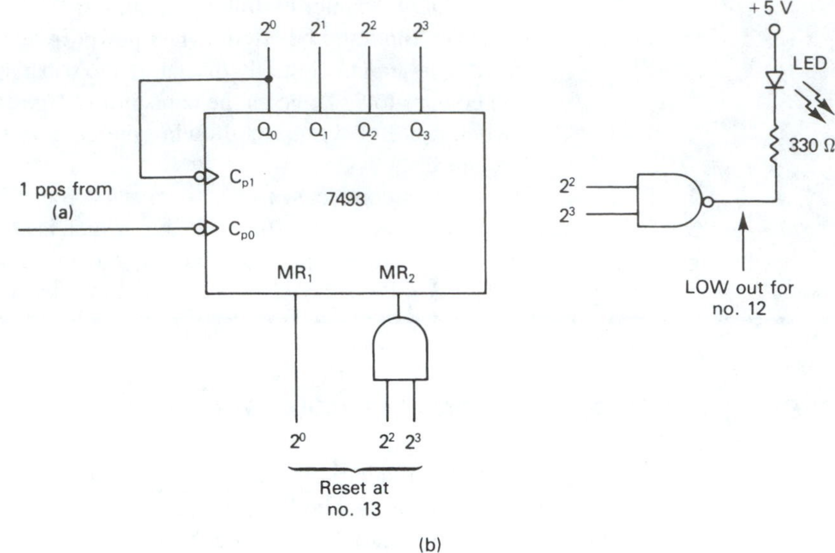

(b)

Figure 6–23 *(Continued)* (b) circuit used to illuminate an LED once every 13 s.

To solve this design problem, we first have to come up with a clock oscillator that produces 1 pulse per second (pps).

The Schmitt circuit in the first part of Figure 6–23a is used to produce the 60-pps clock. To divide the 60 pps down to 1 pps, we can cascade a MOD-10 counter with a MOD-6 counter to create a divide-by-60 circuit.

The 7490 connected as a MOD-10 is chosen for the divide-by-10 section. If you study the output waveforms of a MOD-10 counter, you can see that Q_3 will oscillate at a frequency one-tenth of the frequency at $\overline{C_{p0}}$. Then, if we use Q_3 to trigger the input clock of the divide-by-6 section, the overall effect will be a divide-by-60. (The 7492 is used for the divide-by-6 section simply by using $\overline{C_{p1}}$ as the input and taking the 1-pps output from Q_3 as shown in Figure 6–23a.)

The next step in the system design is to use the 1-pps clock to enable a circuit to turn on an LED for 1 s once every 13 s. It sounds like we need a MOD-13 counter (0 to 12) and a gating scheme that turns on an LED when the count is on the number 12. A 7493 can be used for a MOD-13 counter and a NAND gate can be used to sink the current from an LED when the number 12 ($Q_2 = 1$, $Q_3 = 1$) occurs. Figure 6–23b shows the necessary circuit connections.

Note in Figure 6–23b that a MOD-13 is formed by connecting Q_0 to $\overline{C_{p1}}$ and resetting the counter when the number 13 is reached, resulting in a count of 0 to 12. Also, when the number 12 is reached, the NAND gate's output goes LOW, turning on the LED. [$I_{LED} = (5\text{ V} - 1.7\text{ V})/330\ \Omega = 10\text{ mA}$.]

APPLICATION 6–2

Design a three-digit-decimal counter that can count from 000 to 999.

Solution:

We have already seen that a 7490 is a single-digit-decimal (0 to 9) counter. If we cascade three 7490s together and use the low-order counter to trigger the second digit counter and the second digit counter to trigger the high-order-digit counter, they will count from 000 up to 999. (Keep in mind that the outputs will be binary-coded decimal in groups of 4. In Section 6–5 we will see how we can convert the BCD outputs into actual decimal digits.)

If you review the output waveforms of a 7490 connected as a MOD-10 counter, you can see that at the end of the cycle, when the count changes from 9 (1001) to 0 (0000), the 2^3 output line goes from HIGH to LOW. When cascading counters, you can use that HIGH-to-LOW transition to trigger the input to the next-highest-order counter. That will work out great because we want the next-highest-order decimal digit to increment by 1 each time the lower-order digit has completed its 0-through-9 cycle (i.e., the transition from 009 to 010). The complete circuit diagram for a 000 to 999 BCD counter is shown in Figure 6–24.

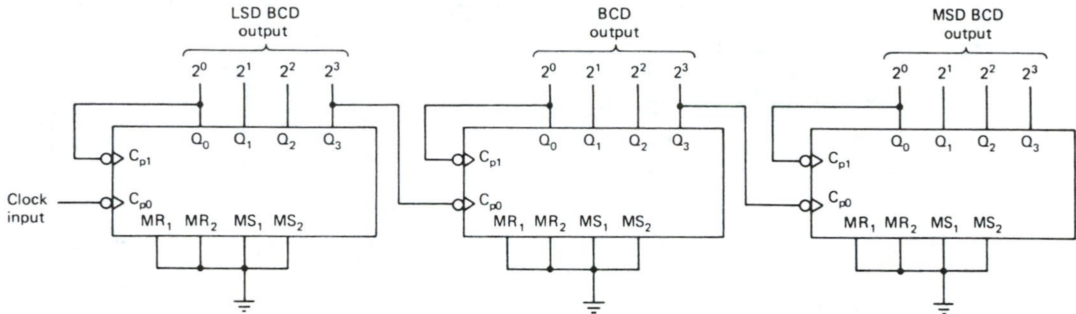

Figure 6–24 Cascading 7490s to form a 000–999 BCD output counter.

APPLICATION 6–3

Design and sketch a block diagram of a digital clock capable of displaying hours, minutes, and seconds.

Solution:

First we have to design a 1-pps clock to feed into the least significant digit of the seconds counter. The seconds will be made up of two cascaded counters that count 00 to 59. When the seconds change from 59 to 00, that transition will be used to trigger the minutes digits to increment by 1. The minutes will also be made up of two cascaded counters that count from 00 to 59. When the minutes change from 59 to 00, that transition will be used to trigger the hours digits to increment by 1. Finally, when the hours reach 12, all counters should be Reset to 0. The digital clock will display the time from 00:00:00 to 11:59:59.

Figure 6–25 is the final circuit that could be used to implement a digital clock. A 1-pps clock (similar to the one shown in Figure 6–23) is used as the initial clock trigger into the least significant digit (LSD) counter of the seconds display. This counter is a MOD-10 constructed from a 7490 IC. Each second that counter will increment. When it changes from 9 to 0, the HIGH-to-LOW edge on the 2^3 line will serve as a clock pulse into the most significant digit (MSD) counter of the seconds display. This counter is a MOD-6 constructed from a 7492 IC.

After 59 seconds, the 2^2 output of that MOD-6 counter will go HIGH to LOW [once each minute (1 ppm)], triggering the MOD-10 of the minutes section. When the minutes exceed 59, the 2^2 output of that MOD-6 counter will trigger the MOD-10 of the hours section.

The MOD-2 of the hours section is just a single toggle flip-flop having a 1 or 0 output. The hours section is set up to count from 0 to 11. When 12 is reached, the NAND gate resets both hours counters. The clock display will therefore be 00:00:00 to 11:59:59.

If you want the clock to display 1:00:00 to 12:59:59 instead, you will have to check for a 13 in the hours section instead of 12. When 13 is reached, we will want to Reset the MOD-2 counter and "Preset" the MOD-10 counter to a 1.

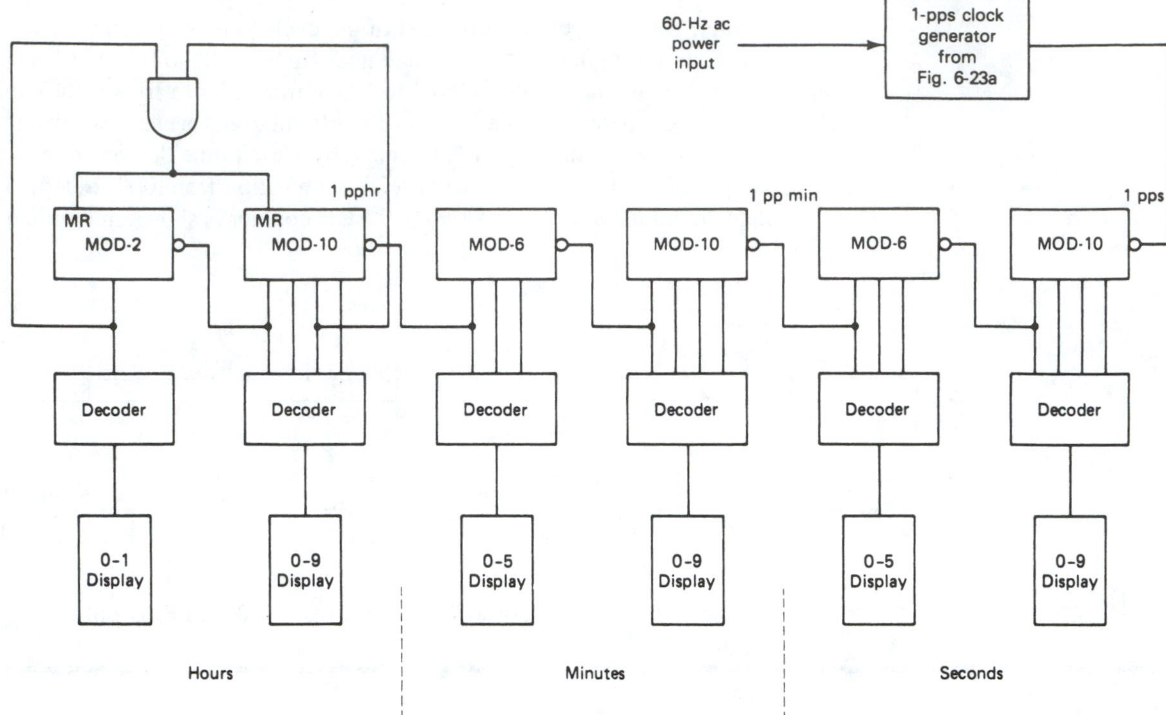

Figure 6–25 Block diagram for a digital clock.

Presettable counters such as the 74192 are used in a case like this. Presettable counters are covered later in this chapter.

The decoders are required to convert the BCD from the counters into a special code that can be used by the actual display device. Digit displays and decoders are discussed in Section 6–5.

APPLICATION 6–4

Design an egg-timer circuit. The timer will be started when you press a pushbutton. After 3 min a 5-V 10-mA dc piezoelectric buzzer will begin buzzing.

Solution:

The first thing to take care of is to divide the 1-pps clock previously designed in Figure 6–23a down to a 1-ppm clock. At 1 ppm when the count reaches 3, the buzzer should be enabled and the input clock disabled. An automatic power-up Reset is required on all the counters so that the minute counter will start at 0. A *D*-latch can be utilized for the pushbutton starter so that after the pushbutton is released, the latch "remembers" and will keep the counting process going.

The circuit of Figure 6–26 can be used to implement this design. When power is first turned on, the automatic Reset circuit will Reset all counter outputs and Reset the 7474, making $\overline{Q} = 1$. With $\overline{Q}$ HIGH, the OR gate will stay HIGH, disabling the clock from getting through to the first 7492.

When the "start" pushbutton is momentarily depressed, $\overline{Q}$ will go LOW, allowing the 1-pps clock to reach $\overline{C_{p1}}$. The first two counters are connected as a MOD-6 and a MOD-10 to yield a divide-by-60 so that we have 1 ppm available for the last counter, which serves as a minute counter. When the count reaches 3 in the last 7492, the AND gate goes HIGH, disabling the clock input. This causes the 7404 to go LOW, providing sink current for the buzzer to operate. The buzzer is turned off by turning off the main power supply.

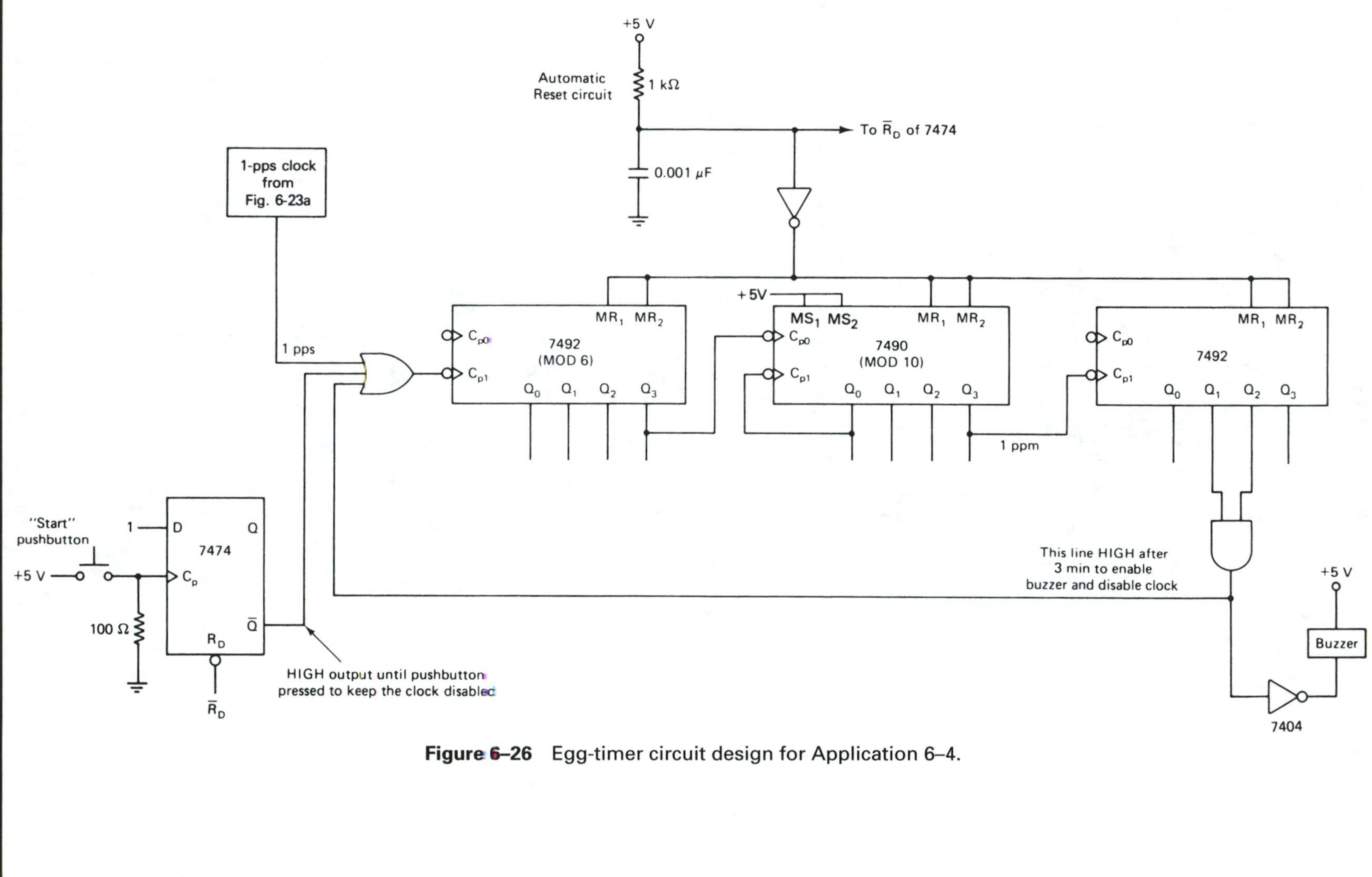

Figure 6–26 Egg-timer circuit design for Application 6–4.

6–5 SEVEN-SEGMENT LED DISPLAY DECODERS

In Section 6–4 we discussed counter circuits that are used to display decimal (0 to 9) numbers. If a counter is to display a decimal number, the count on each 4-bit counter cannot exceed 9 (1001). In other words, the counters must be outputting binary-coded decimal (BCD). As described in Chapter 1, BCD is a 4-bit binary string used to represent the 10 decimal digits. To be useful, however, the BCD must be decoded by a decoder into a format that can be used to drive a decimal numeric display. The most popular display technique is the seven-segment LED display.

A seven-segment LED display is actually made up of seven separate light-emitting diodes in a single package. The LEDs are oriented so as to form an 8. Most seven-segment LEDs have an eighth LED used for a decimal point.

The job of the decoder is to convert the 4-bit BCD code into a "seven-segment code," which will turn on the appropriate LED segments to display the correct decimal digit. For instance, if the BCD is 0111 (7), the decoder must develop a code to turn on the top segment and the two right segments (7).

Common-Anode LED Display

The physical layout of a seven-segment LED display is shown in Figure 6–27. This figure shows that the anode of each LED (segment) is connected to the +5-V supply.

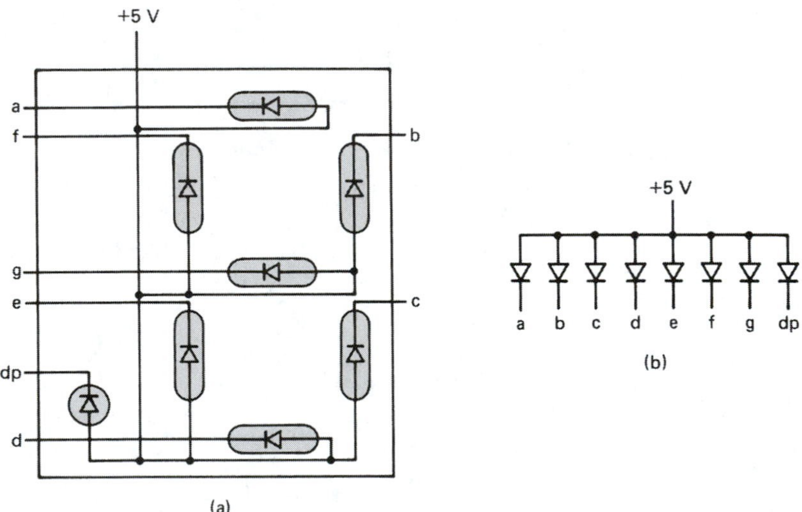

Figure 6–27 Seven-segment common-anode LED display: (a) physical layout; (b) schematic.

Now, to illuminate an LED, its cathode must be grounded through a series-limiting resistor, as shown in Figure 6–28. The value of the limiting resistor can be found by knowing that the voltage drop across an LED is 1.7 V and that it takes approximately 10 mA to illuminate it. Therefore,

$$R_{\text{limit}} = \frac{5.0 - 1.7\text{ V}}{10\text{ mA}} = 330\ \Omega$$

Each segment in the display unit is illuminated in the same way. Figure 6–29 shows the numerical designations for the 10 allowable decimal digits.

Common-anode displays are "active-LOW" (LOW-enable) devices because they take a LOW to turn on (illuminate) a segment. Therefore, the decoder IC used to drive a common-anode LED must have active-LOW outputs.

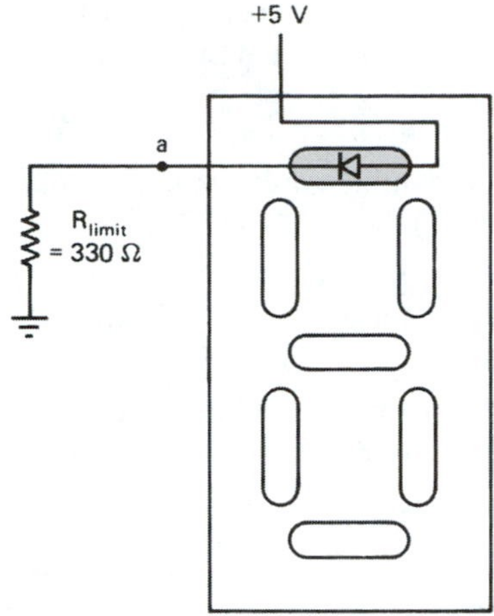

Figure 6–28 Illuminating the *a* segment.

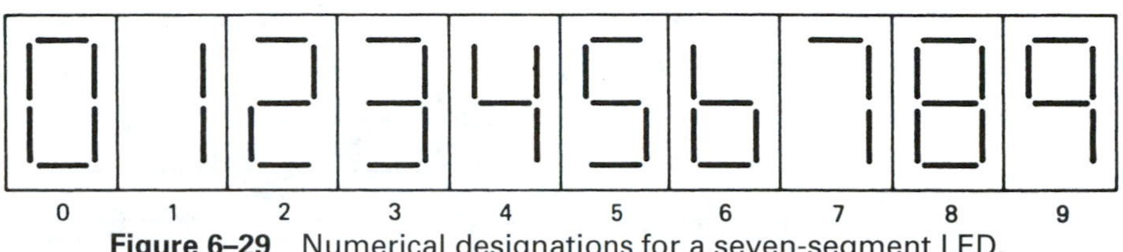

Figure 6–29 Numerical designations for a seven-segment LED.

Common-cathode LEDs and decoders are also available but they are not as popular because they are "active-HIGH," and ICs typically cannot "source" (1 output) as much current as they can "sink" (0 output).

BCD-to-Seven-Segment Decoder/Driver ICs

The 7447 is the most popular common-anode decoder/LED driver. Basically, the 7447 has a 4-bit BCD input and seven individual active-LOW outputs (one for each LED segment). As shown in Figure 6–30, it also has a "lamp test" ($\overline{LT}$) input for testing *all* segments, and it also has ripple blanking input and output.

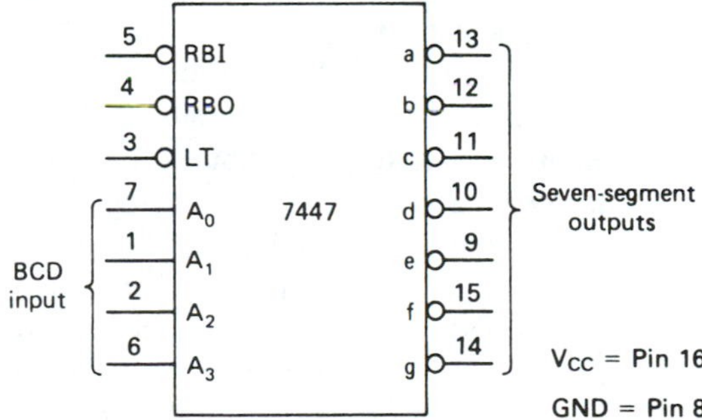

Figure 6–30 Logic symbol for a 7447 decoder.

A very versatile CMOS seven-segment decoder is the 4543, and its high-speed version, the 74HCT4543. The 4543 provides active-HIGH *or* active-LOW outputs and can drive LED displays as well as liquid-crystal displays (LCDs). For the purposes of this chapter, we discuss the 7447 TTL decoder in detail.

To complete the connection between the 7447 and the seven-segment LED, we need seven 330-Ω resistors (eight if the decimal point is included) for current limiting. Dual-in-line-package (DIP) *resistor networks* are available and simplify the wiring process because all seven (or eight) resistors are in a single DIP.

Figure 6–31 shows typical decoder–resistor DIP–LED connections. If a MOD-10 counter's outputs are connected to the BCD input and the count is at six (0110_{BCD}), the following will happen:

1. The decoder will determine that a 0110_{BCD} must send the $\overline{c}$, $\overline{d}$, $\overline{e}$, $\overline{f}$, $\overline{g}$ outputs LOW ($\overline{a}$, $\overline{b}$ will be HIGH for 6).
2. The LOW on those outputs will provide a path for the sink current in the appropriate LED segments via the 330-Ω resistors (the 7447 can sink up to 40 mA at each output).
3. The decimal number 6 will be illuminated together with the decimal point if the dp switch is closed.

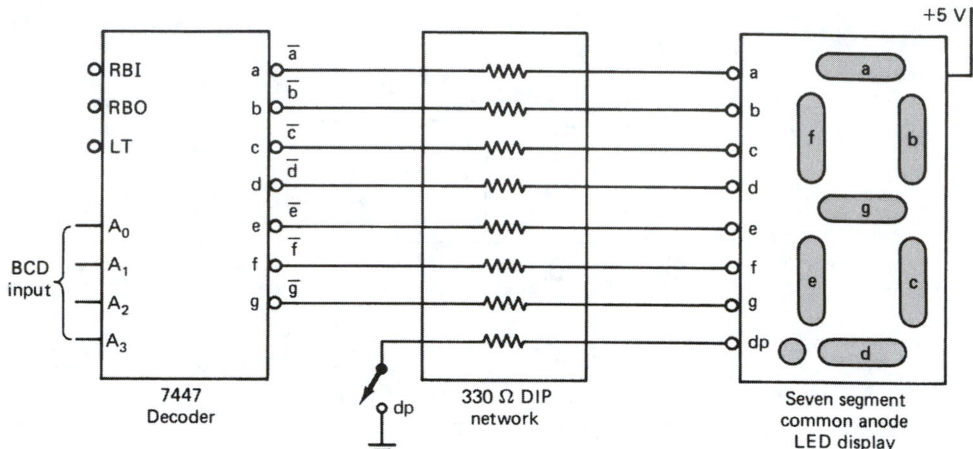

Figure 6–31 Driving a seven-segment LED display.

A complete three-digit decimal display system is shown in Figure 6–32. The three counters in the figure are connected as MOD-10 counters with the input clock oscillator connected to the least significant counter. The three counters are cascaded by connecting the Q_3 output for the first to the $\overline{C_{p0}}$ of the next, and so on.

Note that the decimal point of the LSD (Figure 6–32) is always on, so the counters will therefore count from .0 up to 99.9. If the clock oscillator is set at 10 pps, the LSD will indicate tenths of seconds. Also note that the ripple blanking inputs and outputs ($\overline{\text{RBI}}$ and $\overline{\text{RBO}}$) are used in this design. They are active LOW and are used for *leading-zero suppression.* For example, if the display output is at 1.4, would you like it to read 01.4 or 1.4? To suppress the leading zero and make it a blank, ground the $\overline{\text{RBI}}$ terminal of the MSD decoder. How about if the output is at .6; would you like it to read 00.6, 0.6, or .6? To suppress the second zero when the MSD is blank, simply connect the $\overline{\text{RBO}}$ of the MSD decoder to the $\overline{\text{RBI}}$ of the second digit decoder. The process is: If the MSD is blank (zero suppressed), the MSD decoder puts a LOW out at $\overline{\text{RBO}}$. This LOW is connected to the $\overline{\text{RBI}}$ of the second decoder, which forces a blank output (zero suppression) if its BCD input is zero.

The $\overline{\text{RBI}}$ and $\overline{\text{RBO}}$ can also be used for zero suppression of "trailing" zeros. For example, if you have an eight-digit display, the $\overline{\text{RBI}}$s and $\overline{\text{RBO}}$s could be used to suppress automatically the number 0046.0910 to be displayed as 46.091.

6–6 SYNCHRONOUS COUNTERS

Remember the problems we discussed with ripple counters due to the accumulated propagation delay of the clock from flip-flop to flip-flop? (See Figure 6–5.) Well, synchronous counters eliminate that problem because all the clock inputs ($\overline{C_p}$s) are tied to a common

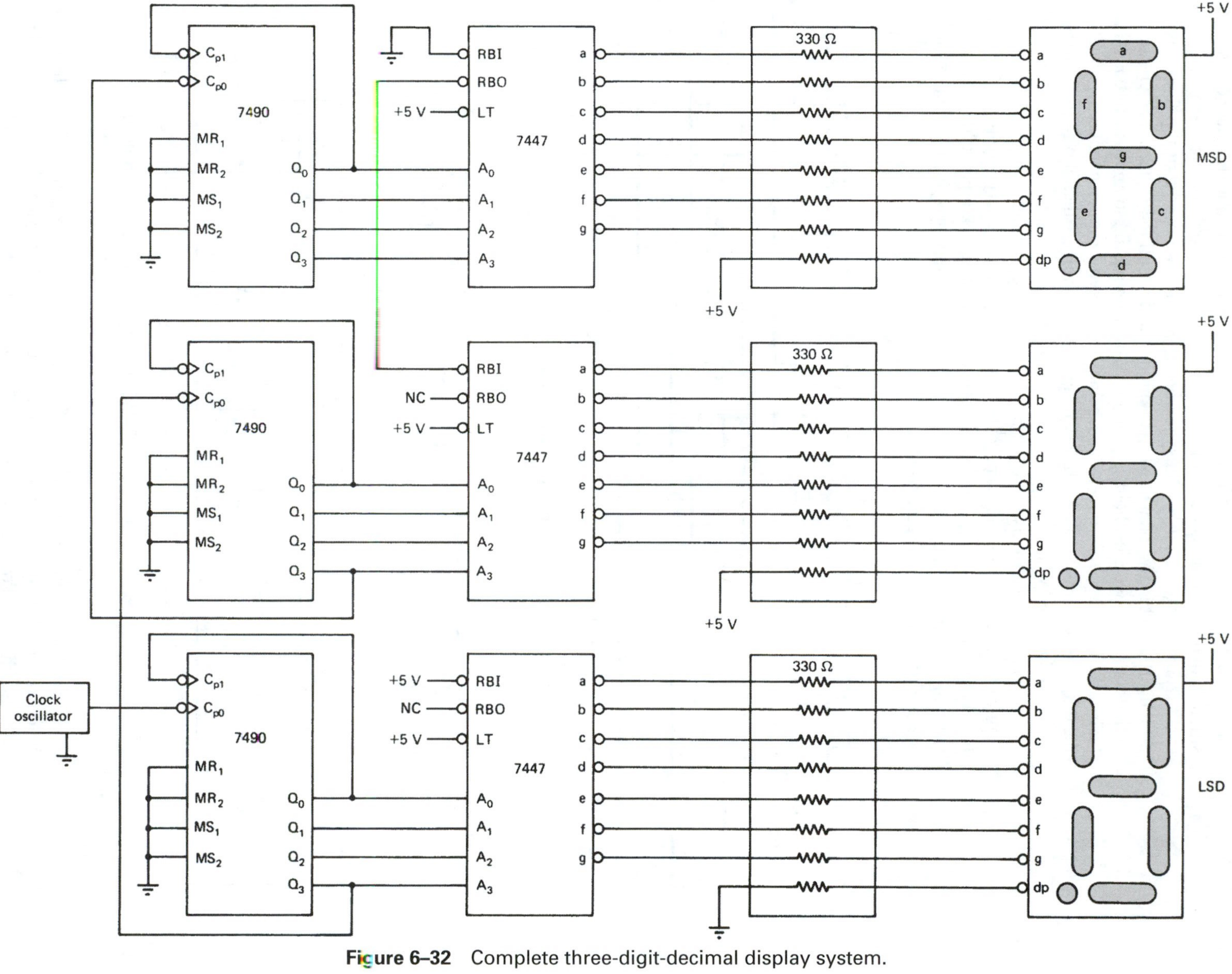

Figure 6–32 Complete three-digit-decimal display system.

clock input line, so each flip-flop will be triggered at the same time (thus any Q output transitions will occur at the same time).

If we want to design a 4-bit synchronous counter, we need four flip-flops, giving us a MOD-16 (2^4) binary counter. Keep in mind that since all the C_p inputs receive a trigger at the same time, we must hold certain flip-flops from making output transitions until it is their turn. To design the connection scheme for the synchronous counter, let's first study the output waveforms of a 4-bit binary counter to determine which flip-flops are to be held from toggling, and when.

From the waveforms in Figure 6–33, we can see that the 2^0 output is a continuous toggle off the clock input line. The 2^1 output line toggles on every negative edge of the 2^0 line, but since the 2^1's $\overline{C_{p0}}$ input is also connected to the clock input, it must be held from toggling until the 2^0 line is HIGH. This can be done simply by tying the J and K inputs to the 2^0 line as shown in Figure 6–34.

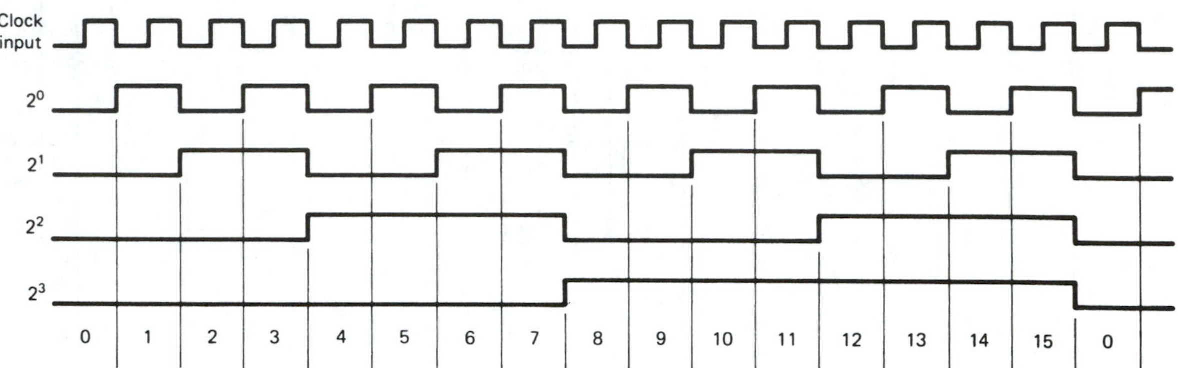

Figure 6–33 Four-bit synchronous binary counter output waveforms.

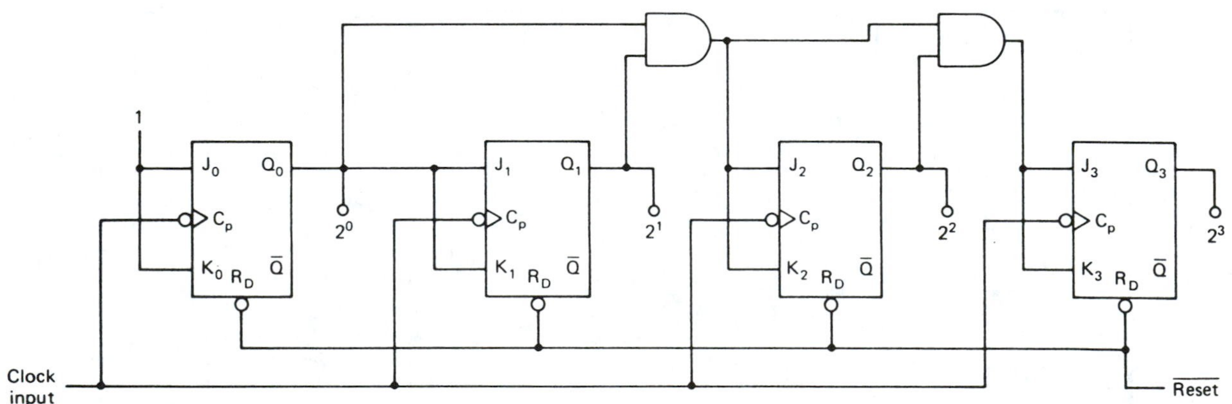

Figure 6–34 Four-bit MOD-16 synchronous up-counter.

The same logic follows through for the 2^2 and 2^3 output lines. The 2^2 line must be held from toggling until the 2^0 *and* 2^1 lines are both HIGH. Also, the 2^3 line must be held from toggling until the 2^0 *and* 2^1 *and* 2^2 lines are all HIGH.

To keep the appropriate flip-flops in the *hold* condition or *toggle* condition, their J and K inputs are tied together and through use of additional AND gates, as shown in Figure 6–34, the J-K inputs will be both 0 or 1, depending on whether they are to be in the hold or toggle mode.

From Figure 6–34 we can see that the same clock input is driving all four flip-flops. The 2^1 flip-flop will be in the hold mode ($J_1 = K_1 = 0$) until the 2^0 output goes HIGH, which will force J_1-K_1 HIGH, allowing the 2^1 flip-flop to toggle when the next negative clock edge comes in.

Now, observe the output waveforms (Figure 6–33) while you look at the circuit design (Figure 6–34) to determine the operation of the last two flip-flops. From the waveforms we see that the 2^2 output must not be allowed to toggle until 2^0 *and* 2^1 are both HIGH. Well, the first AND gate in Figure 6–34 takes care of that by holding J_2-K_2 LOW. The same method is used to keep the 2^3 output from toggling until the 2^0 *and* 2^1 *and* 2^2 outputs are *all* HIGH.

As you can see, the circuit is more complicated, but the cumulative effect of propagation delays through the flip-flops is not a problem as it was in ripple counters because all output transitions will occur at the same time since all flip-flops are triggered from the same input line. (There *is* a propagation delay through the AND gates, but it will not affect the Q outputs of the flip-flops.)

6–7 SYNCHRONOUS UP/DOWN-COUNTER ICs*

Four-bit synchronous binary counters are available in a single integrated-circuit package. Two popular synchronous IC counters are the 74192 and 74193. They both have some features that were not available on the ripple counter ICs. They can count *up or down* and can be *preset* to any count that you desire. The 74192 is a BCD decade up/down-counter and the 74193 is a 4-bit binary up/down-counter. The logic symbol used for both counters is shown in Figure 6–35.

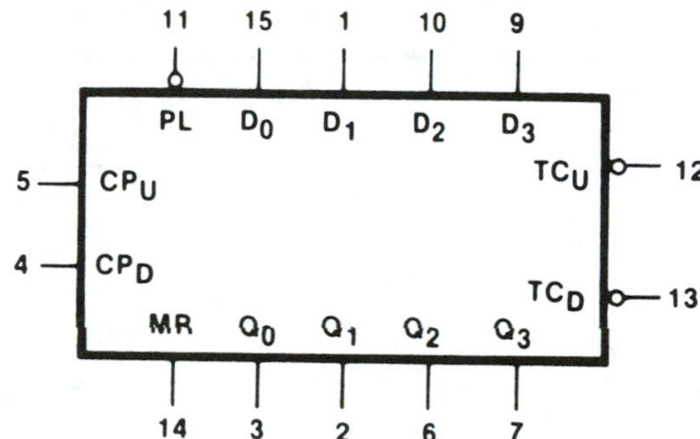

Figure 6–35 Logic symbol for the 74192 and 74193 synchronous counter ICs.

There are two separate clock inputs: C_{pU} for counting up and C_{pD} for counting down. One clock must be held HIGH while counting with the other. The binary output count is taken from Q_0 to Q_3, which are the outputs from four internal J-K flip-flops. The Master Reset (MR) is an active-HIGH Reset for resetting the Q outputs to 0.

The counter can be preset by placing any binary value on the parallel data inputs (D_0 to D_3) and then driving the Parallel Load ($\overline{\text{PL}}$) line LOW. The parallel load operation will change the counter outputs regardless of the conditions of the clock inputs.

The Terminal Count Up ($\overline{\text{TC}_\text{U}}$) and Terminal Count Down ($\overline{\text{TC}_\text{D}}$) are normally HIGH. The $\overline{\text{TC}_\text{U}}$ is used to indicate that the maximum count is reached and the count is about to recycle to 0 (carry condition). The $\overline{\text{TC}_\text{U}}$ line goes LOW for the 74193 when the count reaches 15 *and* the input clock (C_{pU}) goes HIGH to LOW. $\overline{\text{TC}_\text{U}}$ remains LOW until C_{pU} returns HIGH. This LOW pulse at $\overline{\text{TC}_\text{U}}$ can be used as a clock input to the next-higher-order stage of a multistage counter.

The $\overline{\text{TC}_\text{U}}$ output for the 74192 is similar except that it goes LOW at 9 *and* a LOW C_{pU} (see Figure 6-36). The Boolean equations for TC_U, therefore, are as follows:

$$\text{LOW at } \overline{\text{TC}_\text{U}} = Q_0 Q_1 Q_2 Q_3 \overline{C_{pU}} \qquad (74193)$$

$$\text{LOW at } \overline{\text{TC}_\text{U}} = Q_0 Q_3 \overline{C_{pU}} \qquad (74192)$$

*See Application K-9 in Appendix K for a CPLD implementation of a 4-bit counter.

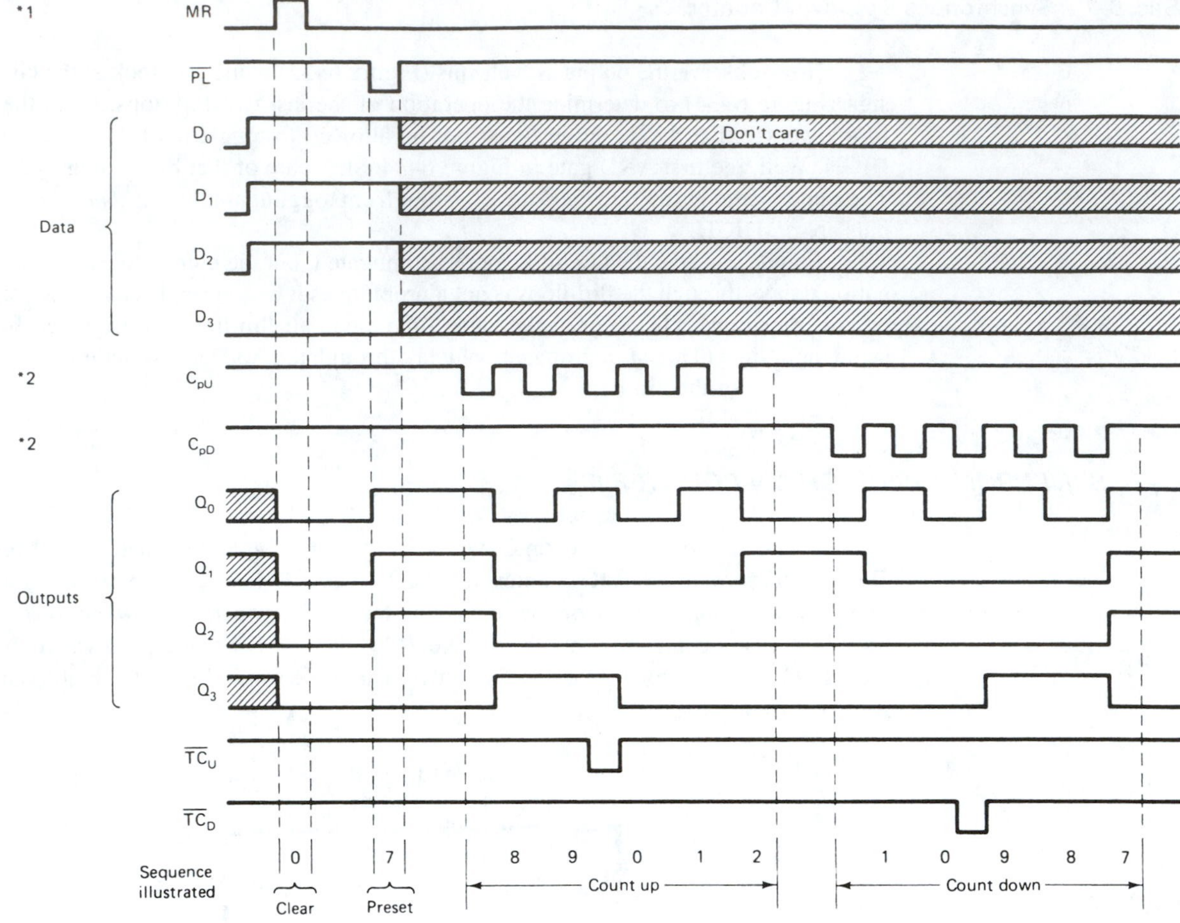

Notes

1. Clear overrides load, data and count inputs.
2. When counting up, count-down input must be HIGH; when counting down, count-up input must be HIGH.

Figure 6–36 Timing waveforms for the 74192 used in Example 6–8.

The Terminal Count Down ($\overline{TC_D}$) is used to indicate that the minimum count is reached and the count is about to recycle to the maximum (15 or 9) count (borrow condition). Therefore, $\overline{TC_D}$ goes LOW when the down-count reaches 0 and the input clock (C_{pD}) goes LOW (see Figure 6–38). The Boolean equation at $\overline{TC_D}$ is

$$\text{LOW at } \overline{TC_D} = \overline{Q_0}\,\overline{Q_1}\,\overline{Q_2}\,\overline{Q_3}\,\overline{C_{pD}} \qquad \text{(74192 and 74193)}$$

TABLE 6–2

Function Table for the 74192/74193 Synchronous Counter IC[a]

Operating mode	*Inputs*								*Outputs*					
	MR	$\overline{PL}$	C_{pU}	C_{pD}	D_0	D_1	D_2	D_3	Q_0	Q_1	Q_2	Q_3	$\overline{TC_U}$	$\overline{TC_p}$
Reset	H	×	×	L	×	×	×	×	L	L	L	L	H	L
	H	×	×	H	×	×	×	×	L	L	L	L	H	H
Parallel Load	L	L	×	L	L	L	L	L	L	L	L	L	H	L
	L	L	×	H	L	L	L	L	L	L	L	L	H	H
	L	L	L	×	H	H	H	H	H	H	H	H	L	H
	L	L	H	×	H	H	H	H	H	H	H	H	H	H
Count up	L	H	↑	H	×	×	×	×	Count up				H	H
Count Down	L	H	H	↑	×	×	×	×	Count down				H	H

[a]H = HIGH voltage level; L = LOW voltage level; × = don't care; ↑ = LOW-to-HIGH clock transition.

The function table shown in Table 6–2 can be used to show the four operating modes (Reset, Load, Count Up, and Count Down) of 74192/74193.

The best way to illustrate how these chips operate is to exercise all its functions and observe the resultant waveforms as shown in the following examples.

EXAMPLE 6–8

Draw the input and output timing waveforms for a 74192 that goes through the following sequence of operation:

1. Reset all outputs to zero.
2. Parallel Load a 7 (0111).
3. Count up five counts.
4. Count down five counts.

Solution:

The timing waveforms are shown in Figure 6–36.

EXAMPLE 6–9

Draw the output waveforms for the 74193 shown in Figure 6–37, given the waveforms shown in Figure 6–38. (Initially, set $D_0 = 1$, $D_1 = 0$, $D_2 = 1$, $D_3 = 1$, and MR = 0.)

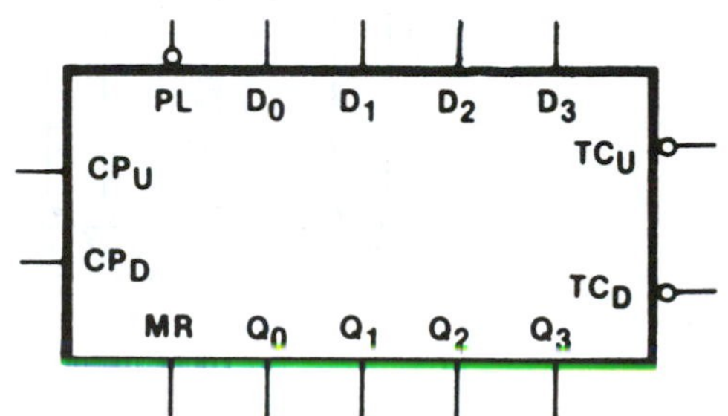

Figure 6–37 Circuit connections for Example 6–9.

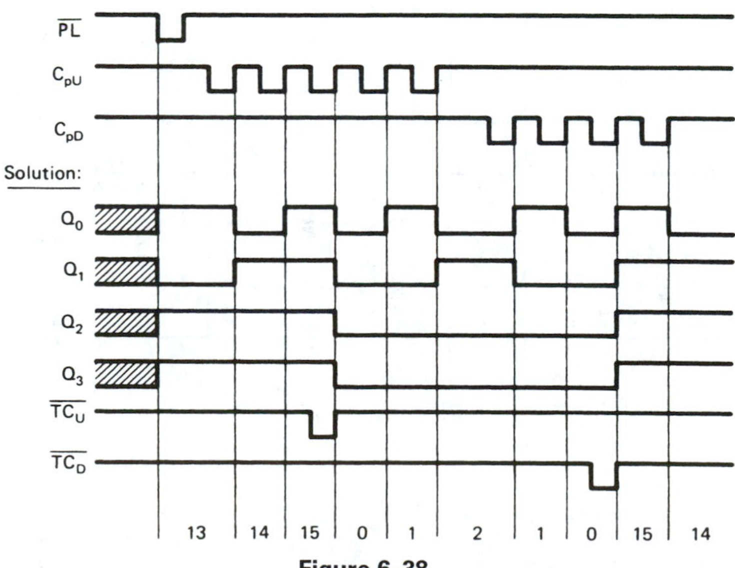

Figure 6–38

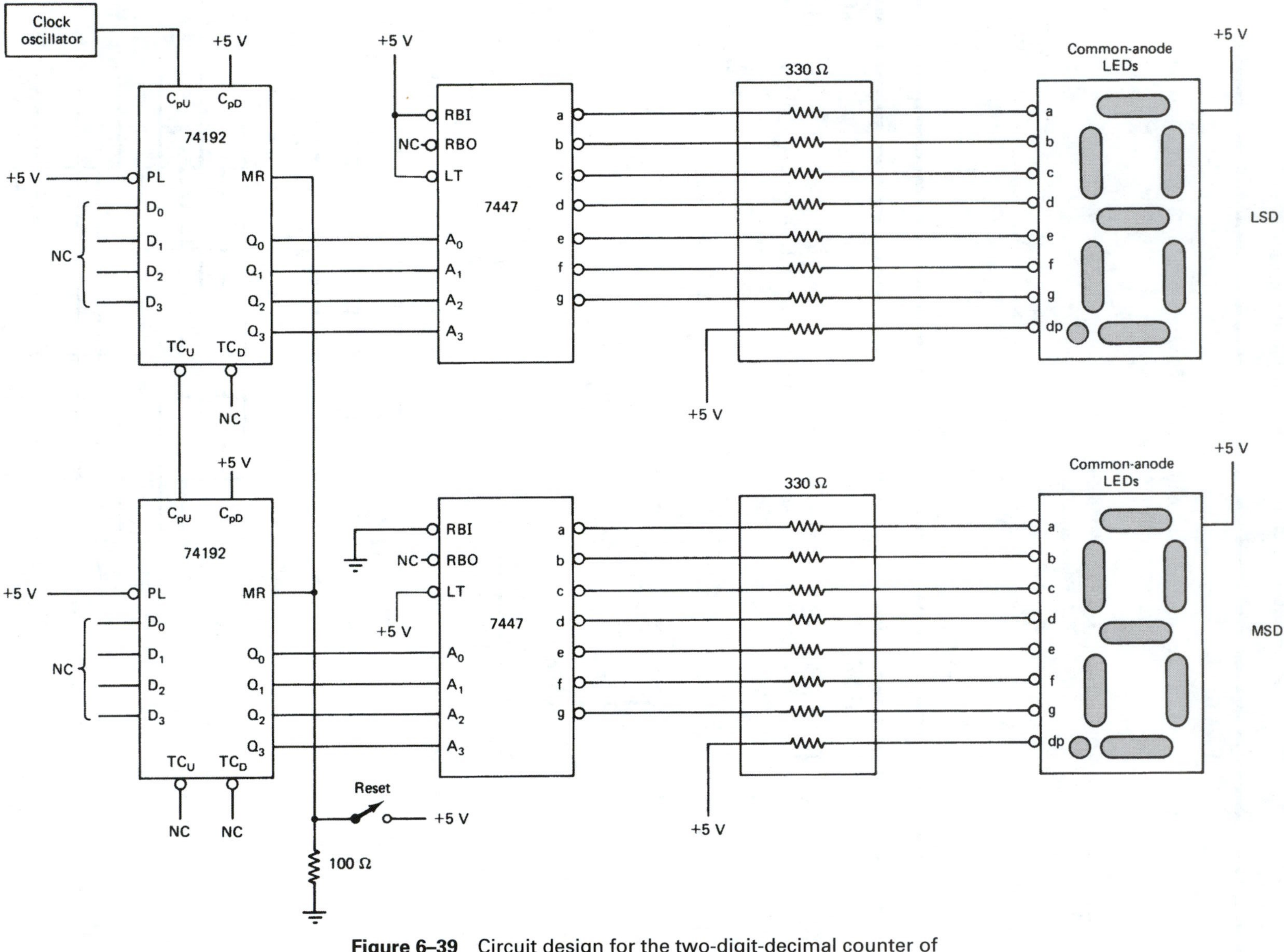

Figure 6–39 Circuit design for the two-digit-decimal counter of Example 6–10.

EXAMPLE 6–10

Design a decimal counter that will count from 00 to 99 using two 74192 counters and the necessary drive circuitry for the two-digit display. (Display circuitry was explained in Section 6–5.)

Solution:

The 74192s can be used to form a multistage counter by connecting the $\overline{\mathrm{TC_U}}$ of the first counter to the C_{pU} of the second counter. $\overline{\mathrm{TC_U}}$ will go LOW, then HIGH, when the first counter goes from 9 to 0 (carry). That LOW-to-HIGH edge can be used as the clock input to the second stage, as shown in Figure 6–39.

EXAMPLE 6–11

Design and sketch the timing waveforms for a divide-by-9 frequency divider using a 74193 counter.

Solution:

We can use the Parallel Load feature of the 74193 to set the counter at some initial value, and then count down to 0. When we reach 0, we will have to Parallel Load the counter to its initial value and count down again, making sure the repetitive cycle repeats once every nine clock periods. Figure 6–40 could be used to implement such a circuit. $\overline{\mathrm{TC_D}}$ is fed back into $\overline{\mathrm{PL}}$. That means that when the Terminal Count is reached, the LOW out of $\overline{\mathrm{TC_D}}$ will enable the Parallel Load, making the outputs equal to the D_0 to D_3 inputs (1001).

The timing waveforms arbitrarily start at 1 and count down. Note that at 0 (Terminal Count), $\overline{\mathrm{TC_D}}$ goes LOW when C_{pD} goes LOW (remember that a LOW at $\overline{\mathrm{TC_D}} = \overline{Q_0}\,\overline{Q_1}\,\overline{Q_2}\,\overline{Q_3}\,\overline{C_{pD}}$). As soon as $\overline{\mathrm{TC_D}}$ goes LOW, the outputs return to 9, thus causing $\overline{\mathrm{TC_D}}$ to go back HIGH again. Therefore, $\overline{\mathrm{TC_D}}$ is a narrow pulse just long enough to perform the Parallel Load operation.

The down counting resumes until 0 is reached again, which causes the Parallel Load of 9 to occur again. The $\overline{\mathrm{TC_D}}$ pulse occurs once every ninth C_{pD} pulse; thus we have a divide-by-9. (A different duty-cycle divide-by-9 can be gotten from the Q_3 or Q_2 outputs.)

6–8 SHIFT REGISTER BASICS

Let's take a look at a 4-bit shift register's contents as it receives 4 bits of parallel data and shifts them to the right four positions into some other digital device. The timing for the shift operations is provided by the input clock and will shift to the right by one position for each input clock pulse, as shown in Figure 6–41.

In the figure, the group of four boxes are four *D* flip-flops comprising the 4-bit shift register. The first step is to parallel load the register with a 1–0–0–0. "Parallel load" means to load all four flip-flops at the same time. This is done by momentarily enabling the appropriate asynchronous Set ($\overline{S_D}$) and Reset ($\overline{R_D}$) inputs.

Next, the first clock pulse causes all bits to shift to the right by one because the input to each flip-flop comes from the Q output of the flip-flop to its left. Each successive pulse causes all data bits to shift one more position to the right.

At the end of the fourth clock pulse, all data bits have been shifted all the way across and now all four original data bits appear, in the correct order, in the serial receiving device. The connections between the fourth flip-flop and the serial receiving device could be a three-conductor serial transmission cable (serial data, clock, and ground).

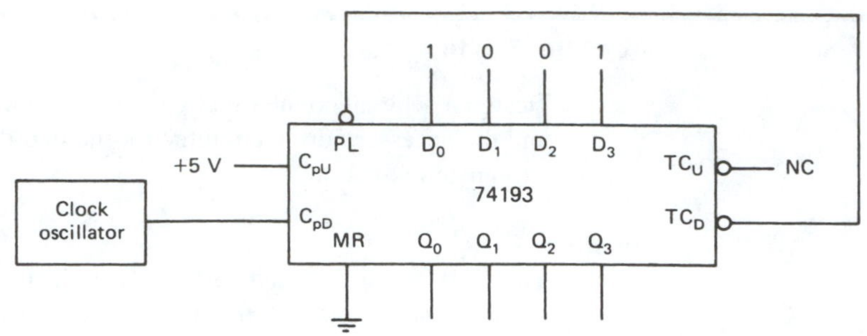

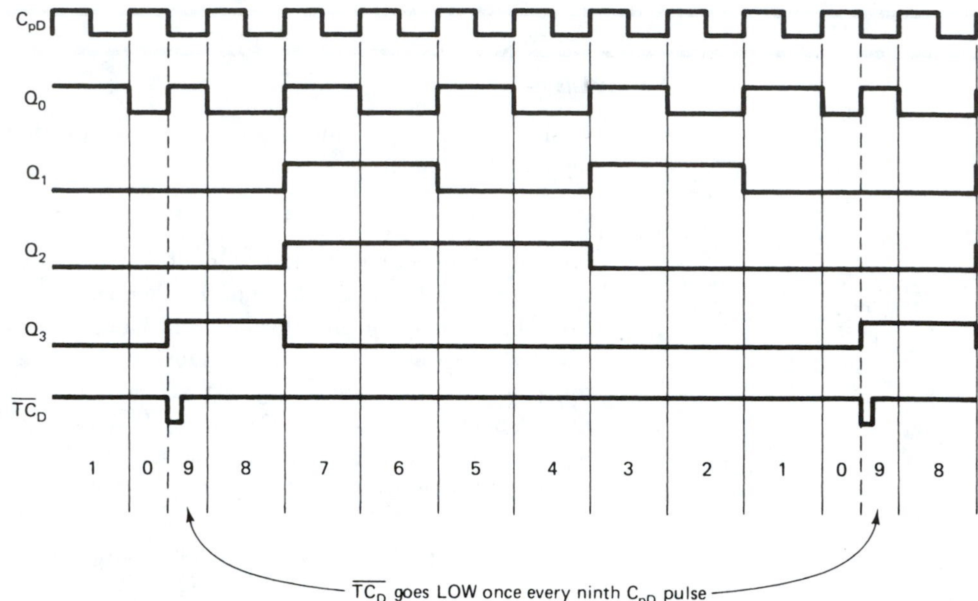

Figure 6–40 Circuit design and timing waveforms for a divide-by-9 frequency divider.

Figure 6–41 illustrates a parallel-to-serial conversion. Shift registers can also be used for serial-to-parallel, parallel-to-parallel, and serial-to-serial as well as shift-right, shift-left operations, as indicated in Figure 6–42. Each of these configurations and an explanation of the need for a "recirculating line" are explained in upcoming sections.

Parallel-to-Serial Conversion

Now let's look at the actual circuit connections for a shift register. The data storage elements can be *D* flip-flops, *S-R* flip-flops, or *J-K* flip-flops. We are pretty familiar with *J-K* flip-flops, so let's stick with them. Most *J-K*s are negative edge triggered (like the 74LS76) and will have an active-LOW asynchronous Set ($\overline{S_D}$) and Reset ($\overline{R_D}$).

Figure 6–43 shows the circuit connections for a 4-bit parallel-in, serial-out shift register that is first Reset, then parallel-loaded with an active-LOW 7 (1000), then shifted right four positions.

Note in Figure 6–43a that all $\overline{C_p}$ inputs are fed from a common clock input. Each flip-flop will respond to its *J-K* inputs at every negative clock input edge. Since every *J-K* input is connected to the preceding stage output, then at each negative clock edge, each flip-flop will change to the state of the flip-flop to its left. In other words, all data bits will be shifted one position to the right.

Now, looking at the timing diagram, in the beginning of period 1, $\overline{R_D}$ goes LOW, resetting Q_0 to Q_3 to 0. Next, the parallel data are input (parallel-loaded) via the D_0 to D_3

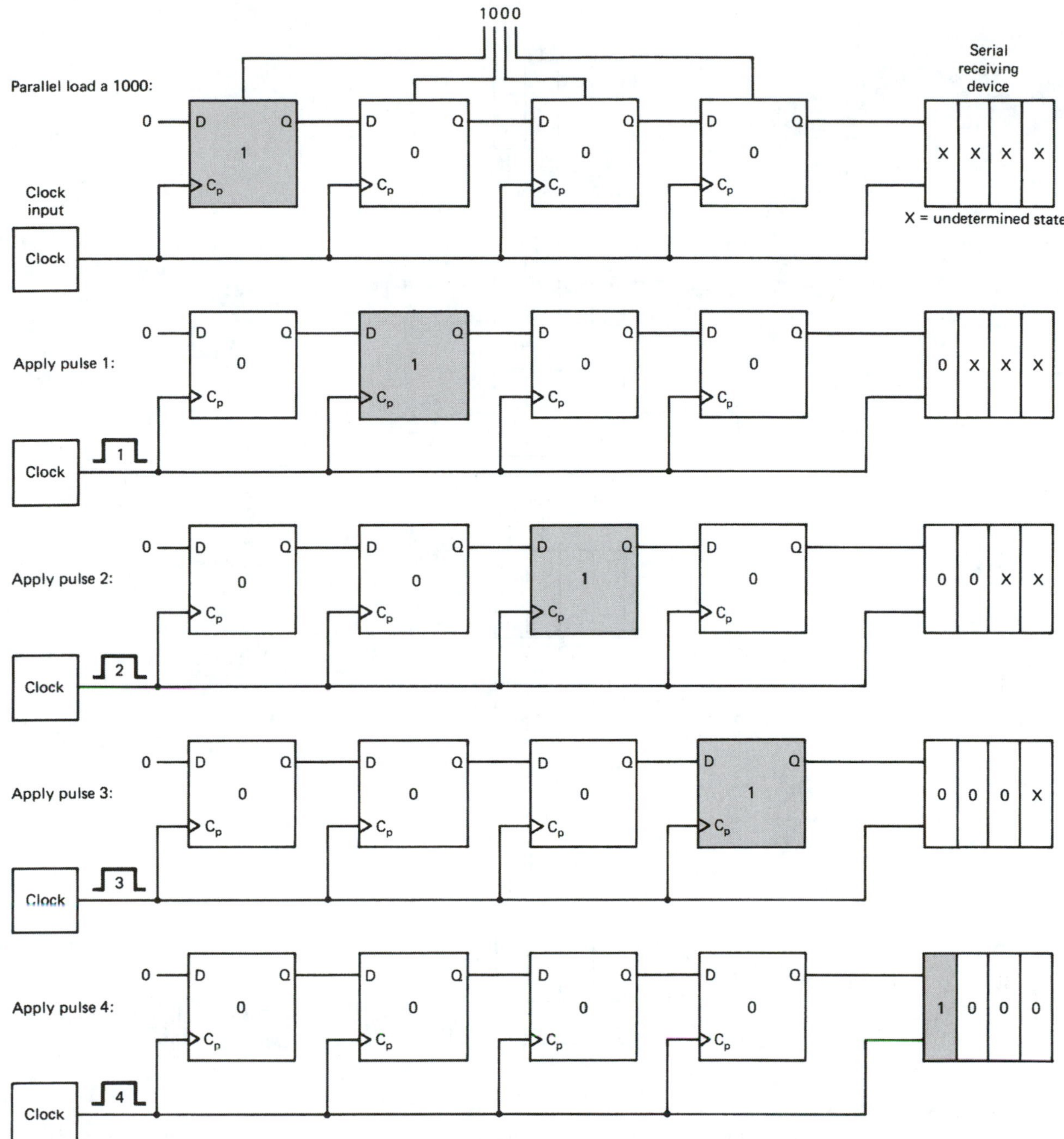

Figure 6–41 Block diagram of a 4-bit shift register used for parallel-serial conversion.

input lines. (Since the $\overline{S_D}$ inputs are active LOW, the complement of the number to be loaded must be used.)

At the first negative clock edge:

Q_0 takes on the value of Q_1

Q_1 takes on the value of Q_2

Q_2 takes on the value of Q_3

Q_3 is Reset by $J = 0$, $K = 1$

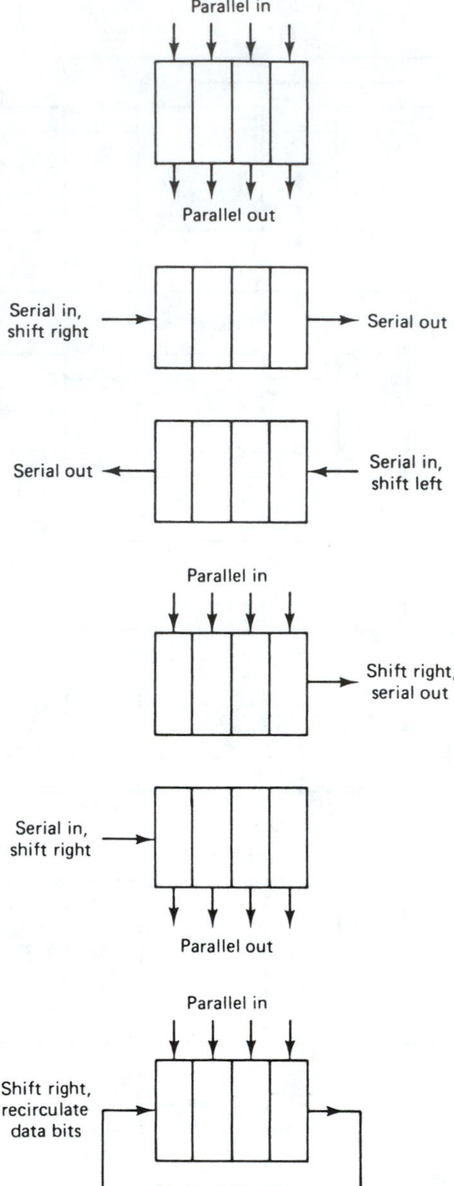

Figure 6–42 Data movement and conversion by shift registers.

In effect, the bits have all shifted one position to the right. Next, the negative edge of periods 2, 3, and 4 will each shift the bits one more position to the right.

The serial output data comes out of the right-end flip-flop (Q_0). Since the LSB was parallel-loaded into the rightmost flip-flop, the LSB will be shifted out first. The order of the parallel input data bits could have been reversed and the MSB would have come out first. Either case is acceptable. It is up to the designer to know which is first, MSB or LSB, and when to sample (or read) the serial output data line.

Recirculating Register

Recirculating the rightmost data bits back into the beginning of the register can be accomplished by connecting Q_0 back to J_3 and $\overline{Q_0}$ back to K_3. That way, the original parallel-loaded data bits will never be lost. After every fourth clock pulse, the Q_3 to Q_0 outputs will contain the original 4 data bits. Therefore, with the addition of the recirculating lines to Figure 6–43a the register becomes a parallel-in, serial, *and* parallel-out.

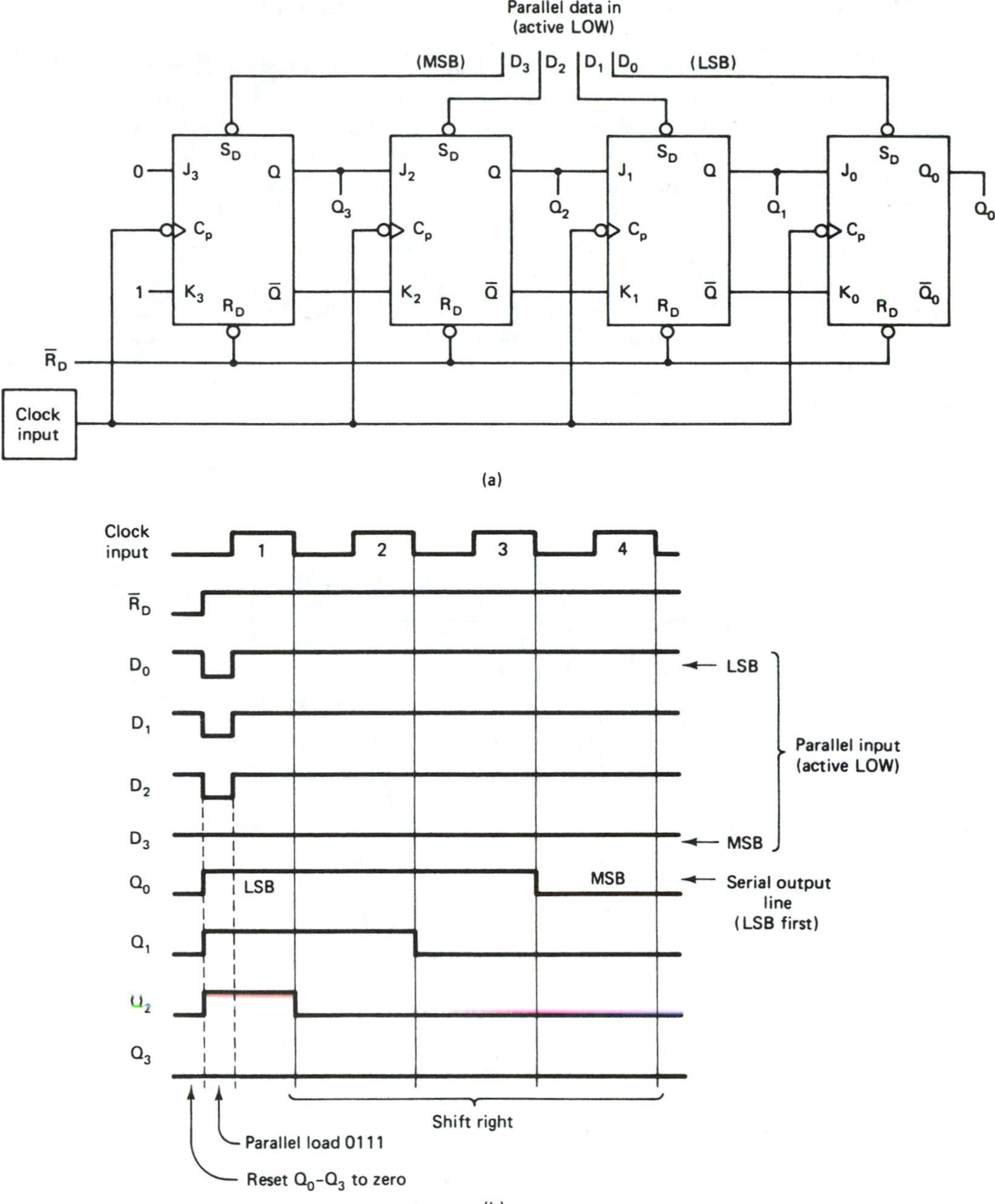

Figure 6–43 (a) Four-bit parallel-in, serial-out shift register using 74LS76 *J-K* flip-flops; (b) waveforms produced by parallel loading a 7 (0111) and shifting right by four clock pulses.

Serial-to-Parallel Conversion

Serial-in, parallel-out shift registers can also be made up of *J-K* flip-flop storage and a shift-right operation. The idea is to put the serial data in on the serial input line (J_3-K_3 in Figure 6–43a) LSB first; "clock" the shift register four times; stop; then read the parallel data from the Q_0 to Q_3 outputs.

6–9 RING SHIFT COUNTER AND JOHNSON SHIFT COUNTER

Two common circuits that are used to create sequential control waveforms for digital systems are the ring and Johnson shift counters. They are similar to a synchronous counter because the clock input to each flip-flop is driven by the same clock input. Their outputs do not count in true binary, but instead provide a repetitive sequence of digital output levels. These shift counters are used to control a sequence of events in a digital system.

In the case of a 4-bit *ring shift counter,* the output at each flip-flop will be HIGH for one clock period, then LOW for the next three, then repeat, as shown in Figure 6–44b. To form the ring shift counter of Figure 6–44a, the Q-$\overline{Q}$ output of each stage is fed to the *J-K* input of the next stage and the Q-$\overline{Q}$ output of the last stage is fed back to the *J-K* input of the first stage. Before applying clock pulses, the shift counter is preset with a 1–0–0–0.

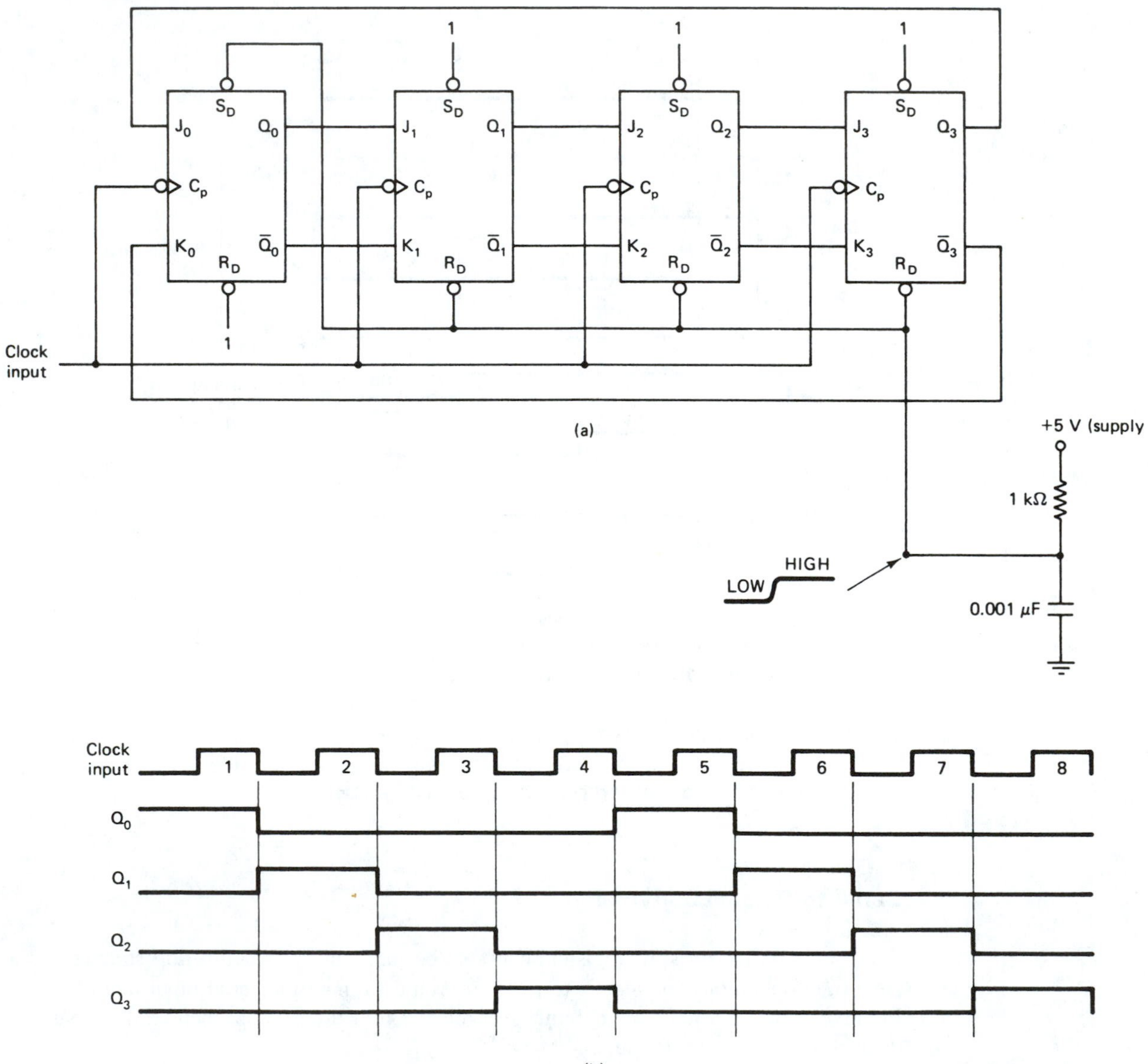

Figure 6–44 Ring shift counter: (a) circuit connections; (b) output waveforms.

Ring Shift Counter Operation

The *RC* circuit connected to the power supply will provide a LOW-then-HIGH as soon as the power is turned on, forcing a HIGH–LOW–LOW–LOW at Q_0–Q_1–Q_2–Q_3, which is the necessary preset condition for a ring shift counter. After the initial preset, the ring shift counter operates just like a recirculating shift-right shift register, producing the waveforms shown.

The *Johnson shift counter* circuit is similar to the ring shift counter except that the output lines of the last flip-flop are crossed (thus an alternative name is "twisted ring counter") before feeding back to the input of the first flip-flop and *all* flip-flops are initially Reset as shown in Figure 6–45a.

Johnson Shift Counter Operation

The *RC* circuit provides an automatic Reset to all four flip-flops so the initial outputs will all be Reset (LOW). At the first negative clock edge, the first flip-flop will Set (HIGH)

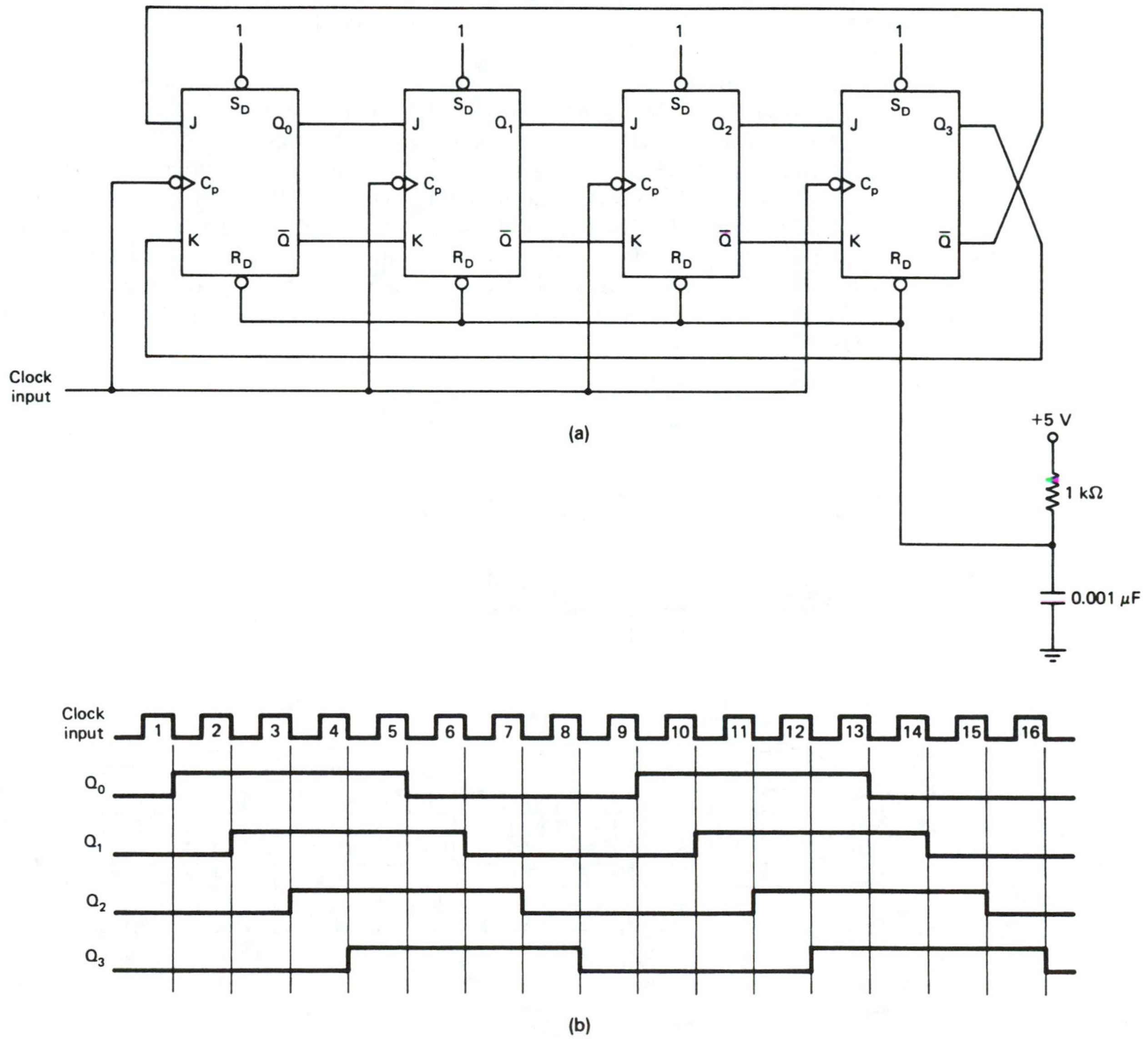

(a)

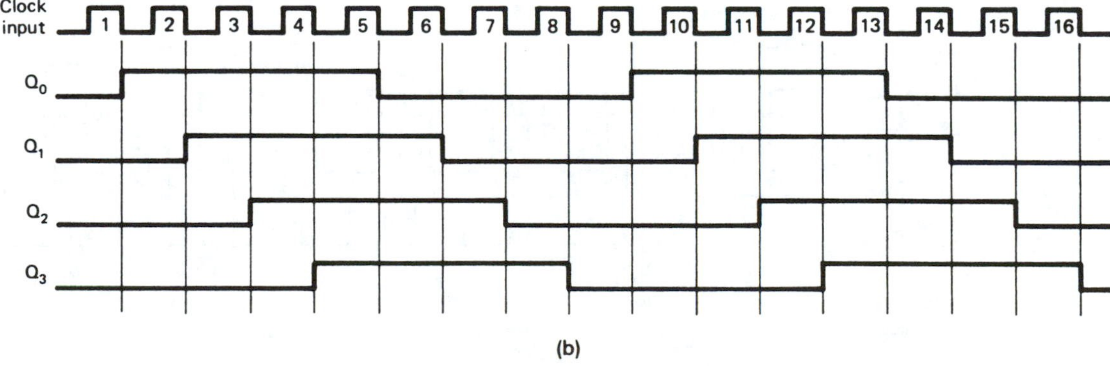

(b)

Figure 6–45 Johnson shift counter: (a) circuit connections; (b) output waveforms.

because J_0 is connected to $\overline{Q_3}$ (HIGH) and K_0 is connected to Q_3 (LOW). The Q_1, Q_2, and Q_3 outputs will follow the state of their preceding flip-flop because of their direct connection J-to-Q. Therefore, during period 2, the output is 1–0–0–0.

At the next negative clock edge, Q_0 remains HIGH because it takes on the *opposite* state of Q_3; Q_1 goes HIGH because it takes on the *same* state as Q_0; Q_2 stays LOW; and Q_3 stays LOW. Now the output is 1–1–0–0.

The sequence continues as shown in Figure 6–45b. Note that during period 5, Q_3 gets Set HIGH. At the end of period 5, Q_0 gets Reset LOW because the outputs of Q_3 are crossed, so Q_0 takes on the opposite state of Q_3.

6–10 SHIFT REGISTER ICS

Four-bit and 8-bit shift registers are commonly available in integrated-circuit packages. Depending on your needs, practically every possible load, shift, and conversion operation is available in a shift register IC.

Let's look at three popular shift register ICs to get familiar with using our data manuals and understanding the terminology and procedure for performing the various operations.

The first shift register to consider is the 74164 8-bit serial-in, parallel-out shift register. By looking at the logic symbol and logic diagram for the 74164 (Figure 6–46) we can see that it saves us the task of wiring together eight D flip-flops. The 74164 has two serial input lines (D_{Sa} and D_{Sb}), synchronously read in by a positive edge-triggered clock (C_p). The logic diagram shows both D_S inputs feeding into an AND gate. Therefore, either input can be used as an active-HIGH enable for data entry through the other input. Each positive edge clock pulse will shift the data bits one position to the right. Therefore, the first data bit entered (either LSB or MSB) will end up in the far right D flip-flop (Q_7) after eight clock pulses. The $\overline{\text{MR}}$ is an active-LOW Master-Reset that resets all eight flip-flops when pulsed LOW.

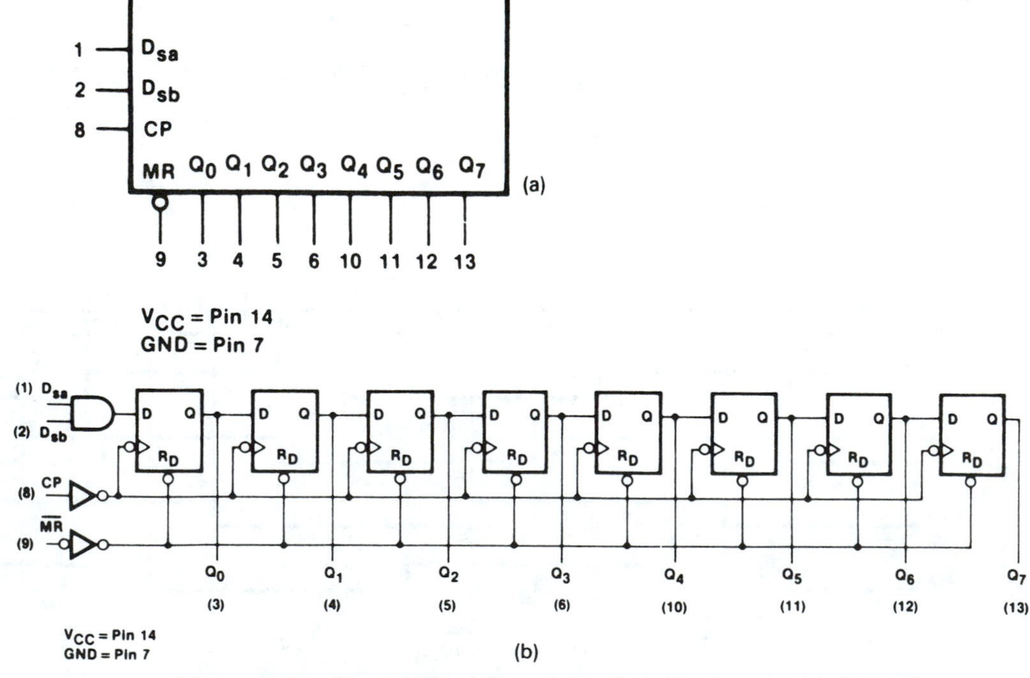

Figure 6–46 The 74164 shift register IC: (a) logic symbol; (b) logic diagram. [(b) Courtesy of Signetics Corporation]

EXAMPLE 6–12

Draw the circuit connections and timing waveforms for the serial-to-parallel conversion of the binary number 11010010 using a 74164 shift register.

Solution:

The serial-to-parallel conversion circuit and waveforms are shown in Figure 6–47. First, the register is cleared by a LOW on $\overline{MR}$, making $Q_0 - Q_7 = 0$. The strobe line is required to make sure that we only get eight clock pulses. The serial data are entered on the D_{Sb} line, MSB first. After eight clock pulses, the 8 data bits can be read at the parallel output pins (MSB at Q_7 and LSB at Q_0).

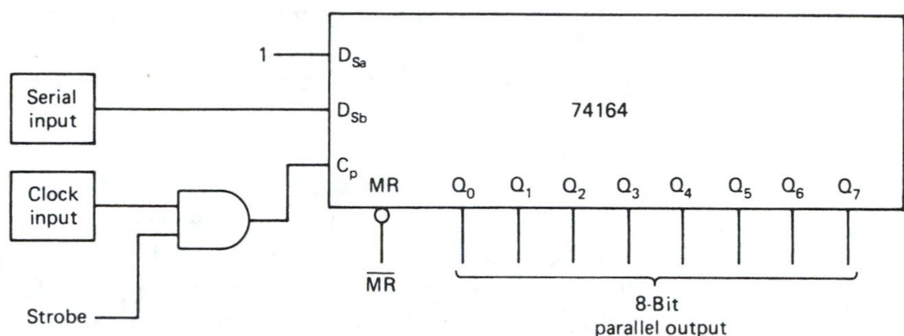

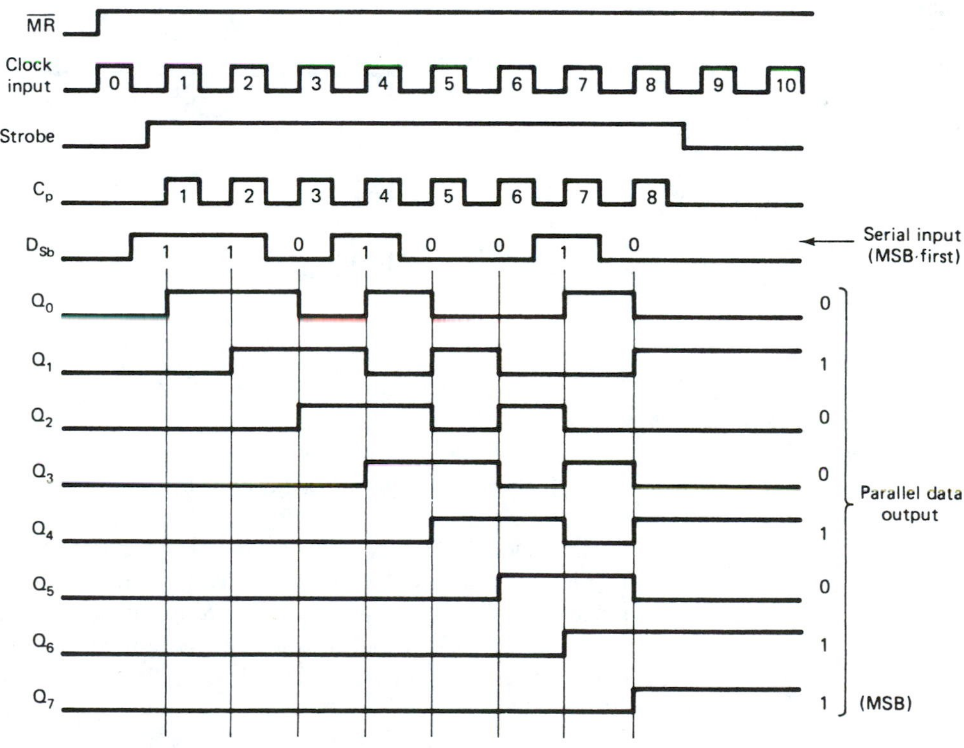

Figure 6–47 Serial-to-parallel conversion using the 74164 shift register IC.

The next IC to consider is the 74165 8-bit serial *or* parallel-in, serial-out shift register. The logic symbol for the 74165 is given in Figure 6–48.

Just by looking at the logic symbol, you should be able to determine the operation of the 74165. The $\overline{PL}$ is an active-LOW terminal for performing a parallel load of the 8 parallel input data bits. The $\overline{CE}$ is an active-LOW clock enable for starting/stopping

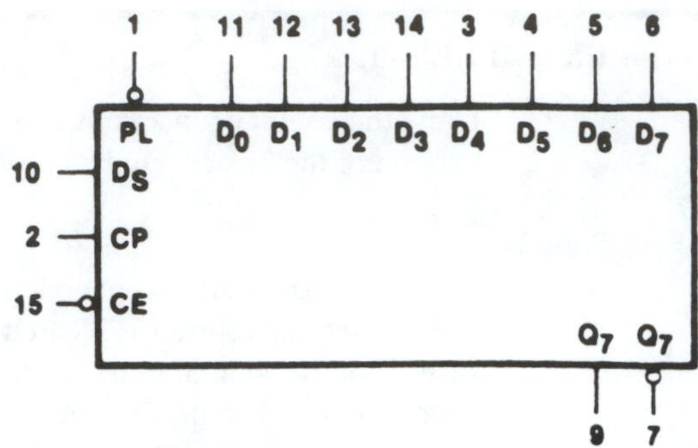

Figure 6–48 Logic symbol for the 74165 8-bit serial or parallel-in, serial-out shift register.

(shifting/holding) the shift operation by enabling/disabling the clock (same function as the "strobe" in Example 6–12).

The clock input (C_p) is positive edge triggered, so after each positive edge, the data bits are shifted one position to the right. The serial output (Q_7) and its complement ($\overline{Q_7}$) are available from the rightmost flip-flop's outputs.

Another shift register IC is the 74194 4-bit bidirectional universal shift register. It is called "universal" because it has a wide range of applications, including serial or parallel input, serial or parallel output, shift left or right, hold, and asynchronous Reset. The logic symbol for the 74194 is shown in Figure 6–49.

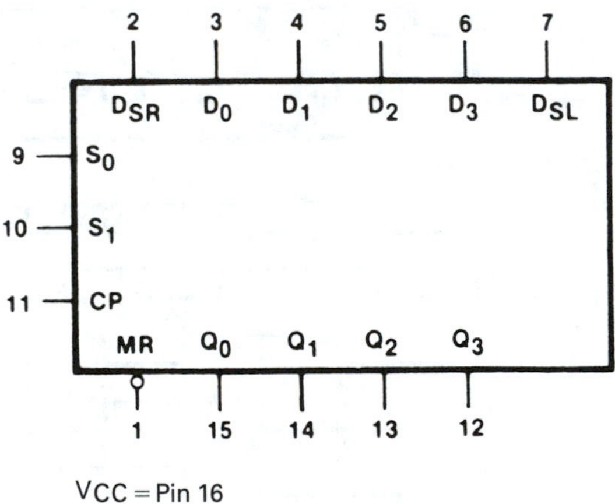

VCC = Pin 16
GND = Pin 8

***Figure 6–49** Logic symbol for the 74194 universal shift register.

The major differences with the 74194 are that there are separate serial inputs for shifting left or shifting right, and the operating mode is determined by the digital states of the mode control inputs, S_0 and S_1. S_0 and S_1 can be thought of as receiving a 2-bit binary code representing one of four possible operating modes ($2^2 = 4$ combinations). The four operating modes are shown in Table 6–3.

TABLE 6–3
Operating Modes of the 74194

Operating mode	S_1	S_0
Hold	0	0
Shift Left	1	0
Shift Right	0	1
Parallel Load	1	1

*See Application K-10 in Appendix K for a CPLD implementation of this shift register.

A complete mode select-function table for the 74194 is shown in Table 6–4. Table 6–4 can be used to determine the procedure and expected outcome of the various shift register operations.

From the function table we can see that a LOW input to the Master Reset ($\overline{MR}$) asynchronously resets Q_0 to Q_3 to 0. A parallel load is accomplished by making S_0, S_1 both HIGH and placing the parallel input data on D_0 to D_3. The register will then be parallel loaded synchronously by the first positive clock (C_p) edge. The 4 data bits can then be shifted to the right or left by making S_0–S_1 1–0 or 0–1 and applying an input clock to C_p.

A recirculating shift-right register can be set up by connecting Q_3 back into D_{SR} and applying a clock input (C_p) with $S_1 = 0$, $S_0 = 1$. Also, a recirculating shift-left

TABLE 6–4
Mode Select-Function Table for the 74194[a]

	Inputs							Outputs			
Operating mode	C_p	$\overline{MR}$	S_1	S_0	D_{SR}	D_{SL}	D_n	Q_0	Q_1	Q_2	Q_3
Reset (clear)	×	L	×	×	×	×	×	L	L	L	L
Hold (do nothing)	×	H	l[b]	l[b]	×	×	×	q_0	q_1	q_2	q_3
Shift Left	↑	H	h	l[b]	×	l	×	q_1	q_2	q_3	L
	↑	H	h	l[b]	×	h	×	q_1	q_2	q_3	H
Shift Right	↑	H	l[b]	h	l	×	×	L	q_0	q_1	q_2
	↑	H	l[b]	h	h	×	×	H	q_0	q_1	q_2
Parallel Load	↑	H	h	h	×	×	d_n	d_0	d_1	d_2	d_3

Courtesy of Philips Components–Signetics

[a]H = HIGH voltage level; h = HIGH voltage level one setup time prior to the LOW-to-HIGH clock transition; L = LOW voltage level; l = LOW voltage level one setup time prior to the LOW-to-HIGH clock transition; $d_n(q_n)$ = lowercase letters indicate the state of the referenced input (or output) one setup time prior to the LOW-to-HIGH clock transition; × = don't care; ↑ = LOW-to-HIGH clock transition.

[b]The HIGH-to-LOW transition of the S_0 and S_1 inputs on the 54/74194 should only take place while C_p is HIGH for conventional operation.

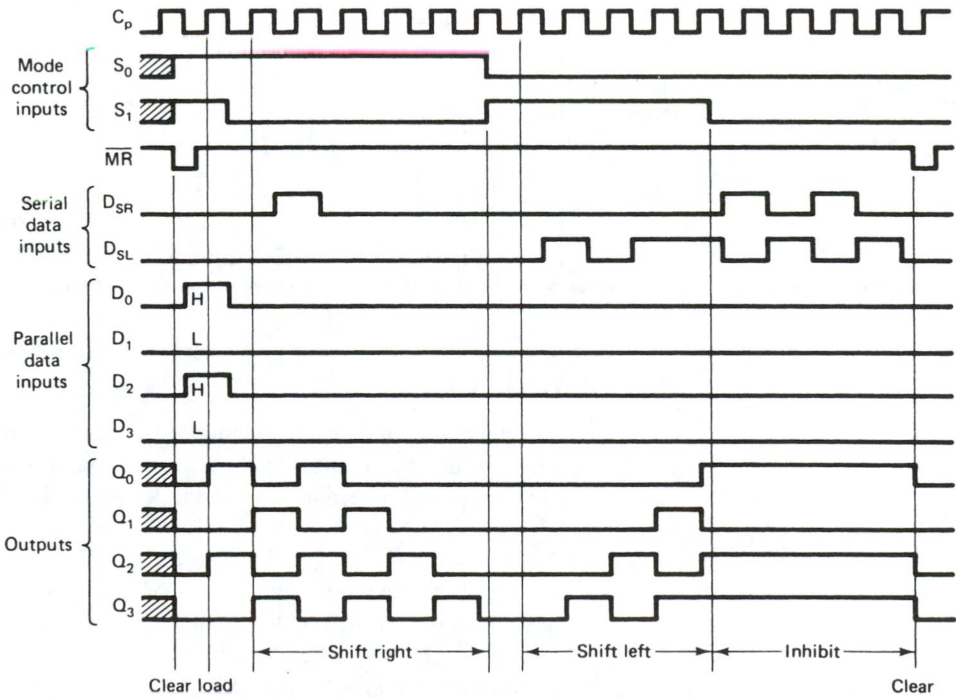

Figure 6–50 Typical clear–load–shift right–shift left–inhibit–clear sequences for a 74194.

register can be set up by connecting Q_0 into D_{SL} and applying a clock input (C_p) with $S_1 = 1$, $S_0 = 0$.

The best way to get a "feel" for the operation of the 74194 is to study the timing waveforms for a typical sequence of operations. Figure 6–50 shows the input data, control waveforms, and the output (Q_0 to Q_3) waveforms generated by a clear–load–shift right–shift left–inhibit–clear sequence. Study these waveforms carefully until you thoroughly understand the setup of the mode controls and the reason for each state change in the Q_0 to Q_3 outputs.

6–11 SYSTEM DESIGN APPLICATIONS FOR SHIFT REGISTERS

Shift registers have many applications in digital sequencing, storage, and transmission of serial and parallel data. The following designs will illustrate some of these applications.

APPLICATION 6–5

Design a 16-bit serial-to-parallel converter.

Solution:

First we have to look through a TTL data manual to see what is available. The 74164 is an 8-bit serial-in, parallel-out shift register. Let's cascade two of them together to form a 16-bit register. Figure 6–51 shows that the Q_7 output is fed into the serial input of the second 8-bit register. That way, as the data bits are shifted through the register, when they reach Q_7 the next shift will pass the data into Q_0 of the second register (via D_{Sa}), making it a 16-bit register. The second serial input (D_{Sb}) of each stage is internally ANDed with the D_{Sa} input so that it serves as an active-HIGH enable input.

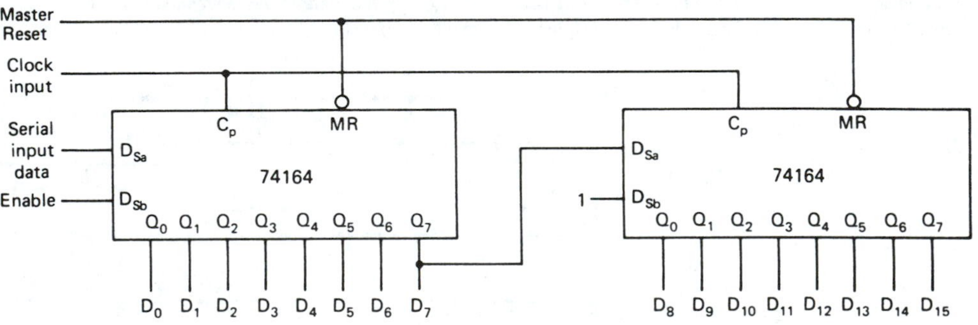

Figure 6–51 Sixteen-bit serial-to-parallel converter.

APPLICATION 6–6

Design a circuit and provide the input waveforms required to perform a parallel-to-serial conversion. Specifically, a hexadecimal B (1011) is to be parallel loaded, then transmitted repeatedly to a serial device LSB first.

Solution:

By controlling the mode control inputs (S_0, S_1) of a 74194 we can perform a parallel load and then shift right repeatedly. The serial output data are taken from Q_3 as shown in Figure 6–52. The 74194 universal shift register is connected as a recirculating parallel-to-serial converter. Each time the Q_3 serial output level is sent to the serial device, it is also recirculated back into the left end of the shift register.

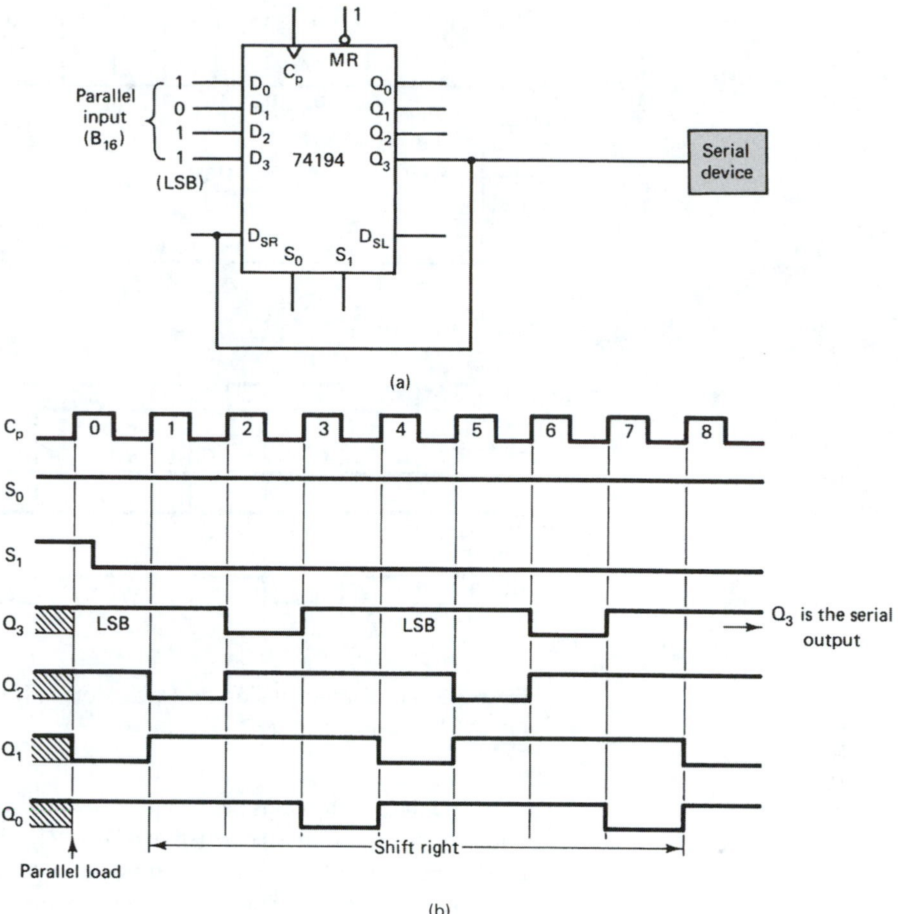

Figure 6–52 Four-bit parallel-to-serial converter: (a) circuit connections; (b) waveforms.

First, at the positive edge of clock pulse 0, the register is parallel loaded with a 1011 (B_{16}) because the mode controls (S_0S_1) are HIGH–HIGH. (D_3 is loaded with the LSB because it will be the first bit out when we shift right.)

Next, the mode controls (S_0S_1) are changed to HIGH–LOW for a shift-right operation. Now, each successive positive clock edge will shift the data bit one position to the right. The Q_3 output will continuously have the levels 1101–1101–1101–etc., which is a backwards hexadecimal B (LSB first).

APPLICATION 6–7

Design an interface to an 8-bit serial printer. Sketch the waveforms required to transmit the single ASCII code for an asterisk (*). *Note:* ASCII is a 7-bit code used to represent alphanumeric data in computer systems. The ASCII code for an asterisk is 010 1010 (see Table 1-4). Let's make the unused eighth bit (MSB) a 0.

Solution:

The circuit design and waveforms are shown in Figure 6–53. The 74165 is chosen for the job because it is an 8-bit register that can be parallel loaded, then shifted synchronously by the clock input to provide the serial output to the printer.

During pulse 0 the register is loaded with the ASCII code for an asterisk (the LSB is put into D_7 because we want it to come out first). The clock input is then enabled by a LOW on $\overline{CE}$. Each positive pulse on C_p from then on will shift the data bits one position to the right. After the eighth clock pulse (0 to 7) the printer will have received all 8 serial data bits. Then the $\overline{CE}$ line is brought HIGH

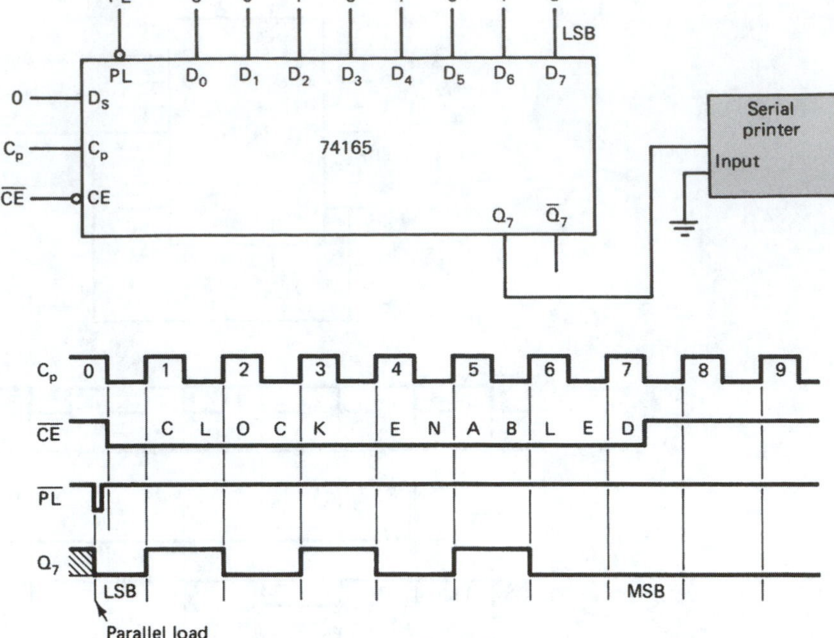

Figure 6–53 Circuit design and waveforms for the transmission of an ASCII character to a serial printer.

to disable the synchronous clock input. To avoid any racing problems, the printer will read the Q_7 line at each negative edge of C_p so that the level will definitely be a stable HIGH or LOW, as shown in Figure 6–53.

At this point you may be wondering how we are practically going to provide electronically the necessary signals on the $\overline{CE}$ and $\overline{PL}$ lines. An exact degree of timing must be provided on these lines to ensure that the register–printer interface communicates properly. These signals will be provided by a microprocessor and are called the *handshaking* signals.

Microprocessor theory and programming is an advanced digital topic and is discussed later in this book. For now, it is important for us to realize that these signals are required and be able to sketch their timing diagrams.

SUMMARY

In this chapter we have learned that

1. Toggle flip-flops can be cascaded end to end to form ripple counters.
2. Ripple counters cannot be used in high-speed circuits because of the problem they have with the accumulation of propagation delay through all of the flip-flops.
3. Any modulus (or divide-by) counter can be formed by resetting the basic ripple counter when a specific count is reached.
4. A glitch is an unwanted level transition that may appear on some of the output bits of a ripple counter.
5. Ripple counter ICs such as the 7490, 7492, and 7493 have four flip-flops integrated into a single package providing four-bit counter operations.

6. Four-bit counter ICs can be cascaded end to end to form counters with higher than MOD-16 capability.
7. Seven-segment LED displays choose among seven separate LEDs (plus a decimal point LED) to form the ten decimal digits. They are constructed with either the anodes or the cathodes connected to a common pin.
8. LED displays require a decoder/driver IC such as the 7447 to decode BCD data into a seven-bit code to activate the appropriate segments to illuminate the correct digit.
9. Synchronous counters eliminate the problem of accumulated propagation delay associated with ripple counters by driving all four flip-flops with a common clock.
10. The 74192 and 74193 are 4-bit synchronous counter ICs. They have a count-up/count-down feature and can accept a 4-bit parallel load of binary data.
11. Shift registers can be used for converting data from serial to parallel and parallel to serial.
12. One common form of digital communication is for a sending computer to convert its data from parallel to serial, then transmit them over a telephone line to a receiving computer, which converts them back from serial to parallel.
13. Simple shift registers can be constructed by connecting the *Q*-outputs of one *J-K* flip-flop into the *J-K* inputs of the next flip-flop. Several flip-flops can be cascaded together this way, driven by a common clock, to form multibit shift registers.
14. The ring and the Johnson shift counters are two specialized shift registers used to create sequential control waveforms.
15. Several multibit shift register ICs are available for the designer to choose from. They generally have 4 or 8 internal flip-flops and are designed to shift either left or right and perform either serial-to-parallel or parallel-to-serial conversions.
16. The 74194 is called a universal 4-bit shift register because it can shift in either direction and can receive and convert to either format.

GLOSSARY

Asynchronous Counter: *See* ripple counter.

Bit String: Two or more binary numbers (bits) that together are used as data bits (a "string" of binary numbers).

Cascade: In multistage systems when the output of one stage is fed directly into the input of the next.

Clock Enable: A separate input pin included on some ICs, used to enable or disable the clock input signal.

Common-Anode LED: A seven-segment LED display whose LED anodes are all connected to a common point and supplied with +5 V. Each LED segment is then turned on by supplying a LOW level (via a limiting resistor) to the appropriate LED cathode.

Data Bit: A single binary representation (0 or 1) of digital information.

Data Conversion: Transformation of digital information from one format to another (e.g., serial-to-parallel conversion).

Data Transmission: The movement of digital information from one location to another.

Digital Sequencer: A system (like a shift counter) that can produce a specific series of digital waveforms to drive another device in a specific sequence.

Divide-by-N: The Q outputs in counter operations will oscillate at a frequency that is at some multiple (N) of the input clock frequency. For example, in a divide-by-8 (MOD-8) counter the output frequency of the highest-order Q (Q_2) is one-eighth the frequency of the input clock.

Glitch: A short-duration-level change in a digital circuit.

Mode Control: Input pins available on some ICs used to control the operating functions of that IC.

Modulus: In a digital counter the modulus is the number of different counter steps.

Oscillate: Change digital states repeatedly (HIGH–LOW–HIGH–LOW–etc.).

Output Enable: An input pin on an IC that can be used to enable or disable the outputs. When disabled, the outputs are in the "float condition."

Parallel Enable: An IC input pin used to enable or disable a synchronous parallel load of data bits.

Parallel Load: A feature on some counters and shift registers that allows you to load all 4 bits at the same time, asynchronously.

Recirculating: In a shift register, instead of letting the shifting data bits "drop" out of the end of the register, a recirculating connection can be made to pass the bits back into the front end of the register.

Ripple Blanking: A feature supplied with display decoders to enable the suppression of leading and trailing zeros.

Ripple Counter: (Asynchronous counter) A multibit counter whose clock input trigger is not connected to each flip-flop but instead has to propagate through each flipflop to reach the input of the next. The fact that the clock has to "ripple" through from stage to stage tends to decrease the maximum operational frequency of the ripple counter.

Sequential: Operations that follow a predetermined sequence of digital states triggered by a timing pulse or clock.

Seven-Segment LED: Seven light-emitting diodes fabricated in a single package. By energizing various combinations of LED segments, the 10 decimal digits can be displayed.

Shift Counter: A special-purpose shift register with modifications to its connections and preloaded with a specific value to enable it to output a special sequence of digital waveforms. It does not count in true binary, but instead is used for special sequential waveform generation.

Shift Register: A storage device containing two or more data bits, capable of moving the data to the left or right and perform conversions between serial and parallel.

Skewed: A "skewed" waveform or pulse is one that is offset to the right or left with respect to the time axis.

Strobe: A connection used in digital circuits to enable or disable a particular function.

Synchronous Counter: A multibit counter whose clock input trigger is connected to each flip-flop, so that each flip-flop will operate in step with the same input clock transition.

Terminal Count: The highest (or lowest) count in a multibit counting sequence.

Up/Down-Counter: A counter that is capable of counting up or counting down.

PROBLEMS

6–1. How are sequential logic circuits different from combinational logic gate circuits?

6–2. What is the modulus of a counter whose output counts from

(a) 0 to 7? **(b)** 0 to 18?
(c) 5 to 0? **(d)** 10 to 0?
(e) 2 to 15? **(f)** 7 to 3?

6–3. How many *J-K* flip-flops are required to construct the following counters?
(a) MOD-7 (b) MOD-8
(c) MOD-2 (d) MOD-20
(e) MOD-33 (f) MOD-15

6–4. If the input frequency to a 6-bit counter is 10 MHz, what is the frequency at the following output terminals?
(a) 2^0 (b) 2^1
(c) 2^2 (d) 2^3
(e) 2^4 (f) 2^5

6–5. Draw the timing waveforms at $\overline{C_p}$, 2^0, 2^1, and 2^2 for a 3-bit binary up-counter for 10 clock pulses.

6–6. Repeat Problem 6–5 for a binary down-counter.

6–7. What is the highest binary number that can be counted using the following number of flip-flops?
(a) 2 (b) 4
(c) 7 (d) 1

6–8. In a 5-bit counter the frequency at the following output terminals is what fraction of the input clock frequency?
(a) 2^0 (b) 2^1
(c) 2^2 (d) 2^3
(e) 2^4

6–9. How many flip-flops are required to form the following divide-by-*N* frequency dividers?
(a) Divide-by-4 (b) Divide-by-15
(c) Divide-by-12 (d) Divide-by-18

6–10. Explain why the propagation delay of a flip-flop affects the maximum frequency at which a ripple counter can operate.

6–11. Sketch the $\overline{C_p}$, 2^0, 2^1, and 2^2 output waveforms for the counter shown in Figure P6–11. (Assume that flip-flops are initially Reset.)

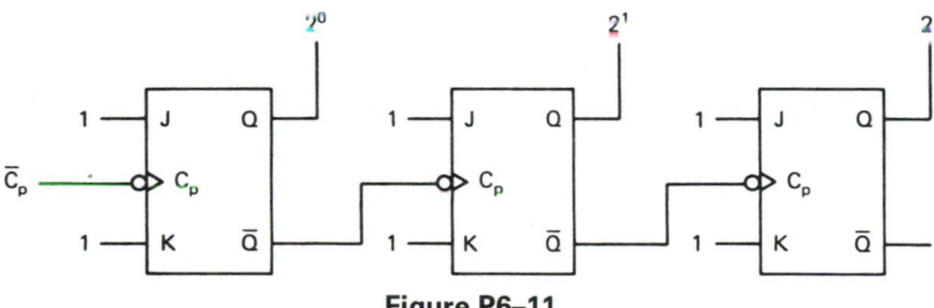

Figure P6–11

6–12. Is the counter of Problem 6–11 an up- or down-counter, and is it a MOD-8 or MOD-16?

6–13. Sketch the connections to a 3-bit ripple up-counter that can be used as a divide-by-6 frequency divider.

6–14. Design a circuit that will convert a 2-MHz input frequency into a 0.4-MHz output frequency.

6–15. Design and sketch a MOD-11 ripple up-counter that can be manually Reset by an external pushbutton.

Counter ICs

6–16. Describe the major differences between the 7490, 7492, and the 7493 TTL ICs.

6–17. Assume that you have one 7490 and one 7492. Show the external connections that are required to form a divide-by-24.

6–18. Repeat Problem 6–17 using two 7492s to form a divide-by-36.

6–19. Using as many 7492s and 7490s as you need, sketch the external connections required to divide a 60-pps clock down to one pulse per day.

6–20. Make the necessary external connections to a 7493 to form a MOD-10 counter.

6–21. Design a ripple counter circuit that will flash an LED ON for 40 ms, OFF for 20 ms (assume that a 100-Hz clock oscillator is available). (*Hint:* Study the output waveforms from a MOD-6 counter.)

6–22. Design a circuit that will turn on an LED 6 s after you press a momentary push-button. (Assume that a 60-pps clock is available.)

6–23. What modification to the egg-timer circuit of Figure 6–26 could be made to allow you to turn off the buzzer without shutting off the power?

6–24. Calculate the size of the series current-limiting resistor that could be used in Figure 6–28 to limit the LED current to 15 mA instead of 10 mA.

6–25. In Figure 6–31 instead of using a resistor dip network, some designers use a single limiting resistor in series with the 5-V supply and connect the 7447 outputs directly to the LED inputs to save money. It works, but the display does not look as good; can you explain why?

Synchronous Counters

6–26. What advantage does a synchronous counter have over a ripple counter?

6–27. Sketch the waveforms at $\overline{C_p}$, 2^0, 2^1, and 2^2 for 10 clock pulses for the 3-bit synchronous counter shown in Figure P6–27.

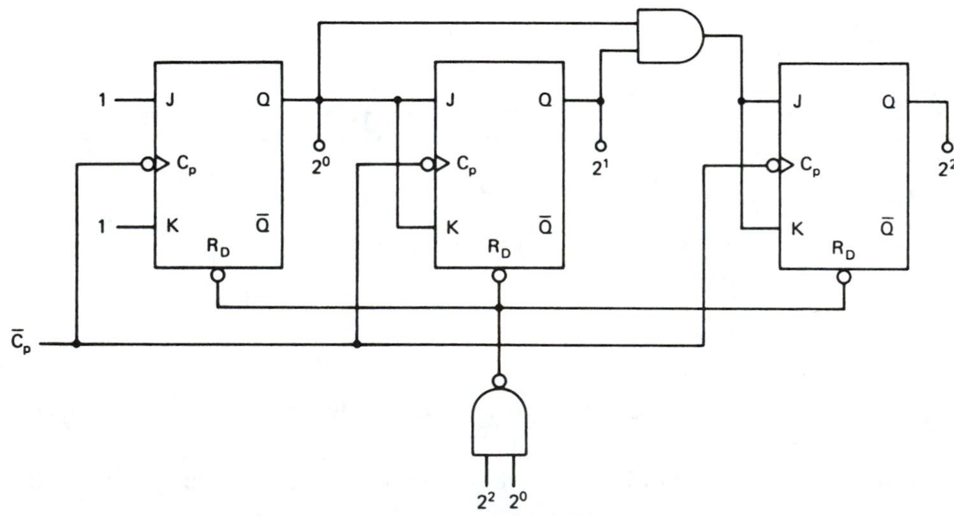

Figure P6–27

6–28. The duty cycle of a square wave is defined as the time the wave is HIGH, divided by the total time for one period. From the waveforms that you sketched for Problem 6–27, find the duty cycle for the 2^2 output wave.

6–29. Sketch the timing waveforms at $\overline{TC_D}$, $\overline{TC_U}$, Q_0, Q_1, Q_2, and Q_3 for the 74192 counter shown in Figure P6–29.

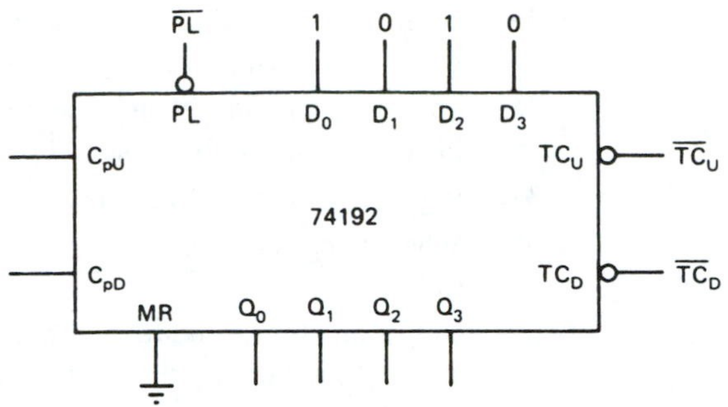

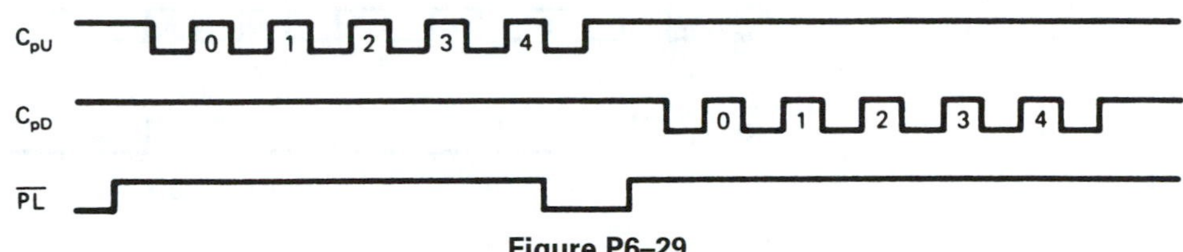

Figure P6–29

6–30. Make all the necessary pin connections to a 74193 without using external gating to form a divide-by-4 frequency divider. Make it an up-counter and show the waveforms at C_{pU}, $\overline{TC_U}$, Q_0, Q_1, Q_2, and Q_3.

Shift Registers

6–31. In Figure P6–31, will the data bits be shifted right or left with each clock pulse? Will they be shifted on the positive or negative clock edge?

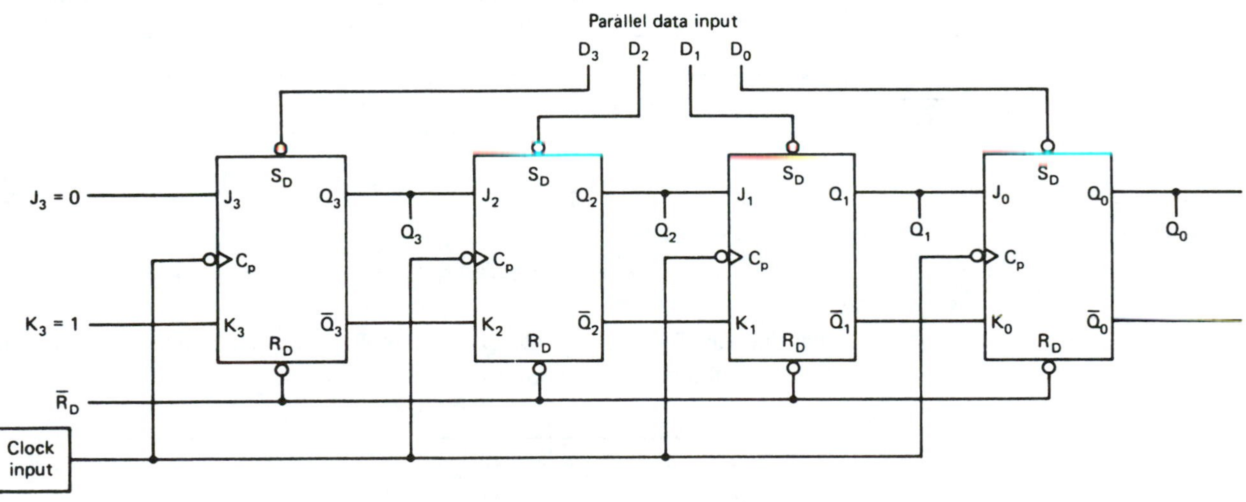

Figure P6–31

6–32. If the register of Figure P6–31 is initially parallel loaded with $D_3 = 0$, $D_2 = 1$, $D_1 = 0$, and $D_0 = 0$, what will the output at Q_3 to Q_0 be after two clock pulses? After four clock pulses?

6–33. Repeat Problem 6–32 for $J_3 = 1$, $K_3 = 0$.

6–34. Change Figure P6–31 to a recirculating shift register by connecting Q_0 back to J_3 and $\overline{Q_0}$ back to K_3. If the register is initially loaded with a 0110, what is the output at Q_3 to Q_0:

(a) After two clock pulses?

(b) After four clock pulses?

6–35. Outline the steps that you would take to parallel load the binary equivalent of a hex B into the register of Figure P6–31.

6–36. To use Figure P6–31 as a parallel-to-serial converter, where are the data input line(s) and data output line(s)?

6–37. Repeat Problem 6–36 for a serial-to-parallel converter.

6–38. What changes have to be made to the circuit of Figure P6–31 to make it a Johnson shift counter?

6–39. How many flip-flops are required to produce the waveform shown in Figure P6–39 at the Q_0 output of a ring shift counter?

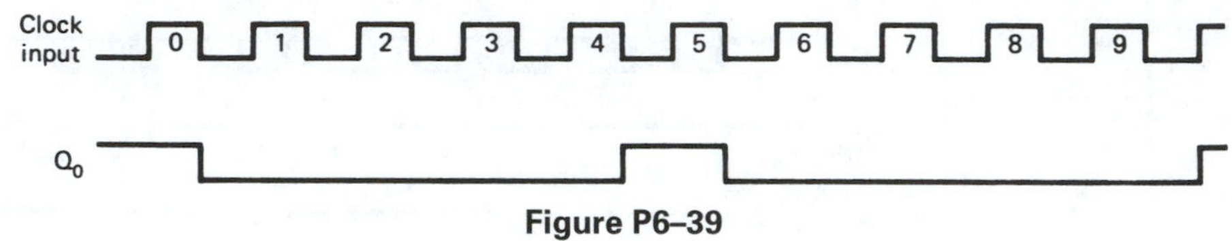

Figure P6–39

6–40. Sketch the waveforms at Q_2 for the first seven clock pulses generated by the circuit shown in Figure P6–40.

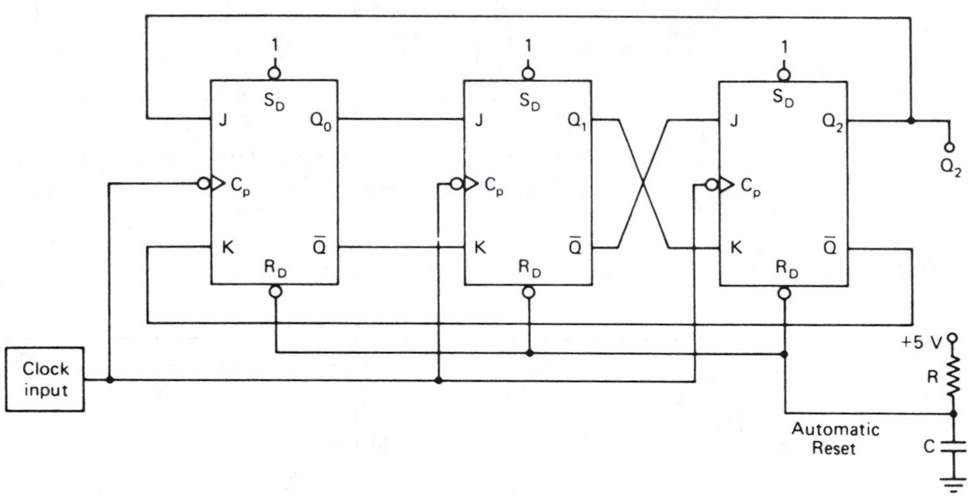

Figure P6–40

6–41. Sketch the waveforms at $\overline{C_p}$, Q_0, Q_1, and Q_2 for seven clock pulses for the ring shift counter shown in Figure P6–41.

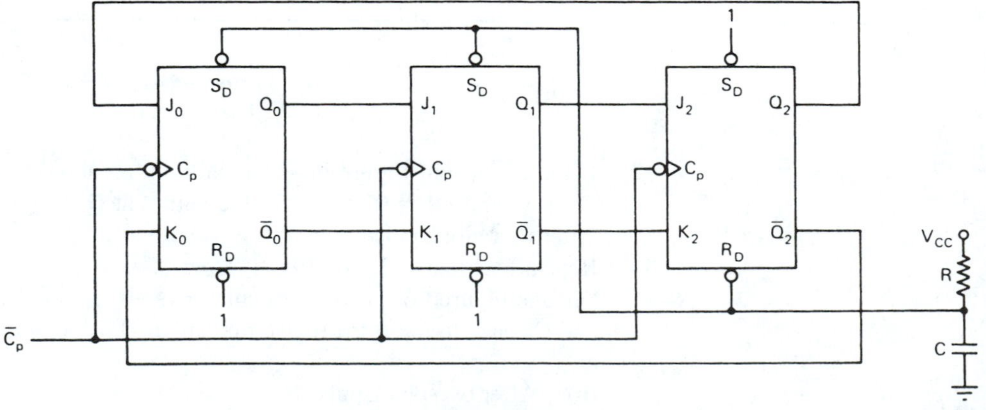

Figure P6–41

6–42. Using the Johnson shift counter output waveforms in Figure 6–45, add some logic gates to produce the waveforms at *X*, *Y*, and *Z* shown in Figure P6–42.

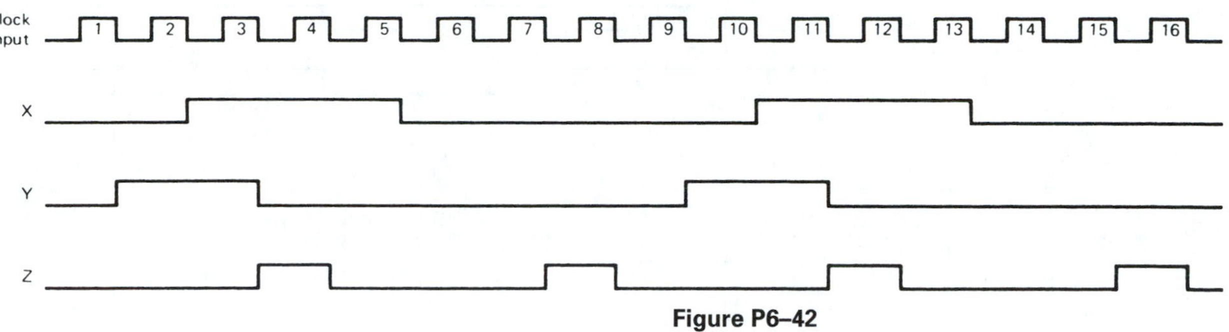

Figure P6–42

Shift Register ICs

6–43. Sketch the output waveforms at Q_0 to Q_3 for the 74194 circuit shown in Figure P6–43. Also, list the operating mode at each positive clock edge.

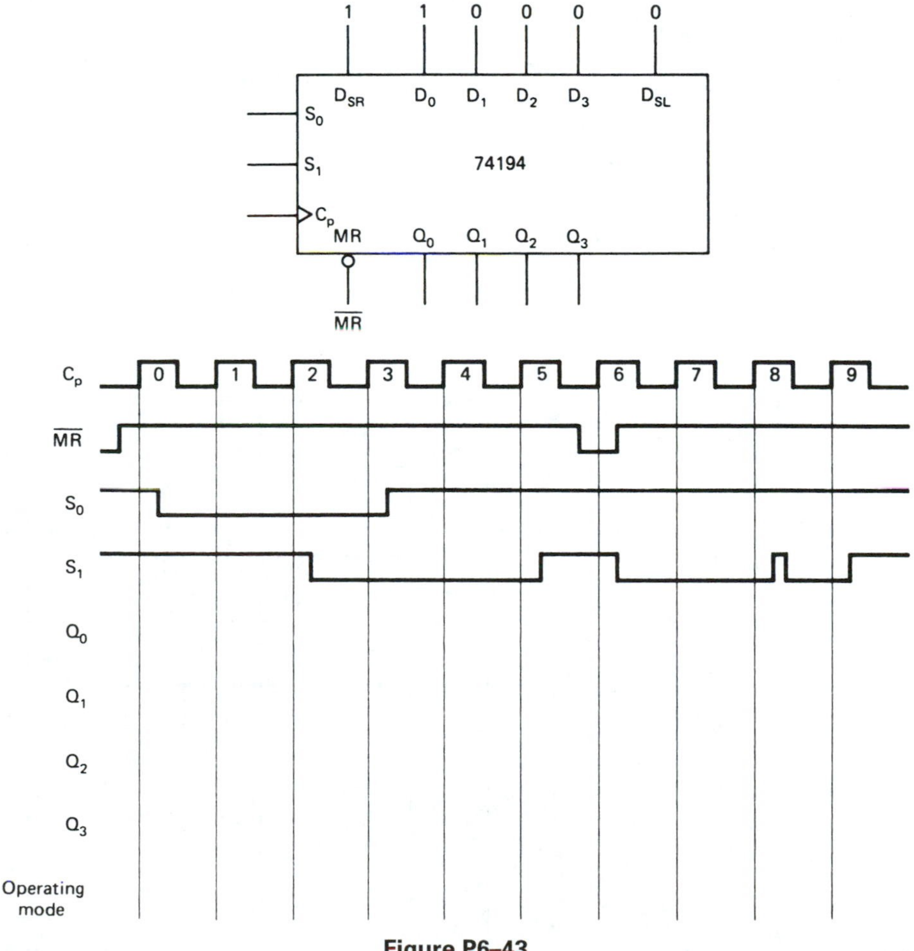

Figure P6–43

6–44. Repeat Problem 6–43 for the input waveforms shown in Figure P6–44.

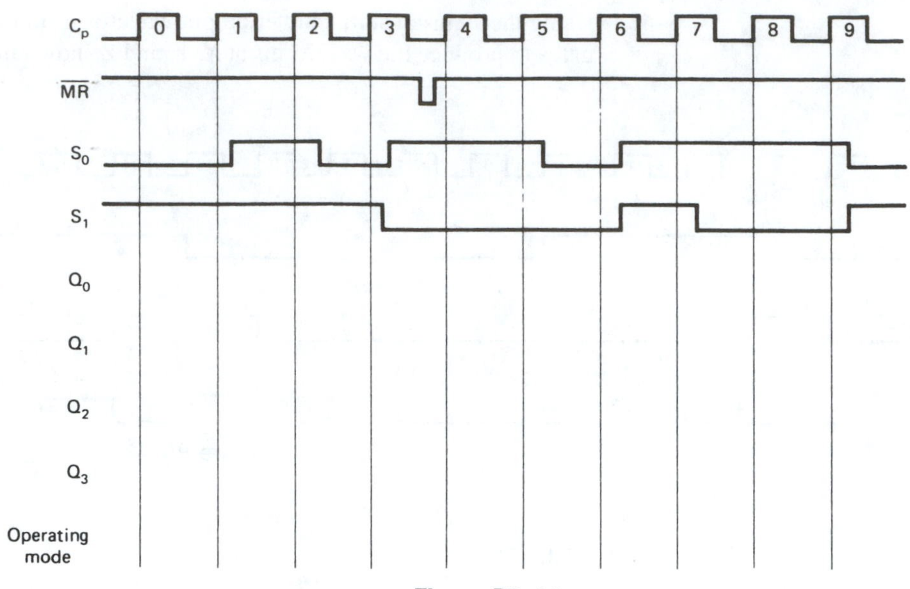

Figure P6–44

6–45. Sketch the output waveforms at Q_0 to Q_3 for the 74194 circuit shown in Figure P6–45. Also, list the operating mode at each positive clock edge.

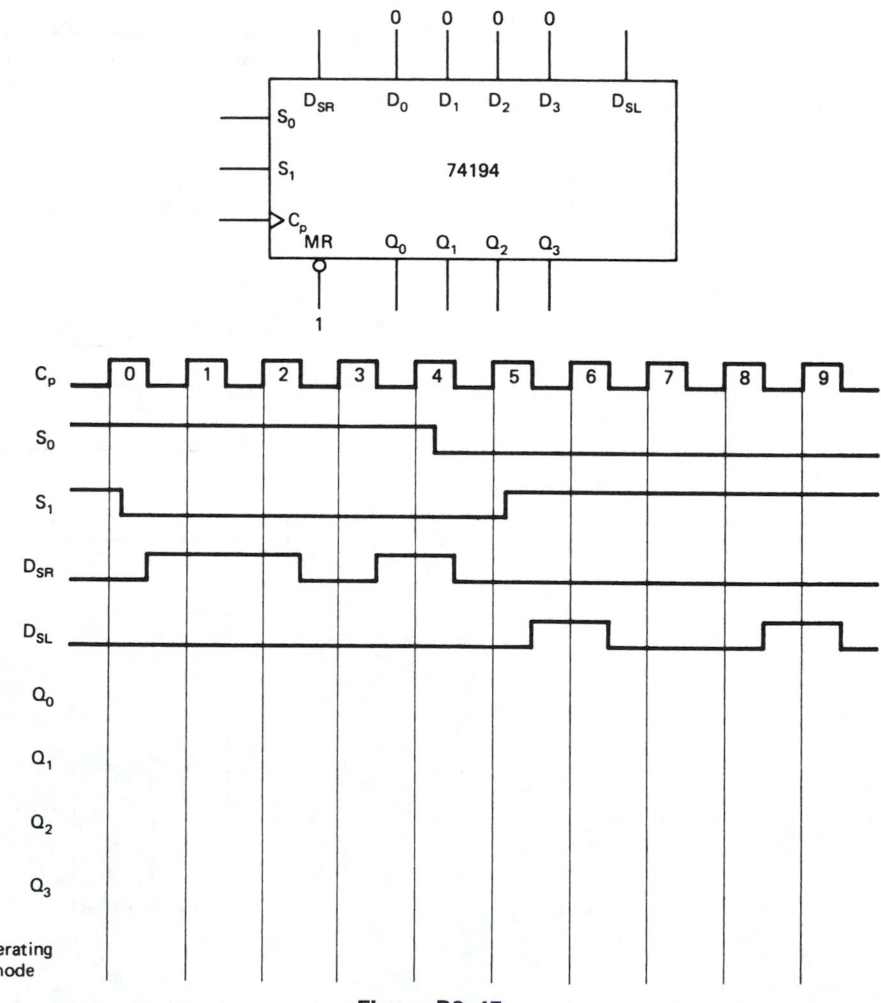

Figure P6–45

6–46. Repeat Problem 6–45 for the waveforms shown in Figure P6–46.

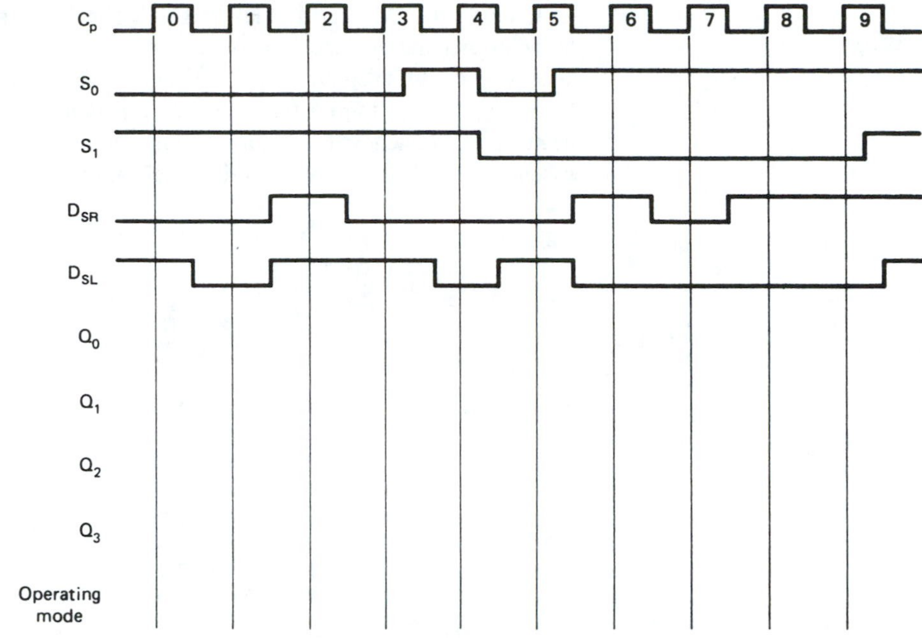

Figure P6–46

6–47. Draw the timing waveforms (similar to Figure 6–47) for a 74164 used to convert the serial binary number 1001 0110 into parallel.

6–48. Draw the circuit connections and timing waveforms for a 74165 used to convert the parallel binary number 1001 0110 into serial MSB first.

6–49. Using your TTL data manual, describe the differences between the 74195 and the 74395A shift registers.

6–50. Using your TTL data manual, describe the differences between the 74164 and the 74165 shift registers.

6–51. Describe how the procedure for parallel loading the 74165 differs from parallel loading a 74166.

6–52. Design a system that can be used to convert an 8-bit serial number LSB first into an 8-bit serial number MSB first. Show the timing waveforms for 16 clock pulses and any control pulses that may be required for the binary number 1011 0100.

SCHEMATIC INTERPRETATION PROBLEMS

6–53. The circuit on the HC11D0 schematic is capable of parallel as well as serial communication via connectors P3 and P2. Which is parallel and which is serial (*hint:* TX stands for transmit, RX stands for receive)?

6–54. Find U9 and U10 of the Watchdog Timer schematic. These are counter ICs that output their 4-bit binary count to Q_0–Q_3. On a separate piece of paper, draw the connections that you would make to add the output of U9 to the output of U10 using 74HC283 4-bit adder IC.

6–55. The 74161s in the Watchdog Timer schematic are used to form an 8-bit counter.

(a) Which is the HIGH-order and which is the LOW-order counter?

(b) Is the parallel-load feature being used on these counters?

(c) How are the counters reset in this circuit?

6–56. On a separate piece of paper redesign the counter section of the Watchdog Timer schematic by replacing the 74161s with 74193s.

6–57. The 68HC11 microcontroller in the HC11D0 master board schematic provides a clock output signal at the pin labeled E. This clock signal is used as the input to the LCD controller, M1 (grid location E-7). The frequency of this signal is 9.8304 MHz, as dictated by the crystal on the 68HC11. To experiment with different clock speeds on the LCD controller, you want to divide that frequency by 2, 4, 8, and 16 before inputting it to pins 6 and 10. Design a circuit using a 4-bit counter IC connected as a frequency divider, and a multiplexer IC to select which counter output is sent to the LCD controller for its clock signal.

7 Interfacing to the Analog World

OBJECTIVES

Upon completion of this chapter, you should be able to:

- Perform the basic calculations involved in the analysis of operational amplifier circuits.
- Explain the operation of binary-weighted and *R*/2*R* digital-to-analog converters.
- Make the external connections to a digital-to-analog IC to convert a numeric binary string into a proportional analog voltage.
- Discuss the meaning of the specifications for converter ICs as given in a manufacturer's data manual.
- Explain the operation of parallel-encoded, counter-ramp, and successive-approximation analog-to-digital converters.
- Make the external connections to an analog-to-digital converter IC to convert an analog voltage to a corresponding binary string.
- Explain the operation and uses of various analog transducers.
- Discuss the operation of a typical data acquisition system.

INTRODUCTION

Most physical quantities that we deal with in this world are *analog* in nature. For example, temperature, pressure, and speed are not simply 1s and 0s but instead take on an infinite number of possible values. To be understood by a digital system, these values must be converted into a binary string representing their value; thus we have the need for *analog-to-digital* conversion. Also, it is important when we need to use a computer to control analog devices to be able to convert from *digital to analog.*

Devices that convert physical quantities into electrical quantities are called *transducers.* Transducers are readily available to convert such quantities as temperature, pressure, velocity, position, and direction into a proportional analog voltage or current. For example, a common transducer for measuring temperature is a thermistor. A thermistor is simply a temperature-sensitive resistor. As its temperature changes, so does its resistance. If we send a constant current through the thermistor, then measure the voltage across it, we can determine its resistance *and* temperature.

7–1 DIGITAL AND ANALOG REPRESENTATIONS

For *analog-to-digital* (A/D) or *digital-to-analog* (D/A) converters to be useful, there has to be a meaningful representation of the analog quantity as a digital representation and the digital quantity as an analog representation. If we choose a convenient range of analog levels such as 0 to 15 V, we could easily represent each 1-V step as a unique digital code, as shown in Figure 7–1.

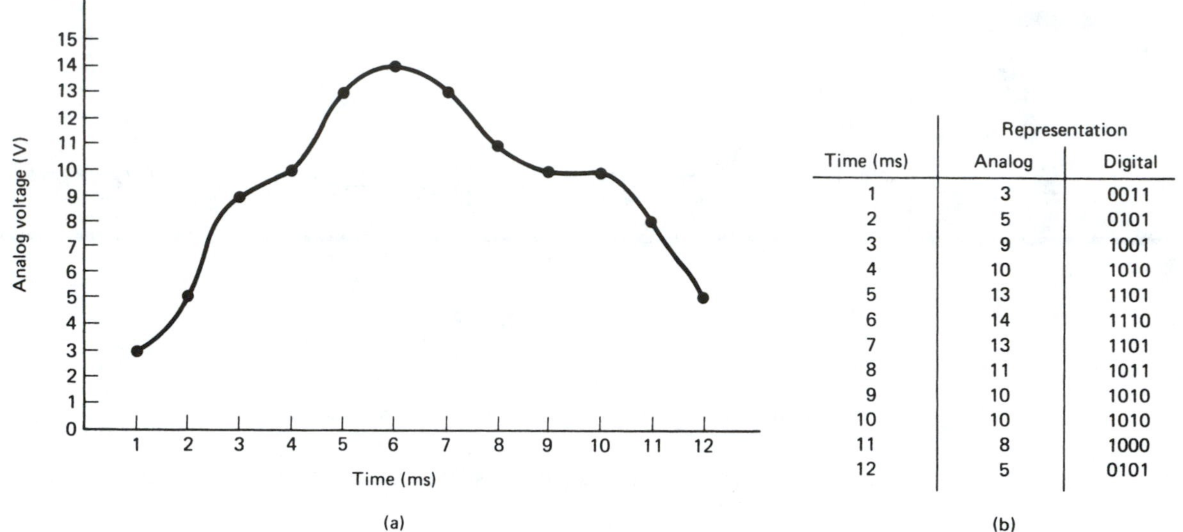

Time (ms)	Representation	
	Analog	Digital
1	3	0011
2	5	0101
3	9	1001
4	10	1010
5	13	1101
6	14	1110
7	13	1101
8	11	1011
9	10	1010
10	10	1010
11	8	1000
12	5	0101

Figure 7–1 Analog and digital representations: (a) voltage versus time; (b) representations at 1-ms intervals.

Figure 7–1 shows that for each analog voltage we can determine an equivalent digital representation. Using four binary positions gives us 4-bit *resolution,* which allows us to develop 16 different representations, with the increment between each being 1 part in 16. If we need to represent more than just 16 different analog levels, we would have to use a digital code with more than four binary positions. For example, a D/A converter with 8-bit resolution will provide increments of 1 part in 256, which provides much more precise representations.

7–2 OPERATIONAL AMPLIFIER BASICS

Most A/D and D/A circuits require the use of an op amp for signal conditioning. There are three characteristics of op amps that make them an almost *ideal amplifier*: (1) very high input impedance, (2) very high voltage gain, and (3) very low output impedance. In this section we gain a basic understanding of how an op amp works, and in future sections we see how it is used in the conversion process. A basic op-amp circuit is shown in Figure 7–2.

The symbol for the op amp is the same as that for a comparator, but when it is connected as shown in Figure 7–2 it provides a much different function. The basic theory involved in the operation of the op-amp circuit in Figure 7–2 is as follows:

1. The impedance looking into the + and − input terminals is assumed to be infinite; therefore, I_{in} (+), (−) = 0 A.

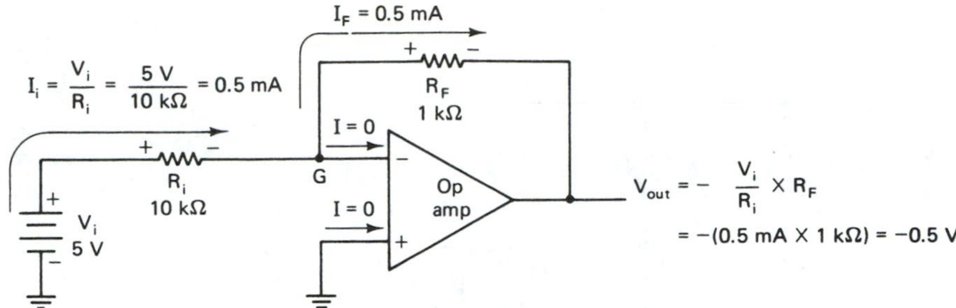

Figure 7–2 Basic op-amp operation.

2. Point G is assumed to be at the same potential as the + input; therefore, point G is at 0 V, called *virtual ground* (virtual means "in effect" but not actual. It is at 0 V, but it cannot sink current).
3. With point G at 0 V, there will be 5 V across the 10-kΩ resistor, causing 0.5 mA to flow.
4. The 0.5 mA cannot flow into the op amp; therefore, it flows up through the 1-kΩ resistor.
5. Since point G is at virtual ground, and since V_{out} is measured with respect to ground, V_{out} is equal to the voltage across the 1-kΩ resistor, which is −0.5 V.

EXAMPLE 7–1

Find V_{out} in Figure 7–3.

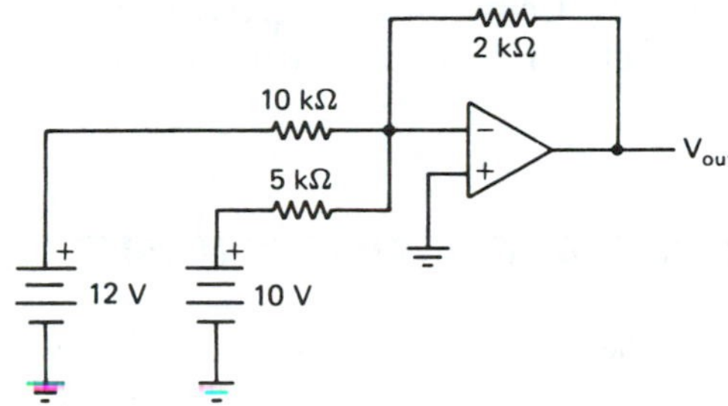

Figure 7–3 Op-amp circuit for Example 7–1.

Solution:

$$I_{10\,k\Omega} = \frac{12\text{ V}}{10\text{ k}\Omega} = 1.2\text{ mA}$$

$$I_{5\,k\Omega} = \frac{10\text{ V}}{5\text{ k}\Omega} = 2\text{ mA}$$

$$I_{2\,k\Omega} = 1.2 + 2 = 3.2\text{ mA}$$

$$V_{out} = -(3.2\text{ mA} \times 2\text{ k}\Omega) = -6.4\text{ V}$$

7–3 BINARY-WEIGHTED DIGITAL-TO-ANALOG CONVERTERS

A basic D/A converter can be built by expanding on the information presented in Section 7–2. Example 7–1 showed us that the 2-kΩ resistor receives the *sum* of the currents heading toward the op amp from the two input resistors. If we scale the input resistors with a binary weighting factor, each input can be made to provide a binary-weighted amount of current, and the output voltage will represent a sum of all the binary-weighted input currents, as shown in Figure 7–4.

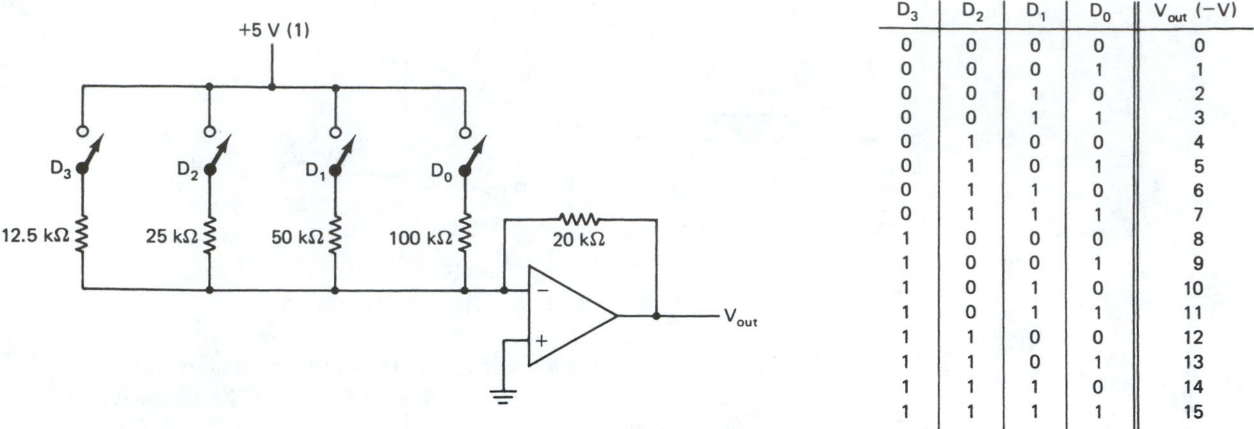

D_3	D_2	D_1	D_0	V_{out} (−V)
0	0	0	0	0
0	0	0	1	1
0	0	1	0	2
0	0	1	1	3
0	1	0	0	4
0	1	0	1	5
0	1	1	0	6
0	1	1	1	7
1	0	0	0	8
1	0	0	1	9
1	0	1	0	10
1	0	1	1	11
1	1	0	0	12
1	1	0	1	13
1	1	1	0	14
1	1	1	1	15

Figure 7–4 Binary-weighted D/A converter.

In Figure 7–4 the 20-kΩ resistor *sums* the currents that are provided by closing any of switches D_0 to D_3. The resistors are scaled in such a way as to provide a binary-weighted amount of current to be summed by the 20-kΩ resistor. Closing D_0 causes 50 μA to flow through the 20 kΩ, creating -1.0 V at V_{out}. Closing each successive switch creates *double* the amount of current of the previous switch. Work through several of the switch combinations presented in Figure 7–4 to prove its operation.

If we were to expand Figure 7–4 to an 8-bit D/A converter, the resistor for D_4 would be one-half of 12.5 kΩ, which is 6.25 kΩ. Each successive resistor is one-half of the previous one. Using this procedure, the resistor for D_7 would be 0.78125 kΩ!

Coming up with accurate resistances over such a large range of values is very difficult. This limits the practical use of this type of D/A converter for any more than 4-bit conversions.

7–4 R/2R LADDER DIGITAL-TO-ANALOG CONVERTERS

The method for D/A conversion that is most often used in integrated-circuit D/A converters is known as the *R*/2*R* ladder circuit. In this circuit, only two resistor values are required, which lends itself nicely to the fabrication of ICs with a resolution of 8, 10, or 12 bits, and higher. Figure 7–5 shows a 4-bit D/A *R*/2*R* converter.

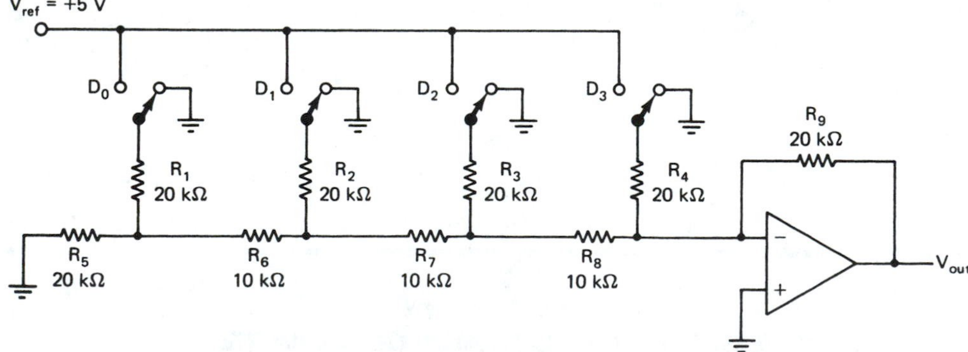

Figure 7–5 The *R*/2*R* ladder D/A converter.

In Figure 7–5, the 4-bit digital information to be converted to analog is entered on the D_0 to D_3 switches. (In an actual IC those switches will be transistor switches.) The arrangement of the circuit is such that as the switches are moved to +5 V or 0 V (1 or 0), they cause a current to flow through R_9 that is proportional to their binary equivalent value. (Each successive switch is worth double the previous one.)

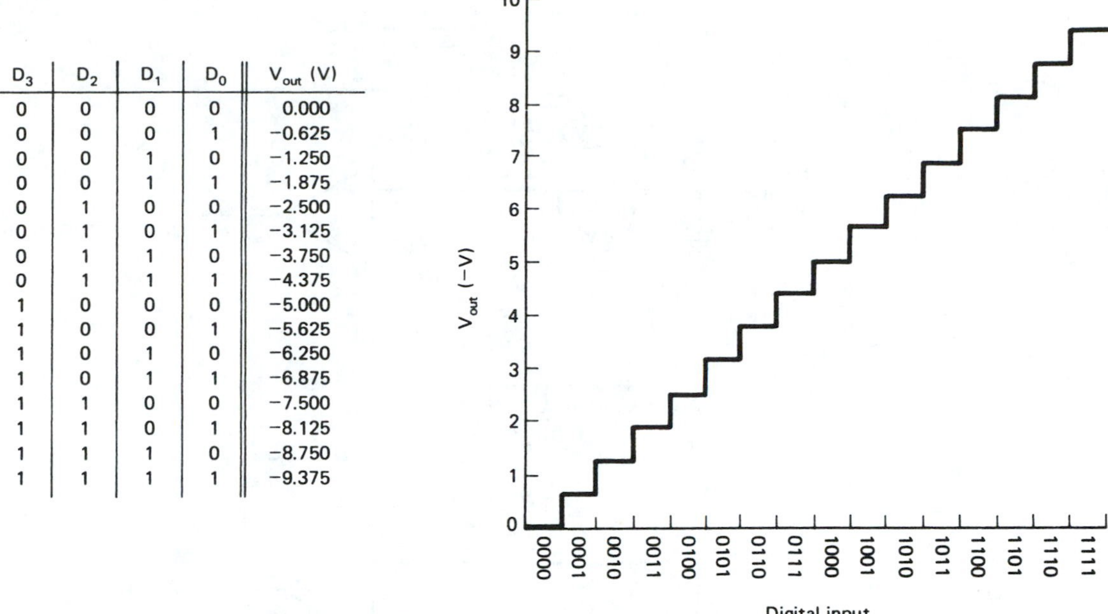

D_3	D_2	D_1	D_0	V_{out} (V)
0	0	0	0	0.000
0	0	0	1	−0.625
0	0	1	0	−1.250
0	0	1	1	−1.875
0	1	0	0	−2.500
0	1	0	1	−3.125
0	1	1	0	−3.750
0	1	1	1	−4.375
1	0	0	0	−5.000
1	0	0	1	−5.625
1	0	1	0	−6.250
1	0	1	1	−6.875
1	1	0	0	−7.500
1	1	0	1	−8.125
1	1	1	0	−8.750
1	1	1	1	−9.375

Figure 7–6 Analog output versus digital input for Figure 7–5.

The analog output voltage from each of the 16 possible switch combinations is shown in Figure 7–6. Let's work through the calculation of V_{out} for three different switch combinations to see how the $R/2R$ method works.

1. For $D_0 = 0$, $D_1 = 0$, $D_2 = 0$, and $D_3 = 1$: R_1 is in parallel with R_5 to equal 10 kΩ, 10 kΩ is in series with R_6 to equal 20 kΩ, 20 kΩ is in parallel with R_2 to equal 10 kΩ, and so on through R_7, R_3, and R_8. The equivalent circuit becomes as shown in Figure 7–7a.
2. For $D_0 = 0$, $D_1 = 0$, $D_2 = 0$, $D_3 = 0$: The equivalent circuit (Figure 7–7b) will be similar to Figure 7–7a except that switch $D_3 = 0$.
3. For $D_0 = 0$, $D_1 = 0$, $D_2 = 1$, $D_3 = 0$: All switches are grounded except D_2, so the equivalent circuit becomes as shown in Figure 7–7c. The circuit in Figure 7–7c is still not simplified far enough to determine V_{out}. The Thévenin equivalent circuit at point A will reduce the circuit to that shown in Figure 7–7d.

As it turns out, no matter which combination of switch positions is used, the magnitude of voltage contributed by closing switch D_3 to 1 is equal to $2V_{ref}/2$; D_2 contributes $2V_{ref}/4$, D_1 contributes $2V_{ref}/8$, and D_0 contributes $2V_{ref}/16$.

7–5 INTEGRATED-CIRCUIT DIGITAL-TO-ANALOG CONVERTERS

One very popular and inexpensive 8-bit D/A converter (DAC) is the DAC0808 and its equivalent, the MC1408. A block diagram, pin configuration, and typical application are shown in Figure 7–8. The circuit in Figure 7–8c is set up to accept an 8-bit digital input and provide a 0- to +10-V analog output. A reference current (I_{ref}) is required for the D/A and is provided by the 10-V, 5-kΩ combination shown. The negative reference (pin 15) is then tied to ground via an equal-size (5-kΩ) resistor.

That 2-mA reference current dictates that the full-scale output current (I_{out}) be approximately 2 mA also. To calculate the *actual* output current, use the formula

$$I_{out} = I_{ref}\left(\frac{A_1}{2} + \frac{A_2}{4} + \ldots + \frac{A_8}{256}\right) \tag{7–1}$$

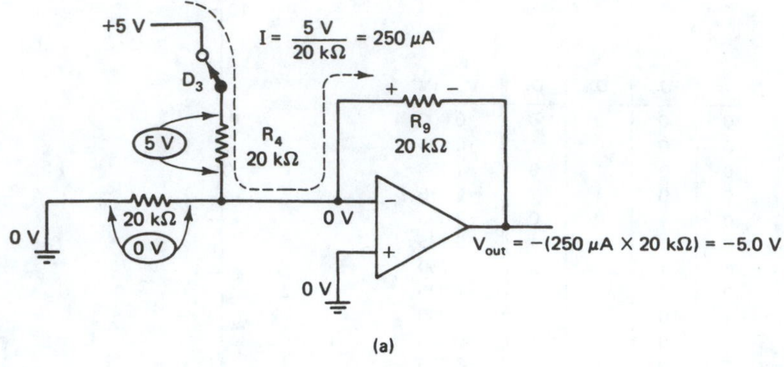

(a)

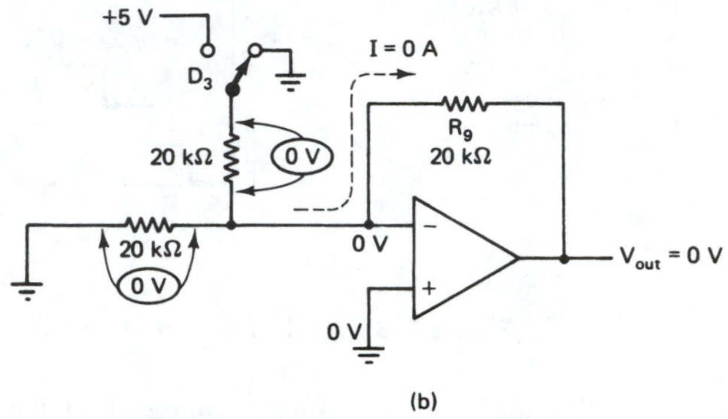

(b)

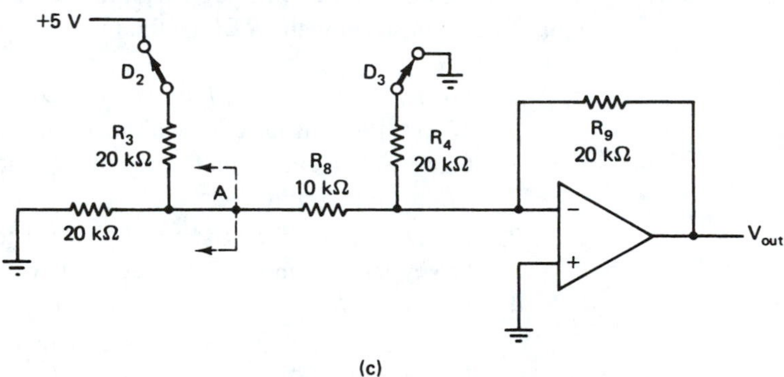

(c)

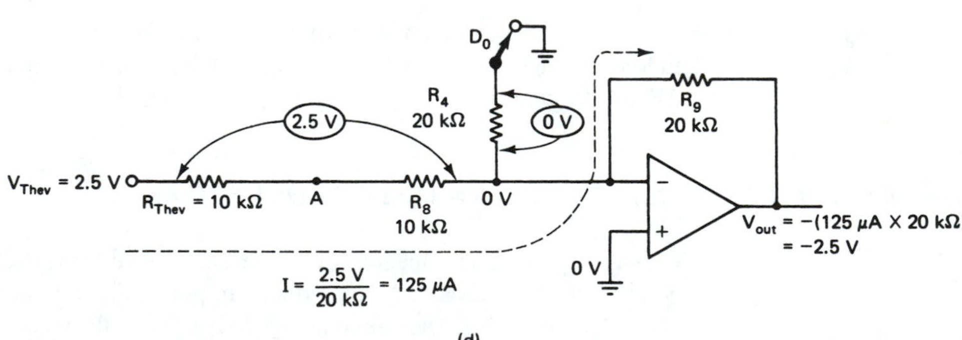

(d)

Figure 7–7 (a) The equivalent $R/2R$ circuit for $D_0 = 0$, $D_1 = 0$, $D_2 = 0$, $D_3 = 1$; (b) the equivalent $R/2R$ circuit for $D_0 = 0$, $D_1 = 0$, $D_2 = 0$, $D_3 = 0$; (c) the equivalent $R/2R$ circuit for $D_0 = 0$, $D_1 = 0$, $D_2 = 1$, $D_3 = 0$; (d) the Thévenin equivalent $R/2R$ circuit for $D_0 = 0$, $D_1 = 0$, $D_2 = 1$, $D_3 = 0$.

[e.g., with all inputs (A_1 to A_8) HIGH, $I_{out} = I_{ref} \times (0.996)$]. To convert an output current to an output voltage, a series resistor could be connected from pin 4 to ground and the output taken across the resistor. That method is simple, but it may cause inaccuracies as various size loads are connected to it.

A more accurate method uses an op amp such as the 741 shown in Figure 7–8c. The output current flows through R_F, which develops an output voltage equal to $I_{out} \times R_F$. The range of output voltage can be changed by changing R_F and is limited only by the specifications of the op amp used.

To test the circuit, an oscillator and an 8-bit counter can be used to drive the digital inputs, and the analog output can be observed on an oscilloscope, as shown in Figure 7–9. In Figure 7–9, as the counters count from 0000 0000 up to 1111 1111, the analog output will go from 0 V up to almost +10 V in 256 steps. The time per step will be equal to the reciprocal of the input clock frequency.

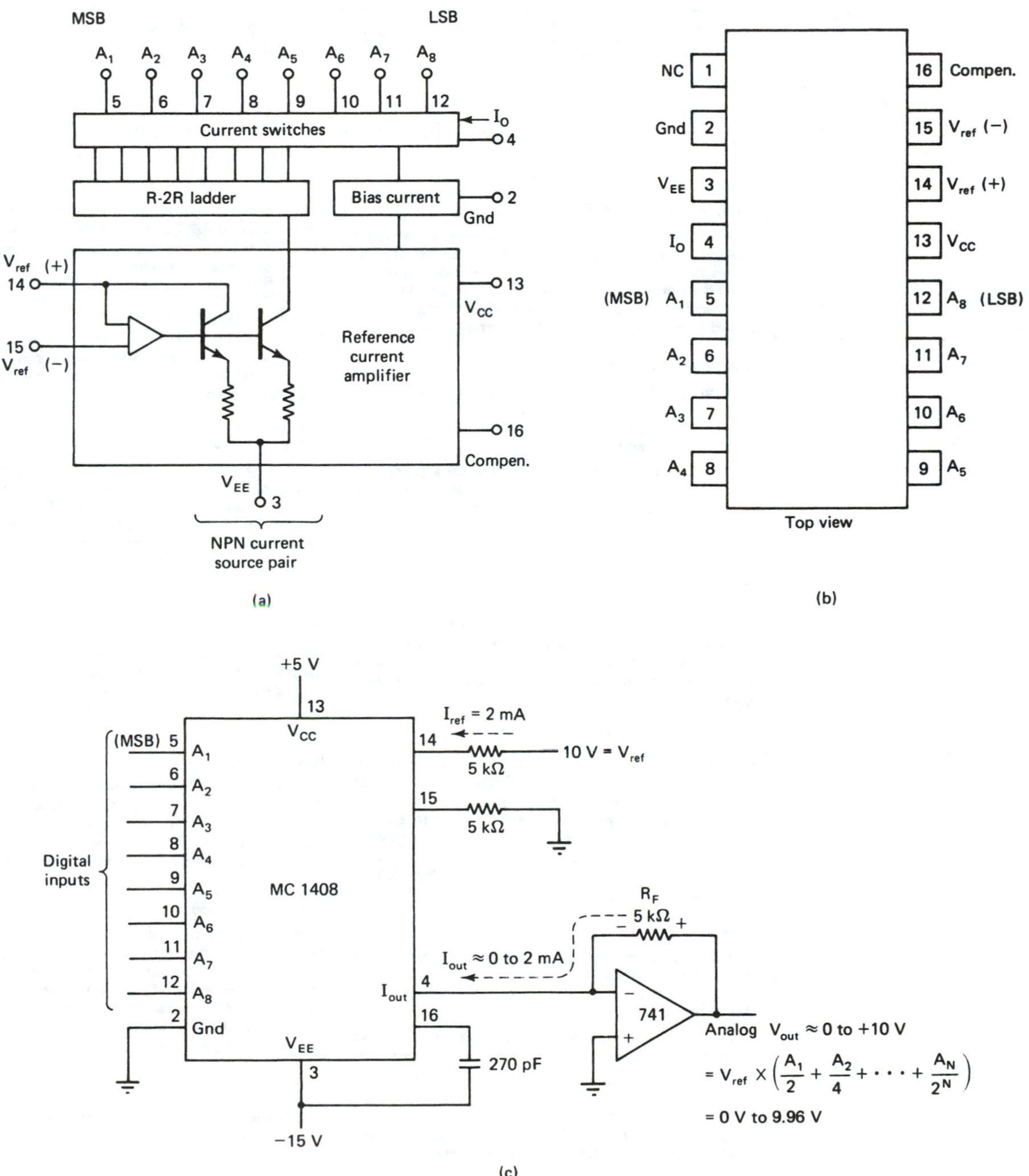

Figure 7–8 The MC1408 D/A converter: (a) block diagram; (b) pin configuration; (c) typical application.

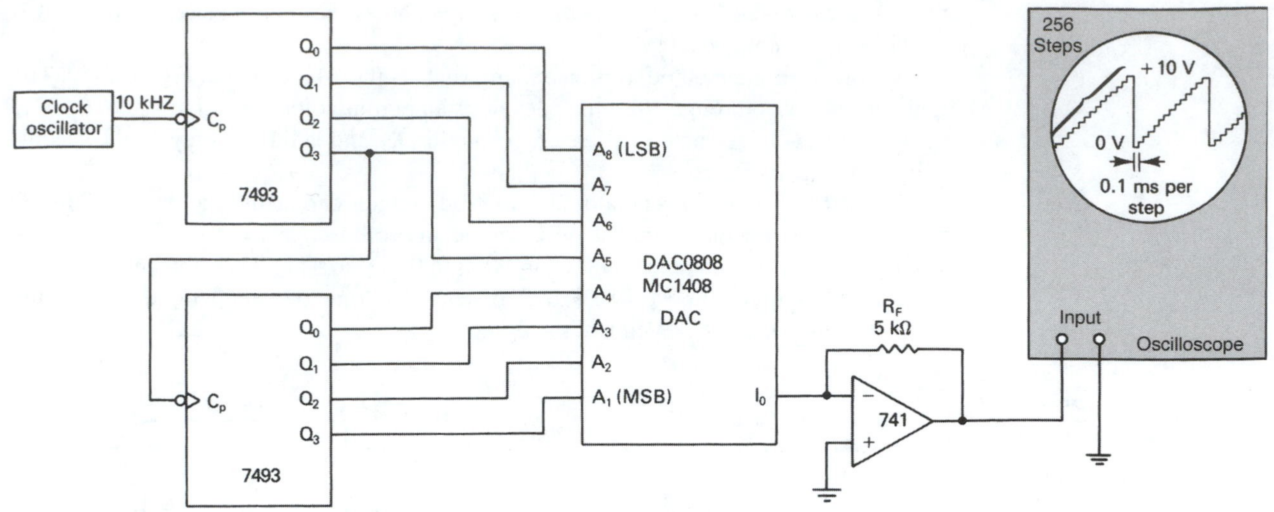

Figure 7–9 Test circuit for a DAC application.

Absolute Accuracy Error

Absolute Accuracy Error is the difference between the theoretical analog input required to produce a given output code and the actual analog input required to produce the same code. The actual input is a range and the error is the midpoint of the measured band and the theoretical band.

Conversion Speed

Conversion Speed is the speed at which a converter can make repetitive conversions.

Conversion Time

Conversion time is the time required for a complete conversion cycle of an ADC. Conversion time is a function of the number of bits and the clock frequency.

Differential Non-Linearity (DNL) (a)

Differential Non-Linearity of a DAC is the deviation of the measured output step size from the ideal step size. In an ADC it is the deviation in the range of inputs from 1 LSB that causes the output to change from one given code to the next code. Excessive DNL gives rise to non-monotonic behavior in a DAC and missing codes in an ADC.

Gain Error (b)

Gain Error is the error of the slope of the line drawn through the midpoints of the steps of the transfer function as compared to the ideal slope. It is usually measured by determining the error of the analog input voltage to cause a full scale output word with the ideal value that should cause this full scale output. This gain error is usually expressed in LSB or in percent of full scale range.

Missing Code (c)

A Missing Code is a code combination that does not appear in the ADC's output range.

Monotonicity (d)

A DAC is monotonic if its output either increases or remains the same when the input code is incremented from any code to the next higher code.

Offset Error (e)

Offset error is the constant error or shift from the ideal transfer characteristic of a converter. In a DAC it is the output obtained when that output should be zero. In an ADC it is the difference between the input level that causes the first code transition and what that input level should be.

Output Voltage Compliance

Output Voltage Compliance of a current output DAC is the range of acceptable voltages at the DAC output for the DAC output current to remain within its specified limits.

Quantizing Error

In an A/D converter there is an infinite number of possible input levels, but only 2^n output codes (n = number of bits). There will, therefore, be an error in the output code that could be as great as $\frac{1}{2}$ LSB because of this quantizing effect. The greatest error occurs at the transition point where the output state changes.

Relative Accuracy (f)

Relative Accuracy is a measure of the difference of the theoretical output value with a given input after any offset and gain errors have been nulled out.

Resolution

Resolution is the number of bits at the input or output of an ADC or DAC. It is the number of discrete steps or states at the output and is equal to 2^n where n is the resolution of the converter. However, n bits of resolution does not guarantee n bits of accuracy.

Setting Time (g)

Setting Time is the delay in a DAC from the 50 percent point on the change in the input digital code to the effected change in the output signal. It is expressed in terms of how long it takes the output to settle to and remain within a certain error band around the final value and is usually specific for full scale range changes.

Transfer Characteristic (h)

The Transfer Characteristic is the relationship of the output to the input, D_{out} vs. A_{in} (ADC) or A_{out} vs. D_{in} (DAC).

Figure 7–10 DAC and ADC specification definitions.

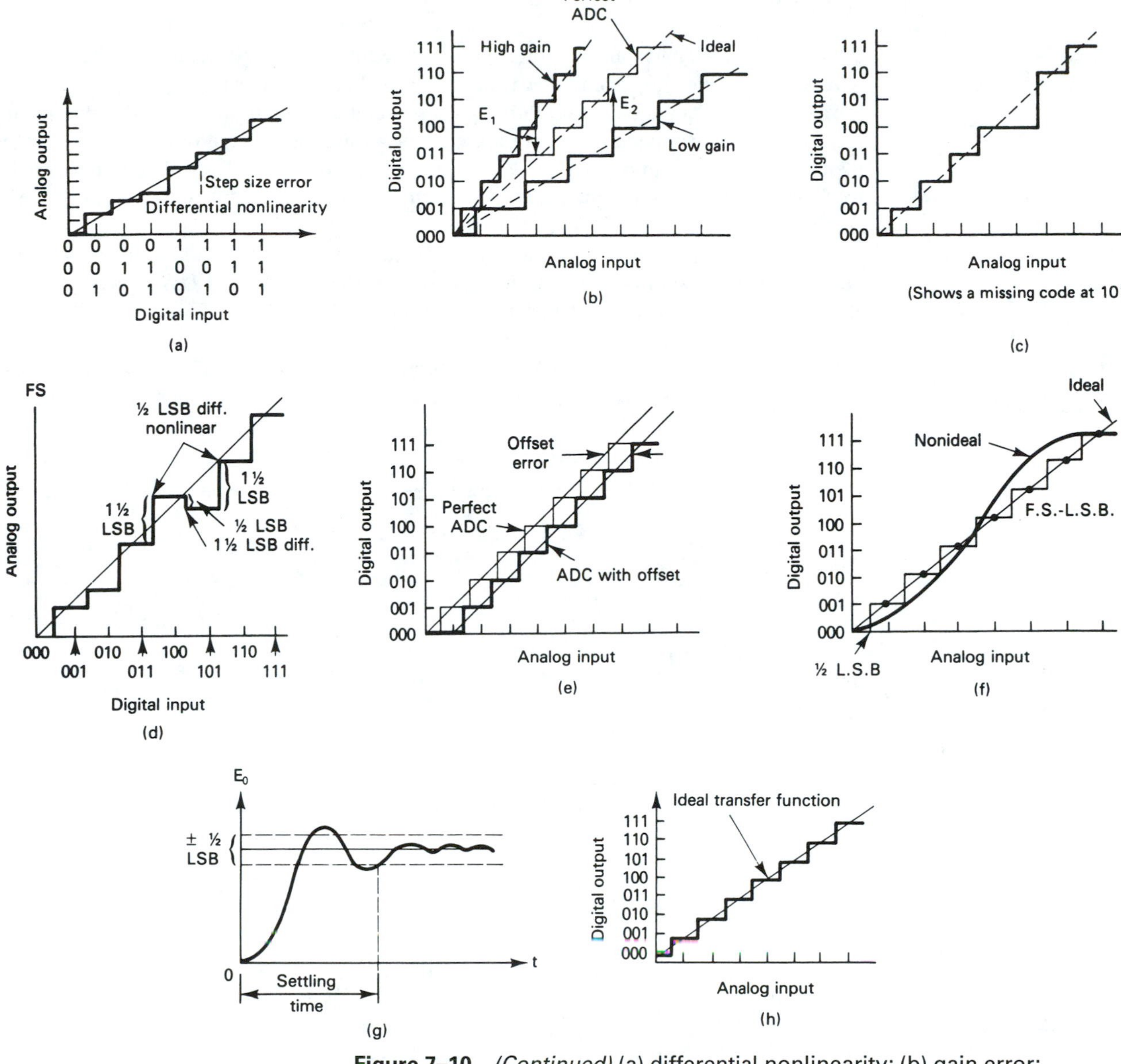

Figure 7–10 *(Continued)* (a) differential nonlinearity; (b) gain error; (c) missing codes; (d) nonmonotonic (must be $> \pm\frac{1}{2}$ LSB nonlinear); (e) offset error; (f) relative accuracy; (g) settling time; (h) 3-bit ADC transfer characteristic. (Courtesy of Philips Components-Signetics)

7–6 IC DATA CONVERTER SPECIFICATIONS

Besides resolution, there are several other specifications that are important in the selection of D/A and A/D converters (DAC and ADC). It is important that the specifications and their definitions given in the manufacturer's data book be studied and understood before selecting a particular DAC or ADC. Figure 7–10 lists some of the more important specifications as presented in the Signetics Linear LSI Data Manual.

7–7 PARALLEL-ENCODED ANALOG-TO-DIGITAL CONVERTERS

The process of taking an analog voltage and converting it to a digital signal can be done in several ways. One simple way that is easy to visualize is by means of parallel encoding (also known as "simultaneous," "multiple comparator," or "flash" converting). In this method, several comparators are set up, each at a different voltage reference level with their outputs driving a priority encoder, as shown in Figure 7–11. The voltage-divider network in Figure 7–11 is designed to drop 1 V across each resistor. This sets up a voltage reference at each comparator input in 1-V steps.

When V_{in} is 0 V, the + input on all seven comparators will be higher than the − input, so they will all output a HIGH. In that case, $\bar{I}_0$ is the only active-LOW input that is enabled, so the 74148 will output an active-LOW binary 0 (111).

When V_{in} exceeds 1.0 V, comparator 1 will output a LOW. Now $\bar{I}_0$ and $\bar{I}_1$ are both enabled, but since it is a *priority* encoder, the output will be a binary 1 (110). As V_{in} increases further, each successive comparator outputs a LOW. The highest input that receives a LOW is encoded into its binary equivalent output.

This particular A/D converter (Figure 7–11) is set up to convert analog voltages in the range 0 to 7 V. The range can be scaled higher or lower, depending on the input voltage

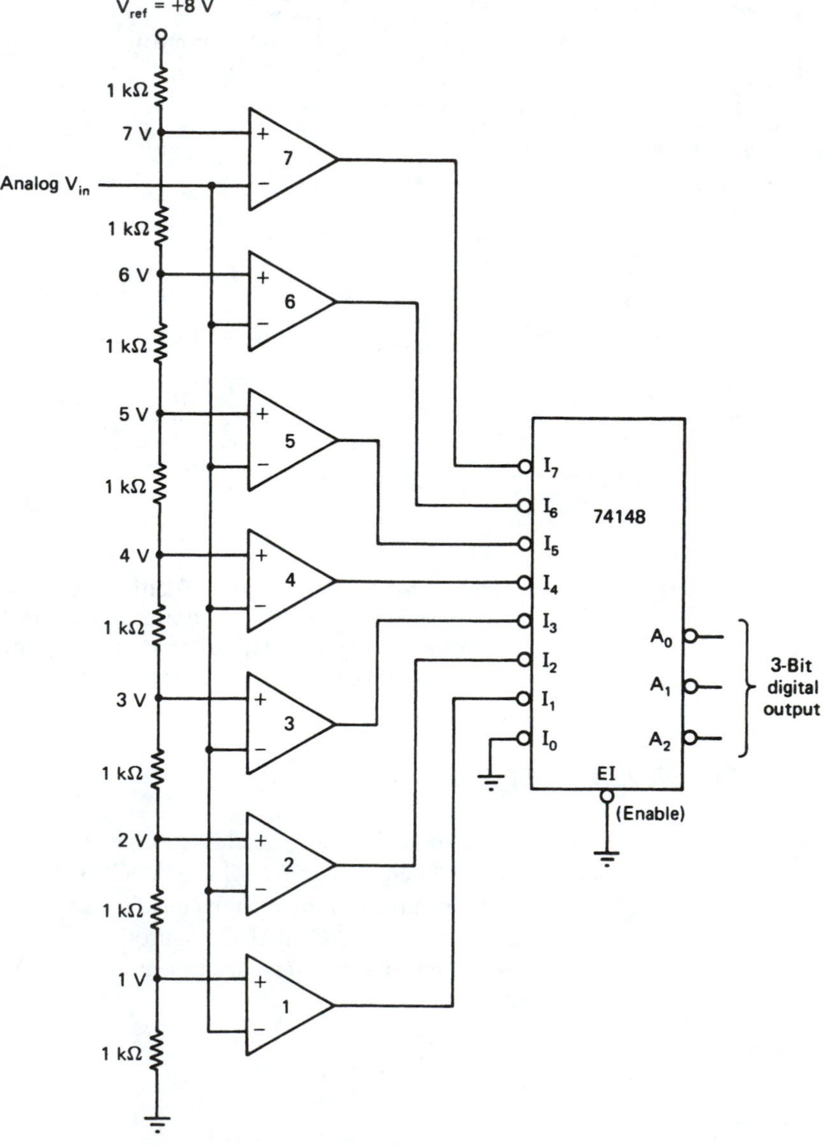

Figure 7–11 Three-bit parallel-encoded ADC.

levels that are expected. The resolution of this converter is only 3 bits, so it can only distinguish between eight different analog input levels. To expand to 4-bit resolution, eight more comparators are required to differentiate the 16 different voltage levels. To expand to 8-bit resolution, 256 comparators would be required! As you can see, circuit complexity becomes a real problem when using parallel encoding for high-resolution conversion. However, a big advantage of using parallel encoding is its high speed. The conversion speed is limited only by the propagation delays of the comparators and encoder (less than 20 ns total).

7–8 COUNTER-RAMP ANALOG-TO-DIGITAL CONVERTERS

The counter-ramp method of A/D conversion uses a counter in conjunction with a D/A converter to determine a digital output that is equivalent to the unknown analog input voltage. In Figure 7–12, depressing the "start conversion" pushbutton clears the counter outputs to 0, which sets the DAC output to 0 V. The (−) input to the comparator is now 0 V, which is less than the positive analog input voltage at the (+) input. Therefore, the comparator outputs a HIGH, which enables the AND gate, allowing the counter to start counting. As the counter's binary output increases, so does the DAC output voltage in the form of a staircase.

When the staircase voltage reaches, then exceeds, the analog input voltage, the comparator output goes LOW, disabling the clock and stopping the counter. The counter output at that point is equal to the binary number that caused the DAC to output a voltage slightly greater than the analog input voltage. Thus we have the binary equivalent of the analog voltage!

The HIGH-to-LOW transition of the comparator is also used to trigger the *D* flip-flop to "latch" on to the binary number at that instant. To perform another conversion, the start pushbutton is depressed again and the process repeats. The result from the previous conversion remains in the *D* flip-flop until the next "end-of-conversion" HIGH-to-LOW edge comes along.

To change the circuit to perform *continuous conversions,* the end-of-conversion line could be tied back to the $\overline{\text{clear}}$ input of the counter. A short delay needs to be inserted into

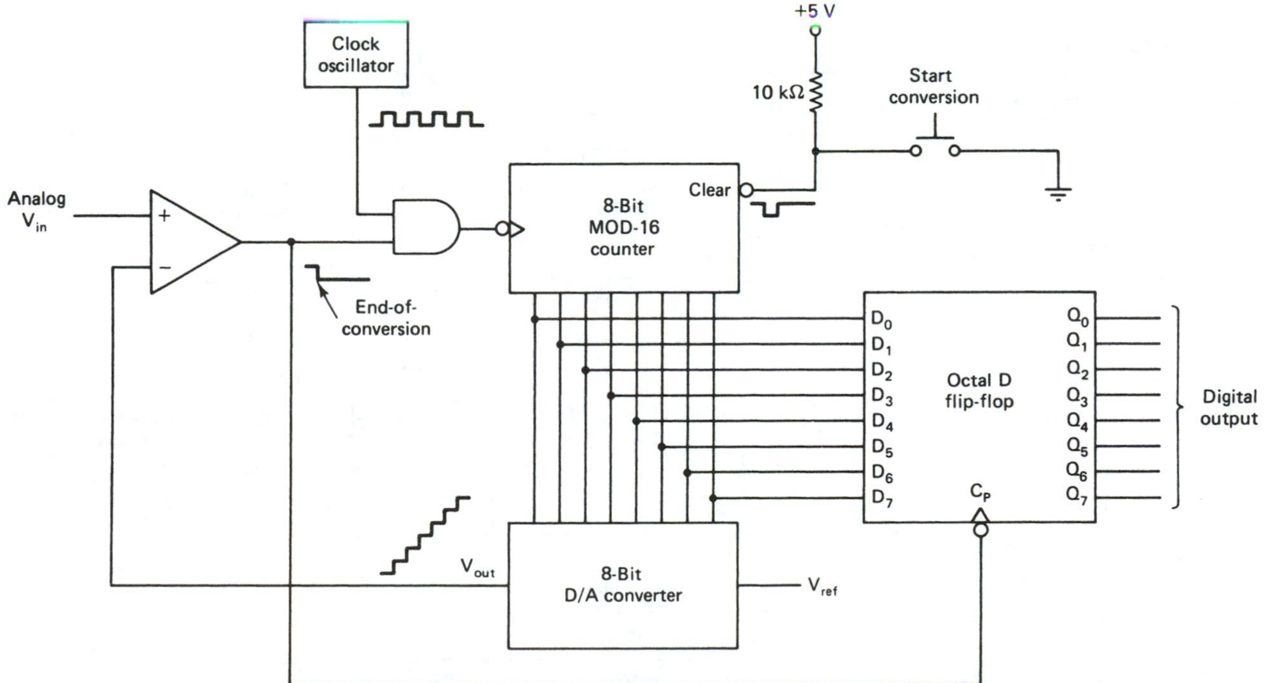

Figure 7–12 Counter-ramp A/D converter.

this new line, however, to allow the *D* flip-flop to read the binary number before the counter is Reset. Two inverters placed end to end in the line will produce a sufficient delay.

The main *disadvantage* of the counter-ramp method of conversion is its slow conversion speed. The worst-case maximum conversion time will occur when the counter has to count all 255 steps before the DAC output voltage matches the analog input voltage.

7–9 SUCCESSIVE-APPROXIMATION ANALOG-TO-DIGITAL CONVERSION

Other methods of A/D conversion employ *up/down-counters* and *integrating slope converters* to "track" the analog input, but the method used in most modern integrated-circuit ADCs is called *successive approximation.* This converter circuit is similar to the counter-ramp ADC circuit except that the method of narrowing in on the unknown analog input voltage is much improved. Instead of counting up from 0 and comparing the DAC output each step of the way, a successive-approximation register (SAR) is used in place of the counter (see Figure 7–13).

In Figure 7–13 the conversion is started by dropping the $\overline{\text{STRT}}$ line LOW. Then the SAR first tries a HIGH on the MSB (D_7) line to the DAC. (Remember, D_7 will cause the DAC to output half of its full-scale output.) If the DAC output is then *higher* than the unknown analog input voltage, the SAR returns the MSB LOW. If the DAC output was still *lower* than the unknown analog input voltage, the SAR leaves the MSB HIGH.

Now, the next lower bit (D_6) is tried. If a HIGH on D_6 causes the DAC output to be higher than the analog V_{in}, it is returned LOW. If not, it is left HIGH. The process continues until all 8 bits, down to the LSB, have been tried. At the end of this eight-step conversion process, the SAR contains a valid 8-bit binary output code that represents the unknown analog input. The $\overline{\text{DR}}$ output now goes LOW, indicating that the *conversion is complete* and the data are ready. That HIGH-to-LOW edge on $\overline{\text{DR}}$ "clocks" the D_0 to D_7 data into the octal *D* flip-flop to make the digital output results available at the Q_0 to Q_7 lines.

The main advantage of the SAR ADC method is its high speed. The ADC in Figure 7–13 takes only eight clock periods to complete a conversion, which is a vast improvement over the counter-ramp method.

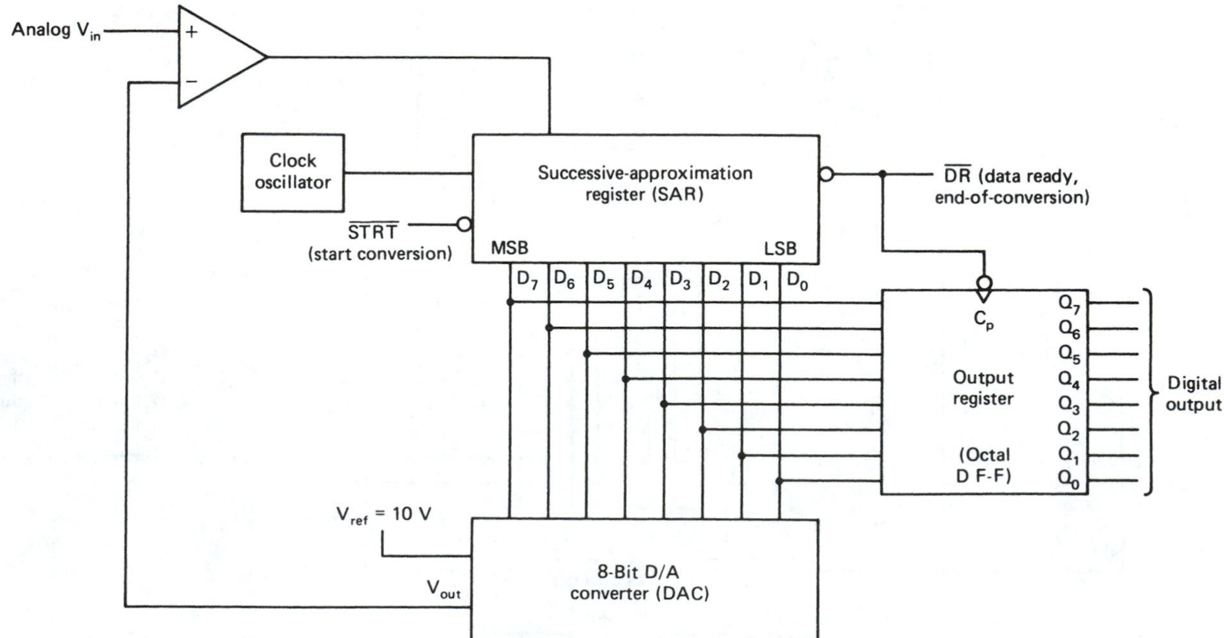

Figure 7–13 Simplified SAR A/D converter.

EXAMPLE 7–2

Show the timing waveforms that would occur in the successive approximation ADC of Figure 7–13 when converting the analog voltage 6.84 V to 8-bit binary, assuming that the full-scale input voltage to the DAC is 10 V ($V_{ref} = 10$ V).

Solution:

Each successive bit, starting with the MSB, will cause the DAC part of the system to output a voltage to be compared. If the V_{ref} of the DAC is 10 V, D_7 will be worth 5 V, D_6 will be worth 2.5 V, D_5 will be worth 1.25 V, and so on, as shown in Table 7–1.

TABLE 7–1
Voltage-Level Contributions by Each Successive Approximation Register Bit

DAC input	*DAC* V_{out}
D_7	5.0000
D_6	2.5000
D_5	1.2500
D_4	0.6250
D_3	0.3125
D_2	0.15625
D_1	0.078125
D_0	0.0390625

Now, when $\overline{\text{STRT}}$ goes LOW, successive bits starting with D_7 will be tried, creating the waveforms shown in Figure 7–14. The HIGH-to-LOW edge on $\overline{\text{DR}}$ "clocks" the final binary number 1010 1111 into the D flip-flop and Q_0 to Q_7 outputs.

Now the Q_0 to Q_7 lines contain the 8-bit binary representation of the analog number 6.8359375, which is an error of only 0.0594% from the target number of 6.84:

$$\% \text{ error} = \frac{\text{actual voltage} - \text{final DAC output}}{\text{actual voltage}} \times 100\%$$

To watch the conversion in progress, an eight-channel oscilloscope or logic analyzer can be connected to the D_0 to D_7 outputs of the SAR.

For *continuous conversions* the $\overline{\text{DR}}$ line can be connected back to the $\overline{\text{STRT}}$ line. That way, as soon as the conversion is complete, the HIGH-to-LOW on $\overline{\text{DR}}$ will issue another start conversion ($\overline{\text{STRT}}$), which forces the data ready ($\overline{\text{DR}}$) line back HIGH for eight clock periods while the new conversion is being made. The latched Q_0 to Q_7 digital outputs will always display the results of the *previous conversion.*

7–10 INTEGRATED-CIRCUIT ANALOG-TO-DIGITAL CONVERTERS

Examples of two popular, commercially available ADCs are the NE5034 and the ADC0801 manufactured by Signetics Corporation.

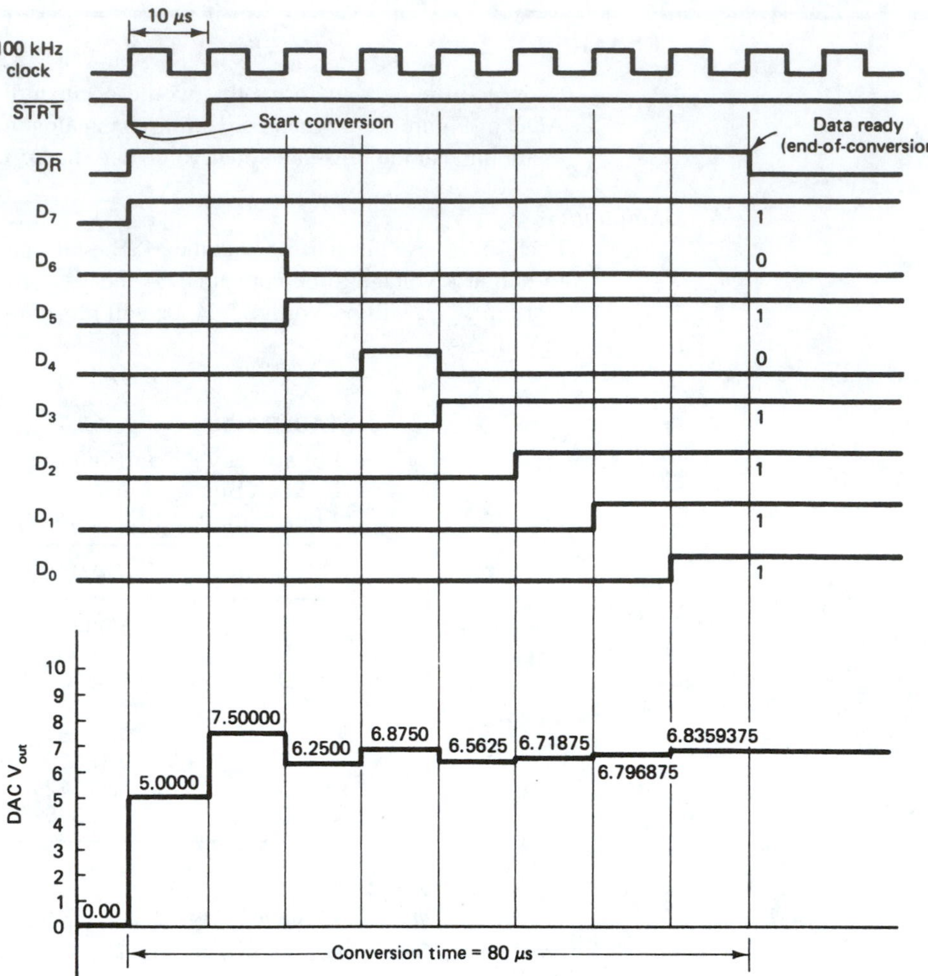

Figure 7–14 Timing waveforms for a successive approximation A/D conversion.

The NE5034

The block diagram and pin configuration for the NE5034 are given in Figure 7–15. Operation of the NE5034 is almost identical to that of the SAR ADC presented in Section 7–9. One difference is that the NE5034 uses a three-state output buffer instead of a *D* flip-flop. With three-state outputs, when $\overline{OE}$ (Output Enable) is LOW, the DB_7 to DB_0 outputs display the continuous status of the eight SAR lines, and when the $\overline{OE}$ line goes HIGH, the DB_7 to DB_0 outputs return to a float or high-impedance state. This way, if the ADC outputs go to a common *data bus* shared by other devices, when DB_7 to DB_0 float, one of the other devices can output information to the data bus without interference.

The NE5034 can provide conversion speeds as high as one per 17 μs, and its three-state outputs make it compatible with bus-oriented microprocessor systems. It also has its own internal clock for providing timing pulses. The frequency is determined by an external capacitor placed between pins 11 and 17. Figure 7–16 shows the frequency and conversion time that can be achieved using the internal clock.

The ADC0801

The pin configuration and block diagram for the ADC0801 are given in Figure 7–17. The ADC0801 uses the successive-approximation method to convert an analog input to an 8-bit binary code. Two analog inputs are provided to allow differential measurements (analog $V_{in} = V_{in(+)} - V_{in(-)}$). It has an internal clock that generates its own timing pulses

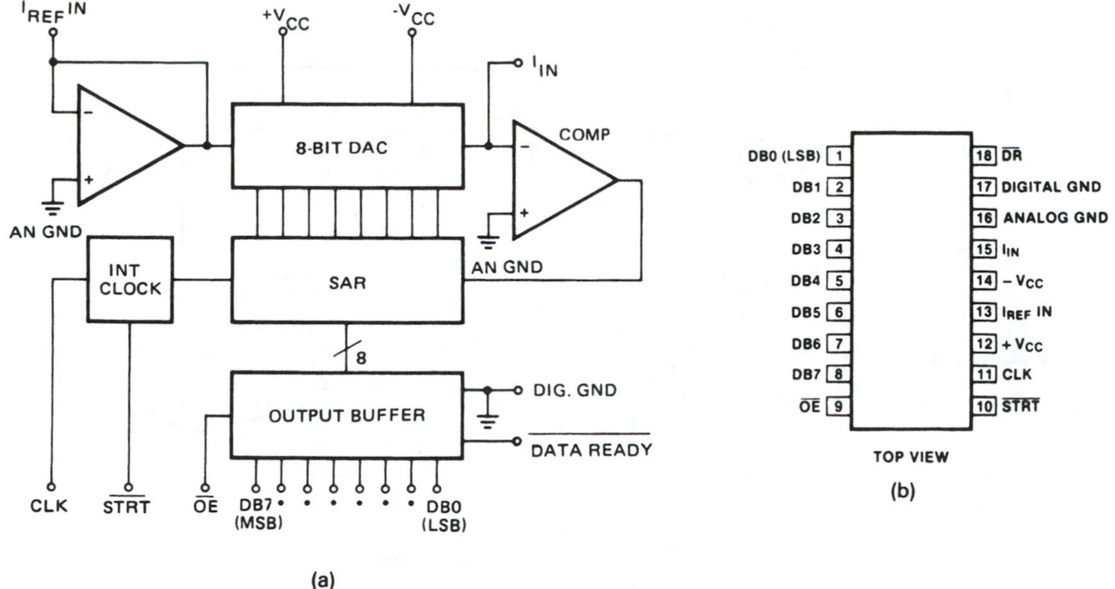

Figure 7–15 The NE5034 A/D converter: (a) block diagram; (b) pin configuration. (Courtesy of Philips Components-Signetics)

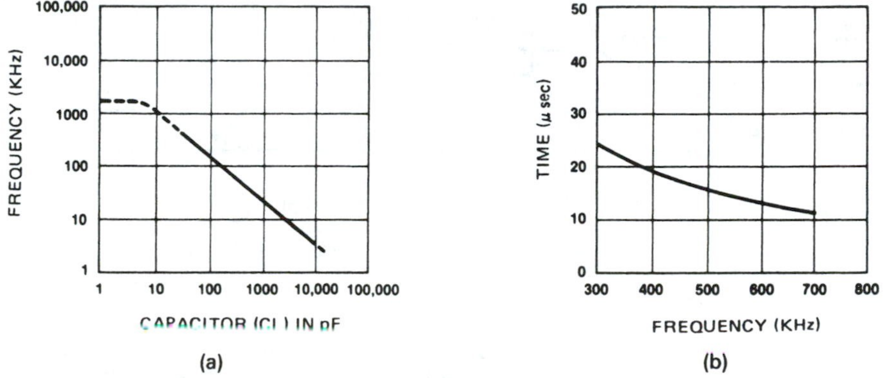

Figure 7–16 The NE5034 internal clock characteristics: (a) internal clock frequency versus external capacitor (CL); (b) conversion time versus clock frequency. (Courtesy of Philips Components–Signetics)

at a frequency equal to $f = 1/(1.1RC)$ (Figure 7–18 shows the connections for the external R and C.) It uses output D latches that are three-stated to facilitate easy bus interfacing.

The convention for naming the ADC0801 pins follows those used by microprocessors to ease interfacing. Basically, the operation of the ADC0801 is similar to that of the NE5034. The ADC0801 pins are defined as follows:

$\overline{\text{CS}}$—active-LOW *Chip Select*

$\overline{\text{RD}}$—active-LOW *Output Enable* (Read)

$\overline{\text{WR}}$—active-LOW *Start Conversion* (Write)

CLK IN—external clock input or capacitor connection point for the internal clock

$\overline{\text{INTR}}$—active-LOW *End-of-Conversion* (Data Ready, Interrupt)

$V_{in(+)}$, $V_{in(-)}$—differential analog inputs (ground one pin for single-ended measurements)

A. Gnd—analog ground

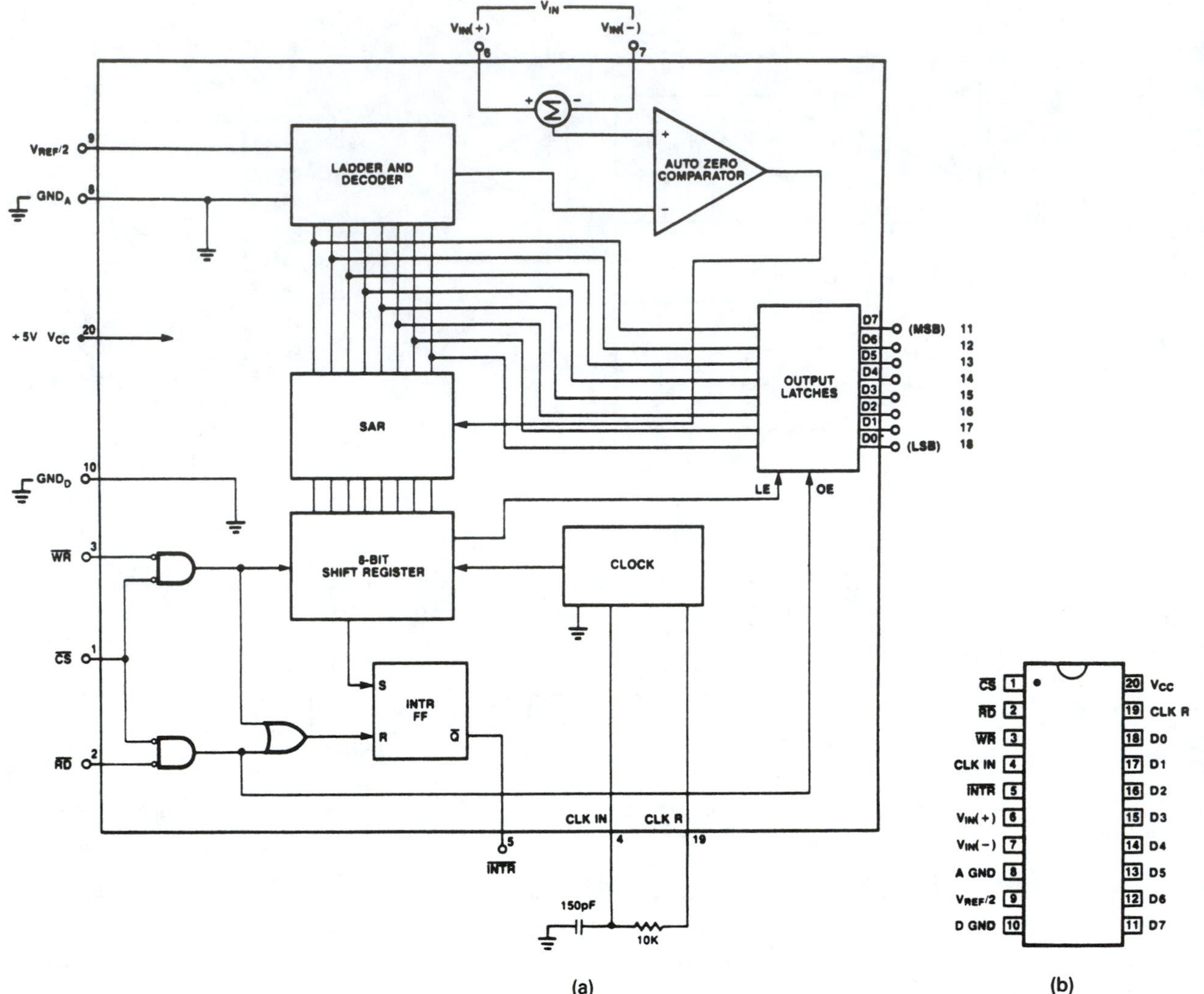

(a) (b)

Figure 7–17 The ADC0801 converter: (a) block diagram; (b) pin configuration. (Courtesy of Philips Components–Signetics)

$V_{ref/2}$—optional reference voltage (used to override the reference voltage assumed at V_{CC})

D. Gnd—digital ground

V_{CC}—5-V power supply and assumed reference voltage

CLK R—resistor connection for the internal clock

D_0 to D_7—digital outputs

To set the ADC0801 up for continuous A/D conversions, the connections shown in Figure 7–18 should be made. The external *RC* will set up a clock frequency of

$$f = \frac{1}{1.1RC} = \frac{1}{1.1(10\ \text{k}\Omega)150\ \text{pF}} = 606\ \text{kHz}$$

The connection from $\overline{\text{INTR}}$ to $\overline{\text{WR}}$ will cause the ADC to start a new conversion each time the $\overline{\text{INTR}}$ (end-of-conversion) line goes LOW. The *RC* circuit with the 7417 open-collector buffer will issue a LOW-to-float pulse at power-up to ensure initial startup. An *open-collector* output gate is required instead of a totem pole output because the $\overline{\text{INTR}}$ is forced LOW by the internal circuitry of the 0801 at the end of each conversion. That

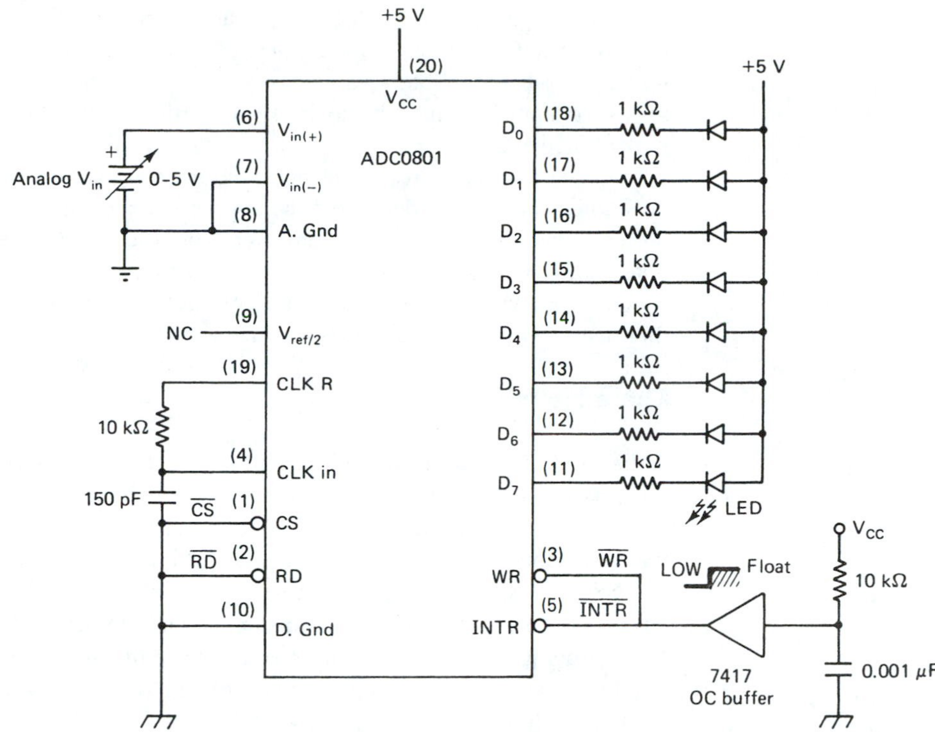

Figure 7–18 Connections for continuous conversions using the ADC0801.

LOW would conflict with the HIGH output level if a totem pole output were used. The $\overline{CS}$ is grounded to enable the ADC chip. $\overline{RD}$ is grounded to enable the D_0 to D_7 outputs. The analog input voltage is positive, 0 to 5 V, so it is connected to $V_{in(+)}$. If it were negative, $V_{in(+)}$ would be grounded and the input voltage would be connected to $V_{in(-)}$. Differential measurements (the difference between two analog voltages) can be made by using both $V_{in(+)}$ and $V_{in(-)}$. The LEDs connected to the digital output will monitor the operation of the ADC outputs. An LED ON indicates a LOW and an LED OFF indicates a HIGH. (In other words, they are displaying the *complement* of the binary output.) To test the circuit operation, you could watch the OFF LEDs count up in binary from 0 to 255 as the analog input voltage is slowly increased from 0 to +5 V.

The analog input voltage range can be changed to values other than 0 to 5 V by using the $V_{ref/2}$ input. This provides the means of encoding small analog voltages to the full 8 bits of resolution. The $V_{ref/2}$ pin is normally not connected and it sits at 2.500 V ($V_{CC}/2$). By connecting 2.00 V to $V_{ref/2}$ the analog input voltage range is changed to 0 to 4 V; 1.5 V would change it to 0 to 3.0 V; and so on. However, the accuracy of the ADC suffers as the input voltage range is decreased.

One final point on the ADC0801. An analog ground *and* a digital ground are both provided to enhance the accuracy of the system. The V_{CC}-to-digital ground lines are inherently noisy due to the switching transients of the digital signals. Using separate analog and digital grounds is not mandatory, but when used it ensures that the analog voltage comparator will not switch falsely due to digital noise and jitter.

7–11 TRANSDUCERS AND SIGNAL CONDITIONING

Hundreds of *transducers* available today are used to convert physical quantities such as heat, light, or force into electrical quantities. The *electrical quantities* (or signal levels) then have to be "conditioned" (or modified) before they can be interpreted by a digital computer.

Signal conditioning is required because transducers each output different ranges and types of electrical signals. Some transducers produce output voltages, some output currents, and others act like a variable resistance. The transducer may have a nonlinear response to input quantities, be inversely proportional, and may output signals that are down in the microvolt range.

The transducers' response specifications are given by the manufacturer and have to be studied carefully to determine the appropriate analog signal conditioning circuitry required to interface it to an A/D converter. After the information is read into a digital computer, software instructions convert that binary input into a meaningful output that can be used for further processing. Let's take a closer look at three commonly used transducers: a thermistor, an IC temperature sensor, and a strain gage.

Thermistors

A thermistor is an electronic component whose resistance is highly dependent on temperature. Its resistance changes by several percent with each degree change in temperature. This makes it a very sensitive temperature-measuring device. One problem, however, is that its response is nonlinear. That means that one degree step changes in temperature will not create equal step changes in resistance. This fact is illustrated in the characteristic curve of the 10-k Ω (at 25°C) thermistor shown in Figure 7–19.

From the characteristic curve you can see that not only is the thermistor nonlinear, but it also has a negative temperature coefficient (i.e., its resistance decreases with increasing temperatures).

To use a thermistor with an A/D converter like the ADC0801, we need to convert the thermistor resistance to a voltage in the range of 0 to 5 V. One way to accomplish this is with the circuit shown in Figure 7–20.

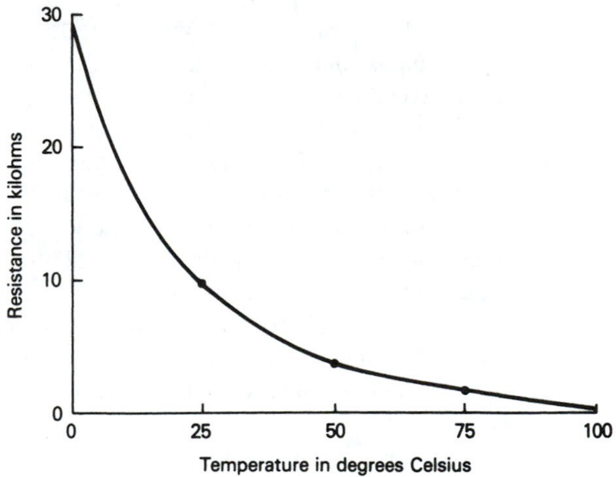

Figure 7–19 Thermistor characteristic curve of resistance versus temperature.

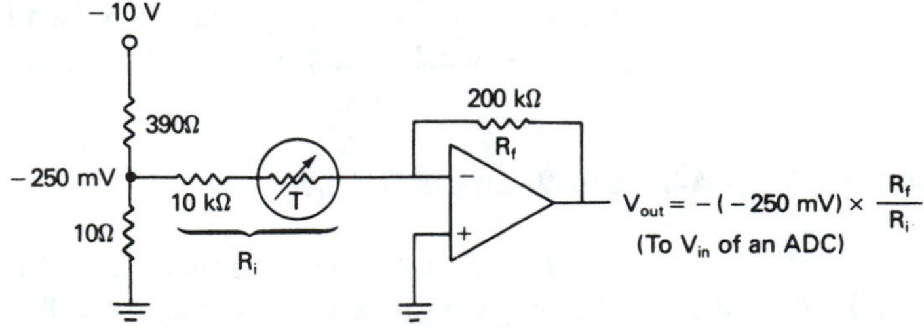

Figure 7–20 Circuit used to convert thermistor ohms to a dc voltage.

This circuit operates similarly to the op-amp circuit explained in Section 7–2. The output of the circuit is found using the following formula: $V_{out} = -V_{in} \times R_f/R_i$ where V_{in} is a fixed reference voltage of -250 mV and R_i is the sum of the thermistor's resistance, R_T, plus 10 kΩ. Using specific values for R_T found in a manufacturer's data manual, we can create Table 7–2, which shows V_{out} as a function of temperature.

TABLE 7–2

Tabulation of Output Voltage Levels for a Temperature Range of 0 to 100°C in Figure 7–20

Temperature (°C)	R_T *(kΩ)*	R_i *(kΩ)*	V_{out} *(V)*
0	29.490	39.490	1.27
25	10.000	20.000	2.50
50	3.893	13.893	3.60
75	1.700	11.700	4.27
100	0.817	10.817	4.62

The output voltage, V_{out}, is fed into an ADC that converts it into an 8-bit binary number. That binary number is then read by a microprocessor that converts it into the corresponding degrees celsius by using software program instructions.

Linear IC Temperature Sensors

The computer software required to convert the output voltages of the previous thermistor circuit is fairly complicated because of the nonlinear characteristics of the device. *Linear temperature sensors* have been developed to simplify the procedure. One such device is the LM35 integrated-circuit temperature sensor. It is fabricated in a three-terminal transistor package and is designed to output 10 mV for each degree celsius above zero. (Another temperature sensor, the LM34 is calibrated in degrees fahrenheit.) This means that at 25°C it will output 250 mV, at 50°C it will output 500 mV, and so on, in linear steps for its entire range. Figure 7–21 shows how we can interface the LM35 to an ADC and microprocessor.

The 1.28-V reference level used in Figure 7–21 is the key to keeping the conversion software programming simple. With 1.28 V at $V_{ref}/2$, the maximum full-scale analog V_{in} is defined as 2.56 V (2560 mV). This corresponds one-for-one with the 256 binary output levels provided by an 8-bit ADC. A one degree rise in temperature increases V_{in} by 10 mV, which increases the binary output by 1. Therefore, if V_{in} equals 0 V, D_7–D_0 equals

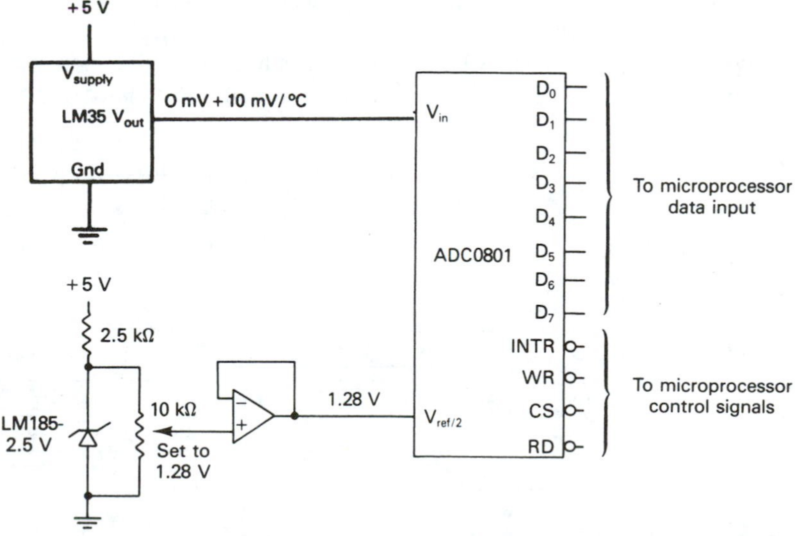

Figure 7–21 Interfacing the LM35 linear temperature sensor to an ADC.

0000 0000, if V_{in} equals 2.55 V, D_7–D_0 equals 1111 1111, and if V_{in} equals 1.00 volt, D_7–D_0 will equal the binary equivalent of 100, which is 0110 0100. Table 7–3 lists some representative values of temperature versus binary output.

TABLE 7–3
Tabulation of Temperature versus Binary Output for a Linear Temperature Sensor and an ADC Set Up for 2560 mV Full Scale

Temperature (°C)	V_{in} *(mV)*	*Binary output* (D_7–D_0)
0	0	0000 0000
1	10	0000 0001
2	20	0000 0010
25	250	0001 1001
50	500	0011 0010
75	750	0100 1011
100	1000	0110 0100

In Figure 7–21, the LM185 is a 2.5-V precision voltage reference diode. This maintains a steady 2.5 V across the 10-kΩ potentiometer even if there are fluctuations in the 5-V power supply line. The 10-kΩ potentiometer must be set to output exactly 1.280 V. The op amp is used as a unity-gain buffer between the potentiometer and ADC and will maintain a steady 1.280 V for the $V_{ref}/2$ pin.

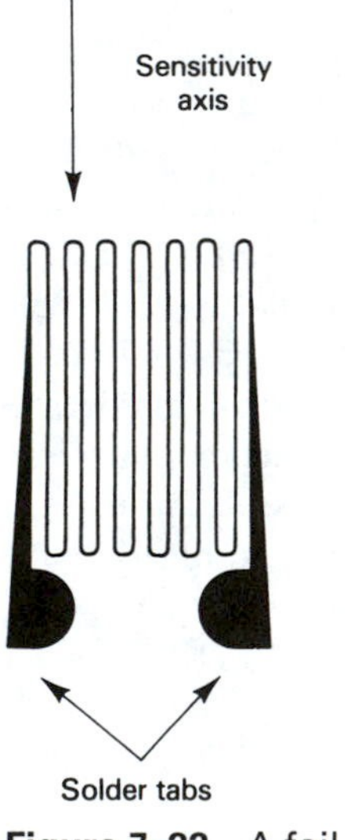

Figure 7–22 A foil-type strain gage.

Strain Gage

The strain gage is a device whose resistance changes when it is stretched. The gage is stretched, or elongated, by its being "strained" by a physical force. This property makes them useful for measuring weight, pressure, flow, and acceleration.

There are several types of strain gages, the most common being the foil type illustrated in Figure 7–22.

The gage is simply a thin electrical conductor that is looped back-and-forth and bonded securely to the piece of material to be strained (see Figure 7–23). Applying a force to the metal beam bends it slightly, which stretches the strain gage in the direction of its sensitivity axis.

As the strain gage conductor is stretched, its cross-sectional area decreases and its length increases, increasing the resistance measured at the solder tabs. The change in resistance is linear with respect to changes in length of the strain gage. However, the change in resistance is very slight, usually milliohms, and has to be converted to a voltage and

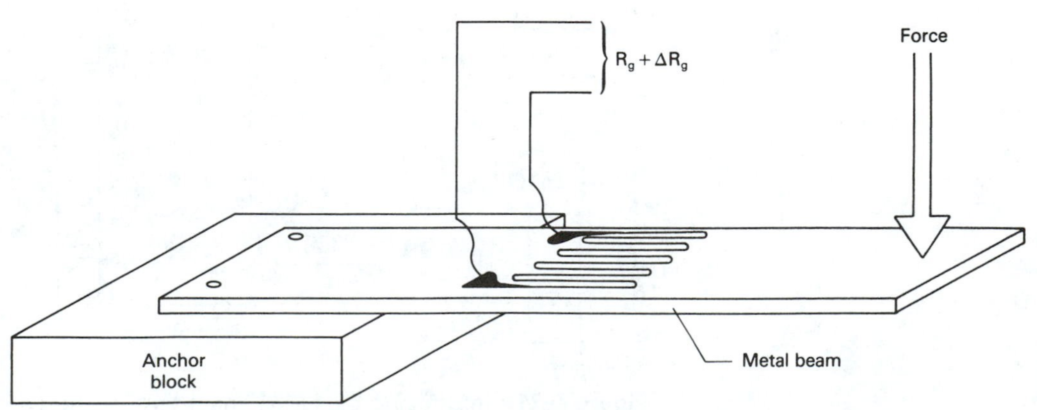

Figure 7–23 Using a strain gage to measure force.

amplified before it is input to an ADC. Figure 7–24 shows the signal conditioning circuitry for a 120-Ω strain gage [R_g (unstrained) = 120 Ω]. The instrumentation amplifier is a special-purpose IC used to amplify small-signal differential voltages such as those at points A–B in the bridge circuit in Figure 7–24.

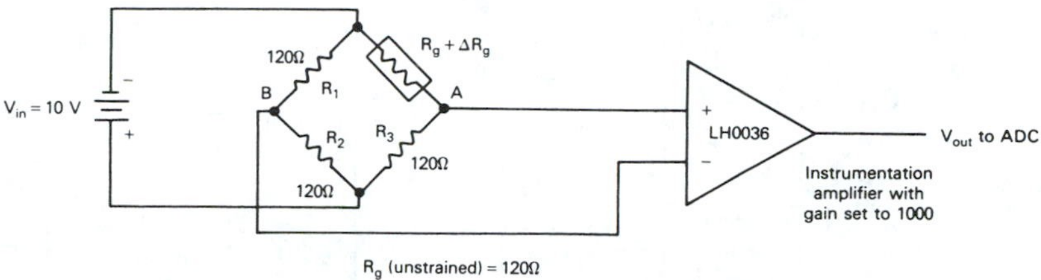

Figure 7–24 Signal conditioning for a strain gage.

With the metal beam unstrained (no force applied) the 120-Ω potentiometer is adjusted so that V_{out} equals 0 V. Then the force is applied to the metal beam, elongating the strain gage, increasing its resistance slightly from its unstrained value of 120 Ω. This causes the bridge circuit to become unbalanced, creating a voltage at points A–B. From basic circuit theory, the voltage is calculated by the following formula:

$$V_{AB} = V_{in}\left[\frac{R_3}{R_3 + (R_g + \Delta R_g)} - \frac{R_2}{R_1 + R_2}\right]$$

For example, if R_2 is set at 120 Ω and ΔR_g is 150 mΩ, the voltage will be

$$V_{AB} = -10\left[\frac{120}{120 + (120.150)} - \frac{120}{120 + 120}\right] = 3.12 \text{ mV}$$

Since the instrumentation amplifier gain is set at 1000, the output voltage sent to the ADC will be 3.12 V.

Through experimentation with several known weights, the relationship of force versus V_{out} can be established and can be programmed into the microprocessor that is reading the ADC output.

7–12 DATA ACQUISITION SYSTEMS

The computerized acquisition of analog quantities is becoming more important than ever in today's automated world. Computer systems are capable of scanning several analog inputs on a particular schedule and sequence to monitor critical quantities and acquire data for future recall. A typical eight-channel computerized data acquisition system (DAS) is shown in Figure 7–25.

The entire system in Figure 7–25 communicates via two common buses, the *data bus* and the *control bus*. The data bus is simply a common set of eight electrical conductors shared by as many devices as necessary to send and receive 8 bits of parallel data to and from anywhere in the system. In this case there are three devices on the data bus: the ADC, the microprocessor, and memory. The control bus passes control signals to and from the various devices for such things as chip select ($\overline{\text{CS}}$), output enable ($\overline{\text{RD}}$), system clock, triggers, and selects.

Each of the eight transducers is set up to output a voltage that is proportional to the analog quantity being measured. The task of the microprocessor is to scan all the quantities at some precise interval and store the digital results in memory for future use.

To do this, the microprocessor must enable and send the proper control signals to each of the devices, in order, starting with the multiplexer and ending with the ADC. This is called *handshaking*, or *polling*, and is all done with software statements. If you are fortunate enough to take a course in microprocessor programming, you will learn how to perform some of these tasks.

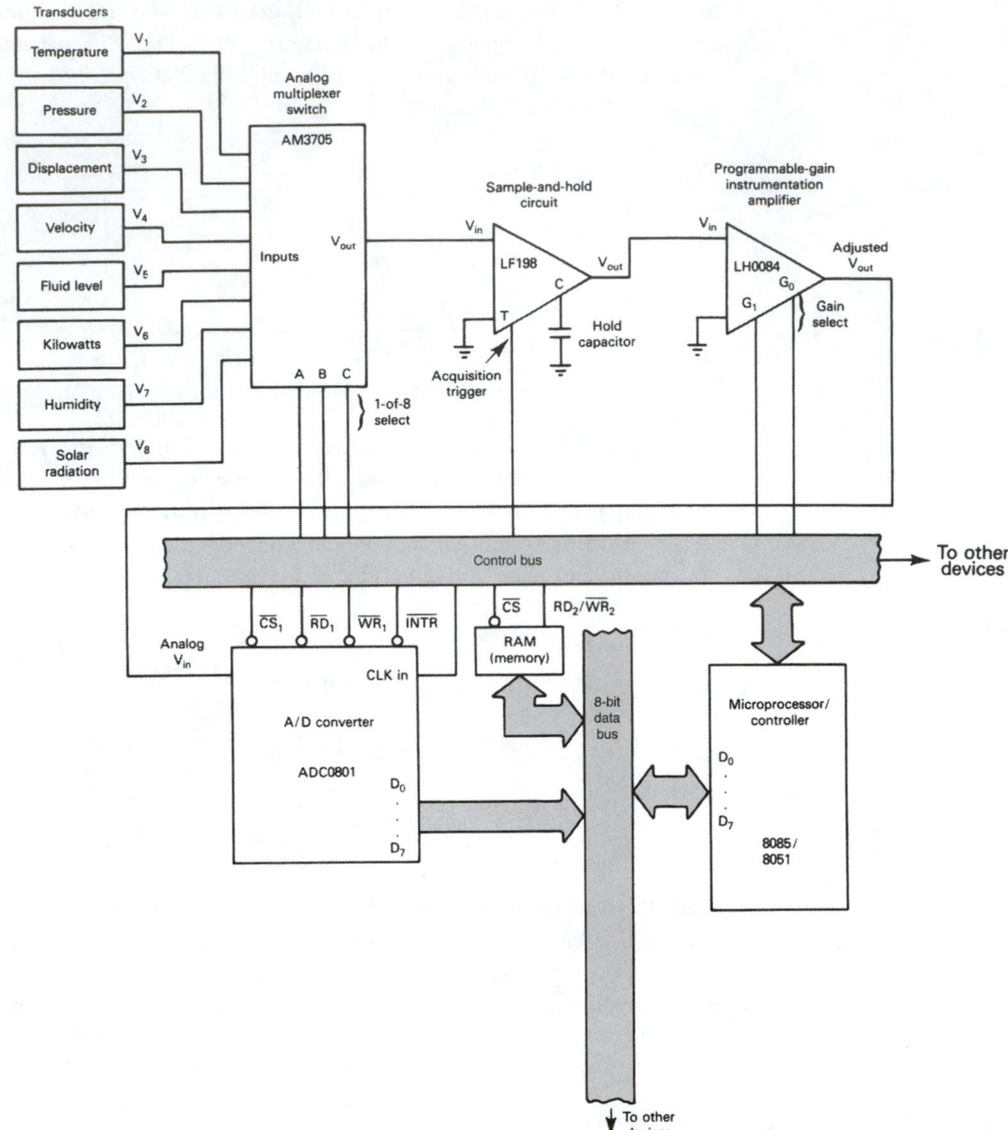

Figure 7–25 Data acquisition system.

All of the hardware interfacing and handshaking that takes place between the microprocessor and the transducers can be explained by taking a closer look at each of the devices in the system.

Analog Multiplexer Switch (AM3705)

The multiplexer reduces circuit complexity and eliminates duplication of circuitry by allowing each of the eight transducer outputs to take turns traveling through the other devices. The microprocessor selects each of the transducers at the appropriate time by setting up the appropriate binary select code on the *A*, *B*, *C* inputs via the control bus. That allows the selected transducer signal to pass through to the next device.

Sample-and-Hold Circuit (LF198)

Since analog quantities can be constantly varying, it is important to be able to select a precise time to take the measurement. The sample-and-hold circuit, with its external *Hold capacitor,* allows the system to take (Sample) *and Hold* an analog value at the precise instant that the microprocessor issues the *acquisition trigger.*

Programmable-Gain Instrumentation Amplifier (LH0084)

Each of the eight transducers have different full-scale output ratings. For instance, the temperature transducer may output in the range 0 to 5 V while the pressure transducer may only output 0 to 500 mV. The LH0084 is capable of being programmed, via the gain select inputs, for gains of 1, 2, 5, or 10. When it is time to read the pressure transducer, the microprocessor will program the gain for 10 so that the range will be 0 to 5 V, to match that of the other transducers. That way the ADC can always operate in its most accurate range, 0 to 5 V.

Analog-to-Digital Converter (ADC0801)

The ADC receives the adjusted analog voltage and converts it to an equivalent 8-bit binary string. To do that, the microprocessor issues chip select ($\overline{CS_1}$) and start conversion ($\overline{WR_1}$) pulses. When the end-of-conversion ($\overline{INTR}$) line goes LOW, the microprocessor issues an output enable ($\overline{RD_1}$) to read the data (D_0 to D_7) that pass, via the data bus, into the microprocessor and then into the random-access-memory (RAM) chip (more on memory in Chapter 8).

This cycle repeats for all eight transducers whenever the microprocessor determines that it is time for the next scan. Other software routines executed by the microprocessor will act on the data that have been gathered. Some possible responses to the measured results might be to sound an alarm, speed up a fan, reduce energy consumption, increase a fluid level, or simply produce a tabular report of the measured quantities.

SUMMARY

In this chapter we have learned that

1. Any analog quantity can be represented by a binary number. Longer binary numbers provide higher resolution, which gives a more accurate representation of the analog quantity.
2. Operational amplifiers are an important building block in analog-to-digital (A/D) and digital-to-analog (D/A) converters. They provide a means for summing currents at the input and converting a current to a voltage at the output of converter circuits.
3. The binary-weighted D/A converter is the simplest to construct but it has practical limitations in resolution (number of input bits).
4. The *R*/2*R* ladder D/A converter uses only two different resistor values no matter how many binary input bits are included. This allows for very high resolution and ease of fabrication in integrated circuit form.
5. The DAC0808 (or MC1408) IC is an 8-bit D/A converter that uses the R/2R ladder method of conversion. It accepts 8 binary input bits and outputs an equivalent analog current. Having eight input bits means that it can resolve up to 256 unique binary values into equivalent analog values.
6. Applying an 8-bit counter to the input of an 8-bit D/A converter will produce a 256-step sawtooth waveform at its output.
7. The simplest way to build an analog-to-digital (A/D) converter is to use the parallel encoding method. The disadvantage is that it is practical only for low resolution applications.
8. The counter-ramp A/D converter employs a counter, a D/A converter, and a comparator to make its conversion. The counter counts from 0 up to a value that causes the D/A output to slightly exceed the analog input value. That binary count is then output as the equivalent to the analog input.

9. The method of A/D conversion used most often is called successive approximation. In this method successive bits are tested to see if they contribute an equivalent analog value that is greater than the analog input to be converted. If they do, they are returned to 0. After all bits are tested, the ones that are left ON are used as the final digital equivalent to the analog input.
10. The NE5034 and the ADC0801 are examples of A/D converter ICs. To make a conversion the *start conversion* pin is made LOW. When the conversion is completed the *end-of-conversion* pin goes LOW. Then to read the digital output the *output enable* pin is made LOW.
11. Data acquisition systems are used to read several different analog inputs, respond to the values read, store the results, and generate reports on the information gathered.
12. Transducers are devices that convert physical quantities such as heat, light, or force into electrical quantities. Those electrical quantities must then be conditioned (or modified) before they can be interpreted by a digital computer.

GLOSSARY

ADC: Analog-to-digital converter.

Binary Weighting: Each binary position in a string is worth double the amount of the bit to its right. By choosing resistors in that same proportion, binary-weighted current levels will flow.

Bridge Circuit: A resistor network used to output a low-level differential voltage that is proportional to the change in resistance in one of its legs.

Bus: A common set of electrical conductors shared by several devices and ICs.

Continuous Conversions: An ADC that is connected to perform repeatedly analog-to-digital conversions by using the end-of-conversion signal to trigger the start-conversion input.

Conversion Time: The length of time between the start of conversion and end of conversion of an ADC.

DAC: Digital-to-analog converter.

Data Acquisition: A term generally used to refer to computer-controlled acquisition and conversion of analog values.

Differential Measurement: The measurement of the difference between two values.

Handshaking: Devices and ICs that are interfaced together must follow a specific protocol, or sequence of control operations, in order to be understood by each other.

Instrumentation Amplifier: A special-purpose operational amplifier capable of high-gain amplification of low-level differential voltages.

Interfacing: The device control and interconnection schemes required for electronic devices and ICs to communicate with each other.

Linearity: Linearity error describes how far the actual transfer function of an ADC or DAC varies from the ideal straight line drawn from 0 up to the full-scale values.

Memory: A storage device capable of holding data that can be read by some other device.

Microprocessor: A large-scale IC capable of performing several functions, including the interpretation and execution of programmed software instructions.

Monotonicity: A monotonic DAC is one in which for every increase in the input digital code, the output level either remains the same or increases.

Op Amp: An amplifier that exhibits almost ideal features (i.e., infinite input impedance, infinite gain, and zero output impedance).

Programmable-Gain Amplifier: An amplifier that has a variable voltage gain that is set by inputting the appropriate digital levels at the gain select inputs.

Reference Voltage: In DAC and ADC circuits, a reference voltage or current is provided to the circuit to set the relative scale of the input and output values.

Resolution: The number of bits in an ADC or DAC. The higher the number, the closer the final representation can be to the actual input quantity.

Sample and Hold: A procedure of taking a reading of a varying analog value at a precise instant and holding that reading.

Signal Conditioning: The process of converting the electrical signal produced by a transducer into a level that is usable by an ADC.

Strain Gage: An electronic component whose resistance increases as it is stretched by a force being applied to it.

Successive Approximation: A method of arriving at a digital equivalent of an analog value by successively trying each of the individual digital bits, starting with the MSB.

Thermistor: An electronic component whose resistance changes with a change in temperature.

Transducer: A device that converts a physical quantity such as heat, light, or force into an electrical quantity such as amperes, volts, or ohms.

Virtual Ground: In certain op-amp circuit configurations, with one input at actual ground potential the other input will be held at a 0-V potential but will not be able to sink or source current.

PROBLEMS

7–1. Describe the function of a transducer.

7–2. How many different digital representations are allowed with:
- **(a)** A 4-bit converter?
- **(b)** A 6-bit converter?
- **(c)** An 8-bit converter?
- **(d)** A 12-bit converter?

7–3. List three characteristics of op amps that make them an almost ideal amplifier.

7–4. Determine V_{out} for the op-amp circuits of Figure P7–4.

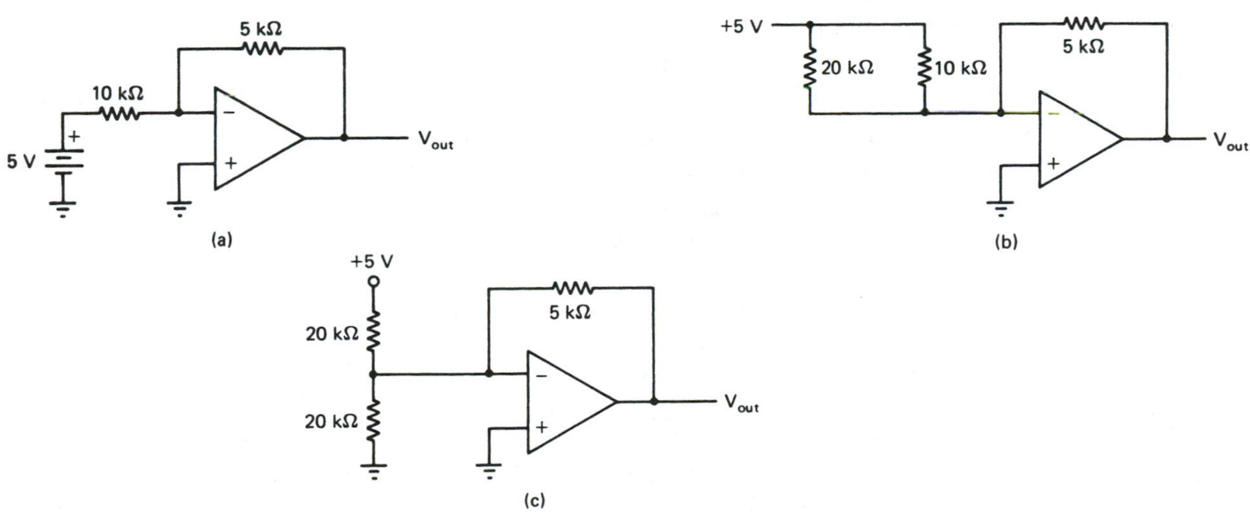

Figure P7–4

7–5. The "virtual ground" concept simplifies the analysis of op-amp circuits by allowing us to assume what?

7–6. **(a)** Change the resistor that is connected to the D_3 switch in Figure 7–4 to 10 kΩ. What values must be used for the other three resistors to ensure the correct binary weighting factors?
(b) Reconstruct the data table in Figure 7–4 with new values for V_{out} using the resistor values found in part (a).

7–7. What effect would doubling the 20-kΩ resistor have on the values for V_{out} in Figure 7–4?

7–8. What effect would changing the reference voltage (V_{ref}) in Figure 7–5 from +5 V to −5 V have on V_{out}?

7–9. Change V_{ref} in Figure 7–5 to +2 V and calculate V_{out} for $D_0 = 0$, $D_1 = 0$, $D_2 = 0$, $D_3 = 1$.

7–10. Reconstruct the data table in Figure 7–6 for a V_{ref} of +2 V instead of +5 V.

7–11. Does the MC1408 DAC use a binary-weighted or an *R*/2*R* method of conversion?

7–12. What is the purpose of the op amp in the DAC application circuit of Figure 7–8c?

7–13. What is the resolution of the DAC0808/MC1408 DAC shown in Figure 7–8?

7–14. Sketch a partial transfer function of analog output versus digital input in Figure 7–8c for digital input values of 0000 0000 through 0000 0111.

7–15. How could the reference current (I_{ref}) in Figure 7–8 be changed to 1.5 mA? What effect would that have on the range of I_{out} and V_{out}?

7–16. In Figure 7–8c, if V_{ref} is changed to 5 V, find V_{out} full-scale (A_1 to A_8 = HIGH).

7–17. Draw a graph of the transfer function (digital output versus analog input) for the parallel-encoded ADC of Figure 7–11.

7–18. What is one advantage and one disadvantage of using the multiple-comparator parallel encoding method of A/D conversion?

7–19. Refer to the counter-ramp ADC of Figure 7–12.
(a) What is the level at the DAC output the *instant after* the "start conversion" pushbutton is pressed?
(b) What is the relationship between the $V(+)$ and $V(-)$ comparator inputs the *instant before* the HIGH-to-LOW edge of "end of conversion"?

7–20. In Figure 7–12, what is the worst-case (longest) conversion time that might be encountered if the clock frequency is 100 kHz?

7–21. Determine the conversion time for an 8-bit ADC that uses a successive-approximation circuit similar to Figure 7–13 if its clock frequency is 50 kHz.

7–22. What connections could be made in the ADC shown in Figure 7–13 to enable it to make continuous conversions?

7–23. Use the SAR ADC of Figure 7–13 to convert the analog voltage 7.28 to 8-bit binary. If $V_{ref} = 10$ V, determine the final binary answer and the percent error.

7–24. Why is the three-state buffer at the output of the NE5034 ADC an important feature?

7–25. Referring to the block diagram of the ADC0801 (Figure 7–17), which inputs are used to enable the three-state output latches? Are they active-LOW or active-HIGH inputs?

7–26. What type of application might require the use of the differential inputs [$V_{in(+)}$, $V_{in(-)}$] on the ADC0801?

7–27. Refer to Figures 7–17 and 7–18.
(a) How would the operation change if $\overline{RD}$ were connected to +5 V instead of ground?
(b) How would the operation change if $\overline{CS}$ were connected to +5 V instead of ground?
(c) What is the purpose of the 10 kΩ–0.001 μF *RC* circuit?
(d) What is the maximum range of the analog V_{in} if $V_{ref/2}$ is changed to 0.5 V?

7–28. Use the thermistor characteristic curve in Figure 7–19 to determine V_{out} in the op-amp circuit of Figure 7–20 at 12.5°C and at 37.5°C.

7–29. Determine the binary output at D_7–D_0 in Figure 7–21 at 20°C and at 80°C.

7–30. The output of the strain gage circuit in Figure 7–24 is initially set at 0 V. Adding a weight to the beam to which the strain gage is attached causes V_{out} to increase to 2.5 V. Determine the change in strain gage resistance (ΔR_g) that would cause that change in V_{out}.

7–31. Briefly describe the flow of the signal from the temperature transducer as it travels through the circuit to the RAM memory in the data acquisition system of Figure 7–25.

SCHEMATIC INTERPRETATION PROBLEMS

7–32. Design a circuit interface that will provide analog output capability to the 4096/4196 control card. Assume that software will be written by a programmer to output the appropriate digital strings to port 1 (P17–P10) of the 8031 microcontroller. Devise an analog output circuit using an MC1408 DAC with a 741 op-amp to output analog voltages in the range of 0 V to 5 V.

7–33. Design a circuit interface that will provide analog input capability to the 4096/4196 control card. The design must be capable of inputting the 8-bit digital results from *two* ADC0801 converters into port 1 (P17–P10) of the 8031 microcontroller. Assume that a single-bit control signal will be output on port 2, bit 0 (P20) to tell which ADC results are to be transmitted. (Assume 1 = ADC#1 and 0 = ADC#2.) Set up the ADCs for continuous conversions and 0-V to 5-V analog input level.

8 Microprocessor and Computer Memory

OBJECTIVES

Upon completion of this chapter, you should be able to:

- Explain the basic concepts involved in memory addressing and data storage.
- Interpret the specific timing requirements given in a manufacturer's data manual for reading or writing to a memory IC.
- Discuss the operation and application for the various types of semiconductor memory ICs.
- Design circuitry to facilitate memory expansion.
- Explain the "refresh" procedure for dynamic RAMs.
- Explain the programming procedure and applications for programmable array ICs.

INTRODUCTION

In digital systems memory circuits provide the means of storing information (data) on a temporary or permanent basis for future recall. The storage medium can be either a semiconductor integrated circuit or a magnetic device such as magnetic tape or disk. Magnetic media generally are capable of storing larger quantities of data than semiconductor memories, but the access time (time it takes to locate, then read or write data) is usually much more for magnetic devices. With magnetic tape or disk it takes time to physically move the read/write mechanism to the exact location to be written to or read from.

With semiconductor memory ICs, electrical signals are used to identify a particular memory location within the integrated circuit and data can be stored in or read from that location in a matter of nanoseconds.

The technology used in the fabrication of memory ICs can be based on either bipolar or MOS transistors. In general, bipolar memories are faster than MOS memories, but MOS can be integrated more densely, providing much more memory locations in the same amount of area.

8–1 MEMORY CONCEPTS

Let's say that you have an application where you must store the digital states of eight binary switches once every hour for 16 hours. This would require 16 *memory locations,* each having a *unique 4-bit address* (0000 to 1111) and each location being capable of containing 8 bits of data. A group of 8 bits is also known as one *byte,* so what we would have is a 16-byte memory, as shown in Figure 8–1.

Location address	Data contents
0000	
0001	
0010	
0011	
0100	
⋮	⋮
1101	
1110	
1111	

16 Bytes

Figure 8–1 Layout for sixteen 8-bit memory locations.

To set up this memory system using actual ICs, we could use sixteen 8-bit flip-flop registers to contain the 16 bytes of data. To identify the correct address, a 4-line-to-16-line decoder can be used to decode the 4-bit location address into an active-LOW chip select to select the appropriate (1-of-16) data register for input/output. Figure 8–2 shows the circuit used to implement this memory application.

The 74LS374s are octal (eight) *D* flip-flops with three-state outputs. To store data in them, 8 bits of data are put on the D_0 to D_7 data inputs via the data bus. Then a LOW-to-HIGH edge on the C_p clock input will cause the data at D_0 to D_7 to be latched into each flip-flop. The value stored in the *D* flip-flops is observed at the Q_0 to Q_7 outputs by making the Output Enable ($\overline{\text{OE}}$) pin LOW.

To select the appropriate (1-of-16) memory location, a 4-bit address is input to the 74LS154 (4-line-to-16-line decoder), which outputs a LOW pulse on one of the output lines when the $\overline{\text{WRITE}}$ enable input is pulsed LOW.

As you can see, the timing of setting up the address bus, data bus, and pulsing the $\overline{\text{WRITE}}$ line is critical. Timing diagrams are necessary for understanding the operation of memory ICs, especially when you are using larger-scale memory ICs. The timing diagram for our 16-byte memory design of Figure 8–2 is given in Figure 8–3.

Figure 8–3 begins to show us some of the standard ways that manufacturers illustrate timing parameters for bus-driven devices. Rather than showing all four address lines and all eight data lines, they group them together and use an X (crossover) to show where any or all of the lines are allowed to change digital levels.

In Figure 8–3 the address and data lines must be set up some time (t_s) before the LOW-to-HIGH edge of $\overline{\text{WRITE}}$. In other words, the address and data lines must *be valid* (be at the appropriate levels) some period of time (t_s) *before* the LOW-to-HIGH edge of $\overline{\text{WRITE}}$ in order for the 74LS374 *D* flip-flop to interpret the input correctly.

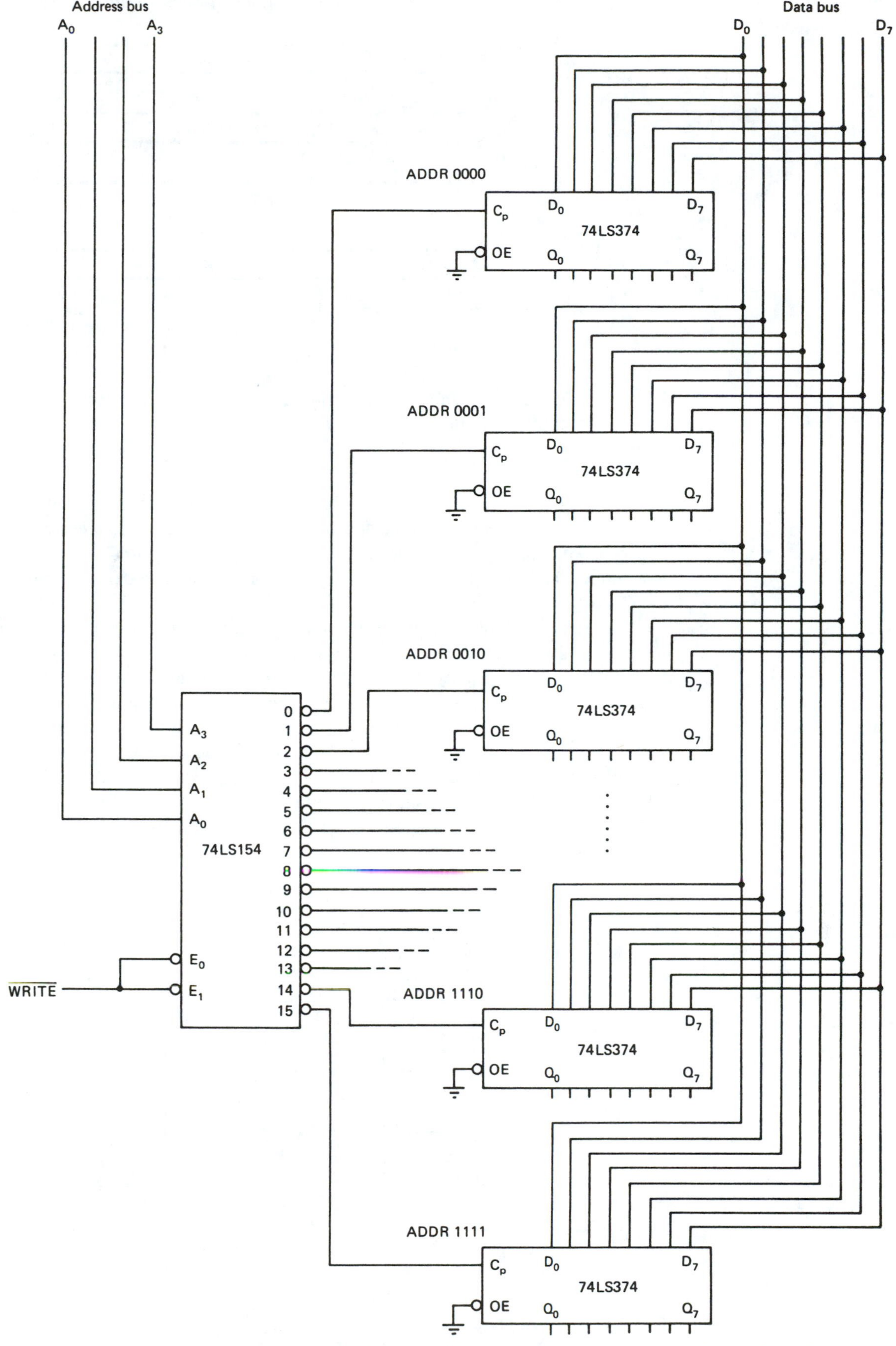

Figure 8–2 Writing to a 16-byte memory constructed from 16 octal *D* flip-flops and a 1-of-16 decoder.

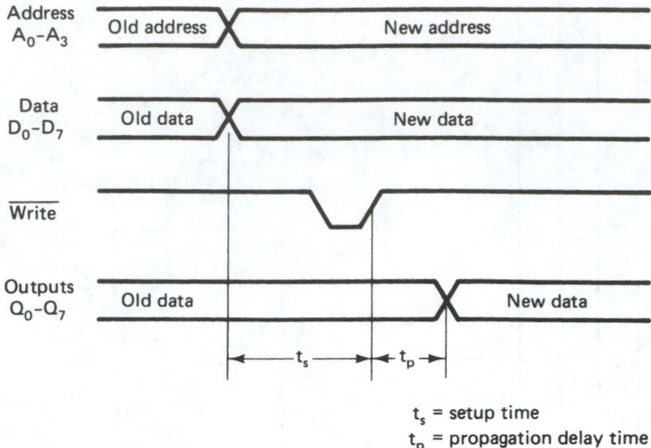

Figure 8–3 Timing requirements for writing data to the 16-byte memory circuit of Figure 8–2.

When the $\overline{\text{WRITE}}$ line is pulsed, the 74LS154 decoder outputs a LOW pulse on one of its 16 outputs, which clocks the appropriate memory location to receive data from the data bus. After the propagation delay (t_p), the data output at Q_0 to Q_7 will be the new data just entered into the D flip-flop; t_p will include the propagation delay of the decoder *and* the C_p-to-Q of the D flip-flop.

In Figure 8–2 all of the three-state outputs are continuously enabled so that their Q outputs are always active. To connect the Q_0 to Q_7 outputs of all 16 memory locations back to the data bus, the $\overline{\text{OE}}$ enables would have to be individually selected, at the appropriate time to avoid a conflict on the data bus, called *bus contention.* Bus contention occurs when two or more devices are trying to send their own digital levels to the shared data bus at the same time. To individually select each group of Q outputs in Figure 8–2, the grounds on the $\overline{\text{OE}}$ enables would be removed, and instead be connected to the output of another 74LS154 1-of-16 decoder, as shown in Figure 8–4.

We have designed in Figures 8–2, 8–3, and 8–4 a small 16-byte (16 × 8) random-access memory (RAM). Commercially available RAM ICs combine all the decoding and storage elements in a single package.

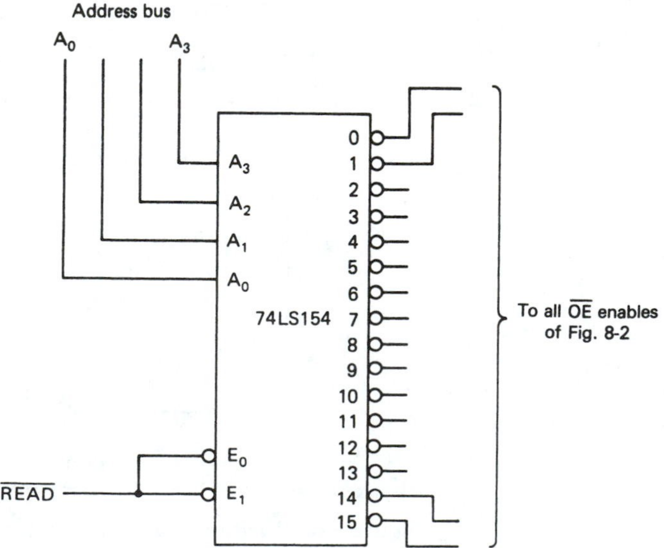

Figure 8–4 Using another decoder to individually select memory locations for Read operations.

8–2 STATIC RAMs

Large-scale *random-access memory* (RAM), also known as *read-write memory,* is used for temporary storage of data and program instructions in microprocessor-based systems. The term *random access* means that the user can access (read or write) data at any location within the entire memory device randomly, without having to sequentially read through several data values until positioned at the desired memory location. [An example of a sequential (nonrandom) memory device is magnetic tape.]

A better term for RAM is *read/write memory* (RWM) because all semiconductor and disk memories have random access. RWM is more specific in that it tells us that data can be *read or written* to any memory location.

RAM is classified as either static or dynamic. *Static* RAMs (SRAMs) use flip-flops as basic storage elements, whereas *dynamic* RAMs (DRAMs) use internal capacitors as basic storage elements. Additional *refresh* circuitry is needed to maintain the charge on the internal capacitors of a dynamic RAM, which makes it more difficult to use. Dynamic RAMs can be packed very densely, however, yielding much more storage capacity per unit area than a static RAM. The cost per bit of dynamic RAMs is also much less than that of the static RAM.

The 2147H Static MOS RAM

The 2147H is a very popular static RAM that uses MOS technology. The 2147H is set up with 4096 (abbreviated 4K, where 1K = 1024) memory locations, with each location containing 1 bit of data. This configuration is called 4096×1.

To develop a unique address for each of the 4096 locations, 12 address lines must be input ($2^{12} = 4096$). The storage locations are set up as a 64×64 array with A_0 to A_5 identifying the row and A_6 to A_{11} identifying the column to pinpoint the specific location to be used. The data sheet for the 2147H is given in Figure 8–5. This figure shows the row and column circuitry used to pinpoint the memory cell within the 64×64 array. The box labeled "Row Select" is actually a 6-to-64 decoder for identifying the appropriate 1-of-64 row. The box labeled "Column Select" is also a 6-to-64 decoder for identifying the appropriate 1-of-64 column. Once the location is selected, the AND gates at the bottom of the block diagram allow the data bit to either pass into (D_{in}) or come out of (D_{out}) the memory location selected. Each memory location, or cell, is actually a configuration of transistors that function like a flip-flop that can be Set (1) or Reset (0).

During *write operations,* in order for D_{in} to pass through its three-state buffer, the Chip Select ($\overline{CS}$) must be LOW *and* the Write Enable ($\overline{WE}$) must also be LOW. During *read operations,* in order for D_{out} to receive data from its three-state buffer, the Chip Select ($\overline{CS}$) must be LOW *and* the Write Enable ($\overline{WE}$) must be HIGH, signifying a *Read operation.*

Read Operation. The circuit connections and waveforms for reading data from a location in a 2147H are given in Figure 8–6. The 12 address lines are brought in from the address bus for address selection. The $\overline{WE}$ input is held HIGH to enable the Read operation.

Referring to the timing diagram, when the new address is entered on the A_0 to A_{11} inputs and the $\overline{CS}$ lines goes LOW, it takes a short period of time, called the *access time,* before the data output is valid. The access time is the length of time from the beginning of the read cycle to the end of t_{ACS} or t_{AA}, whichever ends last. Before the $\overline{CS}$ is brought LOW, D_{out} is in a high-impedance (float) state. The $\overline{CS}$ *and* A_0 to A_{11} inputs must both be held stable for a minimum length of time, t_{RC}, before another Read cycle can be initiated.

After $\overline{CS}$ goes back HIGH, the data out is still valid for a short period of time, t_{HZ}, before returning to its high-impedance state.

Write Operation. A similar set of waveforms are given by the manufacturer for the Write operation. In this case the D_{in} is written into memory while the $\overline{CS}$ and $\overline{WE}$ are

2147H
HIGH SPEED 4096 × 1 BIT STATIC RAM

	2147H-1	2147H-2	2147H-3	2147HL-3	2147H	2147HL
Max. Access Time (ns)	35	45	55	55	70	70
Max. Active Current (mA)	180	180	180	125	160	140
Max. Standby Current (mA)	30	30	30	15	20	10

- **Pinout, Function, and Power Compatible to Industry Standard 2147**
- **HMOS II Technology**
- **Completely Static Memory—No Clock or Timing Strobe Required**
- **Equal Access and Cycle Times**
- **Single +5V Supply**
- **0.8–2.0V Output Timing Reference Levels**
- **Direct Performance Upgrade for 2147**
- **Automatic Power-Down**
- **High Density 18-Pin Package**
- **Directly TTL Compatible—All Inputs and Output**
- **Separate Data Input and Output**
- **Three-State Output**

The Intel® 2147H is a 4096-bit static Random Access Memory organized as 4096 words by 1-bit using HMOS-II, Intel's next generation high-performance MOS technology. It uses a uniquely innovative design approach which provides the ease-of-use features associated with non-clocked static memories and the reduced standby power dissipation associated with clocked static memories. To the user this means low standby power dissipation without the need for clocks, address setup and hold times, nor reduced data rates due to cycle times that are longer than access times.

$\overline{CS}$ controls the power-down feature. In less than a cycle time after $\overline{CS}$ goes high—deselecting the 2147H—the part automatically reduces its power requirements and remains in this low power standby mode as long as $\overline{CS}$ remains high. This device feature results in system power savings as great as 85% in larger systems, where the majority of devices are deselected.

The 2147H is placed in an 18-pin package configured with the industry standard 2147 pinout. It is directly TTL compatible in all respects: inputs, output, and a single +5V supply. The data is read out nondestructively and has the same polarity as the input data. A data input and a separate three-state output are used.

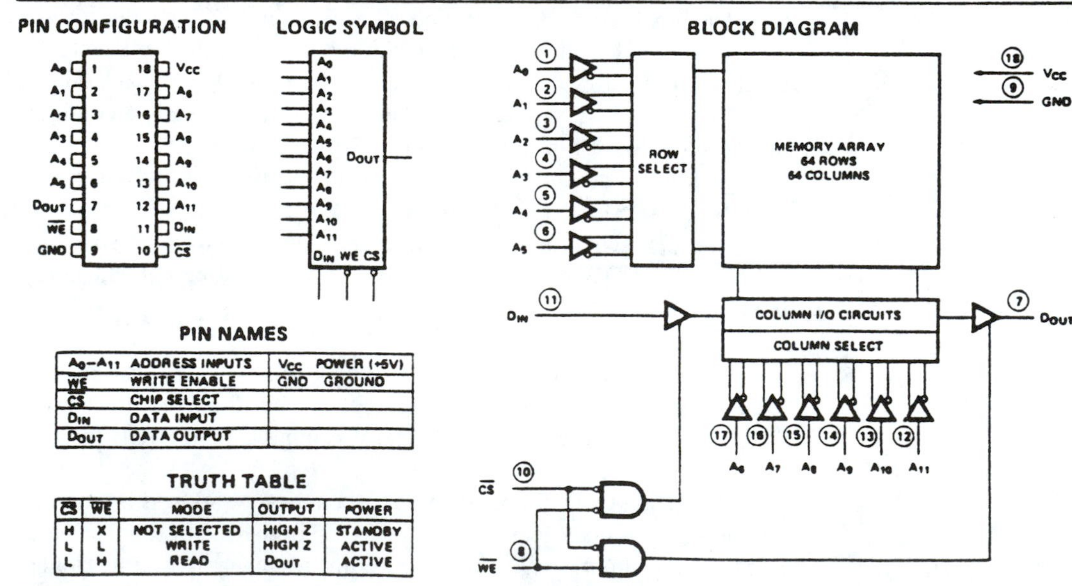

PIN NAMES

A_0–A_{11}	ADDRESS INPUTS	V_{CC}	POWER (+5V)
$\overline{WE}$	WRITE ENABLE	GND	GROUND
$\overline{CS}$	CHIP SELECT		
D_{IN}	DATA INPUT		
D_{OUT}	DATA OUTPUT		

TRUTH TABLE

$\overline{CS}$	$\overline{WE}$	MODE	OUTPUT	POWER
H	X	NOT SELECTED	HIGH Z	STANDBY
L	L	WRITE	HIGH Z	ACTIVE
L	H	READ	D_{OUT}	ACTIVE

Figure 8–5 The 2147H 4K × 1 static RAM. (Courtesy of Intel Corporation)

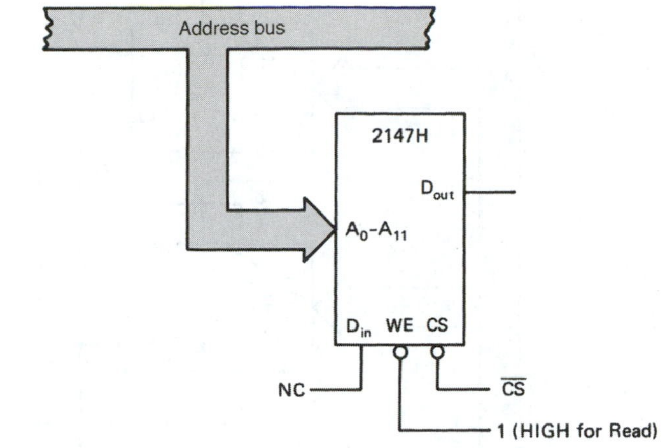

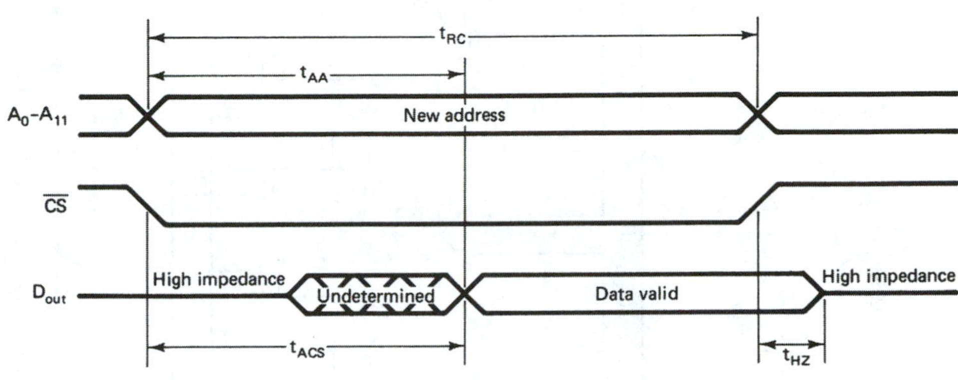

Symbol	Parameter	Min.	Max.	Unit
t_{RC}	Read cycle time	35		ns
t_{AA}	Address access time		35	ns
t_{ACS}	Chip select access time		35	ns
t_{HZ}	Chip deselection to high-Z out	0	30	ns

Figure 8–6 The 2147H Read cycle.

both LOW. The D_{in} must be set up for a length of time *before* either $\overline{CS}$ or $\overline{WE}$ goes back HIGH, and it must also be held for a length of time *after* either $\overline{CS}$ or $\overline{WE}$ goes back HIGH.

Memory Expansion. Since the contents of each memory location in the 2147H is only 1 bit, to be used in an 8-bit computer system, eight 2147Hs must be set up in such a way that when an address is specified, 8 bits of data will be read or written. With eight 2147s we have a 4096 by 8 (4K $\times$ 8) memory system, as shown in Figure 8–7.

The address selection for each 2147H in Figure 8–7 is identical because they are all connected to the same address bus lines. This way, when reading or writing from a specific address, 8 bits, each at the same address, will be sent to, or received from, the data bus simultaneously. The $\overline{WE}$ input determines which internal three-state buffer is enabled, connecting *either* D_{in} or D_{out} to the data bus. The $\overline{WE}$ input is sometimes labeled READ/$\overline{WRITE}$, meaning that it is HIGH for a Read operation, which puts data out to the data bus, via D_{out}, and it is LOW for a Write operation, which writes data into the memory via D_{in}.

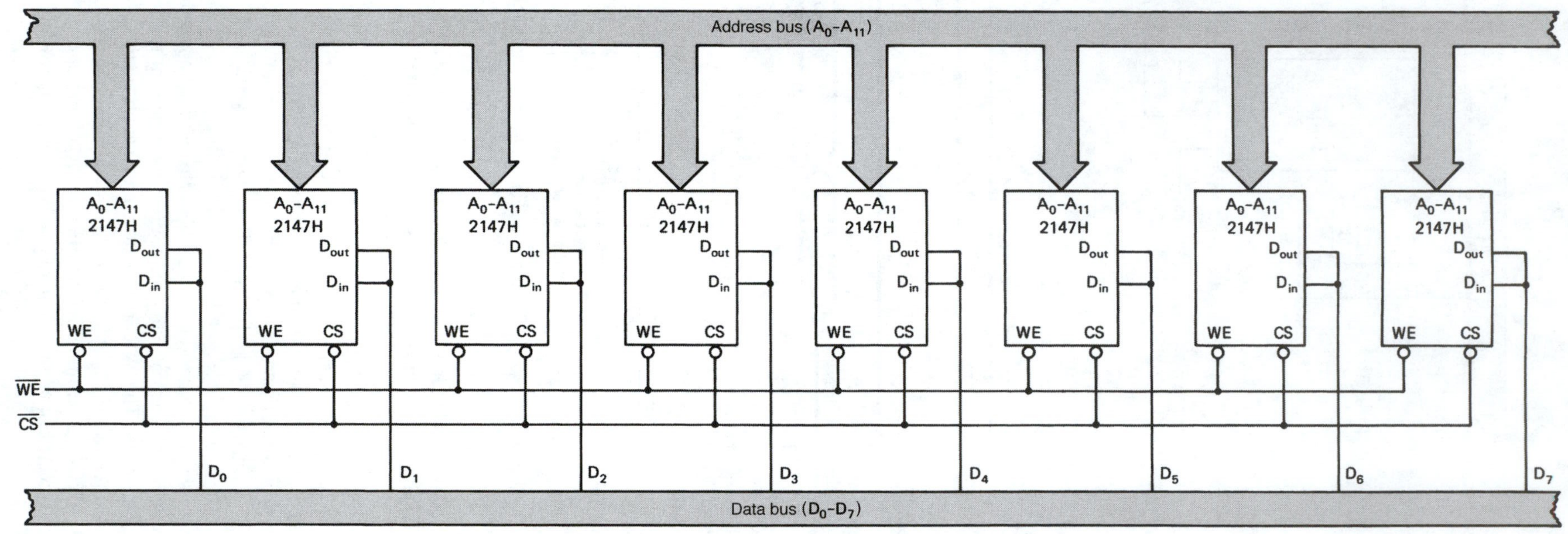

Figure 8–7 4K × 8 memory expansion using eight 2147Hs.

There are several other configurations of RAM memory available. For example, the 2148H is configured as a 1024 × 4-bit (1K × 4) RAM, instead of the 4096 × 1 used by the 2147H. A 1024 × 4-bit RAM will input/output 4 bits at a time for each address specified. This way, interfacing to an 8-bit data bus is simplified by having to use only two 2148Hs, one for the LOW-order data bits (D_0 to D_3) and the other for the HIGH-order data bits (D_4 to D_7).

8–3 DYNAMIC RAMS

Although dynamic RAMs require more support circuitry and are more difficult to use than static RAMs, they are less expensive per bit and have a much higher density, minimizing circuit-board area. Most applications requiring large amounts of read/write memory will use dynamic RAMs instead of static.

Dynamic MOS RAMs store information on a small internal capacitor instead of a flip-flop. All the internal capacitors require recharging, or refreshing, every 2 ms or less to maintain the stored information. An example of a 16K × 1-bit dynamic RAM is the Intel 2118, whose data sheet is shown in Figure 8–8a.

To uniquely address 16,384 locations, 14 address lines are required ($2^{14} = 16{,}384$). However, Figure 8–8a shows only seven address lines (A_0 to A_6). This is because with larger memories such as this, in order to keep the IC pin count to a minimum, the address lines are *multiplexed* into two groups of seven. An external 14-line-to-7-line multiplexer is required in conjunction with the control signals, $\overline{\text{RAS}}$ and $\overline{\text{CAS}}$, in order to access a complete 14-line address.

The controlling device must put the valid 7-bit address of the desired memory array *row* on the A_0 to A_6 inputs, then send the Row Address Strobe ($\overline{\text{RAS}}$) LOW. Next, the controlling device must put the valid 7-bit address of the desired memory array *column* on the *same* A_0 to A_6 inputs, then send the Column Address Strobe ($\overline{\text{CAS}}$) LOW. Each of these 7-bit addresses is latched and will pinpoint the desired 1-bit memory location by its row–column coordinates.

Once the memory location is identified, the $\overline{\text{WE}}$ input is used to direct either a Read or Write cycle similar to the static RAM operation covered in Section 8–2. When $\overline{\text{WE}}$ is LOW, data are written to the RAM via D_{in}; when $\overline{\text{WE}}$ is HIGH, data are read from the RAM via D_{out}.

Read Cycle Timing. (Figure 8–8b)

1. $\overline{\text{WE}}$ is HIGH.
2. A_0 to A_6 are set up with the row address and $\overline{\text{RAS}}$ is sent LOW.
3. A_0 to A_6 are set up with the column address and $\overline{\text{CAS}}$ is sent LOW.
4. After the access time from $\overline{\text{RAS}}$ or $\overline{\text{CAS}}$ (whichever is longer), the D_{out} line will contain valid data.

Write Cycle Timing. (Figure 8–8c)

1. $\overline{\text{WE}}$ is LOW.
2. A_0 to A_6 are set up with the row address and $\overline{\text{RAS}}$ is sent LOW.
3. A_0 to A_6 are set up with the column address and $\overline{\text{CAS}}$ is sent LOW.
4. At the HIGH-to-LOW edge of $\overline{\text{CAS}}$, the level at D_{in} is stored at the specified row–column memory address. D_{in} must be set up prior to, and held after, the HIGH-to-LOW edge of $\overline{\text{CAS}}$ to be interpreted correctly. (There are other setup, hold, and delay times that are not shown. Refer to a memory data book for more complete specifications.)

2118 FAMILY
16,384 x 1 BIT DYNAMIC RAM

	2118-10	2118-12	2118-15
Maximum Access Time (ns)	100	120	150
Read, Write Cycle (ns)	235	270	320
Read-Modify-Write Cycle (ns)	285	320	410

- **Single +5V Supply, ±10% Tolerance**
- **HMOS Technology**
- **Low Power: 150 mW Max. Operating 11 mW Max. Standby**
- **Low V_{DD} Current Transients**
- **All Inputs, Including Clocks, TTL Compatible**
- **$\overline{CAS}$ Controlled Output is Three-State, TTL Compatible**
- **$\overline{RAS}$ Only Refresh**
- **128 Refresh Cycles Required Every 2ms**
- **Page Mode and Hidden Refresh Capability**
- **Allows Negative Overshoot V_{IL} min = -2V**

The Intel® 2118 is a 16,384 word by 1-bit Dynamic MOS RAM designed to operate from a single +5V power supply. The 2118 is fabricated using HMOS — a production proven process for high performance, high reliability, and high storage density.

The 2118 uses a single transistor dynamic storage cell and advanced dynamic circuitry to achieve high speed with low power dissipation. The circuit design minimizes the current transients typical of dynamic RAM operation. These low current transients contribute to the high noise immunity of the 2118 in a system environment.

Multiplexing the 14 address bits into the 7 address input pins allows the 2118 to be packaged in the industry standard 16-pin DIP. The two 7-bit address words are latched into the 2118 by the two TTL clocks, Row Address Strobe ($\overline{RAS}$) and Column Address Strobe ($\overline{CAS}$). Non-critical timing requirements for $\overline{RAS}$ and $\overline{CAS}$ allow use of the address multiplexing technique while maintaining high performance.

The 2118 three-state output is controlled by $\overline{CAS}$, independent of $\overline{RAS}$. After a valid read or read-modify-write cycle, data is latched on the output by holding $\overline{CAS}$ low. The data out pin is returned to the high impedance state by returning $\overline{CAS}$ to a high state. The 2118 hidden refresh feature allows $\overline{CAS}$ to be held low to maintain latched data while $\overline{RAS}$ is used to execute $\overline{RAS}$-only refresh cycles.

The single transistor storage cell requires refreshing for data retention. Refreshing is accomplished by performing $\overline{RAS}$-only refresh cycles, hidden refresh cycles, or normal read or write cycles on the 128 address combinations of A_0 through A_6 during a 2ms period. A write cycle will refresh stored data on all bits of the selected row except the bit which is addressed.

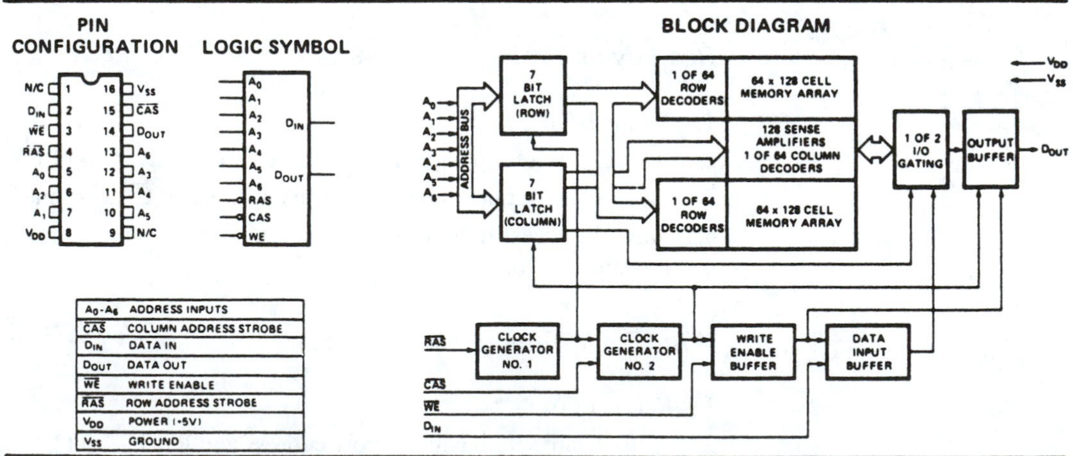

Figure 8–8 (a) The 2118 16K × 1 dynamic RAM.

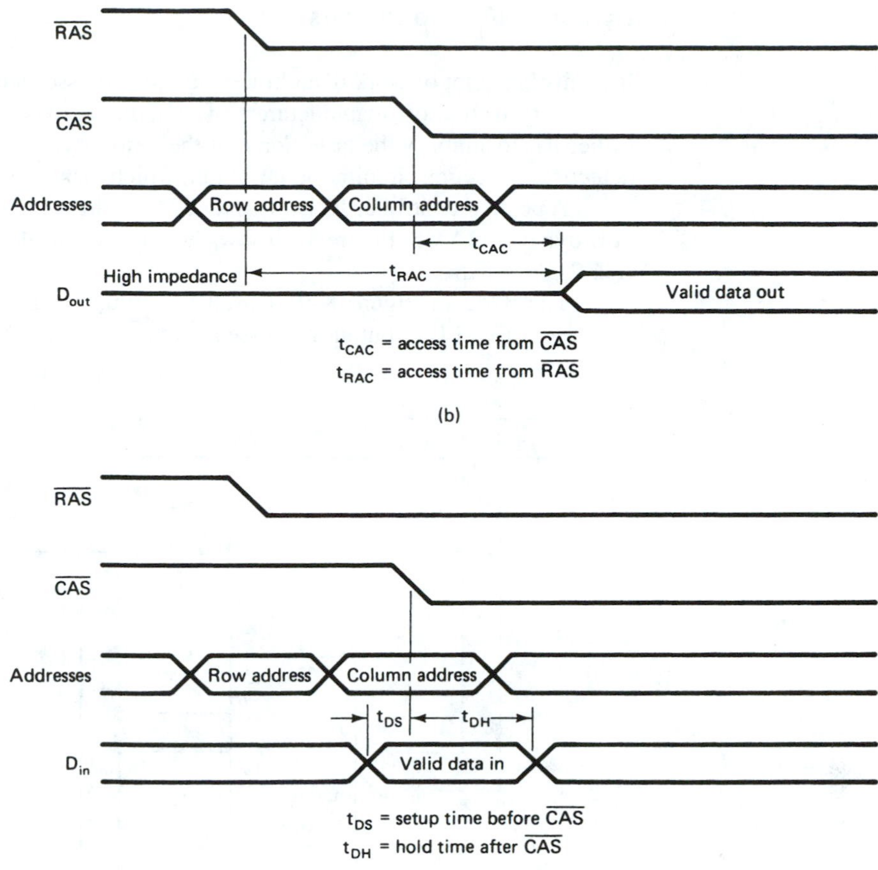

Figure 8–8 (*Continued*) (b) Dynamic RAM Read cycle timing ($\overline{WE}$ = HIGH); (c) dynamic RAM Write cycle timing ($\overline{WE}$ = LOW). [(a) Courtesy of Intel Corporation]

Refresh Cycle Timing. Each of the 128 rows of the 2118 must be *refreshed* every 2 ms or sooner to replenish the charge on the internal capacitors. There are three ways to refresh the memory cells:

1. Read cycle
2. Write cycle
3. $\overline{RAS}$-only cycle

Unless you are reading or writing from all 128 rows every 2 ms, the $\overline{RAS}$-only cycle is the preferred technique to provide data retention. To perform a $\overline{RAS}$-only cycle, the following procedure is used:

1. $\overline{CAS}$ is HIGH.
2. A_0 to A_6 are set up with the row address 000 0000.
3. $\overline{RAS}$ is pulsed LOW.
4. Increment the A_0 to A_6 row address by 1.
5. Repeat steps 3 and 4 until all 128 rows have been accessed.

Dynamic RAM Controllers

It seems like a lot of work demultiplexing the addresses and refreshing the memory cells, doesn't it? Well, most manufacturers of dynamic RAMs (DRAMs) have developed controller ICs to simplify the task. Some of the newer dynamic RAMs have refresh and error detection/correction circuitry built right in, which makes the DRAM look *static* to the user.

A popular controller IC is the Intel 3242 address multiplexer and refresh counter for 16K dynamic RAMS. Figure 8–9 shows how this controller IC is used in conjunction with four 2118 DRAMs.

The 3242 in Figure 8–9 is used to multiplex the 14 input addresses A_0 to A_{13} to seven active-LOW output addresses $\overline{Q}_0$ to $\overline{Q}_6$. When the Row Enable input is HIGH,

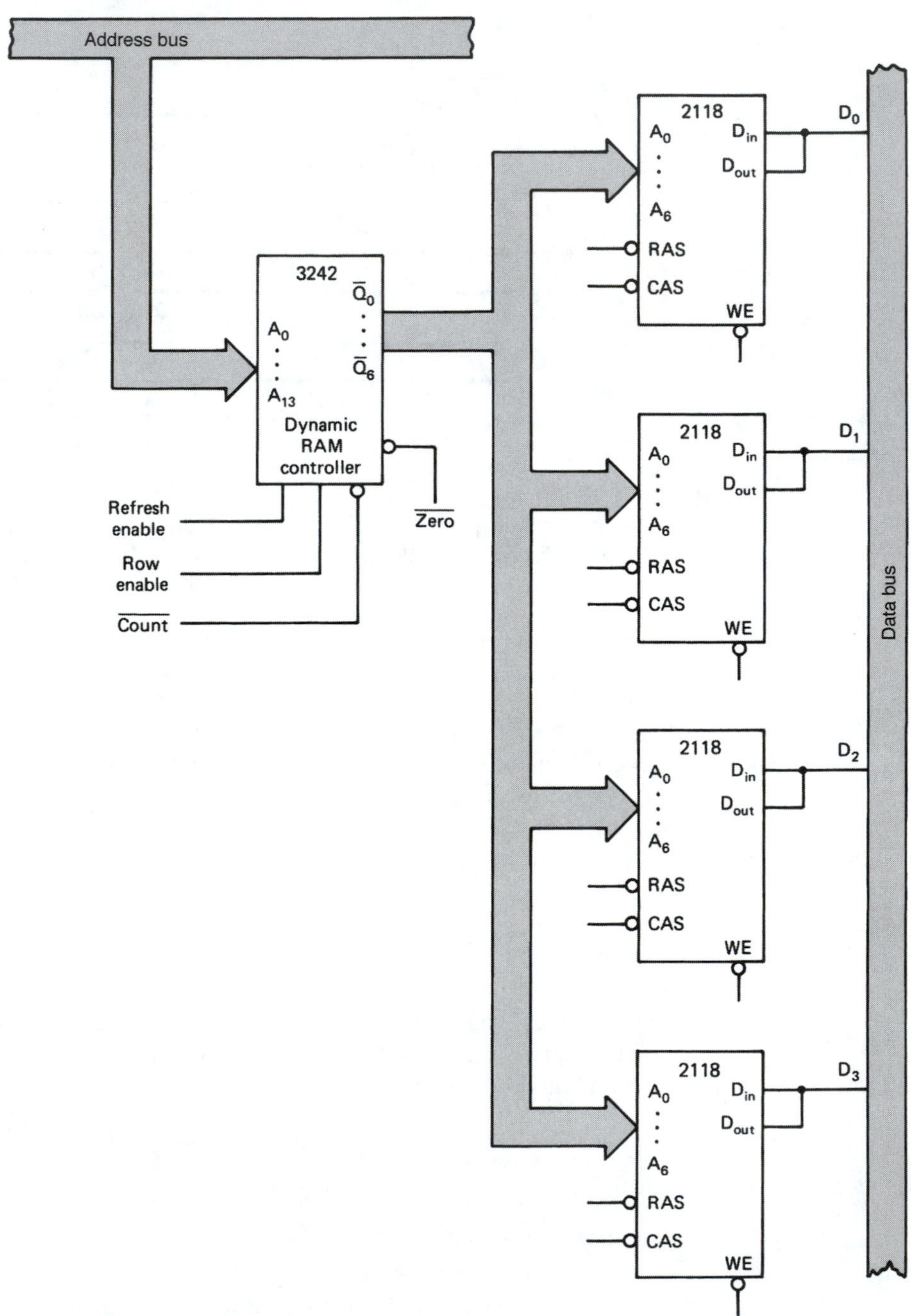

Figure 8–9 Using a 3242 address multiplexer and refresh counter in a 16K × 4 dynamic RAM memory system.

A_0 to A_6 are output inverted to $\overline{Q}_0$ to $\overline{Q}_6$ as the row addresses. When the Row Enable input is LOW, A_7 to A_{13} are output inverted to $\overline{Q}_0$ to $\overline{Q}_6$ as the column address. Of course, the timing of the $\overline{\text{RAS}}$ and $\overline{\text{CAS}}$ on the 2118s must be synchronized with the Row Enable signal.

To provide a "burst" refresh to all 128 rows of the 2118s, the Refresh Enable input in Figure 8–9 is made HIGH. That causes the $\overline{Q}_0$ to $\overline{Q}_6$ outputs to count from 0 to 127 at a rate determined by the $\overline{\text{count}}$ input clock signal. When the first 6 significant bits of the counter sequence to all 0s, the $\overline{\text{zero}}$ output goes LOW, signifying the completion of the first 64 refresh cycles.

One commonly used method of setting up the timing for $\overline{\text{RAS}}$, $\overline{\text{CAS}}$, and Row Enable is with a multitap *delay line,* as shown in Figure 8–10. Basically, the four-tap delay line IC of Figure 8–10a is made up of four inverters with precision *RC*s to develop a 50-ns delay between each inverter. The pulses out of each tap have the same width, but each successive tap is inverted and delayed by 50 ns. (In Figure 8–10b every other tap was used to arrive at noninverted, 100-ns delay pulses.) Delay lines are very useful for circuits requiring sequencing, as dynamic RAM memory systems do.

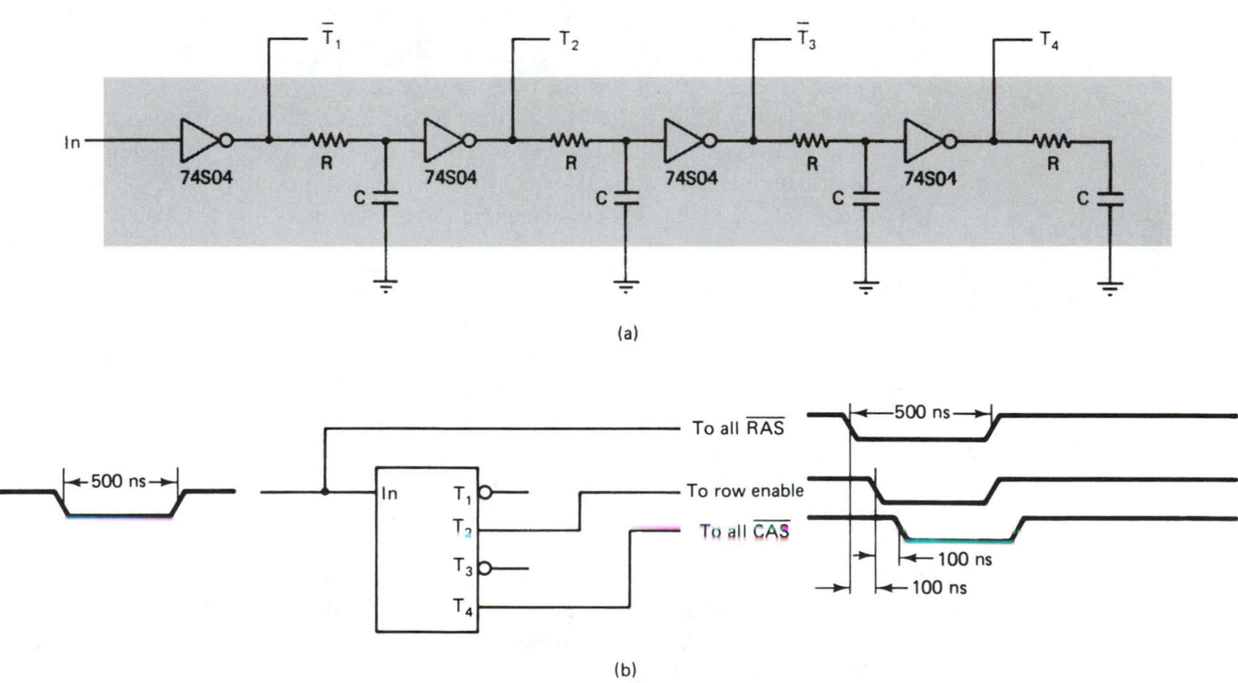

Figure 8–10 Four-tap 50-ns delay line used for dynamic RAM timing: (a) logic diagram; (b) logic symbol and timing.

The waveforms produced by the delay line of Figure 8–10b can be used to drive the control inputs to the 16K × 4 dynamic RAM memory system of Figure 8–9 (a LOW $\overline{\text{RAS}}$ pulse, then a LOW Row Enable pulse, then a LOW $\overline{\text{CAS}}$ pulse). Careful inspection of the data sheets for the 3242 and 2118 is required to determine the maximum and minimum allowable values for pulse widths and delay times. To design the absolute fastest possible memory circuit, all the times would be kept at their minimum value. But it is good practice to design a 10 to 20% margin to be safe.

8–4 READ-ONLY MEMORIES

ROMs are memory ICs used to store data on a permanent basis. They are capable of random access and are *nonvolatile,* meaning that they do not lose their memory contents when power is removed. This makes them very useful for the storage of computer operating systems, software language compilers, table look-ups, specialized code conversion routines, and programs for dedicated microprocessor applications.

ROMs are generally used for read-only operations and are not written to after they are initially programmed. However, there is an erasable variety of ROM called an EPROM (erasable programmable read-only memory) that is very useful because it can be erased and then reprogrammed if desired.

To use a ROM, the user simply specifies the correct address to be read and then enables the Chip Select ($\overline{CS}$). The data contents at that address (usually 8 bits) will then appear at the outputs of the ROM (some ROM outputs will be three-stated, so you will have to enable the output with a LOW on $\overline{OE}$).

Mask Roms

Manufacturers will make a custom mask ROM for users who are absolutely sure of the desired contents of the ROM and have a need for at least 1000 or more chips. To fabricate a custom IC like the mask ROM, the manufacturer charges a one-time fee of more than $1000 for the design of a unique mask that is required in the fabrication of the integrated circuit. After that, each identical ROM that is produced is very inexpensive. In basic terms a mask is a cover placed over the silicon chip during fabrication that determines the permanent logic state to be formed at each memory location. Of course, before the mass production of a quantity of mask ROMs, the user should have thoroughly tested the program or data that will be used as the model for the mask. Most desktop computers use mask ROMs to contain their operating system and for executing procedures that do not change, such as decoding the keyboard and the generation of characters for the CRT.

Fusible-Link PROMs

To avoid the high one-time cost of producing a custom mask, IC manufacturers provide user-programmable ROMs (PROMs). They are available in standard configurations such as 4K × 4, 4K × 8, 8K × 4, and so on.

Initially, every memory cell has a fusible link, keeping its output at 0. A 0 is changed to a 1 by sending a high enough current through the fuse to permanently open it, making the output of that cell a 1. The programming procedure involves addressing each memory location, in turn, and placing the 4-bit or 8-bit data to be programmed at the PROM outputs and then applying a programming pulse (either a HIGH voltage or a constant current to the programming pin). Details for programming are given in the next section.

Once the fusible link is burned open, the data are permanently stored in the PROM and can be read over and over again just by accessing the correct memory address. The process of programming such a large number of locations is best done by a PROM programmer or microprocessor development system (MDS). These systems can copy a good PROM or the data can be input via a computer keyboard or from a magnetic disk.

EPROMs and EEPROMs

When using mask ROMs or PROMs, if you need to make a change in the memory contents or if you make a mistake in the initial programming, you are out of luck! One solution to that problem is to use an erasable PROM (EPROM). These PROMs are erased by exposing an open "window" in the IC to an ultraviolet (UV) light source for a specified length of time. Another type of EPROM is also available, called an electrically erasable PROM (EEPROM or E^2PROM) or electrically alterable PROM (EAROM). By applying a high voltage (about 21 V), a single byte, or the entire chip, can be erased in 10 ms. This is a lot faster than UV erasing and can be done easily while the chip is still in the circuit. One application of the EEPROM is in the tuner of a modern TV set. The EEPROM "remembers" (1) the channel you were watching when you turned off the set, and (2) the volume setting of the audio amplifier.

Examples of two erasable PROMs are the 2716 EPROM and 2816 EEPROM, both manufactured by Intel.

The 2716 EPROM. The data sheet for the 2716 EPROM is given in the Appendices. Referring to the data sheet, note that the 2716 has 16K bits of memory, organized as 2K × 8. 2K locations require 11 address inputs ($2^{11} = 2048$), which are labeled A_0 to A_{10}.

To read a byte (8 bits) of data from the chip, the 11 address lines are set up, then $\overline{\text{CE}}$ and $\overline{\text{OE}}$ are brought LOW to enable the chip and to enable the output. The AC waveforms for the chip show that the data outputs (O_0 to O_7) become valid after a time delay for setting up the addresses (t_{ACC}), or enabling the chip (t_{CE}), or enabling the output (t_{OE}), whichever is completed last. Figure 8–11 shows the circuit connections and waveforms for reading the 2716 EPROM.

In Figure 8–11 the X in the address waveform signifies the point where the address lines must change (1 to 0 or 0 to 1), if they are going to change. The $\overline{\text{CE}}$/PGM line is LOW for Chip Enable and HIGH for programming mode. Outputs O_0 to O_7 are in the high-impedance state (float) until $\overline{\text{OE}}$ goes LOW. The outputs are then undetermined until the delay time t_{OE} has expired, at which time they become the valid levels from the addressed memory contents.

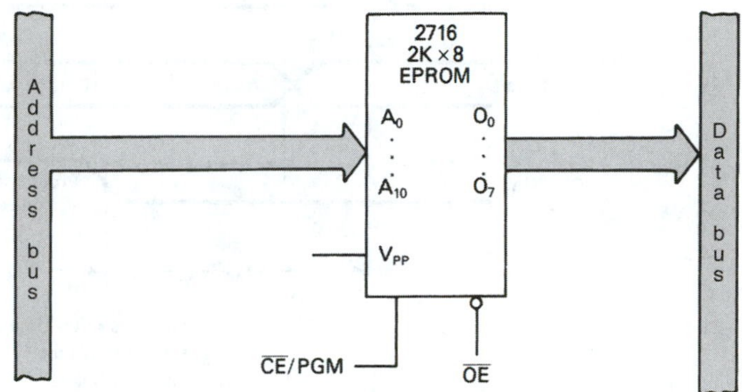

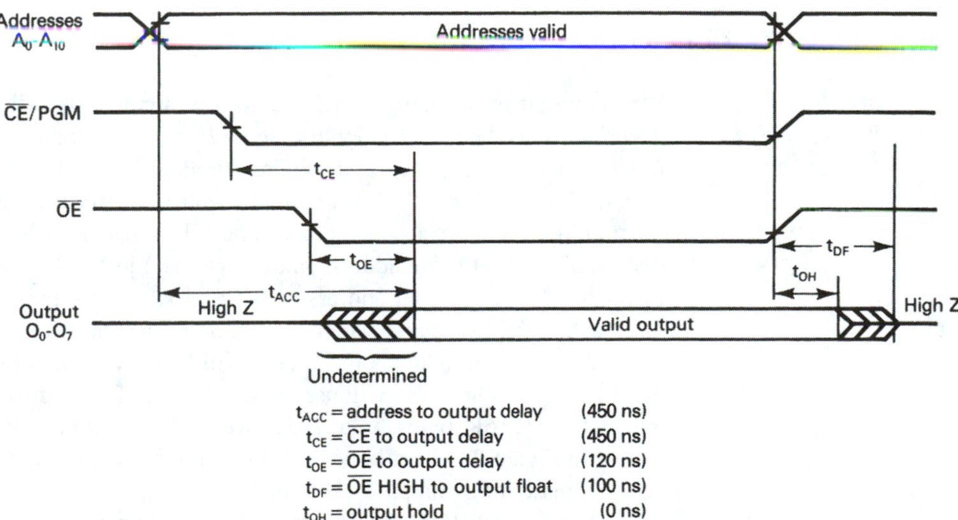

t_{ACC} = address to output delay (450 ns)
t_{CE} = $\overline{\text{CE}}$ to output delay (450 ns)
t_{OE} = $\overline{\text{OE}}$ to output delay (120 ns)
t_{DF} = $\overline{\text{OE}}$ HIGH to output float (100 ns)
t_{OH} = output hold (0 ns)

Figure 8–11 The 2716 EPROM Read cycle.

Programming the 2716. Initially, and after an erasure, all bits in the 2716 are 1s. To program the 2716, the following procedure is used:

1. Set V_{pp} to 25 V and $\overline{\text{OE}}$ = HIGH (5 V).
2. Set up the address of the byte location to be programmed.

3. Set up the 8-bit data to be programmed on the O_0 to O_7 outputs.
4. Apply a 50-ms positive TTL pulse to the $\overline{\text{CE}}$/PGM input.
5. Repeat steps 2, 3, and 4 until all the desired locations have been programmed.

Figure 8–12 shows the circuit connections and waveforms for programming a 2716.

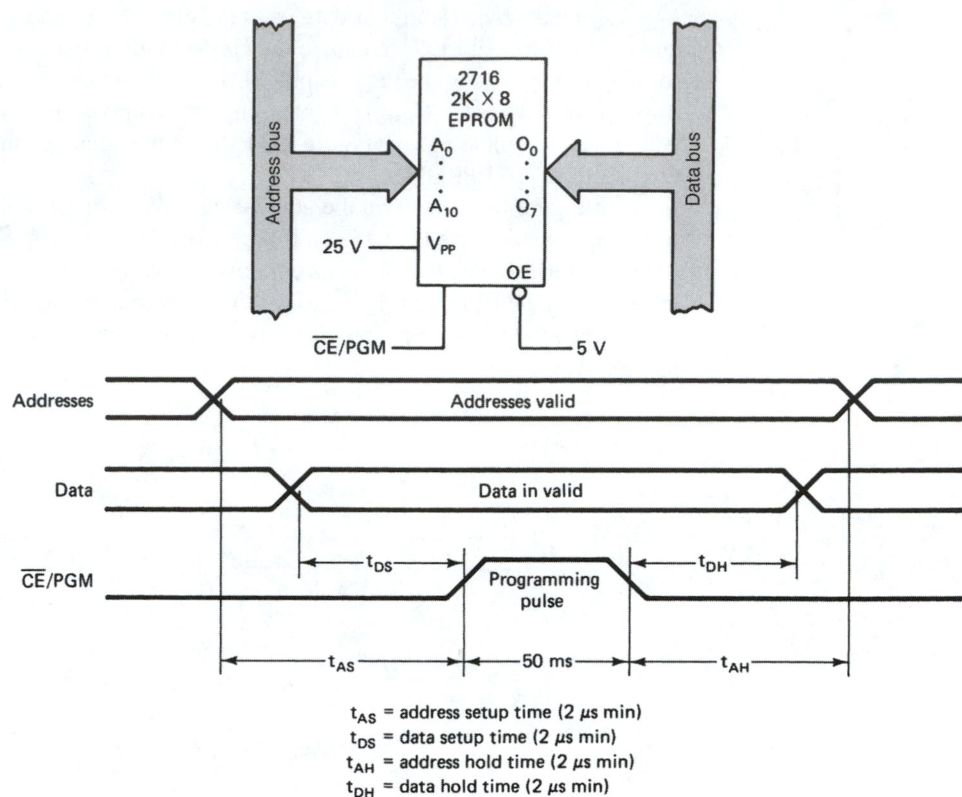

Figure 8–12 The 2716 program cycle.

8–5 MEMORY EXPANSION AND ADDRESS DECODING

When more than one memory IC is used in a circuit, a decoding technique (called *address decoding*) must be used to identify *which IC* is to be read or written to. Most 8-bit microprocessors use 16 separate address lines to identify unique addresses within the computer system. Some of those 16 lines will be used to identify the chip to be accessed, while the others pinpoint the exact memory location. For instance, the 2732 is a 4K × 8 EPROM that requires 12 of those address lines (A_0 to A_{11}) just to locate specific contents within its memory. This leaves four address lines (A_{12} to A_{15}) free for chip address decoding. A_{12} to A_{15} can be used to identify which IC within the system is to be accessed.

With 16 total address lines there will be 64K or 65,536 ($2^{16} = 65{,}536$) unique address locations. One 2732 will use up 4K of those. To design a large EPROM memory system, let's say 16K bytes, four 2732s would be required. The address decoding scheme shown in Figure 8–13 could be used to set up the four EPROMs consecutively in the first 16K addresses of a computer system.

The four EPROMs in Figure 8–13 are set up in consecutive memory locations between 0 to 16K and are individually enabled by the 74LS138 address decoder. The 4K × 8 EPROMs each require 12 address lines for internal memory selection, leaving the four HIGH-order address lines (A_{12} to A_{15}) free for chip selection by the 74LS138.

To read from the EPROMs, the microprocessor first sets up on the address bus the unique 16-bit address that it wants to read from. Then it issues a LOW level on its $\overline{\text{RD}}$ output. This satisfies the three enable inputs for the 74LS138, which then uses A_{12}, A_{13}, and A_{14} to determine which of its outputs is to go LOW, selecting one of the four EPROMs. Once an

EPROM has been selected, it outputs its addressed 8-bit contents to the data bus. (The outputs of the other EPROMs will float because their $\overline{\text{CE}}$s are HIGH.) The microprocessor gives all the chips time to respond, then reads the data that it requested from the data bus.

The address decoding scheme shown in Figure 8–13 is a very common technique used for "mapping" out the memory allocations in microprocessor-based systems (called *memory mapping*). RAM (or RWM) is added to the memory system the same way.

For example, if we wanted to add four 4K × 8 RAMs, their chip enables would be connected to the 4–5–6–7 outputs of the 74LS138, and they would occupy locations 4XXX, 5XXX, 6XXX, and 7XXX. Then, when the microprocessor issues a Read or Write command for, let's say, address 4007, the first RAM would be accessed.

EXAMPLE 8–1

Determine which EPROM and which EPROM address is accessed when the microprocessor of Figure 8–13 issues a Read command for the following hex addresses: **(a)** READ 0007; **(b)** READ 26C4; **(c)** READ 3FFF; **(d)** READ 5007.

Solution:

(a) The HIGH-order hex digit (0) will select the first EPROM. Address 007 (0000 0000 0111) in the first EPROM will be accessed. (Address 007 is actually the *eighth* location in that EPROM.)

(b) The HIGH-order hex digit (2) will select the third EPROM ($A_{15} = 0$, $A_{14} = 0, A_{13} = 1, A_{12} = 1$). Address 6C4 in the third EPROM will be accessed.

(c) The HIGH-order hex digit (3) will select the fourth EPROM ($A_{15} = 0$, $A_{14} = 0, A_{13} = 1, A_{12} = 1$). Address FFF (the last location) in the fourth EPROM will be accessed.

(d) The High-order hex digit (5) will cause the output 5 of the 74LS138 to go LOW. Since no EPROM is connected to it, nothing will be read.

Expansion to 64K

The memory system of Figure 8–13 can be expanded to 64K bytes by utilizing two 74LS138 decoders, as shown in Figure 8–14. Address lines A_0 to A_{12} are not shown in Figure 8–14, but they would go to each 2732 EPROM, just as they did in Figure 8–13. The HIGH-order addresses (A_{12} to A_{15}) are used to select the individual EPROMs. When A_{15} is LOW, the upper decoder in Figure 8–14 is enabled and EPROMs 1 to 8 can be selected. When A_{15} is HIGH, the lower decoder is enabled and EPROMs 9 to 16 can be selected. Using the circuit in Figure 8–14 will allow us to "map-in" sixteen 4K × 8 EPROMs, which will *completely* fill the memory map in a 16-bit address system. Actually, this would not be practical because some room must be set aside for RAM and input/output devices.

One final point on memory and bus operation: Microprocessors and MOS memory ICs are generally designed to drive only a single TTL load. Therefore, when several inputs are being fed from the same bus, a MOS device driving the bus must be buffered. An octal buffer IC such as the 74241, connected between a MOS IC output and the data bus, will provide the current capability to drive a heavily loaded data bus. Bidirectional bus drivers (or transceivers) such as the 74LS640 provide buffering in both directions for use by read/write memories (RAM or RWM). See Section 5–7 for a detailed explanation of buffers, latches, and transceivers.

APPLICATION 8–1: A PROM Look-up Table

Besides being used strictly for memory, ROMs, PROMs, and EPROMs can also be programmed to provide special-purpose functions. One common use is as a look-up table. A simple example is to use a PROM as a 4-bit binary-to-Gray code converter, as shown in Figure 8–15.

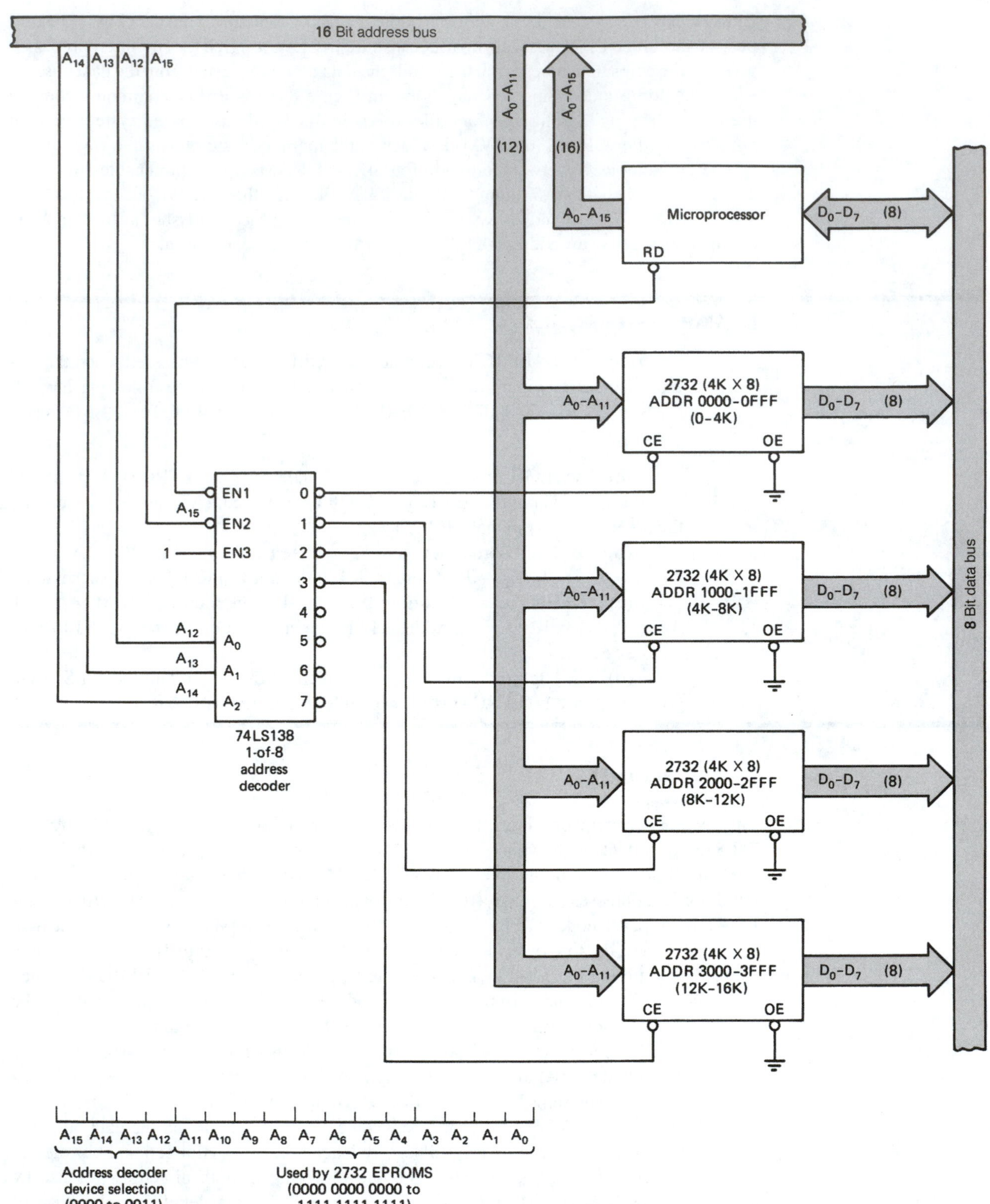

Figure 8–13 Address decoding scheme for a 16K-byte EPROM memory system.

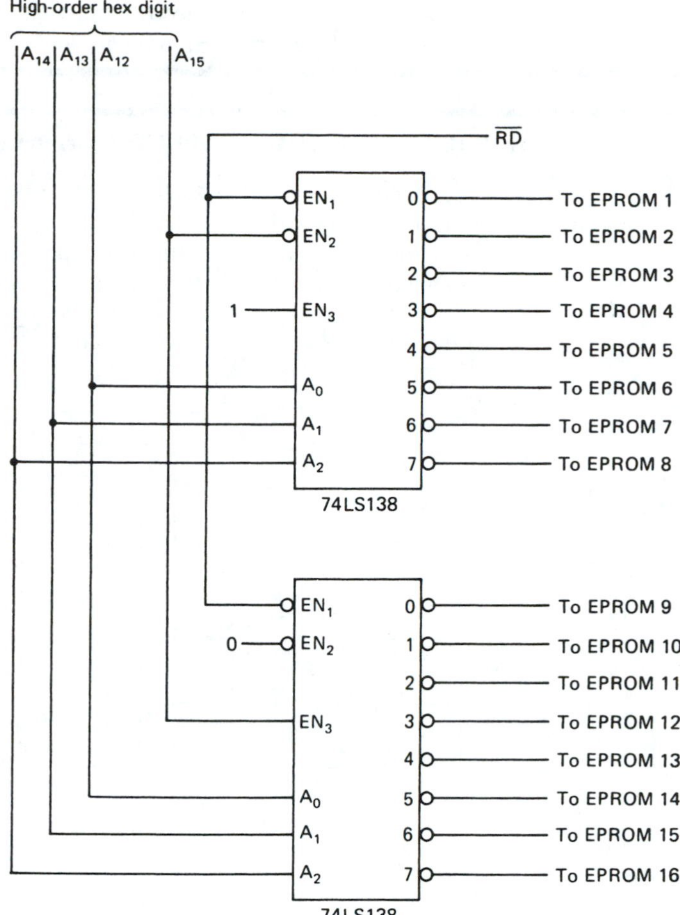

Figure 8–14 Expanding the memory of Figure 8–13 to 64K bytes.

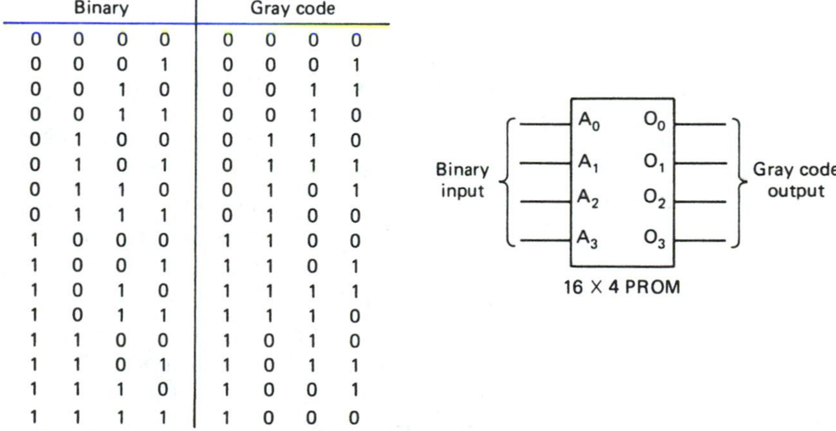

Binary				Gray code			
0	0	0	0	0	0	0	0
0	0	0	1	0	0	0	1
0	0	1	0	0	0	1	1
0	0	1	1	0	0	1	0
0	1	0	0	0	1	1	0
0	1	0	1	0	1	1	1
0	1	1	0	0	1	0	1
0	1	1	1	0	1	0	0
1	0	0	0	1	1	0	0
1	0	0	1	1	1	0	1
1	0	1	0	1	1	1	1
1	0	1	1	1	1	1	0
1	1	0	0	1	0	1	0
1	1	0	1	1	0	1	1
1	1	1	0	1	0	0	1
1	1	1	1	1	0	0	0

Figure 8–15 Using a PROM look-up table to convert binary-to-Gray code.

The PROM chosen for Figure 8–15 must have 16 memory locations, each location containing a 4-bit Gray code. The 4-bit binary string to be converted is used as the address inputs to the PROM. The PROM must be programmed such that each memory contains the equivalent Gray code to be output. For example, address location 0010 will contain 0011, 0100 will contain 0110, and so on, for the complete binary-to-Gray code data table. A more practical application would be to use a PROM to convert 7-bit binary to two BCD digits, which is a very complicated procedure using ordinary logic gates.

APPLICATION 8–2: A Digital LCD Thermometer

Another application, one that covers several topics from within this text, is a digital centigrade thermometer. In this application, using a PROM look-up table simplifies the task of converting meaningless digital strings into decimal digits. Figure 8–16 shows a block diagram of a two-digit centigrade thermometer.

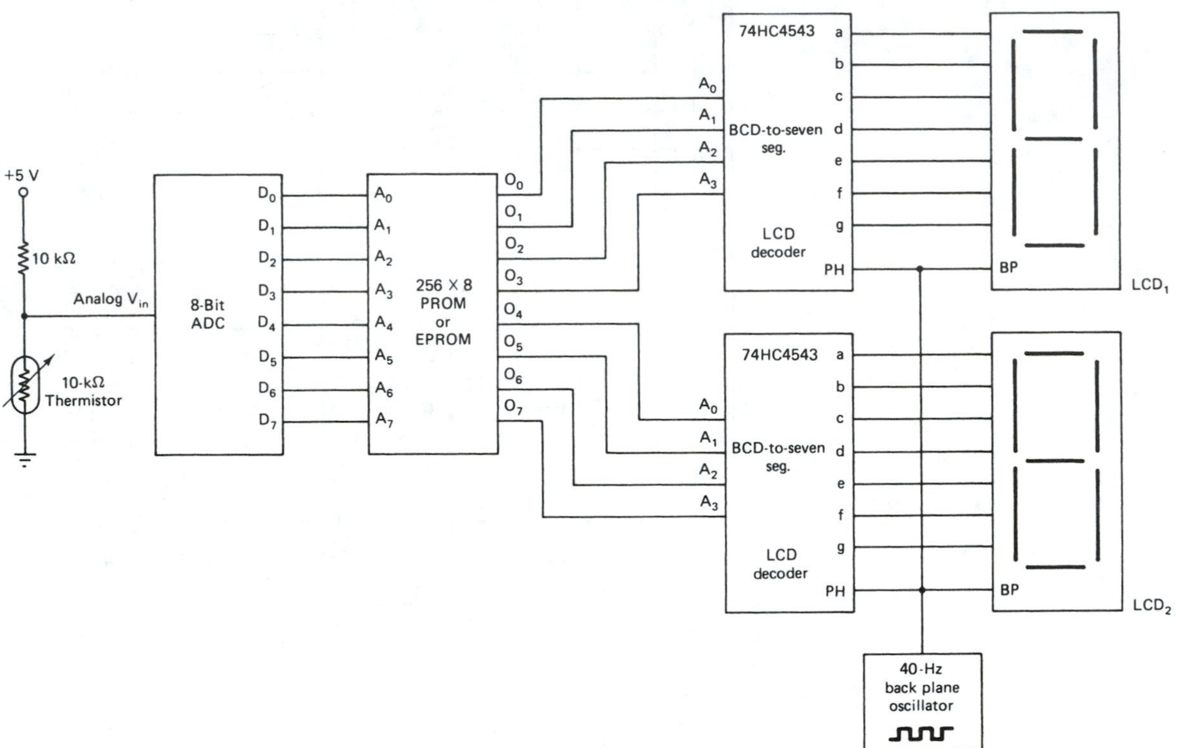

Figure 8–16 Using a PROM as a look-up table for binary-to-BCD conversion for an LCD thermometer.

For the circuit of Figure 8–16 to work, a binary-to-two-digit BCD look-up table has to be programmed into the PROM. Since a standard thermistor is a nonlinear device, as the temperature varies the binary output of the ADC will not change in proportional steps. Programming the PROM with the appropriate codes can compensate for that and can also assure that the output being fed to the two decoders is in the form of two BCD codes, each within the range 0 to 9. The appropriate codes for the PROM contents are best determined through experimentation.

The 74HC4543 will convert its BCD input into a seven-segment code for the liquid-crystal displays (LCDs). Liquid-crystal displays consume significantly less power than LED displays but require a separate square-wave oscillator to drive their backplane. As shown in Figure 8–16, a 40-Hz oscillator is connected to the phase input (PH) of each decoder and the backplane (BP) of each LCD.

8-6 *MAGNETIC AND OPTICAL STORAGE*

The types of memory storage discussed so far have all been based on semiconductor material. They have relied on turning transistors ON or OFF or storing an electrical charge on a capacitor or on a floating gate of a transistor.

The most common magnetic and optical memories are more electromechanical in nature because the memory material containing the 1's and 0's physically spins beneath a read/write head. They are nonvolatile, providing permanent storage and retrieval even when power is removed.

In the case of magnetic memory, the 1 or 0 is represented on the magnetic medium as a tiny north–south or south–north polarity magnet. Optical memory differs in that it uses a laser to optically read the data stored on the medium. An indentation area (called a *pit*) represents 1, and a non-pit (called *land*) represents 0. Because of their electromechanical nature, magnetic and optical storage units are much slower and bulkier than semiconductor memory, but they are also much less expensive and provide much higher storage capacity.

Magnetic Memory: The Floppy Disk and Hard Disk

The most common removable magnetic storage for a PC is the 3.5-inch *floppy disk*. It consists of a magnetizable medium that spins inside a rigid plastic jacket at a speed of 300 rpm (revolutions per minute). The disk drive that holds the floppy disk (or diskette) has two read/write heads, one for each side of the diskette. They are used to record a magnetic charge to, or read a magnetic charge from, the diskette. Up to 1.44MB of data can be stored on the double-sided diskette. The beauty of the diskette is that it can then be removed from its drive unit and used at a different location or filed away for safe-keeping. Additional diskettes can then be purchased for less than one dollar.

The highest-speed magnetic storage is achieved using a *hard-disk drive*. This disk system is not considered a removable medium like the floppy disk but, instead, uses a series of rigid platters mounted in a sealed unit inside a PC. Its recording method is similar to the floppy, except that it uses several two-sided platters (usually three or more) to hold the magnetic data and has one read/write head for each platter surface. Because the platters are rigid and mounted in a permanent housing, their tolerances are more precisely defined, allowing for tighter spacing between bits. This also enables the disks to spin at much higher speeds (thousands of rpm). Storage capacity of hard drives are in the gigabyte range.

The method of storing 1's and 0's magnetically is shown in Figure 8–17. Each disk surface is made of an iron-oxide layer capable of becoming magnetized. This surface is mounted on a substrate (foundation), which is usually a flexible mylar for the floppy disk and aluminum for the hard-drive platter. Initially, the magnetic layer consists of totally nonaligned particles representing no particular magnetic direction. By passing magnetic flux lines through the material, the particles align themselves in a specific north–south or south–north direction.

As the disk surface revolves, the read/write head is moved laterally to the precise track (ring) and bit position as dictated by the PC operating system software. If the software is *writing* data to the disk, the control circuitry places the appropriate polarity on the electromagnet in the read/write head. The example in Figure 8–17(b) shows how to write a 1 onto the disk surface. (Four previously written bits are also shown.) With the + to − polarity shown, magnetic flux lines will flow clockwise through the electromagnet core. As the flux lines pass through the magnetic surface of the disk, they force the particles to align in a specific direction, leaving behind a north–south magnetic charge (south–north as you look at it.) To store a 0, the + to − polarity is reversed, which reverses the flux lines, which in turn reverses the direction of the stored magnetic charge. To *read* the data, the read/write head is used as a magnetic sensor, reading the magnetic polarity as it passes beneath it.

A hard drive achieves higher bit capacity and data access speed because it uses rigid disk platters revolving at a much higher speed within a precision, sealed unit. Typical rotation speed is 3600 rpm. (Some newer drives now exceed 10,000 rpm.) Because the hard drive operates in such a controlled environment, the bits can be packed closer together. Figure 8–18 shows how the bits are laid out on a disk surface. The concentric circles on the disk are called *tracks*. (In the case of a hard drive, the term *cylinder* is used to describe the cylindrical shape that appears from the series of tracks having the same diameter on the entire stack of platters.) The tracks and the bits on the hard-disk drive are packed much closer

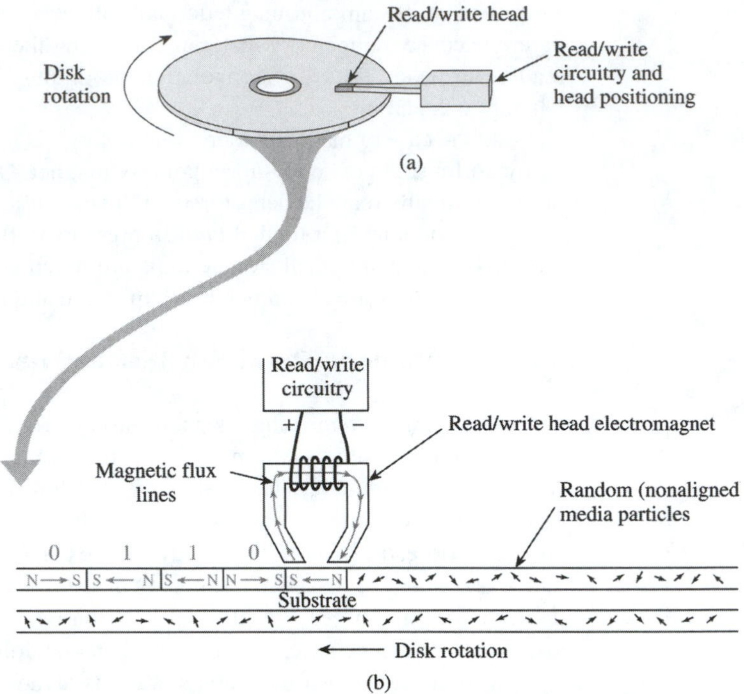

Figure 8–17 Magnetic memory: (a) a floppy disk or hard-drive platter with one of its read/write heads; (b) a cutaway view of Figure 8-17(a), showing how to store a south–north magnetic charge.

together than on a floppy diskette. Newer drives exceed 20,000 tracks per inch and 300 kilobits per inch on each track.

One of the most important specifications on magnetic media is the *data transfer rate*. A floppy can transfer data at 45 kB/s, whereas a hard drive can exceed 30 MB/s. The slow speed of the floppy becomes annoying when you have to wait more than half a minute to transfer a large file that only takes a couple of seconds for a hard drive.

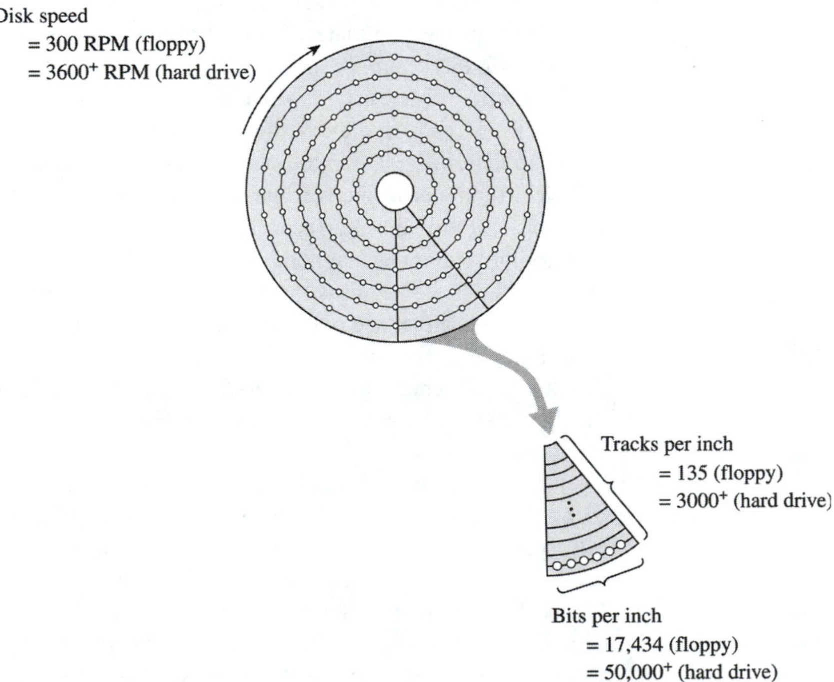

Figure 8–18 Bit density of a floppy disk and a hard drive.

Higher-density, removable magnetic devices are now finding their way into common use with PC users. Two of these are the *Zip disk* and the *Jaz cartridge*. They are removable media like the floppy diskette, but they have much higher storage capacity and faster read/write speed. One reason for the high speed is that the Zip disk spins at 3000 rpm. The magnetic images stored on the Zip disk are spaced much more tightly than on a standard floppy disk, giving it a storage capacity of 100MB. The Jaz cartridge employs two rigid platters and has a capacity as high as 2GB.

Optical Memory: The CD, CD-R, CD-RW, and DVD

Music CDs have been around since the late 1970s as digital medium for the storage and playback of analog music. In the mid-1980s, they were adopted for the storage and retrieval of digital computer data. Their data transfer rate is generally not as fast as a hard-disk drive, but manufacturers are constantly improving the speed. Because it is a removable media and capable of holding up to 650MB of data, most new operating systems and applications software are provided on CDs.

Figure 8–19 illustrates the construction of a CD. It is made of an aluminum alloy coating on the bottom of a rigid polycarbonate wafer. Binary data are stored on the CD by a series of indentations (called *pits*) representing 1's and non-pits (called *lands*) representing 0's. Pits are formed in the CD by stamping tiny indentations into the aluminum alloy. Data are recorded on a CD starting from the center and spiraling outward to the perimeter. The spiral is very tight, having the equivalent of 16,000 tracks per inch. A thin, plastic coating is then used to cover the CD, and a label is placed on top. As the CD spins, each bit position is read optically when a laser beam is reflected off the bottom of the CD surface. A light receptor receives the reflected light and distinguishes between light that is strongly reflected (from the land) versus light that is diffused or absent (from the pit).

User-recordable CDs (CD-Rs) are also available. To make CDs recordable, they are manufactured with a photosensitive dye on a reflective gold layer placed on the bottom of the rigid polycarbonate wafer. To form a 1, the CD recorder superheats a tiny spot on the dye/gold layer, changing its composition so that it will not reflect the laser light in that one spot. This is not actually a pit but, instead, is just an area with reduced reflective properties. These CDs are called WORM (Write Once, Read Many) media because once they are full, they cannot be erased or rewritten.

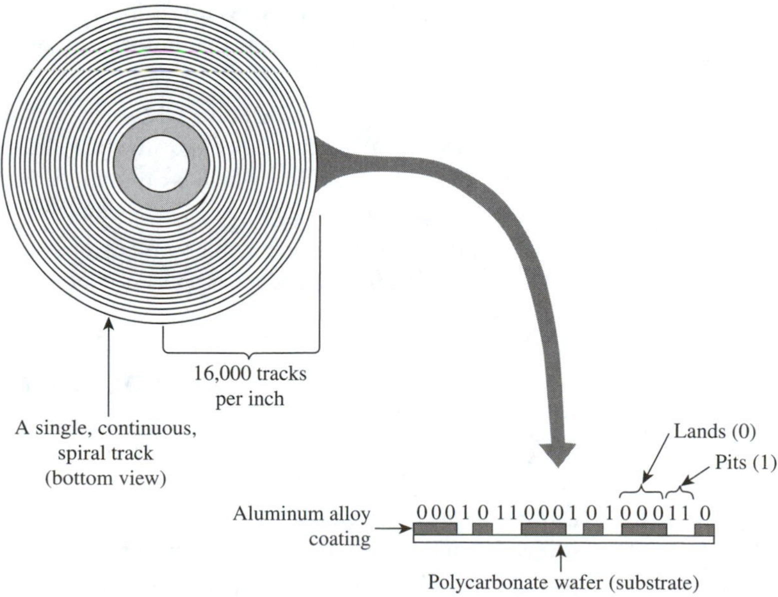

Figure 8–19 CD construction.

The *re-writable* type of CD is the CD-RW. Its surface is a silver alloy crystalline structure that is also converted into a 1 by superheating the specific bit area. The heat turns that area into its amorphous (nonreflective) state. The beauty of the CD-RW technology is that the 1 can be converted back to look like a 0 by reapplying a lower-level heat to that area, which returns the silver alloy back to its crystalline (reflective) state. Again, these are not actual pits and lands but, instead, are reflective and nonreflective areas.

The newest standard for the CD is called DVD (Digital Versatile Disk). DVDs are not only important in the computer industry for data storage but also in home entertainment, where they are replacing prerecorded VCR tapes, offering much higher quality and greater content. They are constructed similar to the CD but have a much higher data capacity, because the pits and tracks are packed much closer on the platter. DVDs can be single- or double-sided and have data capacities of 4.7GB up to 17GB.

SUMMARY

In this chapter we have learned that

1. A simple 16-byte memory circuit can be constructed from 16 octal *D* flip-flops and a decoder. This circuit would have 16 memory locations (addresses) selectable by the decoder, with one byte (8 bits) of data at each location.
2. Static RAM (Random-Access Memory) ICs are also called read/write memory. They are used for the temporary storage of data and program instructions in microprocessor-based systems.
3. A typical RAM IC is the 2147. It is organized as 4K × 1, which means that it has 4K locations with 1 bit of data at each location. (K is the abbreviation for Kilo, which is actually 1024.)
4. Dynamic RAMs are less expensive per bit and have a much higher density than static RAMs. Their basic storage element is an internal capacitor at each memory cell. External circuitry is required to refresh the charge on all capacitors every 2 ms or less.
5. Dynamic RAMs generally multiplex their address bus. This means that the high-order address bits share the same pins as the low-order address bits. They are demultiplexed by the RAS and CAS control signals (Row Address Strobe and Column Address Strobe).
6. Read-Only Memory (ROM) is used to store data on a permanent basis. It is nonvolatile, which means that it does not lose its memory contents when power is removed.
7. The three most common ROMs are: (1) the mask ROM, which is programmed once by a masking process by the manufacturer, (2) the fusible-link programmable ROM (PROM), which is programmed once by the user, and (3) the erasable-programmable ROM (EPROM), which is programmable and erasable by the user.
8. Memory expansion in microprocessor systems is accomplished by using octal or hexadecimal decoders as address decoders to select the appropriate memory IC.

GLOSSARY

Address Decoding: A scheme used to locate and enable the correct IC in a system with several addressable ICs.

Buffer: An IC placed between two other ICs to boost the load-handling capability of the source IC and to provide electrical isolation.

Bus Contention: Bus contention arises when two or more devices are outputting to a common bus at the same time.

Byte: A group of 8 bits.

CAS: Column Address Strobe. An active-LOW signal provided when the address lines contain a valid column address.

Delay Line: An integrated circuit that has a single pulse input and provides a sequence of true and complemented output pulses, with each output delayed from the preceding one by some predetermined time period.

Dynamic: A term used to describe a class of semiconductor memory that uses the charge on an internal capacitor as its basic storage element.

EEPROM: Electrically erasable programmable read-only memory.

EPROM: Erasable programmable read-only memory.

Fusible Link: Used in programmable ICs to determine the logic level at that particular location. Initially, all fuses are intact. Programming the IC either blows the fuse to change the logic state, or leaves it intact.

LCD: Liquid-crystal display. A multisegmented display similar to LED displays except that it uses liquid-crystal technology instead of light-emitting diodes.

Look-Up Table: A table of values that is sometimes programmed into an IC to provide a translation between two quantities.

Magnetic Memory: A storage medium such as tape or disk that holds a magnetic image of large amounts of binary data.

Mask: A material covering the silicon of a masked ROM during the fabrication process. It determines the permanent logic state to be formed at each memory location.

Memory Address: The location of the stored data to be accessed.

Memory Cell: The smallest division of a memory circuit or IC. It contains a single bit of data (1 or 0).

Memory Contents: The binary data quantity stored at a particular memory address.

PROM: Programmable read-only memory.

RAM: Random-access memory (read/write memory).

RAS: Row Address Strobe. An active-LOW signal provided when the address lines contain a valid row address.

ROM: Read-only memory.

Semiconductor Memory: Digital integrated circuits used for the storage of large amounts of binary data. The binary data at each memory cell is stored as the state of a flip-flop, or charge on a capacitor.

Static: A term used to describe a class of semiconductor memory that uses the state on an internal flip-flop as its basic storage element.

PROBLEMS

8–1. **(a)** In general, which type of memory technology is faster, bipolar or MOS?
(b) Which is more dense, bipolar or MOS?

8–2. Design and sketch an 8-byte memory system similar to Figure 8–2, using eight 74LS374s and one 74LS138.

8–3. Briefly describe the difference between static and dynamic RAMs. What are the advantages and disadvantages of each?

8–4. How many address lines are required to select a specific memory location within a RAM having:
(a) 1024 locations? **(b)** 4096 locations? **(c)** 8192 locations?

8–5. How many memory *locations* do the following RAM configurations have?
(a) 2048 × 1 **(b)** 2K × 4
(c) 8192 × 8 **(d)** 1024 × 4
(e) 4K × 8 **(f)** 16K × 1

8–6. What is the total number of *bits* that can be stored in the following RAM configurations?
(a) 1K × 8 **(b)** 4K × 4
(c) 8K × 8 **(d)** 16K × 1

8–7. Design and sketch a 1K × 8 RAM memory system using two 2148Hs.

8–8. Which lines are multiplexed on dynamic RAMs, and why?

8–9. What is the purpose of $\overline{RAS}$ and $\overline{CAS}$ on dynamic RAMs?

8–10. Draw the timing diagrams for a Read cycle and a Write cycle of a dynamic RAM similar to Figures 8–8b and c. Assume that $\overline{CAS}$ is delayed from $\overline{RAS}$ by 100 ns. Also assume that t_{CAC} = 120 ns (max.), t_{RAC} = 180 ns (max.), t_{DS} = 40 ns (min.), and t_{DH} = 30 ns (min.).

8–11. How often does a 2118 dynamic RAM have to be *refreshed,* and why?

8–12. What functions does the 3242 dynamic RAM controller take care of?

8–13. Are the following memory ICs volatile or nonvolatile?
(a) mask ROM **(b)** Static RAM
(c) Dynamic RAM **(d)** EPROM

8–14. Design and sketch an address decoding scheme similar to Figure 8–13 for an 8K × 8 EPROM memory system using 2716 EPROMs. (The 2716 is a 2K × 8 EPROM.)

8–15. What single decoder chip could be used in Figure 8–14 in place of the two 74LS138s?

SCHEMATIC INTERPRETATION PROBLEMS

8–18. Locate the line labeled RAM_SL at location D8 of the HC11D0 schematic. To get a HIGH level on that line, what level must the inputs to U8 be?

8–19. The 62256 (U10) IC in the 4096/4196 schematic is a MOS static RAM. By looking at the number of address lines and data lines, determine the size and configuration of the RAM.

8–20. Repeat Problem 8–19 for U6 of the HC11D0 schematic. Looking at the connections to the address lines, determine how much of the RAM is actually accessible.

8–21. The HC11D0 schematic uses two 27C64 EPROMS.
(a) What are their size and configuration?
(b) What are the labels of the control signals used to determine which EPROM is selected?
(c) Place a jumper from pin 2 to pin 3 of jumper J1 (Grid location D-6). Determine the range of addresses that make SMN_SL active (active-LOW).
(d) Determine the range of addresses that make MON_SL active (active-LOW).

Microprocessor Fundamentals

OBJECTIVES

Upon completion of this chapter you should be able to:

- Describe the benefits that microprocessor design has over hard-wired IC logic design.
- Discuss the functional blocks of a microprocessor-based system having basic input/output capability.
- Describe the function of the address, data, and control buses.
- Discuss the timing sequence on the three buses required to perform a simple input/output operation.
- Explain the role of software program instructions in a microprocessor-based system.
- Understand the software program used to read data from an input port and write it to an output port.
- Discuss the basic function of each of the internal blocks of the 8085A microprocessor.
- Follow the flow of data as it passes through the internal parts of the 8085A microprocessor.

INTRODUCTION

The design applications studied in the previous chapters have all been based on combinational logic gates and sequential logic ICs. One example is a traffic light controller that goes through the sequence green–yellow–red. To implement the circuit using combinational and sequential logic, we would use some counter ICs for the timing, a shift register for sequencing the lights, and a D flip-flop if we want to interrupt the sequence with a pedestrian cross-walk pushbutton. A complete design solution is easily within the realm of SSI and MSI ICs.

On the other hand, think about the complexity of electronic control of a modern automobile. There are several analog quantities to monitor, such as engine speed, manifold pressure, and coolant temperature; and there are several digital control functions to perform, such as spark plug timing, fuel mixture control, and radiator circulation control. The operation is further complicated by the calculations and decisions that have to be made on a continuing basis. This is definitely an application for a *microprocessor-based system.*

A system designer should consider a microprocessor-based solution whenever an application involves making calculations, making decisions based on external stimulus, and maintaining memory of past events. A microprocessor offers several advantages over the "hard-wired" SSI/MSI IC approach. First of all, the microprocessor itself is a general-purpose device. It takes on a unique personality by the software program instructions given by the designer. If you want it to count, you tell it to do so, with software. If you want to shift its output level left, there's an instruction for that. And if you want to add a new quantity to a previous one, there's another instruction for that. Its capacity to perform arithmetic, make comparisons, and update memory make it a very powerful digital problem solver. Making changes to an application can usually be done by changing a few program instructions, unlike the hard-wired system that may have to be totally redesigned and reconstructed.

New microprocessors are introduced every year to fill the needs of the design engineer. However, the theory behind microprocessor technology remains basically the same. It is a general-purpose digital device that is driven by software instructions and communicates with several external "support" chips to perform the necessary input/output of a specific task. Once you have a general understanding of one of the earlier microprocessors that came on the market, such as the Intel 8080/8085, the Motorola 6800, or the Zilog Z80, it is an easy task to teach yourself the necessary information to upgrade to the new microprocessors as they are introduced. Typically, when a new microprocessor is introduced, it will have a few new software instructions available and will have some of the I/O features, previously handled by external support chips, integrated into the microprocessor chip. Learning the basics on these new microprocessor upgrades is more difficult, however, because some of their advanced features tend to hide the actual operation of the microprocessor and may hinder your complete understanding of the system.

This book covers the Intel 8085A microprocessor software, hardware, and support circuitry. A thorough understanding of its architecture and operation will allow you to design and troubleshoot most 8-bit microprocessor-based systems and provide the background for learning the operations of the more highly integrated microprocessors as they are introduced.

9–1 INTRODUCTION TO SYSTEM COMPONENTS AND BUSES

Figure 9–1 shows a microprocessor with the necessary support circuitry to perform basic input and output functions. We will use that figure to illustrate how the microprocessor acts like a general-purpose device, driven by software, to perform a specific task related to the input data switches and output data LEDs. First, let's discuss the components of the system.

Microprocessor

The heart of the system is an 8-bit microprocessor. It could be any of the popular 8-bit microprocessors such as the Intel 8085, the Motorola 6800, or the Zilog Z80. They are called 8-bit microprocessors because external and internal data movement is performed on 8 bits at a time. It will read *program instructions* from memory and execute those instructions that drive the three *external buses* with the proper levels and timing to make the connected

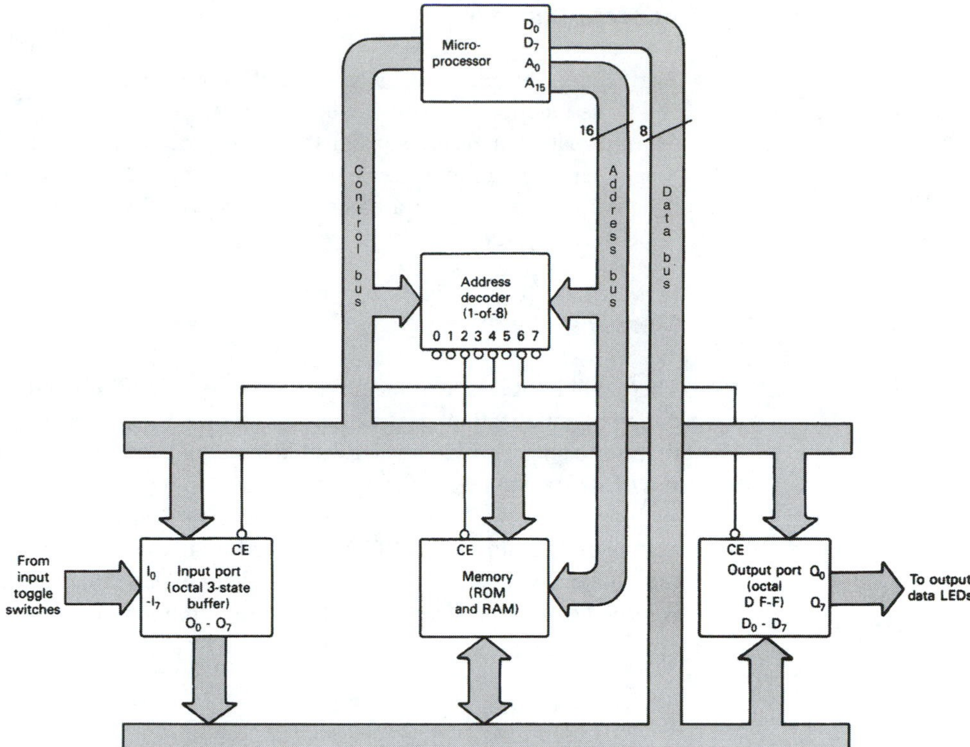

Figure 9–1 An example of a microprocessor-based system used for simple input/output operations.

devices perform specific operations. The buses are simply groups of conductors that are routed throughout the system and tapped into by various devices (or ICs) that need to share the information that is traveling on them.

Address Bus

The address bus is 16 bits wide and is generated by the microprocessor to select a particular location or IC to be active. In the case of a selected memory IC, the low-order bits on the address bus select a particular location within the IC (see Section 8–5). Since the address bus is 16 bits wide, it can actually specify 65,536 (2^{16}) different addresses. The input port is one address, the output port is one address, and the memory in a system of this size may be 4K (4096) addresses. This leaves about 60K addresses available for future expansion.

Data Bus

Once the address bus is set up with the particular address that the microprocessor wants to access,the microprocessor then sends or receives 8 bits of data to or from that address via the *bidirectional* (two-way) data bus.

Control Bus

The control bus is of varying width, depending on the microprocessor being used. It carries control signals that are tapped into by the other ICs to tell what type of operation is being performed. From these signals, the ICs can tell if the operation is a read, a write, an I/O, a memory access, or some other operation.

Address Decoder

The address decoder is usually an octal decoder like the 74LS138 studied in Chapters 4 and 8. Its function is to provide active-LOW Chip Enables ($\overline{CE}$) to the external ICs based on information it receives from the microprocessor via the control and address buses. Since there are multiple ICs on the data bus, the address decoder ensures that only one IC is active at a time to avoid a bus conflict caused by two ICs writing different data to the same bus.

Memory

There will be at least two memory ICs: ROM or EPROM and a RAM. The ROM will contain the "initialization" instructions, telling the microprocessor what to do when power is first turned on. This includes tasks like reading the keyboard and driving the CRT display. It will also contain several subroutines that can be called by the microprocessor to perform such tasks as time delays or input/output data translation. These instructions, which are permanently stored in ROM, are referred to as the "monitor program" or "operating system." The RAM part of memory is volatile, meaning that it loses its contents when power is turned off, and is therefore only used for temporary data storage.

Input Port

The input port provides data to the microprocessor via the data bus. In this case, it is an octal buffer with three-stated outputs. The input to the buffer will be provided by some input device like a keyboard or, as in this case, from eight HIGH–LOW toggle switches. The input port will dump its information to the data bus when it receives a Chip Enable ($\overline{CE}$) from the address decoder and a Read command ($\overline{RD}$) from the control bus.

Output Port

The output port provides a way for the microprocessor to talk to the outside world. It could be sending data to an output device like a printer, or as in this case, it could send data to eight LEDs. An octal *D* flip-flop is used as the interface because after the microprocessor sends data to it, the flip-flop will latch on to the data, allowing the microprocessor to continue with its other tasks.

To load the *D* flip-flop, the microprocessor must first set up the data bus with the data to be output. Then it sets up the address of the output port so that the address decoder will issue a LOW $\overline{CE}$ to it. Finally, it issues a pulse on its $\overline{WR}$ (write) line that travels the control bus to the clock input of the *D* flip-flop. When the *D* flip-flop receives the clock trigger pulse, it latches onto the data that are on the data bus at that time, and drives the LEDs.

9–2 SOFTWARE CONTROL OF MICROPROCESSOR SYSTEMS

The nice thing about microprocessor-based systems is that once you have a working prototype, you can put away the soldering iron because all operational changes can then be made with software. The student of electronics has a big advantage when writing microprocessor software because he or she understands the hardware at work as well as the implications that software will have on the hardware. Areas such as address decoding, chip enables, instruction timing, and hardware interfacing become important when programming microprocessors.

As a brief introduction to microprocessor software, let's refer back to Figure 9–1 and learn the statements required to perform some basic input/output operations. To route the data from the input switches to the output LEDs, the data from the input port must first be read into the microprocessor before they can be sent to the output port. The microprocessor has an 8-bit internal register called the *accumulator* that can be used for that purpose.

The software used to drive microprocessor-based systems is called *assembly language*. The Intel 8080/8085 assembly language statement to load the contents of the input port into the accumulator is LDA *addr.* LDA is called a *mnemonic,* an abbreviation of the operation being performed, which in this case is "Load Accumulator." The suffix *addr* will be replaced with a 16-bit address (4 hex digits) specifying the address of the input port.

After the execution of LDA *addr,* the accumulator will contain the digital value that was on the input switches. Now, to write those data to the output port, we use the command STA *addr.* STA is the mnemonic for "Store Accumulator" and *addr* is the 16-bit address where you want the data stored.

Execution of those two statements is all that is necessary to load the value of the switches into the accumulator and then transfer those data to the output LEDs. The microprocessor takes care of the timing on the three buses, and the address decoder takes care of providing chip enables to the appropriate ICs.

If the system was based on Motorola or Zilog technology, the software in this case will be almost the same. Table 9–1 makes a comparison of the three assembly languages.

TABLE 9–1

Comparison of Input/Output Software on Three Different Microprocessors

Operation	*Intel 8080/8085*	*Motorola 6800*	*Zilog Z80*
Load accumulator with contents of location *addr*	LDA *addr*	LDAA *addr*	LD A, *(addr)*
Store accumulator to location *addr*	STA *addr*	STAA *addr*	LD *(addr),* A

9–3 INTERNAL ARCHITECTURE OF THE 8085A MICROPROCESSOR

The design for the Intel 8085A microprocessor was derived from its predecessor, the 8080A. The 8085A is *software compatible* with the 8080A, meaning that software programs written for the 8080A can run on the 8085A without modification. The 8085A has a few additional features not available on the 8080A. The 8085A also has a higher level of hardware integration, allowing the designer to develop complete microprocessor-based systems with fewer external support ICs than were required by the 8080A. Studying the internal architecture of the 8085A in Figure 9–2 and its pin configuration, Figure 9–3, will give us a better understanding of its operation.

The 8085A is an 8-bit parallel *central processing unit* (CPU). The accumulator discussed in the previous section is connected to an 8-bit *internal data bus.* Six other *general-purpose registers* labeled *B, C, D, E, H,* and *L* are also connected to the same bus.

All arithmetic operations take place in the *arithmetic logic unit* (ALU). The accumulator, along with a temporary register, are used as inputs to all arithmetic operations. The output of the operations is sent to the internal data bus and to five *flag flip-flops* that record the status of the arithmetic operation.

The *instruction register* and *decoder* receive the software instructions from external memory, interpret what is to be done, and then create the necessary timing and control signals required to execute the instruction.

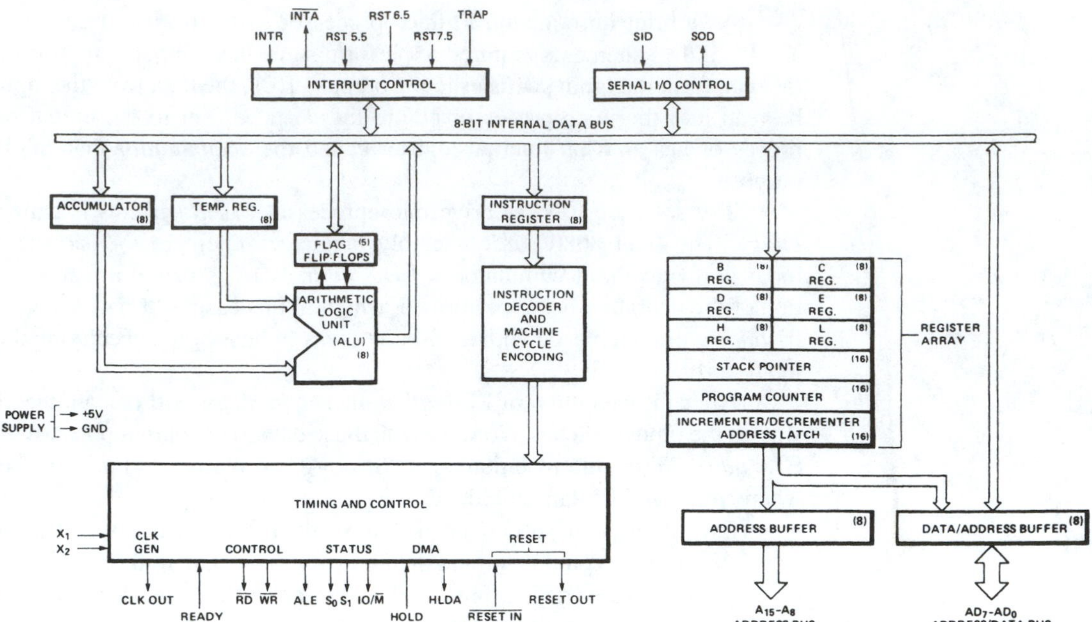

Figure 9–2 8085A CPU functional block diagram. (Courtesy of Intel Corporation)

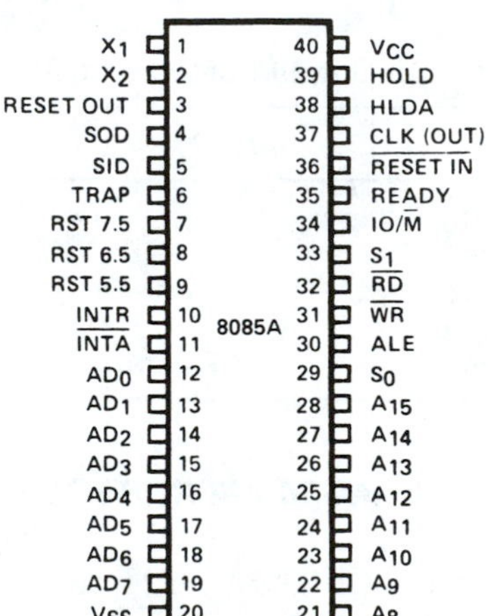

Figure 9–3 8085A pin configuration. (Courtesy of Intel Corporation)

The block diagram also shows *interrupt control,* which provides a way for an external digital signal to interrupt a software program while it is executing. This is accomplished by applying the proper digital signal on one of the interrupt inputs: INTR, RSTx.x, or TRAP. *Serial communication* capabilities are provided via the SID and SOD I/O pins (Serial Input Data, Serial Output Data).

The *register array* contains the six general-purpose 8-bit registers and three 16-bit registers. Sixteen-bit registers are required whenever you need to store addresses. The *stack pointer* stores the address of the last entry on the stack. The stack is a data storage area in RAM used by certain microprocessor operations, which will be covered in a later chapter. The *program counter* contains the 16-bit address of the next software instruction to be executed. The third 16-bit register is the *address latch,* which contains the current 16-bit address that is being sent to the address bus.

The six general-purpose 8-bit registers can also be used in pairs *(B–C, D–E, H–L)* to store addresses or 16-bit data.

9–4 INSTRUCTION EXECUTION WITHIN THE 8085A

Now, referring back to the basic I/O system diagram of Figure 9–1, let's follow the flow of the LDA and STA instructions as they execute in the block diagram of the 8085A. Figure 9–4 shows the 8085A block diagram with numbers indicating the succession of events that occur when executing the LDA instruction.

Remember, LDA *addr* and STA *addr* are assembly language instructions, stored in an external memory IC, that tell the 8085A CPU what to do. LDA *addr* tells the CPU to load its accumulator with the data value that is at address *addr.* STA *addr* tells the CPU to store (or send) the 8-bit value that is in the accumulator to the output port at address *addr.*

The mnemonics LDA and STA cannot be understood by the CPU as they are, they have to be *assembled,* or converted, into a binary string called *machine code.* Binary, or hexadecimal, machine code is what is actually read by the CPU and passed to the instruction register and decoder to be executed. The Intel 8085A Users Manual gives the machine code translation for LDA as $3A_{16}$ (or 3AH) and STA as 32H.

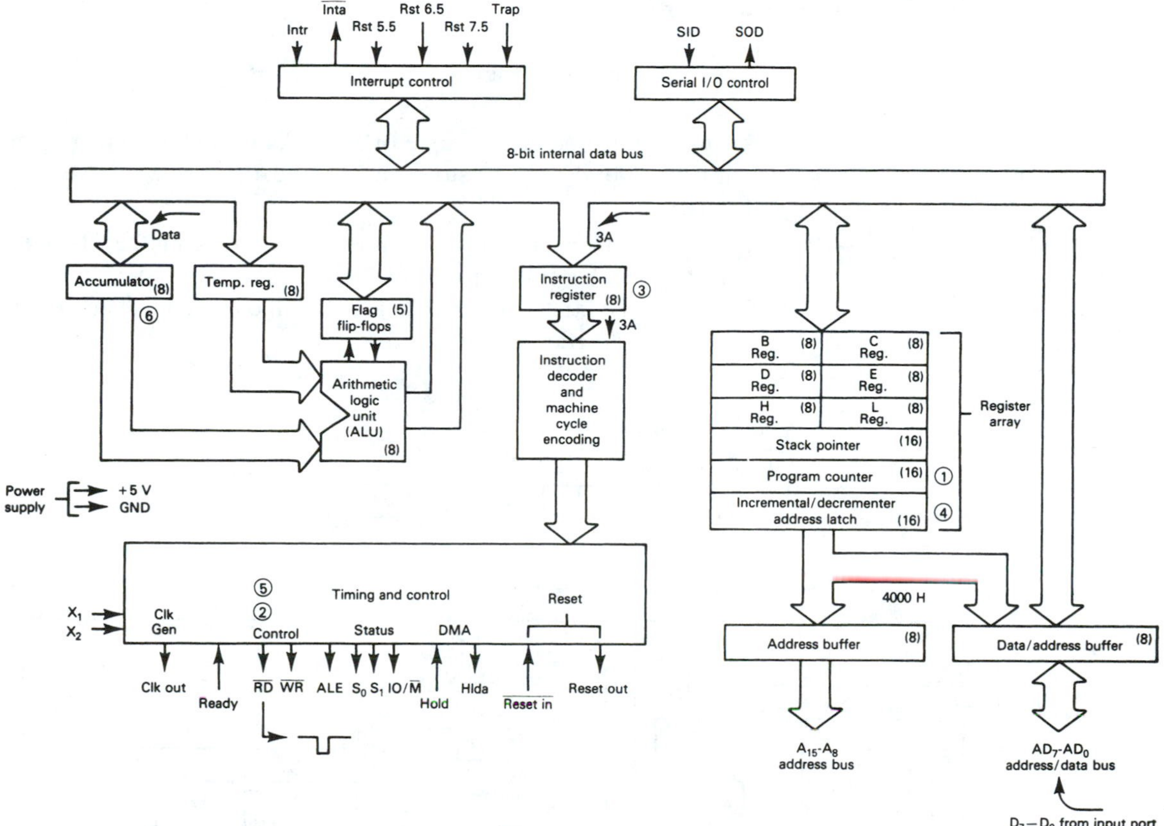

Figure 9–4 Execution of the LDA instruction within the 8085A.

Before studying the flow of execution in Figure 9–4, we need to make a few assumptions. Let's assume that the input port is at address 4000H and the output port is at address 6000H. Let's also assume that the machine code program LDA 4000H, STA 6000H is stored in RAM starting at address 2000H.

Load Accumulator

The sequence of execution of LDA 4000H in Figure 9–4 will be as follows:

1. The program counter will put the address 2000H on the address bus.
2. The timing and control unit will issue a LOW pulse on the $\overline{RD}$ line. This will cause the contents of RAM location 2000H to be put onto the external data bus.

RAM (2000H) has the machine code 3AH, which will travel across the internal data bus to the instruction register.

3. The instruction register passes the 3AH to the instruction decoder, which determines that 3AH is the code for LDA and that a 16-bit (2-byte) address must follow. Since the entire instruction is 3 bytes (one for the 3AH and two for the address 4000H), the instruction decoder increments the program counter two more times so that the address latch register can read and store byte 2 and byte 3 of the instruction.
4. The address latch and address bus now have 4000H on them, which is the address of the input port.
5. The timing and control unit again issues a LOW pulse on the $\overline{\text{RD}}$ line. The data at the input port (4000H) will be put onto the external data bus.
6. That data will travel across the internal data bus to the accumulator where it is now stored. The instruction is complete.

Store Accumulator

Figure 9–5 shows the flow of execution of the STA 6000H instruction.

1. After the execution of the 3-byte LDA 4000H instruction, the program counter will have 2003H in it. (Instruction LDA 4000H resided in locations 2000H, 2001H, 2002H.)
2. The timing and control unit will issue a LOW pulse on the $\overline{\text{RD}}$ line. This will cause the contents of RAM location 20003H to be put onto the external data bus. RAM (2003H) has the machine code 32H, which will travel up the internal data bus to the instruction register.

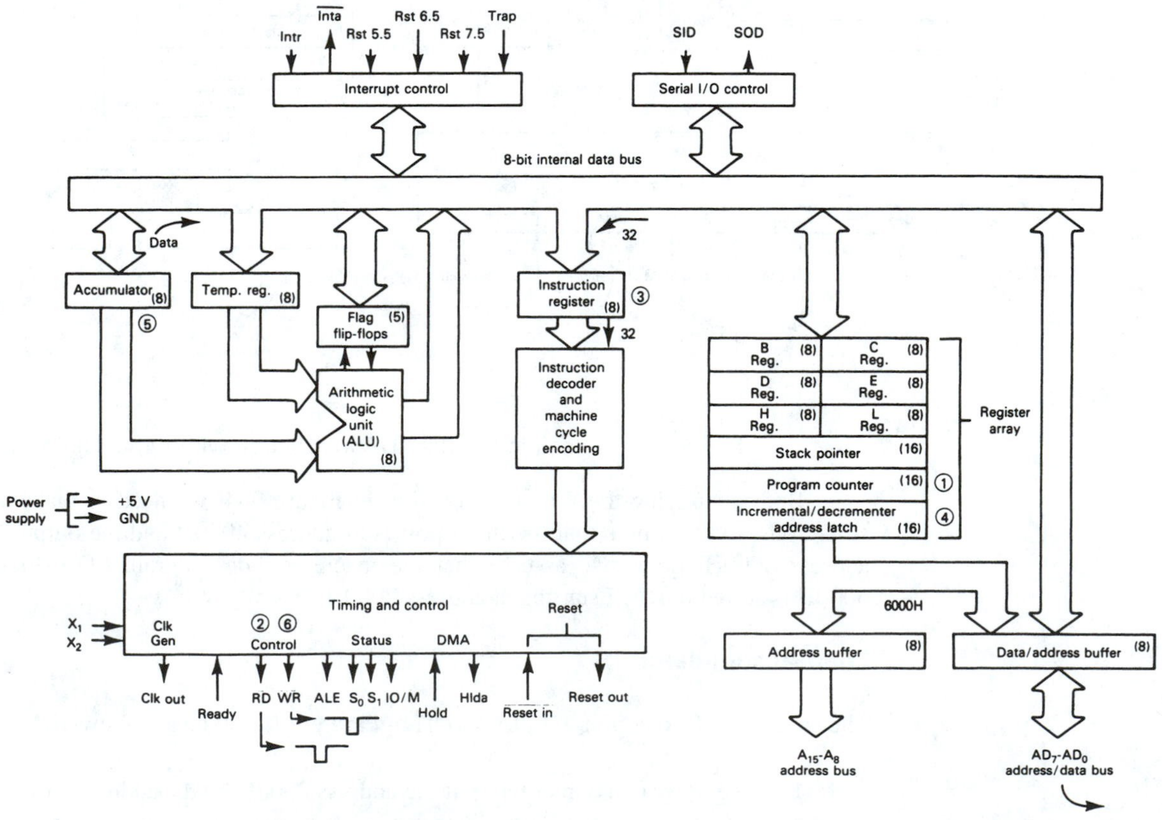

Figure 9–5 Execution of the STA instruction within the 8085A.

3. The instruction register passes the 32H to the instruction decoder, which determines that 32H is the code for STA and that a 2-byte address must follow. The program counter gets incremented two more times, reading and storing byte 2 and byte 3 of the instruction into the address latch.
4. The address latch and address bus now have 6000H on them, which is the address of the output port.
5. The instruction decoder now issues the command to place the contents of the accumulator onto the data bus.
6. The timing and control unit issues a LOW pulse on the $\overline{WR}$ line. Since the $\overline{WR}$ line is used as a clock input to the *D* flip-flop of Figure 9–1, the data from the data bus will be stored and displayed on the LEDs.

The complete assembly language and machine code program for the preceding input/output example is given in Table 9–2.

TABLE 9–2
Assembly Language and Machine Code Listing for the LDA–STA Program

Memory location	*Assembly language*	*Machine code*	
2000H	LDA 4000H	3A	Three-byte instruction to load accumulator with contents from address 4000H
2001H		00	
2002H		40	
2003H	STA 6000H	32	Three-byte instruction to store accumulator out to address 6000H
2004H		00	
2005H		60	

SUMMARY

In this chapter we have learned that

1. A system designer should consider using a microprocessor instead of logic circuitry whenever an application involves making calculations, making decisions based on external stimuli, and maintaining memory of past events.
2. A microprocessor is the heart of a computer system. It reads and acts on program instructions given to it by a programmer.
3. A microprocessor system has three buses: address, data, and control.
4. Microprocessors operate on instructions given to them in the form of machine code (1s and 0s). The machine code is generated by a higher-level language like C or assembly language.
5. The Intel 8085A is an 8-bit microprocessor. It has 7 internal registers, an 8-bit data bus, an arithmetic/logic unit, and several input/output functions.
6. Program instructions are executed inside the microprocessor by the instruction decoder, which issues the machine cycle timing and initiates input/output operations.
7. The microprocessor provides the appropriate logic levels on the data and address buses and takes care of the timing of all control signals output to the connected interface circuitry.
8. Assembly language instructions are written using mnemonic abbreviations and then converted into machine language so that they can be interpreted by the microprocessor.

GLOSSARY

Accumulator: The parallel register in a microprocessor that is the focal point for all arithmetic and logic operations.

Address Bus: A group of conductors that are routed throughout a computer system and is used to select a unique location based on its binary value.

Architecture: The layout and design of a system.

Arithmetic Logic Unit (ALU): The part of a microprocessor that performs all of the arithmetic and digital logic functions.

Assembly Language: A low-level programming language unique to each microprocessor. It is converted, or assembled, into machine code before it can be executed.

Bidirectional: Systems capable of transferring digital information in two directions.

Central Processing Unit (CPU): The "brains" of a computer system. The term is used interchangeably with "microprocessor."

Control Bus: A group of conductors that are routed throughout a computer system and is used to signify special control functions, such as Read, Write, I/O, Memory, and Ready.

Data Bus: A group of conductors that are routed throughout a computer system and contains the binary data used for all arithmetic and I/O operations.

Hardware: The integrated circuits and electronic devices that make up a computer system.

Instruction Decoder: The circuitry inside a microprocessor that interprets the machine code and produces the internal control signals required to execute the instruction.

Instruction Register: A parallel register in a microprocessor that receives the machine code.

Interrupt: A digital control signal that is input to a microprocessor IC pin that suspends current software execution and performs another predefined task.

Machine Code: The binary codes that make up a microprocessor's program instructions.

Microprocessor: An LSI or VLSI integrated circuit that is the fundamental building block of a digital computer. It is controlled by software programs that allow it to do all digital arithmetic, logic, and I/O operations.

Mnemonic: The abbreviated spellings of instructions used in assembly language.

Monitor Program: The computer software program initiated at power-up that supervises system operating tasks, such as reading the keyboard and driving the CRT.

Operating System: *See* Monitor program.

Program Counter: A 16-bit internal register that contains the address of the next program instruction to be executed.

Software: Computer program statements that give step-by-step instructions to a computer to solve a problem.

Stack Pointer: A 16-bit internal register that contains the address of the last entry on the RAM stack.

Support Circuitry: The integrated circuits and electronic devices that assist the microprocessor in performing I/O and other external tasks.

PROBLEMS

9–1. Describe the circumstances that would prompt you to use a microprocessor-based design solution instead of a hard-wired IC logic design.

9–2. In an 8-bit microprocessor system, how many lines are in the data bus? The address bus?

9–3. What is the function of the address bus?

9–4. Use a TTL data manual to find an IC that you could use for the *output port* in Figure 9–1. Draw its logic diagram and external connections.

9–5. Repeat Problem 9–4 for the *input port.*

9–6. Repeat Problem 9–4 for the *address decoder.* Assume that the input port is at address 4000H, the output port is at address 6000H, and memory is at address 2000H. (*Hint:* Use an address decoding scheme similar to that found in Section 8–5.)

9–7. Why does the input port in Figure 9–1 have to have three-stated outputs?

9–8. What two control signals are applied to the input port in Figure 9–1 to cause it to transfer the switch data to the data bus?

9–9. How many different addresses can be accessed using a 16-bit address bus?

9–10. In the assembly language instruction LDA 4000H, what does the LDA signify and what does the 4000H signify?

9–11. Describe what the statement STA 6000H does.

9–12. What are the names of the six internal 8085A general-purpose registers?

9–13. What is the function of the 8085A's instruction register and instruction decoder?

9–14. Why is the program counter register 16 bits instead of 8?

9–15. During the execution of the LDA 4000 instruction in Figure 9–4, the $\overline{RD}$ line goes LOW four times. Describe the activity initiated by each LOW pulse.

9–16. What action does the LOW $\overline{WR}$ pulse initiate during the STA 6000H instruction in Figure 9–5?

SCHEMATIC INTERPRETATION PROBLEMS

9–17. Find the two 4-bit magnitude comparators, U7 and U8, in the Watchdog Timer schematic. Which IC receives the high-order binary data, U7 or U8? [*Hint:* The bold lines in that schematic represent a *bus,* which is a group of conductors that are shared by several ICs. It simplifies the diagram by showing a single bold line instead of several separate lines. When the individual lines are taken off the bus they are labeled appropriately (0–1–2–3 and 4–5–6–7 in this application).]

9–18. Where is the final output of the comparison made by U7, U8 used in the Watchdog Timer schematic?

10 Introduction to 8085A Software

OBJECTIVES

Upon completion of this chapter, you should be able to:

- Make comparisons between assembly language, machine language, and high-level languages.
- Discuss the fundamental circuitry and timing sequence for external microprocessor I/O.
- Hand assemble assembly language programs into machine language code.
- Write a program for a MOD-*n* counter.
- Draw flowcharts and write programs containing compares and conditional branching.
- Move data into, and between, the internal data registers.
- Use the internal register pairs.
- Write time-delay routines using nested loops.
- Calculate program execution time using instruction *T* states.
- Write programs that utilize subroutine calls.
- Write programs that perform external I/O to switches and LEDs.

INTRODUCTION

Chapter 9 gave us a look at how a microprocessor can replace hard-wired logic. The microprocessor is driven by software instructions to perform specific tasks. The instructions are first written in assembly language using mnemonic abbreviations and then converted to machine language so that they can be interpreted by the microprocessor. The conversion from assembly language to machine language involves translating each mnemonic into the appropriate hexadecimal machine code and storing the codes in specific memory addresses. This can be done by a software package called an "assembler," provided by the microprocessor manufacturer, or it can be done by the programmer by looking up the codes and memory addresses (called "hand assembly").

Assembly language is classified as a low-level language because the programmer has to take care of all of the most minute details. High-level languages such as Pascal, FORTRAN, and BASIC are much easier to write but are not as memory efficient or as fast as assembly language. All languages, whether it be Pascal, BASIC, or FORTRAN, get reduced to machine language code before they can be executed by the microprocessor. The conversion from high-level languages to machine code is done by a *compiler.* The compiler makes memory assignments and converts the English-language-type instructions into executable machine code.

Assembly language translates *directly* into machine code without using a compiler. This allows the programmer to write the most streamlined, memory-efficient, and fastest programs possible on the specific hardware configuration that is being used.

Assembly language and its corresponding machine code differs from processor to processor. The fundamentals of the different assembly languages are the same, however, and once you have become proficient on one microprocessor, it is easy to pick it up on another.

The best way to learn assembly language is by studying examples and modifying them to meet your specific needs. Throughout the remainder of this book, we'll do exactly that. The 8085A software instructions will be introduced by writing solutions to specific applications and then covering the advanced instructions by expanding on those solutions.

10–1 HARDWARE REQUIREMENTS FOR BASIC I/O PROGRAMMING

A good way to start out in microprocessor programming is to illustrate program execution by communicating to the outside world. In Chapter 9 we read input switches at memory location 4000H using the LDA instruction and wrote their value to output LEDs at location 6000H using the STA instruction. That was an example of *memory-mapped I/O.* Using that method, the input and output devices were accessed *as if they were memory locations,* by specifying their unique 16-bit address (4000H or 6000H).

The other technique used by the 8085A microprocessor for I/O mapping is called "standard I/O" or "I/O-mapped I/O." *I/O-mapped systems* identify their input and output devices by giving them an 8-bit *port number.* The microprocessor then accesses the I/O ports by using the instructions: OUT *port* and IN *port,* where *port* is 00H to FFH.

Special hardware external to the 8085A is required to provide the source for the IN instruction and the destination for the OUT instruction. Figure 10–1 shows a basic hardware configuration, using standard SSI and MSI ICs, that could be built to input data from eight switches and to output data to eight LEDs using I/O-mapped I/O.

Figure 10–1 is set up to decode the input switches as port FFH and the output LEDs as port FEH. The $IO/\overline{M}$ line from the microprocessor goes HIGH whenever an IN or OUT instruction is being executed (I/O-mapped I/O). All instructions that access memory, and memory-mapped devices, will cause the $IO/\overline{M}$ line to go LOW. The $\overline{RD}$ line from the microprocessor will be pulsed LOW when executing the IN instruction, and the $\overline{WR}$ line will be pulsed LOW when executing the OUT instruction.

IN FFH

The 74LS244 is an octal three-state buffer that is set up to pass the binary value of the input switches over to the data bus as soon as $\overline{OE}_1$ and $\overline{OE}_2$ are brought LOW. To get that LOW, U6a, the inverted-input NAND gate (OR gate), must receive three LOWs at its input. We

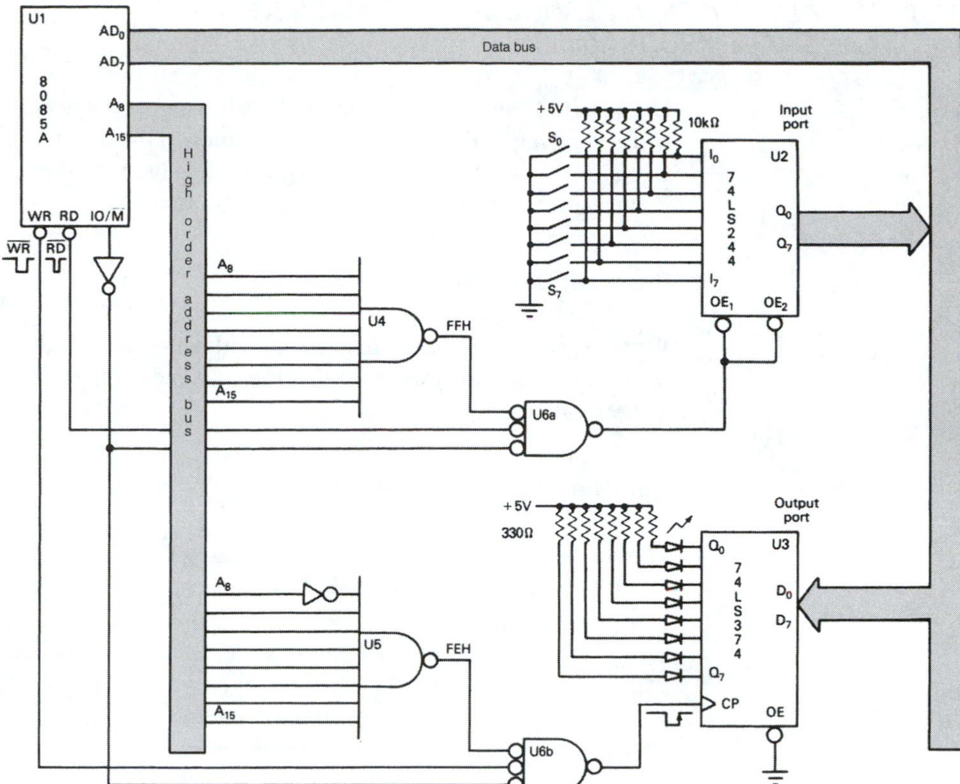

Figure 10–1 Hardware requirements for the IN FFH and OUT FEH instructions.

know that the IN instruction will cause the *inverted* IO/$\overline{\text{M}}$ line to go LOW and the $\overline{\text{RD}}$ line to go LOW. The other input is dependent on the output from the eight-input NAND gate (U4). Gate U4 *will* output a LOW because the binary value of the port number (1111 1111) used in the IN instruction is put onto the high-order address bus during the execution of the IN FFH instruction.

All conditions are now met, U6a will output a LOW pulse (the same width as the LOW $\overline{\text{RD}}$ pulse), which will enable the outputs of U2 to pass to the data bus. After the microprocessor drops the $\overline{\text{RD}}$ line LOW, it waits a short time for external devices (U2 in this case) to respond, then it reads the data bus and raises the $\overline{\text{RD}}$ line back HIGH. The data from the input switches are now stored in the accumulator.

OUT FEH

The 74LS374 is an octal *D* flip-flop set up to sink current to illuminate individual LEDs based on the binary value it receives from the data bus. The outputs at Q_0–Q_7 will latch onto the binary values at D_0–D_7 at the LOW-to-HIGH edge of C_p. U5 and U6b are set up similar to U4 and U6a, except U5's output goes LOW when FEH (1111 1110) is input. Therefore, during the execution of OUT FEH, U6b will output a LOW pulse, the same width as the $\overline{\text{WR}}$ pulse issued by the microprocessor.

The setup time of the 74LS374 latch is accounted for by the microprocessor timing specifications. The microprocessor issues a HIGH-to-LOW edge at $\overline{\text{WR}}$ that makes its way to C_p. At the same time, the microprocessor also sends the value of the accumulator to the data bus. After a time period greater than the setup time for U3, $\overline{\text{WR}}$ goes back HIGH, which applies the LOW-to-HIGH trigger edge for U3, latching the data at Q_0–Q_7.

To summarize, the instruction IN FFH reads the binary value at port FFH into the accumulator. The instruction OUT FEH writes the binary value in the accumulator out to port FEH. Port selection is taken care of by eight-input NAND gates attached to the high-order address bus and by use of the $\overline{\text{RD}}$, $\overline{\text{WR}}$, and IO/$\overline{\text{M}}$ lines.

10–2 WRITING ASSEMBLY LANGUAGE AND MACHINE LANGUAGE PROGRAMS*

Let's start off our software training by studying a completed assembly language program and comparing it to the same program written in the BASIC computer language. BASIC is a high-level language that uses English-language-type commands that are fairly easy to figure out, even by the inexperienced programmer.

Program Definition

Write a program that will function as a down-counter, counting 9 to 0 repeatedly. First draw a flowchart, then write the program statements in the BASIC language, assembly language, and machine language.

Solution

The flowchart in Figure 10–2 is used to show the sequence of program execution, including the branching and looping that takes place.

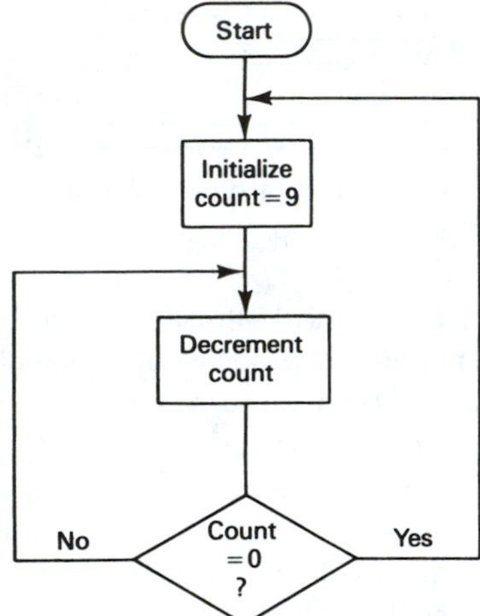

Figure 10–2 Flowchart for Table 10–1.

According to the flowchart, the counter will be decremented repeatedly until 0 is reached, at which time the counter is reinitialized to 9, and the cycle repeats. The instructions used to implement the program are given in Table 10–1.

BASIC

BASIC uses the variable "COUNT" to hold the counter value. Line 30 checks the count. If COUNT is equal to 0, then the program goes back to the beginning. Otherwise it goes back to subtract 1 from COUNT and checks COUNT again.

The 8085A version of the program is first written in assembly language and then it is either hand assembled into machine language or it could be computer assembled using a personal computer with an assembler software package. Throughout this book we will be hand assembling our programs to create machine language.

*All of the programs that follow can be tested on one of the microprocessor trainers described in the Appendix. The SDK-85 by URDA, Inc. is described in Appendix L and the 8085-Primer by EMAC, Inc. is described in Appendix M. Also, free simulation software for the 8085 (SIM8085) is provided on the CD and described in Appendix N.

TABLE 10–1
Down-Counter Program in Three Languages

BASIC language		*8085A Assembly language*		*8085A Machine language*	
Line	*Instruction*	*Label*	*Instruction*	*Address*	*Contents*
10	COUNT=9	START:	MVI A,09H	2000	3E (opcode)
				2001	09 (data)
20	COUNT=COUNT-1	LOOP:	DCR A	2002	3D (opcode)
30	IF COUNT=0 THEN GO TO 10		JZ START	2003	CA (opcode)
				2004	00 } (address)
				2005	20 } (address)
40	GO TO 20		JMP LOOP	2006	C3 (opcode)
				2007	02 } (address)
				2008	20 } (address)

Assembly Language

Assembly language is written using *mnemonics:* MVI, DCR, JZ, etc. The term *mnemonics* is defined as "abbreviations used to assist the memory." The first mnemonic, MVI, stands for "Move Immediate." The instruction MVI A,09H will move the data value 09H into register *A* (register *A* and the accumulator are the same). The next instruction, DCR A, decrements register *A* by 1.

The third instruction, JZ START is called a *conditional jump.* The condition that it is checking for is the *zero* condition. As the *A* register is decremented, if *A* reaches 0, then a flag bit, called the *zero flag,* gets set (a "set" flag is equal to 1). The instruction JZ START is interpreted as "jump to label START if the zero flag is set."

If the condition is not met (zero flag not set), then control passes to the next instruction, JMP LOOP, which is an *unconditional jump.* This instruction is interpreted as "jump to label LOOP regardless of any condition flags."

At this point you should see how the assembly language program functions exactly as the BASIC language program.

Machine Language

Machine language is the final step in creating an executable program for the microprocessor. In this step we must determine the actual hexadecimal codes that will be stored in memory to be read by the microprocessor. First we have to determine what memory locations will be used for our program. This depends on the memory map assignments made in the system hardware design. We have 64K of addressable memory locations (0000H to FFFFH). We'll make an assumption here, and use it throughout the rest of the book, that the user program area was set up in the hardware design to start at location 2000H. The length of the program memory area depends on the size of the ROM or RAM memory IC being used. A 256 × 8 RAM memory is usually sufficient for introductory programming assignments and is commonly used on educational microprocessor trainers. The machine language program listed in Table 10–1 fills up 9 bytes of memory (2000H to 2008H).

The first step in the hand assembly is to determine the code for MVI A. This is known as the opcode (operation code) and is found in the 8085A Assembly Language Reference Chart given in the Appendix. The opcode for MVI A is 3E. The programmer will store the binary equivalent for 3E (0011 1110) into memory location 2000H. Instructions for storing your program into memory are given by the manufacturer of the microprocessor trainer that you are using. If you are using an assembler software package, then the machine code that is generated will usually be saved on a computer disk or used to program an EPROM to be placed in a custom microprocessor hardware design.

The machine language instruction MVI A,09H in Table 10–1 requires 2 bytes to complete. The first byte is the opcode, 3E, which identifies the instruction for the microprocessor. The second byte is the data value, 09H, which is to be moved into register *A*.

The second instruction, DCR A, is a 1-byte instruction. It requires just its opcode, 3D, which is found in the reference chart.

The opcode for the JZ instruction is CA. It must be followed by the 16-bit (2-byte) address to jump to if the condition (zero) is met. This makes it a 3-byte instruction. Byte 2 of the instruction (location 2004H) is the low-order byte of the address, and byte 3 is the high-order byte of the address to jump to. (Be careful to always enter addresses as low-order first, then high-order.)

The opcode for JMP is C3 and must also be followed by a 16-bit (2-byte address specifying the location to jump to. Therefore, this is also a 3-byte instruction where byte 2–byte 3 give a jump address of 2002H.

10–3 COMPARES AND CONDITIONAL BRANCHING

In the previous down-counter program, we watched for the end of the count by checking the zero flag. Let's say that we want to count *up* 0 to 9 instead. In this case we would have to check for a 9 as the terminal count instead of a 0.

CPI data

If we are using the accumulator (*A* register) as our counter, then we can use the CPI *data* (compare) instruction. CPI *data* is a 2-byte instruction that compares the value in the accumulator to the value entered as byte 2 *(data)* of the instruction. Byte 1 of the instruction is the opcode for CPI, which is FE. The function of the CPI instruction is to set the zero flag if the accumulator is equal to byte 2 of the instruction. It is also used to set another flag, called the *carry flag,* if the accumulator is *less than* byte 2. Example 10–1 illustrates the use of the CPI *data* instruction.

EXAMPLE 10–1*

Write a program that will function as an up-counter to count 0 to 9 repeatedly.

Solution:

Machine language		*Assembly language*		
Address	*Contents*	*Label*	*Instruction*	*Comments*
2000	3E	START:	MVI A,00H	; Move 00H into
2001	00			; register A
2002	3C	LOOP:	INR A	; Increment register A
2003	FE		CPI 09H	; Compare reg.A to 09H,
2004	09			; set Z flag if A = 9
2005	CA		JZ START	; Jump to START
2006	00			; if Z flag
2007	20			; is set
2008	C3		JMP LOOP	; Else jump
2009	02			; to LOOP
200A	20			;

Explanation:

This program will continuously loop back to address 2002H, incrementing the accumulator, until the accumulator reaches 9. When A = 9, the CPI instruction

*This example, and several others that follow, are provided on the textbook CD for simulation on the SIM8085 program described in Appendix N.

will set the zero flag. Now the zero condition of the JZ instruction will be met and control will pass to address 2000H, which reinitializes A to 00H. The cycle repeats.

EXAMPLE 10–2

Write a program that will function as a hexadecimal (MOD-16) up-counter. Display the count on the output LEDs provided in the output port design given in Figure 10–1.

Solution:

Machine language		*Assembly language*		
Address	*Contents*	*Label*	*Instruction*	*Comments*
2000	3E	START:	MVI A,00H	; LOAD accumulator
2001	00			; with 00H
2002	2F	LOOP:	CMA	; Complement accumulator
2003	D3		OUT FEH	; Output accumulator
2004	FE			; to port FEH
2005	2F		CMA	; Complement accumulator back
2006	3C		INR A	; Increment accumulator
2007	FE		CPI 10H	; Compare accumulator to 16_{10},
2008	10			; set Z flag if equal
2009	CA		JZ START	; Jump to START
200A	00			; if accumulator
200B	20			; reached 16_{10}
200C	C3		JMP LOOP	; Else jump
200D	02			; to LOOP
200E	20			;

Explanation:

This program is very similar to Example 10–1 with the inclusion of two new instructions: CMA and OUT *port.* Back in Figure 10–1 we developed decoding hardware for port FEH. Note that the output port was driving *active-LOW* LEDs. That is why we had to complement the accumulator before the OUT FEH instruction. Of course the accumulator must be complemented back to its original value before continuing in the counter loop.

Also note that the CPI instruction is checking for a value that is *one greater than* the terminal count. This is because the INR A instruction comes *after* the OUT FEH instruction. If you want to see the terminal count of 15 (0FH), then you have to wait until 16 (10H) before restarting the loop.

10–4 USING THE INTERNAL DATA REGISTERS

As mentioned before, the 8085A has six internal data registers besides the *A* register (accumulator). They are referred to as the *B*, *C*, *D*, *E*, *H*, and *L* registers. Each register is 8 bits wide, but by using certain instructions, they can be grouped together to form *register pairs* (see Figure 10–3). As a register pair they become 16 bits wide, which makes them useful for storing addresses or large numbers.

MVI *r,data*

The easiest way to store data into a register is to use the move immediate command: MVI *r,data.* The *r* is replaced with one of the registers *A*, *B*, *C*, *D*, *E*, *H*, or *L* and *data* is replaced with a 1-byte data value. In machine language, MVI is a 2-byte instruction. Byte 1 is the

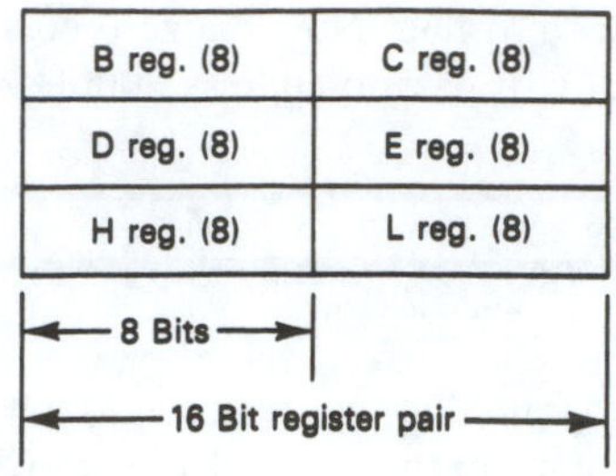

Figure 10–3 The internal data registers.

opcode, which depends on the register being used, and byte 2 is the data. Table 10–2 lists all of the MVI instructions with their opcodes.

TABLE 10–2
MVI Instructions for Internal Registers

Instruction	*Opcode*
MVI A,*data*	3E
MVI B,*data*	06
MVI C,*data*	0E
MVI D,*data*	16
MVI E,*data*	1E
MVI H,*data*	26
MVI L,*data*	2E

EXAMPLE 10–3

List the instructions in assembly language and machine language to meet the following requirements:

(a) Load the *B* register with 20H.
(b) Load the *D* register with 100_{10} (64H).

Solution:

Assembly language	*Machine language*	*Comments*
(a) MVI B,20H	06	B ← 20H
	20	
(b) MVI D,64H	16	D ← 64H
	64	

MOV *r1,r2*

To move data from register to register, we'll use the move register command: MOV *r1,r2*. Register *r2* is the *source* and *r1* is the *destination* of the data. In other words, to move the data from register *E* into register *C*, we would use the instruction MOV C,E. Data from any register can be moved to any other register. The data in the source register (*r2*) remains unchanged. Since there is no data value to be specified as part of the instruction, they are only 1-byte instructions (opcode only). Rather than list all of the opcodes here, take a moment to be sure that you can find the opcodes in the Assembly Language Reference Chart in the Appendix.

EXAMPLE 10–4

List the instructions in assembly language and machine language to meet the following requirements:

(a) Move the contents of the *D* register into the *H* register.

(b) Load the number 44H into both the *C* and *E* registers using only 3 bytes of machine language code.

Solution:

The opcodes are found in the Assembly Language Reference Chart in the Appendix.

	Assembly language	*Machine code*	*Comments*
(a)	MOV H,D	62	H ← D
(b)	MVI C,44H	0E	C ← 44H
		44	
	MOV E,C	59	E ← C

(*Note:* MOV E,C copies register *C* into register *E* and leaves register *C unchanged.* Using MVI E,44H would have made the program 4 bytes long instead of 3.)

LXI *rp,data16*

The LXI *rp,data16* instruction allows us to load all 16 bits of a register pair with one instruction. The *rp* is replaced by a single letter signifying which register pair to be loaded. Use *B* for register pair *B–C*, *D* for register pair *D–E*, and *H* for register pair *H–L*. The *data16* operand is replaced with the 16 bits of data to be loaded into the register pair. This makes LXI a 3-byte instruction: byte 1 is the opcode, byte 2 is the low-order byte of data that goes to the low-order register in the register pair, and byte 3 is the high-order byte of data that goes to the high-order register in the register pair (see Example 10–5).

EXAMPLE 10–5

List the instructions that will perform the following operations:

(a) Load the *B* register with D5 and *C* register with D8.

(b) Redo part (a) using an LXI instruction.

(c) Load the *D–E* register pair with 3800H.

Solution:

	Assembly language	*Machine language*	*Comments*
(a)	MVI B,D5H	06	B ← D5
		D5	
	MVI C,D8H	0E	C ← D8
		D8	
(b)	LXI B,D5D8H	01	
		D8	C ← D8
		D5	B ← D5
(c)	LXI D,3800H	11	
		00	E ← 00
		38	D ← 38

INR *r*, DCR *r* and INX *rp*, DCX *rp*

As you may have guessed, there are commands for incrementing and decrementing registers and register pairs. When working with an 8-bit register, the INR or DCR instruction acts like it is incrementing or decrementing an 8-bit counter. The register pair commands INX and DCX are similar, except they work as a 16-bit counter. (You'll notice that all register pair instructions have an X in their mnemonic.)

If you increment a single register that has an FFH in it, it will "roll over" to 00H. If you decrement a register that has a 00H in it, it will "roll over" to FFH. The same happens to register pairs when they are incremented past FFFFH or decremented below 0000H.

EXAMPLE 10–6

Determine the contents of the *B* and *C* registers after the execution of each of the following programs.

(a) LXI B,24FFH
INX B
(b) LXI B,46FFH
INR C
(c) LXI B,4F88H
DCR B
(d) MVI B,C7H
MVI C,00H
DCX B

Solution:

(a) B = 25H, C = 00H. INX B increments the 16-bit quantity of the B–C register pair.

(b) B = 46H, C = 00H. INR C increments the 8-bit C register only. C rolls over to 00H without affecting the quantity in the B register.

(c) B = 4EH, C = 88H. DCR B decrements the B register only.

(d) B = C6H, C = FFH. Even though B and C were loaded using MVIs, the DCX B command treats them like a 16-bit register pair. In order to decrement, the C register borrows 1 from the B register.

10–5 WRITING TIME-DELAY ROUTINES

Time delays are used in microprocessor applications to insert delays that are required between processes. One good application is the hexadecimal counter that we designed in Example 10–2 to drive output LEDs. In that program, the counter is incrementing and driving the LEDs from 0 to 15 at microprocessor speeds. This is much too fast for a human to see. A time delay, implemented with software, could be inserted within the loop to slow the process down to any speed that we want.

One simple way to create a time delay is to set a register to FFH and count down until it reaches 0, as shown in Table 10–3.

This program will continue to loop back to address 2052 until register C finally reaches 0. After that, the program will continue to the statement following the JNZ instruction.

To determine how long this delay will be in seconds, we need to refer to the Instruction Set Timing Index in the Appendix. This index gives the number of *T states* for each instruction. One T state is the length of one microprocessor clock period. The internal clock frequency of the microprocessor is one-half the frequency of the crystal used to drive the 8085A. Therefore, if for example the 8085A is using a 4-MHz crystal, then one T state will be 0.5 μs. [One T state $= 1/(0.5 \times 4\text{ MHz}) = 0.5\ \mu$s.]

TABLE 10–3
Short Delay Routine Using a Single Register

Address	*Contents*	*Label*	*Instruction*	*Comments*
2050	0E	DELAY:	MVI C,FFH	; C ← FFH
2051	FF			;
2052	0D	LOOP1:	DCR C	; C = C − 1
2053	C2		JNZ LOOP1	; Keeping looping
2054	52			; until
2055	20			; C = 0

The MVI C,FFH instruction takes seven *T* states, DCR C takes four, and JNZ LOOP1 takes 7/10. The 7/10 means that the JNZ instruction takes 10 *T* states *if the condition is met* (if *C* is not 0), and 7 *T* states *if the condition is not met.* It takes the extra time if the condition *is* met because it has to read byte 2–byte 3 of the instruction in order to determine where to jump to.

Now you add up all of the *T* states required to complete the delay, as shown in Table 10–4. Since *C* starts out at FFH (255), then it must be decremented 255 times before setting the zero flag. The 255th time it is decremented, *C* will be 0 so the JNZ instruction will only take seven *T* states. For the first 254 times in the loop, it will take (4 + 10) *T* states for each pass. The last time it will take (4 + 7) *T* states. This gives us a total of 3574 *T* states, and a delay time of 1.787 ms (0.5 μs × 3574 *T* states).

TABLE 10–4
Determining the *T* States of a Delay Routine

	Instruction	*T states*
DELAY:	MVI C,FFH	7
LOOP1:	DCR C	4 ← Loop (254 + 1) times
	JNZ LOOP1	7/10
T states = 7 + 254 × (4 + 10) + 1 × (4 + 7) = 3574		

EXAMPLE 10-7

Determine the length of time of the following delay loop. Assume that the 8085A is driven with a 5-MHz crystal.

```
DELAY:  MVI   A,00H
LOOP:   INR   A
        CPI   6DH
        JNZ   LOOP
```

Solution:

The microprocessor clock period (1 *T* state) = 1/(0.5 × 5 MHz) = 0.4 μs. The *T* states per instruction (found in the Instruction Set Timing Index) are as follows:

MVI A = 7 *T* states

INR A = 4 *T* states

CPI = 7 *T* states

JNZ = 7/10 *T* states

The number of times around the loop is 6DH or 109 times. Therefore, the total number of *T* states is:

$$T \text{ states} = 7 + 108 \times (4 + 7 + 10) + 1 \times (4 + 7 + 7)$$
$$= 2293$$

And the time delay is

$$T = \text{clock period} \times T \text{ states}$$

$$T = 0.4\ \mu s \times 2293 = 917.2\ \mu s$$

Nested Loops

The previous delay of 1.787 ms (Table 10–3) would not be long enough to use as a pause between counts on an LED display for us to see the count changing. We need at least one-quarter of a second (250 ms) between counts.

One way to get a 250-ms delay is to execute the previous delay instructions (Table 10–3) 140 times (140 × 1.787 ms ≈250 ms). This is called a *nested loop.* It is implemented using the flowchart shown in Figure 10–4 and the program statements in Table 10–5.

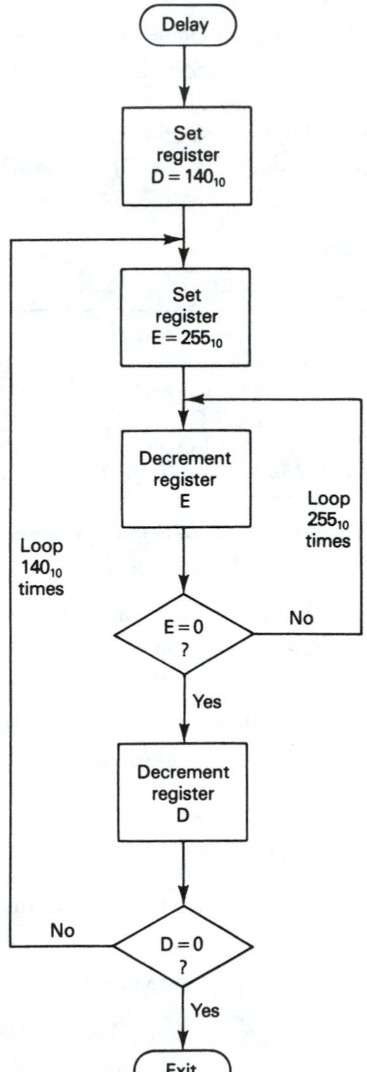

Figure 10–4 Flowchart for a nested-loop delay.

The flowchart is a good means of visualizing the flow of the program. It is recommended that one be drawn before writing any program that has branching or looping. The program instructions in Table 10–5 correspond directly to the flowchart. The inner loop (LOOP1) is the same as the previous delay program. It will decrement register *E* 255 times before setting the zero flag and dropping out. Then, when the DCR D instruction is

TABLE 10–5
A Nested-Loop Program to Create a Delay of ≈ 250 ms

Address	*Contents*	*Label*	*Instruction*	*Comments*
2050	16	DELAY:	MVI D,8CH	; D ← 8CH (140)
2051	8C			;
2052	1E	LOOP2:	MVI E,FFH	; E ← FFH (255)
2053	FF			;
2054	1D	LOOP1:	DCR E	; E ← E-1
2055	C2		JNZ LOOP1	; Jump to
2056	54			; LOOP1
2057	20			; 255 times
2058	15		DCR D	; D ← D-1
2059	C2		JNZ LOOP2	; Jump to
205A	52			; LOOP2
205B	20			; 140 times

executed, register *D* becomes 139 and the zero flag is *reset.* The outer loop (LOOP2) causes the inner loop to be executed repeatedly as register *D* is decremented from 140 down to 0.

For even longer delays you could insert dummy instructions within the loop or you could form a third level of nesting by using a third register. A commonly used dummy or "do nothing" instruction is the NOP (no-operation) instruction, which takes four *T* states.

Register pairs can also be used as our loop counter, giving us a maximum count of FFFFH instead of just FFH provided by single registers. The problem with register pairs, however, is that the DCX *rp* instruction does not set the zero flag when the register pair reaches 0000H. In the previous delays, we were using the zero flag to get us out of the loop. You must be careful when writing programs that depend on the condition of the flags, that the particular instruction *does* affect the flags. For example, DCR *r* will set the zero flag, whereas DCX *rp* will not. This information is provided for each instruction in the 8085A Instruction Set Reference Encyclopedia in the Appendix.

There *are* ways to use register pairs as loop counters, however. To utilize them, we need to use other instructions to set the zero flag when the terminal count is reached. The CMP *r* or the ORA *r* instructions, which are covered in a later chapter, could be used for that purpose.

10–6 USING A TIME-DELAY SUBROUTINE WITH I/O OPERATIONS

Now that we have a one-quarter second delay routine, we can set it up as a *subroutine* and "call" it whenever we need a delay. Subroutines are covered in more detail in a later chapter, but for now all we need to know is that the delay program will be a stand-alone module that will be called by the main program whenever a delay is needed.

To make the delay program of Table 10–5 a subroutine, all we need to do is to put a *return statement* (RET) at the end of it. Then, whenever another program calls it by using a CALL *addr* instruction, it will be executed. When execution of the delay subroutine is complete, the program control returns to the instruction following the CALL *addr* instruction. The operand, *addr,* is the address in memory of the subroutine, which in this case will be 2050H.

Now let's insert a delay within the hexadecimal counter loop of Example 10–2 to slow the count down so that we can see the count change on the output LEDs (see Figure 10–5 and Table 10–6).

The program in Table 10–6 uses the LED circuit that was designed back in Figure 10–1, which was addressed at port FEH. After sending the count to the LEDs, the delay subroutine is called. The opcode for CALL is CD, which is followed by the 2-byte address of the delay subroutine (2050H). Note that the address is entered low order (50) first, then high order (20), the same as with jump addresses.

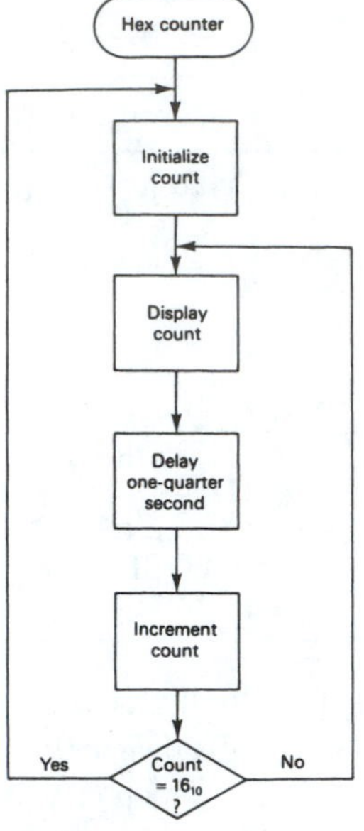

Figure 10–5 Flowchart for a hex counter with a delay.

TABLE 10–6

Hexadecimal Counter Program with a Delay Subroutine* Called between Counts

Address	*Contents*	*Label*	*Instruction*	*Comments*
2000	3E	START:	MVI A,00H	; Initialize A to 0
2001	00			;
2002	2F	DISPLAY:	CMA	; Complement A
2003	D3		OUT FEH	; Output to LEDs
2004	FE			; at port FEH
2005	2F		CMA	; Complement A back
2006	CD		CALL DELAY	; Call the DELAY
2007	50			; subroutine
2008	20			; at 2050H
2009	3C		INR A	; Increment A
200A	FE		CPI 10H	; Compare A to 10H,
200B	10			; set Z flag if equal
200C	CA		JZ START	; Jump to start
200D	00			; if Z flag
200E	20			; set
200F	C3		JMP DISPLAY	; Else jump
2010	02			; to
2011	20			; DISPLAY
—				
—				
—				
2050	16	DELAY:	MVI D,8CH	; D ← 8CH (140)
2051	8C			;
2052	1E	LOOP2:	MVI E,FFH	; E ← FFH (255)
2053	FF			;
2054	1D	LOOP1:	DCR E	; E ← E-1
2055	C2		JNZ LOOP1	; Jump to
2056	54			; LOOP1
2057	20			; 255 times
2058	15		DCR D	; D ← D-1
2059	C2		JNZ LOOP2	; Jump to
205A	52			; LOOP2
205B	20			; 140 times
205C	C9		RET	; Return to main program

*This program assumes that the stack pointer has been initialized (see Section 12–4).

After the execution of the CALL instruction, program control passes to the delay subroutine at address 2050H. At the end of the delay subroutine, we have added a RETURN that will pass control back to the main program at the next instruction, past the CALL instruction.

The delay subroutine was stored, starting at address 2050H, but actually could have started at any memory address past the end of the main program. The addresses between 2011H and 2050H are never executed, and therefore can contain any data without affecting our program.

Now that we know how to use the output port at address FEH, let's concentrate on using the *input port* at address FFH. Executing the instruction IN FFH will load the accumulator with the value of the HIGH/LOW toggle switches shown back in Figure 10–1. The following examples illustrate the use of the input port.

EXAMPLE 10–8

Assume that the switches at input port FFH are connected to eight temperature limit sensors in an industrial process facility. Write a program that will produce a continuous warning flash on the output LEDs if any of the temperature limits go HIGH.

Solution:

Address	*Contents*	*Label*	*Instruction*	*Comments*
2000	DB	READIN:	IN FFH	; Read input switches
2001	FF			; into accumulator
2002	FE		CPI 00H	; Compare A to 00H,
2003	00			; set Z if equal
2004	CA		JZ READIN	; Keep rereading as
2005	00			; long as all
2006	20			; switches are LOW
2007	3E	FLASH:	MVI A,00H	; Active-LOW LEDs
2008	00			; will come on with 00H
2009	D3	LOOP:	OUT FEH	; Activate
200A	FE			; LEDs
200B	CD		CALL DELAY	; Wait one-
200C	50			; quarter
200D	20			; second
200E	2F		CMA	; Complement accumulator
200F	C3		JMP LOOP	; Loop continuously
2010	09			;
2011	20			;

Explanation:

The program will continuously loop around the first three instructions as long as all input switches are LOW. As soon as one temperature limit switch goes HIGH, the CPI 00H instruction will *reset* the zero flag to 0, allowing program flow to drop down to the MVI A,00H instruction.

The first time that the OUT FEH instruction is executed, the active-LOW LEDs will turn on because *A* was set to 00H. We then use the same delay subroutine that we wrote before (at address 2050H) to hold the LEDs on for one-quarter second. The CMA instruction then complements the accumulator to FFH before looping back to the OUT FEH instruction. Sending FFH to the LEDs will turn them all off. The delay will then hold them off for one-quarter second. The effect of this loop is to cause the LEDs to flash. This is used for a warning that a high temperature exists somewhere in the process facility.

EXAMPLE 10–9

Assume that the binary value on the input switches represents a trouble code in an automobile's electronic control. Write a program that flashes the LEDs the number of times equivalent to the trouble code.

Solution:

Address	*Contents*	*Label*	*Instruction*	*Comments*
2000	DB	READCD:	IN FFH	; Read trouble
2001	FF			; code
2002	47		MOV B,A	; Move code to reg. B
2003	3E	LOOP:	MVI A,00H	; A ← 00H
2004	00			;
2005	D3		OUT FEH	; Turn LEDs on
2006	FE			;
2007	CD		CALL DELAY	; Delay one-
2008	50			; quarter
2009	20			; second
200A	2F		CMA	; A ← $\overline{A}$

Solution: (continued)

Address	Contents	Label	Instruction	Comments
200B	D3		OUT FEH	; Turn LEDs off
200C	FE			;
200D	CD		CALL DELAY	; Delay one-
200E	50			; quarter
200F	20			; second
2010	05		DCR B	; B ← B-1
2011	C2		JNZ LOOP	; Flash again
2012	03			; until B = 0
2013	20			;
2014	76		HLT	; Halt

Explanation:

We first read in the trouble code and then transfer it to register *B*. We have to transfer it to register *B* because register *A* is about to be used to flash the LEDs, and we need to remember what the trouble code is. After flashing the LEDs on then off once, we then decrement the trouble code (register *B*) once and check to see if it is 0. If not, then repeat the on/off sequence. This continues until register *B* is decremented to 0. The LEDs will flash the number of times equivalent to the trouble code, then drop to the last statement, HLT, which halts microprocessor execution.

SUMMARY OF INSTRUCTIONS

LDA addr: (Load Accumulator Direct) Load the accumulator with the contents of memory whose address (*addr*) is specified in byte 2–byte 3 of the instruction.

STA addr: (Store Accumulator Direct) Store the contents of the accumulator to memory whose address (*addr*) is specified in byte 2–byte 3 of the instruction.

IN port: (Input) Load the accumulator with the contents of the specified port.

OUT port: (Output) Move the contents of the accumulator to the specified port.

MVI r,data: (Move Immediate) Move into register *r*, the data specified in byte 2 of the instruction.

DCR r: (Decrement Register) Decrement the value in register *r* by 1.

INR r: Increment the value in register *r* by 1.

JMP addr: (Jump) Transfer control to address *addr,* specified in byte 2–byte 3 of the instruction.

JZ addr: (Jump If Zero) Transfer control to address *addr,* if the zero flag is set.

JNZ addr: (Jump If Not Zero) Transfer control to address *addr,* if the zero flag is not set.

CPI data: (Compare Immediate) Compare the accumulator to the data in byte 2 of the instruction. The zero flag is set if $A =$ byte 2. The carry flag is set if $A <$ byte 2.

CMA: (Complement Accumulator) Complement the contents of the accumulator.

MOV r1,r2: (Move Register) Copy the contents of register *r2* into register *r1*.

LXI rp,data16: (Load Register Pair Immediate) Load register pair *rp* with the contents of byte 2–byte 3 of the instruction.

INX rp: (Increment Register Pair) Increment the 16-bit value in register pair *rp* by 1.

DCX rp: (Decrement Register Pair) Decrement the 16-bit value in register pair *rp* by 1.

CALL addr: (Call) Transfer control to address *addr,* specified in byte 2–byte 3 of the instruction.

RET: (Return) Transfer control back to the "calling" program, to the instruction following the last CALL instruction.

HLT: (HALT) Stop the microprocessor.

SUMMARY

In this chapter we have learned that

1. Assembly language converts directly into machine code using a software program called an *assembler.* It can also be converted by hand if the programmer looks up all of the opcodes and addresses required for the machine code translation.
2. Memory-mapped I/O treats every input and output location as if it were a memory address. I/O-mapped I/O (or standard I/O) accesses every input and output location by an assigned port number.
3. I/O-mapped I/O uses the assembly language instructions IN *port* and OUT *port* to read and write data.
4. A flow chart is a useful means of documenting the sequence of program execution.
5. Assembly language is written using mnemonics which are later converted into machine code before being used by the microprocessor.
6. A single microprocessor program instruction occupies 1, 2, or 3 bytes of memory. It always starts with a 1-byte opcode and can be followed by 1 byte of data or a 2-byte address.
7. The looping in a down-counter is redirected to the start when 0 is reached by using the JZ instruction. The looping in an up-counter is redirected to the start when the accumulator reaches a certain count by using the CPI and the JZ instructions.
8. Besides the accumulator, there are 6 other registers labeled: *B*, *C*, *D*, *E*, *H* and *L*. Register pairs are formed by combining two 8-bit registers.
9. Data can be loaded into a register by using the MVI instruction and into a register pair by using the LXI instruction. Data movement between registers is done by using the MOV instruction.
10. Incrementing and decrementing registers and register pairs is done with the INR, DCR, INX, and DCX instructions.
11. Time-delay routines are used by microprocessors to introduce a time delay between operations. The length of the delay can be determined by adding up the *T*-states of all of the instructions executed in the delay and multiplying that times the microprocessor clock period.
12. Subroutines are used to define a program module that is to be executed several times within the main program. The subroutine is executed by using a CALL statement and is ended when a RET statement is encountered.

GLOSSARY

Assembler: A software package that is used to convert assembly language into machine language.

BASIC Language: A high-level computer programming language that uses English-language-type instructions that are converted to executable machine code.

Compiler: A software package that converts a high-level language program into machine language code.

Flowchart: A diagram used by the programmer to map out the looping and conditional branching that a program must make. It becomes the "blueprint" for the program.

Hand Assembly: The act of converting assembly language instructions into machine language codes by hand, using a reference chart.

I/O-Mapped I/O: A method of input/output that addresses each I/O device as a port selected by an 8-bit port number.

Memory-Mapped I/O: A method of input/output that addresses each I/O device as a memory location selected by a 16-bit address.

Nested Loop: A loop embedded within another loop.

Opcode: Operation code. It is the unique 1-byte code given to identify each instruction to the microprocessor.

Operand: The parameters that follow the assembly language mnemonic to complete the specification of the instruction.

Port Number: An 8-bit number used to select a particular I/O port.

Register Pair: In the 8085A microprocessor, six of the internal data registers are paired up to form three register pairs. They are useful for storing 16-bit quantities like addresses and large data values.

Statement Label: A meaningful name given to certain assembly language program lines so that they can be referred to from different parts of the program, using statements like JUMP or CALL.

Subroutine: A reusable group of instructions, ending with a return instruction (RET) and called from another part of the program using a CALL instruction.

Time Delay: A program segment written using repetitive loops to waste time or to slow down a process.

T State: Timing state. It is the length of time of one microprocessor clock period.

Zero Flag: A bit internal to the microprocessor that, when set (1), signifies that the last arithmetic or logic operation had a result of 0.

PROBLEMS

10–1. Describe one advantage and one disadvantage of writing programs in a high-level language instead of assembly language.

10–2. Are the following instructions used for memory-mapped I/O or for I/O-mapped I/O?

(a) LDA *addr* **(b)** STA *addr*
(c) IN *port* **(d)** OUT *port*

10–3. What is the digital level on the microprocessor's IO/$\overline{M}$ line for each of the following instructions?

(a) LDA *addr* **(b)** STA *addr*
(c) IN *port* **(d)** OUT *port*

10–4. List the new IN and OUT instructions that would be used to I/O to the switches and LEDs if the following changes to U4 and U5 were made to Figure 10–1.

(a) Add inverters to inputs A_8 and A_9 of U4 and to A_9 and A_{10} of U5.
(b) Add inverters to inputs A_{14} and A_{15} of U4 and to A_{14} and A_{15} of U5.

10–5. U6a and U6b in Figure 10–1 are OR gates. Why are they drawn as inverted-input NAND gates?

10–6. Are the LEDs in Figure 10–1 active-HIGH or active-LOW?

10–7. Is the $\overline{RD}$ line or the $\overline{WR}$ line pulsed LOW by the microprocessor during the:

(a) IN instruction? **(b)** OUT instruction?

10–8. What three conditions must be met in order to satisfy the output enables of U2 in Figure 10–1?

10–9. What three conditions must be met in order to provide a pulse to the C_p input of U3 in Figure 10–1?

10–10. Which internal data register is used for the IN and OUT instructions?

10–11. Write the assembly language instruction that would initialize the accumulator to 4FH.

10–12. Describe in words what the instruction JZ LOOP does.

10–13. Look up the opcodes for the following instructions:
(a) MVI D,*data* **(b)** INR C
(c) JNZ *addr* **(d)** DCR B

10–14. For each of the instructions in Problem 10–13, determine if they are 1-, 2-, or 3-byte instructions.

10–15. **(a)** Write the machine language code for the following assembly language program. (Start the machine code at address 2010H.)

```
INIT: MVI A,04H
  X1: DCR A
      JZ INIT
      JMP X1
```

(b) Rewrite the program using a JNZ instruction in place of the JZ instruction. Keep the function of the program the same, however.

10–16. Draw a flowchart and write the assembly language and machine language code for a MOD-12 down-counter program. (Start the machine code at address 2000H.)

10–17. Describe in words what the instruction CPI 0DH does.

10–18. Draw a flowchart and write the assembly language and machine language code for a MOD-12 up-counter. (Start the machine language code at address 2030H.)

10–19. Why are there *two* CMA instructions required in Example 10–2?

10–20. Modify your program in Problem 10–18 to output its count to the output LEDs at port FEH.

10–21. List the six internal data registers and the three register pairs inside the 8085A.

10–22. List the instructions in assembly language and machine language to meet the following requirements:
(a) Load the *E* register with 4FH.
(b) Load the *C* register with 12_{10}.
(c) Load the *H–L* register pair with the address 2051H.
(d) Load the *B–C* register pair with 1000_{10}.

10–23. What is the source register and what is the destination register for the instruction: MOV B,D?

10–24. Use the 8085A Instruction Set Reference Encyclopedia in the Appendix to determine which of the following instructions are capable of changing the zero flag:
(a) INR *r* **(b)** INX *rp*
(c) DCR *r* **(d)** DCX *rp*

10–25. Determine the contents of the *D* and *E* registers after the execution of each of the following programs.

(a)
```
MVI D,FFH
INR D
MVI E,00H
DCR E
```
(b)
```
LXI D,4050H
INX D
```
(c)
```
LXI D,40FFH
INX D
```
(d)
```
LXI D,40FFH
INR E
```
(e)
```
LXI D,A000H
DCX D
```
(f)
```
LXI D,E000H
DCR D
```

10–26. Determine the length of time of the following time-delay routines. (Assume that the 8085A is using a 4-MHz crystal.)

(a)
```
DELAY:  MVI B,05H
LOOP:   DCR B
        JNZ LOOP
```
(b)
```
DELAY:  MVI B,C0H
LOOP:   DCR B
        JNZ LOOP
```
(c)
```
DELAY:  MVI A,00H
LOOP:   INR A
        CPI 05H
        JNZ LOOP
```
(d)
```
DELAY:  MVI B,00H
LOOP:   INR B
        MOV A,B
        CPI D0H
        JNZ LOOP
```

10–27. Modify the first instruction in the nested-loop program shown in Table 10–5 to yield a delay of approximately one-tenth of a second.

10–28. Modify one instruction in the counter program given in Table 10–6 to change it to a MOD-12 counter.

10–29. Add the instructions to the DELAY subroutine in Table 10–6 for a third level of nesting to provide a time delay of approximately 1 s.

10–30. Is the CPI 00H instruction in Example 10–8 really necessary for the temperature limit program to work? Why?

10–31. What modification to the temperature limit program in Example 10–8 could be made so that it will only flash the LED that corresponds to the temperature limit that went HIGH?

10–32. Rewrite the program in Example 10–8 so that it will flash all LEDs 10 times and then halt as soon as any of the temperature limit switches go HIGH. (*Hint:* Use the *B* register as a counter.)

10–33. There is the potential of a problem with the automobile electronic control program in Example 10–9. That is, how many times will the LEDs flash if the trouble code is 00H? Fix the program so that a trouble code 00H will mean "no trouble" and keep rereading until there is a nonzero trouble code.

SCHEMATIC INTERPRETATION PROBLEMS

10–34. Find U9 in the HC11D0 schematic. LCD_SL and KEY_SL are active-LOW outputs that signify that either the LCD is selected or the keyboard is selected. Add a logic gate to this schematic that outputs a LOW level called I/O_SEL whenever either the LCD or the keyboard is selected.

10–35. U5 and U6 are octal *D* flip-flops in the Watchdog Timer schematic. They provide two stages of latching for the 8-bit data bus labeled D(7:0).

(a) How are they initially Reset? (*Hint:* CLR is the abbreviation for CLEAR, which is the same as Master Reset.)

(b) What has to happen for the *Q*-outputs of U5 to receive the value of the data bus?

(c) What has to happen for the *Q*-outputs of U6 to receive the value of the U5 outputs?

10–36. S2 in grid location B-1 in the HC11D0 schematic is a set of seven 10-kΩ pull-up resistors contained in a single DIP. They all have a common connection to V_{cc} as shown. Explain their purpose as they relate to the U12 DIP-switch package and the MODA, MODB inputs to the 68HC11 microcontroller.

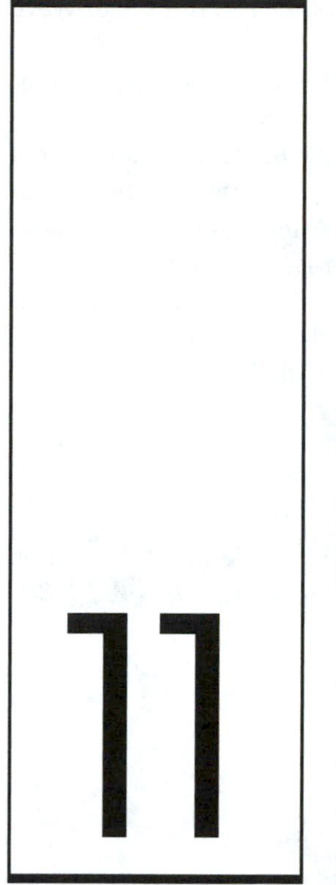

Introduction to 8085A System Hardware

OBJECTIVES

Upon completion of this chapter, you should be able to:

- Describe the differences between memory-mapped I/O and standard I/O.
- Understand the function of each pin on the 8085A microprocessor.
- Describe how the ALE signal is used to demultiplex the address/data bus.
- Discuss Read and Write machine cycle timing.
- Interface octal buffers and latches to microprocessor buses for performing I/O operations.
- Discuss the hardware design and timing of a complete 8085A system using memory-mapped I/O and standard memories.
- Discuss the hardware design and timing of a complete 8085A system using I/O-mapped I/O with 8085A-compatible support chips.
- Develop a memory map for a microprocessor-based system.
- Explain microprocessor instruction cycle and machine cycle timing.
- Draw Read cycle and Write cycle timing waveforms.
- Set up the 8355/8755A support chip for I/O and write the instructions to use its ports for inputting and outputting data.
- Write the instructions to use the I/O ports on the 8155/8156 IC and to use its timer section for timing and waveform generation.

INTRODUCTION

We saw a little bit of microprocessor hardware in Chapter 9 and were introduced to software in Chapter 10. In this chapter we will study the 8085A hardware and timing waveforms as they are used in a complete microprocessor-based system design. We will focus on the two basic types of system configuration: memory-mapped I/O using standard memory ICs and I/O-mapped I/O using special 8085A-compatible memory and I/O ICs.

Using the 8085A-compatible ICs makes life very easy for the hardware designer, but it hides many of the intricacies of the microprocessor chip that are important for the beginner to understand. For that reason, we'll start our discussion by using ICs that we've covered in previous chapters, like the 2716 EPROM, static RAMs, octal buffers, and *D* latches. Once we have a good understanding of the hardware interface and timing considerations in this mode, we'll switch over to the simplified design that uses memory and I/O chips that were specifically designed for interface compatibility with the 8085A.

11–1 8085A PIN DEFINITIONS

Figure 11–1 gives the pin configuration for the 8085A, and Table 11–1 defines the function of each pin, as presented in the Intel publication *Embedded Controller Handbook, 1988.*

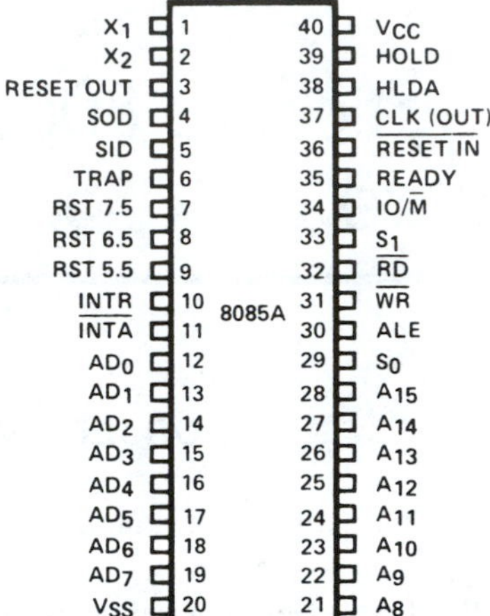

Figure 11–1 8085A pin configuration.

TABLE 11–1
8085A Functional Pin Definitions

Symbol	*Type*	*Name and function*
A_8-A_{15}	O	**Address bus:** The most significant 8 bits of memory address or the 8 bits of the I/O address, 3-stated during Hold and Halt modes and during Reset.
AD_{0-7}	I/O	**Multiplexed address/data bus:** Lower 8 bits of the memory address (or I/O address) appear on the bus during the first clock cycle (*T* state) of a machine cycle. It then becomes the data bus during the second and third clock cycles.
ALE	O	**Address latch enable:** It occurs during the first clock state of a machine cycle and enables the address to get latched into the on-chip latch of peripherals. The falling edge of ALE is set to guarantee setup and hold times for the address information. The falling edge of ALE can also be used to strobe the status information. ALE is never 3-stated.
S_0, S_1, and $IO/\overline{M}$	O	**Machine cycle status:** **$IO/\overline{M}$ S_1 S_0 Status** 0 0 1 Memory write 0 1 0 Memory read 1 0 1 I/O write 1 1 0 I/O read 0 1 1 Opcode fetch 1 1 1 Interrupt Acknowledge * 0 0 Halt

TABLE 11–1
8085A Functional Pin Definitions *(continued)*

Symbol	*Type*	*Name and function*
		* X X Hold * X X Reset * = 3-state (high impedance) X = unspecified S_1 can be used as an advanced R/$\overline{W}$ status. IO/$\overline{M}$, S_0 and S_1 become valid at the beginning of a machine cycle and remain stable throughout the cycle. The falling edge of ALE may be used to latch the state of these lines.
$\overline{RD}$	O	**Read control:** A low level on $\overline{RD}$ indicates the selected memory or I/O device is to be read and that the data bus is available for the data transfer, 3-stated during Hold and Halt modes and during RESET.
$\overline{WR}$	O	**Write control:** A low level on $\overline{WR}$ indicates the data on the data bus are to be written into the selected memory or I/O location. Data are set up at the trailing edge of $\overline{WR}$. 3-stated during Hold and Halt modes and during RESET.
READY	I	**Ready:** If READY is high during a Read or Write cycle, it indicates that the memory or peripheral is ready to send or receive data. If READY is low, the CPU will wait an integral number of clock cycles for READY to go high before completing the Read or Write cycle. READY must conform to specified setup and hold times.
HOLD	I	**Hold:** Indicates that another master is requesting the use of the address and data buses. The CPU, upon receiving the Hold request, will relinquish the use of the bus as soon as the completion of the current bus transfer. Internal processing can continue. The processor can regain the bus only after the HOLD is removed. When the HOLD is acknowledged, the address, data $\overline{RD}$, $\overline{WR}$, and IO/$\overline{M}$ lines are 3-stated.
HLDA	O	**Hold acknowledge:** Indicates that the CPU has received the HOLD request and that it will relinquish the bus in the next clock cycle. HLDA goes low after the HOLD request is removed. The CPU takes the bus one-half clock cycle after HLDA goes low.
INTR	I	**Interrupt request:** Is used as a general-purpose interrupt. It is sampled only during the next to the last clock cycle of an instruction and during Hold and Halt states. If it is active, the Program Counter (PC) will be inhibited from incrementing and an INTA will be issued. During this cycle a RESTART or CALL instruction can be inserted to jump to the interrupt service routine. The INTR is enabled and disabled by software. It is disabled by Reset and immediately after an interrupt is accepted.
$\overline{INTA}$	O	**Interrupt acknowledge:** Is used instead of (and has the same timing as) $\overline{RD}$ during the Instruction cycle after an INTR is accepted. It can be used to activate an 8259A Interrupt chip or some other interrupt port.
RST 5.5 RST 6.5 RST 7.5	I	**Restart interrupts:** These three inputs have the same timing as INTR except they cause an internal RESTART to be automatically inserted. These interrupts have a higher priority than INTR. In addition, they may be individually masked out using the SIM instruction.
TRAP	I	**Trap:** Trap interrupt is a nonmaskable RESTART interrupt. It is recognized at the same time as INTR or RST 5.5–7.5. It is unaffected by any mask or Interrupt Enable. It has the highest priority of any interrupt.

Continued

TABLE 11–1
8085A Functional Pin Definitions *(continued)*

Symbol	*Type*	*Name and function*
$\overline{\text{RESET IN}}$	I	**Reset In:** Sets the Program Counter to zero and resets the Interrupt Enable and HLDA flip-flops. The data and address buses and the control lines are 3-stated during RESET and because of the asynchronous nature of RESET, the processor's internal registers and flags may be altered by RESET with unpredictable results. $\overline{\text{RESET IN}}$ is a Schmitt-triggered input, allowing connection to an *RC* network for power-on RESET delay. Upon power-up, $\overline{\text{RESET IN}}$ must remain low for at least 10 ms after minimum V_{CC} has been reached. For proper reset operation after the power-up duration, $\overline{\text{RESET IN}}$ should be kept low a minimum of three clock periods. The CPU is held in the reset condition as long as $\overline{\text{RESET IN}}$ is applied.
RESET OUT	O	**Reset Out:** Reset Out indicates CPU is being reset. Can be used as a system reset. The signal is synchronized to the processor clock and lasts an integral number of clock periods.
X_1, X_2	I	**X_1 and X_2:** Are connected to a crystal, *LC,* or *RC* network to drive the internal clock generator. X_1 can also be an external clock input from a logic gate. The input frequency is divided by 2 to give the processor's internal operating frequency.
CLK	O	**Clock:** Clock output for use as a system clock. The period of CLK is twice the X_1, X_2 input period.
SID	I	**Serial input data line:** The data on this line are loaded into accumulator bit 7 whenever a RIM instruction is executed.
SOD	O	**Serial output data line:** The output SOD is set or reset as specified by the SIM instruction.
V_{CC}		**Power:** +5-V supply.
V_{SS}		**Ground:** Reference.

11–2 THE MULTIPLEXED BUS AND READ/WRITE TIMING

To keep the IC pin count to a minimum and thus reduce printed-circuit-board area, most manufacturers have used a *multiplexing* scheme on their address/data bus to allow two functions to "share" the same group of pins.

Pins 12 through 19 on the 8085A are multiplexed, using the same pins for both the low-order address lines (A_0–A_7) and the eight data lines (D_0–D_7). A special output control signal, ALE (Address Latch Enable), provides a signal to the external hardware to tell it when the multiplexed lines, AD_0–AD_7, contain addresses and when they contain data.

If we are using standard memories such as the 2716 EPROM and the 2114 RAM, and using standard octal devices like the 74LS244 buffer and the 74LS374 latch, then we must *demultiplex* the AD_0–AD_7 lines before they can be used. Figure 11–2 on p. 341 shows the most common method for doing that.

The A_8–A_{15} high-order address lines in Figure 11–2 are always valid, and are routed directly to the address bus. For example, the instruction STA 38FFH will place 38H on the A_8–A_{15} lines. The low-order part of the address, FFH, will appear on the AD_0–AD_7 lines early in the instruction cycle, and then disappear to allow the data or the STA instruction to pass out to the data bus via the same AD_0–AD_7 lines.

ALE

The function of the 74LS373 transparent *D* latch is to "grab" hold of the low-order address information before it disappears. The ALE signal shown in the microprocessor timing diagram in Figure 11–3 on p. 341 is the key to the demultiplexing process.

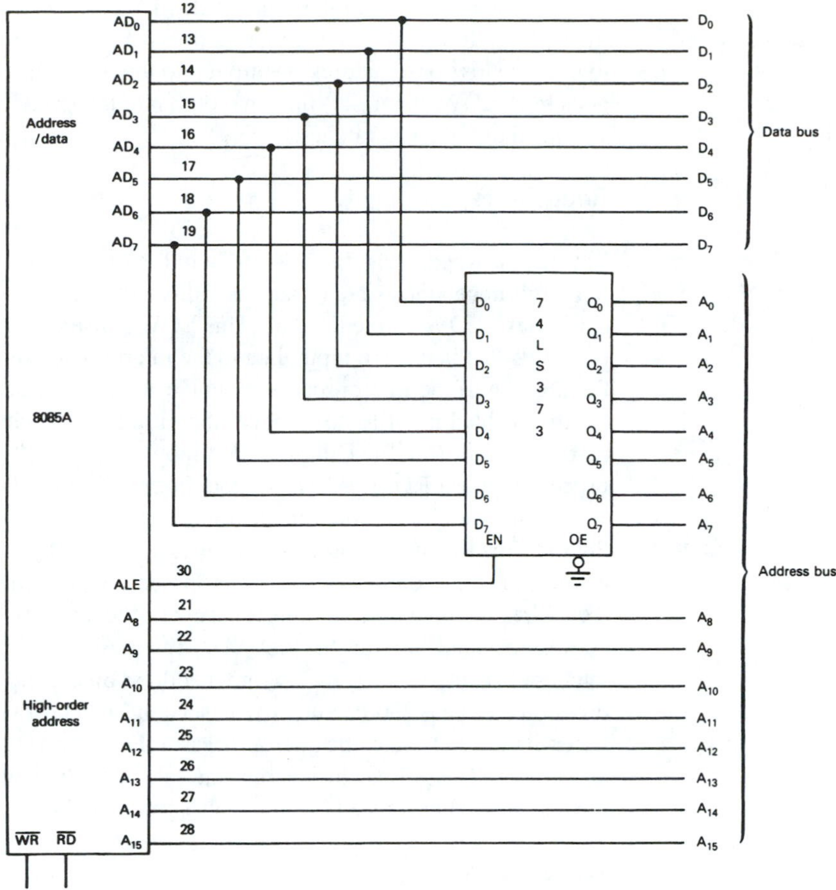

Figure 11–2 Using ALE and a D latch to demultiplex the AD_0–AD_7 lines.

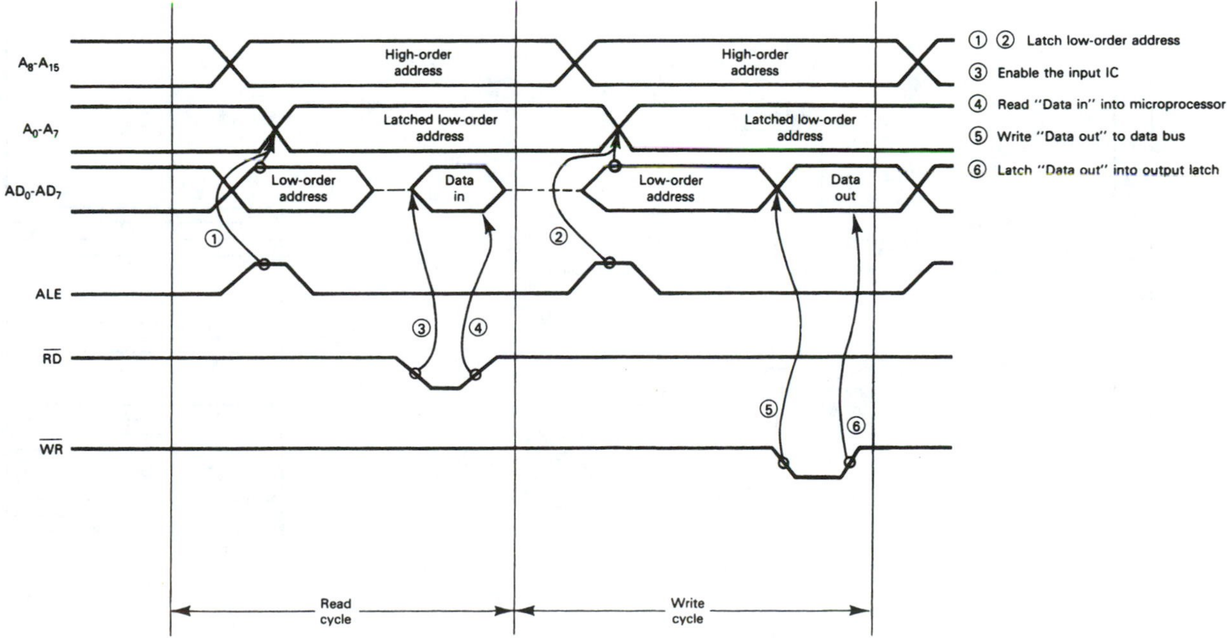

Figure 11–3 ALE and Read/Write cycle timing for the 8085A.

In the beginning of the Read cycle or Write cycle, the 8085A sends out all 16 bits of the address. It also issues a HIGH on the ALE line. The HIGH ALE enables the 74LS373 to pass the low-order addresses through to the A_0–A_7 lines (① and ②). When the ALE line goes back LOW, the A_0–A_7 information *remains latched* in the 74LS373. (See Chapter 5 if you need to review *D* latches.)

Read Cycle

Later in the Read cycle, the low-order address is removed from AD_0–AD_7, and the lines go to a high-impedance state (float), as indicated by the dashed lines. Next, the microprocessor issues a LOW on the $\overline{RD}$ line. This LOW is used to drive an active-LOW output enable on the IC that is used to input data to the microprocessor's data bus. This IC now provides the Data In to the microprocessor via the AD_0–AD_7 lines (③). The microprocessor waits for the enabled input IC to respond, then it raises $\overline{RD}$ back HIGH, and at the same time it reads the Data In (④). This so-called input IC is any IC used to input data to the microprocessor, like a ROM, RAM, or input buffer. (See Figure 11–4.)

An example of a software instruction that reads from a memory-mapped input is LDA 2800H (Load accumulator from 2800H). After the 3-byte LDA instruction is read in and decoded by the microprocessor, the microprocessor performs a fourth Read cycle to read from the addressed input port. For this Read cycle, the microprocessor puts 28H on the high-order address bus and 00H on the AD_0–AD_7 lines. An ALE pulse immediately follows, which latches 00H out to the A_0–A_7 low-order address bus. An address decoder connected to the address bus sees the 2800H and provides a chip select (or $\overline{OE}$ in Figure 11–4) to port 2800H. Later in the Read cycle, the microprocessor floats the AD_0–AD_7 lines (A_0–A_7 is still latched), and issues a $\overline{RD}$ pulse. This enables the selected port at address 2800H to output its data to the data bus, which is read into the microprocessor on the rising edge of $\overline{RD}$.

Write Cycle

In Figure 11–3 you can see that the Write cycle starts out the same way as the Read cycle does. It sends out 16 address bits and an ALE, but then, instead of floating the AD_0–AD_7 lines, it immediately puts the Data Out onto them. The microprocessor drops the $\overline{WR}$ line

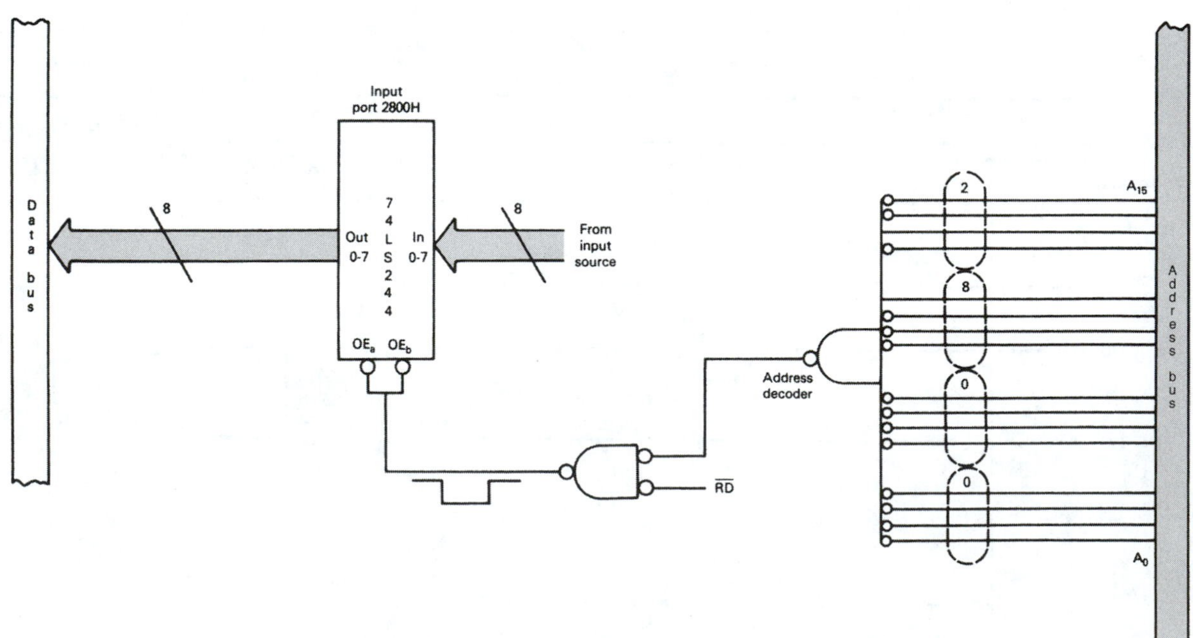

Figure 11–4 The 74LS244 octal buffer used as a memory-mapped input at address 2800H.

LOW at the same time that it sends the data out on AD_0–AD_7 (⑤). The LOW $\overline{WR}$ tells the addressed output or memory that data are available for it on the data bus. Often the output device being written to is a positive edge-triggered *D* flip-flop (as shown in Figure 11–5), which is triggered when $\overline{WR}$ returns LOW to HIGH (⑥).

An example of a software instruction that writes to a memory-mapped output is STA 3800H (Store accumulator to 3800H). The Write cycle for this instruction sets up A_8–A_{15} with 38H, AD_0–AD_7 with 00H, and drives ALE HIGH. A_0–A_7 latches onto 00H, which together with the 38H, selects the output at 3800H. The microprocessor next places the data from the accumulator on the AD_0–AD_7 lines (data bus) and pulses the $\overline{WR}$ line. The LOW-to-HIGH edge of the $\overline{WR}$ pulse is used as a clock trigger to the *D* flip-flop connected at address 3800H (see Figure 11–5). At the end of this cycle, the *Q* outputs of the *D* flip-flop contain the value that was in the accumulator.

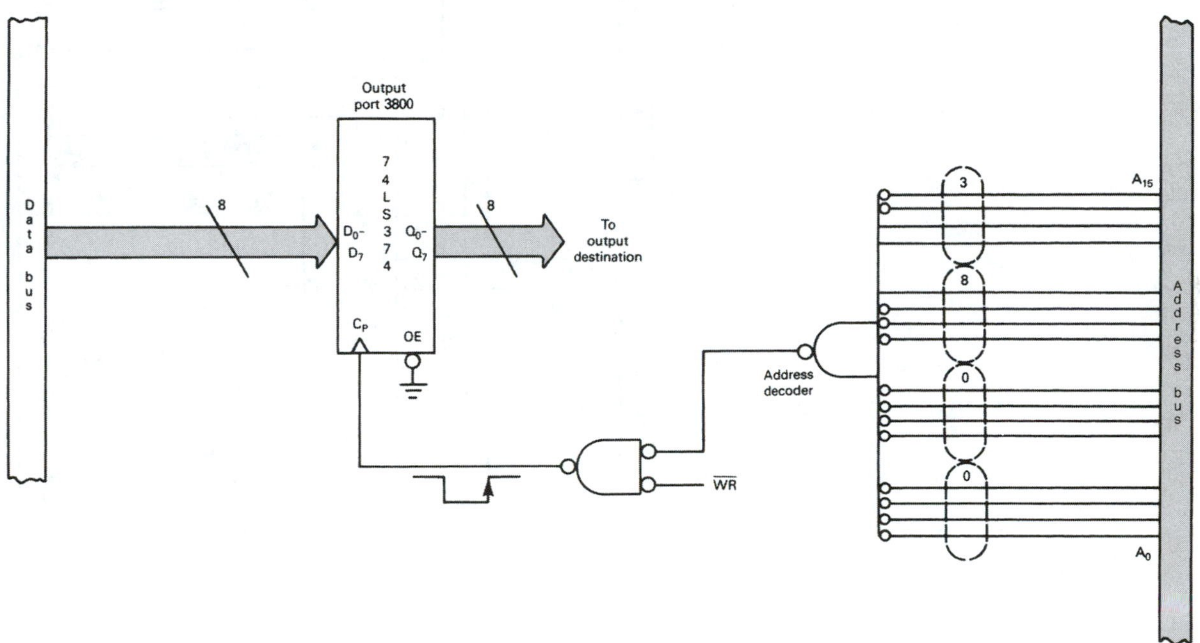

Figure 11–5 The 74LS374 octal *D* flip-flop used as a memory-mapped output at address 3800H.

11–3 MICROPROCESSOR SYSTEM DESIGN USING MEMORY-MAPPED I/O AND STANDARD MEMORIES

Now, if we put together the pieces from the previous section and from Chapter 8 (memories), we can develop a complete system design. We'll design a memory-mapped I/O system, which means that each I/O and memory is treated as a unique 16-bit memory location. Using this mapping scheme, we can read and write to the I/O ICs as if they were memory locations. This allows us to use all of the software instructions related to memory operations to perform input and output operations.

The fact that we are using *standard memories* means that they are general-purpose, universal memories used by any type of microprocessor. Intel also has a family of memory and I/O chips that are designed to interface directly to the 8085A's multiplexed bus, eliminating the need for an address latch. We'll use those in the next section after we have a better understanding of demultiplexing and memory-mapped I/O.

The complete system with I/O and memory is shown in Figure 11–6.

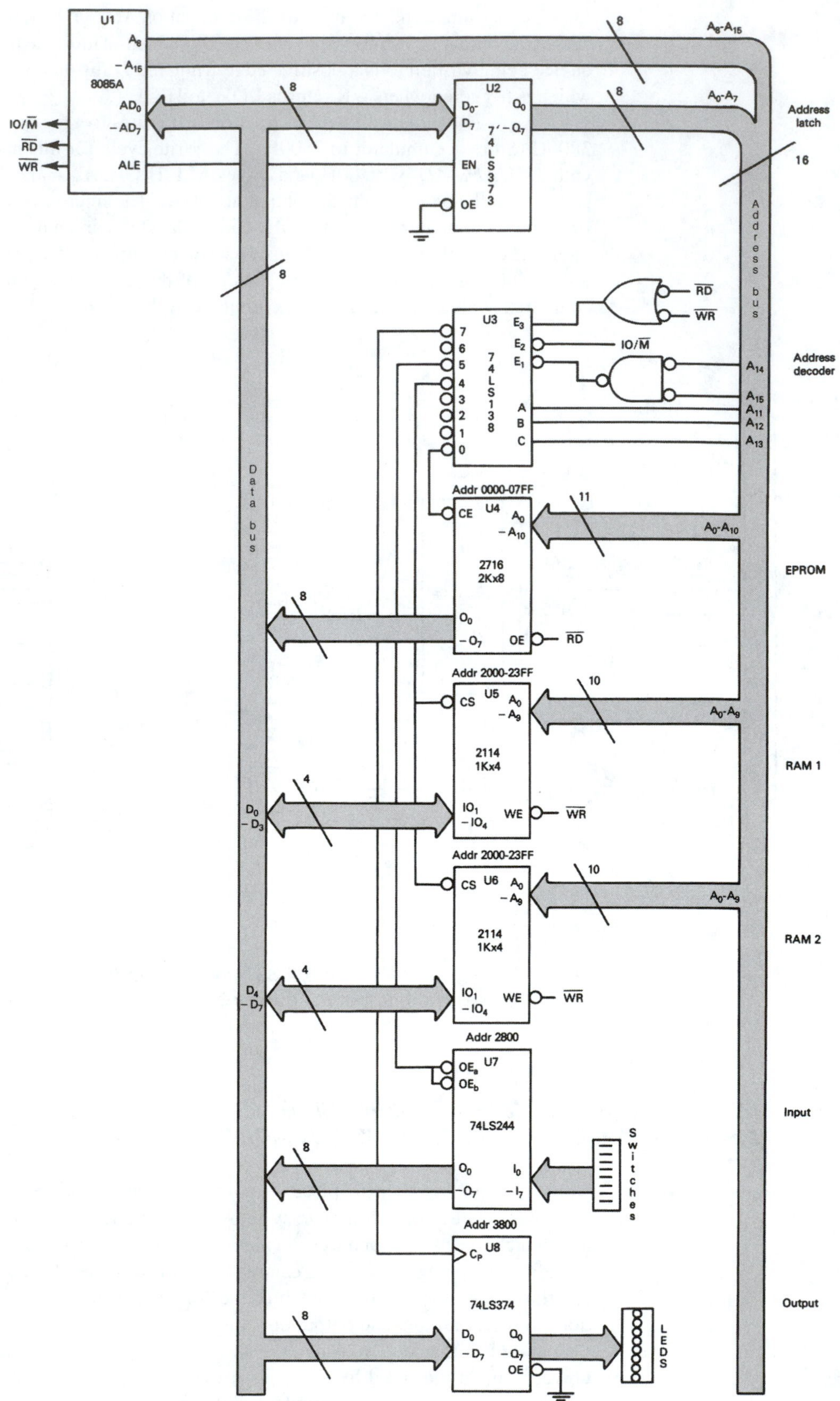

Figure 11–6 An 8085A system design using memory-mapped I/O and standard memories.

Address Latch—U2

Starting at the top of the diagram, you should recognize the 74LS373 octal latch used as the address latch. The multiplexed AD_0–AD_7 lines are input at D_0–D_7. When the ALE line is pulsed, the Q_0–Q_7 outputs will hold the low-order address byte (A_0–A_7).

Also note that the multiplexed AD_0–AD_7 lines are used as the *data bus.* We must be careful to only use the data bus *after* the low-order address byte has been removed. This safeguard is built into our system because all memory transfers and I/O to the data bus are enabled via the 74LS138 address decoder, which is only enabled when we are doing a Read ($\overline{RD}$) or a Write ($\overline{WR}$). If you review the timing waveform in Figure 11–3, you'll see that $\overline{RD}$ and $\overline{WR}$ are active *only* after the low-order address byte has been removed from the multiplexed bus.

Address Decoder—U3

Address decoding using the 74LS138 was covered back in Chapter 8, Figure 8–13. The 138 provides an active-LOW at one of its eight outputs, based on the binary input at *A*, *B*, and *C* (where *A* = LSB). The one output that is active will select the memory IC, input IC, or output IC that is to have access to the data bus.

Before any IC is selected, the 138 enables (E1, $\overline{E2}$, and $\overline{E3}$) have to be satisfied. This means that it must be receiving a LOW $\overline{RD}$ or $\overline{WR}$ pulse; the IO/$\overline{M}$ line must be LOW (which it is for all memory-related software instructions); and A_{15}–A_{14} must both be LOW.

From there, to determine which IC is selected, we have to further analyze the *A*, *B*, and *C* inputs. At this point, we want to develop a *memory map* of address allocations, as shown in Table 11–2 and Figure 11–7.

TABLE 11–2
Developing a Memory Map for Figure 11–6

Given: $\overline{RD}$ or $\overline{WR}$ = 0
IO/$\overline{M}$ = 0
A_{14} = 0
A_{15} = 0
A(LSB) ← A_{11}
B ← A_{12}
C(MSB) ← A_{13}
16-bit address

$A_{15}A_{14}A_{13}A_{12}$	$A_{11}A_{10}A_9A_8$	$A_7A_6A_5A_4$	$A_3A_2A_1A_0$
0 0 C B	A X X X	X X X X	X X X X

where X = don't care

IC Selection:
EPROM (U4) = #0 (A = 0, B = 0, C = 0)
∴ $A_{15} - A_0$ = 0000 0XXX XXXX XXXX
Address range = 0000H to 07FFH

RAMs (U5, U6) = #4 (A = 0, B = 0, C = 1)
∴ $A_{15} - A_0$ = 0010 0XXX XXXX XXXX
Address range = 2000H to 27FFH
Usable range = 2000H to 23FFH

Input Buffer (U7) = #5 (A = 1, B = 0, C = 1)
∴ $A_{15} - A_0$ = 0010 1XXX XXXX XXXX
Address range = 2800H to 2FFFH
Address used = 2800H

Output Latch (U8) = #7 (A = 1, B = 1, C = 1)
∴ $A_{15} - A_0$ = 0011 1XXX XXXX XXXX
Address range = 3800H to 3FFFH
Address used = 3800H

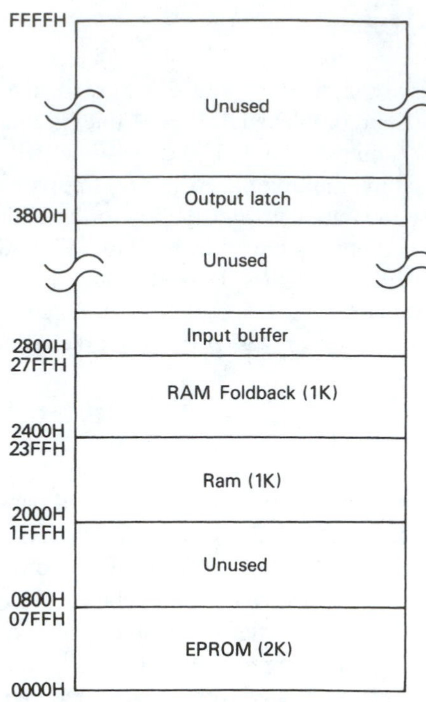

Figure 11–7 Memory map for Figure 11–6.

Using the procedure outlined in Table 11–2, we're able to develop a pictorial "map" of the placement of I/O and memory as shown in Figure 11–7. As the memory map shows, the decoder has placed each IC at a unique set of addresses.

EPROM—U4

The EPROM will be selected whenever the microprocessor reads from any address from 0000H to 07FFH. Address bits A_{11}, A_{12}, A_{13}, A_{14}, and A_{15} were used to select the EPROM, which leaves bits A_0–A_{10} to select the specific address within the EPROM. Having those 11 address lines allows us to specify 2048 (2^{11}) unique addresses, which happens to be the size of the 2716 EPROM. The selected address will place its 8-bit contents out to the data bus via O_0–O_7. The $\overline{\text{OE}}$ line is connected to $\overline{\text{RD}}$ to prevent a bus conflict if the programmer mistakenly tries to write to the EPROM's output.

When the 8085A is first powered up and reset, its program counter register points to address 0000H and starts executing from there. Therefore, in our system we'll store the main application's program in EPROM starting at 0000H. This way, when we power up, the software program that we write will automatically take over and perform whatever task we define.

RAMs—U5, U6

Two RAMs are required in our design because we need 8 bits at each location and one RAM location is only 4 bits wide. Note that both RAMs use the same address bits (A_0–A_9) for internal memory selection. Therefore, if you read from location 0, both RAMs will place 4 bits from their own location 0 onto the data bus. RAM1 provides D_0–D_3 and RAM2 provides D_4–D_7. Together they give us a full byte of data. If we were using 1K × 1 RAM, we would set up eight RAMs to provide 1 byte at each location.

The pair of RAMs are selected whenever reading or writing to address 2000H to 23FFH or 2400H to 27FFH. Address bits A_{11}–A_{15} are used to select the RAMs, and bits A_0–A_9 are used to select the specific location within the RAMs. With 10 address location bits (A_0–A_9) we can pinpoint 1024 (2^{10}) unique locations. This will take us from location 2000H to 23FFH.

After that, you would think that we are beyond the bounds of the RAMs, but we are not. Address bit A_{10} is not used by the address decoder, nor by the RAMs. Therefore, it is a don't care. If we try to read from address 2400H, A_{10} will be HIGH, but it makes no difference to the RAMs. The RAMs receive all 0s on A_0–A_9 and will output its first memory location. As you try to read locations 2400H through 27FFH, the RAMs are still selected, and you will access all 1024 locations all over again. This is called the *fold-back* area. The RAMs are intended to be accessed at locations 2000H to 23FFH, but the same locations can be accessed at locations 2400H to 27FFH. If you need 2K of RAM, you would put the additional 1K at location 2400H to 27FFH by using the A_{10} bit to select between the two pairs of RAMs.

The Write Enable ($\overline{\text{WE}}$) input is used by the RAM to determine which direction the data will flow. We use the $\overline{\text{WR}}$ line from the microprocessor to drive it. If $\overline{\text{WE}}$ is driven LOW, the addressed RAM location will be written to (Write). If $\overline{\text{WE}}$ is driven HIGH, the addressed RAM location will be sent to the data bus (Read).

Input Buffer—U7

The input buffer is accessed at any address from 2800H to 2FFFH. This is because only A_{11}–A_{15} are used in its selection. If we wanted to narrow it down to a single location, we could use a decoding scheme that checks all 16 address bits like the one shown in Figure 11–4. But why bother? We have plenty of address spaces to waste, and using the method given in Figure 11–6 is simpler. We'll assume that the input buffer is at address 2800H, and read from it using the instruction LDA 2800H.

Output Latch—U8

The output latch also has a *range* of addresses that access it (3800H–3FFFH). It could have been narrowed down to a single address using a decoding scheme such as the one shown in Figure 11–5. But instead, we'll use the circuit of Figure 11–6 and write to the LEDs using STA 3800H.

The fact that U8 is a *latch* allows us to write data to the LEDs and then go perform other program instructions while the LEDs remain latched.

11–4 CPU INSTRUCTION TIMING

All software instructions used by the 8085A are made up of a sequence of *Read* and *Write* operations. Each *Read* or *Write* is called a *machine cycle*. A machine cycle takes between three to six CPU clock periods (T states) to complete. We have already seen that the total number of T states varies, depending on the complexity of the instruction.

Now we'll see that the number of T states required for an instruction depends on the number of Read and Write operations it takes to complete. Take for example the instruction STA 3800H. The function of STA is to "store the accumulator to memory whose address is specified in byte 2–byte 3 of the instruction." It takes four machine cycles to execute STA 3800H.

The 8085A Instruction Set Timing Index, in the Appendix, lists the machine cycles and number of T states for each instruction. For the STA instruction it shows machine cycles: F, R, R, W. Machine cycle F is defined as a four T-state *opcode fetch,* R is defined as a three T-state *memory read,* and W is defined as a three T-state *memory write.* Figure 11–8 shows the four machine cycles required for the STA 3800 instruction. (ALE is omitted for clarity but would pulse HIGH at the beginning of each machine cycle.)

During machine cycle M1, the microprocessor performs a four T-state *opcode fetch* to determine what instruction to perform. (The M1 cycle is *always* an *opcode fetch.*) In order to do this, the processor must first set up the address bus with the address of the machine language code (we'll assume 2000H). It then issues a LOW pulse on $\overline{\text{RD}}$, which allows the accessed memory at address 2000H to put the opcode, 32H, onto the data bus. The

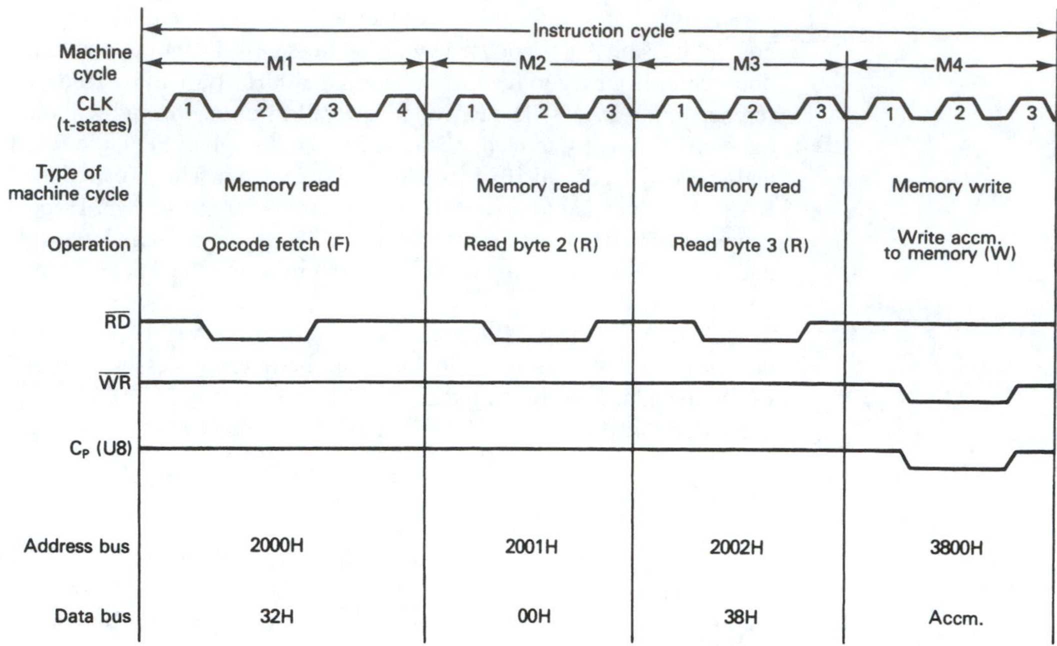

Figure 11–8 Microprocessor timing for the STA 3800H instruction.

microprocessor's instruction decoder determines that a 32H is the opcode for STA and that two more bytes for the instruction must follow.

Machine cycle M2 is a *memory read* for byte 2 of the instruction. The program counter is incremented by 1 (2001H) and put on the address bus. The $\overline{RD}$ pulse causes the accessed memory at 2001H to be put on the data bus (00H). Then machine cycle M3 is a *memory read* for the third byte of the instruction (38H).

The microprocessor now has read the complete instruction and must act on it. The M4 machine cycle is a *memory write,* which completes the instruction by storing the accumulator to the specified address. During M4, the processor sets up the address bus with the specified address (3800H), puts the accumulator's contents on the data bus, and then pulses the $\overline{WR}$ line. Referring back to the memory-mapped design of Figure 11–6, you'll see that the $\overline{WR}$ pulse will be duplicated on the C_p input to U8. This provides the positive trigger to the 74LS374, which latches the accumulator contents from the data bus into the latch and drives the LEDs.

EXAMPLE 11–1

Sketch the timing waveforms at $\overline{RD}$, $\overline{WR}$, C_p (U8), and $\overline{OE}$ (U7) for the following program, which is executing on the microprocessor circuit of Figure 11–6. [Assume that the microprocessor clock frequency is 2 MHz (4-MHz crystal).]

Address	*Contents*	*Label*	*Instruction*	*Comments*
2000	3A	LOOP:	LDA 2800H	; A ← Memory(2800H)
2001	00			;
2002	28			;
2003	32		STA 3800H	; Memory(3800H) ← A
2004	00			;
2005	38			;
2006	C3		JMP LOOP	; Jump to LOOP
2007	00			;
2008	20			;

Solution:

The waveforms that are generated will repeat each time around the loop and could be observed on a logic analyzer or a four-trace oscilloscope. The machine cycles and T states from the Instruction Set Timing Index Appendix are as follows:

$$\text{LDA} = \text{FRRR } (13\ T \text{ states})$$

$$\text{STA} = \text{FRRW } (13\ T \text{ states})$$

$$\text{JMP} = \text{FRR } (10\ T \text{ states})$$

The waveforms are shown in Figure 11–9.

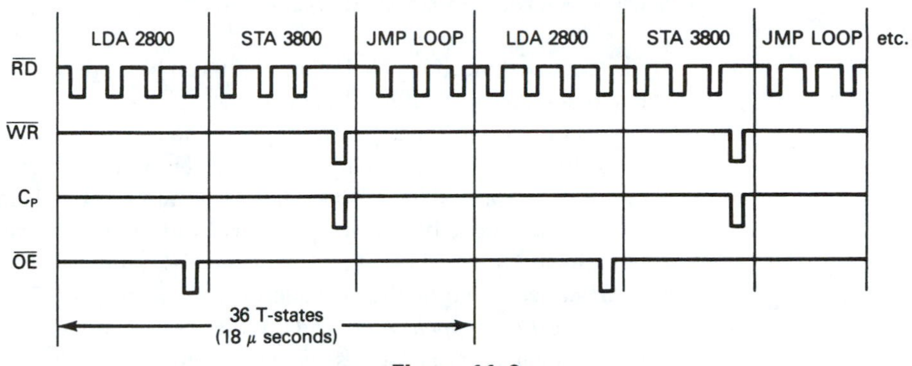

Figure 11–9

Explanation:

The individual CLK T states are not drawn out but would total 36. The time for a complete loop is: $36 \times [1/(2\text{ MHz})] = 18\ \mu\text{s}$.

The LDA instruction requires four machine cycles: opcode fetch, memory read (byte 2), memory read (byte 3), and memory read (from 2800H). During the last machine cycle, the address bus is set up with 2800H so that when $\overline{RD}$ goes LOW, $\overline{OE}$ also goes LOW, passing the input data from the switches to the data bus.

The STA instruction requires four machine cycles: opcode fetch, memory read (byte 2), memory read (byte 3), and memory write (to 3800H). During the last machine cycle, the address bus is set up with 3800H so that when $\overline{WR}$ is pulsed LOW, C_p receives a pulse, latching the accumulator output data from the data bus to the LEDs.

The JMP instruction requires three machine cycles: opcode fetch, memory read (byte 2), and memory read (byte 3). During the third machine cycle, the program counter is loaded with the address specified in byte 2–byte 3, and control passes to that location (2000H).

11–5 A MINIMUM COMPONENT 8085A-BASED SYSTEM USING I/O-MAPPED I/O

Intel Corporation has made it possible to reduce the chip count of an 8085A-based system to three ICs. Using specifically designed 8085A-compatible memory and I/O chips enables us to eliminate several of the ICs that were needed to fulfill the design requirements of the system shown in Figure 11–6. These 8085A-compatible *support chips* accept the multiplexed AD_0–AD_7 lines directly, and use the ALE signal to demultiplex the address internally, eliminating the need for the address latch IC. They also integrate input/output ports on the same chip with RAM or ROM memory. Figure

11–10 shows a three-chip microprocessor system made possible by the 8156 and 8355/8755A support ICs.

The 8355/8755A is the read-only memory for the system. The 8355 is the mask-ROM version, and the 8755A is the EPROM version. The 8755A is used in the early stages of system development and then the pin-compatible 8355 would be used for final mass production. They are both organized as 2K $\times$ 8, and each have two 8-bit I/O ports built in.

The 8156 provides the random-access (Read/Write) memory for the system. It is organized as 256 $\times$ 8 and has two 8-bit I/O ports, one 6-bit I/O port, and a 14-bit timer/counter.

Memory Mapping versus I/O Mapping

The circuit connections for Figure 11–10 can be made to form a memory-mapped I/O system or an I/O-mapped I/O system.

For a *memory-mapped I/O* system the A_{15} address line has to be used to drive the $IO/\overline{M}$ line on the support chips. (Use the connection indicated by the dashed line.) To access the I/O ports on the support chips, A_{15} must be HIGH. To access memory, A_{15} must be LOW. This is done by specifying A_{15} as 1 or 0 in the address field of the software instruction. Since A_{15} is dedicated to driving the $IO/\overline{M}$ line, then it cannot be used as the 16th bit of an address, limiting the total addressable memory to 32K instead of 64K.

For *I/O-mapped I/O* (standard I/O) the $IO/\overline{M}$ line is connected directly to the $IO/\overline{M}$ line of the microprocessor as shown in Figure 11–10. This way, the A_{15} line is free to be used on the address bus. The only limitation with standard I/O, however, is that all input and output to the ports must be made with the IN *port* or OUT *port* instructions; whereas with memory-mapped I/O, you can use any of the memory-directed instructions such as STA, LDA, MOV, and MVI. For most of our applications we'll use the standard I/O method for input/output.

Chip Decoding

The chip decoding (or selection) used in Figure 11–10 couldn't be any simpler. If A_{13} is HIGH, the 8156 is enabled; if A_{13} is LOW, the 8355/8755A is enabled. Also, since the 8355/8755A has 2048 memory locations, it needs 11 lines to pinpoint each internal location. Therefore, to enable the 8355/8755A we'll make A_{11} through A_{15} all LOW, and use AD_0–AD_7 and A_8–A_{10} to select specific memory locations. The range of addresses for the 8355/8755A will therefore be 0000H to 07FFH.

The 8156 RAM is accessed when A_{13} is HIGH; and memory is pinpointed by AD_0–AD_7 (256 locations). The range of addresses for the 8156 memory will therefore be 2000H to 20FFH.

The other control signals, $IO/\overline{M}$, $\overline{RD}$, and $\overline{WR}$, brought into both chips, tell whether it is an I/O port Read or Write or whether it is a memory Read or Write.

I/O Ports

Since we will be using the standard I/O mapped system, we'll use the IN *port* and OUT *port* instructions for I/O operations. When using IN or OUT, the 8-bit port number is duplicated on the high-order, and the low-order, address bus.

For example, IN 23H will set up the address bus with 2323H. With 2323H on the address bus, bit A_{13} will be HIGH, enabling the 8156. The three low-order bits of the port number are used to determine which port to use. The 3 in the number 23H specifies port *C* in the 8156. (More on that in Section 11–6.)

To access the I/O ports on the 8355/8755A you need to make A_{13} LOW. To read port *A* you would use IN 00H. To read port *B* you would use IN 01H. (See Section 11–6.)

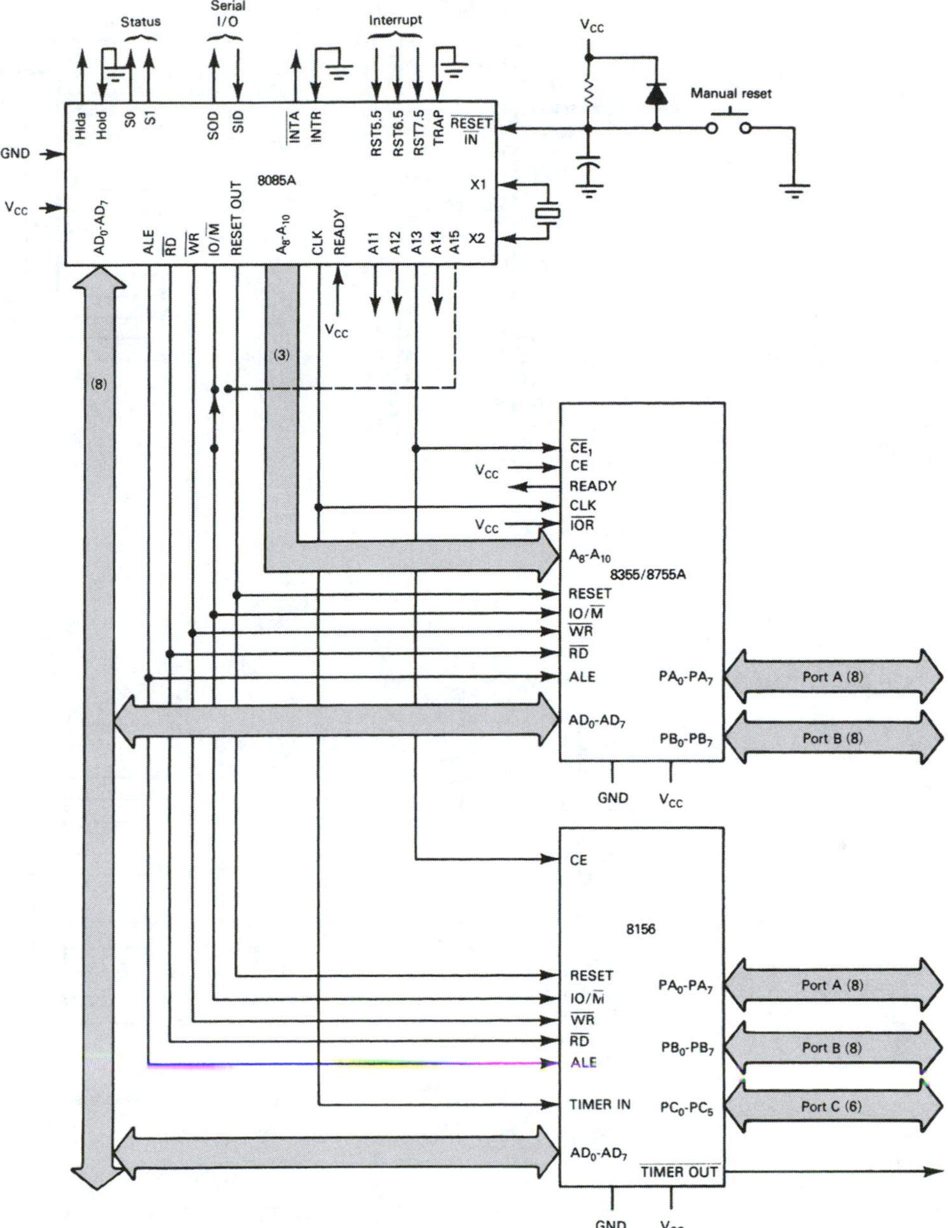

Figure 11–10 A minimum component 8085A-based system.

11–6 THE 8355/8755A AND 8155/8156 PROGRAMMABLE SUPPORT ICs

As mentioned earlier, the 8355/8755A and 8155/8156 ICs are specifically designed to simplify system design with the 8085A microprocessor (and also the 8088 microprocessor). These memory chips accept the multiplexed address/data bus directly and have built-in I/O ports.

The 8355/8755A

The block diagrams for the 8355 and 8755A are given in Figures 11–11a and b. The two ICs are pin compatible with each other so that after program development and debugging is complete using the EPROM version, the mask-ROM version can be directly substituted. The 2K ROM (or EPROM) section will contain program or operating system instructions

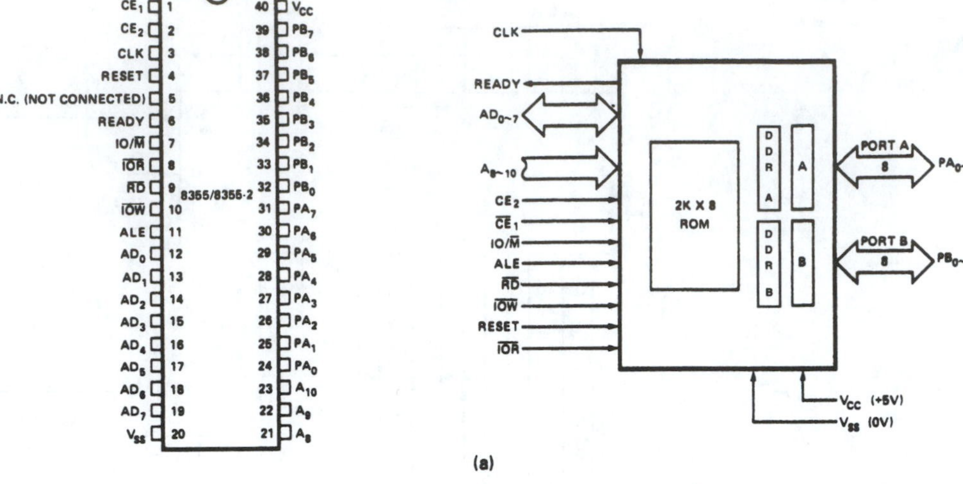

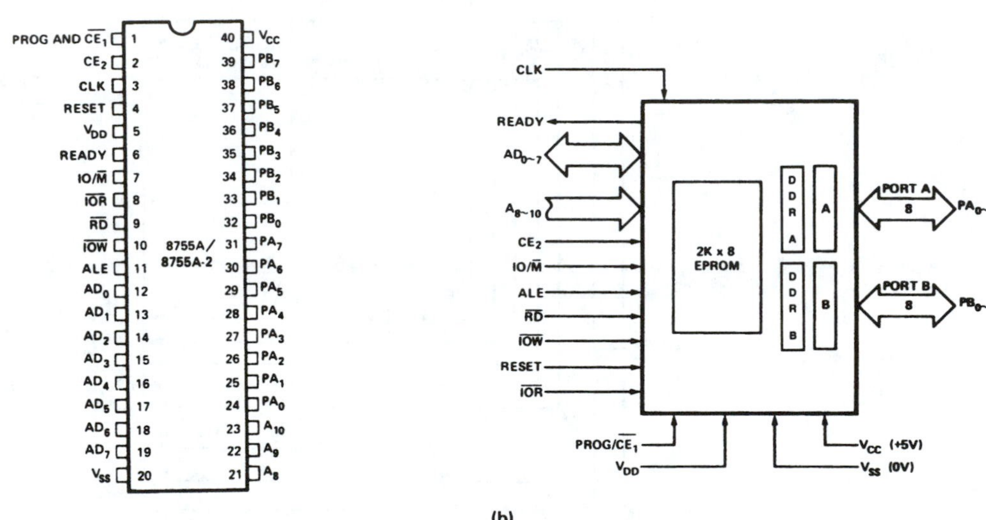

Figure 11–11 Pin configurations and block diagrams: (a) 8355 ROM with I/O; (b) 8755A EPROM with I/O. (Courtesy of Intel Corporation)

and is read like any other ROM. The ALE signal is used in conjunction with AD_0–AD_7 to latch the low-order address, A_0–A_7, internally.

Each of the 16 I/O lines are individually programmable as input or output. To use the I/O ports, they must first be programmed as either input or output. Each 8-bit port has what is called a *Data Direction Register* (DDR) associated with it. Outputting 1s to the DDR designate it as an output and 0s designate it as an input. The DDR for port *A* is accessed as port number XXXX XX10, and the DDR for port *B* as XXXX XX11. The X positions are used for chip decoding (selection). An IN *port* or OUT *port* instruction will send or receive data from the selected 8355/8755. The 8355/8755 then looks at AD_1 and AD_0 to determine which port or DDR is being referred to. Table 11–3 summarizes port and DDR designations. Therefore when developing a port number for the IN and OUT instructions, we'll use Table 11–3 to determine the least significant bits (AD_1, AD_0).

TABLE 11–3
Summary of Port and DDR Designations

AD_1	AD_0	*Selection*
0	0	Port A
0	1	Port B
1	0	Port A Data Direction Register (DDR A)
1	1	Port B Data Direction Register (DDR B)

We need to refer back to Figure 11–10 to determine the other six bits. To provide a Chip Enable to the 8355 (or 8755A), A_{13} must be LOW. Remember that the low-order address byte is duplicated on the high-order address byte. Therefore a LOW on AD_5 will show up on A_{13}, providing an active-LOW for $\overline{\text{CE}}_1$. Now we can build Table 11–4, which will show the actual port numbers to be used to access the ports and DDRs for the 8355/8755A in Figure 11–10.

TABLE 11–4
Port Numbers Used for the 8355/8755A in Figure 11–10

AD_7–AD_0 *7 6 5 4 3 2 1 0*	*Selection*	*Port numbers used*
X X 0 X X X 0 0	Port A	00H
X X 0 X X X 0 1	Port B	01H
X X 0 X X X 1 0	DDR A	02H
X X 0 X X X 1 1	DDR B	03H

The following examples will illustrate the use of the I/O ports on the 8355/8755A.

EXAMPLE 11–2

Write the instructions that would be used to designate port *A* of the 8355 in Figure 11–10 as input and port *B* as output.

Solution:

```
MVI A,00H    ; Send 0s to DDR A
OUT 02H      ;  to designate "Input"
MVI A,FFH    ; Send 1s to DDR B
OUT 03H      ;  to designate "Output"
```

EXAMPLE 11–3

Write the instructions that would be used to designate port *A*'s least significant 4 bits as input, and most significant 4 bits as output, for the 8355 in Figure 11–10.

Solution:

```
MVI A,F0H    ; LSBs = IN, MSBs = OUT
OUT 02H      ; DDR A ← Acc
```

EXAMPLE 11–4

In Figure 11–10, assume that input switches are connected to port *A* of the 8355, and output LEDs are connected to port *B*. Write a program to read the input switches and write their value to the output LEDs.

Solution:

```
MVI A,00H    ; Define Port A
OUT 02H      ;   as input
MVI A,FFH    ; Define Port B
OUT 03H      ;   as output
IN 00H       ; Read Port A into Acc
OUT 01H      ; Write Acc to Port B
```

The 8155/8156

The 8155/8156 is a RAM with I/O ports and a timer. It is used in a minimum-component 8085A-based system. The difference between the two versions is that the 8155 uses an active-LOW Chip Enable ($\overline{CE}$), while the 8156 uses an active-HIGH Chip Enable (CE). The pin configuration and block diagram for the 8155/8156 are given in Figure 11–12.

The RAM section provides 256 bytes of volatile program or data storage. Ports *A* and *B* each provide 8 bits, and port *C* provides 6 bits, of I/O capability. The IO/$\overline{M}$ line is used to select either the I/O ports or the RAM memory. The ALE line is used to internally demultiplex the AD_0–AD_7 lines.

Programming the Command Register

Before using the I/O and timer features, we must program an 8-bit latch called the *command register.* The command register, shown in Figure 11–13, defines how the ports and timer section are to be used. Bits 0 and 1 define the 8-bits of ports *A* and *B* as input or output. (You do not have the capability to define individual bits within the port as you could with the 8355/8755A.)

Bits 2 and 3 define the 6-bit port *C* as input (ALT 1), output (ALT 2), or control (ALT 3, ALT 4). The "Control" feature of port *C* provides special handshaking capability for 8085A-based systems. For our purposes, we'll only use ALT 1 and 2.

Command register bits 4 and 5 are used when in the ALT 3 and 4 mode, and bits 6 and 7 are used to start/stop the internal timer.

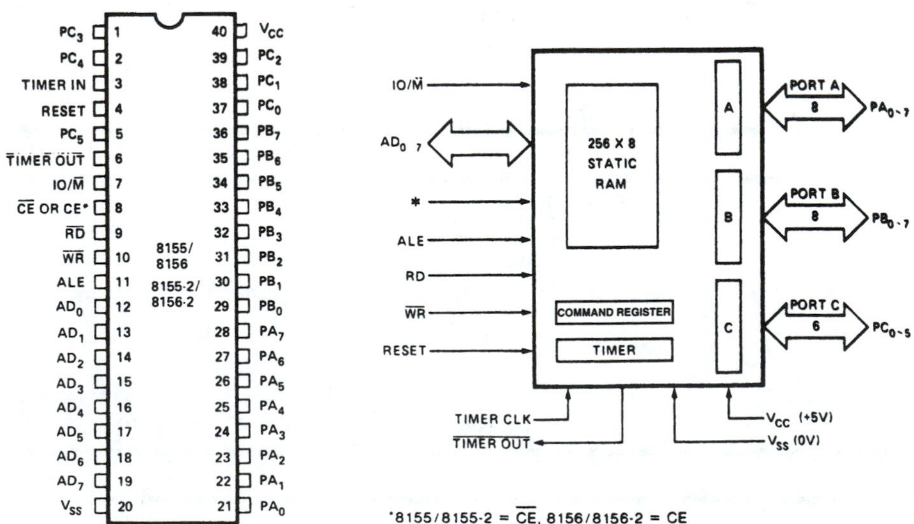

Figure 11–12 Pin configuration and block diagram for the 8155/8156 RAM with I/O ports and timer. (Courtesy of Intel Corporation)

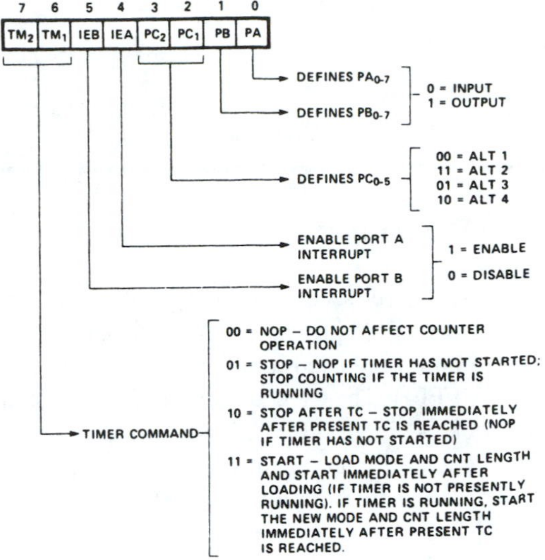

Figure 11–13 Command register bit assignment. (Courtesy of Intel Corporation)

To program the command register, you must first load the accumulator with the bit string that you want sent to the command register, and then use the OUT *port* instruction. The *port* number for the command registers, I/O ports, and timer are determined based on the information presented in Table 11–5.

TABLE 11–5
I/O and Timer Port Numbers

AD_7–AD_0 7 6 5 4 3 2 1 0	*Selection*	*Port numbers used*[a]
X X 1 X X 0 0 0	Command/Status Register	20
X X 1 X X 0 0 1	Port A	21
X X 1 X X 0 1 0	Port B	22
X X 1 X X 0 1 1	Port C	23
X X 1 X X 1 0 0	Low-order timer	24
X X 1 X X 1 0 1	High-order timer, timer mode	25

X: Don't care.
[a] *Port number* is based on bit AD_5 being HIGH to enable the 8156 in the system design of Figure 11–10.

EXAMPLE 11–5

Write the assembly language to program the command register of the 8156 in Figure 11–10 so that port *A* is defined as input, port *B* as output, and port *C* as input.

Solution:

```
MVI   A,02H ; A ← 0000 0010
            ;  (see Figure 11–13)
OUT   20H   ; Output to Command Register
               (see Table 11–5)
```

EXAMPLE 11–6

Write the assembly language to program the command register of the 8156 in Figure 11–10 so that all 22 I/O port bits are defined as outputs.

Solution:

```
        MVI    A,0FH  ; A ← 0000 1111
                      ;   (see Figure 11–13)
        OUT    20H    ; Output to Command Register
                      ;   (see Table 11–5)
```

Timer Operation

The timer section of the 8155/8156 is a 14-bit down-counter that can be used to produce varying frequency square waves or timing pulses at a time period determined by the initial count preloaded into the counter. Since it is 14 bits, the counter length can be as high as 3FFFH. The CLK output pin from the 8085A microprocessor is usually used to drive the clock input (TIMER IN) to the counter. The nice thing about using an external counter/timer like this is that once you have programmed and started it, the microprocessor is then free to return to performing other tasks as the counter ticks away. Figure 11–14 shows the format of the timer register. The initial starting value for the down-counter is loaded in two steps. The least significant 8 bits of the count length are output to port 24, as was determined in Table 11–5. The most significant 6 bits of the count length are output to port 25 along with the 2-bit *timer mode* as shown in Figure 11–14.

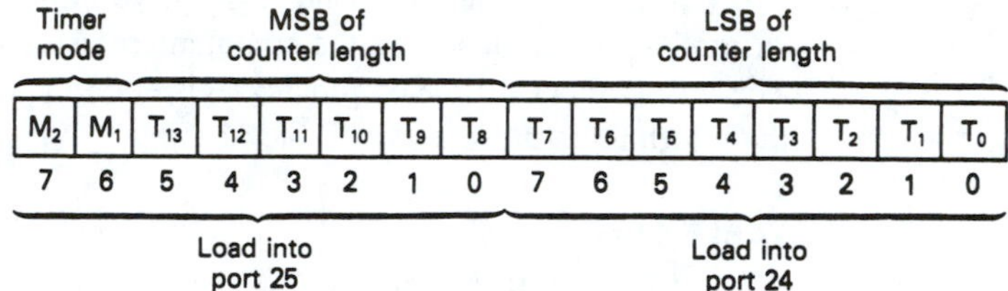

Figure 11–14 Timer register format.

Figure 11–15 lists the four timer modes and the resultant waveform produced at the $\overline{\text{TIMER OUT}}$ pin of the 8155/8156. There are four possible waveforms output at the $\overline{\text{TIMER OUT}}$ pin: a single square wave, a continuous square wave, a single pulse, and continuous pulses.

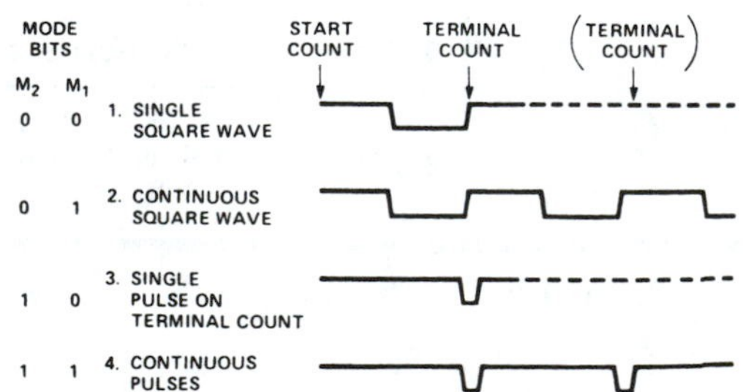

Figure 11–15 $\overline{\text{TIMER OUT}}$ operation modes and waveforms. (Courtesy of Intel Corporation)

The length of time from the point labeled "START COUNT" to the point labeled "TERMINAL COUNT" is the length of time it takes for the counter to count from its initial value down to 0. If we call this time the "$\overline{\text{TIMER OUT}}$ period," we can write the following equation:

$$\overline{\text{TIMER OUT}} \text{ period} = \text{TIMER IN period} \times \text{count length}$$

For example, if the microprocessor clock, connected to TIMER IN, has a period of 0.5 μs (2 MHz), and the initial counter value is 00FFH, then the period of a single square wave would be

$$\overline{\text{TIMER OUT}}\text{ period} = \text{TIMER IN period} \times \text{count length}$$
$$= 0.5\ \mu\text{s} \times 255$$
$$= 127.5\ \mu\text{s}$$

Once we have loaded the initial count and timer mode into the timer register, we then have to start the counter. If you refer back to Figure 11–13, you'll see that bit 6 and bit 7 of the command register are used to start and stop the counter (1–1 starts the counter). The following examples illustrate the use of the timer section.

EXAMPLE 11–7

Write the assembly language instructions that would program the 8156 to continuously output LOW pulses at $\overline{\text{TIMER OUT}}$ every 100 μs. Use the port assignments developed in Table 11–5, which were based on the microprocessor system shown in Figure 11–10. (Assume a microprocessor CLK period of 0.5 μs.)

Solution:

To find the count length:

$$\overline{\text{TIMER OUT}}\text{ period} = \text{TIMER IN period} \times \text{count length}$$
$$100\ \mu\text{s} = 0.5\ \mu\text{s} \times \text{count length}$$
$$\text{Count length} = 200 \text{ or } 00\text{C8H}$$

Timer register:

```
1100   0000          1100   1000
\/\_____/            \________/
Timer  MSB of         LSB of
mode   count          count
       length         length
\__________/          \______/
    To                   To
  port 25              port 24
```

Command register:

```
11XX   XXXX
\/\______/
Start  Don't
count  care
\_________/
    To
  port 20
```

Assembly language:

```
MVI A,C0H     ; Load Timer Mode
OUT 25H       ;  and MSB of Count
MVI A,C8H     ; Load
OUT 24H       ;  LSB of Count
MVI A,C0H     ; Start timer
OUT 20H       ;  running
HLT           ; End
```

EXAMPLE 11–8

Write a program that will produce a continuous 50-kHz square wave.

Solution:

To find count length:

$$\overline{\text{TIMER OUT}}\text{ period} = 1/50\text{ kHz}$$
$$= 20\ \mu s$$
$$\text{Count length} = \overline{\text{TIMER OUT}}\text{ period/TIMER IN period}$$
$$= 20\ \mu s/0.5\ \mu s$$
$$= 40\text{ or 0028H}$$

Timer register:

01	00 0000	0010 1000
Timer mode	MSB of count length	LSB of count length
To port 25		To port 24

Command register:

11	XX XXXX
Start count	Don't care
To port 20	

Assembly language:

```
MVI A,40H    ; Load Timer Mode
OUT 25H      ;  and MSB of Count
MVI A,28H    ; Load
OUT 24H      ;  LSB of Count
MVI A,C0H    ; Start timer
OUT 20H      ;  running
HLT          ; End
```

SUMMARY

In this chapter we have learned that

1. The low-order address bus is multiplexed with the data bus to reduce the pin count on the 8085A.
2. The 8085A pulses the address latch enable (ALE) line HIGH when there is a valid address on the multiplexed address/data bus.
3. The address/data bus can be demultiplexed by a *D* latch like the 74LS373.
4. During the READ cycle, data to be input is placed on the data bus by an external device and the $\overline{RD}$ line is pulsed LOW.
5. During the WRITE cycle, the microprocessor places data on the data bus and then pulses the $\overline{WR}$ line LOW.

6. The 74LS138 is a common address decoder found in microprocessor systems. Its outputs are used to enable a single device to be active on the data bus.
7. The 74LS244 octal buffer is commonly used as the interface chip for input devices. The 74LS374 octal *D* flip-flop is commonly used as the interface chip to provide latched outputs.
8. Each machine cycle is made up of a READ or a WRITE cycle and can take from three to six CPU clock periods.
9. The first machine cycle in every instruction is an opcode fetch (READ) to determine what instruction is to be executed.
10. A minimum-component microprocessor system can be constructed with an 8085A, an 8355/8755 ROM, and an 8156 RAM.
11. The 8355 (ROM) and the 8755(EPROM) contain a 2K × 8 memory, two I/O ports, and were designed to interface directly to the 8085A.
12. The data direction registers (DDR A and DDR B) of the 8355/8755 must be loaded with an input or output designation before the two I/O ports can be used.
13. The 8155/8156 interface chip has RAM, I/O ports, and a timer. Its command register must be loaded before using its I/O ports or timer.

GLOSSARY

Address Latch Enable (ALE): A signal output by the microprocessor to demultiplex the data bus and low-order address bus. A HIGH ALE signifies that the multiplexed bus currently contains address information.

Command Register: A register inside a microprocessor support IC programmed to designate the function and operation of the IC.

Data Direction Register (DDR): A register inside a microprocessor support IC programmed to designate the direction of data flow (in or out).

Fold-Back: An area of memory that duplicates a memory area at a different address. This occurs when an address decoding scheme has "don't-care" address bits. By changing the state of the "don't-care" bits, you will access the same memory using different addresses.

Instruction Cycle: The instruction cycle is made up of a number of Read or Write machine cycles required to complete the execution of a single microprocessor instruction.

IO/$\overline{M}$: A control signal issued by the microprocessor to signify whether the I/O of data is to a memory device (LOW) or an I/O port (HIGH).

Machine Cycle: The execution of each microprocessor instruction consists of several Read or Write operations to external devices. Each Read or Write operation is a machine cycle.

Memory Map: A table developed for each microprocessor system design that lists the range of addresses that access each device or IC in the system.

Multiplexed Bus: Two or more signals sharing the same bus. The data bus and the low-order address bus share the same pins on the 8085A microprocessor.

Read Cycle: The microprocessor machine cycle that issues a LOW $\overline{RD}$ pulse and reads 8 bits from the addressed memory or I/O port.

Support Chips: ICs that interface to microprocessor buses to perform the specific tasks required of a microprocessor-based system.

Write Cycle: The microprocessor machine cycle that issues a LOW $\overline{WR}$ pulse and writes 8 bits to the addressed memory or I/O port.

PROBLEMS

11–1. For what are the SID and SOD pins on the 8085A used?

11–2. Which pins on the 8085A are multiplexed?

11–3. Does the IO/$\overline{\text{M}}$ line go HIGH or LOW during:
(a) An opcode fetch? **(b)** A memory write? **(c)** An I/O read?

11–4. Describe the relationship between the X1–X2 pins and the CLK pin of the 8085A microprocessor.

11–5. What is one advantage and one disadvantage of the multiplexed bus?

11–6. When the ALE line goes HIGH, does AD_0–AD_7 contain an address or data?

11–7. Within a Read cycle or a Write cycle, what information is output to the AD_0–AD_7 lines first: address or data?

11–8. Assume that the waveforms in Figure P11–8 are observed in the circuit of Figure 11–2. Sketch the waveform that will result at A_0.

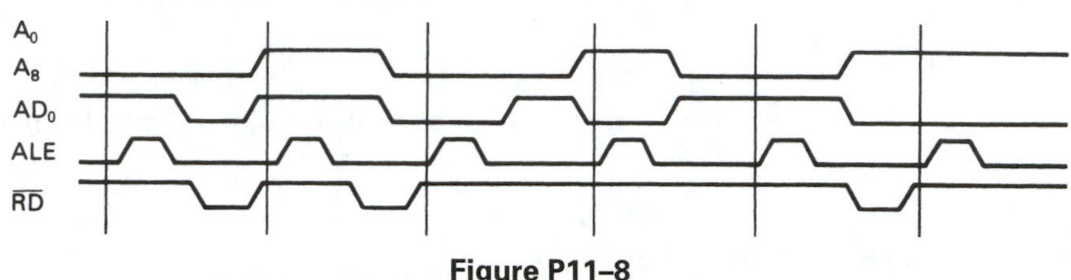

Figure P11–8

11–9. During what transition on the $\overline{\text{RD}}$ line does the microprocessor read data from the data bus (positive edge or negative edge)?

11–10. During what transition on the $\overline{\text{WR}}$ line does the microprocessor write data to the data bus (positive edge or negative edge)?

11–11. Modify the 16-input NAND gate in Figure 11–4 so that it decodes the address 6400H instead of 2800H.

11–12. Our 8085A system design in Figure 11–6 uses AD_0–AD_7 as the data bus, D_0–D_7. What safeguard is built into the circuit to ensure that we only use the data bus *after* the low-order address is removed?

11–13. List the conditions that must be met to *enable* the 74LS138 in Figure 11–6.

11–14. Develop a new memory map for Figure 11–6 using the new 74LS138 connections shown in Figure P11–14.

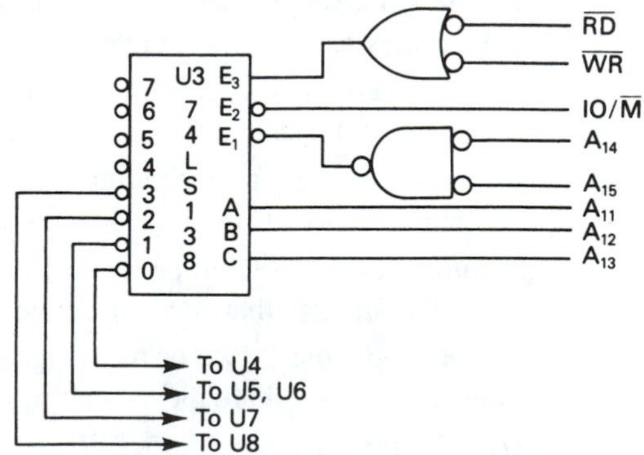

Figure P11–14

11–15. How many address locations does the 2114 RAM have? How many data bits are at each location?

11–16. Why does the design in Figure 11–6 use a *pair* of 2114s?

11–17. What is the range of addresses of the *fold-back* area for the 2114s in Figure 11–6? Why does the fold-back area exist?

11–18. Use the 8085A Instruction Set Timing Index in the Appendix to list the machine cycles for the following instructions:
(a) JMP *label* **(b)** IN *port*
(c) MVI A,*data* **(d)** RET

11–19. How long does each instruction in Problem 11–18 take if you are using a 4-MHz crystal?

11–20. The first machine cycle in *every* instruction is an opcode fetch. (True or False?)

11–21. Briefly describe the action that is taking place during each of the four machine cycles of the instruction LDA 2800H.

11–22. The following program is being executed on the microprocessor circuit of Figure 11–6. Sketch the waveforms at $\overline{RD}$, $\overline{WR}$, C_p (U8), and $\overline{OE}$ (U7) for two executions of the loop. (Assume that a 4-MHz crystal is being used.)

```
LOOP: LDA 2800H
      CMA
      STA 3800H
      JMP LOOP
```

11–23. What is the length of time from the first LOW pulse on $\overline{OE}$ to the second LOW pulse on $\overline{OE}$ in Problem 11–22?

11–24. Why is there no need for an address latch IC when using the 8355 and 8155 support ICs?

11–25. At what point would the 8755A IC be used in place of the 8355 in Figure 11–10?

11–26. The I/O-mapped microprocessor circuit of Figure 11–10 can be changed to memory-mapped I/O by changing the switch connection to A_{15}. List one advantage and one disadvantage of doing this.

11–27. What range of addresses would access the 8156 RAM in Figure 11–10 if CE were connected to A_{11} instead of A_{13}?

11–28. **(a)** Write the assembly language instructions to set up port *A* of the 8355 in Figure 11–10 as an output port and port *B* as an input port.
(b) Write the instructions to read the input port and write those data to the output port.

11–29. What is the port address of the command register in the 8156 of Figure 11–10?

11–30. Rebuild Table 11–5 with the new port numbers that would be used if CE were connected to A_{11} instead of A_{13} in Figure 11–10.

11–31. Write the assembly language to program the command register of the 8156 in Figure 11–10 so that port *A* is output, port *B* is output, and port *C* is input.

11–32. Write the assembly language program that will generate a pulse every 20 μs at $\overline{\text{TIMER OUT}}$ in Figure 11–10. Use a 4-MHz crystal.

11–33. Write the assembly language program that will create a continuous 10-KHz square wave at $\overline{\text{TIMER OUT}}$ of Figure 11–10. Use a 4-MHz crystal.

SCHEMATIC INTERPRETATION PROBLEMS

11–34. Locate the output pins labeled E and R/W on U1 of the HC11D0 schematic. During certain operations line E goes HIGH and line R/W is then used to signify a READ operation if it is HIGH or a WRITE operation if it is LOW. For a READ operation, which line goes LOW: WE_B or OE_B?

11–35. Find the octal decoder U5 in the HC11D0 schematic. Determine the levels on AS, AD13, AD14, and AD15 required to provide an active-LOW signal on the line labeled MON_SL.

11–36. The octal decoder U9 in the HC11D0 schematic is used to determine if the LCD (LCD_SL) or the keyboard (KEY_SL) is to be active.

(a) Determine the levels on AD3-5, AD11-15, and AS required to select the LCD.

(b) Repeat for selecting the keyboard.

11–37. Locate the 68HC11 microcontroller in the HC11D0 schematic. (A microcontroller is a microprocessor with built-in RAM, ROM, and I/O ports.) Pins 31–38 are the low-order address bus (*A*0–*A*7) multiplexed with (shared with) the data bus (*D*0–*D*7). Pins 9–16 are the high-order address bus (*A*8–*A*15). The low-order address bus is demultiplexed (selected and latched) from the shared address/data lines by U2 and the AS (Address Strobe) line.

(a) Which ICs are connected to the data bus (*DB*0–*DB7*)?

(b) Which ICs are connected to the address bus (*AD*0–*AD*15)?

11–38. U9 and U5 in the HC11D0 schematic are used for address decoding. Determine the levels on *AD*11–*AD*15 and *AD*3–*AD*5 to select:

(a) the LCD (LCD_SL) and

(b) the keyboard (KEY_SL).

The 8085A Software Instruction Set

OBJECTIVES

Upon completion of this chapter, you should be able to:

- Write intermediate-level applications programs.
- Use the indirect-addressing instructions for data transfer.
- Use logic instructions to perform Boolean operations.
- Perform multidigit BCD addition.
- Determine the status of the flag byte after the execution of arithmetic and logical instructions.
- Use logical instructions for masking off unwanted data.
- Write structured programs using subroutine modules.
- Describe stack operations associated with the CALL, RETurn, PUSH, and POP instructions.
- Use the interrupt capability of the 8085A microprocessor.

INTRODUCTION

A microprocessor only does exactly what it is told to do by the program instructions. We have covered several of the 8085A instructions in the previous chapter and will now proceed to cover most of the remaining instructions in this chapter.

The 8085A microprocessor has 80 different instructions, totalling 246 different opcodes when all of the instruction variations are included. For example, the instruction MOV *r1*,*r2* has 49 unique opcodes, depending on which destination register (*r1*) and which source register (*r2*) are used.

The complete instruction set is divided into five different functional groups as follows:

1. **Data transfer group:** Copies data between registers, or between registers and memory locations.
2. **Arithmetic group:** Performs addition, subtraction, incrementing, and decrementing with registers and memory locations.
3. **Logical group:** Performs logical (Boolean) operations on data in registers and memory.
4. **Branch group:** Modifies program flow by performing conditional or unconditional jumps, calls, or returns.
5. **Stack, I/O, and machine control group:** Performs stack operations, input/output, and interrupt control.

As you become more familiar with 8085A programming, you'll find that there are usually several different ways to get the same job done. Some ways may be more time efficient, others may be more memory efficient, and still others may make more sense logically and be easier to transfer as a "module" to another user application.

As a beginning programmer, you will find that much of your time is spent searching through the instruction set trying to find an instruction that will perform the specific task that you have in mind, not knowing if such an instruction even exists! That's why I suggest that you spend time skimming through the Instruction Set Reference Encyclopedia in the Appendix, over and over again. Even though you won't totally understand what you are reading, you will become familiar with the various operations that are available and the format of the reference encyclopedia.

A key to learning assembly and machine languages is to write and "debug" your programs on a microprocessor trainer. There are several of these trainers available, the Intel SDK-85 being one of the more popular ones. The SDK-85, and most other trainers, use support ICs like the 8355 and 8156, and have other circuitry to ease program development and testing.

The programs in this book can be tested on any 8085A or 8080-based trainer. Several of the programs read an input port and write to an output port. We will assume that we are using an 8085A-based system similar to the circuit presented back in Figure 11–10. That circuit has a 256-byte RAM area for program and data storage at 2000H to 20FFH, and I/O ports at 00H, 01H, 21H, 22H, and 23H. If you are familiar with the SDK-85 trainer, you will notice a great deal of similarities.

12–1 THE DATA TRANSFER INSTRUCTION GROUP

The instructions in this group are used to *transfer,* or to *move,* data around the system. Data can be moved into a register (MVI), into a register pair (LXI), from register to register (MOV), to and from memory directly (LDA,STA), and to and from memory indirectly. All of the data transfer instructions have been covered in previous chapters except for the *indirect* memory transfers.

Indirect Addressing of Memory

The LDA *addr* and STA *addr* instructions were examples of *direct* addressing of memory. For example, the instruction STA 20A0H will store the accumulator data to memory location 20A0H. Direct memory addressing is limited to data transfers to and from memory and the accumulator. If you want to move one of the other registers, or a data byte, to memory, then you must use *indirect memory addressing.*

Indirect addressing uses the *H–L* register pair as a *memory address pointer.* Before using one of the indirect-addressing instructions, you must first load the *H–L* register pair with the 16-bit address of the memory location that you want to access. You can then use one of the MVI or MOV instructions that refer to memory moves (instructions having a capital letter *M* in the operand field).

MVI M,*data*

For example, to move the data byte F8H to memory location 20A0H you would use the program given in Table 12–1.

TABLE 12–1
Using Indirect Addressing to Load F8H into Memory Location 20A0H

Address	*Contents*	*Label*	*Instruction*	*Comments*
2000	21	START:	LXI H,20A0H	; HL ← 20A0H
2001	A0			;
2002	20			;
2003	36		MVI M,F8H	; M(20A0H) ← F8H
2004	F8			;

The instruction MVI M,*data,* used in Table 12–1, is interpreted as follows: "Move the *data* given in byte 2 of the instruction to memory (M) whose address is pointed to by the *H–L* register pair." Figure 12–1 illustrates this data transfer.

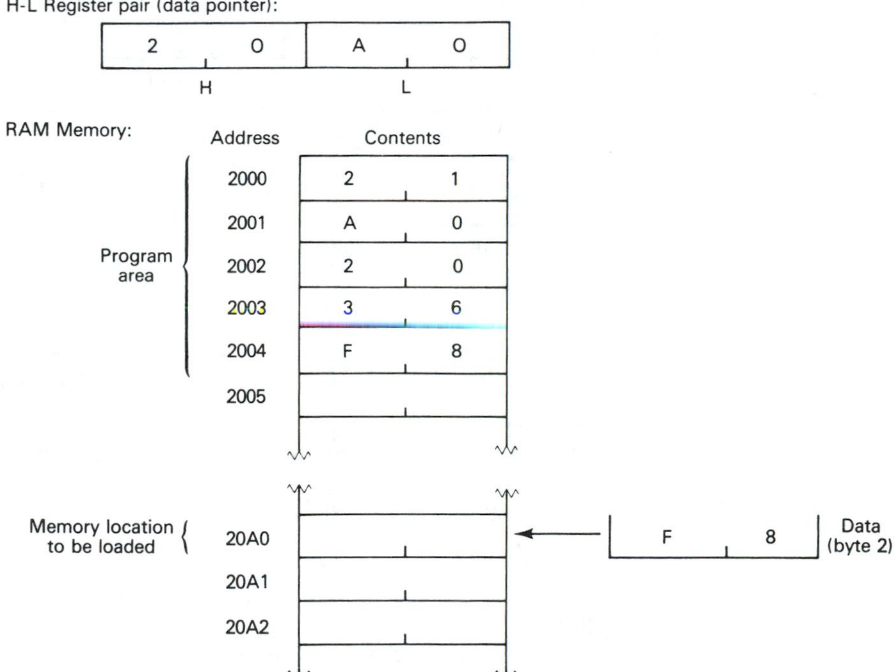

Figure 12–1 Illustration of the MVI M,F8H where *H–L* is 20A0H.

MOV M,*r* and MOV *r*,M

To use indirect memory addressing with the registers, we use the instructions MOV M,*r* and MOV *r*,M where *r* is one of the registers *A*, *B*, *C*, *D*, *E*, *H*, or *L*. Again, we use the *H–L* register pair as the memory pointer. For example, to move the contents of memory location 20B0H into register *D* and then transfer it to location 20C0H, we would use the program given in Table 12–2.

If you skim through the data transfer instructions in the Appendix, you won't find a move instruction that allows you to move memory to memory. That is why we had to use

TABLE 12–2
Program Using Indirect Addressing to Move Data from 20B0H, to Register *D*, to 20C0H

Address	*Contents*	*Label*	*Instruction*	*Comments*
2000	21	START:	LXI H,20B0H	; HL ← 20B0H
2001	B0			;
2002	20			;
2003	56		MOV D,M	; D ← M(20B0H)
2004	21		LXI H,20C0H	; HL ← 20C0H
2005	C0			;
2006	20			;
2007	72		MOV M,D	; M(20C0H) ← D

the *D* register in Table 12–2 as a temporary holding area, before moving the data to memory location 20C0H.

The following examples further illustrate the use of indirect addressing for data movement.

EXAMPLE 12–1

Write a program that will load memory locations 20C0H to 20CFH with even numbers, starting with the number 00H.

Solution:

Address	*Contents*	*Label*	*Instruction*	*Comments*
2000	06	START:	MVI B,00H	; Initialize
2001	00			; B ← 00H
2002	21		LXI H,20C0H	; Initialize
2003	C0			; HL ← 20C0H
2004	20			;
2005	70	LOOP:	MOV M,B	; M(HL) ← B
2006	04		INR B	; B ← B + 1
2007	04		INR B	; B ← B + 1
2008	23		INX H	; HL ← HL + 1
2009	3E		MVI A,D0H	; } Set Z flag
200A	D0			; } if L has
200B	BD		CMP L	; } reached D0H
200C	C2		JNZ LOOP	; If not, loop back
200D	05			; to load next
200E	20			; even number
200F	76		HLT	; Stop

Explanation:

We initialize the *B* register to 0 and use it for the even-number counter. The *H–L* memory pointer is initialized to 20C0H. Each time through the loop, register *B* is incremented twice, keeping it even, and *H–L* is incremented once, pointing to the next memory location. Program lines 2009 through 200B are used to determine if the low-order part of the *H–L* pointer has gone past the last location to be loaded (20CFH).

EXAMPLE 12–2

The 8355/8755A ROM/EPROM IC in the microprocessor system of Figure 11–10 is to be used for I/O operations. Figure 12–2 shows the connections for reading switches at port 00 and writing to LEDs at port 01. Use indirect memory

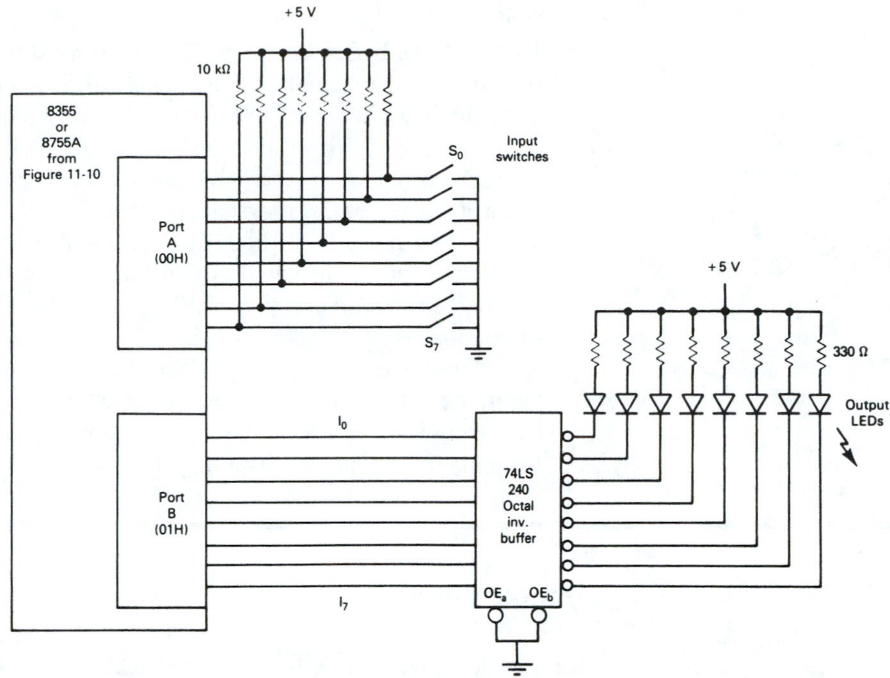

Figure 12–2 Interfacing input switches and output LEDs to an 8355/8755A microprocessor support IC.

addressing to send the contents of memory locations 20C0H through 20CFH to the LEDs, repeatedly. Use the one-quarter second time-delay program given in Table 10–6 to slow the display down.

Solution:

Address	*Contents*	*Label*	*Instruction*	*Comments*
2000	3E	START:	MVI A,FFH	; Designate port B
2001	FF			; as output
2002	D3		OUT 03H	; Program
2003	03			; DDR B
2004	21	LOOPA:	LXI H,20C0H	; Starting address
2005	C0			; of data
2006	20			; for display
2007	7E	LOOPB:	MOV A,M	; A ← M(HL)
2008	D3		OUT 01H	; Send contents of A
2009	01			; to output LEDs
200A	CD		CALL DELAY	; Delay
200B	50			; one-quarter
200C	20			; second at 2050H
200D	23		INX H	; Increment HL
200E	3E		MVI A,D0H	; } Set Z flag
200F	D0			; } if L has
2010	BD		CMP L	; } reached D0H
2011	C2		JNZ LOOPB	; Display next
2012	07			; memory
2013	20			; location
2014	C3		JMP LOOPA	; Restart at
2015	04			; beginning of
2016	20			; display memory
.				
.				
.				
2050		DELAY: (See Table 10–6)		

Explanation:

The 74LS240 buffer in Figure 12–2 is required because the 8355/8755A cannot sink the 10 mA required to illuminate an LED. An *inverting* buffer was chosen so that placing a 1 at pin I_0 through I_7 will turn the corresponding LED ON.

In Chapter 11 we determined that the data direction register for port *B* (DDR B) is at port number 03H, and to access port *B* we use port number 01H. Lines 2000H to 2003H program port *B* as *output.* Line 2008H writes the accumulator to port *B* (LEDs). The DELAY subroutine that we wrote in Chapter 10 is called to slow the display down so that we can see it. The output ports on the 8355/8755A are *latches,* so the LEDs will remain ON while we are in the DELAY subroutine.

Once we have displayed the data from all memory locations, 20C0H through 20CFH, then the CMP L instruction sets the zero flag, which forces us back to LOOPA to repeat the same displays again. (If this program is run *after* Example 12–1, then the data displayed will be the even numbers starting with 00H.)

Load/Store Data Transfer

Another way to move data is by using the LDA *addr* and STA *addr* instructions. You should remember from previous examples that LDA *addr* will "load the accumulator with the contents of memory at address *addr*." Also from previous examples, STA *addr* will "store the accumulator to memory at address *addr*." The operand *addr* is a 16-bit address given in byte 2–byte 3 of the instruction.

Another form of Load/Store data transfer is available using indirect addressing. The LDAX B, LDAX D, STAX B, and STAX D instructions use the 16-bit address stored in the *B–C* or *D–E* register pairs as the memory pointer. For example, the instruction LDAX D is interpreted as "load the accumulator with the contents of memory pointed to by *D–E*."

EXAMPLE 12–3

The following program is used to transfer data from one place in memory to another:

Instruction	*Comments*
LDA 20A0H	; A ← M(20A0H)
STA 20B0H	; M(20B0H) ← A

Rewrite that program using LDAX B and STAX D instructions.

Solution:

Instruction	*Comments*
LXI B,20A0H	; BC ← 20A0H (SOURCE)
LXI D,20B0H	; DE ← 20B0H (DESTINATION)
LDAX B	; A ← M(BC)
STAX D	; M(DE) ← A

Explanation:

In order to use the LDAX B instruction, we must first load the *B–C* register pair with the address 20A0H. (This is the address of the data *source.*) The LDAX B instruction will load the accumulator with the contents of memory location

20A0H. The *D–E* register pair is loaded with the *destination* address, 20B0H, and the STAX D instruction stores the accumulator to that location.

The advantage of using LDAX and STAX instead of LDA and STA is that the memory address becomes a *variable* that can be changed during program execution by changing the contents of the *B–C* or *D–E* register pairs.

12–2 THE ARITHMETIC INSTRUCTION GROUP

The arithmetic group includes instructions to increment, decrement, add, and subtract. We have used the increment and decrement instructions with registers and register pairs in previous examples. Instructions in the arithmetic group have an effect on the various flags that are used by the 8085A.

The Flag Byte

When a register is decremented to 00H, a flag bit, called the zero flag, is set to 1. There are several other flag bits besides the zero flag that can be affected as a result of arithmetic and logical instructions. These flags are useful for determining the status of a register after an arithmetic or logical instruction. These flags can then be used by the "conditional branching" instructions to Jump, Call, or Return, only if a specific flag bit is set or reset. (A "set" flag equals 1; a "reset" flag equals 0.)

The five flags used by the 8085A are stored in the flag byte shown in Figure 12–3. You must read the definition given in the Instruction Set Reference Encyclopedia in the Appendix for each instruction used to determine what effect that instruction has on each flag bit. If an instruction states that a particular flag is affected, then that flag will be forced to a 0 (reset) or a 1 (set) after completion of that instruction.

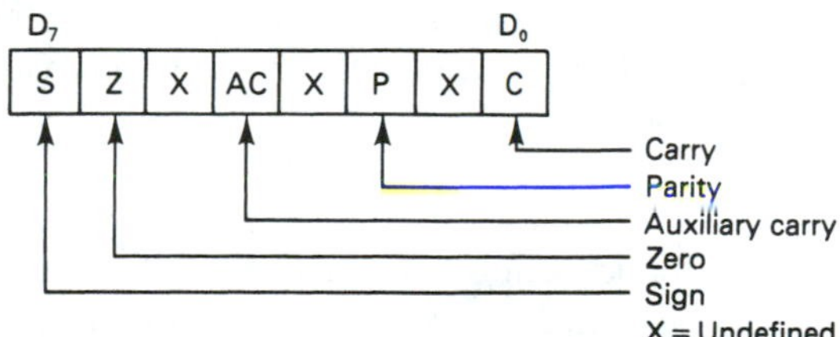

Figure 12–3 The 8085A flag byte.

The *carry flag* is set when the result of an addition is greater than FFH. Subtraction operations use the carry bit to indicate a "borrow." It is also affected by the logical instructions.

The *parity flag* is set if the number of 1 bits in the result of the operation is even (i.e., if the result has even parity). It is reset if the number of 1 bits is odd.

The *auxiliary carry flag* is set when there is a carry (overflow) from bit 3 to bit 4, as a result of the operation. This flag is used by the DAA instruction to form BCD numbers.

The *zero flag* is set when the result of certain instructions is 0.

The *sign flag* is used to indicate a negative result from an arithmetic or logical operation. It assumes that two's-complement notation is being used, and sets the sign flag if bit 7 of the accumulator is 1 (negative).

EXAMPLE 12–4

Determine the value of the flag byte upon completion of the following groups of instructions. (Assume that all flag bits are initially reset.)

(a) MVI A,55H
INR A
(b) MVI C,FFH
INR C

Solution: **(a)**

```
          A = 0101 0101
                    + 1
          -------------
          A = 0101 0110
          C = Unaffected by INR A
          P = 1
         AC = 0
          Z = 0
          S = 0
  Flag Byte = 00X0 X1X0 = 04H
```

(b)

```
          C = 1111 1111
                    + 1
          -------------
          C = 0000 0000
          C = Unaffected by INR C
          P = 1
         AC = 1
          Z = 1
          S = 0
  Flag Byte = 01X1 X1X0 = 54H
```

ADD r, ADI data, SUB r, and SUI data

These instructions are used to add or subtract a register, a memory location, or a data byte to or from the accumulator. The result of all arithmetic operations is placed in the accumulator.

For example, instruction ADD *r* adds the value in register *r* to the accumulator and places the result in the accumulator. (SUB *r* is a similar instruction used to subtract a register value from the accumulator.) The instruction ADI *data* adds the data value entered in byte 2 of the instruction to the accumulator. (SUI *data* is a similar instruction used to subtract a data value from the accumulator.) When performing multibyte addition and subtraction, there are four other instructions: ADC *r*, SBB *r*, ACI *data,* and SBI *data* that use the carry (or borrow) flag within the arithmetic operation. The following examples illustrate the use of the arithmetic instructions.

EXAMPLE 12–5

Determine the contents of the accumulator after completion of the following program instructions.

(a)

Instruction	*Comments*
MVI A,52H	; A ← 52H
MVI B,28H	; B ← 28H
ADD B	; A ← A + B
Answer: A = 7AH	

(b)

Instruction	Comments
MVI A,74H	; A ← 74H
MVI D,6BH	; D ← 6BH
SUB D	; A ← A − D
Answer: A = 09H	

(c)

Instruction	Comments
MVI A,0CH	; A ← 0CH
ADI 22H	; A ← A + 22H
Answer: A = 2EH	

(d)

Instruction	Comments
MVI A,6DH	; A ← 6DH
SUI 1FH	; A ← A − 1FH
Answer: A = 4EH	

(e)

Instruction	Comments
LXI H,20C0H	; HL ← 20C0H
MVI M,20H	; M(HL) ← 20H
MVI A,2AH	; A ← 2AH
ADD M	; A ← A + M(HL)
Answer: A = 4AH	

EXAMPLE 12–6

Write the assembly language instructions to store 10 numbers in RAM starting at address 20B0H. Use the ADI instruction to make each number stored 5 larger than the previous. The first location should contain the data byte 00H.

Solution:

	Instruction	Comments
	LXI H,20B0H	; Starting location of RAM
	MVI B,0AH	; B = counter for ten numbers
	MVI A,00H	; A holds numbers to be stored
LOOP:	MOV M,A	; Move number to memory
	ADI 05H	; Add 5 to A
	INX H	; Next memory location
	DCR B	; Decrement counter
	JNZ LOOP	; Loop back to store next number
	HLT	; End

DAA

The DAA (decimal adjust accumulator) instruction is required whenever you are using BCD numbers in arithmetic operations. The reason for this is that the microprocessor naturally performs all arithmetic in binary. You would get incorrect answers if you enter and add BCD numbers and expect to get BCD answers.

For example, if we want to add the base 10 decimal numbers 44 + 28, we would enter them as BCD, and the operation would look as follows:

$$\begin{array}{r} 44 = 0100\ 0100 \\ +\ 28 = 0010\ 1000 \\ \hline 72 \neq 0110\ 1100\ (6CH) \end{array}$$

The microprocessor adds the two BCD numbers and gets 6CH for an answer. We wanted to get the answer 72 (0111 0010), however. To correct the problem, we need to execute the DAA instruction. The DAA instruction uses the following rules to form the correct BCD result.

1. If the value of the least significant 4 bits of the accumulator is greater than 9 *or* if the AC flag is set, 6 is added to the accumulator.
2. If the value of the most significant 4 bits of the accumulator is now greater than 9 *or* if the CY flag is set, 6 is added to the most significant 4 bits of the accumulator.

Applying those rules to the previous addition will result in the following operations:

$$\begin{array}{r} 44 = 0100\ 0100 \\ +\ 28 = 0010\ 1000 \\ \hline 0110\ 1100 \\ {}^*+\quad 0110 \\ \hline 0111\ 0010 = 72\ \text{(correct)} \end{array}$$

*The least significant 4 bits are greater than 9, so add 6.

EXAMPLE 12–7

Add the decimal numbers 39 + 29 and adjust the result to a valid BCD answer by applying the rules of the DAA instruction.

Solution:

$$\begin{array}{rl} 39 = 0011\ 1001 & \text{BCD} \\ +\ 29 = 0010\ 1001 & \text{BCD} \\ \hline 0110\ 0010 & \\ \text{AC} & \\ \text{Add 6} \qquad +\ 0110 & \\ \hline 0110\ 1000\ \text{BCD} = 68 & \textit{Answer} \end{array}$$

EXAMPLE 12–8

Write an assembly language program to store the following five decimal numbers in RAM, in BCD form, starting at location 20B0H: 42, 51, 77, 32, and 63. Add the numbers and store the three-digit BCD answer in the *B–C* register pair (*B* = hundreds digit, *C* = tens digit and ones digit).

Solution:

	Instruction	Comments
MAIN:	CALL STORE	; Store the numbers
	CALL ADD	; Add the numbers
	MOV C,A	; Transfer tens and ones to C
	HLT	; End
		;
		; Subroutine to store 5 numbers
STORE:	MVI A,42H	; A ← 42H
	STA 20B0H	; Store 1st number
	MVI A,51H	; A ← 51H
	STA 20B1H	; Store 2nd number
	MVI A,77H	; A ← 77H
	STA 20B2H	; Store 3rd number
	MVI A,32H	; A ← 32H
	STA 20B3H	; Store 4th number
	MVI A,63H	; A ← 63H
	STA 20B4H	; Store 5th number
	RET	; Return to MAIN
		;
		; Subroutine to add 5 numbers
ADD:	LXI H,20B0H	; Address of 1st number
	MVI B,00H	; Initialize hundreds digit to zero
	MVI D,04H	; Numbers counter
	MOV A,M	; Move 1st number to A
	INX H	; Increment address to 2nd number
LOOP:	ADD M	; Add each successive number to A
	CC HNDRDS	; Call HNDRDS if carry set
	DAA	; Adjust A to valid BCD
	CC HNDRDS	; Call HNDRDS if carry set
	DCR D	; Decrement numbers counter
	RZ	; Return if zero
	INX H	; Increment address to next number
	JMP LOOP	; Loop back
		;
		; Subroutine to increment hundreds counter
HNDRDS:	INR B	; Increment hundreds counter
	RET	; Return to ADD subroutine

Explanation:

This program is a good example of using a *"structured programming"* technique. The program is broken up into three *modules,* or subroutines. Each module has a specific function and can be written and tested on its own. This is very helpful for program development and debugging. Each module would be entered in a different block of memory. For example, MAIN could start at address 2000H, STORE could start at 2010H, ADD could start at 2030H, and HNDRDS could start at 2050H.

The MAIN program does nothing but CALL two subroutines and perform a move. The STORE subroutine uses MVI and STA instructions to store the five numbers to be added. These numbers are entered in hexadecimal, which when translated to binary for the microprocessor will be the same as BCD.

The ADD subroutine has two new instructions in it: a conditional call (CC HNDRDS) and a conditional return (RZ). Calls and returns (as well as jumps) can be based on the condition of any of the following flags: Z, C, P, or S. The CC HNDRDS instruction is necessary to increment the hundreds counter if there is an overflow (carry) due to the ADD M or DAA instruction (result > 99). The RZ instruction returns control to the MAIN program when D is decremented to 0.

12–3 THE LOGICAL INSTRUCTION GROUP

The logical group provides a way to perform the Boolean operations AND, OR, and exclusive-OR between registers or between memory and a register. This group also provides instructions to rotate (shift) the bits in the accumulator, compare data, and complement data. The compares (CPI, CMP) and complement (CMA) have been explained in previous ex.amples.

ANA r, ANI data, XRA r, XRI data, ORA r, and ORI data

These instructions perform Boolean operations between the accumulator and a register or memory. The result of the operation is placed in the accumulator. For example, Figure 12–4 illustrates the operation that takes place when executing the instruction ANA B (A ← A AND B). Bit A_0 is ANDed with bit B_0, and the result is put back into bit A_0. This process is repeated for all 8 bits. The XRA and ORA instructions operate the same way except exclusive-OR gates or OR gates are used in place of the AND gates.

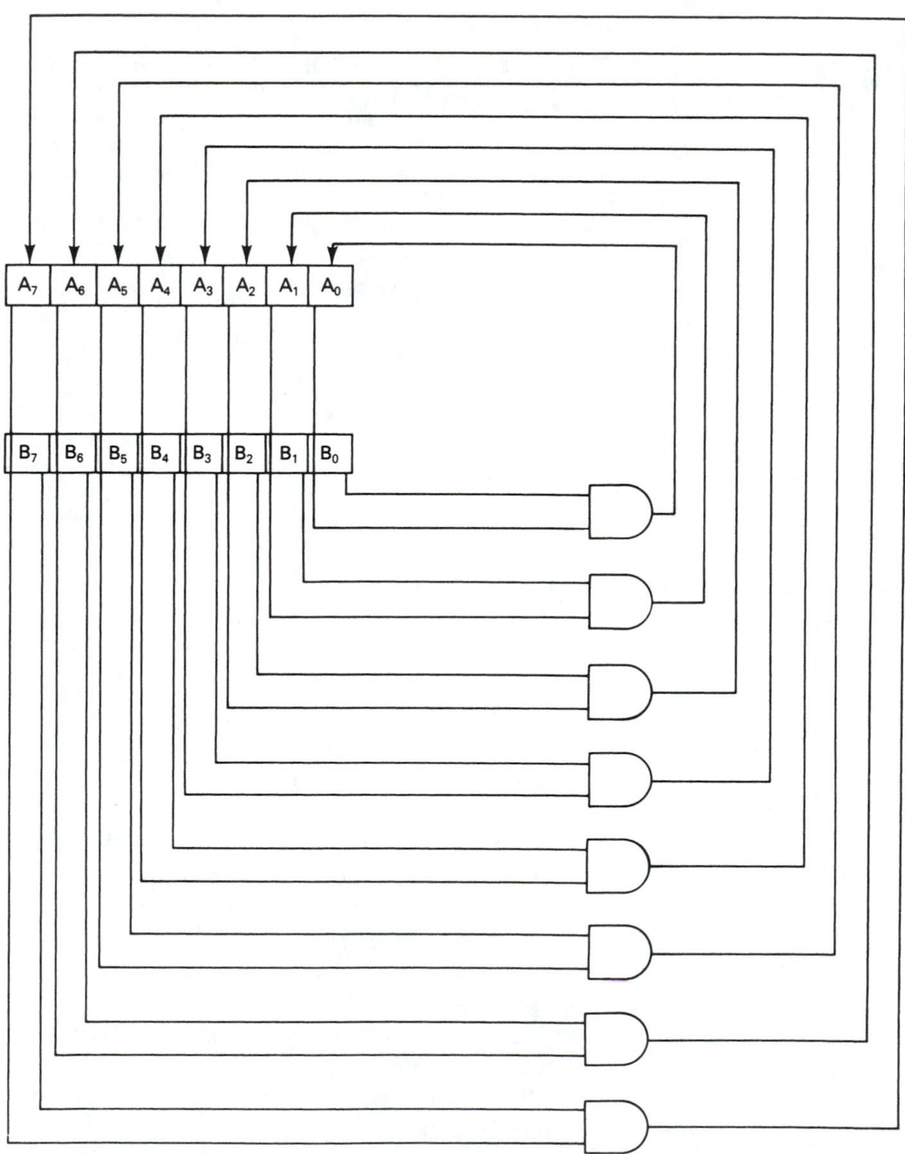

Figure 12–4 Illustration of the instruction ANA B.

The logic instructions ending with an *I* (ANI, XRI, and ORI) operate with an *immediate* data byte entered as byte 2 of the instruction. For example, ORI 35H would OR the accumulator with the bit string 0011 0101.

EXAMPLE 12–9

Determine the contents of the accumulator after the completion of each of the following groups of instructions.

(a)
```
MVI A,42H
MVI D,15H
ORA D
```
(b)
```
MVI A,5CH
XRI FEH
```
(c)
```
MVI A,F0H
XRA A
```
(d)
```
MVI A,3CH
ANI 87H
```

Solution: **(a)**

A = 0100 0010
D = 0001 0101
A OR D = 0101 0111 = 57H *Answer*

(b)

A = 0101 1100
data = 1111 1110
A ex-OR *data* = 1010 0010 = A2H *Answer*

(c)

A = 1111 0000
A = 1111 0000
A ex-OR A = 0000 0000 = 00H *Answer*

Note: XRA A is often used as a 1-byte instruction to zero the accumulator.

(d)

A = 0011 1100
data = 1000 0111
A AND *data* = 0000 0100 = 04H *Answer*

Masking

Occasionally in your application program you have a need to read the status of a single bit. Let's say for example that you have eight level-sensing switches connected to input port 00H as shown in Figure 12–5. When the temperature or pressure exceeds a certain limit, the digital level on that switch goes HIGH. Let's say that temperature *A* and pressure *A* are critical monitoring points in a chemical processing plant. The other temperatures and pressures are just "nice to know" levels, not critical to the operation.

Your job is to flash the warning LEDs if temperature *A* or pressure *A* go HIGH, ignoring the other temperatures and pressures. Unfortunately you cannot read just bit 0 and bit 4. When you execute the instruction IN 00H, the accumulator is loaded with the status of *all eight* switches. If you then want to check if bit 0 or bit 4 is HIGH, you would have to have several CPI *data* compares to check for all of the accumulator values that could arise from varying digital levels on the don't-care inputs.

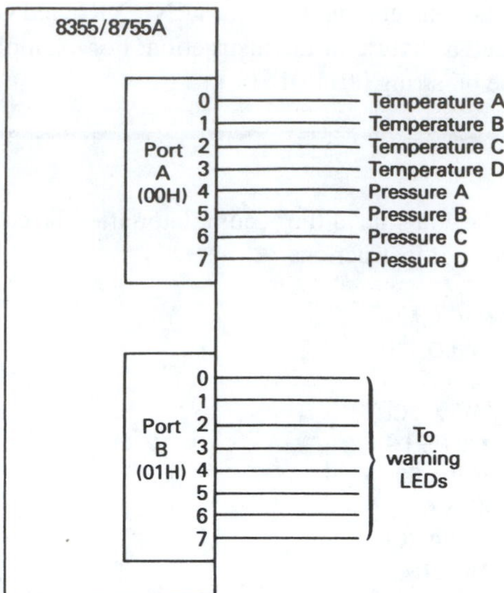

Figure 12–5 Level-sensing switches and warning LEDs connected to the I/O ports of an 8355/8755A.

You want to ignore all of the don't-care inputs. This is where *masking* is very helpful. After you read the input switches into the accumulator (IN 00H), you then AND the accumulator with 0001 0001 (ANI 11H). This is called "masking off" bits 1, 2, 3, 5, 6, and 7. They are forced to 0 while preserving the level of bits 0 and 4. Now if the accumulator is not 0, then bit 0 or bit 4 must be HIGH; flash the LEDs (see Example 12–10).

EXAMPLE 12–10

Assume that the connections shown in Figure 12–5 are used for monitoring temperatures and pressures in a chemical processing plant. Write an assembly language program that will flash the warning LEDs continuously if temperature *A* or pressure *A* goes HIGH, regardless of the other inputs.

Solution:

	Instruction	*Comments*
START:	MVI A,00H	; Send 0s to DDR A to
	OUT 02H	; designate "Input" (Chp 11)
	MVI A,FFH	; Send 1s to DDR B to
	OUT 03H	; designate "Output" (Chp 11)
READSW:	IN 00H	; Read limit switches
	ANI 11H	; Mask OFF all but bits 0 and 4
	CNZ FLASH	; If A is not 0, then CALL FLASH
	JMP READSW	; Else keep reading
		;
		; Subroutine FLASH
FLASH:	MVI A,FFH	; LEDs ON
LOOPA:	OUT 01H	; Outputs to LEDs
	CALL DELAY	; Use Delay from Table 10–6
	CMA	; Complement A
	JMP LOOPA	; Repeat

EXAMPLE 12–11

Modify the program in Example 12–10 to flash the LEDs if *all temperatures* are HIGH, regardless of the pressures.

Solution:

Modified instruction	*Comments*
READSW: IN 00H	; Read limit switches
ANI 0FH	; Mask OFF all pressure switches
CPI 0FH	; Set Z flag if all temperatures HIGH
CZ FLASH	; If zero, call FLASH
JMP READSW	; Else keep reading

Rotates: RLC, RRC, RAL, and RAR

These instructions treat the accumulator like a recirculating shift register. Each bit can be shifted left or right, and the carry flag can be included as part of the shift register. Figure 12–6 illustrates the effect that each instruction has on the accumulator and carry flag.

As the illustration shows, RLC and RRC are 8-bit rotates, and RAL and RAR are 9-bit rotates. The following example will help you understand the effect that the rotates have on the accumulator and carry flag.

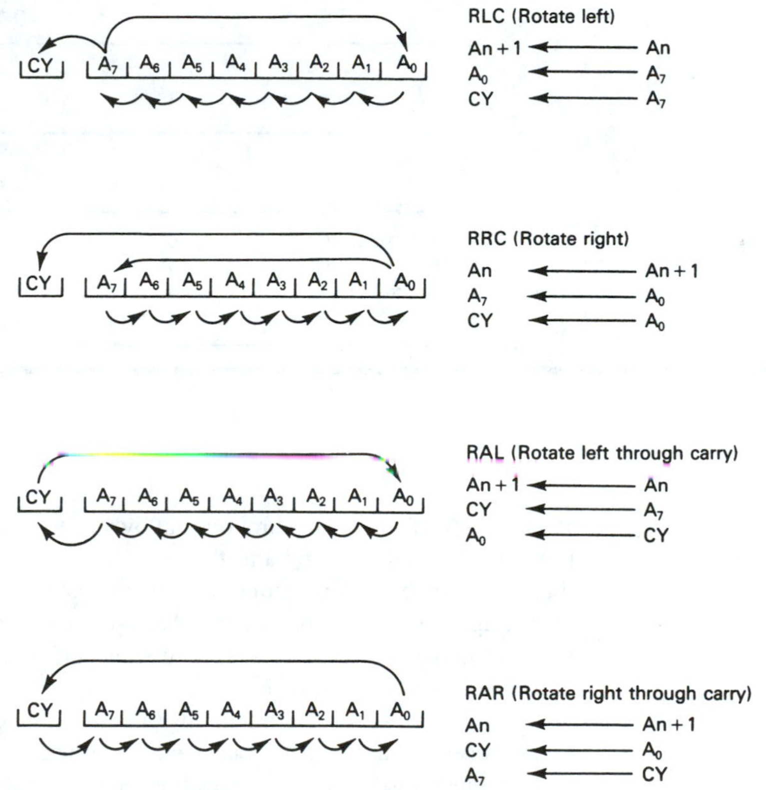

Figure 12–6 The rotate instructions.

EXAMPLE 12–12

Determine the contents of the accumulator and carry flag after each of the following groups of instructions is executed. (Assume CY = 0 initially.)

(a)
```
MVI A,C3H
RLC
```
(b)
```
MVI A,73H
RRC
RRC
RRC
```

(c)
```
MVI A,FFH
RAR
RAR
RAR
```

Solution: (a)

	CY	A_7–A_0 7654 3210	
Before RLC	0	1100 0011	
After RLC	1	1000 0111	*Answer*

(b)

	CY	A_7–A_0 7654 3210	
Before RRC	0	0111 0011	
1st RRC	1	1011 1001	
2nd RRC	1	1101 1100	
3rd RRC	0	0110 1110	*Answer*

(c)

	CY	A_7–A_0 7654 3210	
Before RAR	0	1111 1111	
1st RAR	1	0111 1111	
2nd RAR	1	1011 1111	
3rd RAR	1	1101 1111	*Answer*

12–4 SUBROUTINES AND THE STACK

We have seen that using subroutines allows us to develop a modular approach to solving program applications. Programs that need to execute the same group of instructions more than once can reduce program size by using that group of instructions as a subroutine and CALLing it whenever required. Another use for subroutines is to divide large applications programs into several modules, each module performing a specific function that can be developed and tested on its own.

An example of a structured (modular) program application might be for a microprocessor-based energy management system. The program has to read several area temperature sensors, convert the readings to centigrade degrees, and provide ON/OFF control to the area heating units. The flowchart for that application is shown in Figure 12–7.

Each subroutine can be written and tested by a different programmer and then linked together to form a complete application solution. The machine language and assembly language for the main program is given in Table 12–3.

The Stack

Each subroutine must end with a return or a conditional return. In Table 12–3, upon completion of the READTEMP routine, we want program control to pass to the instruction following the call, which is at address 2003H. This is where the *stack* becomes important. The stack is an area in RAM that is used by certain program instructions to automatically store addresses and data. In the case of the CALL and RETurn instructions, when a CALL is

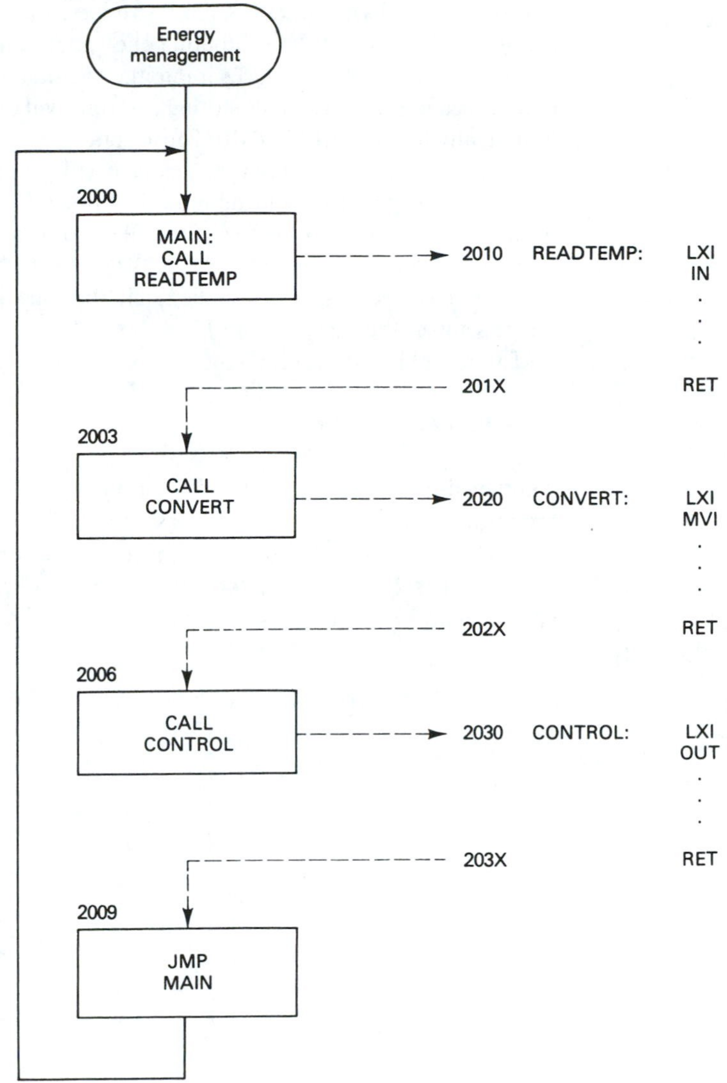

Figure 12–7 A modular approach to solving a microprocessor-based energy management application.

TABLE 12–3

Listing for the Main Program of the Energy Management Application

Address	*Contents*	*Label*	*Instruction*	*Comments*
2000	CD	MAIN:	CALL READTEMP	; Read area temperatures
2001	10			;
2002	20			;
2003	CD		CALL CONVERT	; Convert to centigrade
2004	20			;
2005	20			;
2006	CD		CALL CONTROL	; Control heating units
2007	30			;
2008	20			;
2009	C3		JMP MAIN	; Repeat
200A	00			;
200B	20			;

executed, the 2-byte address of the instruction following the CALL (2003H) is placed on the stack. Then, when the subroutine encounters a RETurn, the microprocessor takes the address on the stack, and places it into the program counter, which forces control to return to that location (2003H). This storing and retrieval of the return address is taken care of automatically by the CALL and RET instructions.

The default location of the stack is usually at the very end of the available RAM area. It can also be specified by loading the 16-bit *stack pointer* (SP) with the address of the top of the stack. If we want the top of the stack to be at 20C0H, then we would start our program with the instruction: LXI SP,20C0H.

When an address is put on the stack, the stack pointer is decremented by 2 to point to the new top of the stack. Figure 12–8 shows the stack operations that take place due to the CALL READTEMP and RET instructions, assuming the stack pointer starts at 20C0H.

PUSH *rp* and POP *rp*

Another use for the stack is for the temporary storage of the registers and flags. As your programs get more complex, you may need to use certain registers for more than one purpose. For instance, the accumulator must be used for the IN and OUT instructions, but it is also used to receive the results of all arithmetic operations. Therefore, if the accumulator has some important data in it, you can temporarily save it out on the stack before executing

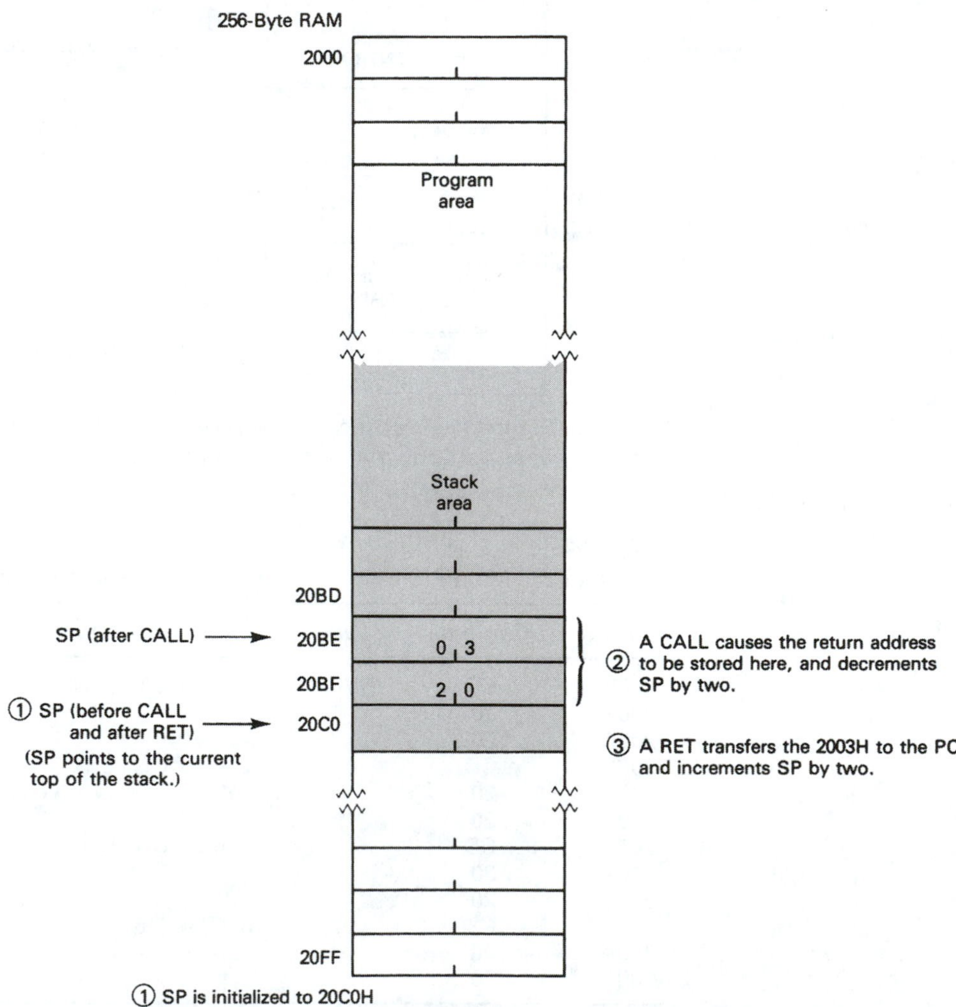

Figure 12–8 Stack operations due to CALL READTEMP and RET.

an IN instruction. After you are done working with the data received from the IN instruction, you can retrieve the accumulator data back from the stack.

The saving and retrieval of registers from the stack is accomplished by using the PUSH *rp* and POP *rp* instructions. The PUSH *rp* instruction saves the register pair *rp* on the top of the stack, and POP *rp* retrieves the register pair. The high-order register of the register pair is saved at stack address SP-1, and the low-order register is saved at address SP-2.

The register pairs are *B–C*, *D–E*, *H–L*, and a new pair called PSW. PSW stands for *Program Status Word,* and is made up of the accumulator byte, followed by the flag byte.

More than one register pair can be saved on the stack by using successive PUSH instructions. However, when doing so, you must remember to POP the registers back off the stack in the reverse order (last on, first off). For each PUSH the SP is decremented by 2, and for each POP the SP is incremented by 2. Table 12–4 and Figure 12–9 illustrate a program that uses the stack for both PUSH/POPs and CALL/RETs. The 2-byte stack pointer shown in Figure 12–9 is initialized at 20C0H and then decremented by 2 for the CALL and each PUSH instruction. (SP always points to the top of the stack.) Then SP is incremented by 2 for each POP and the RET instruction. Upon returning to the INR instruction in the MAIN program, SP is back to its original value, 20C0H. Even though the subroutine "CONTROL" may have altered the registers, they are restored to their original values before returning to MAIN.

TABLE 12–4

Program Using the Stack for the Instructions CALL, RET, PUSH, and POP

	Instruction	*Comments*
MAIN:	LXI SP,20C0H	; Initialize SP to top of stack
	MVI	; ⎫ Program instructions
	ADD	; ⎬ which use all
	DCR	; ⎭ registers.
	etc.	;
	.	
	.	
	CALL CONTROL	; Sub CONTROL needs registers for other
		; purposes.
	INR	; Remainder of MAIN
	MOV	; ⎫ program which needs
	RAR	; ⎬ to use previously
	etc.	; ⎭ defined data registers.
	.	
	.	
	.	
	HLT	; End
CONTROL:	PUSH B	; Save register
	PUSH D	; ⎫ values
	PUSH H	; ⎬ defined
	PUSH PSW	; ⎭ in MAIN program.
	MVI	; Use
	IN	; ⎫ registers for
	RAR	; ⎬ other
	etc.	; ⎭ purposes.
	.	
	.	
	.	
	POP PSW	; Retrieve previous
	POP H	; ⎫ register values (in
	POP D	; ⎬ reverse order)
	POP B	; ⎭ before returning.
	RET	; Return to MAIN (INR instruction)

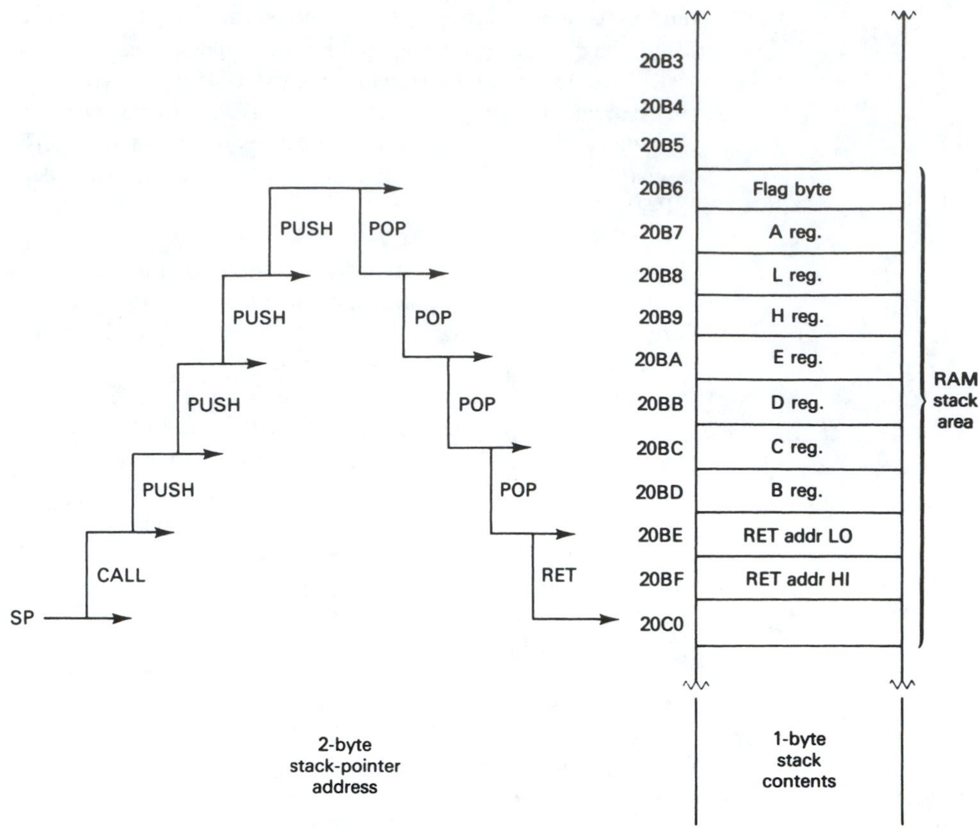

Figure 12–9 Stack operations resulting from the program in Table 12–4.

The values that were temporarily stored on the stack remain there, but you don't have access to them unless you alter the stack pointer, which is now 20C0H. The next time that you use the stack, those values will be overwritten by the new ones.

Note that when loading the stack, the addresses are heading toward your program area. You must be careful to leave room for both. Also, all PUSHes need corresponding POPs, and CALLs need RETurns to restore the stack pointer to its initial value. When you become an advanced programmer, there are some fancy ways to work around that by using instructions such as INX SP, XTHL, and SPHL. Those instructions will also allow you to return to a location other than the instruction following the CALL. It *is* okay to use nested CALLs as long as each subroutine has an equal number of PUSH and POPs and ends with a RETurn (or conditional return).

EXAMPLE 12–13

Draw a diagram of the stack contents upon completion of the following program. (Solution is given in Figure 12–10.)

Address	*Contents*	*Label*	*Instruction*	*Comments*
2000	31	MAIN:	LXI SP,20C0H	; Initialize stack pointer
2001	C0			;
2002	20			;
2003	01		LXI B,44FFH	; B ← 44H, C ← FFH
2004	FF			;
2005	44			;
2006	11		LXI D,AA77H	; D ← AAH, E ← 77H
2007	77			;
2008	AA			;

(continued)

Address	Contents	Label	Instruction	Comments
2009	CD		CALL X1	; CALL subroutine X1
200A	10			;
200B	20			;
200C	76		HLT	; End
200D	00		NOP	; No operation
200E	00		NOP	; No operation
200F	00		NOP	; No operation
2010	C5	X1:	PUSH B	; Save BC on stack
2011	D5		PUSH D	; Save DE on stack
2012	1E		MVI E,00H	; E ← 00H
2013	00			;
2014	DB		IN 00H	; A ← port 00H
2015	00			;
2016	67		MOV H,A	; H ← A
2017	D1		POP D	; Retrieve DE
2018	C1		POP B	; Retrieve BC
2019	C9		RET	; Return to MAIN

Solution:

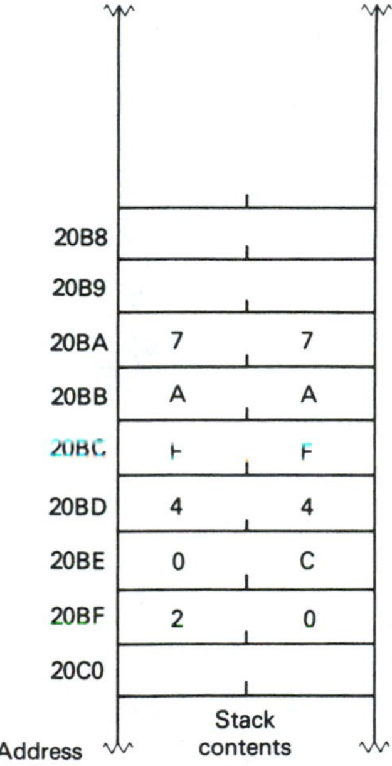

Figure 12–10 Solution to Example 12–13.

12–5 INTERRUPTS

The 8085A microprocessor provides several ways to *interrupt* program execution by means of an external digital signal. This allows you to break into the execution of your main applications program to perform some special operation that is required whenever the interrupt signal is provided. Depending on which interrupt input pin is used on the 8085A, the interrupt is initiated by providing either a rising edge or a HIGH-level input signal or both. Acknowledgment of the interrupt signal can be either enabled or disabled by software instructions. If a particular interrupt pin is enabled and it receives the proper input trigger, program execution will branch to a new address just as if a CALL was made. The five interrupt inputs and resulting branch addresses are given in Table 12–5.

TABLE 12–5
The 8085A Interrupts

Name	*Priority*	*Address[a] branched to when interrupt occurs*	*Type trigger*
TRAP	1	0024H	Rising edge AND high level until sampled
RST 7.5	2	003CH	Rising edge (latched)
RST 6.5	3	0034H	High level until sampled
RST 5.5	4	002CH	High level until sampled
INTR	5	b	High level until sampled

[a]In the case of TRAP and RST 5.5–7.5, the contents of the Program Counter are pushed onto the stack before the branch occurs.

[b]Depends on the instruction that is provided to the 8085A by the 8259 or other circuitry when the interrupt is acknowledged.

For the purposes of this book, we'll use the RST 7.5 pin for interrupts. We will initiate an interrupt by pressing a pushbutton to simulate an interrupt signal from an external device. The circuit in Figure 12–11 could be used to provide the rising-edge trigger signal required at the RST 7.5 interrupt input. (*Note:* Due to switch bounce, this circuit will provide multiple rising edges when the pushbutton is pressed *and* released.)

For example, if the 8085A receives a rising-edge signal on its pin labeled RST 7.5, then the microprocessor will go through the steps of executing a CALL 003CH. Address 003CH is within the system ROM (or EPROM) and must contain the instructions for the action to be taken upon receiving an interrupt.

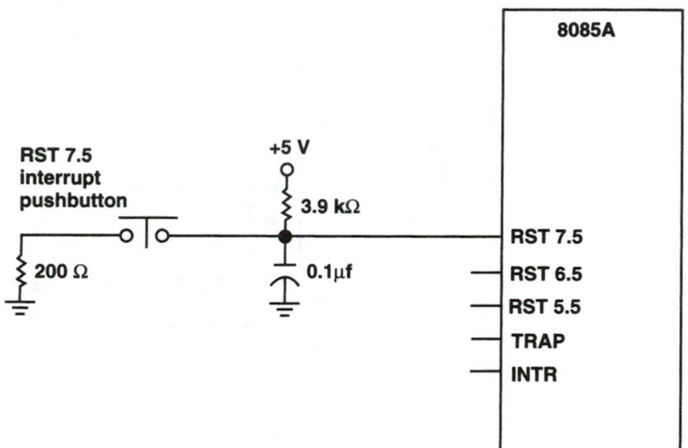

Figure 12–11 A pushbutton circuit used to provide an interrupt signal for the RST 7.5 input.

All five interrupt branch addresses chosen by Intel are packed into a small area (0024H to 003CH), leaving only eight addresses between each one for interrupt instructions. We only have room at each of these branch addresses to put in a JMP *addr* instruction to jump to another location that has more room for us to write our *Interrupt Service Routine* (ISR). The interrupt service routine is a subroutine that performs the action that is required whenever the interrupt condition is initiated.

Before applying an interrupt signal, we have to define which interrupts will be used, and then enable the interrupts to be acknowledged. This requires two new instructions, SIM and EI.

SIM and EI

The SIM (Set Interrupt Mask) instruction is used to tell which interrupts RST 5.5, 6.5, and 7.5 are to be enabled by the execution of the EI (Enable Interrupt) instruction. Before executing SIM, you must load the accumulator with the data for the interrupt mask. The definition of the accumulator bits required before execution of SIM is shown in Figure 12–12.

Bits 6 and 7 are used for outputting serial data to the SOD pin on the microprocessor. Since we are not using SIM for serial I/O, then make bits 6 and 7 equal to 0. Bit 5 is a don't care. Making bit 4 of the accumulator a 1 before executing SIM will reset the RST 7.5 edge-triggered flip-flop.

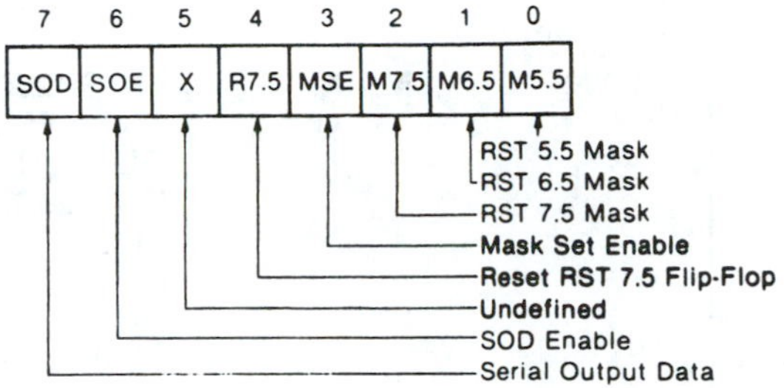

Figure 12–12 Accumulator contents before executing SIM.

To set up the interrupt masks, you first set bit 3 of the accumulator to 1 and put a 0 in bits 2, 1, and 0 to enable interrupts RST 7.5, 6.5, and 5.5, respectively. You then execute SIM.

The interrupts are not actually enabled to be acknowledged by the microprocessor until the EI instruction is executed. When executing the EI instruction, the microprocessor uses the interrupt mask defined by the SIM instruction to determine which interrupts to enable. Each time an interrupt signal is applied and acknowledged by the microprocessor, the interrupts are *disabled* and must be reenabled by another EI instruction if they are to be acknowledged again later in the program.

EXAMPLE 12–14

Write the program instructions to enable the RST 6.5 interrupt.

Solution:

```
MVI A,0DH    ; A ← 0000 1101
SIM          ; Set Interrupt Mask
EI           ; Enable Interrupts
```

EXAMPLE 12–15

Write the program instructions to enable the RST 7.5 and the RST 6.5 interrupts.

Solution:

```
MVI A,09H    ; A ← 0000 1001
SIM          ; Set Interrupt Mask
EI           ; Enable Interrupts
```

If we want to use the interrupt pushbutton circuit of Figure 12–11, we need to first enable the RST 7.5 interrupt within our MAIN program. Then as the MAIN program is executing other tasks, if the pushbutton is pressed, the microprocessor will execute a CALL 003CH.

Location 003CH is within the monitor ROM (operating system). The contents at 003CH must be predetermined before manufacturing the ROM or before programming the EPROM. Since we want to be in an area where we can write an interrupt service routine, let's assume that the instruction JMP 20CEH is at location 003CH. We then jump to 2080H because that is a location in RAM that provides room for us to write

TABLE 12–6
Program Execution Using the RST 7.5 Interrupt Pushbutton to Turn On the LEDs

Address	*Label*	*Instruction*	*Comments*
			; Pressing the RST 7.5 pushbutton
			; executes a "CALL 003CH"
003C	RST7.5:	JMP 20CEH	; Jump to ISR in RAM
.			
.			
.			
2000	MAIN:	LXI SP,20A0H	; Initalize stack pointer
		MVI A,FFH	; Define port 01
		OUT 03H	; DDR B = output
		MVI A,0BH	; RST7.5 to be enabled
		SIM	; Set interrupt mask
		EI	; Enable interrupt
	LOOP:	NOP	; Endless loop ←
		JMP LOOP	; awaiting interrupt
.			
.			
.			
→ 2080	ISR:	MVI A,FFH	; A ← 1111 1111
		OUT 01H	; Turn ON the LEDs
		RET	; Return to LOOP
.			
.			
.			
→ 20CE		JMP 2080H	; Jump to different RAM area

our ISR to service our interrupt request. The program in Table 12–6 lists the program instructions required to turn on the LEDs at port 01 if the RST 7.5 pushbutton is pressed.

The main program contains an endless loop that is executed continuously until the RST 7.5 pushbutton is pressed. The RST 7.5 interrupt causes the microprocessor to execute "CALL 003CH." At 003CH is a jump to 20CEH, the location of our interrupt service routine. The ISR turns on the LEDs and executes a return (RET). RET causes program control to return to the instruction that was just about to be executed before the interrupt occurred. Now we are back in the loop and the LEDs remain latched on. Since the microprocessor has acknowledged an interrupt, all interrupts are disabled. We will remain within the loop even if RST 7.5 is pressed again.

EXAMPLE 12–16

Write a program that will rotate a single LED at port 01 continuously to the left if the RST 7.5 interrupt is pressed.

Solution:

Address	*Label*	*Instruction*	*Comments*
003C	RST7.5:	JMP 20CEH	; Jump to ISR in RAM*
.			
.			
.			
2000	MAIN:	LXI SP,20A0H	; Initialize stack pointer
		MVI A,FFH	; Define port 01
		OUT 03H	; DDR B = output
		MVI A,0BH	; RST7.5 to be enabled
		SIM	; Set interrupt mask
		EI	; Enable interrupt

(continued)

Address	Label	Instruction	Comments
	LOOP:	NOP	; Endless loop
		JMP LOOP	; awaiting interrupt
. . .			
2050	DELAY:	MVI D,8CH	; Delay one-quarter second
	LOOP2:	MVI E,FFH	; (from Table 10–6)
	LOOP1:	DCR E	;
		JNZ LOOP1	;
		DCR D	;
		JNZ LOOP2	;
		RET	;
. . .			
2080	ISR:	MVI A,01H	; Make LSB HIGH
	DISP:	OUT 01H	; Turn ON right-most LED
		CALL DELAY	; Wait one-quarter second
		RLC	; Rotate the ON bit
		JMP DISP	; Do it again
. . .			
20CE		JMP 2080H	; Jump to different RAM area

Explanation:

The MAIN program enables the RST 7.5 interrupt and then waits in an endless loop for the RST 7.5 pushbutton to be pressed. When it is pressed, the microprocessor executes a "CALL 003CH." Location 003CH instructs the microprocessor to jump to location 20CEH. Since there are only 3 bytes available at 20CEH, we need to jump to another area of RAM (2080H), which will have enough room for our Interrupt Service Route (ISR). The ISR turns on the right-most LED and then rotates it slowly to the left, continuously.

*(*Note:* To determine the jump address for the RST 7.5 interrupt, a programmer must look at the ROM listing for the particular microprocessor trainer that is being used. This example assumes that we are using the SDK-85 trainer, which uses 20CEH for the jump address of the RST 7.5 interrupt. Also, the SDK-85 reserves only 3 bytes of memory at that location, so we need to jump immediately to another area of RAM that has enough room for our ISR.)

EXAMPLE 12–17

Write a program that will flash the LEDs at port 01H ON—then—OFF the number of times indicated on the binary input switches at port 00H as soon as the RST 7.5 interrupt is pressed.

Solution:

Address	Label	Instruction	Comments
003C	RST7.5:	JMP 20CEH	; Jump to ISR in RAM
. . .			
2000	MAIN:	LXI SP,20A0H	; Initialize stack pointer
		MVI A,00H	; Define port 00

Solution: (continued)

Address	Label	Instruction	Comments
		OUT 02H	; DDR A = input
		MVI A,FFH	; Define port 01
		OUT 03H	; DDR B = output
		MVI A,0BH	; RST7.5 to be enabled
		SIM	; Set interrupt mask
		EI	; Enable interrupt
	LOOP:	NOP	; Endless loop
		JMP LOOP	; awaiting interrupt
. . .			
2050	DELAY:	MVI D,8CH	; Delay one-quarter second
	LOOP2:	MVI E,FFH	; (from Table 10–6)
	LOOP1:	DCR E	;
		JNZ LOOP1	;
		DCR D	;
		JNZ LOOP2	;
		RET	;
. . .			
2080	ISR:	IN 00H	; Read the binary input switches
		MOV B,A	; Move switch value to Reg B
	FLASH:	MVI A,00H	; Turn LEDs
		OUT 01H	; OFF
		CALL DELAY	; Delay one-quarter second
		CMA	; Turn LEDs
		OUT 01H	; ON
		CALL DELAY	; Delay one-quarter second
		DCR B	; Decrement switch counter
		JNZ FLASH	; Keep looping until B = 0
		RET	; Return
. .			
20CE		JMP 2080H	; Jump to different RAM area

Explanation:

The MAIN program defines both I/O ports, enables the RST 7.5 interrupt, and then waits in an endless loop for the RST 7.5 pushbutton to be pressed. The ISR is placed in RAM starting at 2080H because there is room for only three bytes at 20CEH (again, assuming we are using the SDK-85 trainer). The first statement in the ISR is to read the binary input switches into the accumulator. This value is then moved to the *B* register because the accumulator will later be used to flash the LEDs. The LEDs are flashed OFF then ON once. The *B* register is then decremented and checked for 0. If it is not 0 the LEDs are flashed again and again until *B* is decremented to 0.

SUMMARY OF INSTRUCTIONS

CMP r: (Compare register) Compare the accumulator to register *r*. The Z flag is set if $A = r$. The CY flag is set if $A < r$.

MVI, M,data: (Move to memory immediate) Move the *data* in byte 2 of the instruction to the memory location whose address is pointed to by *H–L*.

MOV r,M: (Move from memory) Move the data in the memory location whose address is pointed to by *H–L* to register *r*.

LDAX rp: (Load accumulator indirect) Load the accumulator with the contents of the memory location whose address is pointed to by register pair *rp*.

STAX rp: (Store accumulator indirect) Store the contents of the accumulator to the memory location whose address is pointed to by register pair *rp*.

ADD r: (Add register) Add the contents of register *r* to the accumulator.

ADI data: (Add immediate) Add the contents of byte 2 of the instruction to the accumulator.

SUB r: (subtract register) Subtract register *r* from the accumulator.

SUI data: (Subtract immediate) Subtract the contents of byte 2 of the instruction from the accumulator.

DAA: (Decimal adjust accumulator) Adjust the 8-bit number in the accumulator to form two BCD digits.

CC addr: (Call if carry set) Transfer control to the program statement whose address, *addr,* is specified in byte 2–byte 3 of the instruction, if the carry flag is set.

RZ: (return if zero) Return to the calling program if the zero flag is set.

ANA r: (AND register) Logically AND the contents of register *r* with the accumulator.

ANI data: (AND immediate) Logically AND the *data* in byte 2 of the instruction with the accumulator.

XRA r: (Ex-OR register) Exclusive-OR the contents of register *r* with the accumulator.

XRI data: (Ex-OR immediate) Exclusive-OR the *data* in byte 2 of the instruction with the accumulator.

ORA r: (OR register) Logically OR the contents of register *r* with the accumulator.

ORI data: (OR immediate) Logically OR the *data* in byte 2 of the instruction with the accumulator.

RLC: (Rotate left) Rotate the 8 bits of the accumulator one position to the left.

RRC: (Rotate right) Rotate the 8 bits of the accumulator one position to the right.

RAL: (Rotate left through carry) Rotate the 9 bits of the accumulator plus carry, one position to the left.

RAR: (Rotate right through carry) Rotate the 9 bits of the accumulator plus carry, one position to the right.

PUSH rp: (Push register pair) Store the 16-bit contents of register pair *rp* on the top of the stack.

POP rp: (Pop register pair) Move the 16-bits in the two top positions of the stack into register pair *rp*.

SIM: (Set interrupt mask) Program the interrupt mask for the RST 7.5, 6.5, and 5.5 hardware interrupts.

EI: (Enable Interrupt) Enable the interrupt system.

SUMMARY

In this chapter we have learned that

1. The move instructions (MOV, MVI) allow the programmer to move data from register to register, register to memory, and memory to register.
2. Indirect addressing gives the programmer access to the data that are stored in the memory location specified in byte2–byte3 of the instruction.

3. The *H–L* register pair is often used as the memory address pointer by indirect address instructions.
4. The I/O ports on the 8355/8755A can be used to read data switches and write to LED indicators using the IN *port* and OUT *port* instructions.
5. The LDA *addr* and STA *addr* instructions are like the IN *port* and OUT *port* except they use indirect addresses *(addr)* to specify the source and destination of the data.
6. The flag byte can be examined to tell whether the following flags are set or reset; sign, zero, auxiliary carry, parity, and carry.
7. Flags are set or reset based on the result of particular program instructions. Conditional branching is often dictated by the status of these flags.
8. Several different addition and subtraction instructions are available. The accumulator is used to receive the results of all additions and subtractions.
9. The DAA instruction is used after performing arithmetic operations on BCD numbers. This ensures that the resulting number is always a valid BCD number.
10. Several logical instructions are available to perform AND, OR, complement, and Exclusive-OR operations.
11. Masking is a technique used to force unwanted bits in the accumulator to the 1-state or the 0-state.
12. Four different rotate instructions are available for shifting bits to the left or right. The value in the carry flag can be included or excluded in the rotation.
13. A subroutine is a group of instructions that is used to form a *module* that can be executed whenever it is called by the main program.
14. The stack is an area set aside in RAM and used by certain program instructions to temporarily store addresses and data.
15. The PUSH and POP instructions are used to store and retrieve register pairs to and from the stack.
16. The *Program Status Work* (PSW) is a 16-bit register-pair that consists of the accumulator plus the flag byte.
17. As the stack is being loaded and unloaded with data and addresses, the *stack pointer* keeps track of the address of the current top of the stack.
18. Interrupt inputs are provided by the 8085A as a means to interrupt normal program execution by inputting an external electrical pulse.
19. An interrupt is like a CALL. When the interrupt signal is acknowledged, program control passes to a subroutine that was previously defined to service that interrupt request.

GLOSSARY

Auxiliary Carry Flag: Tells if there was a carry from bit 3 to bit 4 as a result of the previous arithmetic or logical operation.

Carry Flag: Tells if there was a carry or borrow out of bit 7 due to the previous arithmetic or logical operation.

Conditional Branching: Program branching, or rerouting, due to CALLs, RETurns, and JMPs can be made based on the "condition" of any one of the flag bits.

Flag Byte: The status of the five 8085A flags are stored in the "flag byte."

Immediate Data: The data that is byte 2 of the instructions that end with the letter *I*. The data are used in the execution of the instruction.

Indirect Addressing: A means of addressing a memory location by using the contents of a register pair (usually *H–L*) as a memory pointer to specify the memory location.

Interrupt: A way to use an external digital signal to break into normal program execution to initiate a branch to a special service subroutine.

Interrupt Mask: A data string that is used to determine which interrupts are to be enabled and which are to be disabled.

Interrupt Service Routine (ISR): The subroutine that is executed when the microprocessor receives an interrupt.

Logical Instructions: Microprocessor instructions that deal with the Boolean functions: AND, OR, and exclusive-OR.

Masking: Covering up or nullifying unwanted bits in a data byte.

Parity Flag: Tells if the result of the previous arithmetic or logical operation has even or odd parity.

Program Status Word (PSW): The 16 bits composed of the accumulator plus the flag byte.

Sign Flag: Tells whether the result of the previous arithmetic or logical operation is positive or negative.

Stack: An area set aside in RAM and used by certain instructions for the temporary storage of data and addresses.

Stack Pointer: A 16-bit register containing an address that is used to point to the top of the stack.

Structured Programming: A programming method that emphasizes breaking the application into several subroutine modules, each performing a specific function.

Top of Stack: The address of the last entry on the stack.

PROBLEMS

12–1. Rewrite the following instructions using indirect addressing to perform the same function.

```
LDA 20B0H
STA 20B1H
```

12–2. Write the instructions that use indirect addressing to load memory location 20B0H with the contents of register *C*.

12–3. Write the instructions that use indirect addressing to move the contents of memory location 20C5H to location 20C6H.

12–4. Rewrite the solution to Problem 12–3 using the LDA and STA instructions.

12–5. Write a program that transfers the contents of memory address 20XXH to the output LEDs at port 01. (Use the I/O interface circuit in Figure 12–2.) The low-order address, XX, will be read in from the input switches.

12–6. Why is the 74LS240 buffer required in Figure 12–2? Why use an *inverting* buffer?

12–7. Change one statement in the solution to Example 12–2 so that LEDs will display memory contents 20C0H to 20DFH instead of 20C0H to 20CFH.

12–8. Most indirect-addressing instructions use the *H–L* register pair as a pointer. What register pairs do the LDAX and STAX instructions use?

12–9. Expand the solution to Example 12–3 so that it transfers the memory contents at 20A0H through 20AFH to locations 20B0H through 20BFH.

12–10. Determine which flag bits (S, Z, AC, P, C) are set if the flag byte equals:

(a) 44H

(b) 95H

12-11. Use the Instruction Set Reference Encyclopedia in the Appendix to determine which flags are affected by the following instructions.

(a)	INR *r*	**(b)**	ANA *r*
(c)	DCX *rp*	**(d)**	CMA
(e)	STC	**(f)**	RLC
(g)	MVI *r*	**(h)**	CMP *r*

12–12. Determine the value of the registers and flag byte (F) upon completion of the following instructions. (Assume all flag bits are initially reset.)

(a)
```
MVI A,4FH
INR A
```
A = _____
F = _____

(b)
```
LLXI B,027EH
INR C
INX B
```
B = _____
C = _____
F = _____

(c)
```
LXI B,05FFH
INR C
MVI C,2AH
```
B = _____
C = _____
F = _____

12–13. Repeat Problem 12–12 for the following instructions:

(a)
```
LXI B,253AH
MVIA,52H
ADD B
MOV B,A
ORA C
```
A = _____
B = _____
F = _____

(b)
```
MVI A,4FH
ADI 1AH
```
A = _____
F = _____

(c)
```
MVI A,2FH
XRI A2H
```
A = _____
F = _____

(d)
```
MVI A,29H
ADI 38H
DAA
```
A = _____
F = _____

(e)
```
XRA A
LXI B, 7267H
ADD B
ADD C
DAA
```
A = _____
F = _____

12–14. The solution to Example 12–8 is an example of "structured" programming. Why is it important to write large programs using this technique?

12–15. Assume that you are reading temperatures and pressures at input port 00, as shown in Figure 12–5. You are only interested in the pressures. What instruction could be used to mask off (ignore) all temperatures?

12–16. Modify the solution to Example 12–10 to flash the LEDs only if all temperatures are HIGH and all pressures are LOW by using an XRI *data* instruction.

12–17. Determine the contents of the accumulator and carry flag after the completion of the following instructions. (Assume CY = 0 initially.)

(a)
```
MVI A,0FH
RAR
RAR
```
CY = ____
A = ____

(b)
```
MVI A,FFH
INR A
RAL
```
CY = ____
A = ____

(c)
```
MVI A,AFH
RRC
RRC
```
CY = ____
A = ____

12–18. Write a program that will continuously rotate a single ON LED at port 01, left to right, with a one-quarter second stop at each position.

12–19. Write a program that will continuously bounce a single ON LED left to right ($2^7 - 2^6 \ldots 2^0$), then right to left, then left to right, etc., at one-quarter second for each position. (*Hint:* Use a conditional branch on carry.)

12–20. How is the stack used when making a subroutine CALL and RETurn?

12–21. The stack pointer should be initially set at the (beginning, middle, or end) of the available RAM area?

12–22. What does the instruction PUSH PSW do? What happens to the stack pointer due to that instruction?

12–23. When using multiple PUSH instructions in a subroutine, the registers must be POPped in the *same* order. (True or false?)

12–24. What would happen if a RETurn instruction is encountered before all registers are POPped off the stack?

12–25. The one-quarter second DELAY subroutine that we have been using uses the *D* and *E* registers for counters. Modify the subroutine so that it is "transparent," which means that it has no effect on the other programs that also need to use the *D* and *E* registers.

12–26. Draw a diagram of the stack contents upon completion of the following program.

Address	*Label*	*Instruction*
2000	MAIN:	LXI SP,20B0H
2003		LXI B,3344H
2006		PUSH B
2007		CALL SUB1
200A		POP B
200B		HLT
200C	SUB1:	PUSH B
200D		NOP
200E		POP B
200F		RET

12–27. How is an 8085A interrupt like a subroutine? How is it different?

12–28. When an interrupt is made, program control first branches to (ROM or RAM)?

12–29. Write the program instructions to enable the RST 6.5 and RST 5.5 interrupts.

12–30. Why is a second EI instruction sometimes required at the end of the interrupt service routine?

12–31. Write a program that will display the status of the temperature and pressure limit switches of Figure 12–5 each time the RST 7.5 interrupt pushbutton is pressed.

12–32. Write a program that will display the number of times that the RST 7.5 interrupt pushbutton is pressed. (Each time the pushbutton is pressed, increase the count on the LEDs.) Assume the LEDs are connected at port 01.

SCHEMATIC INTERPRETATION PROBLEMS

12–33. Describe the operation of U6 in the 4096/4196 schematic. Use the names of the input/output labels provided on the IC for your discussion.

12–34. Refer to sheet 2 of the 4096/4196 schematic. Describe the sequence of operations that must take place to load the 8-bit data string labeled IA0-IA7 and the 8-bit data string labeled ID0-ID7. Include reference to U30, U32, U23, U13 : A, U1 : F, and U33.

Interfacing and Applications

OBJECTIVES

Upon completion of this chapter you should be able to:

- Interface a digital-to-analog converter (DAC) to the I/O ports of an 8085A microprocessor system.
- Write software to use a DAC as a programmable voltage source and waveform generator.
- Interface an analog-to-digital converter (ADC) to the I/O ports of an 8085A microprocessor system.
- Write the software for handshaking between an ADC and the microprocessor.
- Describe the operation of the hardware and software required for a digital thermometer application.
- Understand how a look-up table is used for data translation.
- Describe the operation of the hardware and software required to drive a multiplexed display.
- Understand the software required to read a matrix keyboard.
- Describe the construction and operation of a stepper motor.
- Write the software to drive a stepper motor at a particular speed and number of revolutions.

INTRODUCTION

So far we've covered 90% of the instructions that you'll ever need to solve most 8085A-based system applications. Up to now our I/O has dealt with reading eight input switches and writing to eight output LEDs.

Practical I/O interfacing involves much more than just switches and LEDs. Take, for example, a microprocessor-based microwave oven. It has to read and interpret a matrix keypad, control a high-wattage microwave element, sense analog temperatures, and drive a multidigit display. Sounds complex doesn't it? Well, we have all of the fundamental building blocks involved in such an application: multiplexing, demultiplexing, A/D conversion, load buffering, and D/A conversion.

The complete solution to such a comprehensive application as a microwave oven is beyond the scope of this book. However, the hardware interfacing and software driver routines for each component within the system *is* within our grasp. In this chapter we'll draw on the knowledge gained in previous chapters so that we can develop practical interfaces to the analog world, as well as expand on our digital I/O capability.

13–1 INTERFACING TO A DIGITAL-TO-ANALOG CONVERTER

For microprocessor D/A applications, we can use the DAC0808 IC discussed in Chapter 7. (See Figure 7–8.) The DAC0808 is an 8-bit DAC that produces a current output that is proportional to the binary value applied at its inputs. (See Equation 7–1.) Using a microprocessor to drive the binary inputs gives us a tremendous capability for producing *programmable current and voltage sources* as well as creating *specialized waveforms.* Interfacing the DAC to the microprocessor system is done simply by connecting to any of the 8-bit output ports provided by one of the microprocessor support chips.

Figure 13–1 shows the connections and calculated V_{out} that will result from outputting a C5H (1100 0101) to the DAC0808. See Chapter 7 if you need a review of the calculations or circuit theory.

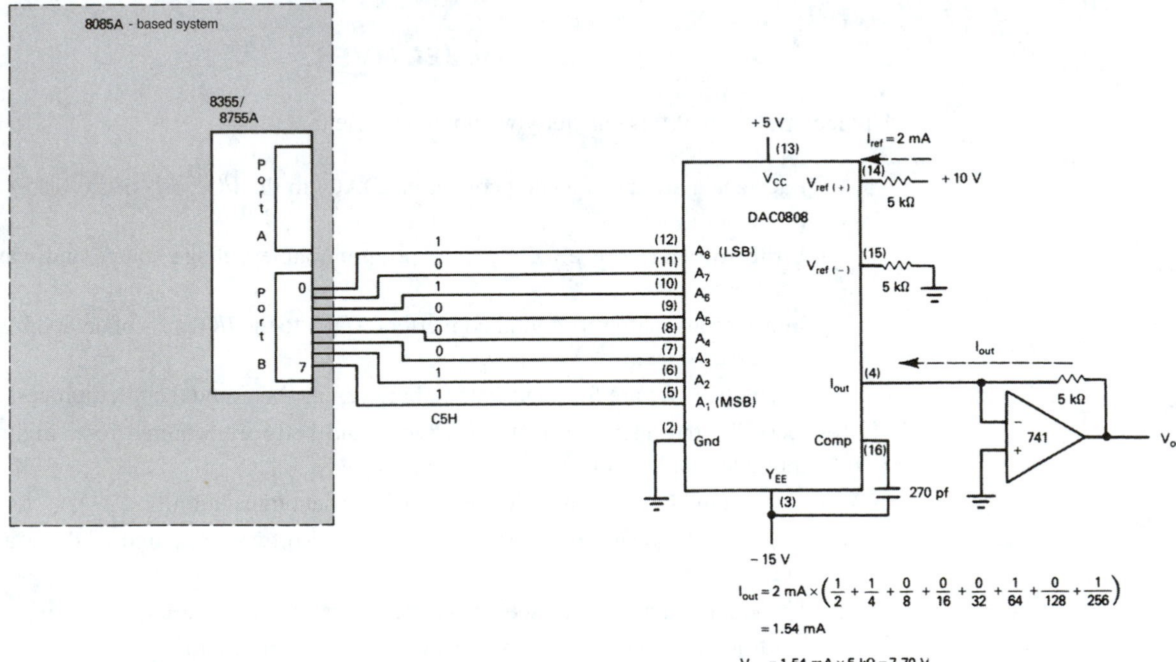

Figure 13–1 Interfacing the DAC0808 to a microprocessor system.

EXAMPLE 13–1

Write an assembly language program for the DAC circuit of Figure 13–1 to produce an output voltage of 4.50 V.

Solution:

V_{out} is proportional to the binary input. The largest binary input, 1111 1111, produces an output voltage of 9.96 V. The following algebraic ratio can be set up to solve for the required binary input (Req B_{in}):

$$\frac{\text{Max } B_{in}}{\text{Max } V_{out}} = \frac{\text{Req } B_{in}}{\text{Req } V_{out}}$$

$$\frac{1111\ 1111}{9.96} = \frac{\text{Req } B_{in}}{4.50}$$

$$255 \times 4.50 = 9.96 \times \text{Req } B_{in}$$

$$\text{Req } B_{in} = 115_{10}\ (73\text{H or } 0111\ 0011_2)$$

Assembly language program:

Label	*Instruction*	*Comments*
Start:	MVI A,FFH	; Program DDR B for
	OUT 03H	; Port B = Output
	MVI A,73H	; Output 0111 0011
	OUT 01H	; to Port B
	HLT	; End

13–2 USING A DAC FOR WAVEFORM GENERATION

Besides using the DAC as a programmable voltage or current source, it can also be used to produce specialized waveforms. You can create a *square wave* by outputting a high voltage, then a low voltage, repeatedly. The frequency and duty cycle can be set by inserting appropriate delays within the loop.

A 256-step *sawtooth wave* can be created by counting 00H to FFH repeatedly. Its frequency can be adjusted by adding a small delay to each step.

More exotic repetitive waveforms can also be created. For example, to create a *sine wave* with, let's say, 24 steps, we would have to determine the binary value that will yield the correct output voltage at each of the 15° increments along the sine wave. These values would then be stored in a data table (a look-up table) and then used as input to the DAC to reproduce the sine wave. The resolution, or smoothness, of the sine wave will improve as you increase the number of data points in the table. The following examples illustrate the capability of the DAC as a waveform generator.

EXAMPLE 13–2

The following program is used to create a square wave at V_{out} in Figure 13–1. Sketch the resultant waveform and label the voltage levels and times. (Assume that a 6.144-MHz crystal is used. The solution is given in Figure 13–2.)

Label	*Instruction*	*Comments*	*T states*
	MVI A,FFH	; Program	7
	OUT 03H	; DDR B = Output	10
LOOP:	MVI A,80H	; Output 1000 0000	7
	OUT 01H	; to Port B DAC	10
	NOP	; No operation*	4
	MVI A,00H	; Output 0000 0000	7
	OUT 01H	; to Port B DAC	10
	NOP	; No operation*	4
	JMP LOOP	; Repeat	10

**Note:* The NOPs were added to increase the pulse widths slightly. The HIGH pulse width could be increased to 10.1 μs (50% duty cycle) by inserting a 10 *T* state "dummy" instruction like LXI after the first NOP.

Solution:

Voltage levels:

$$V_{out} \text{ for 80H:}$$

$$V_{out} = 10\text{ V} \times (\tfrac{1}{2}) = 5.0\text{ V}$$

$$V_{out} \text{ for 00H:}$$

$$V_{out} = 10\text{ V} \times (0) = 0.0\text{ V}$$

Time periods:

$$1\ T\text{ state} = 2 \times \text{crystal period}$$

$$= 2 \times 1/6.144\text{ MHz}$$

$$= 0.326\ \mu\text{s}$$

V_{out} = +5 V from the end of the first OUT 01H to the end of the second OUT 01H.

$$T\text{ states} = 4 + 7 + 10 = 21$$

$$T_{HIGH} = 21 \times 0.326\ \mu\text{s} = 6.84\ \mu\text{s}$$

V_{out} = 0 V from the end of the second OUT 01H to the end of the first OUT 01H.

$$T\text{ states} = 4 + 10 + 7 + 10 = 31$$

$$T_{LOW} = 31 \times 0.326\ \mu\text{s} = 10.1\ \mu\text{s}$$

$$\text{Frequency} = 1/(6.84\ \mu\text{s} + 10.1\ \mu\text{s}) = 59\text{ kHz}$$

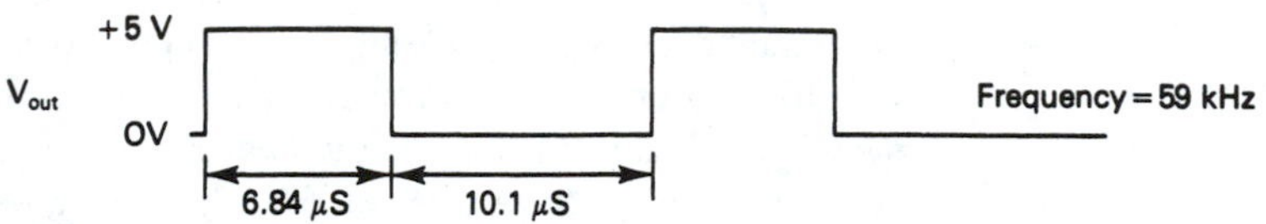

Figure 13–2 Solution to Example 13–2.

EXAMPLE 13–3

The following program is used to create a sawtooth waveform at V_{out} in Figure 13–1. Sketch the resultant waveform and label the voltage levels and times. (Assume that a 6.144-MHz crystal is used. The solution is given in Figure 13–3.)

Label	*Instructions*	*Comments*	*T states*
	MVI A, FFH	; Program Port B	7
	OUT 03H	; DDR B = Output	10
	MVI A,00H	; Start V_{out} at zero	7
LOOP:	OUT 01H	; Output to DAC	10
	INR A	; Increase one step	4
	JMP LOOP	; Next step	10

Solution:

Voltage level ranges from 0 V (0000 0000) to 9.96 V (1111 1111).

$$1\ T\text{ state} = 2 \times 1/6.144\text{ MHz} = 0.326\ \mu\text{s}$$

$$\text{Each step} = 4 + 10 + 10 = 24\ T\text{ states}$$

$$\text{Time per step} = 24\ T \text{ states} \times 0.326\ \mu\text{s} = 7.82\ \mu\text{s}$$

$$\text{Total time} = 256 \text{ steps} \times 7.82\ \mu\text{s} = 2.00 \text{ ms}$$

$$\text{Frequency} = 1/(2 \text{ ms}) = 500 \text{ Hz}$$

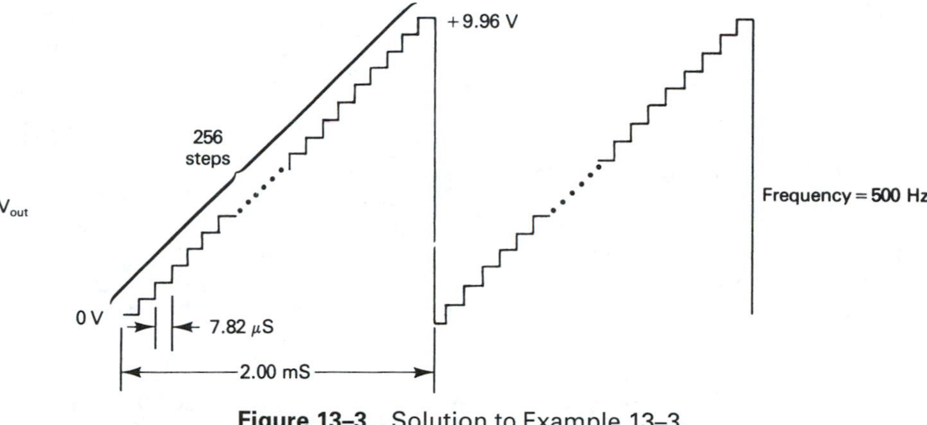

Figure 13–3 Solution to Example 13–3.

13–3 INTERFACING TO AN ANALOG-TO-DIGITAL CONVERTER

Analog quantities such as temperature, pressure, and strain need to be converted to an equivalent digital value if they are to be interpreted by a microprocessor system. To perform an analog-to-digital conversion, we can use the ADC0801 circuit that we studied back in Chapter 7. (See Section 7–10.) We can modify that circuit to interface to our 8085A system by connecting the 8-bit digital output of the ADC0801 to an input port on our ROM IC and connect start-conversion/end-conversion control signals to a second I/O port on our ROM. Figure 13–4 shows the connections that we'll use to interface the ADC0801 to our 8085A-based microprocessor system.

Port *A* will be programmed as an input port to receive the 8 bits of data from the ADC. The conversion process is initiated when bit 0 of port *B* outputs the LOW pulse for the start-conversion ($\overline{SC}$) signal. The microprocessor will now continuously read bit 7 of

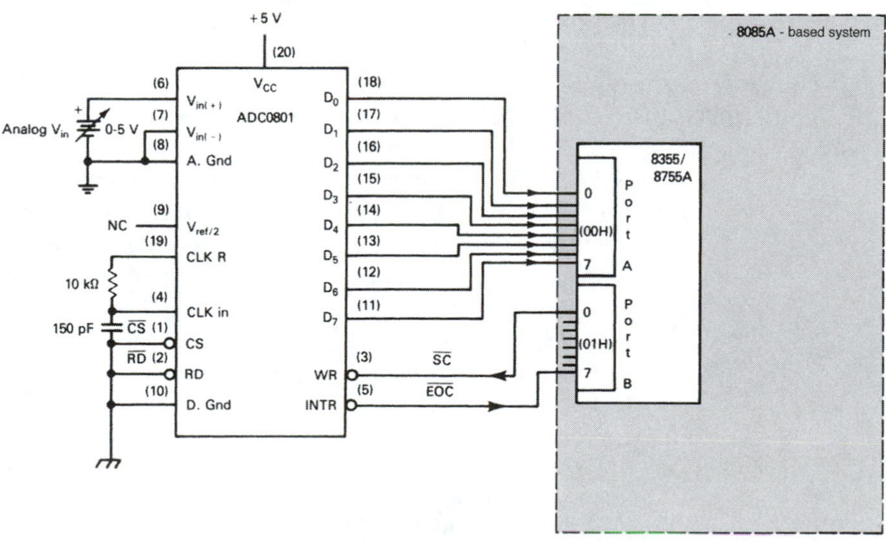

Figure 13–4 Interfacing an ADC0801 to an 8085A microprocessor system.

TABLE 13–1
Software Handshaking Required to Perform an A/D Conversion in Figure 13–4

Label	*Instruction*	*Comments*
INIT:	MVI A,00H	; Program Port A
	OUT 02H	; DDR A = Input
	MVI A,01H	; Program Port B
	OUT 03H	; Bit 7 = Input, Bit 0 = Output
SC:	MVI A,00H	; Output a LOW on Bit 0
	OUT 01H	; for start conversion ($\overline{SC}$)
	MVI A,01H	; Output a HIGH on Bit 0
	OUT 01H	; to return line HIGH
WAIT:	IN 01H	; Read Port B
	ANI 80H	; Mask off all but Bit 7
	JNZ WAIT	; Reread until $\overline{EOC}$ goes LOW
DONE:	IN 00H	; Read digital result into accumulator
	HLT	; End

port *B*, waiting for the end-of-conversion ($\overline{EOC}$) line to drop LOW. When $\overline{EOC}$ does drop LOW, the conversion is complete, and the microprocessor reads the 8 bits of data on port *A*. The program listed in Table 13–1 shows the "handshaking" that takes place between the I/O ports and the ADC when making an A/D conversion.

The first four lines of the program define the data direction registers for port *A* and port *B* (DDR A and DDR B). Port *B* is to be used for both input (bit 7) *and* output (bit 0).

The start-conversion signal, $\overline{SC}$, is generated by the instructions following the "SC:" label. The A/D conversion starts when pin 3 ($\overline{WR}$) on the ADC receives a LOW-to-HIGH signal. That LOW to HIGH is made by outputting a LOW on bit 0, followed by a HIGH.

After starting the conversion, we have to wait several clock periods for the conversion to be complete (66 to 73 clock periods according to manufacturer specifications). There are several ways to do this. One way is to insert a time delay that is longer than 73 clock periods. Another way is to connect the $\overline{EOC}$ line into an interrupt pin on the microprocessor and wait for an interrupt. The method used in this program uses the three instructions following the "WAIT:" label. In this method, we keep reading port *B* and check to see if bit 7 is LOW. A LOW on bit 7 tells us that $\overline{EOC}$ is LOW and the digital data are available at D_0–D_7.

The last two instructions read the 8-bit result into the accumulator (IN 00H) and halt (HLT).

13–4 DESIGNING A DIGITAL THERMOMETER USING AN ADC

A good way to illustrate the versatility of a microprocessor in A/D applications is by working through the design of a *digital thermometer.* In Chapter 7 we discussed the theory of interfacing a *linear temperature sensor* to an ADC. (See Figure 7–21.) By choosing an appropriate reference voltage, the binary output of the ADC will increase in linear steps numerically equal to the temperature in degrees centigrade. (See Table 7–3.)

In order to display the temperature as a two-digit decimal number, we need to convert the 8-bit binary output to two BCD digits and output them to a pair of seven-segment LED displays. Figure 13–5 shows the complete circuit required for a two-digit microprocessor-based thermometer.

The thermometer circuit is centered around the three-chip minimum-component 8085A microprocessor system introduced in Figure 11–10. We'll use the two I/O ports on the 8355/8755A for the A/D handshaking as we did before.

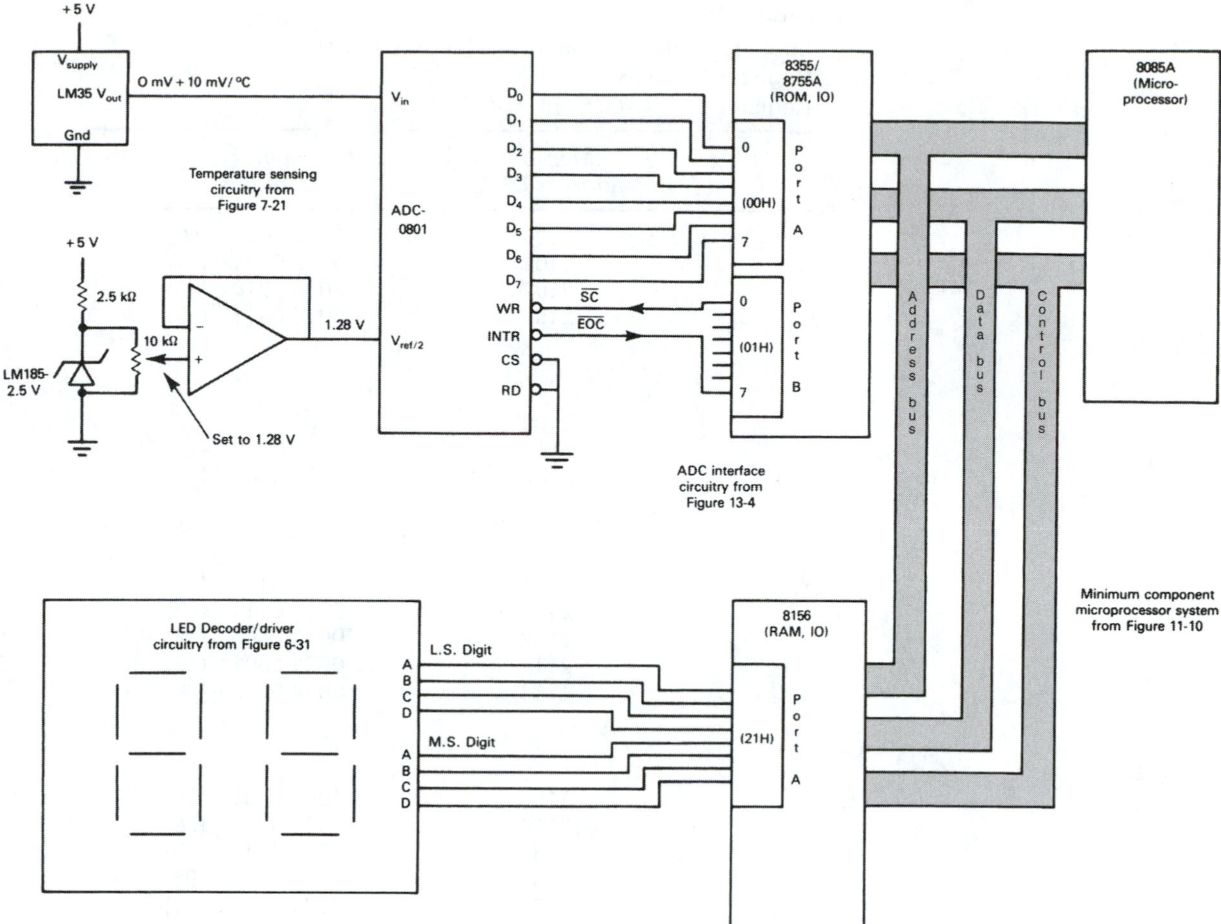

Figure 13–5 An 8085A-based centigrade thermometer.

The 8 bit port A on the 8156 will be used to drive the two-digit display. The decoder/driver circuitry accepts BCD data, converts them to seven-segment code, and turns on the appropriate segments. Port *A* (21H) will output two digits worth of BCD data. The least significant BCD data are output on bits 0–3 and the most significant BCD data are output on bits 4–7.

The binary data that the microprocessor receives from the ADC are not in a form that can be used by the two-digit display. For example, if the temperature is 20°, the ADC will output 0001 0100_2. This has to be converted to 0010 0000_{BCD} before being output to the display.

This conversion could be done with MSI ICs specifically designed for binary-to-BCD conversion. Another way (the method that we'll use) is to write a software subroutine to perform the conversion. A third way is to use a *look-up table.*

To use a look-up table for this application, we would need 100 additional RAM locations to hold the table values. A second 8156 could be placed in the foldback area, 2100H to 21FFH, to provide the required memory space. Indirect memory addressing would be used to access the BCD data to be output. The *H* register would contain 21H to point to the HIGH-order address in the look-up table. The *L* register would hold the binary output of the ADC.

For example, if the temperature is 20°, the ADC will output 0001 0100_2 (14H), making the *H–L* register pair 2114H. The contents of RAM location 2114H is 0010 0000_{BCD}. (See Table 13–2.) This value can be output directly to the LED display. In a sense, we have

TABLE 13–2
Look-up Table for Binary-to-BCD Conversion
Register H = 21H
Register L = D_7–D_0 from ADC

Memory address (H–L)	Memory contents
2100	0000 0000 (0)
2101	0000 0001 (1)
2102	0000 0010 (2)
2103	0000 0011 (3)
.	.
.	.
.	.
2109	0000 1001 (9)
210A	0001 0000 (10)
210B	0001 0001 (11)
.	.
.	.
.	.
2110	0001 0110 (16)
2111	0001 0111 (17)
2112	0001 1000 (18)
2113	0001 1001 (19)
2114	0010 0000 (20)
.	.
.	.
.	.
215F	1001 0101 (95)
2160	1001 0110 (96)
2161	1001 0111 (97)
2162	1001 1000 (98)
2163	1001 1001 (99)
D_7–D_0 from ADC ↑	Equivalent BCD (63H = 99_{10})

"looked up" the BCD value to be output by using the binary value placed in the *L* register as a pointer to the RAM table.

The table look-up technique is very useful for nonlinear and other complex data conversion. Its disadvantage is that it uses a lot of valuable memory to hold the table entries.

In our temperature application, since the results are linear and have a one-to-one correlation, a simple *software algorithm* can be written to do the conversion. A DAA instruction by itself is not enough to convert binary to BCD. (It would have no effect on the binary string 0001 0100, for instance.) What we could do, however, is count from 0 up to the numeric value that was read from the ADC. We will execute a DAA instruction for each count, keeping the result a valid BCD number for each increment. The end result will be the two-digit BCD equivalent of the binary ADC output. Table 13–3 lists the complete program solution for the thermometer application.

This program is another example of a *structured, modular* program. The MAIN program does nothing but CALL the three subroutines, which do all of the work. The first subroutine, INIT, programs the data direction registers of the 8355/8755A, and the command register of the 8156, for I/O.

The second subroutine, ADC, is the same as the A/D program listed in Table 13–1. At the completion of this subroutine, the accumulator contains the binary equivalent of the temperature.

The last subroutine, CONVERT, converts the binary value into an equivalent two-digit BCD result. The BCD answer is then sent to output port 21H, which drives the seven-segment LED display circuitry.

TABLE 13–3
Program Listing for the 0–99°C Thermometer Circuit of Figure 13–5

Label	*Instruction*	*Comments*
MAIN:	LXI SP,20C0H	; Initialize stack pointer
	CALL INIT	; Initialize I/O ports
LOOP:	CALL ADC	; Perform A/D conversion
	CALL CONVRT	; Convert bin-to-BCD and display
	JMP LOOP	; Repeat continuously
		;
		;
		;
INIT:	MVI A,00H	; Program Port A
	OUT 02H	; DDR A = Input
	MVI A,01H	; Program Port B
	OUT 03H	; Bit 7 = Input, Bit 0 = Output
	MVI A,01H	; Program Port A of the 8156
	OUT 20H	; as an Output port
	RET	; Return to MAIN
		;
		;
		;
ADC:	MVI A,00H	; Output a
	OUT 01H	; LOW-then-HIGH
	MVI A,01H	; on Bit 0 to
	OUT 01H	; Start conversion ($\overline{SC}$)
WAIT:	IN 01H	; Keep rereading Bit 7 of
	ANI 80H	; Port B ($\overline{EOC}$) until
	JNZ WAIT	; it goes LOW
	IN 00H	; Read digital result into accumulator
	RET	; Return to MAIN
		;
		;
		;
CONVRT:	MOV D,A	; Move binary value to D
	XRA A	; Zero out A
COUNT:	INR A	; Count up
	DAA	; in BCD
	DCR D	; Decrement binary value
	JNZ COUNT	; Keep counting until D = 0
	OUT 21H	; Output BCD to seven segment display
	RET	; Return to MAIN

13–5 *DRIVING A MULTIPLEXED DISPLAY*

Multidigit LED or LCD displays are commonly used in microprocessor systems. The theory behind decoding and driving seven-segment displays was covered in Chapter 6.

To drive each digit of a six-digit display using separate, dedicated drivers would require six 8-bit I/O ports. Instead, a *multiplexing* scheme is usually used. Using the multiplexing technique, up to eight digits can be driven by using only two output ports. One output port is used to select which *digit* is to be active, while the other port is used to drive the appropriate *segments* within the selected digit. Figure 13–6 shows how the two I/O ports of an 8355/8755A can be used to drive a six-digit multiplexed display.

The displays used in Figure 13–6 are common-cathode LEDs. To enable a digit to work, the connection labeled COM must be grounded. The individual segments are then illuminated by supplying +5 V, via a 150-Ω limiting resistor, to the appropriate segment.

It takes about 10 mA to illuminate a single segment. If all segments in one digit are on, as with the number 8, the current in the COM line will be 70 mA. The output ports of the 8355/8755A can only sink 2 mA. This is why we need the *PNP* transistors set up as current buffers. When port *A* outputs a 0 on bit 0, the first *PNP* turns on, shorting the emitter

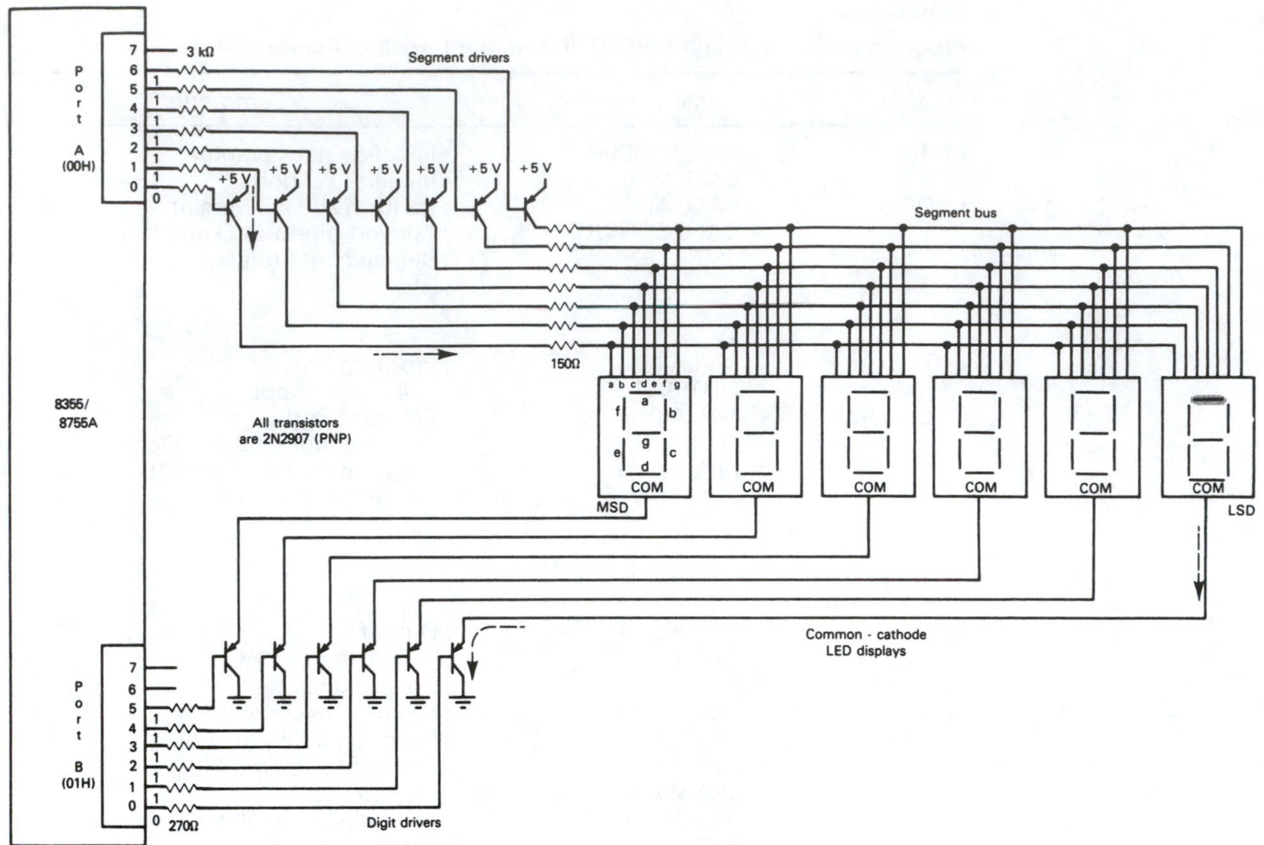

Figure 13–6 A multiplexed six-digit display with the *a* segment of the LSD illuminated.

to collector. This allows current to flow from the +5-V supply through the 150-Ω limiting resistor, to the *a* segments. None of the *a* segments will illuminate unless one of the digits' COM lines is brought LOW. To enable the LSD, port *B* will output a 0 on bit 0, which shorts the emitter to collector of that transistor. That provides a path for current to flow from the COM on the LSD, to ground. Figure 13–7 shows the bit assignments for each of the segments and each of the digits.

Note that enabling both the segment and the digit requires an active-LOW signal. To drive all six digits, we have to *scan* the entire display repeatedly with the appropriate numbers to be displayed. For example, to display the number 123456, we need to turn on the segments for the number 1 (*b* and *c*), then turn on the MSD. We then turn off all digits, turn on the segments for the number 2 (*a*, *b*, *g*, *e*, and *d*) and turn on the next digit. We then turn off all digits, turn on the segments for the number 3, and turn on the next digit. This process repeats until all six digits have been flashed on once. At that point, the MSD is cycled back

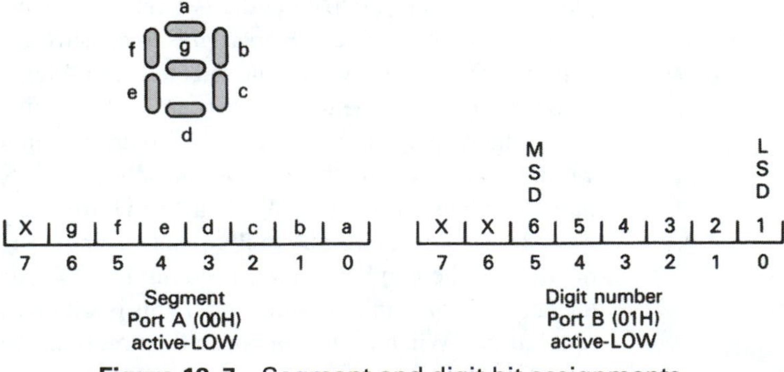

Figure 13–7 Segment and digit bit assignments.

on, followed by each of the next digits. By repeating this cycle over and over again, the number 123456 appears to be on all of the time. The following example shows the instructions that are required to carry out this task.

EXAMPLE 13–4

Write a program to display the number 123456 in the multiplexed display of Figure 13–6.

Solution:

Label	*Instructions*	*Comments*
MAIN:	LXI SP,20C0H	; Initialize stack pointer
	MVI A,FFH	; Program Ports A and B
	OUT 02H	; DDR A = Output
	OUT 03H	; DDR B = Output
LOOP:	MVI B,F9H	; B ← #1 segments
	MVI C,DFH	; C ← 6th digit (MSD)
	CALL DISP	; Display #1
	MVI B,A4H	; B ← #2 segments
	MVI C,EFH	; C ← 5th digit
	CALL DISP	; Display #2
	MVI B,B0H	; B ← #3 segments
	MVI C,F7H	; C ← 4th digit
	CALL DISP	; Display #3
	MVI B,99H	; B ← #4 segments
	MVI C,FBH	; C ← 3rd digit
	CALL DISP	; Display #4
	MVI B,92H	; B ← #5 segments
	MVI C,FDH	; C ← 2nd digit
	CALL DISP	; Display #5
	MVI B,82H	; B ← #6 segments
	MVI C,FEH	; C ← 1st digit (LSD)
	CALL DISP	; Display #6
	JMP LOOP	; Repeat
DISP:	MVI A,FFH	; Turn off
	OUT 01H	; all digits
	MOV A,D	; Move segment data to A
	OUT 00H	; Output A to segment bus
	MOV A,C	; Move digit data to A
	OUT 01H	; Turn on selected digit
	RET	; Return for next digit

Explanation:

Figures 13–8 and 13–9 give the segment assignments and digit position assignments that are used in our program. The LOOP instructions load the *B* register with the data required for the active-LOW segment port, and load the *C* register with the data required for the active-LOW digit port. With *B* and *C* loaded with the data for the MSD, the subroutine DISP is called.

Digit	x	g	f	e	d	c	b	a	Hex code
6	1	0	0	0	0	0	1	0	82H
5	1	0	0	1	0	0	1	0	92H
4	1	0	0	1	1	0	0	1	99H
3	1	0	1	1	0	0	0	0	B0H
2	1	0	1	0	0	1	0	0	A4H
1	1	1	1	1	1	0	0	1	F9H

Segments

Figure 13–8 Segment assignments for number 1 2 3 4 5 and 6 in Example 13–4.

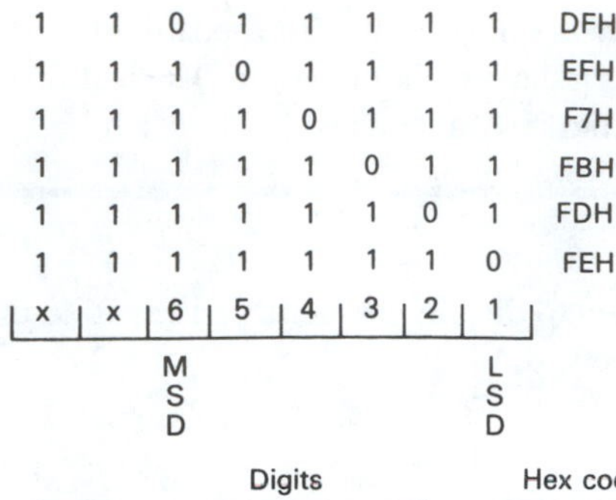

Figure 13–9 Digit assignments for all six-digit positions in Example 13–4.

DISP starts by turning off all digits. This is necessary because the next step is to drive the segment bus, not knowing which digit is to be turned on. OUT 00H drives the appropriate lines on the segment bus LOW. Since all digits are disabled, no segments are illuminated yet. The last two instructions in DISP turn on the digit that goes with the segment data currently on the segment bus.

The first digit to be illuminated is the MSD. The next group of three instructions in LOOP turns on digit 5, the next group turns on digit 4, and so on, down to the LSD. This cycle repeats continuously, illuminating each digit one-sixth of the time.

EXAMPLE 13–5

Write a program that will continuously rotate the number 0 left to right in the six-digit display given in Figure 13–6. Pause at each digit position for one-quarter second.

Solution:

Label	*Instruction*	*Comments*
MAIN:	LXI SP,20C0H	; Initialize stack pointer
	MVI A,FFH	; Program Ports A and B
	OUT 02H	; DDR A = Output
	OUT 03H	; DDR B = Output
	MVI A,C0H	; Load segment bus with
	OUT 00H	; data for number 0
MSD:	MVI A,DFH	; Data string to enable MSD
LOOP:	OUT 01H	; Drive digit bus
	CALL DELAY	; One-quarter second delay (Table 10–5)
	RRC	; Rotate right
	JNC MSD	; If CY now equals 0 then restart at MSD
	JMP LOOP	; Else display next position

Explanation:

After programming ports *A* and *B* as outputs, we send the segment code (C0H) for 0 to the segment bus. This information is latched in port *A* (00H) and will remain constant on the segment bus. Next, the MSD is enabled by sending DFH (1101 1111) to the digits. The 0 now appears on the MSD. The delay will cause the display to stay on for one-quarter second. The RRC instruction is used to rotate the accumulator, which has a single zero in it, to enable just one of the digits at a time. The display–delay–rotate continues to the LSD. The next rotate sends the zero bit of the accumulator into the carry bit, resetting CY. The JNC condition will be met, forcing program control back to label MSD, which sets up the accumulator to point to the leftmost digit (MSD). The cycle repeats.

13–6 SCANNING A KEYBOARD

A keyboard is one of the most common means of entering data into a microprocessor or computer system. With the large number of different keys on a typical keyboard, it is impossible to assign a separate input pin for each key. A 64-key keyboard would require 64 separate input pins, or eight, 8-bit input ports.

Instead, most keyboards are connected up as an *X-Y matrix.* For example, a 64-key keyboard would be wired up with eight rows and eight columns. A software program is then written to drive each of the rows active, one at a time, while reading the columns to see if any of the keys in the active row were depressed. When a key is pressed, the program decodes the particular key by knowing which row and which column were active.

The software for decoding on 8-by-8 keyboard is fairly complex, but we can get the general idea of the decoding process by studying the hardware and software solution to the hexadecimal keyboard shown in Figure 13–10.

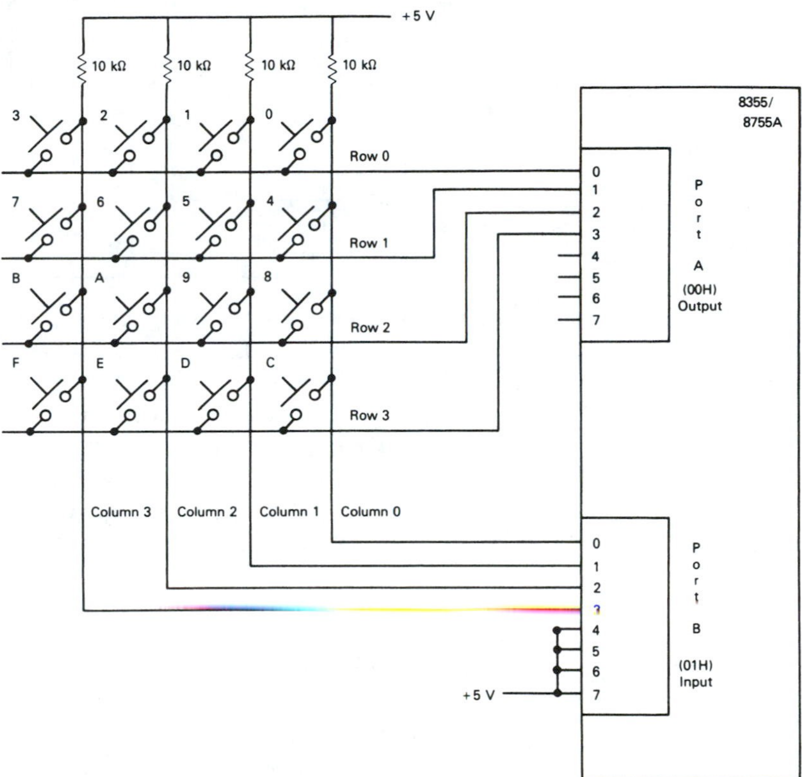

Figure 13–10 Hardware requirements for decoding a 4-by-4 hexadecimal keyboard.

Instead of using 16 individual lines for the 16 keys, a 4-by-4 matrix is set up. When a key is pressed, the adjacent row and column are connected electrically. To determine which key is pressed, the program must know which row is active-LOW when a LOW is encountered on one of the columns.

The scanning process is initiated by outputting a 0 on row 0, and 1s on the other three rows (port 00H). Next, the microprocessor reads the columns (port 01H). If one of the keys in row 0 (0, 1, 2, or 3) are depressed at the time, then the data read at port 01H will have a 0 in bit 0, 1, 2, or 3. If so, the program will be able to pinpoint which key was pressed.

If none of the columns are LOW when row 0 is active, then row 1 is made LOW. Again, the program reads the columns (port 01H) and checks for a LOW bit. The program continues to loop around all four rows, checking the columns each time, until one of the column bits is LOW.

To convert the row–column value into an actual hexadecimal number, refer to Table 13–4. For example, Table 13–4 shows that if row 1 was made LOW and column 2 was read

TABLE 13–4

Determining the Key Pressed by Applying Weighting Factors to the Active-LOW Row and Column Indicators

Row 3 2 1 0	Column 3 2 1 0	Key pressed
1 1 1 0	1 1 1 0	0
1 1 1 0	1 1 0 1	1
1 1 1 0	1 0 1 1	2
1 1 1 0	0 1 1 1	3
1 1 0 1	1 1 1 0	4
1 1 0 1	1 1 0 1	5
1 1 0 1	1 0 1 1	6
1 1 0 1	0 1 1 1	7
1 0 1 1	1 1 1 0	8
1 0 1 1	1 1 0 1	9
1 0 1 1	1 0 1 1	A
1 0 1 1	0 1 1 1	B
0 1 1 1	1 1 1 0	C
0 1 1 1	1 1 0 1	D
0 1 1 1	1 0 1 1	E
0 1 1 1	0 1 1 1	F

Row weighting (bits 3, 2, 1, 0): Add 12, Add 8, Add 4, Add 0

Column weighting (bits 0, 1, 2, 3): Add 0, Add 1, Add 2, Add 3

as a LOW, then the number 6 key is being pressed. This holds true if you trace through the schematic given in Figure 13–10. Trace through several other keys, just to be sure that you can see how Table 13–4 is built from Figure 13–10.

The weighting factors given at the bottom of the table are used to establish the value of the key pressed. For example, whenever row 1 is LOW, we are looking for the numbers 4, 5, 6, or 7. If a column position goes LOW, we add 4 to the column number to get the key value. In the case of the number 6 key, we are in row 1 and column 2, so we add 4 plus 2 and get a value of 6. Try adding the weighting factors for several of the other row–column combinations to prove to yourself that this technique works.

A final note on keyboard scanning. As with any other mechanical switch, there will be *switch bounce* that has to be accounted for. One simple way to avoid erroneous results is to wait for one-quarter second after encountering a key closure, to give the user time to release the key and for the key to stop bouncing.

Table 13–5 lists a program that will scan the keyboard until a key is pressed, and then place the numeric value of that key into the accumulator.

TABLE 13–5

Program Used to Scan and Decode the Hex Keyboard of Figure 13–10

Label	Instruction	Comments
INIT:	MVI A,FFH	; Program Port A
	OUT 02H	; DDR A = Output
	CMA	; Program Port B
	OUT 03H	; DDR B = Input
		;
ROW0:	MVI B,00H	; B = Row weighting = 0
	MVI A,FEH	; A ← 1111 1110
	OUT 00H	; Drive Row 0 LOW
	IN 01H	; Read columns
	CPI FFH	; Set Z flag if all columns HIGH
	JNZ COL	; Jump to COL if any column LOW
		;
ROW1:	MVI B,04H	; Row weighting = 4
	MVI A,FDH	; A ← 1111 1101
	OUT 00H	; Drive Row 1 LOW

TABLE 13–5
Program Used to Scan and Decode the Hex Keyboard of Figure 13–10 *(continued)*

Label	*Instruction*	*Comments*
	IN 01H	; Read columns
	CPI FFH	; Check for a
	JNZ COL	; LOW column
		;
ROW2:	MVI B,08H	; Row weighting = 8
	MVI A,FBH	; A ← 1111 1011
	OUT 00H	; Drive Row 2 LOW
	IN 01H	; Read columns
	CPI FFH	; Check for a
	JNZ COL	; LOW column
		;
ROW3:	MVI B,0CH	; Row weighting = 12
	MVI A,F7H	; A ← 1111 0111
	OUT 00H	; Drive Row 3 LOW
	IN 01H	; Read columns
	CPI FFH	; Check for a
	JNZ COL	; LOW column
	JMP ROW0	; No key pressed, scan again
		;
COL:	MVI C,00H	; C = column counter = 0
	RRC	; Rotate LOW column
	JNC DONE	; If LOW is now in CY, then done
	INR C	; C = 1
	RRC	; Rotate LOW column again
	JNC DONE	; If LOW is now in CY, then done
	INR C	; C = 2
	RRC	; Rotate LOW column again
	JNC DONE	; If LOW is now in CY, then done
	INR C	; Else C = 3
		;
DONE:	XRA A	; Clear A
	ADD B	; Add row weighting plus
	ADD C	; column weighting to A
	HLT	; A = Value of key pressed

13–7 DRIVING A STEPPER MOTOR

A stepper motor makes its rotation in steps instead of a smooth continuous motion as with conventional motors. Typical stepping angles are 15° or 7.5° per step, requiring 24 or 48 steps, respectively, to complete one revolution. The stepping action is controlled by digital levels that energize magnetic coils within the motor.

Because they are driven by digital signals, they are well suited for microprocessor-based control applications. For example, a program could be written to cause the stepper motor to rotate at 100 rpm, for 32 revolutions, and then stop. This is useful for applications requiring exact positioning control without the use of closed-loop feedback circuitry to monitor the position. Typical applications are floppy disk Read/Write head positioning, printer type head and line feed control, and robotics.

There are several ways to construct a motor to achieve this digitally controlled stepping action. One such way is illustrated in Figure 13–11.

This particular stepper motor construction uses four *stator* (stationary) *coils* set up as four pole-pairs. Each stator pole is offset from the previous one by 45°. The directions of the windings are such that energizing any one coil will develop a "north" field at one pole and a "south" field at the opposite pole. The north and south poles created by energizing coil 1 are shown in Figure 13–11. The rotating part of the motor (the *rotor*) is designed with three ferromagnetic pairs, spaced 60° apart from each other. (A ferromagnetic material is one that is attracted to magnetic fields.) Since the stator poles are spaced 45° apart, this makes the next stator-to-rotor 15° out of alignment.

In Figure 13–11 the rotor has aligned itself with the flux lines created by the north–south stator poles of coil 1. To step the rotor 15° clockwise, coil 1 is deenergized and coil 2 is energized. The rotor pair closest to coil 2 will now line up with stator pole pair 2's flux lines. The next 15° steps are made by energizing coil 3, then 4, then 1, then 2, etc., for as many steps as you require. Figure 13–12 shows the stepping action achieved by energizing each successive coil six times. Table 13–6 shows the digital

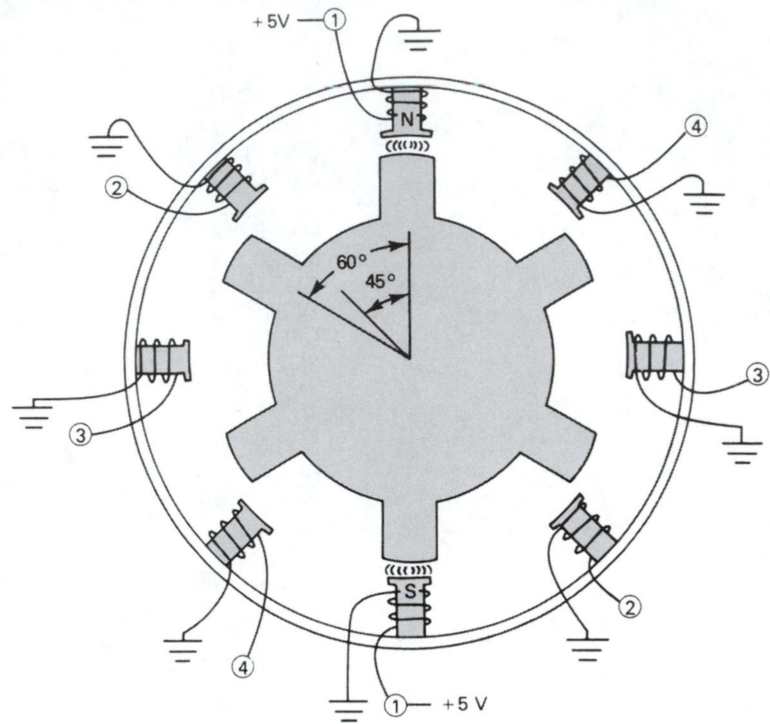

Figure 13–11 A four-coil stepper motor with stator coil 1 energized.

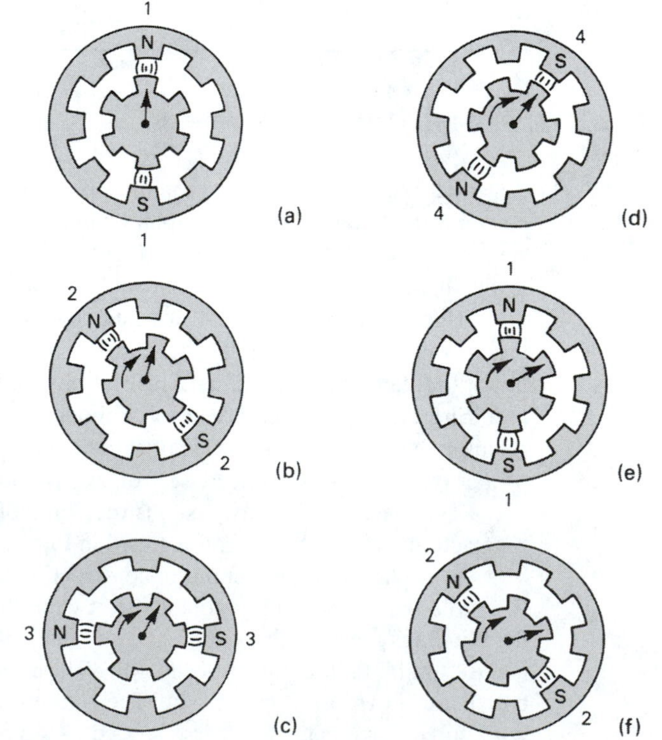

Figure 13–12 Coil energizing sequence for 15° clockwise steps.

TABLE 13–6
Digital Codes for 15° Clockwise and Counterclockwise Rotation

Clockwise coil 1 2 3 4	*Counterclockwise coil 1 2 3 4*
1 0 0 0	0 0 0 1
0 1 0 0	0 0 1 0
0 0 1 0	0 1 0 0
0 0 0 1	1 0 0 0
1 0 0 0	0 0 0 1
0 1 0 0	0 0 1 0
etc.	etc.

codes that are applied to the stator coils for 15° clockwise and 15° counterclockwise rotation.

The amount of current required to energize a coil pair is much higher than the capability of a parallel I/O port, so we will need some current-buffering circuitry similar to that shown in Figure 13–13.

The output of the upper 7406 inverting buffer in Figure 13–13 is LOW, forward biasing the base–emitter of the MJ2955 *PNP* power transistor. This causes the collector–emitter to short, allowing the large current to flow through the number 1 coils to ground. The IN4001 diodes protect the coils from arcing over when the current is stopped.

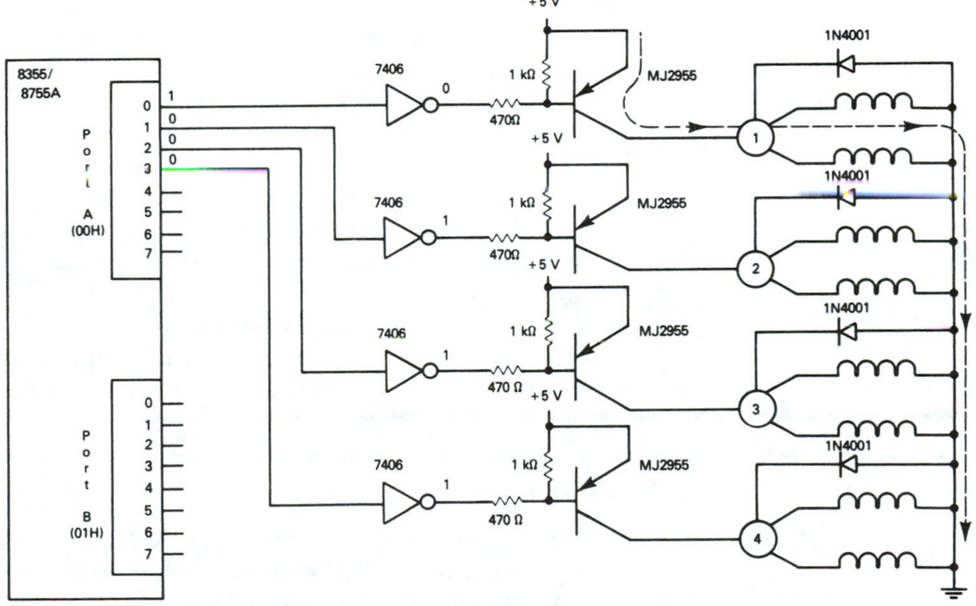

Figure 13–13 Drive circuitry for a four-coil stepper motor showing the number 1 coils energized.

The software program to drive the stepper motor involves rotating a single ON bit left to right, or right to left, repeatedly. The speed of rotation will be dictated by the delay period that is inserted between rotates. One complete revolution will be made by executing the RRC or RLC instruction 24 times (15° per step). The following examples illustrate the software requirements for driving a stepper motor.

EXAMPLE 13–6

Connect eight input switches to port *B* of the stepper motor circuit in Figure 13–13. Rotate the stepper motor clockwise, continuously, at a speed dictated by the value read in on the input switches (00H = fastest, FFH = slowest).

Solution:

Label	Instruction	Comments
INIT:	LXI SP,20C0H	; Initialize stack pointer
	MVI A,FFH	; Program Port A
	OUT 02H	; DDR A = Output
	DMA	; Program Port B
	OUT 03H	; DDR B = Input
	MVI A,11H	; A ← 0001 0001
STEP:	OUT 00H	; Send low-order nibble to stepper
	PUSH PSW	; Save A
	CALL DELAY	; Pause at that step
	POP PSW	; Restore A
	RLC	; Rotate A to energize next coil
	JMP STEP	; Repeat
DELAY:	IN 01H	; Read input switches for delay
	MOV B,A	; Move delay value to B
	MVI C,FFH	; C ← FFH
LOOP:	DCX B	; Decrement B-C Register pair
	MOV A,C	; Set Z flag if
	ORA B	; C = 0 and B = 0
	RZ	; Return if B-C reached 0000H
	JMP LOOP	; Else decrement B-C again

Explanation:

The stepper motor is connected to the low-order 4 bits (nibble) of port *A*. Since RLC is an 8-bit rotate, we have to repeat the low-order information on the high-order bits so that after the fourth rotate, the A_0 bit will be HIGH again to continue the rotation (A = 0001 0001). The PUSH and POP instructions are required because the DELAY subroutine changes *A* when it reads the input switches. The DELAY subroutine is used to insert a pause between each step in the rotation. The pause is proportional to the length of time that it takes to decrement the *B–C* register pair from XXFFH, down to 0, where XX is the value read from the input switches. Since the DCX instruction affects *no* flags, the ORA B instruction is used to set the zero flag when both *B* and *C* equal 0. The maximum decrement is FFFFH to 0000H (65,536) and the minimum is 00FFH to 0000H (256). This gives us a very broad range of stepper motor speeds.

EXAMPLE 13–7

Write a program to drive the stepper motor five revolutions clockwise, then five revolutions counterclockwise, then five revolutions clockwise, etc., continuously. Use the same subroutine that was developed in Example 13–6 to set the speed of rotation based on the input switch settings.

Solution:

Label	Instructions	Comments
INIT:	LXI SP,20C0H	; Initialize stack pointer
	MVI A,FFH	; Program Port A
	OUT 02H	; DDR A = Output
	CMA	; Program Port B

Solution: (continued)

Label	*Instructions*	*Comments*
	OUT 03H	; DDR B = Input
	MVI A,11H	; A ← 0001 0001
LOOP:	MVI D,78H	; D steps = 5 rev × 24 = 78H
CWISE:	OUT 00H	; Output to stepper coils
	PUSH PSW	; Save A
	CALL DELAY	; Pause at that step (Ex. 13–6)
	POP PSW	; Restore A
	RLC	; Rotate one step clockwise
	DCR D	; Decrement step counter
	JNZ CWISE	; If not zero, step again
	MVI D,78H	; Else reset counter for 5 revs
CCWISE:	OUT 00H	; Output to stepper coils
	PUSH PSW	; Save A
	CALL DELAY	; Pause at that step (Ex. 13–6)
	POP PSW	; Restore A
	RRC	; Rotate one step counterclockwise
	DCR D	; Decrement step counter
	JNZ CCWISE	; If not zero, step again
	JMP LOOP	; Else restart with clockwise

Explanation:

This is similar to Example 13–6 except that we have to count the number of revolutions. The step angle of our motor is 15°, requiring 24 steps for one revolution. Five revolutions will therefore require 120 steps (5 × 24 = 120). And 120 converted to hexadecimal is 78H. Clockwise rotation is made by using the RLC instruction, which will energize coil 1, then 2, and 3, etc. Counterclockwise rotation is made by using the RRC instruction, which will energize coil 1, then 4, then 3, then 2, etc.

SUMMARY

In this chapter we have learned that

1. A digital-to-analog converter can be driven by an output port of a microprocessor to create specialized square, sawtooth, and triangle waveforms. The period of the waveform is controlled by a delay inserted between the steps.
2. Two ports are required on a microprocessor system to interface to an analog-to-digital converter. One port is used to receive the digital result and the other port uses one bit to issue a *start-conversion* signal and then monitors another bit for the *end-of-conversion* response.
3. An LM35 linear temperature sensor can be input to an A-to-D converter to provide an accurate means of monitoring temperatures with a microprocessor.
4. A *look-up table* is a simple way to convert between digital codes but it can use up a lot of memory locations to implement a complete table.
5. Driving multiplexed displays can be accomplished using two ports: one to drive the segments of all digits, and one to select which digit is to be active.
6. Scanning a keyboard is accomplished by setting it up as a row–column matrix and then activating individual rows and seeing which column receives the signal. Once the row–column combination is known, the individual key that is pressed can be determined.
7. Stepper motors revolve by taking small angular steps. Their speed and number of revolutions can be accurately controlled by software. To rotate the motor, one out of the four stator coils is energized, then each adjacent coil is energized in succession.

GLOSSARY

Algorithm: A procedure or formula used to solve a problem.

Closed-Loop Feedback: A system that sends information about an output device back to the device that is driving the output device, to keep track of the particular activity.

Common-Cathode: A seven-segment display having the cathodes of each segment connected to a common ground point. Individual segments are then illuminated by supplying +5 V via a series-limiting resistor.

Ferromagnetic: A material in which magnetic flux lines can easily pass (high permeability).

Flux Lines: The north-to-south magnetic field set up by magnets is made up of flux lines.

Handshaking: The communication between a data sending device and receiving device that is necessary to determine the status of the transmitted data.

Look-Up Table: A data table that is used in software programs to translate a particular binary string into a meaningful binary output. The translation is made by "looking up" the output value by using the input string as an address pointer.

Matrix Keyboard: A multikey keyboard whose keys are electrically connected up in an *X-Y* matrix configuration with each key located by its row–column address.

Multiplexed Display: A multidigit display whose digits share a common segment bus and are enabled one at a time.

Pole-Pair: Two opposing magnetic poles situated opposite each other in a motor housing and energized concurrently.

Programmable Voltage Source: A device whose output voltage level is controlled by software program instructions.

Rotor: The rotating part of the stepper motor.

Sawtooth Wave: A repetitive waveform characterized by a linearly rising voltage level that reaches a high point and then drops back to its initial value.

Segment Bus: A common set of seven conductors (eight if the decimal point is used) shared by each digit of a multiplexed display.

Stator Coil: A stationary coil, mounted on the inside of the motor housing.

Step Angle: The number of degrees that a stepper motor rotates for each change in the digital input signal (usually 15° or 7.5°).

Stepper Motor: A motor whose rotation is made in steps that are controlled by a digital input signal.

PROBLEMS

13–1. Determine the value of V_{out} in the DAC circuit of Figure 13–1 that would result from the execution of the following program segments:

(a) MVI A,01H
OUT 01H

(b) MVI A,80H
OUT 01H

(c) MVI A,FFH
OUT 01H

13–2. Write the programs that will output the following voltage in the DAC circuit in Figure 13–1.

(a) $V_{out} \cong 1.2$ V

(b) $V_{out} \cong 5.0$ V

(c) $V_{out} \cong 0.2$ V

13–3. Sketch and label the voltage and time for the output waveform that will result from executing the following program with the DAC circuit of Figure 13–1. (Assume that a 6.144-MHz crystal is being used.)

```
       MVI A,FFH
       OUT 03H
LOOP:  MVI A,40H
       OUT 01H
       MVI A,04H
       OUT 01H
       JMP LOOP
```

13–4. What change could be made to the program in Problem 13–3 to make the duty cycle of the output wave 50%?

13–5. Write a program that uses the DAC circuit of Figure 13–1 to create a triangle wave of any frequency and any amplitude.

13–6. Describe the handshaking that takes place between the ADC0801 and the I/O ports in Figure 13–4 when making an A/D conversion.

13–7. Which instructions in Table 13–1 are responsible for issuing the LOW-to-HIGH start-conversion signal to the ADC in Figure 13–4?

13–8. What is the 8-bit binary output of the ADC in Figure 13–5 if the temperature is 25° C?

13–9. In Figure 13–5 what conversion must be made to the 8-bit data string received from the ADC before it can be output to the two-digit display?

13–10. What is one advantage and one disadvantage of using a look-up table for the translation of data values?

13–11. Briefly describe the operation of the multiplexed display circuit of Figure 13–6.

13–12. Why are transistors required for segment drivers and digit drivers in Figure 13–6?

13–13. Why is it necessary to turn off all digits in the beginning of the DISP subroutine in Example 13–4?

13–14. Write a program that will display the number 7 in at least significant digit position in Figure 13–6.

13–15. Modify the program in Example 13–4 so that it displays the word HELP in the four rightmost positions.

13–16. Write a program that will continuously flash the number 4, ON one-quarter second, OFF one-quarter second, in the LSD of Figure 13–6.

13–17. Briefly describe the technique used to decode the matrix keyboard shown in Figure 13–10.

13–18. Assume that an active-LOW LED circuit is connected to bit 7 of port *A* in Figure 13–10. Determine which key must be pressed in order for the following program to turn on the LED.

```
INIT:  MVI A,FFH
       OUT 02H
       CMA
       OUT 03H
ROW:   MVI A,FBH
       OUT 00H
COL:   IN 01H
       CPI FDH
       JNZ COL
LED:   MVI A,7FH
       OUT 00H
       JMP LED
```

13–19. Rewrite the program given in Problem 13–18 so that it turns on the LED if the number 6 key is pressed.

13–20. Briefly describe how the instructions following label COL: in Table 13–5 convert the key pressed into the actual hexadecimal value.

13–21. Describe how a stepper motor differs from other motors.

13–22. Which direction will the stepper motor rotor turn in Figure 13–11 if the coil sequence is 1–4–3–2–1?

13–23. The stepper motor coils in Figure 13–11 are activated by (a HIGH, or a LOW)?

13–24. What is the purpose of the MJ2955 *PNP* transistors in Figure 13–13?

13–25. **(a)** Why are the PUSH and POP instructions required in the solution to Example 13–6?

(b) Could they be placed *inside* the DELAY subroutine instead of where they are?

13–26. Assume that there are input switches connected to port *B* of the 8355/8755A in Figure 13–13. Write a program that drives the stepper motor counterclockwise at a slow speed. The motor will make XX revolutions, where XX is the value read in from the input switches.

SCHEMATIC INTERPRETATION PROBLEMS

13–27. On a separate piece of paper draw the circuit connections to add a bank of eight LEDs with current limiting resistors to the octal *D* flip-flop, U5, in the 4096/4196 schematic.

13–28. Identify the following ICs on the 4096/4196 schematic (sheets 1 and 2):

(a) The three-state octal buffers
(b) The three-state octal *D* flip-flops
(c) The three-state octal transceivers
(d) The three-state octal latches

13–29. Design a "missing pulse detector." It will be used to monitor the DAV input line in the 4096/4196 schematic. Assume that the DAV line is supposed to provide a 2-μS HIGH pulse every 100 μS. Monitor the DAV line with a 74123 monostable multivibrator. Have the 74123 output a HIGH to port 1, bit 7(P1.7) of the 8031 microcontroller if a missing pulse is detected.

14 The 8051 Microcontroller*

OBJECTIVES

Upon completion of this chapter you should be able to:

- Describe the advantages that a microcontroller has over a microprocessor for control applications.
- Describe the functional blocks within the 8051 microcontroller.
- Make comparisons between the different microcontrollers available within the 8051 family.
- Write the software instructions to read and write data to and from the I/O ports.
- Describe the use of the alternate functions on the I/O ports.
- Make the distinction between the internal and external data and code memory spaces.
- Use various addressing modes to access internal data memory and the Special Function Registers (SFRs).
- Interface an external EPROM and a RAM to the 8051.
- Use some of the more commonly used instructions of the 8051 instruction set.
- Write simple 8051 I/O programs.
- Use the bit operations of the 8051.
- Understand program solutions to applications such as a keyboard decoder and analog-to-digital converter.

INTRODUCTION

In the previous chapters you should have recognized several common components that are incorporated into most microprocessor-based system applications. These are the microprocessor, RAM, ROM (or EPROM), and parallel I/O ports. Intel Corporation has

*Visit Appendix A for a list of web sites for downloadable 8051 software tools.

recognized this and over the years since the introduction of the 8085A, has been developing a new line of microprocessors specifically designed for control applications. They are called *microcontrollers.*

The microcontroller has the CPU, RAM, ROM, timer/counter, and parallel and serial I/O ports fabricated into a single IC. It is often called "a computer on a chip." The CPU's instruction set is improved for control applications and offers bit-oriented data manipulation, branching, and I/O, as well as multiply and divide instructions.

The microcontroller is most efficiently used in systems that have a fixed program for a dedicated application. A microcontroller is used in a keyboard for a personal computer to scan and decode the keys. It is used in an automobile for sensing and controlling engine operation. Other dedicated applications such as a microwave oven, a video cassette recorder, a gas pump, and an automated teller machine use the microcontroller IC also.

The previous chapters have provided a good background for understanding the microcontroller. The theory behind microprocessor buses, external memory, and I/O ports is necessary to utilize the features available on a microcontroller. In this chapter we'll be introduced to one of the most widely used microcontrollers available today: the Intel 8051. You'll see several similarities between it and the 8085A-based systems previously discussed. You should appreciate the way that you were eased into microprocessor operational theory using the 8085A, but now you will see why the 8051 can be a much better solution to dedicated control applications.

Most control applications require extensive I/O and need to work with individual bits. The 8051 addresses both of these needs by having 32 I/O lines and a CPU instruction set that handles single-bit I/O, bit manipulation, and bit checking.

14–1 THE 8051 FAMILY OF MICROCONTROLLERS

The basic architectural structure of the 8051 is given in Figure 14–1. The block diagram gives us a good picture of the hardware included in the 8051 IC. For internal memory it has a 4K × 8 ROM and a 128 × 8 RAM. It has two 16-bit counter/timers and interrupt control for five interrupt sources. Serial I/O is provided by TXD and RXD (transmit and receive), and it also has four 8-bit parallel I/O ports (P0, P1, P2, P3). There is also an 8052 series of microcontrollers available that has an 8K ROM, 256 RAM, and three counter/timers.

Other versions of the 8051 are the 8751, which has an internal EPROM for program storage in place of the ROM, and the 8031, which has *no* internal ROM, but instead

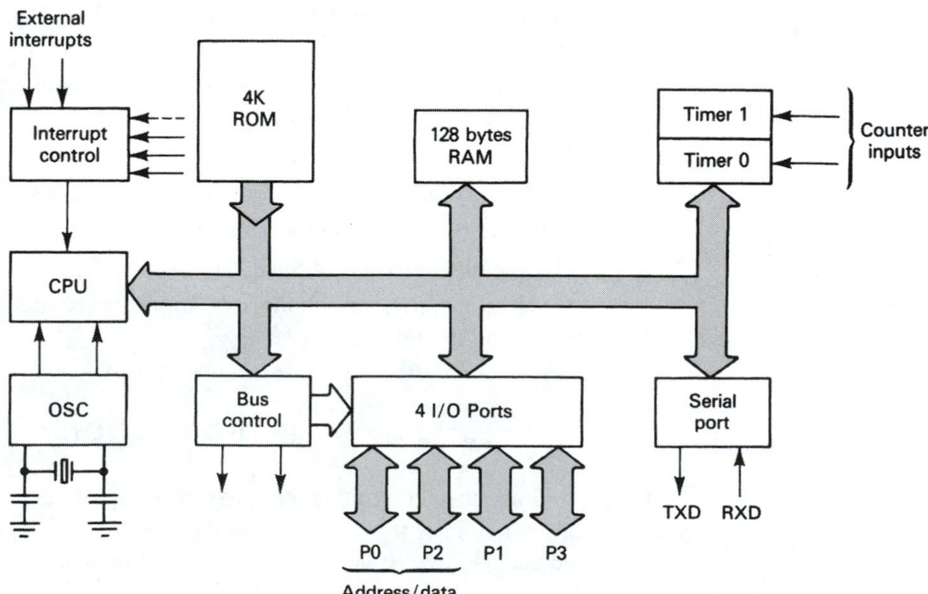

Figure 14–1 Block diagram of the Intel 8051 microcontroller.

accesses an external ROM or EPROM for program instructions. Table 14–1 summarizes the 8051 family of microcontrollers. Note that the 8052/8752/8032 series has an extra 4K of program space (except the 8032), double the RAM area, an extra timer/counter, and one additional interrupt source. All parts use the same CPU instruction set. The ROMless versions (8031 and 8032) are the least expensive parts but require an external ROM or EPROM like a 2732 or 2764 for program storage.

TABLE 14–1
The 8051 Family of Microcontrollers

Device number	Internal memory: Program	Internal memory: Data	Timers/ event counters	Interrupt sources
8051	4K × 8 ROM	128 × 8 RAM	2 × 16-bit	5
8751H	4K × 8 EPROM	128 × 8 RAM	2 × 16-bit	5
8031	None	128 × 8 RAM	2 × 16-bit	5
8052AH	8K × 8 ROM	256 × 8 RAM	3 × 16-bit	6
8752BH	8K × 8 EPROM	256 × 8 RAM	3 × 16-bit	6
8032AH	None	256 × 8 RAM	3 × 16-bit	6

14–2 8051 ARCHITECTURE

In order to squeeze so many functions on a single chip, the designers had to develop an architecture that uses the same internal address spaces and external pins for more than one function. This technique is similar to that used by the 8085A for the multiplexed AD_0–AD_7 lines.

The 8051 is a 40-pin IC. Thirty-two pins are needed for the four I/O ports. In order to provide for the other microcontroller control signals, most of those pins have alternate functions, which will be described in this section. Also in this section, we'll see how the 8051 handles the overlapping address spaces used by internal memory, external memory, and the special function registers.

The pin configuration for the 8051 is given in Figure 14–2.

Port 0

Port 0 is dual purpose, serving as either an 8-bit bidirectional I/O port (P0.0–P0.7) or the low-order multiplexed address/data bus (AD_0–AD_7). As an I/O port, it can sink up to 8 LS TTL loads in the LOW condition and is a float for the HIGH condition (I_{OL} = 3.2 mA). The alternate port designations, AD_0–AD_7, are used to access external memory. They are activated automatically whenever reference is made to external memory. The AD lines are demultiplexed into A_0–A_7 and D_0–D_7 by using the ALE signal, the same way that it was done with the 8085A.

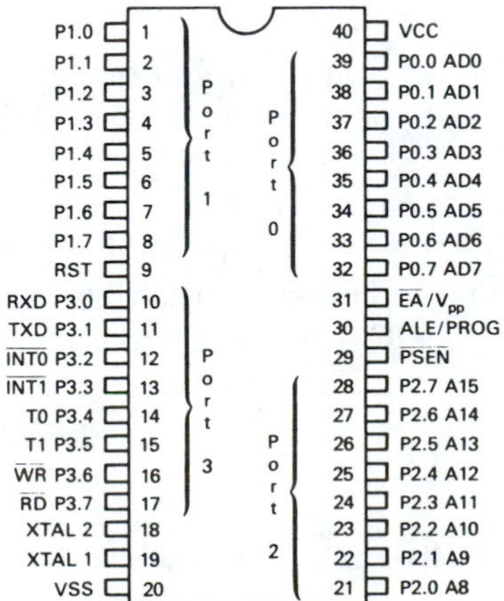

Figure 14–2 The 8051 pin configuration.

Port 1

Port 1 is an 8-bit bidirectional I/O port that can sink or source up to 4 LS TTL loads. ($I_{OL} =$ 1.6 mA, $I_{OH} = -80\ \mu$A.)

Port 2

Port 2 is dual purpose, serving as either an 8-bit bidirectional I/O port (P2.0–P2.7), or as the high-order address bus (A_8–A_{15}) for access to external memory. As an I/O port it can sink or source up to 4 LS TTL loads. The port becomes active as the high-order address bus whenever reference to external memory is made.

Port 3

Port 3 is dual purpose, serving as an 8-bit bidirectional I/O port that can sink or source up to 4 LS TTL loads or as special-purpose I/O to provide the functions listed in Table 14–2.

TABLE 14–2
Alternative Functions of Port 3

Port pin	*Alternative function*
P3.0	RXD (serial input port)
P3.1	TXD (serial output port)
P3.2	$\overline{\text{INT0}}$ (external interrupt 0)
P3.3	$\overline{\text{INT1}}$ (external interrupt 1)
P3.4	T0 (Timer 0 external input)
P3.5	T1 (Timer 1 external input)
P3.6	$\overline{\text{WR}}$ (external data memory write strobe)
P3.7	$\overline{\text{RD}}$ (external data memory read strobe)

RST

Reset input. A HIGH on this pin resets the microcontroller.

ALE/$\overline{\text{PROG}}$

Address Latch Enable output pulse for latching the low-order byte of the address during accesses to external memory. This pin is also the program pulse input ($\overline{\text{PROG}}$) during programming of the EPROM parts.

$\overline{\text{PSEN}}$

Program Store Enable is a read strobe for external program memory. It will be connected to the Output Enable ($\overline{\text{OE}}$) of an external ROM or EPROM.

$\overline{\text{EA}}$/VPP

External Access ($\overline{\text{EA}}$) is tied LOW to enable the microcontroller to fetch its program code from an external memory IC. This pin also receives the 21-V programming supply voltage (VPP) for programming the EPROM parts.

XTAL1,XTAL2

Connections for a crystal or an external oscillator.

Address Spaces

The address spaces of the 8051 are divided into four distinct areas: internal data memory, external data memory, internal code memory, and external code memory (see Figure 14–3).

The 8051 allows for up to 64K of external data memory (RAM) and 64K of external code memory (ROM/EPROM). The only disadvantage of external memory, besides the additional circuitry, is that ports 0 and 2 get tied up for the address and data bus. The actual hardware interfacing for external memory will be given later in this chapter.

When the 8051 is first reset, the program counter starts at 0000H. This points to the first program instruction in the *internal code memory* unless $\overline{EA}$ (External Access) is tied LOW. If $\overline{EA}$ is tied LOW, the CPU issues a LOW on $\overline{PSEN}$ (Program Store Enable), enabling the *external code memory* ROM/EPROM instead. $\overline{EA}$ *must* be tied LOW if using the ROMless (8031) version.

With $\overline{EA}$ tied HIGH, the first 4K of program instruction fetches are made from the internal code memory. Any fetches from code memory above 4K (1000H to FFFFH) will automatically be made from the external code memory.

If the application has a need for large amounts of data memory, then *external data memory* (RAM) can be used. I/O to external RAM can only be made by using one of the MOVX instructions. When executing a MOVX instruction, the 8051 knows that you are referring to external RAM and issues the appropriate $\overline{WR}$ or $\overline{RD}$ signal.

If 128 bytes (256 for the 8052) of RAM storage is enough, then the *internal data memory* is better to use because it has a much faster response time and a broad spectrum of instructions to access it. Figure 14–3 shows two blocks of internal data memory. The first is the 128 byte RAM at address 00H to 7FH. The second is made up of the Special Function Registers (SFRs) at addresses 80H to FFH. Note that *all addresses are only 1 byte wide,* which allows for more efficient use of code space and faster access time. The use of the address space in the 128-byte RAM is detailed in Figure 14–4.

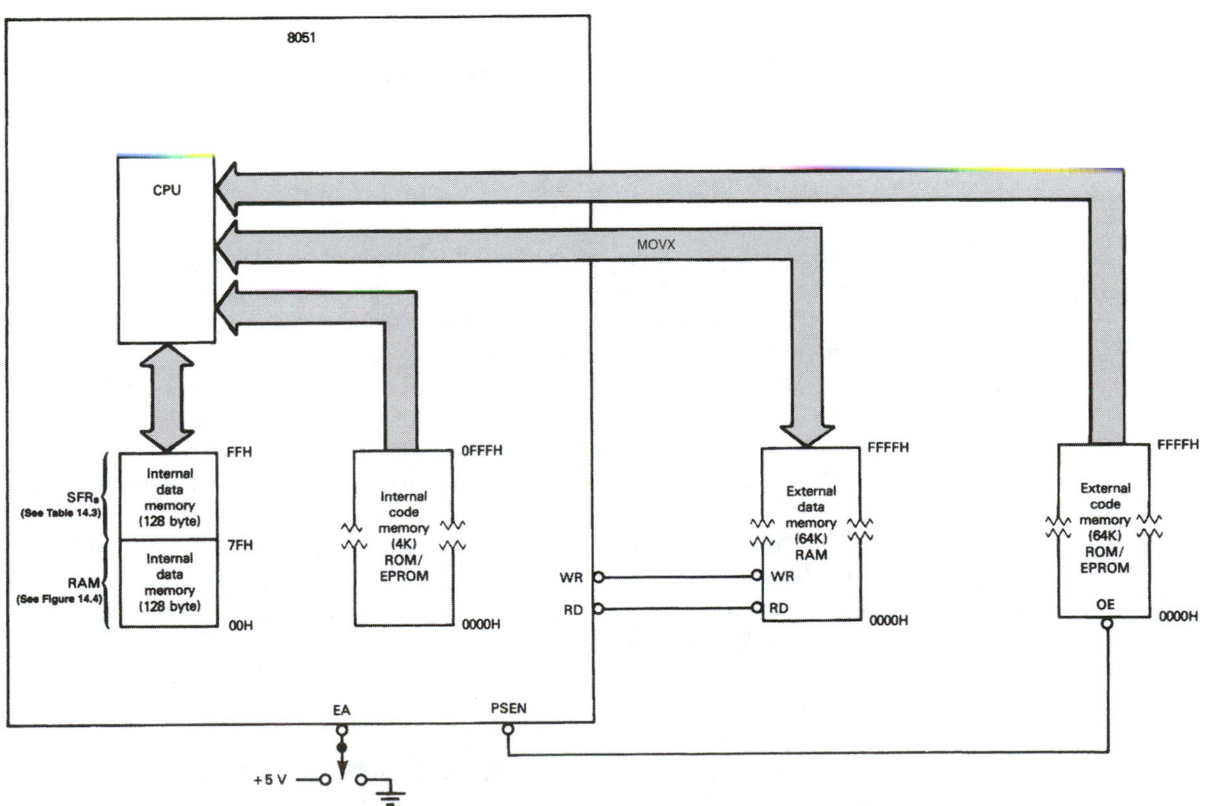

Figure 14–3 8051 address spaces.

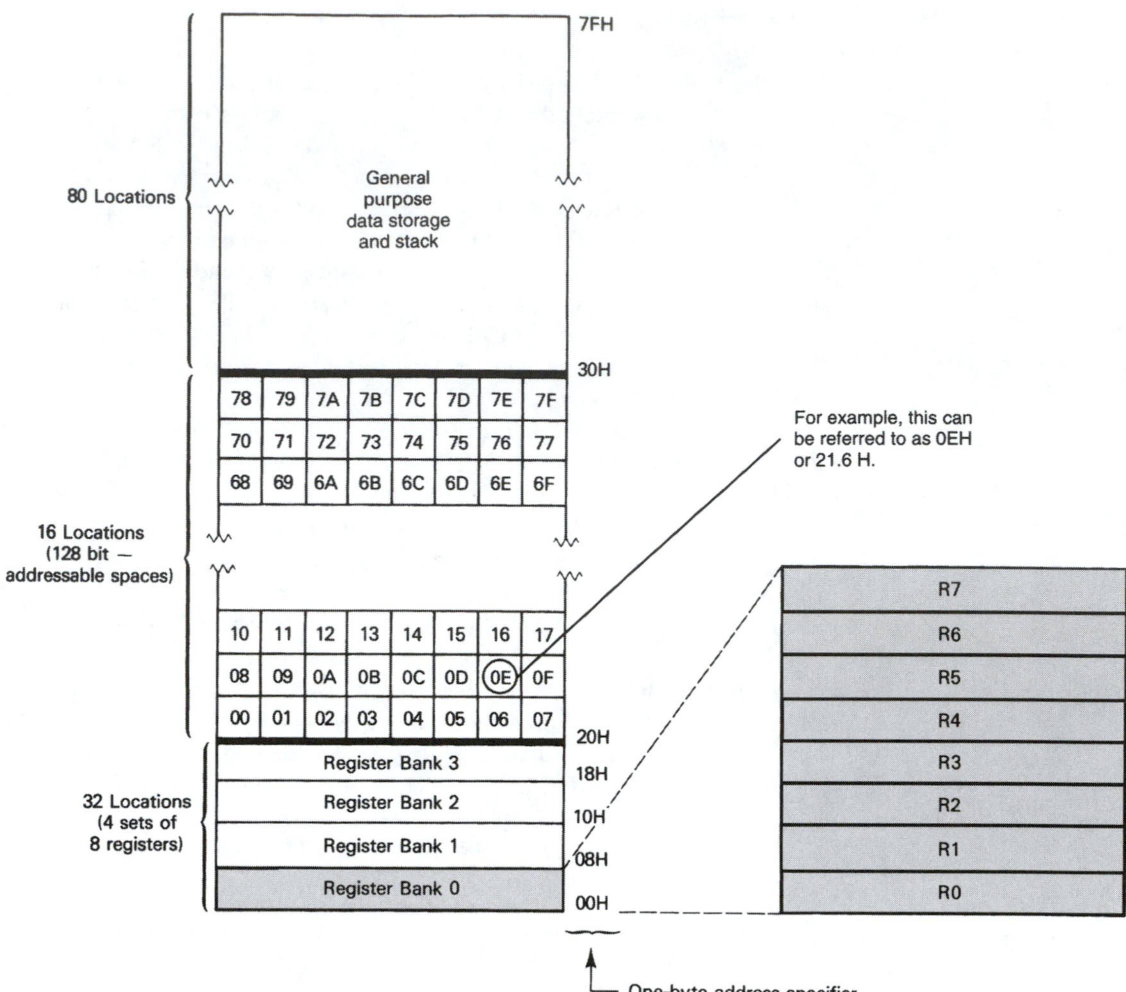

Figure 14–4 Address space in the 128-byte internal data memory RAM.

The first 32 locations are set aside for four banks of data registers, giving the programmer the use of 32 separate registers. The registers within each bank are labeled R0 through R7. The bank that is presently in use is defined by the setting of the two "bank select bits" in the PSW (Program Status Word).

RAM addresses 20H through 2FH are designated as bit-addressable memory locations. This meets the needs of control applications for a large number of ON/OFF bit flags, by providing 128 uniquely addressable bit locations.

The last 80 locations are set aside for general-purpose data storage and stack operations.

The Special Function Registers (SFRs) are maintained in the next 128 addresses in the internal data memory of the 8051 (address 80H to FFH). It contains registers that are required for software instruction execution as well as those used to service the special hardware features built into the 8051. Table 14–3 lists the SFRs and their addresses.

The address of an SFR is also only 1-byte wide. However, instead of specifying the 1-byte hex address, you also have the option of addressing any SFR location by simply specifying its register name, like P0 or SP. Several of the SFRs are bit addressable. For example, to address bit 3 of the accumulator, you would use the address E0H.3 (or you could use ACC.3). The PSW is also bit addressable, having the bit addresses listed in Table 14–4.

For example, to address the parity bit in the PSW, you would use the address D0H.0 (or you could use PSW.0).

TABLE 14–3
The Special Function Registers (SFRs) [Internal Data Memory (80H to FFH)]

Register	*Address*	*Function*
P0	80H[a]	Port 0
SP	81H	Stack pointer
DPL	82H	Data pointer (Low)
DPH	83H	Data pointer (High)
TCON	88H[a]	Timer register
TMOD	89H	Timer mode register
TL0	8AH	Timer 0 low byte
TL1	8BH	Timer 1 low byte
TH0	8CH	Timer 0 high byte
TH1	8DH	Timer 1 high byte
P1	90H[a]	Port 1
SCON	98H[a]	Serial port control register
SBUF	99H	Serial port data buffer
P2	A0H[a]	Port 2
IE	A8H[a]	Interrupt enable register
P3	B0H[a]	Port 3
IP	B8H[a]	Interrupt priority register
PSW	D0H[a]	Program status word
ACC	E0H[a]	Accumulator (direct address)
B	F0H[a]	B register

[a]Bit-addressable register.

TABLE 14–4
The PSW Bit Addresses

Symbol	*Address*	*Function*
CY	D0H.7	Carry flag
AC	D0H.6	Auxiliary carry flag
F0	D0H.5	General-purpose flag
RS1	D0H.4	Register bank select (MSB)
RS0	D0H.3	Register bank select (LSB)
OV	D0H.2	Overflow flag
—	D0H.1	User-definable flag
P	D0H.0	Parity flag

14–3 INTERFACING TO EXTERNAL MEMORY

Up to 64K of code memory (ROM/EPROM) and 64K of data memory (RAM) can be added to any of the 8051 family members. If you are using the 8031 (ROMless) part, then you *have* to use external code memory for storing your program instructions. As mentioned earlier, the alternate function of port 2 is to provide the high-order address byte (A_8–A_{15}), and the alternate function of port 0 is to provide the multiplexed low-order address/data byte (AD_0–AD_7).

If you are interfacing to a general-purpose EPROM like the 2732, then the ALE signal provided by the 8031 is used to demultiplex the AD_0–AD_7 lines via an address latch, as shown in Figure 14–5.

As you can see, two of the I/O ports are used up in order to provide the address and data buses. The AD_0–AD_7 lines that are output on port 0 are demultiplexed by the ALE signal and the address latch, the same way as they were in the 8085A circuits studied earlier. The $\overline{\text{PSEN}}$ signal is asserted at the end of each instruction fetch cycle to enable the EPROM outputs to put the addressed code byte on the data bus to be read by port 0. The $\overline{\text{EA}}$ line is tied LOW so that the 8031 knows to fetch all program code from external memory.

If extra *data memory* is needed, then the interface circuit given in Figure 14–6 can be used. The 8155 RAM accepts the AD_0–AD_7 lines directly and uses the ALE signal to

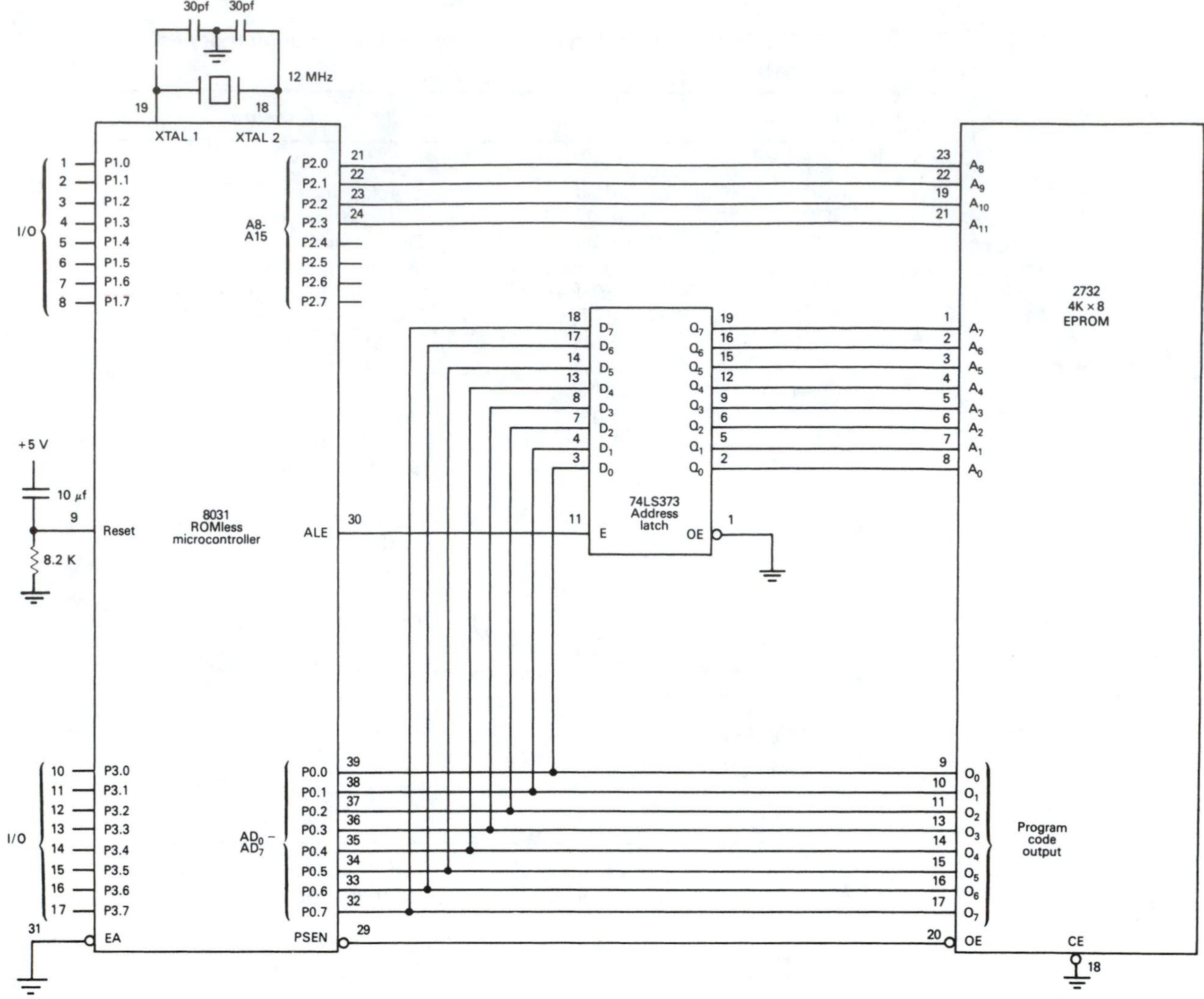

Figure 14–5 Interfacing a 2732 EPROM to the ROMless 8031 microcontroller.

internally demultiplex, eliminating the need for an address latch IC. We have to use 11 of the 8051 pins to interface to the 8155, but the 8155 provides an additional 22 I/O lines, giving us a net gain of 11.

The addresses of the external RAM locations are 0000H to 00FFH, overlapping the addresses of the internal data memory. There is not a conflict, however, because all I/O to the external data memory is made using the MOVX instruction. The MOVX instruction ignores internal memory and instead, activates the appropriate control signal, $\overline{RD}$ or $\overline{WR}$ via port 3. The LOW $\overline{RD}$ or $\overline{WR}$ signal allows the 8155 to send or receive data to or from port 0 of the 8051.

The I/O ports on the 8155 are accessed by making the IO/$\overline{M}$ line HIGH. This is done by using memory-mapped I/O and specifying an address whose bit A_{15} is HIGH (8000H or higher).

14–4 THE 8051 INSTRUCTION SET

All the members of the 8051 family use the same instruction set. (The complete instruction set is given in the 8051 Instruction Set Summary in Appendix F.) Several new instructions in the 8051 make it especially well suited for control applications. The discussion that follows assumes that you are using a commercial assembler software package like ASM51,

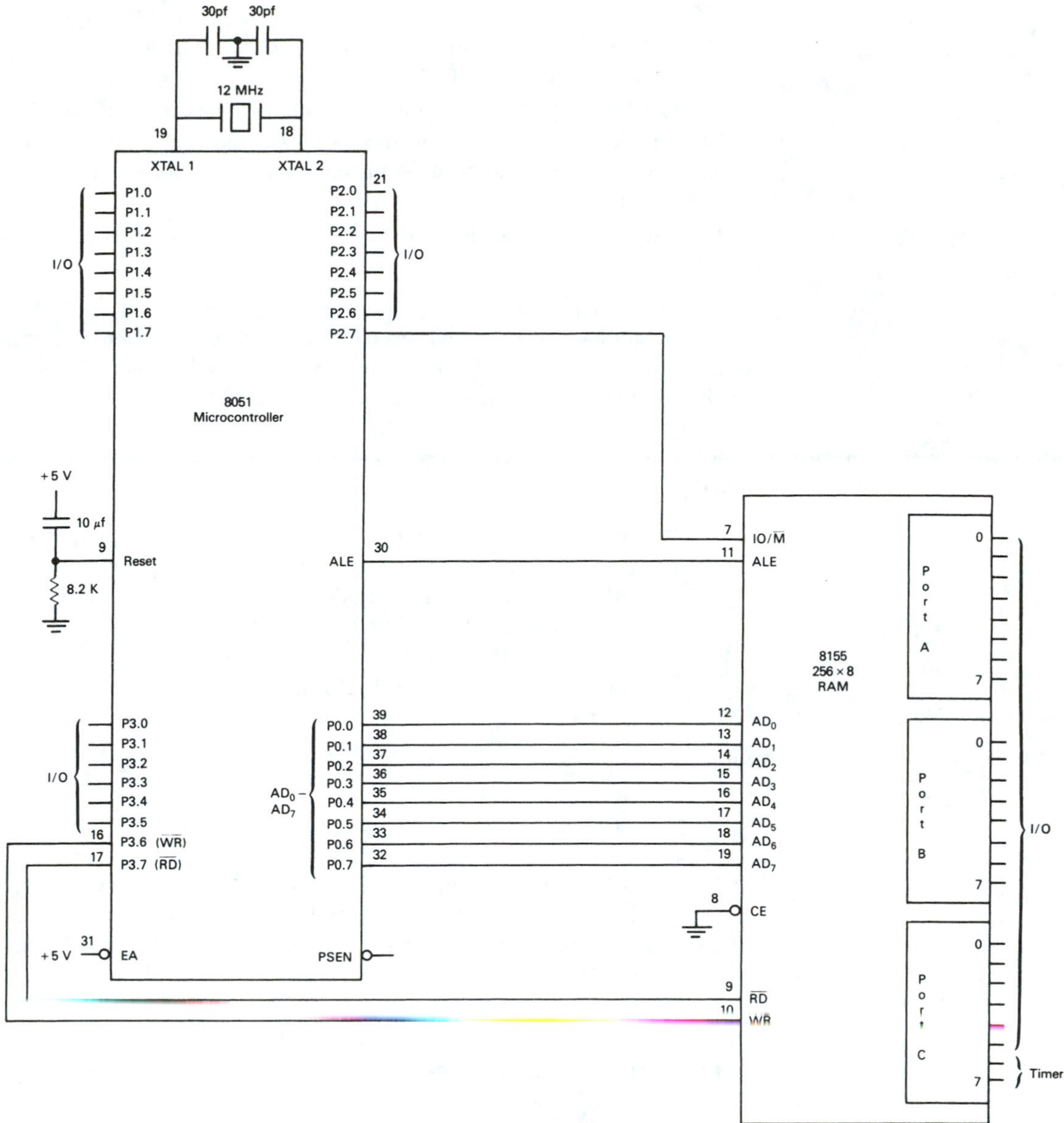

Figure 14–6 Interfacing an 8155 RAM/IO/Timer to the 8051 microcontroller.

available from Intel. Hand assembly of the 8051 instructions into executable machine code is very difficult and misses out on several of the very useful features available to the ASM51 programmer.

Addressing Modes

The instruction set provides several different means to address data memory locations. We'll use the MOV instruction to illustrate several common addressing modes. For example, to move data into the accumulator, any of the following instructions could be used:

MOV A,*Rn*: *Register addressing,* the contents of register *Rn* (where $n = 0–7$) is moved to the accumulator.

MOV A,@*Ri*: *Indirect addressing,* the contents of memory whose address is in *Ri* (where $i = 0$ or 1) is moved to the accumulator. (Note, only registers R_0 and R_1 can be used to hold addresses for the indirect-addressing instructions.)

MOV A,20H: *Direct addressing,* move the contents of RAM location 20H to the accumulator. I/O ports can also be accessed as a direct address as shown in the following instruction.

MOV A,P3: *Direct addressing,* move the contents of port 3 to the accumulator. Direct addressing allows you to specify the address by giving its actual hex address (e.g., B0H) or by giving its abbreviated name (e.g., P3), which is listed in the SFR table (Table 14–3).

MOV A,#64H: *Immediate constant,* move the number 64H into the accumulator.

In each of the previous instructions, the *destination* of the move was the accumulator. The destination in any of those instructions could also have been a register, a direct address location, or an indirect address location. You should already be realizing the extended flexibility that this instruction set has over the 8085A.

EXAMPLE 14–1

Write the 8051 instruction to perform each of the following operations.

(a) Move the contents of the accumulator to register 5.
(b) Move the contents of RAM memory location 42H to port 1.
(c) Move the value at port 2 to register 3.
(d) Send FFH to port 0.
(e) Send the contents of RAM memory, whose address is in register 1, to port 3.

Solution:

(a) MOV R5,A
(b) MOV P1,42H
(c) MOV R3,P2
(d) MOV P0,#0FFH (Note the addition of the leading 0 to the FFH. This is because the ASM51 assembler requires that the first position of any hexadecimal number must be a numeric digit from 0 to 9.)
(e) MOV P3,@R1

Program Branching Instructions

8051 assembly language provides several different ways to branch (jump) to various program segments within the 64K of code memory area. Below are some of the most useful jump instructions.

JMP *label* *(Unconditional jump):* Program control passes to location *label.* The JMP instruction is converted by the ASM51 assembler into an absolute jump, AJMP; a short jump, SJMP; or a long jump, LJMP, depending on the destination pointed to by *label.*

JZ *label* *(Jump if accumulator zero):* Program control passes to location *label* if the accumulator equals zero.

JNZ *label* *(Jump if accumulator not zero):* Program control passes to location *label* if the accumulator is not equal to zero.

JB *bit,label* *(Jump if bit set):* Program control passes to location *label* if the specified bit, *bit,* is set.

JNB *bit,label* *(Jump if bit not set):* Program control passes to location *label* if the specified bit, *bit,* is not set.

DJNZ ***Rn,label*** *(Decrement register and jump if not zero):* Program control passes to location *label* if, after decrementing register *Rn*, the register is not equal to zero (n = 0–7).

CJNE ***Rn,#data,label*** *(Compare immediate data to register and jump if not equal):* Program control passes to location *label* if register *Rn* is not equal to the immediate data *#data*. The compare can also be made to the accumulator by specifying *A* instead of *Rn*.

CALL ***label*** *(Call subroutine):* Program control passes to location *label*. The return address is stored on the stack. The CALL instruction is converted by the ASM51 assembler into an absolute call, ACALL, or a long call, LCALL, depending on the destination pointed to by *label*.

RET *(Return):* Program control is returned back to the instruction following the CALL instruction.

EXAMPLE 14–2

Write a program that continuously reads a byte from port 1 and writes it to port 0 until the byte read equals zero.

Solution:

```
READ:    MOV A,P1        ; A ← P1
         MOV P0,A        ; P0 ← A
         JNZ READ        ; Repeat until A = 0
         NOP             ; Remainder of program
         etc.
```

EXAMPLE 14–3

Repeat Example 14–2, except stop the looping when the number 77H is read.

Solution:

```
READ:    MOV A,P1             ; A ← P1
         MOV P0,A             ; P0 ← A
         CJNE A,#77H,READ     ; Repeat until A = 77H
         NOP                  ; Remainder of program
         etc.
```

EXAMPLE 14–4

Repeat Example 14–2, except stop the looping when bit 3 of port 2 is set.

Solution:

```
READ:    MOV A,P1          ; A ← P1
         MOV P0,A          ; P0 ← A
         JNB P2.3,READ     ; Repeat until bit 3 of port 2 is set
         NOP               ; Remainder of program
         etc.              ;
```

EXAMPLE 14–5

Write a program that will produce an output at port 0 that counts down from 80H to 00H.

Solution:

```
        MOV R0,#80H        ; R0 ← 80H
COUNT:  MOV P0,R0          ; P0 ← R0
        DJNZ R0,COUNT      ; Decrement R0, jump to COUNT if not 0
        NOP                ; Remainder of program
        etc.               ;
```

Logical and Bit Operations

The 8051 instruction set provides the basic logical operations (OR, AND, Ex-OR, and NOT), rotates (left or right, with or without carry), and bit operations (clear, set, and complement). The following list shows some of the most commonly used of those operations.

ANL A,*Rn* *(AND register to accumulator):* Logically AND register *Rn,* bit by bit, with the accumulator and store the result in the accumulator.

ANL A,*#data* *(AND data byte to accumulator):* Logically AND data byte *#data,* bit by bit, with the accumulator and store the result in the accumulator.

The descriptions of logical ORs and exclusive-ORs are similar to the AND, and use the following instructions:

ORL A,*Rn* *(OR register to accumulator)*

ORL A,*#data* *(OR data byte to accumulator)*

XRL A,*Rn* *(Ex-OR register to accumulator)*

XRL A,*#data* *(Ex-OR data byte to accumulator)*

The most commonly used instructions to operate on individual bits are as follows:

CLR *bit* *(Clear bit):* Clear (reset to 0) the value of the bit located at address *bit.*

SETB *bit* *(Set bit):* Set (set to 1) the value of the bit located at address *bit.*

CPL *bit* *(Complement bit):* Complement the value of the bit located at address *bit.*

The rotate commands available to the 8051 programmer are as follows:

RL A *(Rotate accumulator left):* Rotate the 8 bits of the accumulator one position to the left.

RLC A *(Rotate accumulator left through carry):* Rotate the 9 bits of the accumulator including carry one position to the left.

RR A *(Rotate accumulator right):* Rotate the 8 bits of the accumulator one position to the right.

RRC A *(Rotate accumulator right through carry):* Rotate the 9 bits of the accumulator including carry one position to the right.

The following examples illustrate the use of the logical and bit operations.

EXAMPLE 14–6

Determine the contents of the accumulator after the execution of the following program segments.

(a)
```
MOV A,#3CH      ; A ← 0011 1100
MOV R4,#66H     ; R4 ← 0110 0110
ANL A,R4        ; A ← A AND R4
```

Solution:

A = 0011 1100
R4 = 0110 0110
A AND R4 = 0010 0100 = 24H *Answer*

(b)
```
MOV A,#3FH      ; A ← 0011 1111
XRL A,#7CH      ; A ← EX-OR 7CH
```

Solution:

A = 0011 1111
7CH = 0111 1100
A EX-OR 7CH = 0100 0011 = 43H *Answer*

(c)
```
MOV A,#0A3H     ; A ← 1010 0011
RR A            ; Rotate right
```

Solution:

A = 1010 0011
Rotate right
A = 1101 0001 *Answer*

(d)
```
MOV A,#0C3H     ; A ← 1100 0011
RLC A           ; Rotate left through carry
```

Solution:

Assume carry = 0 initially
A = 1100 0011
Rotate left through carry
A = 1000 0110, carry = 1 *Answer*

EXAMPLE 14–7

Use the SETB, CLR, and CPL instructions to do the following operations:

(a) Clear bit 7 of the accumulator.
(b) Output a 1 on bit 0 of port 3.
(c) Complement the parity flag (bit 0 of the PSW).

Solution:

(a) CLR ACC.7 (*Note:* According to Table 14–3, the symbol for the direct address of the accumulator is ACC.)
(b) SETB P3.0
(c) CPL PSW.0

EXAMPLE 14–8

Describe the activity at the output of port 0 during the execution of the following program segment:

```
        MOV R7,#0AH
        MOV P0,#00H
LOOP:   CPL P0.7
        DJNZ R7,LOOP
```

Solution:

Register 7 is used as a loop counter with the initial value of 10 (0AH). Each time that the complement instruction (CPL) is executed, bit 7 will toggle to its opposite state. Toggling bit 7 ten times will create a waveform with five positive pulses.

Arithmetic Operations

The 8051 is capable of all the basic arithmetic functions: addition, subtraction, incrementing, decrementing, multiplication, and division. The following list outlines the most commonly used forms of the arithmetic instructions.

ADD A,*Rn* *(Add register to accumulator):* Add the contents of register *Rn* (where $n = 0–7$) to the accumulator and place the result in the accumulator.

ADD A,*#data* *(Add immediate data to accumulator):* Add the value of *#data* to the accumulator and place the result in the accumulator.

SUBB A,*Rn* *(Subtract register from accumulator with borrow):* Subtract the contents of register *Rn* and borrow (carry flag) from the accumulator and place the result in the accumulator.

SUBB A,*#data* *(Subtract immediate data from accumulator with borrow):* Subtract the value of *#data* and borrow (carry flag) from the accumulator and place the result in the accumulator.

INC A *(Increment accumulator)*

INC *Rn* *(Increment register)*

DEC A *(Decrement accumulator)*

DEC *Rn* *(Decrement register)*

MUL AB *(Multiply A times B):* Multiply the value in the accumulator times the value in the *B* register. The low-order byte of the 16-bit product is left in the accumulator and the high-order byte is placed in register *B*.

DIV AB *(Divide A by B):* Divide the value in the accumulator by the value in the *B* register. The accumulator receives the quotient and register *B* receives the remainder.

DA A *(Decimal adjust accumulator):* Adjust the value in the accumulator, resulting from an addition, into two BCD digits.

The following examples illustrate the use of arithmetic instructions.

EXAMPLE 14–9

Add the value being input at port 1 to the value at port 2 and send the result to port 3.

Solution:

```
MOV R0,P1     ; R0 ← P1
MOV A,P2      ; A ← P2
ADD A,R0      ; A ← A + R0
MOV P3,A      ; P3 ← A
```

EXAMPLE 14–10

Multiply the value being input at port 0 times the value at port 1 and send the result to ports 3 and 2 (high order, low order).

Solution:

```
MOV A,P0        ; A ← P0
MOV B,P1        ; B ← P1
MUL AB          ; A × B
MOV P2,A        ; P2 ← A (low order)
MOV P3,B        ; P3 ← B (high order)
```

14–5 8051 APPLICATIONS

Having bit-handling instructions and built-in I/O makes the 8051 a good choice for data acquisition and control applications. In this section we'll look at a few applications that illustrate how we can utilize these new features to simplify our program solutions.

Instruction Timing

The 8051 circuitry that we looked at earlier was driven with a 12-MHz crystal. The 12 MHz is a convenient choice because each instruction machine cycle takes 12 clock periods to complete. This means that one machine cycle takes 1 μs. All 8051 instructions are completed in one or two machine cycles (12 or 24 oscillator periods), except MUL and DIV, which take four. The oscillator periods for each instruction are given in the 8051 Instruction Set Summary Appendix. For example, the MOV A,*Rn* instruction requires 12 oscillator periods to complete, which will take 1 μs [12 × (1/12 MHz)].

Time Delay

Knowing the time duration of each instruction, we can write accurate time delays for our applications programs. Writing counter/loop programs is much easier now, with the introduction of the DJNZ and the CJNE instructions. Table 14–5 and Figure 14–7 show a time-delay program that we could call as a subroutine from an applications program.

TABLE 14–5
Time-Delay Subroutine

```
DELAY:   MOV R1,#0F3H     ; Outer-loop counter (F3H = 243)
         MOV R0,#00H      ; Inner-loop counter
LOOP:    DJNZ R0,LOOP     ; Loop 256 times
         DJNZ R1,LOOP     ; Loop 243 times
         RET              ; Return
```

The DJNZ instruction takes 24 oscillator periods, or 2 μs. By initializing register 0 at 00H, the first time DJNZ R0,LOOP is executed, R0 will become FFH. The program will loop within that same instruction 255 more times until R0 equals 0. At that point, control drops down to the next DJNZ, which functions as an outer loop as shown in the flowchart.

The R0 (inner) loop will take 512 μs to complete. The R1 (outer) loop executes the inner loop 243 times for a total time delay of approximately one-eighth second. A third loop, using R2, could be added to increase the time by another factor of 256.

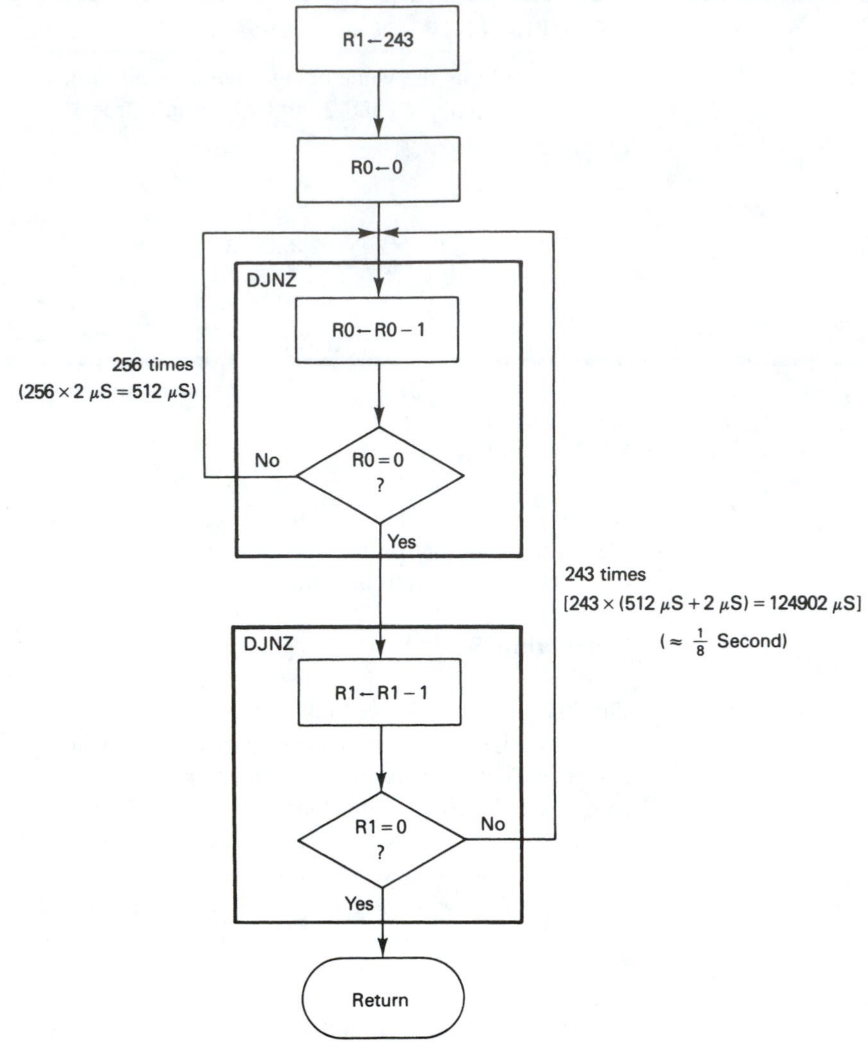

Figure 14–7 Flowchart for the time-delay subroutine of Table 14–5.

EXAMPLE 14–11

Assume that there are input switches connected to port 0 and output LEDs connected to port 1 of an 8051. Write a program that will flash the LEDs ON one second, OFF one second, the number of times indicated on the input switches.

Solution:

Label	*Instruction*	*Comments*
READ:	MOV A,P0	; Keep reading port 0 switches
	JZ READ	; into A until A ≠ 0
	MOV R0,A	; Transfer A to register 0
ON:	MOV P1,#0FFH	; Turn ON port 1 LEDs
	CALL DELAY	; Delay 1 second
OFF:	MOV P1,#00H	; Turn OFF port 1 LEDs
	CALL DELAY	; Delay 1 second
	DJNZ R0,ON	; Loop back number of times on switches
STOP:	JMP STOP	; Suspend operation
		;

Solution: (continued)

Label	Instruction	Comments
		; Delay 1 second subroutine
DELAY:	MOV R7,#08H	; Outer loop counter
	MOV R5,#00H	; Inner loop counter
LOOP2:	MOV R6,#0F3H	; Middle loop counter
LOOP1:	DJNZ R5,LOOP1	; LOOP1 delays for
	DJNZ R6,LOOP1	; one-eighth second
	DJNZ R7,LOOP2	; LOOP2 executes LOOP1 eight times
	RET	; Return

EXAMPLE 14–12

Write a program that will decode the hexadecimal keyboard shown in Figure 14–8.

Solution:

Label	Instruction	Comments
KEYSCAN:	CALL ROWRD	; Determine row of key pressed
	CALL COLRD	; Determine column of key pressed
	CALL CONVRT	; Convert row/column to key value
STOP:	JMP STOP	; Suspend operation
		;
ROWRD:	MOV P0,#0FH	; Output 0s to all columns
	MOV R0,#00H	; Row = 0
	JNB P0.0,RET1	; Return if row 0 is LOW
	MOV R0,#01H	; Row = 1
	JNB P0.1,RET1	; Return if row 1 is LOW
	MOV R0,#02H	; Row = 2
	JNB P0.2,RET1	; Return if row 2 is LOW
	MOV R0,#03H	; Row = 3
	JNB P0.3,RET1	; Return if row 3 is LOW
	JMP ROWRD	; Else keep reading
RET1:	RET	; Return
		;
COLRD:	MOV P0,#0F0H	; Output 0s to all rows
	MOV R1,#00H	; Column = 0
	JNB P0.4,RET2	; Return if column 0 is LOW
	MOV R1,#01H	; Column = 1
	JNB P0.5,RET2	; Return if column 1 is LOW
	MOV R1,#02H	; Column = 2
	JNB P0.6,RET2	; Return if column 2 is LOW
	MOV R1,#03H	; Column = 3
	JNB P0.7,RET2	; Return if column 3 is LOW
	JMP COLRD	; Else keep reading
RET2:	RET	; Return
		;
CONVRT:	MOV B,#04H	; B = Multiplication factor
	MOV A,R0	; Move row number to A
	MUL AB	; A = row × 4
	ADD A,R1	; A = row × 4 + column
	RET	; A now contains value of key pressed

Explanation:

The keyboard is wired as a 4 × 4 row–column matrix. The low-order nibble of port 0 is connected to the rows, and the high order is connected to the columns. The rows *and* columns are held HIGH with the 10-kΩ pull-up resistors. Since the I/O ports of the 8051 can be both read *and* written to (bidirectional), we can use

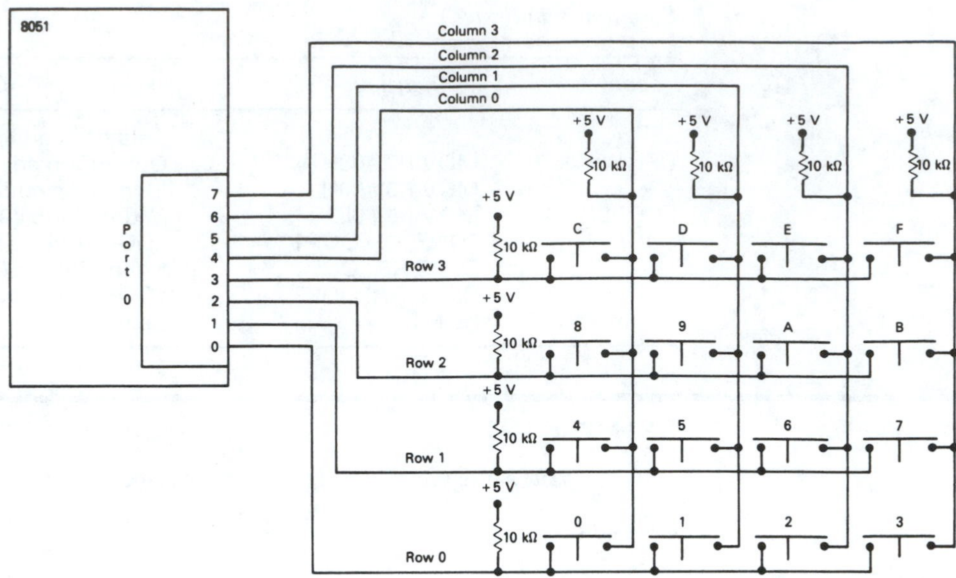

Figure 14–8 Keyboard interface to an 8051.

a different technique to scan a keyboard than we used with the 8085A. To determine the row of a depressed key, we drive all of the columns LOW and all of the rows HIGH by executing the instruction MOV P0,#0FH. The HIGH on the rows is actually a *float* state, allowing the rows to be read. The next series of instructions in the ROWRD subroutine read each row to determine the row number that is LOW, if any. If, for example, key 5 is depressed, then row 1 will be LOW and the other three rows will be HIGH.

Now that we know the row, we must next determine the column. The instruction MOV P0,#0F0H will drive the rows LOW and float the columns. If the number 5 key is still depressed, column 1 will be LOW.

The last subroutine converts the row–column combination to the numeric value of the key pressed. Rows 0, 1, 2, and 3 have weighting factors of 0, 4, 8, and 12, respectively, and columns 0, 1, 2, and 3 have weighting factors of 0, 1, 2, and 3, respectively. Knowing that, the CONVRT subroutine uses the following formula to determine the numeric value of the key pressed:

$$\text{Key pressed} = \text{row} \times 4 + \text{column}$$

EXAMPLE 14–13

Figure 14–9 shows how an ADC0801 is interfaced to an 8051. Write a program that takes care of the handshaking requirements for $\overline{SC}$ and $\overline{EOC}$ to complete an analog-to-digital conversion.

Solution:

Label	*Instruction*	*Comments*
FLOAT:	MOV P1,#0FFH	; Write 1s to port 1
	MOV P0,#0FFH	; Write 1s to port 0
SC:	CLR P0.6	; Output LOW-then-HIGH
	SETB P0.6	; on $\overline{SC}$
WAIT:	JB P0.7,WAIT	; Wait here until $\overline{EOC}$ goes LOW
DONE:	MOV R0,P1	; Transfer ADC result to register 0

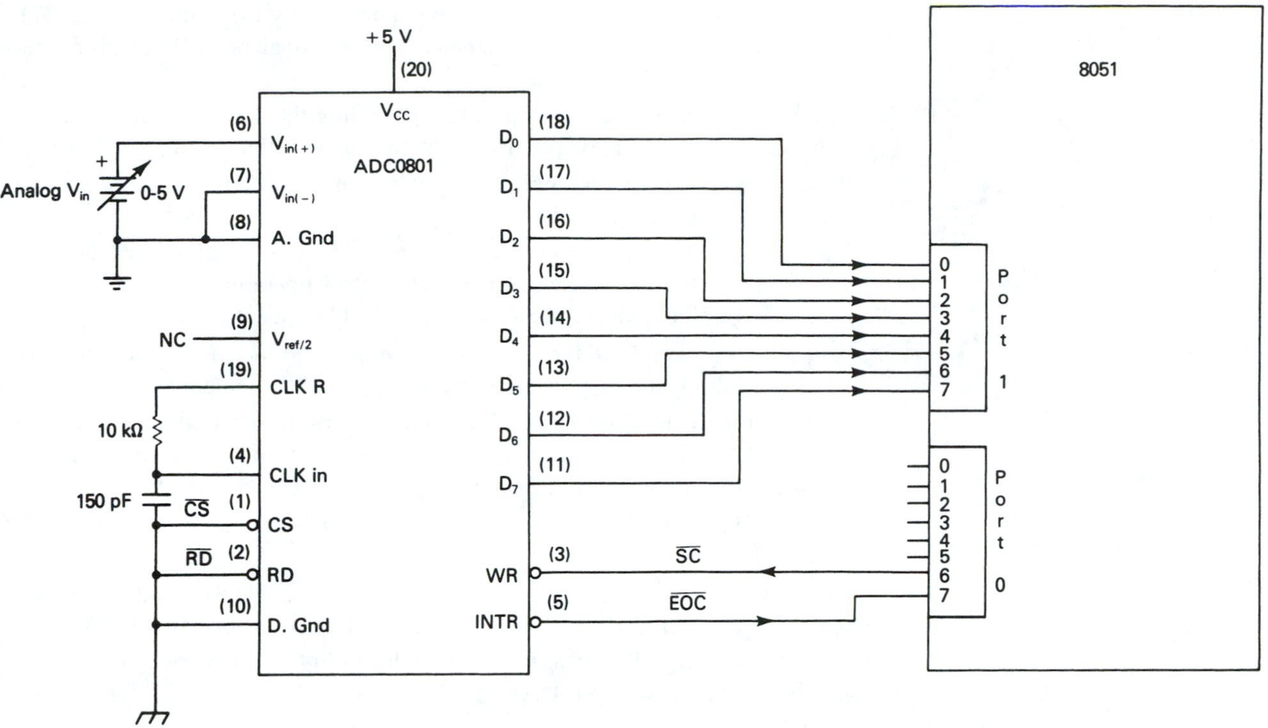

Figure 14–9 Interfacing an analog-to-digital converter to an 8051.

Explanation:

The I/O ports are in the float condition after the initial reset of the 8051. They must be floating in order to be used as inputs. The first two instructions in this program write 1s to ports 1 and 0, which make them float just in case they had 0s in them from a previous step in the program. To start the conversion process, bit 6 of port 0 ($\overline{SC}$) is pulsed LOW then HIGH using the clear (CLR) and set bit (SETB) instructions. The jump-if-bit-set instruction (JB) then monitors bit 7 of port 0 ($\overline{EOC}$) and remains in a WAIT loop until it goes LOW. Finally, at the end of conversion, port 1 is read into register 0, which is used to hold the ADC result.

SUMMARY

In this chapter we have learned that

1. The 8051 microcontroller is different from a microprocessor because it has the CPU, ROM, RAM, timer/counter, and parallel and serial ports fabricated into a single IC.
2. Thirty-two of the forty pins of the 8051 are used for the four 8-bit parallel I/O ports. Three of the ports share their function with the address, data, and control buses.
3. The address spaces of the 8051 are divided into four distinct areas: internal data memory, external data memory, internal code memory, and external code memory. The internal data memory is further divided into user RAM and Special Function Registers (SFRs).
4. To interface to an external EPROM like the 2732, an octal *D*-latch is required to demultiplex the address/data bus, which is shared with port 0. The External Access ($\overline{EA}$) pin is tied LOW and the ($\overline{PSEN}$) output is used to enable the output of the EPROM.

5. Extra data memory and I/O ports can be interfaced by using the 8155 IC. The 8155 demultiplexes the address/data bus internally so an octal *D*-latch is not required.
6. The MOV instruction is very powerful, providing the ability to move data almost anywhere internal or external to the microcontroller and to the I/O ports.
7. Program branching is accomplished by use of many different conditional and unconditional jumps and calls.
8. The 8051 instruction set provides the ability to work with individual bits, which makes it very efficient for on/off control operations. Instructions are available for all the logic functions, rotates, and bit manipulations.
9. Instructions are provided for all the basic arithmetic instructions: addition, subtraction, multiplication, division, incrementing, and decrementing.
10. Each instruction machine cycle takes 12 clock periods to complete. This means that if a 12 MHz crystal is used, each machine cycle takes 1 microsecond to complete. One complete instruction takes 1, 2, or 4 machine cycles.
11. A 4-by-4 matrix keyboard can be scanned using bit operations on a single I/O port.
12. Interfacing an 8-bit analog-to-digital converter to an 8051 is accomplished with one port and two bits on a second port. The *start-conversion* LOW pulse is issued with bit-setting instructions and the *end-of-conversion* signal is monitored with bit-checking instructions.

GLOSSARY

Assembler: A software package used to convert assembly language programs into executable machine code.

Bidirectional: An I/O port that can be used for both input and output.

Bit Addressable: Memory spaces in the internal RAM and SFR areas that allow for the addressing of individual bits.

External Access ($\overline{EA}$): An active-LOW input that when forced LOW, tells the processor that all program memory is external

Microcontroller: A computer on a chip. It contains a CPU, ROM, RAM, counter/timers, and I/O ports. It is especially well suited for control applications.

Program Store Enable ($\overline{PSEN}$): An active-Low output control signal issued by the microcontroller as the read strobe for external program memory.

Register Bank: A group of eight registers. The 8051 has four sets of banks that are selected by writing to the two bank-select bits in the PSW.

Special Function Register (SFR): A data register that has a dedicated space within the microcontroller. It is used to maintain the data required for microcontroller operations.

PROBLEMS

14–1. Why is a microcontroller sometimes referred to as a "computer on a chip"?

14–2. Describe the difference between the 8031, 8051, and the 8751.

14–3. What additional features does the 8052 have over the 8051?

14–4. Port 2 has a dual purpose. One is as a bidirectional I/O port. What is its other purpose?

14–5. If you are using the internal ROM in the 8051 for your program code memory, should the $\overline{EA}$ be tied HIGH or LOW?

14–6. What is the maximum size EPROM and RAM that can be interfaced to the 8051?

14–7. What is the address range of the Special Function Registers (SFRs)?

14–8. What are the SFR addresses of the four I/O ports?

14–9. The 8051 has four register banks with eight registers in each bank. How does the CPU know which bank is currently in use?

14–10. Why is there a need for the 74LS373 address latch when interfacing the 2732 EPROM in Figure 14–5 but not when interfacing the 8155 RAM in Figure 14–6?

14–11. Determine the value of the accumulator after the execution of instruction A:, B:, C:, and D:

```
        MOV 40H,#88H
        MOV R0,#40H
A:      MOV A,R0        ; A = _____
B:      MOV A,@R0       ; A = _____
C:      MOV A,40H       ; A = _____
D:      MOV A,#40H      ; A = _____
```

14–12. Repeat Problem 14–11 for the following instructions:

```
    MOV 50H,#33H
    MOV 40H,#22H
    MOV R0,#30H
    MOV 30H,50H
    MOV R1,#40H
A:  MOV A,40H       ; A = _____
B:  MOV A,@R0       ; A = _____
C:  MOV A,R1        ; A = _____
D:  MOV A,30H       ; A = _____
```

14–13. Write the instructions to perform each of the following operations:

(a) Output a C7H to port 3.
(b) Load port 1 into register 7.
(c) Load the accumulator with the number 55H.
(d) Send the contents of RAM memory, whose address is in register 0, to the accumulator.
(e) Output the contents of register 1 to port 0.

14–14. Write a program that continuously reads port 0 until the byte read equals A7H. At that time, turn on the output LEDs connected at port 1.

14–15. Write a program that will produce an output at port 1 that counts up from 20H to 90H repeatedly.

14–16. Determine the value of the accumulator after the execution of instruction A:, B:, and C:

```
    MOV A,#00H
    MOV R0,#36H
A:  XRL A,R0        ; A = _____
B:  ORL A,#71H      ; A = _____
C:  ANL A,#0F6H     ; A = _____
```

14–17. Repeat Problem 14–16 for the following program:

```
    MOV A,#77H
A:  CLR ACC.1       ; A = _____
B:  SETB ACC.7      ; A = _____
C:  RL A            ; A = _____
```

14–18. Modify one instruction in Example 14–8 so that it will output eight pulses at port 0 instead of five.

14–19. Write a program that adds 05H to the number read at port 0 and outputs the result to port 1.

14–20. What is the length of time of the following time-delay subroutine? (Assume a 12-MHz crystal is being used.)

```
DELAY:  MOV R1,#00H
        MOV R0,#00H
LOOP:   DJNZ R0,LOOP
        DJNZ R1,LOOP
        RET
```

SCHEMATIC INTERPRETATION PROBLEMS

14–21. Find Port 2 (P2.7–P2.0) of U8 in the 4096/4196 schematic. This port outputs the high-order address bits for the system (*A*8–*A*15). On a separate piece of paper, draw a binary comparator that compares the four bits *A*8–*A*11 to the four bits *A*12–*A*15. The HIGH output for an equal comparison is to be input to P3.4 (pin 14).

14–22. Locate the microcontroller in the 4096/4196 schematic.

(a) What is its grid location and part number?

(b) Its low-order address is multiplexed like the 68HC11 in Problem 11–37. What IC and control signal is used to demultiplex the address/data bus (*AD*0–*AD*7) into the low-order address bus (*A*0–*A*7)?

(c) What IC and control signal is used to demultiplex the address/data bus (*AD*0–*AD*7) into the data bus (*D*0–*D*7)?

WWW Sites

Useful WWW sites for this book:

www.acebus.com (AceBus) Downloadable 8051 microcontroller editor, assembler, and simulator

www.ahinc.com/scsi.htm (Advanced Horizons, Inc.) SCSI, CD, and DVD definitions and standards

www.altera.com (Altera Corporation) CPLDs and FPGAs

www.amanb.net/8085.html (ABCreations) 8085 simulator

www.amd.com (Advanced Micro Devices, Inc.) Microprocessors, EPROMs, and Flash memory

www.analog.com (Analog Devices, Inc.) Analog ICs, ADCs, and DACs

www.atmel.com (ATMEL Corporation) microcontrollers

www.bipom.com (BiPom Electronics, Inc.) Downloadable 8051 microcontroller editor, assembler, and simulator

www.chips.ibm.com (IBM Microelectronics Corporation) Microprocessors, SRAMs, DRAMs, and custom logic

www.cyrix.com (Cyrix Corporation) Microprocessors

www.digikey.com Sales catalog of electronic products and components

www.emacinc.com (EMAC, Inc.) Microprocessor and microcontroller trainers and single-board computers

www.elexp.com (Electronics Express, Inc.) Sales of electronic products, CPLD programmer boards, microprocessor trainers and components

www.fairchildsemi.com (Fairchild Semiconductor, Inc.) Digital Logic ICs, Analog ICs, discrete semiconductors, EEPROMs, and microcontrollers

www.fujitsumicro.com (Fujitsu, Inc.) Microcontrollers, DRAMs, and Flash memory

www.hitachi.com (Hitachi, LTD.) Microprocessors, DRAMs, SRAMs, Flash memories, EEPROMs, digital logic ICs, analog ICs, and discrete semiconductors www.idt.com (Integrated Device Technology, Inc.) SRAMs and digital logic ICs

www.insoluz.com (Insoluz Corporation) 8085 simulator

www.intel.com (Intel Corporation) Microprocessors, microcontrollers, and Flash memory

www.intersil.com (Intersil Corporation) Analog ICs, ADCs, DACs, SRAMs, microprocessors, microcontrollers, and discrete semiconductors

www.jameco.com Sales catalog of electronic products and components

www.jdr.com Sales catalog of electronic products and components

www.latticesemi.com (Lattice Semiconductor Corporation) CPLDs

www.linear-tech.com (Linear Technology Corporation) ADCs and DACs

www.microchip.com (Microchip Technology, Inc.) EEPROMs and microcontrollers

www.micronsemi.com (Micron Semiconductor Products, Inc.) DRAMs, SRAMs, and Flash memory

www.mot-sps.com (Motorola Semiconductors, Inc.) Analog ICs, digital logic ICs, microcontrollers, microprocessors, SRAMs, and discrete semiconductors

www.mp3-tech.org (MP3 Tech) MP3 definitions and standards

www.national.com (National Semiconductor Corp.) Analog ICs, digital logic ICs, and microcontrollers

www.onsemi.com (ON Semiconductor, Inc.) Analog ICs, digital logic ICs, and discrete semiconductors

www.semiconductors.philips.com (Philips Semiconductors, Inc.) Analog ICs, digital logic ICs, microcontrollers, and discrete semiconductors

www.sharpmeg.com (Sharp Microelectronics Corporation) Microprocessors, microcontrollers, SRAMs, and Flash memory

www.ti.com (Texas Instruments, Inc.) Analog ICs, digital logic ICs, and microcontrollers

www.toshiba.com (Toshiba America, Inc.) Analog ICs, digital logic ICs, microcontrollers, DRAMs, SRAMs, and Flash memory

www.urda.com (URDA, Inc.) Microprocessor trainers

www.usb.org (Universal Serial Bus) USB definitions and standards

www.xess.org (XESS Corporation) CPLD programmer boards

www.xilinx.com (Xilinx, Inc.) CPLDs and FPGAs

www.zilog.com (Zilog Inc.) Microprocessors and microcontrollers

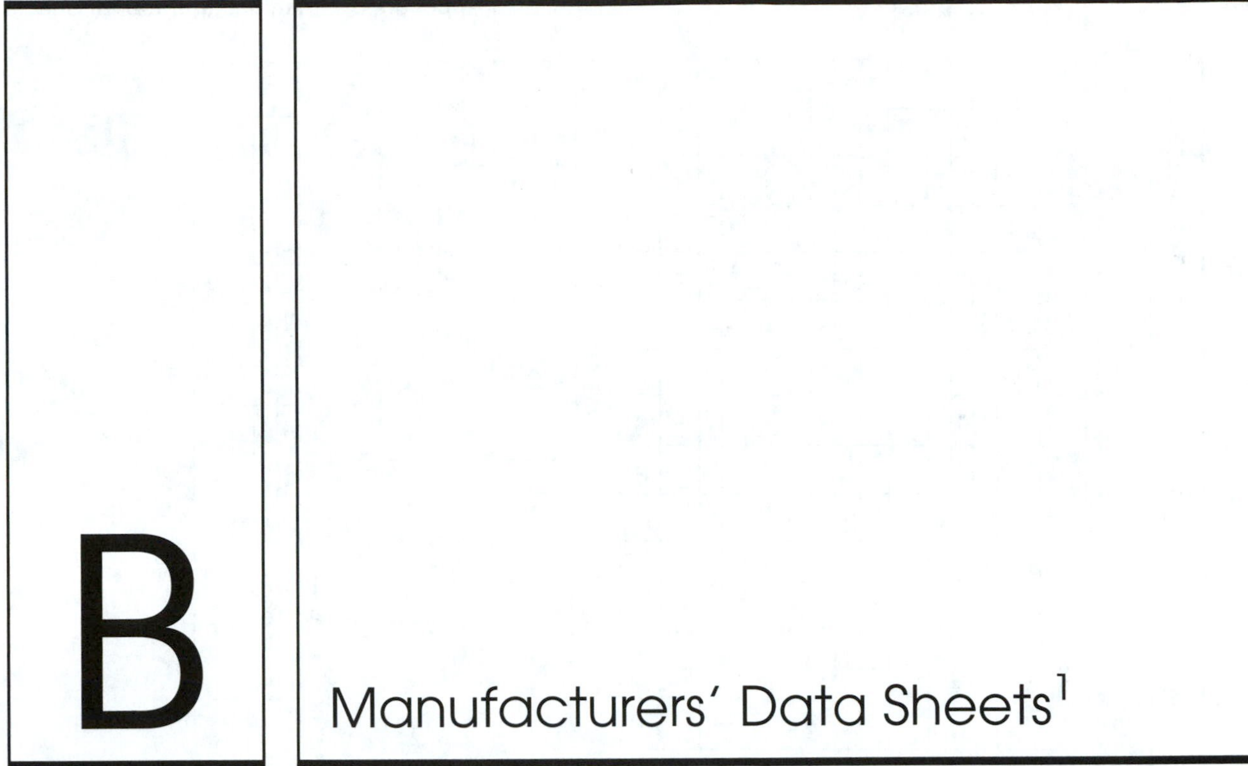

IC NUMBERS

7400
2732
ADC0801
MC1508/1408
μA741
NE555

[1]Courtesy of Philips Components—Signetics and Intel Corporation.

GATES — 54/7400, LS00, S00

Quad Two-Input NAND Gate

TYPE	TYPICAL PROPAGATION DELAY	TYPICAL SUPPLY CURRENT (Total)
7400	9ns	8mA
74LS00	9.5ns	1.6mA
74S00	3ns	15mA

ORDERING CODE

PACKAGES	COMMERCIAL RANGES $V_{CC} = 5V \pm 5\%$; $T_A = 0°C$ to $+70°C$	MILITARY RANGES $V_{CC} = 5V \pm 10\%$; $T_A = -55°C$ to $+125°C$
Plastic DIP	N7400N • N74LS00N N74S00N	
Plastic SO	N74LS00D N74S00D	
Ceramic DIP		S5400F • S54LS00F S54S00F
Flatpack		S5400W • S54LS00W S54S00W
LLCC		S54LS00G

FUNCTION TABLE

INPUTS		OUTPUT
A	B	Y
L	L	H
L	H	H
H	L	H
H	H	L

H = HIGH voltage level
L = LOW voltage level

INPUT AND OUTPUT LOADING AND FAN-OUT TABLE

PINS	DESCRIPTION	54/74	54/74S	54/74LS
A, B	Inputs	1ul	1Sul	1LSul
Y	Output	10ul	10Sul	10LSul

NOTE
Where a 54/74 unit load (ul) is understood to be 40µA I_{IH} and −1.6mA I_{IL}, a 54/74S unit load (Sul) is 50µA I_{IH} and −2.0mA I_{IL}, and 54/74LS unit load (LSul) is 20µA I_{IH} and −0.4mA I_{IL}

PIN CONFIGURATION

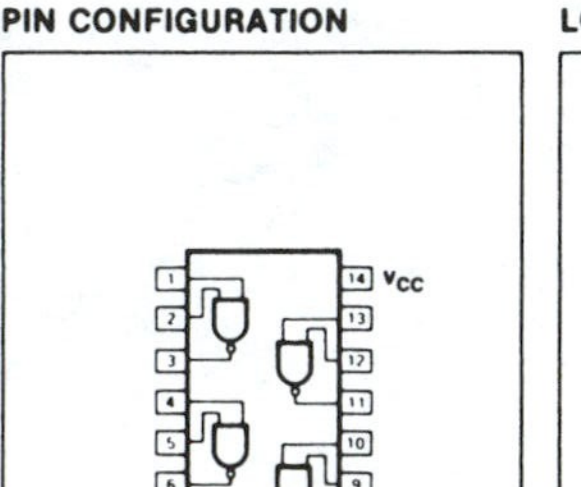

LOGIC SYMBOL

LOGIC SYMBOL (IEEE/IEC)

GATES — 54/7400, LS00, S00

DC ELECTRICAL CHARACTERISTICS (Over recommended operating free-air temperature range unless otherwise noted.)

PARAMETER		TEST CONDITIONS[1]			54/7400 Min	54/7400 Typ[2]	54/7400 Max	54/74LS00 Min	54/74LS00 Typ[2]	54/74LS00 Max	54/74S00 Min	54/74S00 Typ[2]	54/74S00 Max	UNIT
V_{OH}	HIGH-level output voltage	V_{CC} = MIN, V_{IH} = MIN, V_{IL} = MAX, I_{OH} = MAX		Mil	2.4	3.4		2.5	3.4		2.5	3.4		V
				Com'l	2.4	3.4		2.7	3.4		2.7	3.4		V
V_{OL}	LOW-level output voltage	V_{CC} = MIN, V_{IH} = MIN	I_{OL} = MAX	Mil		0.2	0.4		0.25	0.4			0.5[4]	V
				Com'l		0.2	0.4		0.35	0.5			0.5	V
			I_{OL} = 4mA	74LS					0.25	0.4				V
V_{IK}	Input clamp voltage	V_{CC} = MIN, $I_I = I_{IK}$					−1.5			−1.5			−1.2	V
I_I	Input current at maximum input voltage	V_{CC} = MAX	V_I = 5.5V				1.0						1.0	mA
			V_I = 7.0V							0.1				mA
I_{IH}	HIGH-level input current	V_{CC} = MAX	V_I = 2.4V				40							µA
			V_I = 2.7V							20			50	µA
I_{IL}	LOW-level input current	V_{CC} = MAX	V_I = 0.4V				−1.6			−0.4				mA
			V_I = 0.5V										−2.0	mA
I_{OS}	Short-circuit output current[3]	V_{CC} = MAX		Mil	−20		−55	−20		−100	−40		−100	mA
				Com'l	−18		−55	−20		−100	−40		−100	mA
I_{CC}	Supply current (total)	V_{CC} = MAX	I_{CCH}	Outputs HIGH		4	8		0.8	1.6		10	16	mA
			I_{CCL}	Outputs LOW		12	22		2.4	4.4		20	36	mA

NOTES
1. For conditions shown as MIN or MAX, use the appropriate value specified under recommended operating conditions for the applicable type.
2. All typical values are at $V_{CC} = 5V$, $T_A = 25°C$.
3. I_{OS} is tested with $V_{OUT} = +0.5V$ and $V_{CC} = V_{CC}$ MAX + 0.5V. Not more than one output should be shorted at a time and duration of the short circuit should not exceed one second.
4. $V_{OL} = +0.45V$ MAX for 54S at $T_A = +125°C$ only.

AC CHARACTERISTICS $T_A = 25°C$, $V_{CC} = 5.0V$

PARAMETER		TEST CONDITIONS	54/74 $C_L = 15pF$, $R_L = 400\Omega$ Min	54/74 Max	54/74LS $C_L = 15pF$, $R_L = 2k\Omega$ Min	54/74LS Max	54/74S $C_L = 15pF$, $R_L = 280\Omega$ Min	54/74S Max	UNIT
t_{PLH} t_{PHL}	Propagation delay	Waveform 1		22 15		15 15		4.5 5.0	ns

AC WAVEFORM

WAVEFORM FOR INVERTING OUTPUTS

V_{IN}, V_{OUT}, V_M, t_{PHL}, t_{PLH}

V_M = 1.3V for 54LS/74LS; V_M = 1.5V for all other TTL families.

Waveform 1

GATES — 54/7400, LS00, S00

ABSOLUTE MAXIMUM RATINGS (Over operating free-air temperature range unless otherwise noted.)

	PARAMETER	54	54LS	54S	74	74LS	74S	UNIT
V_{CC}	Supply voltage	7.0	7.0	7.0	7.0	7.0	7.0	V
V_{IN}	Input voltage	− 0.5 to + 5.5	− 0.5 to + 7.0	− 0.5 to + 5.5	− 0.5 to + 5.5	− 0.5 to + 7.0	− 0.5 to + 5.5	V
I_{IN}	Input current	− 30 to + 5	− 30 to + 1	− 30 to + 5	− 30 to + 5	− 30 to + 1	− 30 to + 5	mA
V_{OUT}	Voltage applied to output in HIGH output state	− 0.5 to + V_{CC}	− 0.5 to + V_{CC}	− 0.5 to + V_{CC}	− 0.5 to + V_{CC}	− 0.5 to + V_{CC}	− 0.5 to + V_{CC}	V
T_A	Operating free-air temperature range	− 55 to + 125			0 to 70			°C

RECOMMENDED OPERATING CONDITIONS

PARAMETER			54/74			54/74LS			54/74S			UNIT
			Min	Nom	Max	Min	Nom	Max	Min	Nom	Max	
V_{CC}	Supply voltage	Mil	4.5	5.0	5.5	4.5	5.0	5.5	4.5	5.0	5.5	V
		Com'l	4.75	5.0	5.25	4.75	5.0	5.25	4.75	5.0	5.25	V
V_{IH}	HIGH-level input voltage		2.0			2.0			2.0			V
V_{IL}	LOW-level input voltage	Mil			+ 0.8			+ 0.7			+ 0.8	V
		Com'l			+ 0.8			+ 0.8			+ 0.8	V
I_{IK}	Input clamp current				− 12			− 18			− 18	mA
I_{OH}	HIGH-level output current				− 400			− 400			− 1000	μA
I_{OL}	LOW-level output current	Mil			16			4			20	mA
		Com'l			16			8			20	mA
T_A	Operating free-air temperature	Mil	− 55		+ 125	− 55		+ 125	− 55		+ 125	°C
		Com'l	0		70	0		70	0		70	°C

NOTE

V_{IL} = + 0.7V MAX for 54S at T_A = + 125°C only

TEST CIRCUITS AND WAVEFORMS

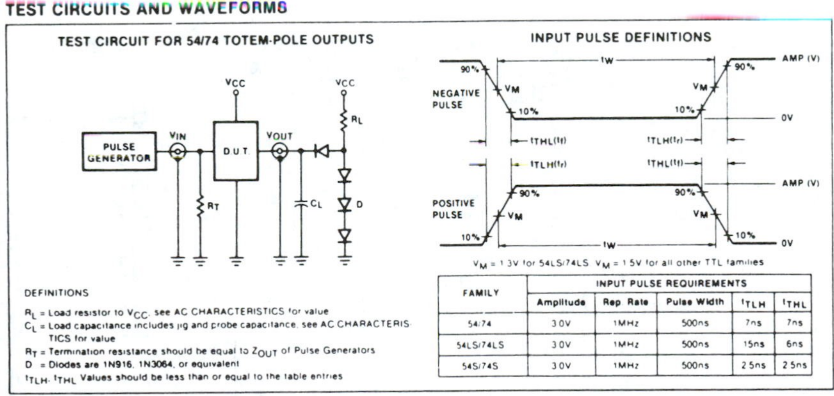

DEFINITIONS

R_L = Load resistor to V_{CC}. see AC CHARACTERISTICS for value

C_L = Load capacitance includes jig and probe capacitance. see AC CHARACTERISTICS for value

R_T = Termination resistance should be equal to Z_{OUT} of Pulse Generators

D = Diodes are 1N916, 1N3064, or equivalent

t_{TLH}, t_{THL} Values should be less than or equal to the table entries

FAMILY	INPUT PULSE REQUIREMENTS				
	Amplitude	Rep. Rate	Pulse Width	t_{TLH}	t_{THL}
54/74	3.0V	1MHz	500ns	7ns	7ns
54LS/74LS	3.0V	1MHz	500ns	15ns	6ns
54S/74S	3.0V	1MHz	500ns	2.5ns	2.5ns

2732A
32K (4K x 8) PRODUCTION AND UV ERASABLE PROMS

- **200 ns (2732A-2) Maximum Access Time . . . HMOS*-E Technology**
- **Compatible with High-Speed Microcontrollers and Microprocessors . . . Zero WAIT State**
- **Two Line Control**
- **10% V_{CC} Tolerance Available**
- **Low Current Requirement**
  - **100 mA Active**
  - **35 mA Standby**
- **int$_e$ligent Identifier™ Mode**
  - **Automatic Programming Operation**
- **Industry Standard Pinout . . . JEDEC Approved 24 Pin Ceramic Package** (See Packaging Spec. Order #231369)

The Intel 2732A is a 5V-only, 32,768-bit ultraviolet erasable (cerdip) Electrically Programmable Read-Only Memory (EPROM). The standard 2732A access time is 250 ns with speed selection (2732A-2) available at 200 ns. The access time is compatible with high performance microprocessors such as the 8 MHz iAPX 186. In these systems, the 2732A allows the microprocessor to operate without the addition of WAIT states.

An important 2732A feature is Output Enable ($\overline{OE}$) which is separate from the Chip Enable ($\overline{CE}$) control. The $\overline{OE}$ control eliminates bus contention in microprocessor systems. The $\overline{CE}$ is used by the 2732A to place it in a standby mode ($\overline{CE} = V_{IH}$) which reduces power consumption without increasing access time. The standby mode reduces the current requirement by 65%; the maximum active current is reduced from 100 mA to a standby current of 35 mA.

*HMOS is a patented process of Intel Corporation.

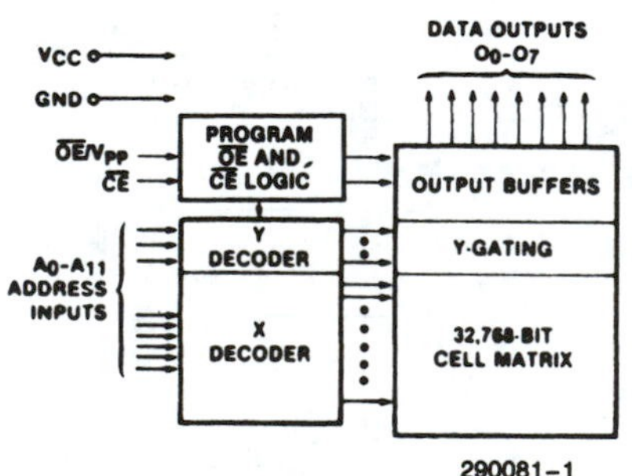

290081-1

Figure 1. Block Diagram

Pin Names

A_0–A_{11}	Addresses
$\overline{CE}$	Chip Enable
$\overline{OE}/V_{PP}$	Output Enable/V_{PP}
O_0–O_7	Outputs

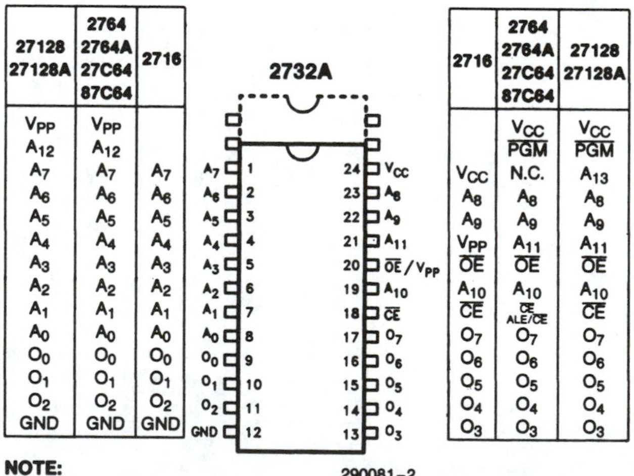

27128 27128A	2764 2764A 27C64 87C64	2716	2732A	Pin	Pin	2732A	2716	2764 2764A 27C64 87C64	27128 27128A
V_{PP}	V_{PP}							V_{CC}	V_{CC}
A_{12}	A_{12}							$\overline{PGM}$	$\overline{PGM}$
A_7	A_7	A_7	A_7	1	24	V_{CC}	V_{CC}	N.C.	A_{13}
A_6	A_6	A_6	A_6	2	23	A_8	A_8	A_8	A_8
A_5	A_5	A_5	A_5	3	22	A_9	A_9	A_9	A_9
A_4	A_4	A_4	A_4	4	21	A_{11}	V_{PP}	A_{11}	A_{11}
A_3	A_3	A_3	A_3	5	20	$\overline{OE}/V_{PP}$	$\overline{OE}$	$\overline{OE}$	$\overline{OE}$
A_2	A_2	A_2	A_2	6	19	A_{10}	A_{10}	A_{10}	A_{10}
A_1	A_1	A_1	A_1	7	18	$\overline{CE}$	$\overline{CE}$	$\overline{CE}$ ALE/$\overline{CE}$	$\overline{CE}$
A_0	A_0	A_0	A_0	8	17	O_7	O_7	O_7	O_7
O_0	O_0	O_0	O_0	9	16	O_6	O_6	O_6	O_6
O_1	O_1	O_1	O_1	10	15	O_5	O_5	O_5	O_5
O_2	O_2	O_2	O_2	11	14	O_4	O_4	O_4	O_4
GND	GND	GND	GND	12	13	O_3	O_3	O_3	O_3

290081-2

NOTE:
Intel "Universal Site" compatible EPROM configurations are shown in the blocks adjacent to the 2732A pins.

Figure 2. Cerdip Pin Configuration

DEVICE OPERATION

The modes of operation of the 2732A are listed in Table 1. A single 5V power supply is required in the read mode. All inputs are TTL levels except for $\overline{OE}/V_{PP}$ during programming and 12V on A_9 for the int$_e$ligent Identifier™ mode. In the program mode the $\overline{OE}/V_{PP}$ input is pulsed from a TTL level to 21V.

Table 1. Mode Selection

Mode \ Pins	$\overline{CE}$	$\overline{OE}/V_{PP}$	A_9	A_0	V_{CC}	Outputs
Read/Program Verify	V_{IL}	V_{IL}	X	X	V_{CC}	D_{OUT}
Output Disable	V_{IL}	V_{IH}	X	X	V_{CC}	High Z
Standby	V_{IH}	X	X	X	V_{CC}	High Z
Program	V_{IL}	V_{PP}	X	X	V_{CC}	D_{IN}
Program Inhibit	V_{IH}	V_{PP}	X	X	V_{CC}	High Z
Int$_e$ligent Identifier(3)						
—Manufacturer	V_{IL}	V_{IL}	V_H	V_{IL}	V_{CC}	89H
—Device	V_{IL}	V_{IL}	V_H	V_{IH}	V_{CC}	01H

NOTES:
1. X can be V_{IH} or V_{IL}.
2. V_H = 12V ±0.5V.
3. A_1–A_8, A_{10}, A_{11} = V_{IL}.

Read Mode

The 2732A has two control functions, both of which must be logically active in order to obtain data at the outputs. Chip Enable ($\overline{CE}$) is the power control and should be used for device selection. Output Enable ($\overline{OE}/V_{PP}$) is the output control and should be used to gate data from the output pins, independent of device selection. Assuming that addresses are stable, address access time (t_{ACC}) is equal to the delay from $\overline{CE}$ to output (t_{CE}). Data is available at the outputs after the falling edge of $\overline{OE}/V_{PP}$, assuming that $\overline{CE}$ has been low and addresses have been stable for at least t_{ACC}–t_{OE}.

Standby Mode

EPROMs can be placed in a standby mode which reduces the maximum active current of the device by applying a TTL-high signal to the $\overline{CE}$ input. When in standby mode, the outputs are in a high impedance state, independent of the $\overline{OE}/V_{PP}$ input.

Two Line Output Control

Because EPROMs are usually used in larger memory arrays, Intel has provided two control lines which accommodate this multiple memory connection. The two control lines allow for:

a) The lowest possible memory power dissipation, and

b) complete assurance that output bus contention will not occur.

To use these two control lines most efficiently, $\overline{CE}$ should be decoded and used as the primary device selecting function, while $\overline{OE}/V_{PP}$ should be made a common connection to all devices in the array and connected to the $\overline{READ}$ line from the system control bus. This assures that all deselected memory devices are in their low power standby mode and that the output pins are active only when data is desired from a particular memory device.

SYSTEM CONSIDERATION

The power switching characteristics of EPROMs require careful decoupling of the devices. The supply current, I_{CC}, has three segments that are of interest to the system designer—the standby current level, the active current level, and the transient current peaks that are produced by the falling and rising edges of Chip Enable. The magnitude of these transient current peaks is dependent on the output capacitive and inductive loading of the device. The associated transient voltage peaks can be suppressed by complying with Intel's two-line control and by use of properly selected decoupling capacitors. It is recommended that a 0.1 μF ceramic capacitor be used on every device between V_{CC} and GND. This should be a high frequency capacitor of low inherent inductance and should be placed as close to the device as possible. In addition, a 4.7 μF bulk electrolytic capacitor should be used between V_{CC} and GND for every eight devices. The bulk capacitor should be located near where the power supply is connected to the array. The purpose of the bulk capacitor is to overcome the voltage droop caused by the inductive effects of PC board traces.

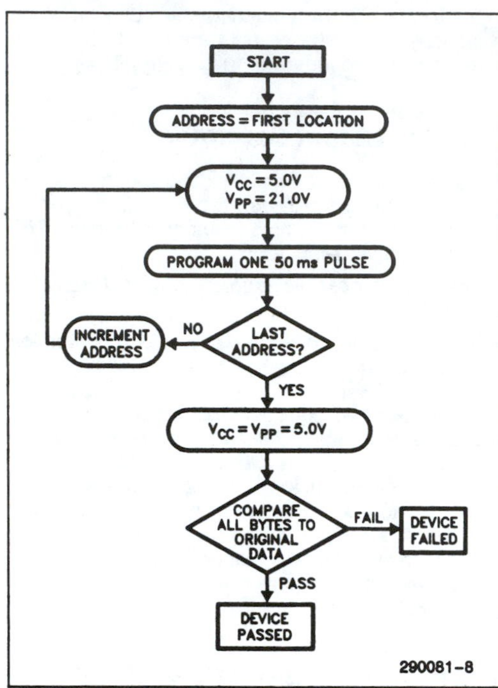

Figure 3. Standard Programming Flowchart

PROGRAMMING MODES

CAUTION: Exceeding 22V on $\overline{OE}/V_{PP}$ will permanently damage the device.

Initially, and after each erasure (cerdip EPROMs), all bits of the EPROM are in the "1" state. Data is introduced by selectively programming "0s" into the bit locations. Although only "0s" will be programmed, both "1s" and "0s" can be present in the data word. The only way to change a "0" to a "1" in cerdip EPROMs is by ultraviolet light erasure.

The device is in the programming mode when the $\overline{OE}/V_{PP}$ input is at 21V. It is required that a 0.1 μF capacitor be placed across $\overline{OE}/V_{PP}$ and ground to suppress spurious voltage transients which may damage the device. The data to be programmed is applied 8 bits in parallel to the data output pins. The levels required for the address and data inputs are TTL.

When the address and data are stable, a 20 ms (50 ms typical) active low, TTL program pulse is applied to the $\overline{CE}$ input. A program pulse must be applied at each address location to be programmed (see Figure 3). Any location can be programmed at any time—either individually, sequentially, or at random. The program pulse has a maximum width of 55 ms. The EPROM must not be programmed with a DC signal applied to the $\overline{CE}$ input.

Programming of multiple 2732As in parallel with the same data can be easily accomplished due to the simplicity of the programming requirements. Like inputs of the paralleled 2732As may be connected together when they are programmed with the same data. A low level TTL pulse applied to the $\overline{CE}$ input programs the paralleled 2732As.

Program Inhibit

Programming of multiple EPROMs in parallel with different data is easily accomplished by using the Program Inhibit mode. A high level $\overline{CE}$ input inhibits the other EPROMs from being programmed. Except for $\overline{CE}$, all like inputs (including $\overline{OE}/V_{PP}$) of the parallel EPROMs may be common. A TTL low level pulse applied to the $\overline{CE}$ input with $\overline{OE}/V_{PP}$ at 21V will program that selected device.

Program Verify

A verify (Read) should be performed on the programmed bits to determine that they have been correctly programmed. The verify is performed with $\overline{OE}/V_{PP}$ and $\overline{CE}$ at V_{IL}. Data should be verified t_{DV} after the falling edge of $\overline{CE}$.

int$_e$ligent Identifier™ Mode

The int$_e$ligent Identifier Mode allows the reading out of a binary code from an EPROM that will identify its manufacturer and type. This mode is intended for use by programming equipment for the purpose of automatically matching the device to be programmed with its corresponding programming algorithm. This mode is functional in the 25°C ±5°C ambient temperature range that is required when programming the device.

To activate this mode, the programming equipment must force 11.5V to 12.5V on address line A9 of the EPROM. Two identifier bytes may then be sequenced from the device outputs by toggling address line A0 from V_{IL} to V_{IH}. All other address lines must be held at V_{IL} during the int$_e$ligent Identifier Mode.

Byte 0 (A0 = V_{IL}) represents the manufacturer code and byte 1 (A0 = V_{IH}) the device identifier code. These two identifier bytes are given in Table 1.

2732A

ERASURE CHARACTERISTICS (FOR CERDIP EPROMS)

The erasure characteristics are such that erasure begins to occur upon exposure to light with wavelengths shorter than aproximately 4000 Angstroms (Å). It should be noted that sunlight and certain types of fluorescent lamps have wavelengths in the 3000–4000Å range. Data shows that constant exposure to room level fluorescent lighting could erase the EPROM in approximately 3 years, while it would take approximately 1 week to cause erasure when exposed to direct sunlight. If the device is to be exposed to these types of lighting conditions for extended periods of time, opaque labels should be placed over the window to prevent unintentional erasure.

The recommended erasure procedure is exposure to shortwave ultraviolet light which has a wavelength of 2537 Angstroms (Å). The integrated dose (i.e., UV intensity × exposure time) for erasure should be a minimum of 15 Wsec/cm^2. The erasure time with this dosage is approximately 15 to 20 minutes using an ultraviolet lamp with a 12000 μW/cm^2 power rating. The EPROM should be placed within 1 inch of the lamp tubes during erasure. The maximum integrated dose an EPROM can be exposed to without damage is 7258 Wsec/cm^2 (1 week @ 12000 μW/cm^2). Exposure of the device to high intensity UV light for longer periods may cause permanent damage.

PROGRAMMING

D.C. PROGRAMMING CHARACTERISTICS

T_A = 25°C ±5°C, V_{CC} = 5V ±5%, V_{PP} = 21V ±0.5V

Symbol	Parameter	Limits			Units	Test Conditions (Note 1)
		Min	Typ[3]	Max		
I_{LI}	Input Current (All Inputs)			10	μA	$V_{IN} = V_{IL}$ or V_{IH}
V_{IL}	Input Low Level (All Inputs)	−0.1		0.8	V	
V_{IH}	Input High Level (All Inputs Except $\overline{OE}/V_{PP}$)	2.0		V_{CC} + 1	V	
V_{OL}	Output Low Voltage During Verify			0.45	V	I_{OL} = 2.1 mA
V_{OH}	Output High Voltage During Verify	2.4			V	I_{OH} = −400 μA
I_{CC_2}[4]	V_{CC} Supply Current (Program and Verify)		85	100	mA	
I_{PP_2}[4]	V_{PP} Supply Current (Program)			30	mA	$\overline{CE} = V_{IL}$, $\overline{OE}/V_{PP} = V_{PP}$
V_{ID}	A_9 int$_e$ligent Identifier Voltage	11.5		12.5	V	

CMOS 8-BIT A/D CONVERTERS — ADC0801/2/3/4/5-1

Preliminary

DESCRIPTION

The ADC0801 family is a series of five CMOS 8-bit successive approximation A/D converters using a resistive ladder and capacitive array together with an auto-zero comparator. These converters are designed to operate with microprocessor controlled buses using a minimum of external circuitry. The three-state output data lines can be connected directly to the data bus.

The differential analog voltage input allows for increased common-mode rejection and provides a means to adjust the zero scale offset. Additionally, the voltage reference input provides a means of encoding small analog voltages to the full 8 bits of resolution.

FEATURES

- **Compatible with most microprocessors**
- **Differential inputs**
- **Three-state outputs**
- **Logic levels TTL and MOS compatible**
- **Can be used with internal or external clock**
- **Analog input range 0V to V_{CC}**
- **Single 5V supply**
- **Guaranteed specification with 1MHz clock**

APPLICATIONS

- **Transducer to microprocessor interface**
- **Digital thermometer**
- **Digitally-controlled thermostat**
- **Microprocessor-based monitoring and control systems**

PIN CONFIGURATION

F,N PACKAGE

Pin	Name	Pin	Name
1	$\overline{CS}$	20	V_{CC}
2	$\overline{RD}$	19	CLK R
3	$\overline{WR}$	18	D0
4	CLK IN	17	D1
5	$\overline{INTR}$	16	D2
6	$V_{IN}(+)$	15	D3
7	$V_{IN}(-)$	14	D4
8	A GND	13	D5
9	V_{REF} 2	12	D6
10	D GND	11	D7

TOP VIEW

ORDER NUMBERS
ADC0801/02-1F
ADC0801/02/03-1 LCF
ADC081/02/03/04/05-1 LCN
ADC0804-1 CN

ABSOLUTE MAXIMUM RATINGS

SYMBOL & PARAMETER		RATING	UNIT
V_{CC}	Supply Voltage	6.5	V
	Logic Control Input Voltages	−0.3 to +16	V
	All Other Input Voltages	−0.3 to (V_{CC} +0.3)	V
T_A	Operating Temperature Range ADC0801/02-1 F	−55 to +125	°C
	ADC0801/02/03-1 LCF	−40 to +85	°C
	ADC0801/02/03/04/05-1 LCN	−40 to +85	°C
	ADC0804-1 CN	0 to +70	°C
T_{STG}	Storage Temperature	−65 to +150	°C
T_{SOLD}	Lead Soldering Temperature (10 seconds)	300	°C
P_D	Package Power Dissipation at T_A = 25°C	875	mW

CMOS 8-BIT A/D CONVERTERS — ADC0801/2/3/4/5-1

Preliminary

BLOCK DIAGRAMS

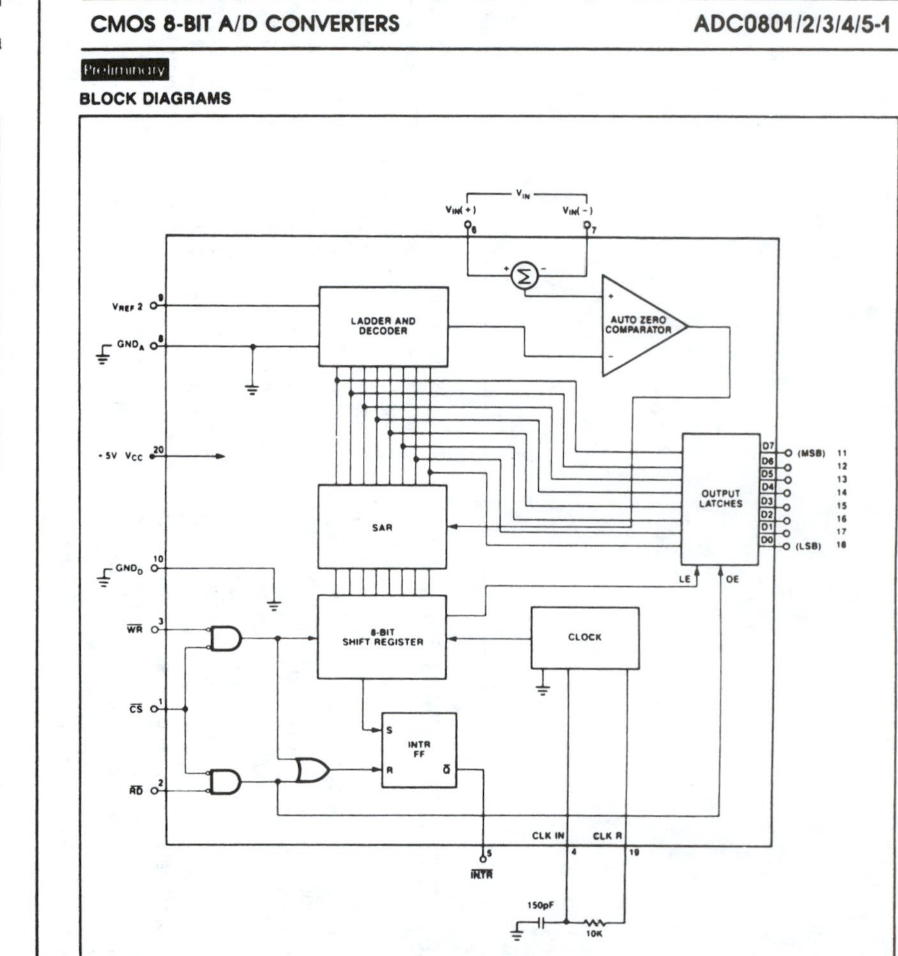

CMOS 8-BIT A/D CONVERTERS — ADC0801/2/3/4/5-1

Preliminary

DC ELECTRICAL CHARACTERISTICS V_{CC} = 5.0V, f_{CLK} = 1MHz, $T_{MIN} \le T_A \le T_{MAX}$, unless otherwise specified.

SYMBOL & PARAMETER	TEST CONDITIONS	ADC0801/2/3/4/5 Min	Typ	Max	UNIT
ADC0801 Relative Accuracy Error (Adjusted)	Full Scale Adjusted			0.25	LSB
ADC0802 Relative Accuracy Error (Unadjusted)	$\frac{V_{REF}}{2}$ = 2.500 V_{DC}			0.50	LSB
ADC0803 Relative Accuracy Error (Adjusted)	Full Scale Adjusted			0.50	LSB
ADC0804 Relative Accuracy Error (Unadjusted)	$\frac{V_{REF}}{2}$ = 2.500 V_{CC}			1	LSB
ADC0805 Relative Accuracy Error (Unadjusted)	$\frac{V_{REF}}{2}$ = has no connection			1	LSB
$\frac{V_{REF}}{2}$ Input Resistance		400	640		Ω
Analog Input Voltage Range		−0.05		V_{CC} +0.05	V
DC Common Mode Error	Over Analog Input Voltage Range		1/16	1/8	LSB
Power Supply Sensitivity	V_{CC} = 5V ± 10%[1]				
CONTROL INPUTS					
V_{IH} Logical "1" Input Voltage	V_{CC} = 5.25V_{DC}	2.0		15	V_{DC}
V_{IL} Logical "0" Input Voltage	V_{CC} = 4.75V_{DC}			0.8	V_{DC}
I_{IH} Logical "1" Input Current	V_{IN} = 5V_{DC}		0.005	1	μA_{DC}
I_{IL} Logical "0" Input Current	V_{IN} = 0V_{DC}	−1	0.005		μA_{DC}
CLOCK IN AND CLOCK R					
V_{T+} Clk In Positive-Going Threshold Voltage		2.7	3.1	3.5	V_{DC}
V_{T-} Clk In Negative-Going Threshold Voltage		1.5	1.8	2.1	V_{DC}
V_H Clk In Hysteresis $(V_{T+}) - (V_{T-})$		0.6	1.3	2.0	V_{DC}
V_{OL} Logical "0" Clk R Output Voltage	I_{OL} = 360μA, V_{CC} = 4.75 V_{DC}			0.4	V_{DC}
V_{OH} Logical "1" Clk R Output Voltage	I_{OH} = −360μA, V_{CC} = 4.75 V_{DC}	2.4			V_{DC}
DATA OUTPUT AND $\overline{INTR}$					
V_{OL} Logical "0" Output Voltage					
Data Outputs	I_{OL} = 1.6mA, V_{CC} = 4.75 V_{DC}			0.4	V_{DC}
$\overline{INTR}$ Outputs	I_{OL} = 1.0mA, V_{CC} = 4.75 V_{DC}			0.4	V_{DC}
V_{OH} Logical "1" Output Voltage	I_{OH} = −360μA, V_{CC} = 4.75 V_{DC}	2.4			V_{DC}
	I_{OH} = −10μA, V_{CC} = 4.75 V_{DC}	4.5			V_{DC}
I_{OZL} 3-State Output Leakage	V_{OUT} = 0V_{DC}, $\overline{CS}$ = Logical "1"	−3			μA_{DC}
I_{OZH} 3-State Output Leakage	V_{OUT} = 5V_{DC}, $\overline{CS}$ = Logical "1"			3	μA_{DC}
I_{SC} + Output Short Circuit Current	V_{OUT} = O_V, T_A = 25°C	4.5	6		mA_{DC}
I_{SC} − Output Short Circuit Current	V_{OUT} = V_{CC}, T_A = 25°C	9.0	16		mA_{DC}
I_{CC} Power Supply Current	f_{CLK} = 1MHz, $V_{REF/2}$ = Open $\overline{CS}$ = Logical "1", T_A = 25°C		3.0	3.5	mA

NOTE:
1. Analog inputs must remain within the range: $-0.05 \le V_{IN} \le V_{CC} + 0.05V$.

CMOS 8-BIT A/D CONVERTERS — ADC0801/2/3/4/5-1

Preliminary

AC ELECTRICAL CHARACTERISTICS

SYMBOL & PARAMETER	TO	FROM	TEST CONDITIONS	ADC0801/2/3/4/5 Min	Typ	Max	UNIT
Conversion Time			f_{CLK} = 1MHz[1]	66		73	μs
f_{CLK} Clock Frequency			See Note 1.	0.1	1.0	3.0	MHz
Clock Duty Cycle			See Note 1.	40		60	%
CR Free-Running Conversion Rate			$\overline{CS}$ = 0, f_{CLK} = 1MHz $\overline{INTR}$ Tied To $\overline{WR}$			13690	conv/s
$t_{W(\overline{WR})L}$ Start Pulse Width			$\overline{CS}$ = 0	30			ns
t_{ACC} Access Time	Output	$\overline{RD}$	$\overline{CS}$ = 0, C_L = 100 pF		75	100	ns
t_{1H}, t_{OH} Three-State Control	Output	$\overline{RD}$	CL = 10 pF, RL = 10K See Three-State Test Circuit		70	100	ns
t_{W1}, t_{R1} $\overline{INTR}$ Delay	$\overline{INTR}$	$\overline{WD}$ or $\overline{RD}$			100	150	ns
C_{IN} Logic Input Capacitance					5	7.5	pF
C_{OUT} Three-State Output Capacitance					5	7.5	pF

NOTE
1. Accuracy is guaranteed at f_{CLK} = 1MHz. Accuracy may degrade at higher clock frequencies.

CMOS 8-BIT A/D CONVERTERS — ADC0801/2/3/4/5-1

Preliminary

FUNCTIONAL DESCRIPTION

The ADC0801 through ADC0805 series of A/D converters are successive approximation devices with 8-bit resolution and no missing codes. The most significant bit is tested first and after 64 clock cycles a digital 8-bit binary word is transferred to an output latch and the $\overline{\text{INTR}}$ pin goes low, indicating that conversion is complete. A conversion in progress can be interrupted by issuing another start command. The device may be operated in a continuous conversion mode by connecting the $\overline{\text{INTR}}$ and $\overline{\text{WR}}$ pins together and holding the $\overline{\text{CS}}$ pin low. To insure start-up when connected this way, an external $\overline{\text{WR}}$ pulse is required at power-up.

As the $\overline{\text{WR}}$ input goes low, when $\overline{\text{CS}}$ is low, the SAR is cleared and remains so as long as these two inputs are low. Conversion begins between 1 and 8 clock periods after at least one of these inputs goes high. As the conversion begins, the $\overline{\text{INTR}}$ line goes high. Note that the $\overline{\text{INTR}}$ line will remain low until 1 to 8 clock cycles after either the $\overline{\text{WR}}$ or the $\overline{\text{CS}}$ input (or both) goes high.

When the $\overline{\text{CS}}$ and $\overline{\text{RD}}$ inputs are both brought low to read the data, the $\overline{\text{INTR}}$ line will go low and the three-state output latches are enabled.

The digital control lines ($\overline{\text{CS}}$, $\overline{\text{RD}}$, and $\overline{\text{WR}}$) operate with standard TTL levels and have been renamed when compared with standard A/D Start and Output Enable labels. For non-microprocessor based applications, the $\overline{\text{CS}}$ pin can be grounded, the $\overline{\text{WR}}$ pin can be interpreted as a $\overline{\text{START}}$ pulse pin, and the $\overline{\text{RD}}$ pin performs the OE (Output Enable) function.

The $V_{IN}(-)$ input can be used to subtract a fixed voltage from the input voltage. Because there is a time interval between sampling the $V_{IN}(+)$ and the $V(-)$ inputs, it is important that these inputs remain constant, during the entire conversion cycle.

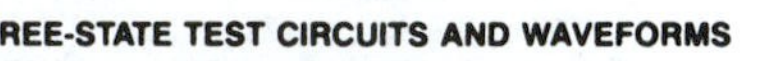

THREE-STATE TEST CIRCUITS AND WAVEFORMS

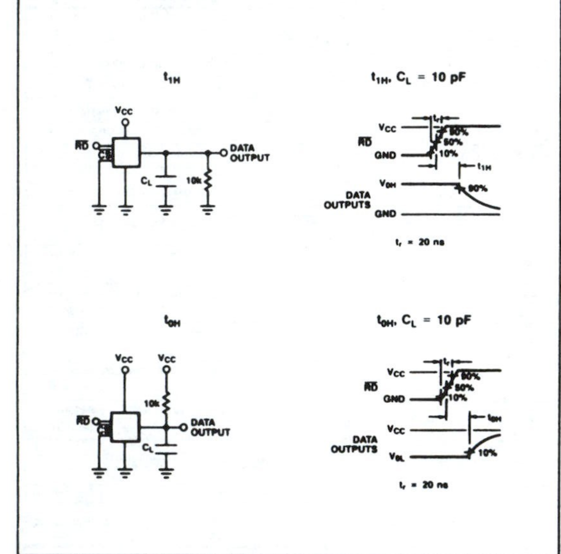

CMOS 8-BIT A/D CONVERTERS — ADC0801/2/3/4/5-1

Preliminary

TIMING DIAGRAMS (All timing is measured from the 50% voltage points)

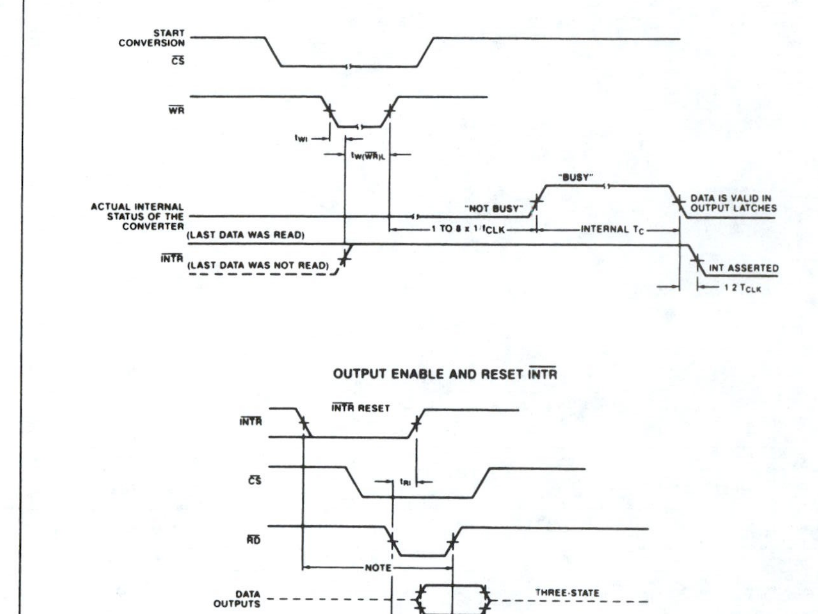

Note: Read strobe must occur 8 clock periods (8 t_{CLK}) after assertion of interrupt to guarantee reset of $\overline{\text{INTR}}$

8-BIT MULTIPLYING D/A CONVERTER — MC1508-8/1408-8/1408-7

DESCRIPTION

The MC1508/MC1408 series of 8-bit monolithic digital-to-analog converters provide high speed performance with low cost. They are designed for use where the output current is a linear product of an 8-bit digital word and an analog reference voltage.

FEATURES

- **Fast settling time — 70ns (typ)**
- **Relative accuracy ±0.19% (max error)**
- **Non-inverting digital inputs are TTL and CMOS compatible**
- **High speed multiplying rate 4.0mA/μs (input slew)**
- **Output voltage swing +.5V to -5.0V**
- **Standard supply voltages +5.0V and -5.0V to -15V**
- **Military qualifications pending**

APPLICATIONS

- Tracking A-to-D converters
- 2½-digit panel meters and DVM's
- Waveform synthesis
- Sample and hold
- Peak detector
- Programmable gain and attenuation
- CRT character generation
- Audio digitizing and decoding
- Programmable power supplies
- Analog-digital multiplication
- Digital-digital multiplication
- Analog-digital division
- Digital addition and subtraction
- Speech compression and expansion
- Stepping motor drive
- Modems
- Servo motor and pen drivers

CIRCUIT DESCRIPTION

The MC1508/MC1408 consists of a reference current amplifier, an R-2R ladder, and 8 high speed current switches. For many applications, only a reference resistor and reference voltage need be added.

The switches are non-inverting in operation, therefore, a high state on the input turns on the specified output current component.

The switch uses current steering for high speed, and a termination amplifier consisting of an active load gain stage with unity gain feedback. The termination amplifier holds the parasitic capacitance of the ladder at a constant voltage during switching, and provides a low impedance termination of equal voltage for all legs of the ladder.

The R-2R ladder divides the reference amplifier current into binarily-related components, which are fed to the switches. Note that there is always a remainder current which is equal to the least significant bit. This current is shunted to ground, and the maximum output current is 255/256 of the reference amplifier current, or 1.992mA for a 2.0mA reference amplifier current if the NPN current source pair is perfectly matched.

PIN CONFIGURATION

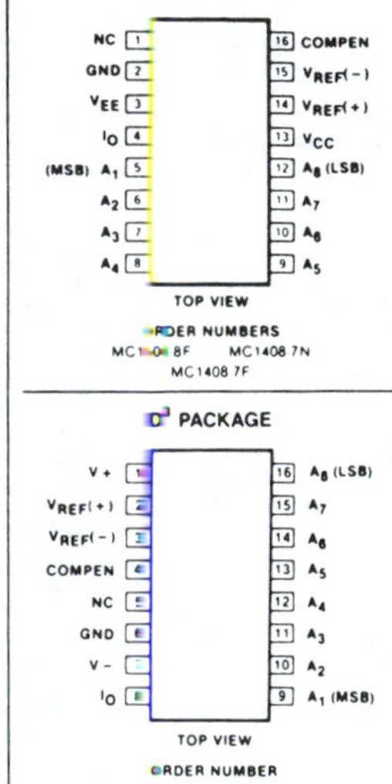

NOTES
1 SOL Released in Large SO package only
2 SOL and non-standard pinout
3 SO and non-standard pinouts

BLOCK DIAGRAM

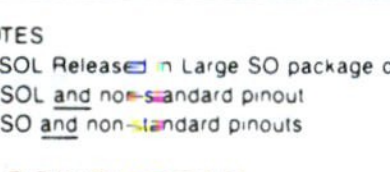

ABSOLUTE MAXIMUM RATINGS $T_A = +25°C$ unless otherwise specified

	PARAMETER	RATING	UNIT
	Power Supply Voltage		
V_{CC}	Positive	+5.5	V
V_{EE}	Negative	-16.5	V
V_5-V_{12}	Digital Input Voltage	0 to V_{CC}	V
V_0	Applied Output Voltage	-5.2 to +18	V
I_{14}	Reference Current	5.0	mA
V_{14}, V_{15}	Reference Amplifier Inputs	V_{EE} to V_{CC}	
P_D	Power Dissipation (Package Limitation)		
	Ceramic Package	1000	mW
	Plastic Package	800	mW
	Lead Soldering Temperature (60 sec)	300	°C
T_A	Operating Temperature Range		
	MC1508	-55 to +125	°C
	MC1408	0 to +75	°C
T_{STG}	Storage Temperature Range	-65 to +150	°C

DC ELECTRICAL CHARACTERISTICS[1]

Pin 3 must be 3V more negative than the potential to which R_{15} is returned
$V_{CC} = +5.0Vdc$, $V_{EE} = -15Vdc$, $\frac{V_{ref}}{R_{14}} = 2.0mA$ unless otherwise specified
MC1508: $T_A = -55°C$ to $125°C$. MC1408: $T_A = 0°C$ to $75°C$ unless otherwise noted.

	PARAMETER	TEST CONDITIONS	MC1508-8 Min	MC1508-8 Typ	MC1508-8 Max	MC1408-8 Min	MC1408-8 Typ	MC1408-8 Max	MC1408-7 Min	MC1408-7 Typ	MC1408-7 Max	UNIT
E_r	Relative accuracy	Error relative to full scale Io, Figure 3			±0.19			±0.19			±0.39	%
t_s	Settling time[1]	To within ½ LSB, includes t'PLH, T'A = +25°C, Figure 4		70			70			70		ns
	Propagation delay time											ns
t_{PLH}	Low-to-high	$T_A = +25°C$,		35	100		35	100		35	100	
t_{PHL}	High-to-low	Figure 4										
TCIo	Output full scale current drift			-20			-20			-20		PPM/°C
	Digital input logic level (MSB)											Vdc
V_{IH}	High	Figure 5	2.0			2.0			2.0			
V_{IL}	Low				0.8			0.8			0.8	
	Digital input current (MSB)	Figure 5										mA
I_{IH}	High	$V_{IH} = 5.0V$		0	0.04		0	0.04		0	0.04	
I_{IL}	Low	$V_{IL} = 0.8V$		-0.4	-0.8		-0.4	-0.8		-0.4	-0.8	
I_{15}	Reference input bias current	Pin 15, Figure 5		-1.0	-5.0		-1.0	-5.0		-1.0	-5.0	μA
I_{OR}	Output current range	Figure 5										mA
		$V_{EE} = -5.0V$	0	2.0	2.1	0	2.0	2.1	0	2.0	2.1	
		$V_{EE} = -7.0V$ to $-15V$	0	2.0	4.2	0	2.0	4.2	0	2.0	4.2	
Io	Output current	Figure 5, $V_{ref} = 2.000V$, R14 = 1000Ω	1.9	1.99	2.1	1.9	1.99	2.1	1.9	1.99	2.1	mA
$I_{O(min)}$	Off-state	All bits low		0	4.0		0	4.0		0	4.0	μA
Vo	Output voltage compliance	$E_r \le 0.19\%$ at $T_A = +25°C$, Figure 5										Vdc
		$V_{EE} = -5V$		-0.6, +10	-0.55, +0.5		-0.6, +10	-0.55, +0.5		-0.6, +10	-0.55, +0.5	
		V_{EE} below -10V		-5.5, +10	-5.0, +0.5		-5.5, +10	-5.0, +0.5		-5.5, +10	-5.0, +0.5	
SRI_{ref}	Reference current slew rate	Figure 6		8.0			8.0			8.0		mA/μs
$PSRR_{(-)}$	Output current power supply sensitivity	$I_{ref} = 1mA$		0.5	2.7		0.5	2.7		0.5	2.7	μA/V
	Power supply current											mA
I_{CC}	Positive	All bits low,		+2.5	+22		+2.5	+22		+2.5	+22	
I_{EE}	Negative	Figure 5		-6.5	-13		-6.5	-13		-6.5	-13	
	Power supply voltage range											Vdc
V_{CCR}	Positive	$T_A = +25°C$,	+4.5	+5.0	+5.5	+4.5	+5.0	+5.5	+4.5	+5.0	+5.5	
V_{EER}	Negative	Figure 5	-4.5	-15	-16.5	-4.5	-15	-16.5	-4.5	-15	-16.5	
P_D	Power dissipation	All bits low, Figure 5										mW
		$V_{EE} = -5.0Vdc$		34	170		34	170		34	170	
		$V_{EE} = -15Vdc$		110	305		110	305		110	305	

NOTES
1 All bits switched

TYPICAL PERFORMANCE CHARACTERISTICS

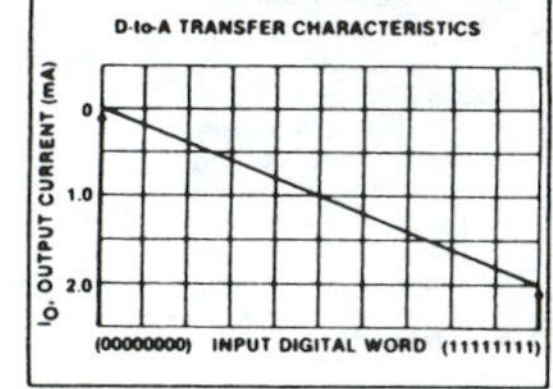

FUNCTIONAL DESCRIPTION

Reference Amplifier Drive and Compensation

The reference amplifier input current must always flow into pin 14 regardless of the setup method or reference supply voltage polarity.

Connections for a positive reference voltage are shown in Figure 1. The reference voltage source supplies the full reference current. For bipolar reference signals, as in the multiplying mode, R_{15} can be tied to a negative voltage corresponding to the minimum input level. R_{15} may be eliminated and pin 15 grounded, with only a small sacrifice in accuracy and temperature drift.

The compensation capacitor value must be increased with increasing values of R_{14} to maintain proper phase margin. For R_{14} values of 1.0, 2.5, and 5.0K ohms, minimum capacitor values are 15, 37, and 75pF. The capacitor may be tied to either V_{EE} or ground, but using V_{EE} increases negative supply rejection. (Fluctuations in the negative supply have more effect on accuracy than do any changes in the positive supply).

A negative reference voltage may be used if R_{14} is grounded and the reference voltage is applied to R_{15}, as shown in Figure 2. A high input impedance is the main advantage of this method. The negative reference voltage must be at least 3.0V above the V_{EE} supply. Bipolar input signals may be handled by connecting R_{14} to a positive reference voltage equal to the peak positive input level at pin 15.

Capacitive bypass to ground is recommended when a DC reference voltage is used. The 5.0V logic supply is not recommended as a reference voltage, but if a well regulated 5.0V supply which drives logic is to be used as the reference, R_{14} should be formed of two series resistors and the junction of the two resistors bypassed with 0.1μF to ground. For reference voltages greater than 5.0V, a clamp diode is recommended between pin 14 and ground.

If pin 14 is driven by a high impedance such as a transistor current source, none of the above compensation methods apply and the amplifier must be heavily compensated, decreasing the overall bandwidth.

Output Voltage Range

The voltage at pin 4 must always be at least 4.5 volts more positive than the voltage of the negative supply (pin 3) when the reference current is 2mA or less, and at least 8 volts more positive than the negative supply when the reference current is between 2mA and 4mA. This is necessary to avoid saturation of the output transistors, which would cause serious degradation of accuracy.

Signetics' MC1508/MC1408 does not need a range control because the design extends the compliance range down to 4.5 volts (or 8 volts—see above) above the negative supply voltage without significant degradation of accuracy. Signetics' MC1508/MC1408 can be used in sockets designed for other manufacturers' MC1508/MC1408 without circuit modification.

Output Current Range

Any time the full scale current exceeds 2mA, the negative supply must be at least 8 volts more negative than the output voltage. This is due to the increased internal voltage drops between the negative supply and the outputs with higher reference currents.

Accuracy

Absolute accuracy is the measure of each output current level with respect to its intended value, and is dependent upon relative accuracy, full scale accuracy and full scale current drift. Relative accuracy is the measure of each output current level as a fraction of the full scale current after zero scale current has been nulled out. The relative accuracy of the MC1508/MC1408 is essentially constant over the operating temperature range because of the excellent temperature tracking of the monolithic resistor ladder. The reference current may drift with temperature, causing a change in the absolute accuracy of output current; however, the MC1508/MC1408 has a very low full scale current drift over the operating temperature range.

The MC1508/MC1408 series is guaranteed accurate to within ±1/2 LSB at +25°C at a full scale output current of 1.99mA. The relative accuracy test circuit is shown in Figure 3. The 12-bit converter is calibrated to a full scale output current of 1.99219mA; then the MC1508/MC1408's full scale current is trimmed to the same value with R_{14} so that a zero value appears at the error amplifier output. The counter is activated and the error band may be displayed on the oscilloscope, detected by comparators, or stored in a peak detector.

Two 8-bit D-to-A converters may not be used to construct a 16-bit accurate D-to-A converter. Sixteen-bit accuracy implies a total of ±1/2 part in 65,536, or ±0.00076%, which is much more accurate than the ±0.19% specification of the MC1508/MC1408.

Monotonicity

A monotonic converter is one which always provides an analog output greater than or equal to the preceding value for a corresponding increment in the digital input code. The MC1508/MC1408 is monotonic for all values of reference current above 0.5mA. The recommended range for operation is a DC reference current between 0.5mA and 4.0mA.

Settling Time

The worst case switching condition occurs when all bits are switched on, which corresponds to a low-to-high transition for all input bits. This time is typically 70ns for settling to within 1/2 LSB for 8-bit accuracy. This time applies when RL < 500 ohms and C_O < 25pF. The slowest single switch is the least significant bit, which typically turns on and settles in 65ns. In applications where the D-to-A converter functions in a positive going ramp mode, the worst case condition does not occur and settling times less than 70ns may be realized.

Extra care must be taken in board layout since this usually is the dominant factor in satisfactory test results when measuring settling time. Short leads, 100μF supply bypassing for low frequencies, minimum scope lead length, good ground planes, and avoidance of ground loops are all mandatory.

8-BIT MULTIPLYING D/A CONVERTER MC1508-8/1408-8/1408-7

TEST CIRCUITS

Figure 1. Positive V_{REF}

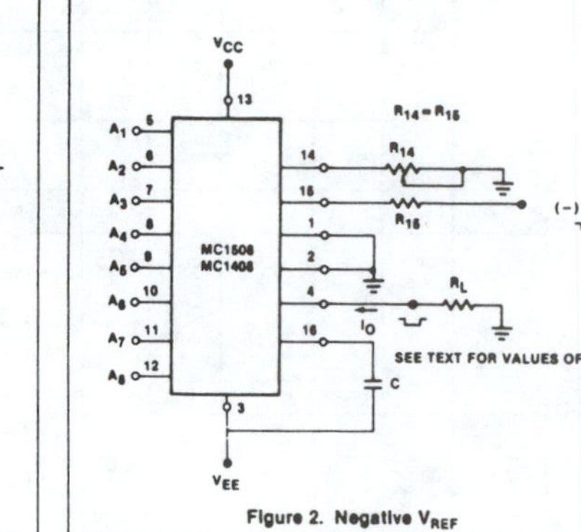

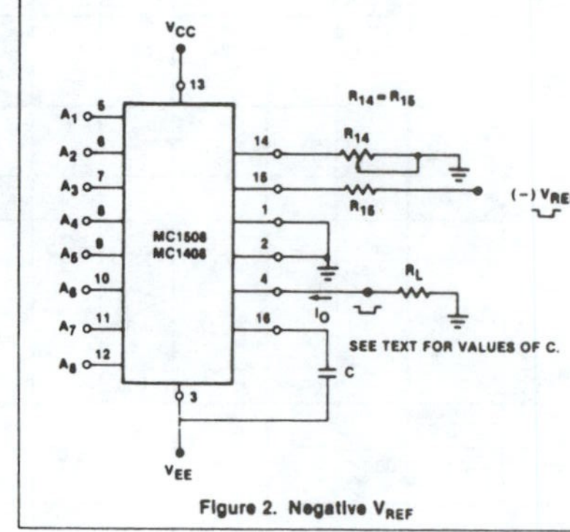

Figure 2. Negative V_{REF}

8-BIT MULTIPLYING D/A CONVERTER MC1508-8/1408-8/1408-7

TEST CIRCUITS (Cont'd)

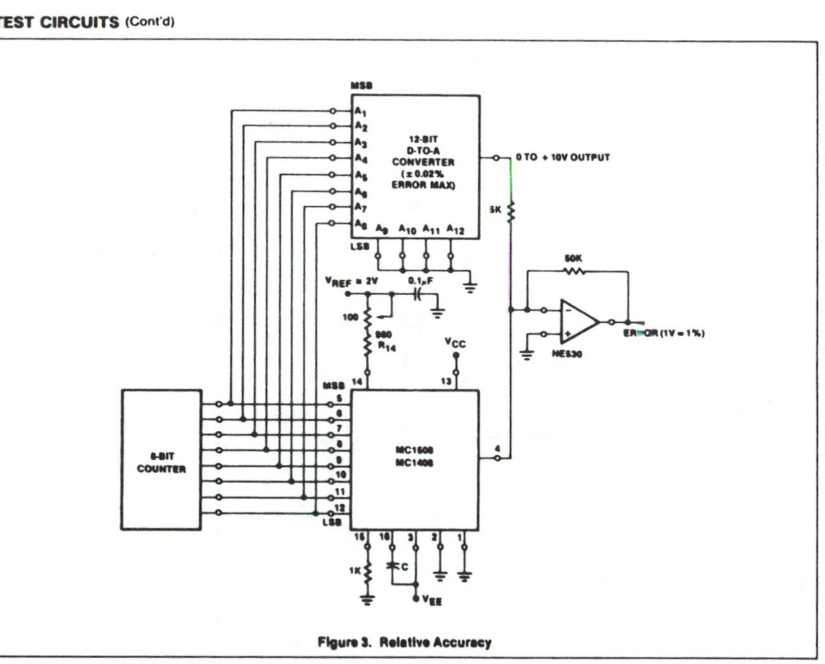

Figure 3. Relative Accuracy

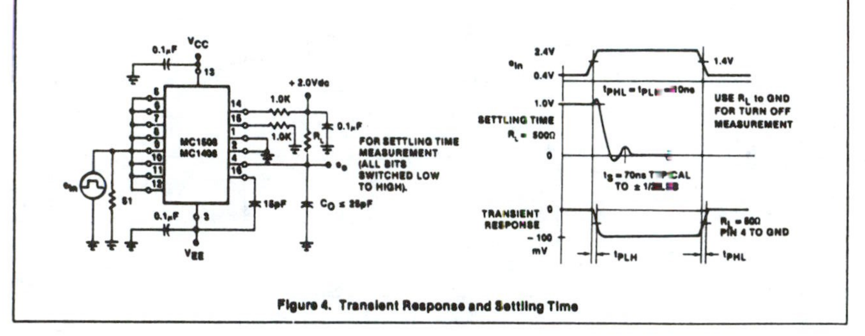

Figure 4. Transient Response and Settling Time

8-BIT MULTIPLYING D/A CONVERTER MC1508-8/1408-8/1408-7

TEST CIRCUITS (Cont'd)

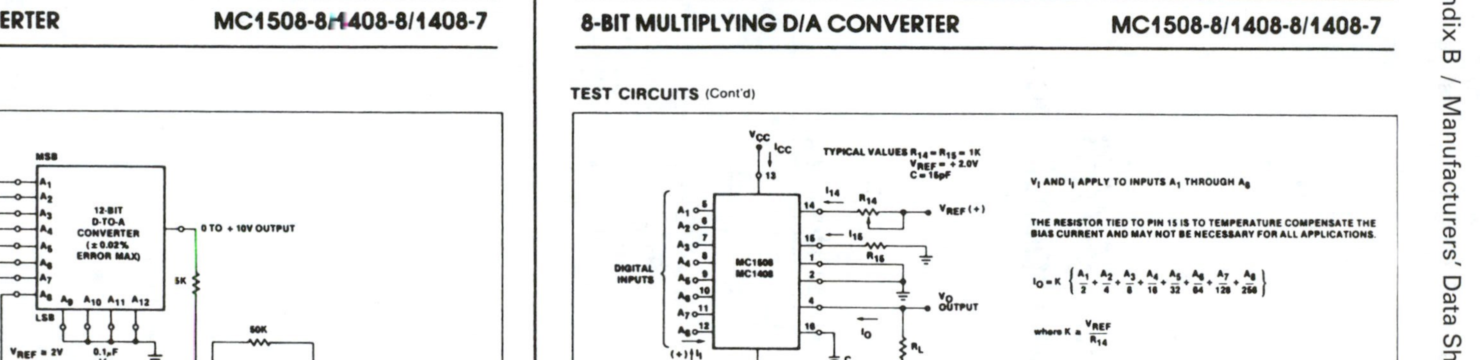

Figure 5. Notation Definitions

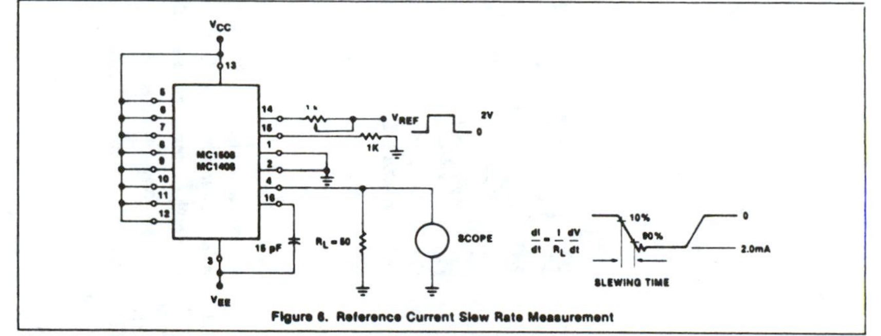

Figure 6. Reference Current Slew Rate Measurement

Signetics

μA741/μA741C/SA741C
General Purpose Operational Amplifier

Product Specification

Linear Products

DESCRIPTION

The μA741 is a high performance operational amplifier with high open-loop gain, internal compensation, high common mode range and exceptional temperature stability. The μA741 is short-circuit-protected and allows for nulling of offset voltage.

FEATURES

- **Internal frequency compensation**
- **Short circuit protection**
- **Excellent temperature stability**
- **High input voltage range**

PIN CONFIGURATION

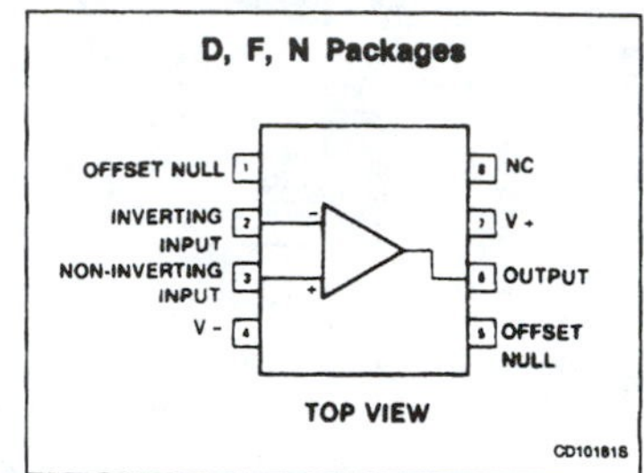

ORDERING INFORMATION

DESCRIPTION	TEMPERATURE RANGE	ORDER CODE
8-Pin Plastic DIP	−55°C to +125°C	μA741N
8-Pin Plastic DIP	0 to +70°C	μA741CN
8-Pin Plastic DIP	−40°C to +85°C	SA741CN
8-Pin Cerdip	−55°C to +125°C	μA741F
8-Pin Cerdip	0 to +70°C	μA741CF
8-Pin SO	0 to +70°C	μA741CD

EQUIVALENT SCHEMATIC

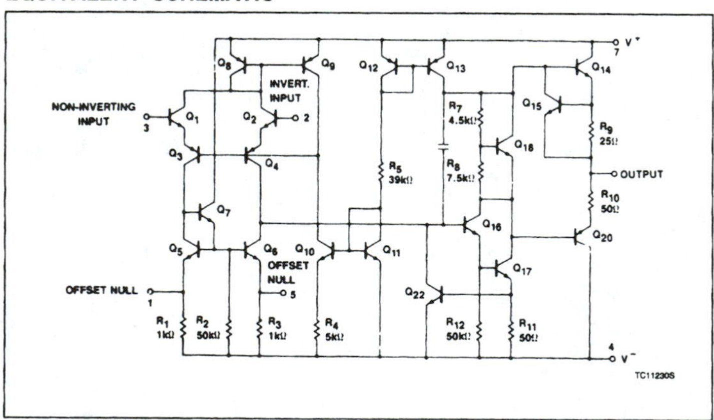

General Purpose Operational Amplifier μA741/μA741C/SA741C

ABSOLUTE MAXIMUM RATINGS

SYMBOL	PARAMETER	RATING	UNIT
V_S	Supply voltage μA741C μA741	 ±18 ±22	 V V
P_D	Internal power dissipation D package N package F package	 500 1000 1000	 mW mW mW
V_{IN}	Differential input voltage	±30	V
V_{IN}	Input voltage[1]	±15	V
I_{SC}	Output short-circuit duration	Continuous	
T_A	Operating temperature range μA741C SA741C μA741	 0 to +70 −40 to +85 −55 to +125	 °C °C °C
T_{STG}	Storage temperature range	−65 to +150	°C
T_{SOLD}	Lead soldering temperature (10sec max)	300	°C

NOTE:
1. For supply voltages less than ±15V, the absolute maximum input voltage is equal to the supply voltage.

DC ELECTRICAL CHARACTERISTICS (μA741, μA741C) $T_A = 25°C$, $V_S = \pm15V$, unless otherwise specified.

SYMBOL	PARAMETER	TEST CONDITIONS	μA741			μA741C			UNIT
			Min	Typ	Max	Min	Typ	Max	
V_{OS} $\Delta V_{OS}/\Delta T$	Offset voltage	$R_S = 10k\Omega$ $R_S = 10k\Omega$, over temp.		1.0 1.0 10	5.0 6.0		2.0 10	6.0 7.5	mV mV μV/°C
I_{OS} $\Delta I_{OS}/\Delta T$	Offset current	 Over temp. $T_A = +125°C$ $T_A = -55°C$		20 7.0 20 200	200 200 500		20 200	200 300	nA nA nA nA pA/°C
I_{BIAS} $\Delta I_B/\Delta T$	Input bias current	 Over temp. $T_A = +125°C$ $T_A = -55°C$		80 30 300 1	500 500 1500		80 1	500 800	nA nA nA nA nA/°C
V_{OUT}	Output voltage swing	$R_L = 10k\Omega$ $R_L = 2k\Omega$, over temp.	±12 ±10	±14 ±13		±12 ±10	±14 ±13		V V
A_{VOL}	Large-signal voltage gain	$R_L = 2k\Omega$, $V_O = \pm10V$ $R_L = 2k\Omega$, $V_O = \pm10V$, over temp.	50 25	200		20 15	200		V/mV V/mV
	Offset voltage adjustment range			±30			±30		mV
PSRR	Supply voltage rejection ratio	$R_S \leq 10k\Omega$ $R_S \leq 10k\Omega$, over temp.		10	150		10	150	μV/V μV/V
CMRR	Common-mode rejection ratio	Over temp.	70	90					dB dB
I_{CC}	Supply current	 $T_A = +125°C$ $T_A = -55°C$		1.4 1.5 2.0	2.8 2.5 3.3		1.4	2.8	mA mA mA

General Purpose Operational Amplifier μA741/μA741C/SA741C

DC ELECTRICAL CHARACTERISTICS (Continued) (μA741, μA741C) $T_A = 25°C$, $V_S = \pm 15V$, unless otherwise specified.

SYMBOL	PARAMETER	TEST CONDITIONS	μA741			μA741C			UNIT
			Min	Typ	Max	Min	Typ	Max	
V_{IN} R_{IN}	Input voltage range Input resistance	(μA741, over temp.)	±12 0.3	±13 2.0		±12 0.3	±13 2.0		V MΩ
P_D	Power consumption	 $T_A = +125°C$ $T_A = -55°C$		50 45 45	80 75 100		50	85	mW mW mW
R_{OUT}	Output resistance			75			75		Ω
I_{SC}	Output short-circuit current		10	25	60	10	25	60	mA

DC ELECTRICAL CHARACTERISTICS (SA741C) $T_A = 25°C$, $V_S = \pm 15V$, unless otherwise specified.

SYMBOL	PARAMETER	TEST CONDITIONS	SA741C			UNIT
			Min	Typ	Max	
V_{OS} $\Delta V_{OS}/\Delta T$	Offset voltage	$R_S = 10k\Omega$ $R_S = 10k\Omega$, over temp.		2.0 10	6.0 7.5	mV mV μV/°C
I_{OS} $\Delta I_{OS}/\Delta T$	Offset current	Over temp.		20 200	200 500	nA nA pA/°C
I_{BIAS} $\Delta I_B/\Delta T$	Input bias current	Over temp.		80 1	500 1500	nA nA nA/°C
V_{OUT}	Output voltage swing	$R_L = 10k\Omega$ $R_L = 2k\Omega$, over temp.	±12 ±10	±14 ±13		V V
A_{VOL}	Large-signal voltage gain	$R_L = 2k\Omega$, $V_O = \pm 10V$ $R_L = 2k\Omega$, $V_O = \pm 10V$, over temp.	20 15	200		V/mV V/mV
	Offset voltage adjustment range			±30		mV
PSRR	Supply voltage rejection ratio	$R_S \leq 10k\Omega$		10	150	μV/V
V_{IN}	Input voltage range	(μA741, over temp.)	±12	±13		V
R_{IN}	Input resistance		0.3	2.0		MΩ
P_d	Power consumption			50	85	mW
R_{OUT}	Output resistance			75		Ω
I_{SC}	Output short-circuit current			25		mA

AC ELECTRICAL CHARACTERISTICS $T_A = 25°C$, $V_S = \pm 15V$, unless otherwise specified.

SYMBOL	PARAMETER	TEST CONDITIONS	μA741, μA741C			UNIT
			Min	Typ	Max	
C_{IN}	Parallel input capacitance	Open-loop, f = 20Hz		1.4		pF
	Unity gain crossover frequency	Open-loop		1.0		MHz
 t_R SR	Transient response unity gain Rise time Overshoot Slew rate	$V_{IN} = 20mV$, $R_L = 2k\Omega$, $C_L \leq 100pF$ $C \leq 100pF$, $R_L \geq 2k\Omega$, $V_{IN} = \pm 10V$		 0.3 5.0 0.5		 μs % V/μs

Signetics

NE/SE555/SE555C
Timer

Product Specification

Linear Products

DESCRIPTION

The 555 monolithic timing circuit is a highly stable controller capable of producing accurate time delays, or oscillation. In the time delay mode of operation, the time is precisely controlled by one external resistor and capacitor. For a stable operation as an oscillator, the free running frequency and the duty cycle are both accurately controlled with two external resistors and one capacitor. The circuit may be triggered and reset on falling waveforms, and the output structure can source or sink up to 200mA.

FEATURES

- **Turn-off time less than 2μs**
- **Max. operating frequency greater than 500kHz**
- **Timing from microseconds to hours**
- **Operates in both astable and monostable modes**
- **High output current**
- **Adjustable duty cycle**
- **TTL compatible**
- **Temperature stability of 0.005% per °C**

APPLICATIONS

- **Precision timing**
- **Pulse generation**
- **Sequential timing**
- **Time delay generation**
- **Pulse width modulation**
- **Pulse position modulation**
- **Missing pulse detector**

PIN CONFIGURATIONS

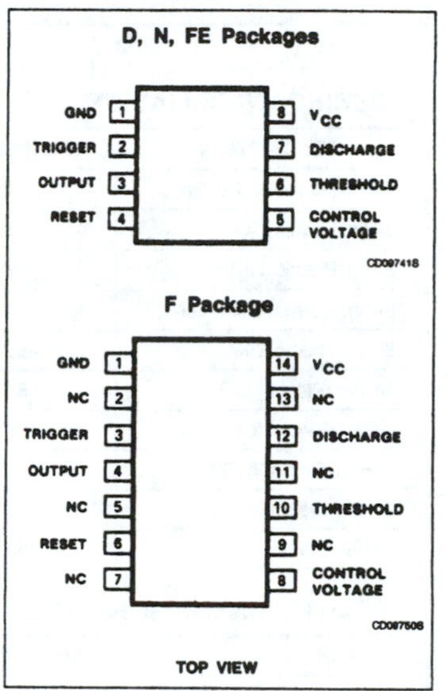

EQUIVALENT SCHEMATIC

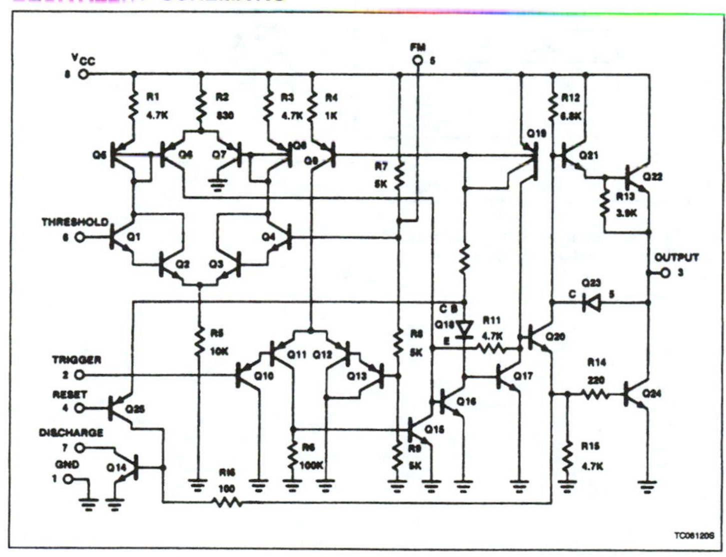

Signetics Linear Products | Product Specification

Timer

NE/SE555/SE555C

ORDERING INFORMATION

DESCRIPTION	TEMPERATURE RANGE	ORDER CODE
8-Pin Hermetic Cerdip	0 to +70°C	NE555FE
8-Pin Plastic SO	0 to +70°C	NE555D
8-Pin Plastic DIP	0 to +70°C	NE555N
8-Pin Hermetic Cerdip	−55°C to +125°C	SE555CFE
8-Pin Plastic DIP	−55°C to +125°C	SE555CN
14-Pin Plastic DIP	−55°C to +125°C	SE555N
8-Pin Hermetic Cerdip	−55°C to +125°C	SE555FE
14-Pin Ceramic DIP	0 to +70°C	NE555F
14-Pin Ceramic DIP	−55°C to +125°C	SE555F
14-Pin Ceramic DIP	−55°C to +125°C	SE555CF

BLOCK DIAGRAM

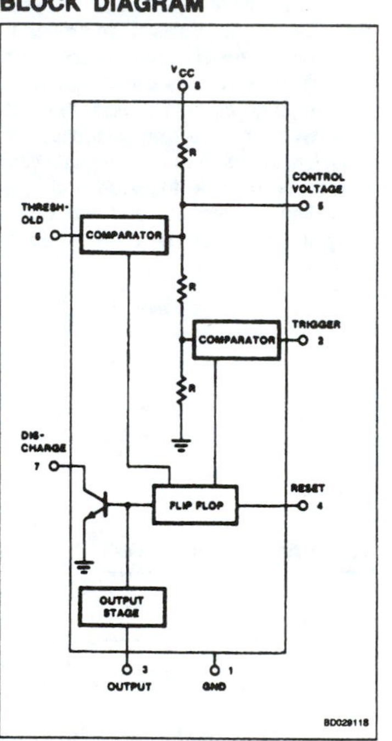

ABSOLUTE MAXIMUM RATINGS

SYMBOL	PARAMETER	RATING	UNIT
V_{CC}	Supply voltage SE555 NE555, SE555C	 +18 +16	 V V
P_D	Maximum allowable power dissipation[1]	600	mW
T_A	Operating ambient temperature range NE555 SE555, SE555C	 0 to +70 −55 to +125	 °C °C
T_{STG}	Storage temperature range	−65 to +150	°C
T_{SOLD}	Lead soldering temperature (10sec max)	+300	°C

NOTE:

1. The junction temperature must be kept below 125°C for the D package and below 150°C for the FE, N and F packages. At ambient temperatures above 25°C, where this limit would be derated by the following factors:
   D package 160 °C/W
   FE package 150 °C/W
   N package 100 °C/W
   F package 105 °C/W

Timer NE/SE555/SE555C

DC AND AC ELECTRICAL CHARACTERISTICS $T_A = 25°C$, $V_{CC} = +5V$ to $+15$ unless otherwise specified.

SYMBOL	PARAMETER	TEST CONDITIONS	SE555			NE555/SE555C			UNIT
			Min	Typ	Max	Min	Typ	Max	
V_{CC}	Supply voltage		4.5		18	4.5		16	V
I_{CC}	Supply current (low state)[1]	$V_{CC} = 5V$, $R_L = \infty$ $V_{CC} = 15V$, $R_L = \infty$		3 10	5 12		3 10	6 15	mA mA
 t_M $\Delta t_M/\Delta T$ $\Delta t_M/\Delta V_S$	Timing error (monostable) Initial accuracy[2] Drift with temperature Drift with supply voltage	$R_A = 2k\Omega$ to $100k\Omega$ $C = 0.1\mu F$		 0.5 30 0.05	 2.0 100 0.2		 1.0 50 0.1	 3.0 150 0.5	 % ppm/°C %/V
 t_A $\Delta t_A/\Delta T$ $\Delta t_A/\Delta V_S$	Timing error (astable) Initial accuracy[2] Drift with temperature Drift with supply voltage	R_A, $R_B = 1k\Omega$ to $100k\Omega$ $C = 0.1\mu F$ $V_{CC} = 15V$		 4 0.15	 6 500 0.6		 5 0.3	 13 500 1	 % ppm/°C %/V
V_C	Control voltage level	$V_{CC} = 15V$ $V_{CC} = 5V$	9.6 2.9	10.0 3.33	10.4 3.8	9.0 2.6	10.0 3.33	11.0 4.0	V V
V_{TH}	Threshold voltage	$V_{CC} = 15V$ $V_{CC} = 5V$	9.4 2.7	10.0 3.33	10.6 4.0	8.8 2.4	10.0 3.33	11.2 4.2	V V
I_{TH}	Threshold current[3]			0.1	0.25		0.1	0.25	µA
V_{TRIG}	Trigger voltage	$V_{CC} = 15V$ $V_{CC} = 5V$	4.8 1.45	5.0 1.67	5.2 1.9	4.5 1.1	5.0 1.67	5.6 2.2	V V
I_{TRIG}	Trigger current	$V_{TRIG} = 0V$		0.5	0.9		0.5	2.0	µA
V_{RESET}	Reset voltage[4]		0.3		1.0	0.3		1.0	V
I_{RESET}	Reset current Reset current	 $V_{RESET} = 0V$		0.1 0.4	0.4 1.0		0.1 0.4	0.4 1.5	mA mA
V_{OL}	Output voltage (low)	$V_{CC} = 15V$ $I_{SINK} = 10mA$ $I_{SINK} = 50mA$ $I_{SINK} = 100mA$ $I_{SINK} = 200mA$ $V_{CC} = 5V$ $I_{SINK} = 8mA$ $I_{SINK} = 5mA$		 0.1 0.4 2.0 2.5 0.1 0.05	 0.15 0.5 2.2 0.25 0.2		 0.1 0.4 2.0 2.5 0.3 0.25	 0.25 0.75 2.5 0.4 0.35	 V V V V V V
V_{OH}	Output voltage (high)	$V_{CC} = 15V$ $I_{SOURCE} = 200mA$ $I_{SOURCE} = 100mA$ $V_{CC} = 5V$ $I_{SOURCE} = 100mA$	 13.0 3.0	 12.5 13.3 3.3		 12.75 2.75	 12.5 13.3 3.3		 V V V
t_{OFF}	Turn-off time[5]	$V_{RESET} = V_{CC}$		0.5	2.0		0.5	2.0	µs
t_R	Rise time of output			100	200		100	300	ns
t_F	Fall time of output			100	200		100	300	ns
	Discharge leakage current			20	100		20	100	ns

NOTES:
1. Supply current when output high typically 1mA less.
2. Tested at $V_{CC} = 5V$ and $V_{CC} = 15V$.
3. This will determine the max value of $R_A + R_B$, for 15V operation, the max total $R = 10M\Omega$, and for 5V operation, the max. total $R = 3.4M\Omega$.
4. Specified with trigger input high.
5. Time measured from a positive going input pulse from 0 to $0.8 \times V_{CC}$ into the threshold to the drop from high to low of the output. Trigger is tied to threshold.

8085A Assembly Language Reference Chart[1] and Alphabetized Mnemonics

[1]Courtesy of Intel Corporation.

DATA TRANSFER GROUP

Move

Mnemonic	Operand	Code
MOV	A,A	7F
MOV	A,B	78
MOV	A,C	79
MOV	A,D	7A
MOV	A,E	7B
MOV	A,H	7C
MOV	A,L	7D
MOV	A,M	7E
MOV	B,A	47
MOV	B,B	40
MOV	B,C	41
MOV	B,D	42
MOV	B,E	43
MOV	B,H	44
MOV	B,L	45
MOV	B,M	46
MOV	C,A	4F
MOV	C,B	48
MOV	C,C	49
MOV	C,D	4A
MOV	C,E	4B
MOV	C,H	4C
MOV	C,L	4D
MOV	C,M	4E
MOV	D,A	57
MOV	D,B	50
MOV	D,C	51
MOV	D,D	52
MOV	D,E	53
MOV	D,H	54
MOV	D,L	55
MOV	D,M	56

Move (cont)

Mnemonic	Operand	Code
MOV	E,A	5F
MOV	E,B	58
MOV	E,C	59
MOV	E,D	5A
MOV	E,E	5B
MOV	E,H	5C
MOV	E,L	5D
MOV	E,M	5E
MOV	H,A	67
MOV	H,B	60
MOV	H,C	61
MOV	H,D	62
MOV	H,E	63
MOV	H,H	64
MOV	H,L	65
MOV	H,M	66
MOV	L,A	6F
MOV	L,B	68
MOV	L,C	69
MOV	L,D	6A
MOV	L,E	6B
MOV	L,H	6C
MOV	L,L	6D
MOV	L,M	6E
MOV	M,A	77
MOV	M,B	70
MOV	M,C	71
MOV	M,D	72
MOV	M,E	73
MOV	M,H	74
MOV	M,L	75
XCHG		EB

Move Immediate

Mnemonic	Operand	Code
MVI	A, byte	3E
MVI	B, byte	06
MVI	C, byte	0E
MVI	D, byte	16
MVI	E, byte	1E
MVI	H, byte	26
MVI	L, byte	2E
MVI	M, byte	36

Load Immediate

Mnemonic	Operand	Code
LXI	B, dble	01
LXI	D, dble	11
LXI	H, dble	21
LXI	SP, dble	31

Load/Store

Instruction	Code
LDAX B	0A
LDAX D	1A
LHLD adr	2A
LDA adr	3A
STAX B	02
STAX D	12
SHLD adr	22
STA adr	32

byte = constant, or logical/arithmetic expression that evaluates to an 8-bit data quantity (Second byte of 2-byte instructions)

dble = constant, or logical/arithmetic expression that evaluates to a 16-bit data quantity (Second and Third bytes of 3-byte instructions)

adr = 16-bit address (Second and Third bytes of 3-byte instructions)

* = all flags (C, Z, S, P, AC) affected

** = all flags except CARRY affected (exception INX and DCX affect no flags)

† = only CARRY affected

ARITHMETIC AND LOGICAL GROUP

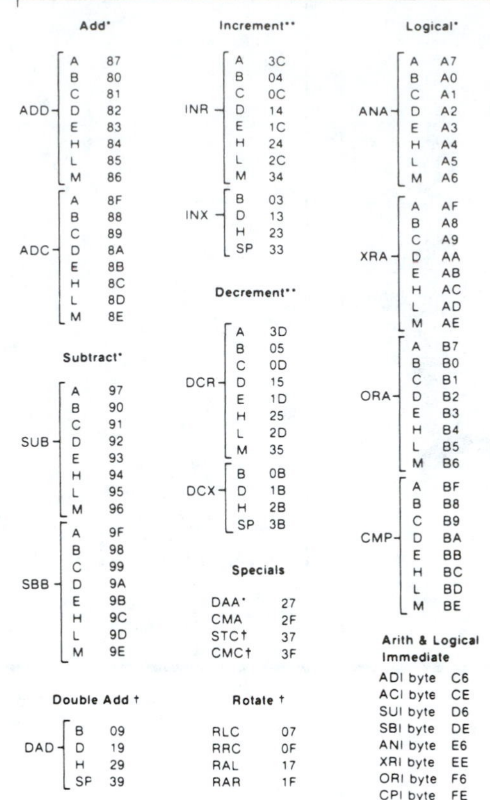

Add*

Mnemonic	Operand	Code
ADD	A	87
ADD	B	80
ADD	C	81
ADD	D	82
ADD	E	83
ADD	H	84
ADD	L	85
ADD	M	86
ADC	A	8F
ADC	B	88
ADC	C	89
ADC	D	8A
ADC	E	8B
ADC	H	8C
ADC	L	8D
ADC	M	8E

Subtract*

Mnemonic	Operand	Code
SUB	A	97
SUB	B	90
SUB	C	91
SUB	D	92
SUB	E	93
SUB	H	94
SUB	L	95
SUB	M	96
SBB	A	9F
SBB	B	98
SBB	C	99
SBB	D	9A
SBB	E	9B
SBB	H	9C
SBB	L	9D
SBB	M	9E

Double Add †

Mnemonic	Operand	Code
DAD	B	09
DAD	D	19
DAD	H	29
DAD	SP	39

Increment**

Mnemonic	Operand	Code
INR	A	3C
INR	B	04
INR	C	0C
INR	D	14
INR	E	1C
INR	H	24
INR	L	2C
INR	M	34
INX	B	03
INX	D	13
INX	H	23
INX	SP	33

Decrement**

Mnemonic	Operand	Code
DCR	A	3D
DCR	B	05
DCR	C	0D
DCR	D	15
DCR	E	1D
DCR	H	25
DCR	L	2D
DCR	M	35
DCX	B	0B
DCX	D	1B
DCX	H	2B
DCX	SP	3B

Specials

Instruction	Code
DAA*	27
CMA	2F
STC†	37
CMC†	3F

Rotate †

Instruction	Code
RLC	07
RRC	0F
RAL	17
RAR	1F

Logical*

Mnemonic	Operand	Code
ANA	A	A7
ANA	B	A0
ANA	C	A1
ANA	D	A2
ANA	E	A3
ANA	H	A4
ANA	L	A5
ANA	M	A6
XRA	A	AF
XRA	B	A8
XRA	C	A9
XRA	D	AA
XRA	E	AB
XRA	H	AC
XRA	L	AD
XRA	M	AE
ORA	A	B7
ORA	B	B0
ORA	C	B1
ORA	D	B2
ORA	E	B3
ORA	H	B4
ORA	L	B5
ORA	M	B6
CMP	A	BF
CMP	B	B8
CMP	C	B9
CMP	D	BA
CMP	E	BB
CMP	H	BC
CMP	L	BD
CMP	M	BE

Arith & Logical Immediate

Instruction	Code
ADI byte	C6
ACI byte	CE
SUI byte	D6
SBI byte	DE
ANI byte	E6
XRI byte	EE
ORI byte	F6
CPI byte	FE

BRANCH CONTROL GROUP

Jump

Instruction	Code
JMP adr	C3
JNZ adr	C2
JZ adr	CA
JNC adr	D2
JC adr	DA
JPO adr	E2
JPE adr	EA
JP adr	F2
JM adr	FA
PCHL	E9

Call

Instruction	Code
CALL adr	CD
CNZ adr	C4
CZ adr	CC
CNC adr	D4
CC adr	DC
CPO adr	E4
CPE adr	EC
CP adr	F4
CM adr	FC

Return

Instruction	Code
RET	C9
RNZ	C0
RZ	C8
RNC	D0
RC	D8
RPO	E0
RPE	E8
RP	F0
RM	F8

Restart

Mnemonic	Operand	Code
RST	0	C7
RST	1	CF
RST	2	D7
RST	3	DF
RST	4	E7
RST	5	EF
RST	6	F7
RST	7	FF

I/O AND MACHINE CONTROL

Stack Ops

Mnemonic	Operand	Code
PUSH	B	C5
PUSH	D	D5
PUSH	H	E5
PUSH	PSW	F5
POP	B	C1
POP	D	D1
POP	H	E1
POP	PSW*	F1
XTHL		E3
SPHL		F9

Input/Output

Instruction	Code
OUT byte	D3
IN byte	DB

Control

Instruction	Code
DI	F3
EI	FB
NOP	00
HLT	76

New Instructions (8085 Only)

Instruction	Code
RIM	20
SIM	30

RESTART TABLE

Name	Code	Restart Address
RST 0	C7	0000_{16}
RST 1	CF	0008_{16}
RST 2	D7	0010_{16}
RST 3	DF	0018_{16}
RST 4	E7	0020_{16}
TRAP	Hardware* Function	0024_{16}
RST 5	EF	0028_{16}
RST 5.5	Hardware* Function	$002C_{16}$
RST 6	F7	0030_{16}
RST 6.5	Hardware* Function	0034_{16}
RST 7	FF	0038_{16}
RST 7.5	Hardware* Function	$003C_{16}$

*NOTE The hardware functions refer to the on-chip Interrupt feature of the 8085 only

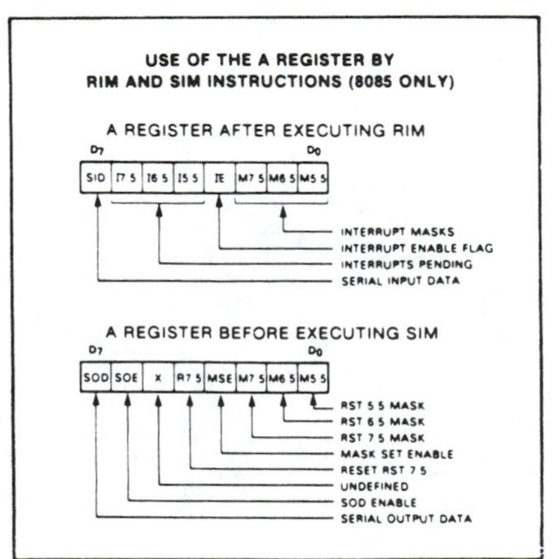

Code	Mnemonic	Code	Mnemonic	Code	Mnemonic	Code	Mnemonic	Code	Mnemonic	Code	Mnemonic
00	NOP	2B	DCX H	56	MOV D,M	81	ADD C	AC	XRA H	D7	RST 2
01	LXI B,dble	2C	INR L	57	MOV D,A	82	ADD D	AD	XRA L	D8	RC
02	STAX B	2D	DCR L	58	MOV E,B	83	ADD E	AE	XRA M	D9	· · ·
03	INX B	2E	MVI L,byte	59	MOV E,C	84	ADD H	AF	XRA A	DA	JC adr
04	INR B	2F	CMA	5A	MOV E,D	85	ADD L	B0	ORA B	DB	IN byte
05	DCR B	30	SIM*	5B	MOV E,E	86	ADD M	B1	ORA C	DC	CC adr
06	MVI B,byte	31	LXI SP,dble	5C	MOV E,H	87	ADD A	B2	ORA D	DD	· · ·
07	RLC	32	STA adr	5D	MOV E,L	88	ADC B	B3	ORA E	DE	SBI byte
08	· · ·	33	INX SP	5E	MOV E,M	89	ADC C	B4	ORA H	DF	RST 3
09	DAD B	34	INR M	5F	MOV E,A	8A	ADC D	B5	ORA L	E0	RPO
0A	LDAX B	35	DCR M	60	MOV H,B	8B	ADC E	B6	ORA M	E1	POP H
0B	DCX B	36	MVI M,byte	61	MOV H,C	8C	ADC H	B7	ORA A	E2	JPO adr
0C	INR C	37	STC	62	MOV H,D	8D	ADC L	B8	CMP B	E3	XTHL
0D	DCR C	38	· · ·	63	MOV H,E	8E	ADC M	B9	CMP C	E4	CPO adr
0E	MVI C,byte	39	DAD SP	64	MOV H,H	8F	ADC A	BA	CMP D	E5	PUSH H
0F	RRC	3A	LDA adr	65	MOV H,L	90	SUB B	BB	CMP E	E6	ANI byte
10	· · ·	3B	DCX SP	66	MOV H,M	91	SUB C	BC	CMP H	E7	RST 4
11	LXI D,dble	3C	INR A	67	MOV H,A	92	SUB D	BD	CMP L	E8	RPE
12	STAX D	3D	DCR A	68	MOV L,B	93	SUB E	BE	CMP M	E9	PCHL
13	INX D	3E	MVI A,byte	69	MOV L,C	94	SUB H	BF	CMP A	EA	JPE adr
14	INR D	3F	CMC	6A	MOV L,D	95	SUB L	C0	RNZ	EB	XCHG
15	DCR D	40	MOV B,B	6B	MOV L,E	96	SUB M	C1	POP B	EC	CPE adr
16	MVI D,byte	41	MOV B,C	6C	MOV L,H	97	SUB A	C2	JNZ adr	ED	· · ·
17	RAL	42	MOV B,D	6D	MOV L,L	98	SBB B	C3	JMP adr	EE	XRI byte
18	· · ·	43	MOV B,E	6E	MOV L,M	99	SBB C	C4	CNZ adr	EF	RST 5
19	DAD D	44	MOV B,H	6F	MOV L,A	9A	SBB D	C5	PUSH B	F0	RP
1A	LDAX D	45	MOV B,L	70	MOV M,B	9B	SBB E	C6	ADI byte	F1	POP PSW
1B	DCX D	46	MOV B,M	71	MOV M,C	9C	SBB H	C7	RST 0	F2	JP adr
1C	INR E	47	MOV B,A	72	MOV M,D	9D	SBB L	C8	RZ	F3	DI
1D	DCR E	48	MOV C,B	73	MOV M,E	9E	SBB M	C9	RET	F4	CP adr
1E	MVI E,byte	49	MOV C,C	74	MOV M,H	9F	SBB A	CA	JZ adr	F5	PUSH PSW
1F	RAR	4A	MOV C,D	75	MOV M,L	A0	ANA B	CB	· · ·	F6	ORI byte
20	RIM*	4B	MOV C,E	76	HLT	A1	ANA C	CC	CZ adr	F7	RST 6
21	LXI H,dble	4C	MOV C,H	77	MOV M,A	A2	ANA D	CD	CALL adr	F8	RM
22	SHLD adr	4D	MOV C,L	78	MOV A,B	A3	ANA E	CE	ACI byte	F9	SPHL
23	INX H	4E	MOV C,M	79	MOV A,C	A4	ANA H	CF	RST 1	FA	JM adr
24	INR H	4F	MOV C,A	7A	MOV A,D	A5	ANA L	D0	RNC	FB	EI
25	DCR H	50	MOV D,B	7B	MOV A,E	A6	ANA M	D1	POP D	FC	CM adr
26	MVI H,byte	51	MOV D,C	7C	MOV A,H	A7	ANA A	D2	JNC adr	FD	· · ·
27	DAA	52	MOV D,D	7D	MOV A,L	A8	XRA B	D3	OUT byte	FE	CPI byte
28	· · ·	53	MOV D,E	7E	MOV A,M	A9	XRA C	D4	CNC adr	FF	RST 7
29	DAD H	54	MOV D,H	7F	MOV A,A	AA	XRA D	D5	PUSH D		
2A	LHLD adr	55	MOV D,L	80	ADD B	AB	XRA E	D6	SUI byte		

*8085 Only

INTEL® 8080/8085 INSTRUCTION SET REFERENCE TABLES

INTERNAL REGISTER ORGANIZATION

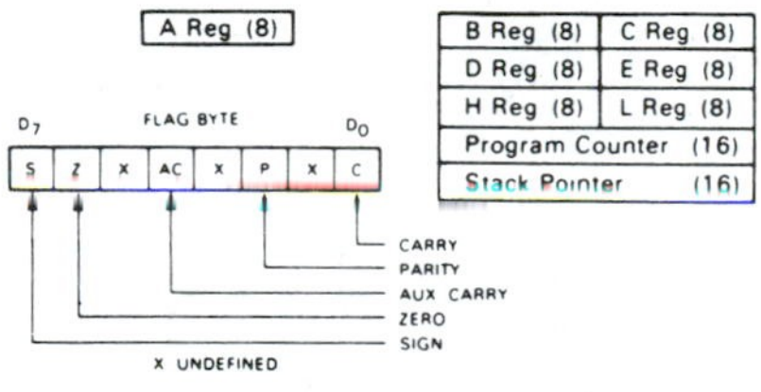

REGISTER-PAIR ORGANIZATION

PSW

A (8)	FLAGS (8)

NOTE Leftmost Byte is high-order byte for arithmetic operations and addressing Left byte is pushed on stack first Right byte is popped first

B	(B/C) (16)
D	(D/E) (16)
H	(H/L) (16)
Prog Ctr	(16)
Stack Ptr	(16)

BRANCH CONTROL INSTRUCTIONS

Flag Condition	Jump		Call		Return	
Zero=True	JZ	CA	CZ	CC	RZ	C8
Zero=False	JNZ	C2	CNZ	C4	RNZ	C0
Carry=True	JC	DA	CC	DC	RC	D8
Carry=False	JNC	D2	CNC	D4	RNC	D0
Sign=Positive	JP	F2	CP	F4	RP	F0
Sign=Negative	JM	FA	CM	FC	RM	F8
Parity=Even	JPE	EA	CPE	EC	RPE	E8
Parity=Odd	JPO	E2	CPO	E4	RPO	E0
Unconditional	JMP	C3	CALL	CD	RET	C9

ACCUMULATOR OPERATIONS

	Code	Function
XRA A	AF	Clear A and Clear Carry
ORA A	B7	Clear Carry
CMC	3F	Complement Carry
CMA	2F	Complement Accumulator
STC	37	Set Carry
RLC	07	Rotate Left
RRC	0F	Rotate Right
RAL	17	Rotate Left Thru Carry
RAR	1F	Rotate Right Thru Carry
DAA	27	Decimal Adjust Accum

REGISTER PAIR AND STACK OPERATIONS

		Register Pair					
	PSW (A/F)	B (B/C)	D (D/E)	H (H/L)	SP	PC	Function
INX		03	13	23	33		Increment Register Pair
DCX		0B	1B	2B	3B		Decrement Register Pair
LDAX		0A	1A	7E(1)			Load A Indirect (Reg Pair holds Adrs)
STAX		02	12	77(2)			Store A Indirect (Reg Pair holds Adrs)
LHLD				2A			Load H L Direct (Bytes 2 and 3 hold Adrs)
SHLD				22			Store H L Direct (Bytes 2 and 3 hold Adrs)
LXI		01	11	21	31	C3(3)	Load Reg Pair Immediate (Bytes 2 and 3 hold immediate data)
PCHL						E9	Load PC with H L (Branch to Adrs in H L)
XCHG			EB				Exchange Reg Pairs D E and H L
DAD		09	19	29	39		Add Reg Pair to H L
PUSH	F5	C5	D5	E5			Push Reg Pair on Stack
POP	F1	C1	D1	E1			Pop Reg Pair off Stack
XTHL				E3			Exchange H L with Top of Stack
SPHL					F9		Load SP with H L

Notes 1 This is MOV A,M 2 This is MOV M,A 3 This is JMP

Alphabetized 8085A Mnemonics with Hex Opcodes

Hex	*Mnemonic*		*Hex*	*Mnemonic*		*Hex*	*Mnemonic*		*Hex*	*Mnemonic*	
CE	ACI	8-Bit	2B	DCX	H	52	MOV	D,D	E5	PUSH	H
8F	ADC	A	3B	DCX	SP	53	MOV	D,E	F5	PUSH	PSW
88	ADC	B	F3	DI		54	MOV	D,H	17	RAL	
89	ADC	C	FB	EI		55	MOV	D,L	1F	RAR	
8A	ADC	D	76	HLT		56	MOV	D,M	D8	RC	
8B	ADC	E	DB	IN	8-bit	5F	MOV	E,A	C9	RET	
8C	ADC	H	3C	INR	A	58	MOV	E,B	20	RIM	
8D	ADC	L	04	INR	B	59	MOV	E,C	07	RLC	
8E	ADC	M	0C	INR	C	5A	MOV	E,D	F8	RM	
87	ADD	A	14	INR	D	5B	MOV	E,E	D0	RNC	
80	ADD	B	IC	INR	E	5C	MOV	E,H	C0	RNZ	
81	ADD	C	24	INR	H	5D	MOV	E,L	F0	RP	
82	ADD	D	2C	INR	L	5E	MOV	E,M	38	RPE	
83	ADD	E	34	INR	M	67	MOV	H,A	E0	RPO	
84	ADD	H	03	INX	B	60	MOV	H,B	0F	RRC	
85	ADD	L	13	INX	D	61	MOV	H,C	C7	RST	0
86	ADD	M	23	INX	H	62	MOV	H,D	CF	RST	1
C6	ADI	8-Bit	33	INX	SP	63	MOV	H,E	D7	RST	2
A7	ANA	A	DA	JC	16-Bit	64	MOV	H,H	DF	RST	3
A0	ANA	B	FA	JM	16-Bit	65	MOV	H,L	E7	RST	4
A1	ANA	C	C3	JMP	16-Bit	66	MOV	H,M	EF	RST	5
A2	ANA	D	D2	JNC	16-Bit	6F	MOV	L,A	F7	RST	6
A3	ANA	E	C2	JNZ	16-Bit	68	MOV	L,B	FF	RST	7
A4	ANA	H	F2	JP	16-Bit	69	MOV	L,C	C8	RZ	
A5	ANA	L	EA	JPE	16-Bit	6A	MOV	L,D	9F	SBB	A
A6	ANA	M	E2	JPO	16-Bit	6B	MOV	L,E	98	SBB	B
E6	ANI	8-Bit	CA	JZ	16-Bit	6C	MOV	L,H	99	SBB	C
CD	CALL	16-Bit	3A	LDA	16-Bit	6D	MOV	L,L	9A	SBB	D
DC	CC	16-Bit	0A	LDAX	B	6E	MOV	L,M	9B	SBB	E
FC	CM	16-Bit	1A	LDAX	D	77	MOV	M,A	9C	SBB	H
2F	CMA		2A	LHLD	16-Bit	70	MOV	M,B	9D	SBB	L
3F	CMC		01	LXI	B,16-Bit	71	MOV	M,C	9E	SBB	M
BF	CMP	A	11	LXI	D,16-Bit	72	MOV	M,D	DE	SBI	8-Bit
B8	CMP	B	21	LXI	H,16-Bit	73	MOV	M,E	22	SHLD,	16-Bit
B9	CMP	C	31	LXI	SP,16-Bit	74	MOV	M,H	30	SIM	
BA	CMP	D	7F	MOV	A,A	75	MOV	M,L	F9	SPHL	
BB	CMP	E	78	MOV	A,B	3E	MVI	A, 8-Bit	32	STA	16-Bit
BC	CMP	H	79	MOV	A,C	06	MVI	B, 8-bit	02	STAX	B
BD	CMP	L	7A	MOV	A,D	0E	MVI	C, 8-Bit	12	STAX	D
BE	CMP	M	7B	MOV	A,E	16	MVI	D, 8-Bit	37	STC	
D4	CNC	16-Bit	7C	MOV	A,H	1E	MVI	E, 8-Bit	97	SUB	A
C4	CNZ	16-Bit	7D	MOV	A,L	26	MVI	H, 8-Bit	90	SUB	B
F4	CP	16-Bit	7E	MOV	A,M	2E	MVI	L, 8-Bit	91	SUB	C
EC	CPE	16-Bit	47	MOV	B,A	36	MVI	M, 8-Bit	92	SUB	D
FE	CPI	8-Bit	40	MOV	B,B	00	NOP		93	SUB	E
E4	CPO	16-Bit	41	MOV	B,C	B7	ORA	A	94	SUB	H
CC	CZ	16-Bit	42	MOV	B,D	B0	ORA	B	95	SUB	L
27	DAA		43	MOV	B,E	B1	ORA	C	96	SUB	M
09	DAD	B	44	MOV	B,H	B2	ORA	D	D6	SUI	8-Bit
19	DAD	D	45	MOV	B,L	B3	ORA	E	EB	XCHG	
29	DAD	H	46	MOV	B,M	B4	ORA	H	AF	XRA	A
39	DAD	SP	4F	MOV	C,A	B5	ORA	L	A8	XRA	B
3D	DCR	A	48	MOV	C,B	B6	ORA	M	A9	XRA	C
05	DCR	B	49	MOV	C,C	F6	ORI	8-Bit	AA	XRA	D
0D	DCR	C	4A	MOV	C,D	D3	OUT	8-Bit	AB	XRA	E
15	DCR	D	4B	MOV	C,E	E9	PCHL		AC	XRA	H
1D	DCR	E	4C	MOV	C,H	C1	POP	B	AD	XRA	L
25	DCR	H	4D	MOV	C,L	D1	POP	D	AE	XRA	M
2D	DCR	L	4E	MOV	C,M	E1	POP	H	EE	XRI	8-Bit
35	DCR	M	57	MOV	D,A	F1	POP	PSW	E3	XTHL	
0B	DCX	B	50	MOV	D,B	C5	PUSH	B			
1B	DCX	D	51	MOV	D,C	D5	PUSH	D			

D

8085A Instruction Set Reference Encyclopedia[1]

[1]Courtesy of Intel Corporation.

THE INSTRUCTION SET

INSTRUCTION SET ENCYCLOPEDIA

In the ensuing dozen pages, the complete 8085A instruction set is described, grouped in order under five different functional headings, as follows:

1. **Data Transfer Group** — Moves data between registers or between memory locations and registers. Includes moves, loads, stores, and exchanges.
2. **Arithmetic Group** — Adds, subtracts, increments, or decrements data in registers or memory.
3. **Logic Group** — ANDs, ORs, XORs, compares, rotates, or complements data in registers or between memory and a register.
4. **Branch Group** — Initiates conditional or unconditional jumps, calls, returns, and restarts.
5. **Stack, I/O, and Machine Control Group** — Includes instructions for maintaining the stack, reading from input ports, writing to output ports, setting and reading interrupt masks, and setting and clearing flags.

The formats described in the encyclopedia reflect the assembly language processed by Intel-supplied assembler, used with the Intellec® development systems.

Data Transfer Group

This group of instructions transfers data to and from registers and memory. **Condition flags are not affected by any instruction in this group.**

MOV r1, r2 (Move Register)
(r1) ← (r2)
The content of register r2 is moved to register r1.

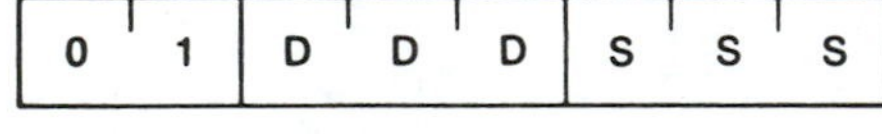

Cycles: 1
States: 4 (8085), 5 (8080)
Addressing: register
Flags: none

*All mnemonics copyrighted © Intel Corporation 1976.

MOV r, M (Move from memory)
(r) ← ((H) (L))
The content of the memory location, whose address is in registers H and L, is moved to register r.

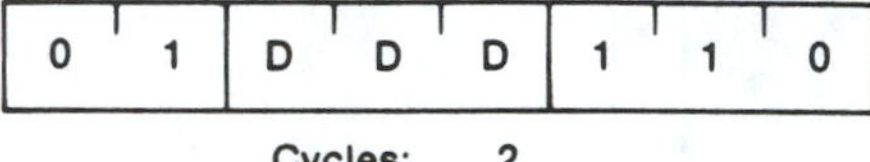

Cycles: 2
States: 7
Addressing: reg. indirect
Flags: none

MOV M, r (Move to memory)
((H)) (L)) ← (r)
The content of register r is moved to the memory location whose address is in registers H and L.

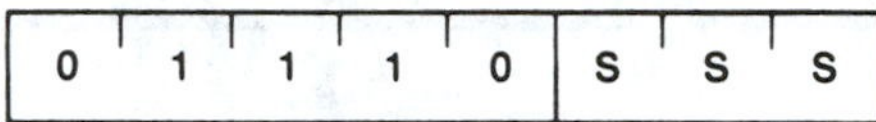

Cycles: 2
States: 7
Addressing: reg. indirect
Flags: none

MVI r, data (Move Immediate)
(r) ← (byte 2)
The content of byte 2 of the instruction is moved to register r.

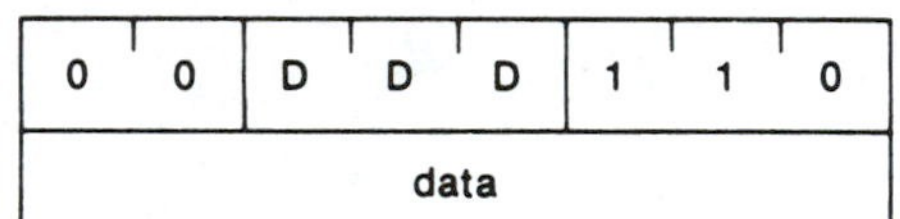

Cycles: 2
States: 7
Addressing: immediate
Flags: none

MVI M, data (Move to memory immediate)
((H) (L)) ← (byte 2)
The content of byte 2 of the instruction is moved to the memory location whose address is in registers H and L.

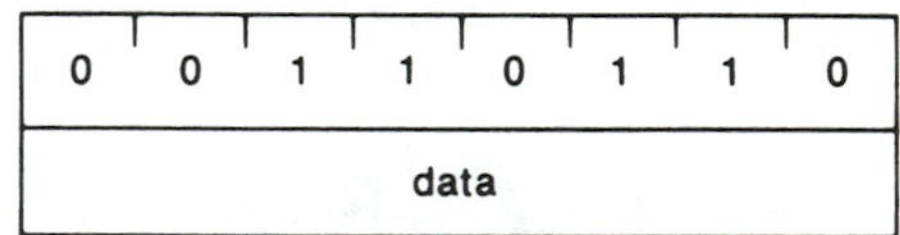

Cycles: 3
States: 10
Addressing: immed./reg. indirect
Flags: none

THE INSTRUCTION SET

LXI rp, data 16 (Load register pair immediate)
(rh) ← (byte 3),
(rl) ← (byte 2)
Byte 3 of the instruction is moved into the high-order register (rh) of the register pair rp. Byte 2 of the instruction is moved into the low-order register (rl) of the register pair rp.

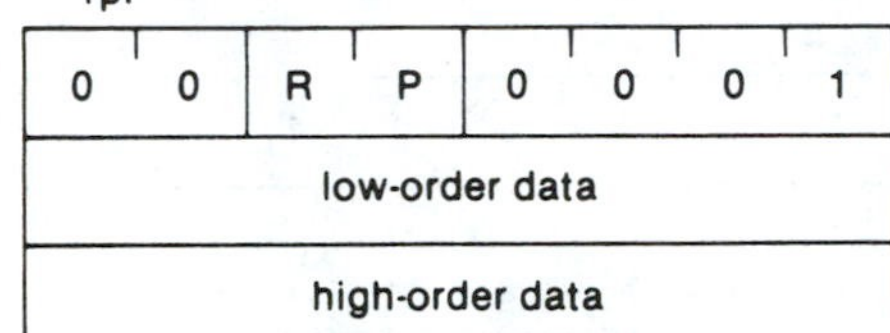

Cycles: 3
States: 10
Addressing: immediate
Flags: none

LDA addr (Load Accumulator direct)
(A) ← ((byte 3)(byte 2))
The content of the memory location, whose address is specified in byte 2 and byte 3 of the instruction, is moved to register A.

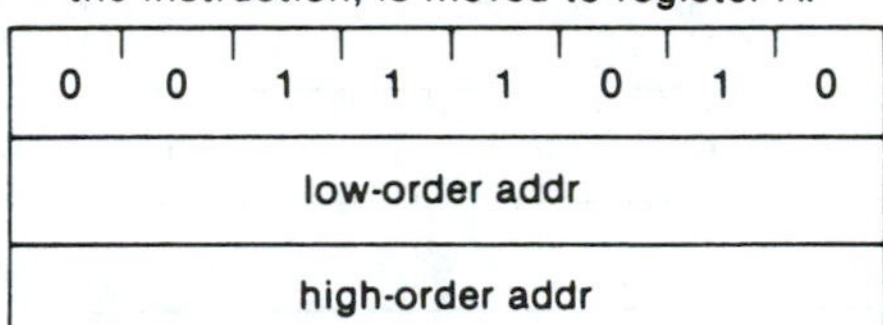

Cycles: 4
States: 13
Addressing: direct
Flags: none

STA addr (Store Accumulator direct)
((byte 3)(byte 2)) ← (A)
The content of the accumulator is moved to the memory location whose address is specified in byte 2 and byte 3 of the instruction.

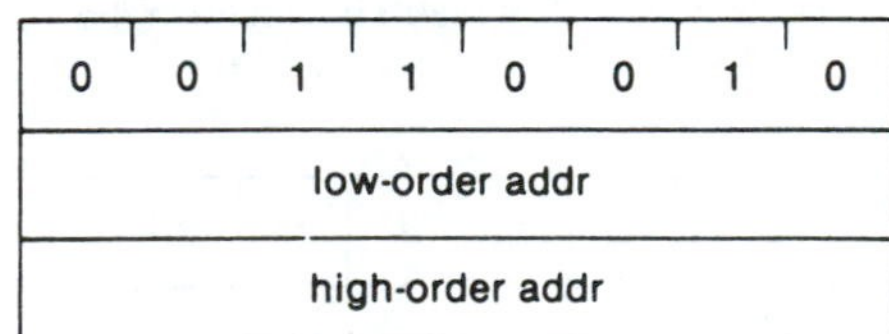

Cycles: 4
States: 13
Addressing: direct
Flags: none

LHLD addr (Load H and L direct)
(L) ← ((byte 3)(byte 2))
(H) ← ((byte 3)(byte 2) + 1)
The content of the memory location, whose address is specified in byte 2 and byte 3 of the instruction, is moved to register L. The content of the memory location at the succeeding address is moved to register H.

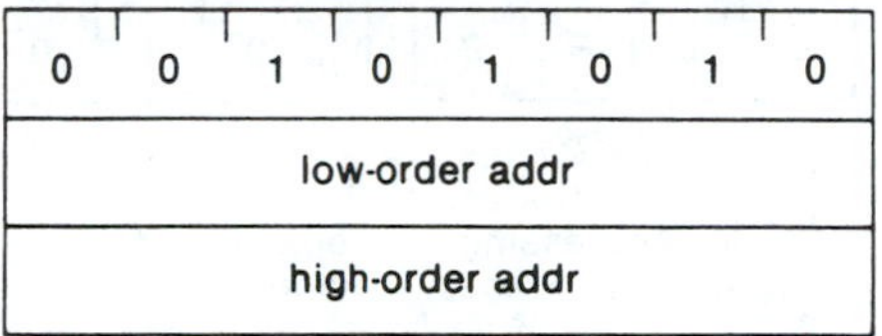

Cycles: 5
States: 16
Addressing: direct
Flags: none

SHLD addr (Store H and L direct)
((byte 3)(byte 2)) ← (L)
((byte 3)(byte 2) + 1) ← (H)
The content of register L is moved to the memory location whose address is specified in byte 2 and byte 3. The content of register H is moved to the succeeding memory location.

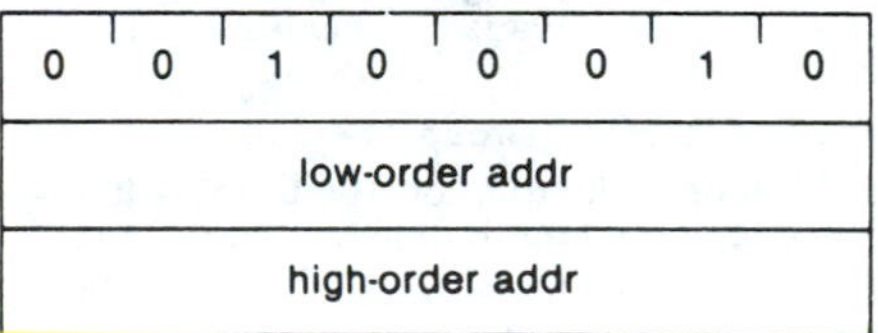

Cycles: 5
States: 16
Addressing: direct
Flags: none

LDAX rp (Load accumulator indirect)
(A) ← ((rp))
The content of the memory location, whose address is in the register pair rp, is moved to register A. Note: only register pairs rp = B (registers B and C) or rp = D (registers D and E) may be specified.

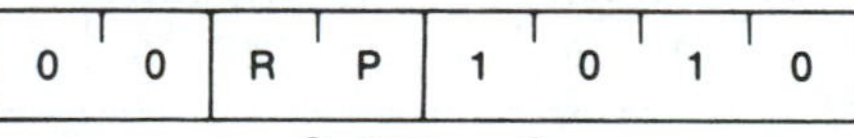

Cycles: 2
States: 7
Addressing: reg. indirect
Flags: none

THE INSTRUCTION SET

STAX rp (Store accumulator indirect)
((rp)) ← (A)
The content of register A is moved to the memory location whose address is in the register pair rp. Note: only register pairs rp = B (registers B and C) or rp = D (registers D and E) may be specified.

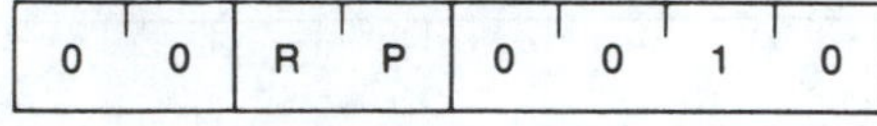

Cycles: 2
States: 7
Addressing: reg. indirect
Flags: none

XCHG (Exchange H and L with D and E)
(H) ↔ (D)
(L) ↔ (E)
The contents of registers H and L are exchanged with the contents of registers D and E.

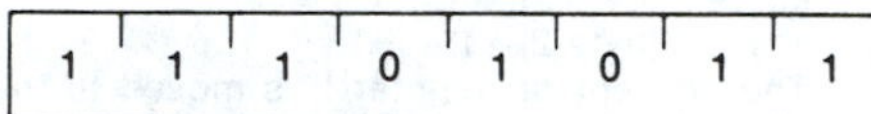

Cycles: 1
States: 4
Addressing: register
Flags: none

Arithmetic Group

This group of instructions performs arithmetic operations on data in registers and memory.

Unless indicated otherwise, all instructions in this group affect the Zero, Sign, Parity, Carry, and Auxiliary Carry flags according to the standard rules.

All subtraction operations are performed via two's complement arithmetic and set the carry flag to one to indicate a borrow and clear it to indicate no borrow.

ADD r (Add Register)
(A) ← (A) + (r)
The content of register r is added to the content of the accumulator. The result is placed in the accumulator.

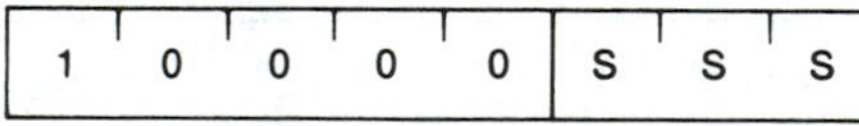

Cycles: 1
States: 4
Addressing: register
Flags: Z,S,P,CY,AC

ADD M (Add memory)
(A) ← (A) + ((H) (L))
The content of the memory location whose address is contained in the H and L registers is added to the content of the accumulator. The result is placed in the accumulator.

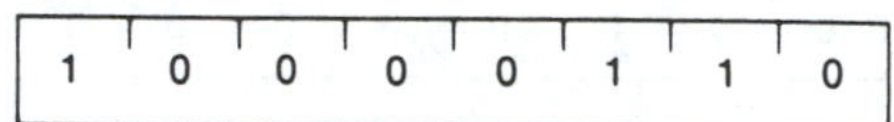

Cycles: 2
States: 7
Addressing: reg. indirect
Flags: Z,S,P,CY,AC

ADI data (Add immediate)
(A) ← (A) + (byte 2)
The content of the second byte of the instruction is added to the content of the accumulator. The result is placed in the accumulator.

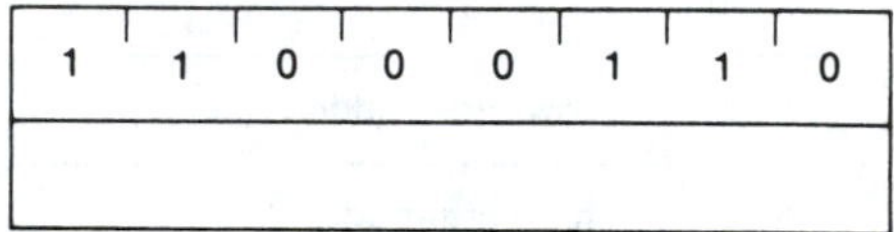

Cycles: 2
States: 7
Addressing: immediate
Flags: Z,S,P,CY,AC

ADC r (Add Register with carry)
(A) ← (A) + (r) + (CY)
The content of register r and the content of the carry bit are added to the content of the accumulator. The result is placed in the accumulator.

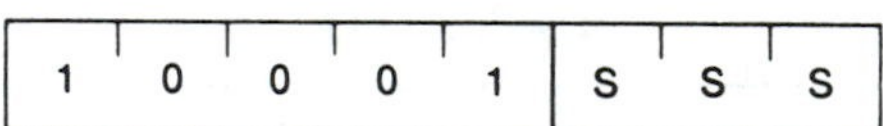

Cycles: 1
States: 4
Addressing: register
Flags: Z,S,P,CY,AC

THE INSTRUCTION SET

ADC M (Add memory with carry)
(A) ← (A) + ((H) (L)) + (CY)
The content of the memory location whose address is contained in the H and L registers and the content of the CY flag are added to the accumulator. The result is placed in the accumulator.

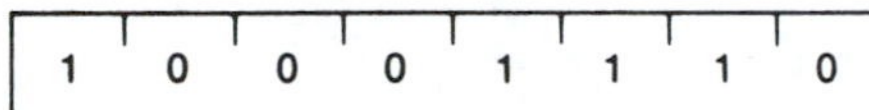

Cycles: 2
States: 7
Addressing: reg. indirect
Flags: Z,S,P,CY,AC

SUB M (Subtract memory)
(A) ← (A) − ((H) (L))
The content of the memory location whose address is contained in the H and L registers is subtracted from the content of the accumulator. The result is placed in the accumulator.

1 0 0 1 0 1 1 0

Cycles: 2
States: 7
Addressing: reg. indirect
Flags: Z,S,P,CY,AC

ACI data (Add immediate with carry)
(A) ← (A) + (byte 2) + (CY)
The content of the second byte of the instruction and the content of the CY flag are added to the contents of the accumulator. The result is placed in the accumulator.

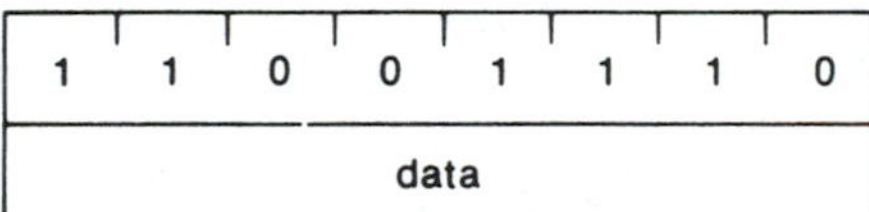

Cycles: 2
States: 7
Addressing: immediate
Flags: Z,S,P,CY,AC

SUI data (Subtract immediate)
(A) ← (A) − (byte 2)
The content of the second byte of the instruction is subtracted from the content of the accumulator. The result is placed in the accumulator.

1 1 0 1 0 1 1 0
data

Cycles: 2
States: 7
Addressing: immediate
Flags: Z,S,P,CY,AC

SUB r (Subtract Register)
(A) ← (A) − (r)
The content of register r is subtracted from the content of the accumulator. The result is placed in the accumulator.

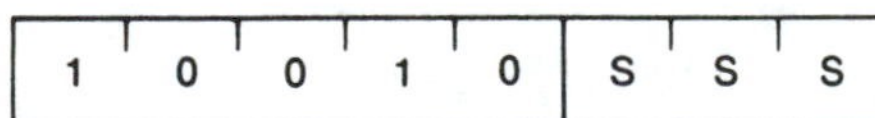

Cycles: 1
States: 4
Addressing: register
Flags: Z,S,P,CY,AC

SBB r (Subtract Register with borrow)
(A) ← (A) − (r) − (CY)
The content of register r and the content of the CY flag are both subtracted from the accumulator. The result is placed in the accumulator.

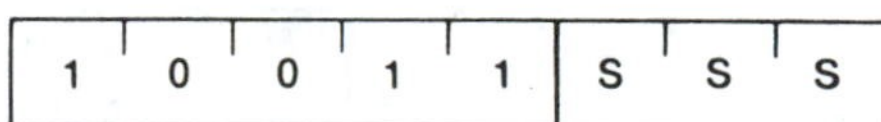

Cycles: 1
States: 4
Addressing: register
Flags: Z,S,P,CY,AC

THE INSTRUCTION SET

SBB M (Subtract memory with borrow)
(A) ← (A) − ((H) (L)) − (CY)
The content of the memory location whose address is contained in the H and L registers and the content of the CY flag are both subtracted from the accumulator. The result is placed in the accumulator.

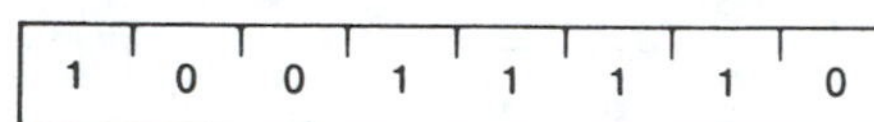

Cycles: 2
States: 7
Addressing: reg. indirect
Flags: Z,S,P,CY,AC

SBI data (Subtract immediate with borrow)
(A) ← (A) − (byte 2) − (CY)
The contents of the second byte of the instruction and the contents of the CY flag are both subtracted from the accumulator. The result is placed in the accumulator.

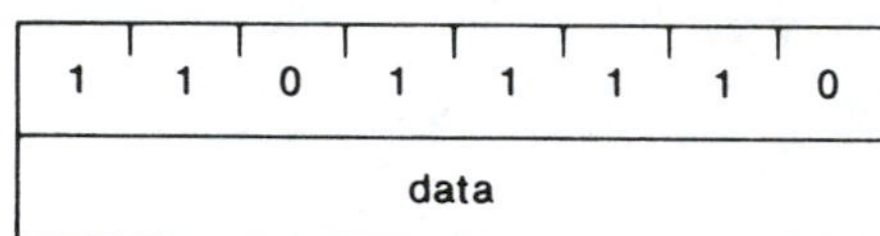

Cycles: 2
States: 7
Addressing: immediate
Flags: Z,S,P,CY,AC

INR r (Increment Register)
(r) ← (r) + 1
The content of register r is incremented by one. Note: All condition flags **except CY** are affected.

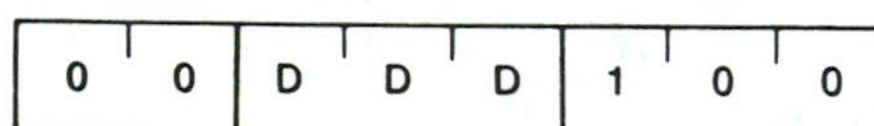

Cycles: 1
States: 4 (8085), 5 (8080)
Addressing: register
Flags: Z,S,P,AC

INR M (Increment memory)
((H) (L) ← ((H) (L)) + 1
The content of the memory location whose address is contained in the H and L registers is incremented by one. Note: All condition flags **except CY** are affected.

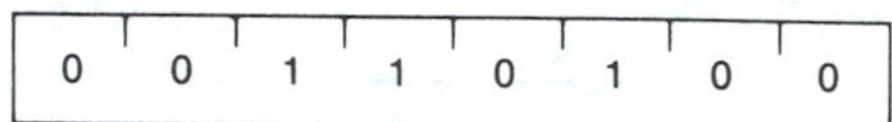

Cycles: 3
States: 10
Addressing: reg. indirect
Flags: Z,S,P,AC

DCR r (Decrement Register)
(r) ← (r) − 1
The content of register r is decremented by one. Note: All condition flags **except CY** are affected.

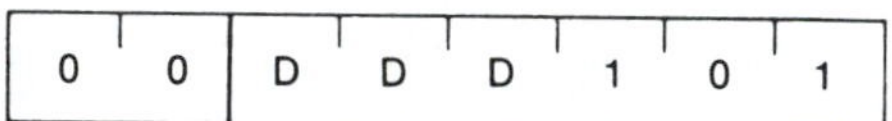

Cycles: 1
States: 4 (8085), 5 (8080)
Addressing: register
Flags: Z,S,P,AC

DCR M (Decrement memory)
((H) (L)) ← ((H) (L)) − 1
The content of the memory location whose address is contained in the H and L registers is decremented by one. Note: All condition flags **except CY** are affected.

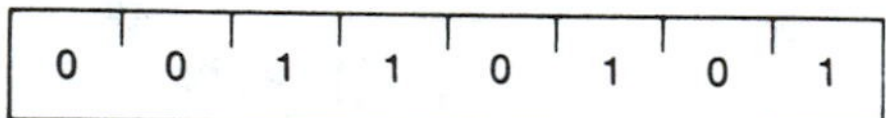

Cycles: 3
States: 10
Addressing: reg. indirect
Flags: Z,S,P,AC

THE INSTRUCTION SET

INX rp (Increment register pair)
(rh) (rl) ← (rh) (rl) + 1
The content of the register pair rp is incremented by one. Note: **No condition flags are affected.**

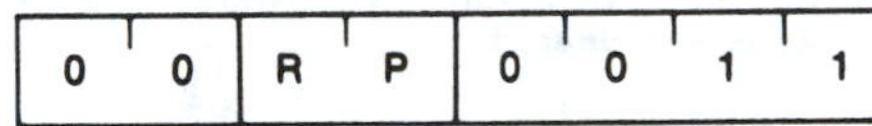
0 0 R P 0 0 1 1

Cycles: 1
States: 6 (8085), 5 (8080)
Addressing: register
Flags: none

DCX rp (Decrement register pair)
(rh) (rl) ← (rh) (rl) − 1
The content of the register pair rp is decremented by one. Note: **No condition flags are affected.**

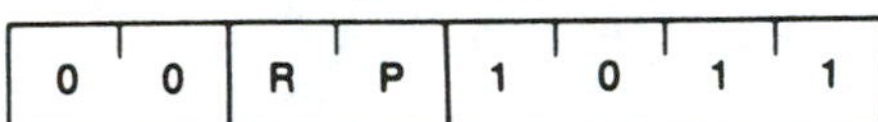
0 0 R P 1 0 1 1

Cycles: 1
States: 6 (8085), 5 (8080)
Addressing: register
Flags: none

DAD rp (Add register pair to H and L)
(H) (L) ← (H) (L) + (rh) (rl)
The content of the register pair rp is added to the content of the register pair H and L. The result is placed in the register pair H and L. Note: **Only the CY flag is affected.** It is set if there is a carry out of the double precision add; otherwise it is reset.

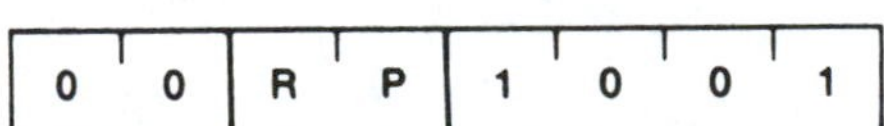
0 0 R P 1 0 0 1

Cycles: 3
States: 10
Addressing: register
Flags: CY

DAA (Decimal Adjust Accumulator)
The eight-bit number in the accumulator is adjusted to form two four-bit Binary-Coded-Decimal digits by the following process:

1. If the value of the least significant 4 bits of the accumulator is greater than 9 **or** if the AC flag is set, 6 is added to the accumulator.
2. If the value of the most significant 4 bits of the accumulator is now greater than 9, **or** if the CY flag is set, 6 is added to the most significant 4 bits of the accumulator.

NOTE: All flags are affected.

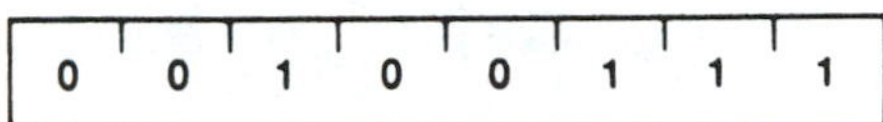
0 0 1 0 0 1 1 1

Cycles: 1
States: 4
Flags: Z,S,P,CY,AC

Logical Group

This group of instructions performs logical (Boolean) operations on data in registers and memory and on condition flags.

Unless indicated otherwise, all instructions in this group affect the Zero, Sign, Parity, Auxiliary Carry, and Carry flags according to the standard rules.

ANA r (AND Register)
(A) ← (A) ∧ (r)
The content of register r is logically ANDed with the content of the accumulator. The result is placed in the accumulator. **The CY flag is cleared and AC is set (8085). The CY flag is cleared and AC is set to the OR'ing of bits 3 of the operands (8080).**

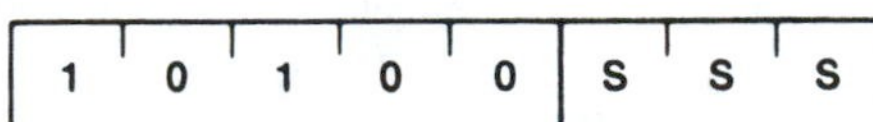
1 0 1 0 0 S S S

Cycles: 1
States: 4
Addressing: register
Flags: Z,S,P,CY,AC

THE INSTRUCTION SET

ANA M (AND memory)

(A) ← (A) ∧ ((H) (L))

The contents of the memory location whose address is contained in the H and L registers is logically ANDed with the content of the accumulator. The result is placed in the accumulator. **The CY flag is cleared and AC is set (8085). The CY flag is cleared and AC is set to the OR'ing of bits 3 of the operands (8080).**

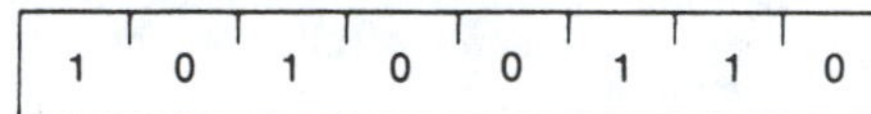

Cycles: 2
States: 7
Addressing: reg. indirect
Flags: Z,S,P,CY,AC

ANI data (AND immediate)

(A) ← (A) ∧ (byte 2)

The content of the second byte of the instruction is logically ANDed with the contents of the accumulator. The result is placed in the accumulator. **The CY flag is cleared and AC is set (8085). The CY flag is cleared and AC is set to the OR'ing of bits 3 of the operands (8080).**

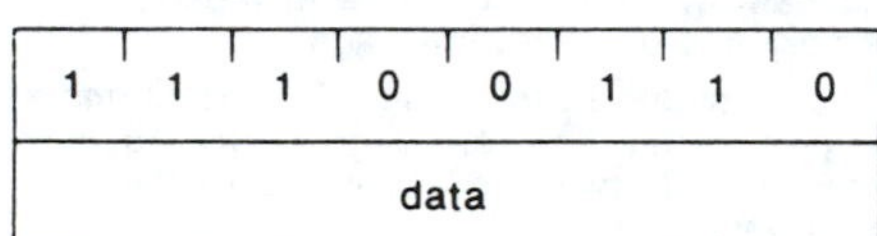

Cycles: 2
States: 7
Addressing: immediate
Flags: Z,S,P,CY,AC

XRA r (Exclusive OR Register)

(A) ← (A) ⊻ (r)

The content of register r is exclusive-OR'd with the content of the accumulator. The result is placed in the accumulator. **The CY and AC flags are cleared.**

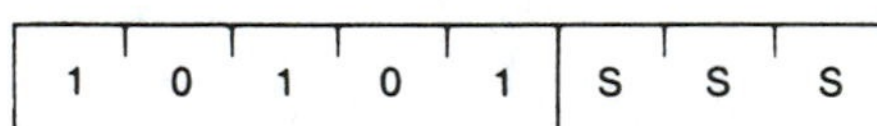

Cycles: 1
States: 4
Addressing: register
Flags: Z,S,P,CY,AC

XRA M (Exclusive OR Memory)

(A) ← (A) ⊻ ((H) (L))

The content of the memory location whose address is contained in the H and L registers is exclusive-OR'd with the content of the accumulator. The result is placed in the accumulator. **The CY and AC flags are cleared.**

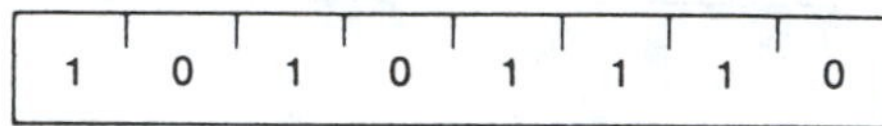

Cycles: 2
States: 7
Addressing: reg. indirect
Flags: Z,S,P,CY,AC

XRI data (Exclusive OR immediate)

(A) ← (A) ⊻ (byte 2)

The content of the second byte of the instruction is exclusive-OR'd with the content of the accumulator. The result is placed in the accumulator. **The CY and AC flags are cleared.**

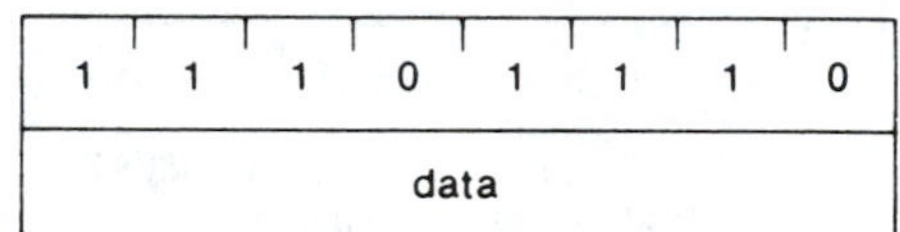

Cycles: 2
States: 7
Addressing: immediate
Flags: Z,S,P,CY,AC

ORA r (OR Register)

(A) ← (A) V (r)

The content of register r is inclusive-OR'd with the content of the accumulator. The result is placed in the accumulator. **The CY and AC flags are cleared.**

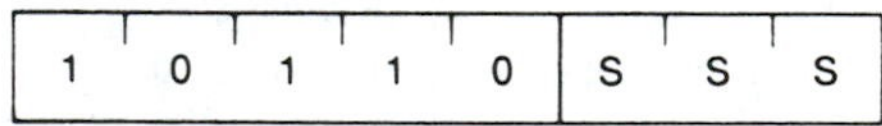

Cycles: 1
States: 4
Addressing: register
Flags: Z,S,P,CY,AC

THE INSTRUCTION SET

ORA M (OR memory)

(A) ← (A) V ((H) (L))

The content of the memory location whose address is contained in the H and L registers is inclusive-OR'd with the content of the accumulator. The result is placed in the accumulator. **The CY and AC flags are cleared.**

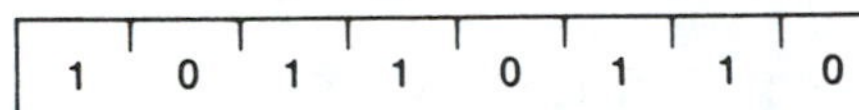

Cycles:	2
States:	7
Addressing:	reg. indirect
Flags:	Z,S,P,CY,AC

ORI data (OR Immediate)

(A) ← (A) V (byte 2)

The content of the second byte of the instruction is inclusive-OR'd with the content of the accumulator. The result is placed in the accumulator. **The CY and AC flags are cleared.**.

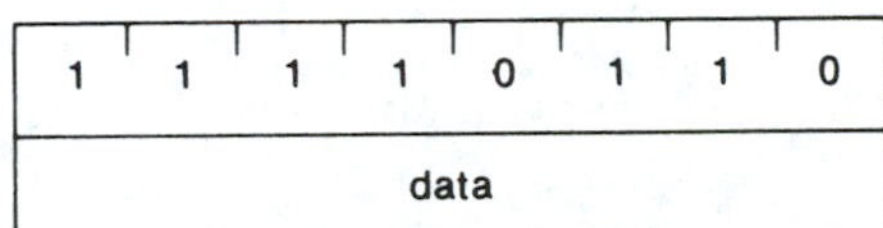

Cycles:	2
States:	7
Addressing:	immediate
Flags:	Z,S,P,CY,AC

CMP r (Compare Register)

(A) – (r)

The content of register r is subtracted from the accumulator. The accumulator remains unchanged. The condition flags are set as a result of the subtraction. **The Z flag is set to 1 if (A) = (r). The CY flag is set to 1 if (A) < (r).**

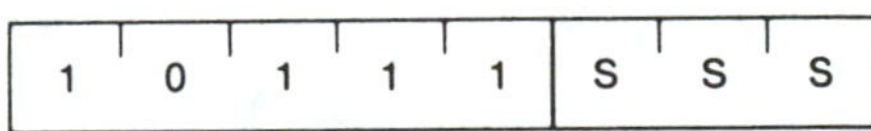

Cycles:	1
States:	4
Addressing:	register
Flags:	Z,S,P,CY,AC

CMP M (Compare memory)

(A) – ((H) (L))

The content of the memory location whose address is contained in the H and L registers is subtracted from the accumulator. The accumulator remains unchanged. The condition flags are set as a result of the subtraction. **The Z flag is set to 1 if (A)=((H) (L)). The CY flag is set to 1 if (A)<((H) (L)).**

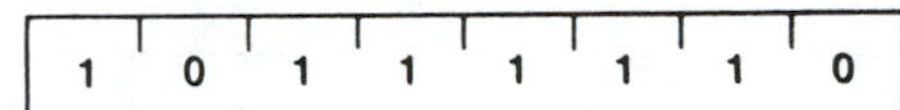

Cycles:	2
States:	7
Addressing:	reg. indirect
Flags:	Z,S,P,CY,AC

CPI data (Compare immediate)

(A) – (byte 2)

The content of the second byte of the instruction is subtracted from the accumulator. The condition flags are set by the result of the subtraction. **The Z flag is set to 1 if (A)=(byte 2). The CY flag is set to 1 if (A)<(byte 2).**

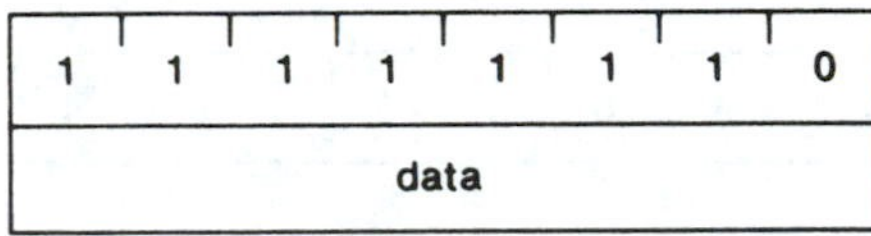

Cycles:	2
States:	7
Addressing:	immediate
Flags:	Z,S,P,CY,AC

RLC (Rotate left)

$(A_{n+1}) \leftarrow (A_n)$;$(A_0) \leftarrow (A_7)$

$(CY) \leftarrow (A_7)$

The content of the accumulator is rotated left one position. The low order bit and the CY flag are both set to the value shifted out of the high order bit position. **Only the CY flag is affected.**

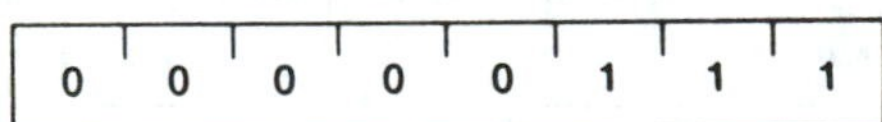

Cycles:	1
States:	4
Flags:	CY

THE INSTRUCTION SET

RRC (Rotate right)
$(A_n) \leftarrow (A_{n+1})$; $(A_7) \leftarrow (A_0)$
$(CY) \leftarrow (A_0)$
The content of the accumulator is rotated right one position. The high order bit and the CY flag are both set to the value shifted out of the low order bit position. **Only the CY flag is affected.**

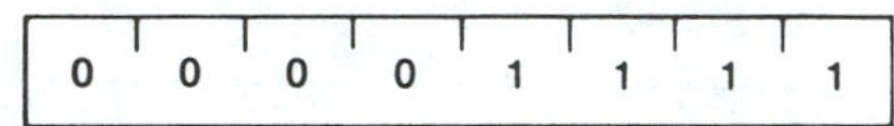

0	0	0	0	1	1	1	1

Cycles: 1
States: 4
Flags: CY

RAL (Rotate left through carry)
$(A_{n+1}) \leftarrow (A_n)$; $(CY) \leftarrow (A_7)$
$(A_0) \leftarrow (CY)$
The content of the accumulator is rotated left one position through the CY flag. The low order bit is set equal to the CY flag and the CY flag is set to the value shifted out of the high order bit. **Only the CY flag is affected.**

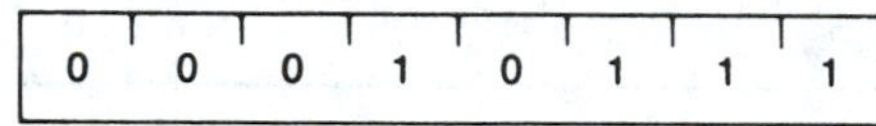

0	0	0	1	0	1	1	1

Cycles: 1
States: 4
Flags: CY

RAR (Rotate right through carry)
$(A_n) \leftarrow (A_{n+1})$; $(CY) \leftarrow (A_0)$
$(A_7) \leftarrow (CY)$
The content of the accumulator is rotated right one position through the CY flag. The high order bit is set to the CY flag and the CY flag is set to the value shifted out of the low order bit. **Only the CY flag is affected.**

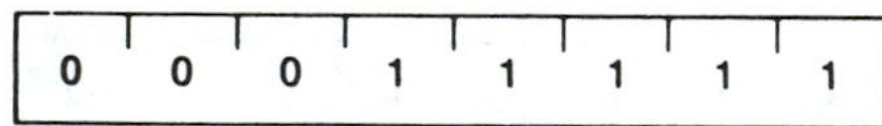

0	0	0	1	1	1	1	1

Cycles: 1
States: 4
Flags: CY

CMA (Complement accumulator)
$(A) \leftarrow (\overline{A})$
The contents of the accumulator are complemented (zero bits become 1, one bits become 0). **No flags are affected.**

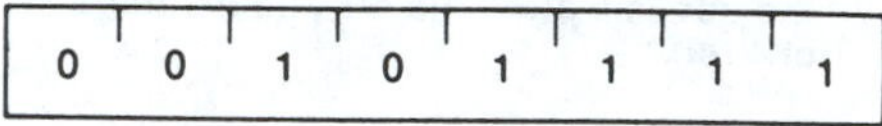

0	0	1	0	1	1	1	1

Cycles: 1
States: 4
Flags: none

CMC (Complement carry)
$(CY) \leftarrow (\overline{CY})$
The CY flag is complemented. **No other flags are affected.**

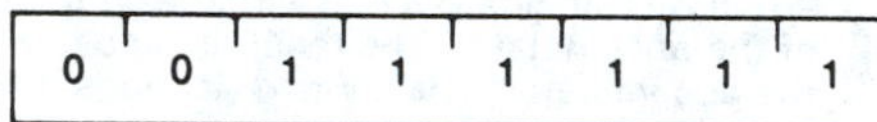

0	0	1	1	1	1	1	1

Cycles: 1
States: 4
Flags: CY

STC (Set carry)
$(CY) \leftarrow 1$
The CY flag is set to 1. **No other flags are affected.**

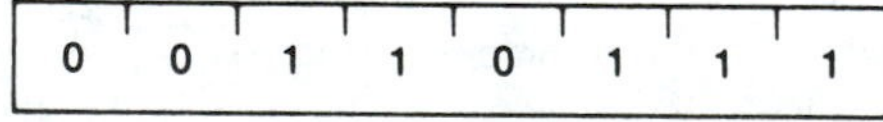

0	0	1	1	0	1	1	1

Cycles: 1
States: 4
Flags: CY

THE INSTRUCTION SET

Branch Group

This group of instructions alter normal sequential program flow.

Condition flags are not affected by any instruction in this group.

The two types of branch instructions are unconditional and conditional. Unconditional transfers simply perform the specified operation on register PC (the program counter). Conditional transfers examine the status of one of the four processor flags to determine if the specified branch is to be executed. The conditions that may be specified are as follows:

CONDITION		CCC
NZ —	not zero (Z = 0)	000
Z —	zero (Z = 1)	001
NC —	no carry (CY = 0)	010
C —	carry (CY = 1)	011
PO —	parity odd (P = 0)	100
PE —	parity even (P = 1)	101
P —	plus (S = 0)	110
M —	minus (S = 1)	111

JMP addr (Jump)

(PC) ← (byte 3) (byte 2)

Control is transferred to the instruction whose address is specified in byte 3 and byte 2 of the current instruction.

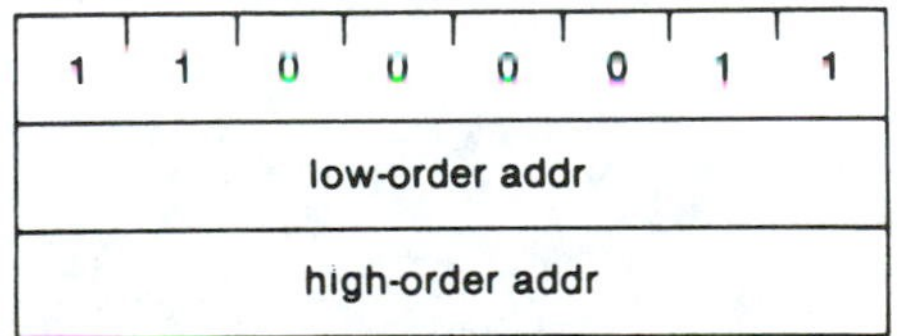

1	1	0	0	0	0	1	1
low-order addr							
high-order addr							

Cycles: 3
States: 10
Addressing: immediate
Flags: none

Jcondition addr (Conditional jump)

If (CCC),

(PC) ← (byte 3) (byte 2)

If the specified condition is true, control is transferred to the instruction whose address is specified in byte 3 and byte 2 of the current instruction; otherwise, control continues sequentially.

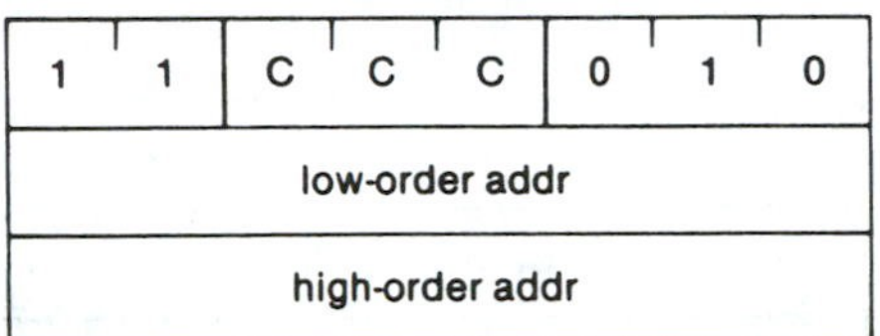

1	1	C	C	C	0	1	0
low-order addr							
high-order addr							

Cycles: 2/3 (8085), 3 (8080)
States: 7/10 (8085), 10 (8080)
Addressing: immediate
Flags: none

CALL addr (Call)

((SP) − 1) ← (PCH)
((SP) − 2) ← (PCL)
(SP) ← (SP) − 2
(PC) ← (byte 3) (byte 2)

The high-order eight bits of the next instruction address are moved to the memory location whose address is one less than the content of register SP. The low-order eight bits of the next instruction address are moved to the memory location whose address is two less than the content of register SP. The content of register SP is decremented by 2. Control is transferred to the instruction whose address is specified in byte 3 and byte 2 of the current instruction.

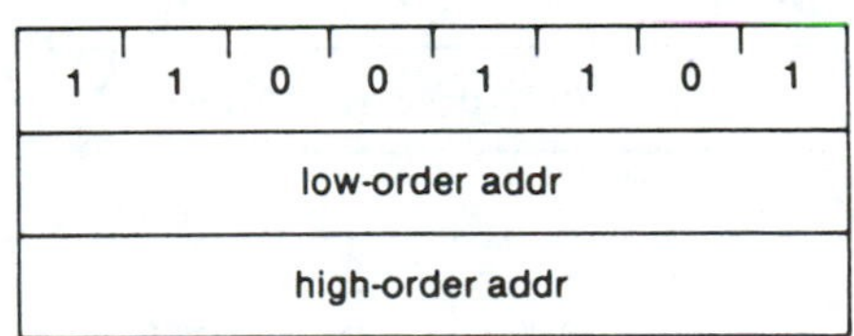

1	1	0	0	1	1	0	1
low-order addr							
high-order addr							

Cycles: 5
States: 18 (8085), 17 (8080)
Addressing: immediate/reg. indirect
Flags: none

THE INSTRUCTION SET

Ccondition addr (Condition call)
If (CCC),
((SP) − 1) ← (PCH)
((SP) − 2) ← (PCL)
(SP) ← (SP) − 2
(PC) ← (byte 3) (byte 2)
If the specified condition is true, the actions specified in the CALL instruction (see above) are performed; otherwise, control continues sequentially.

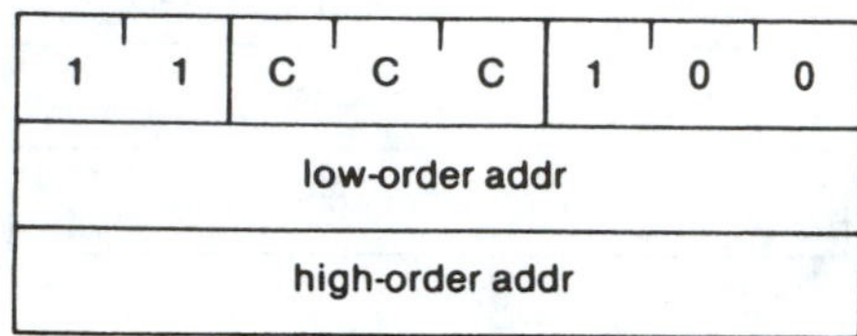

Cycles: 2/5 (8085), 3/5 (8080)
States: 9/18 (8085), 11/17 (8080)
Addressing: immediate/ reg. indirect
Flags: none

RET (Return)
(PCL) ← ((SP));
(PCH) ← ((SP) + 1);
(SP) ← (SP) + 2;
The content of the memory location whose address is specified in register SP is moved to the low-order eight bits of register PC. The content of the memory location whose address is one more than the content of register SP is moved to the high-order eight bits of register PC. The content of register SP is incremented by 2.

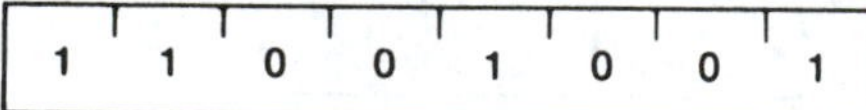

Cycles: 3
States: 10
Addressing: reg. indirect
Flags: none

Rcondition (Conditional return)
If (CCC),
(PCL) ← ((SP))
(PCH) ← ((SP) + 1)
(SP) ← (SP) + 2
If the specified condition is true, the actions specified in the RET instruction (see above) are performed; otherwise, control continues sequentially.

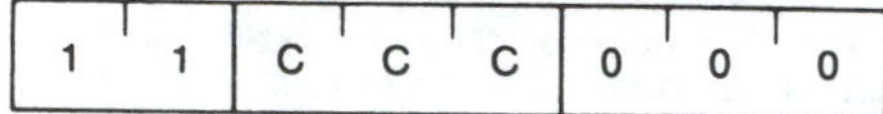

Cycles: 1/3
States: 6/12 (8085), 5/11 (8080)
Addressing: reg. indirect
Flags: none

RST n (Restart)
((SP) − 1) ← (PCH)
((SP) − 2) ← (PCL)
(SP) ← (SP) − 2
(PC) ← 8 * (NNN)
The high-order eight bits of the next instruction address are moved to the memory location whose address is one less than the content of register SP. The low-order eight bits of the next instruction address are moved to the memory location whose address is two less than the content of register SP. The content of register SP is decremented by two. Control is transferred to the instruction whose address is eight times the content of NNN.

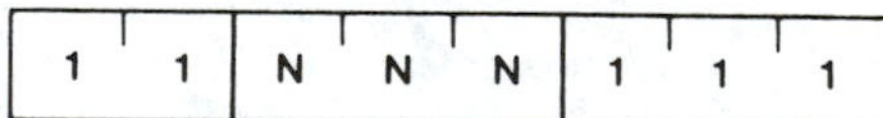

Cycles: 3
States: 12 (8085), 11 (8080)
Addressing: reg. indirect
Flags: none

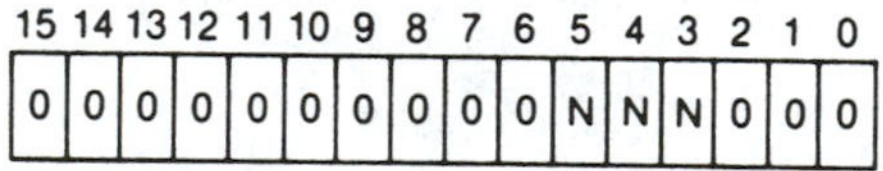

Program Counter After Restart

THE INSTRUCTION SET

PCHL (Jump H and L indirect — move H and L to PC)

(PCH) ← (H)
(PCL) ← (L)

The content of register H is moved to the high-order eight bits of register PC. The content of register L is moved to the low-order eight bits of register PC.

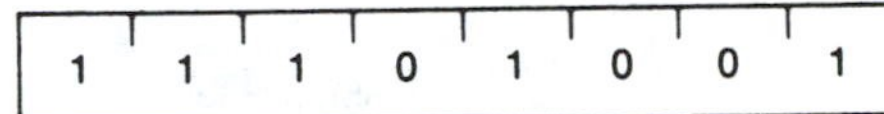

1	1	1	0	1	0	0	1

Cycles: 1
States: 6 (8085), 5 (8080)
Addressing: register
Flags: none

Stack, I/O, and Machine Control Group

This group of instructions performs I/O, manipulates the Stack, and alters internal control flags.

Unless otherwise specified, **condition flags are not affected by any instructions in this group.**

PUSH rp (Push)

((SP) − 1) ← (rh)
((SP) − 2) ← (rl)
((SP) ← (SP) − 2

The content of the high-order register of register pair rp is moved to the memory location whose address is one less than the content of register SP. The content of the low-order register of register pair rp is moved to the memory location whose address is two less than the content of register SP. The content of register SP is decremented by 2. **Note: Register pair rp = SP may not be specified.**

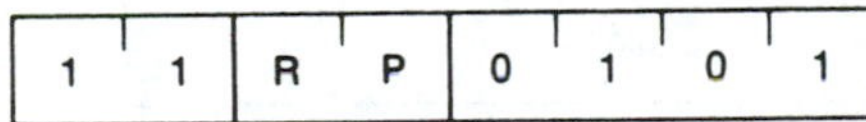

1	1	R	P	0	1	0	1

Cycles: 3
States: 12 (8085), 11 (8080)
Addressing: reg. indirect
Flags: none

PUSH PSW (Push processor status word)

((SP) − 1) ← (A)
((SP) − $2)_0$ ← (CY), ((SP) − $2)_1$ ← X
((SP) − $2)_2$ ← (P), ((SP) − $2)_3$ ← X
((SP) − $2)_4$ ← (AC), ((SP) − $2)_5$ ← X
((SP) − $2)_6$ ← (Z), ((SP) − $2)_7$ ← (S)
(SP) ← (SP) − 2 X: Undefined.

The content of register A is moved to the memory location whose address is one less than register SP. The contents of the condition flags are assembled into a processor status word and the word is moved to the memory location whose address is two less than the content of register SP. The content of register SP is decremented by two.

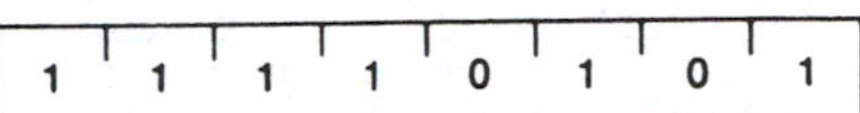

1	1	1	1	0	1	0	1

Cycles: 3
States: 12 (8085), 11 (8080)
Addressing: reg. indirect
Flags: none

FLAG WORD

D_7	D_6	D_5	D_4	D_3	D_2	D_1	D_0
S	Z	X	AC	X	P	X	CY

X: undefined

POP rp (Pop)

(rl) ← ((SP))
(rh) ← ((SP) + 1)
(SP) ← (SP) + 2

The content of the memory location, whose address is specified by the content of register SP, is moved to the low-order register of register pair rp. The content of the memory location, whose address is one more than the content of register SP, is moved to the high-order register of register rp. The content of register SP is incremented by 2. **Note: Register pair rp = SP may not be specified.**

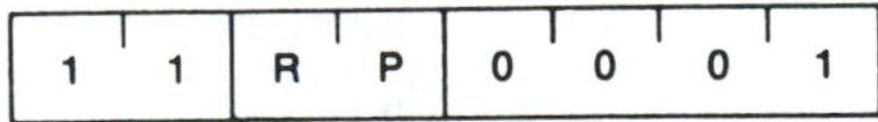

1	1	R	P	0	0	0	1

Cycles: 3
States: 10
Addressing: reg.indirect
Flags: none

THE INSTRUCTION SET

POP PSW (Pop processor status word)

(CY) ← $((SP))_0$
(P) ← $((SP))_2$
(AC) ← $((SP))_4$
(Z) ← $((SP))_6$
(S) ← $((SP))_7$
(A) ← ((SP) + 1)
(SP) ← (SP) + 2

The content of the memory location whose address is specified by the content of register SP is used to restore the condition flags. The content of the memory location whose address is one more than the content of register SP is moved to register A. The content of register SP is incremented by 2.

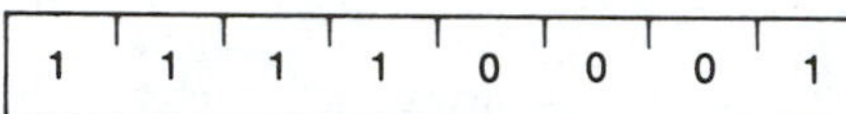

Cycles: 3
States: 10
Addressing: reg. indirect
Flags: Z,S,P,CY,AC

XTHL (Exchange stack top with H and L)

(L) ↔ ((SP))
(H) ↔ ((SP) + 1)

The content of the L register is exchanged with the content of the memory location whose address is specified by the content of register SP. The content of the H register is exchanged with the content of the memory location whose address is one more than the content of register SP.

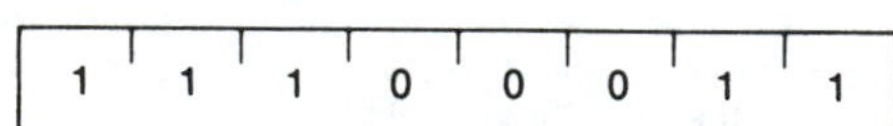

Cycles: 5
States: 16 (8085), 18 (8080)
Addressing: reg. indirect
Flags: none

SPHL (Move HL to SP)

(SP) ← (H) (L)

The contents of registers H and L (16 bits) are moved to register SP.

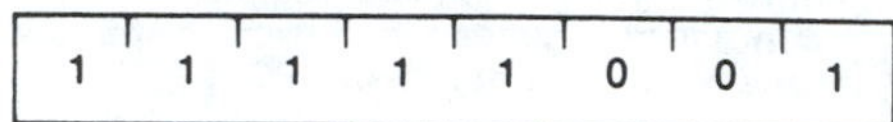

Cycles: 1
States: 6 (8085), 5 (8080)
Addressing: register
Flags: none

IN port (Input)

(A) ← (data)

The data placed on the eight bit bi-directional data bus by the specified port is moved to register A.

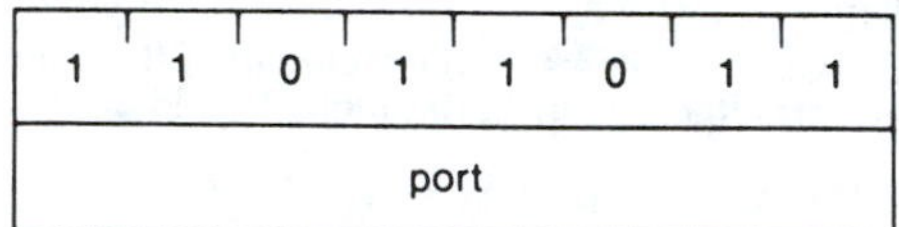

Cycles: 3
States: 10
Addressing: direct
Flags: none

OUT port (Output)

(data) ← (A)

The content of register A is placed on the eight bit bi-directional data bus for transmission to the specified port.

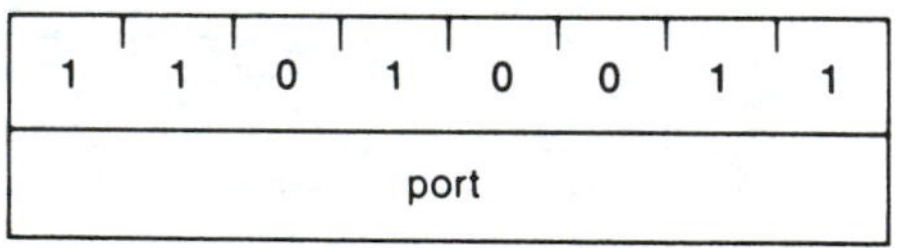

Cycles: 3
States: 10
Addressing: direct
Flags: none

THE INSTRUCTION SET

EI (Enable interrupts)
The interrupt system is enabled **following the execution of the next instruction. Interrupts are not recognized during the EI instruction.**

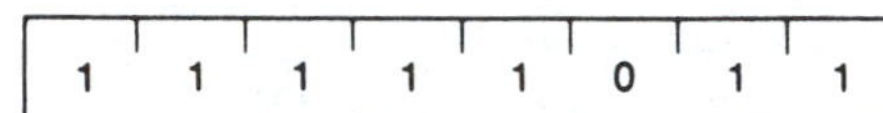

Cycles: 1
States: 4
Flags: none

NOTE: Placing an EI instruction on the bus in response to $\overline{INTA}$ during an INA cycle is prohibited. (8085)

DI (Disable interrupts)
The interrupt system is disabled **immediately following the execution of the DI instruction. Interrupts are not recognized during the DI instruction.**

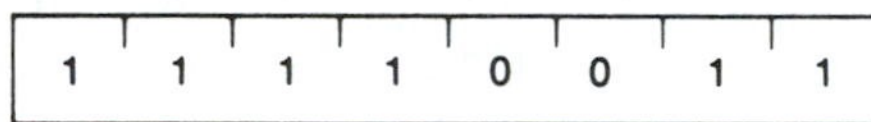

Cycles: 1
States: 4
Flags: none

NOTE: Placing a DI instruction on the bus in response to $\overline{INTA}$ during an INA cycle is prohibited. (8085)

HLT (Halt)
The processor is stopped. The registers and flags are unaffected. (8080) A second ALE is generated during the execution of HLT to strobe out the Halt cycle status information. (8085)

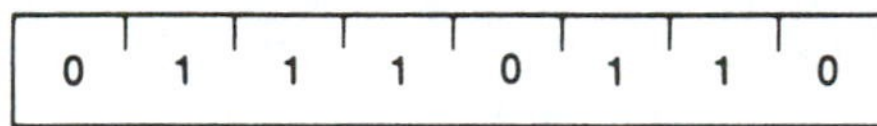

Cycles: 1+ (8085), 1 (8080)
States: 5 (8085), 7 (8080)
Flags: none

NOP (No op)
No operation is performed. The registers and flags are unaffected.

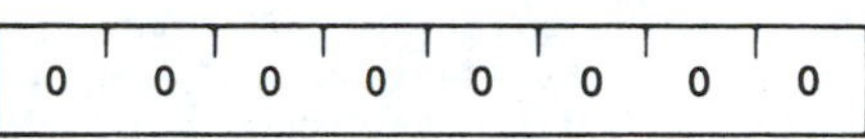

Cycles: 1
States: 4
Flags: none

RIM (Read Interrupt Masks) (8085 only)
The RIM instruction loads data into the accumulator relating to interrupts and the serial input. This data contains the following information:

- Current interrupt mask status for the RST 5.5, 6.5, and 7.5 hardware interrupts (1 = mask disabled)
- Current interrupt enable flag status (1 = interrupts enabled) except immediately following a TRAP interrupt. (See below.)
- Hardware interrupts pending (i.e., signal received but not yet serviced), on the RST 5.5, 6.5, and 7.5 lines.
- Serial input data.

Immediately following a TRAP interrupt, the RIM instruction must be executed as a part of the service routine if you need to retrieve current interrupt status later. Bit 3 of the accumulator is (in this special case only) loaded with the interrupt enable (IE) flag status that existed prior to the TRAP interrupt. Following an RST 5.5, 6.5, 7.5, or INTR interrupt, the interrupt flag flip-flop reflects the current interrupt enable status. Bit 6 of the accumulator (I7.5) is loaded with the status of the RST 7.5 flip-flop, which is always set (edge-triggered) by an input on the RST 7.5 input line, even when that interrupt has been previously masked. (See SIM Instruction.)

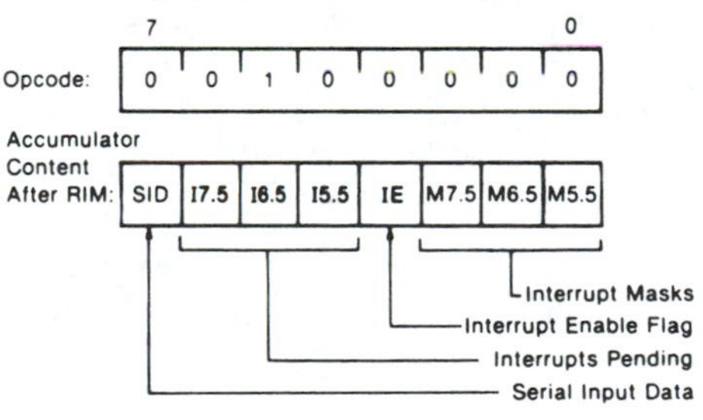

Cycles: 1
States: 4
Flags: none

THE INSTRUCTION SET

SIM (Set Interrupt Masks) (8085 only)

The execution of the SIM instruction uses the contents of the accumulator (which must be previously loaded) to perform the following functions:

- Program the interrupt mask for the RST 5.5, 6.5, and 7.5 hardware interrupts.
- Reset the edge-triggered RST 7.5 input latch.
- Load the SOD output latch.

To program the interrupt masks, first set accumulator bit 3 to 1 and set to 1 any bits 0, 1, and 2, which disable interrupts RST 5.5, 6.5, and 7.5, respectively. Then do a SIM instruction. If accumulator bit 3 is 0 when the SIM instruction is executed, the interrupt mask register will not change. If accumulator bit 4 is 1 when the SIM instruction is executed, the RST 7.5 latch is then reset. RST 7.5 is distinguished by the fact that its latch is always set by a rising edge on the RST 7.5 input pin, even if the jump to service routine is inhibited by masking. This latch remains high until cleared by a $\overline{\text{RESET IN}}$, by a SIM Instruction with accumulator bit 4 high, or by an internal processor acknowledge to an RST 7.5 interrupt subsequent to the removal of the mask (by a SIM instruction). The $\overline{\text{RESET IN}}$ signal always sets all three RST mask bits.

If accumulator bit 6 is at the 1 level when the SIM instruction is executed, the state of accumulator bit 7 is loaded into the SOD latch and thus becomes available for interface to an external device. The SOD latch is unaffected by the SIM instruction if bit 6 is 0. SOD is always reset by the $\overline{\text{RESET IN}}$ signal.

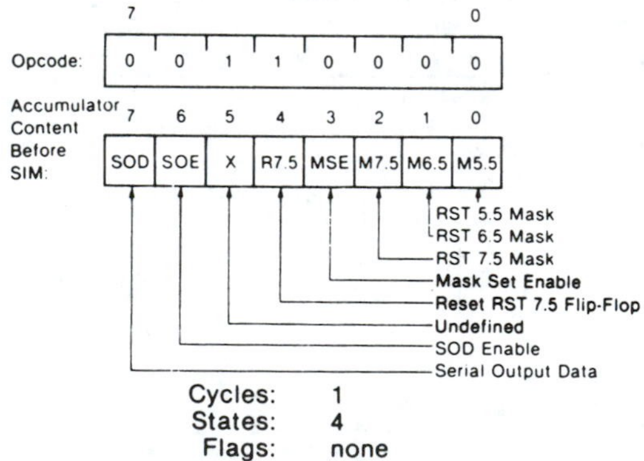

Cycles: 1
States: 4
Flags: none

E

8085A Instruction Set Timing Index[1]

[1]Courtesy of Intel Corporation.

8085A

8080A/8085A INSTRUCTION SET INDEX

Instruction		Code	Bytes	T States		Machine Cycles
				8085A	8080A	
ACI	DATA	CE data	2	7	7	F R
ADC	REG	1000 1SSS	1	4	4	F
ADC	M	8E	1	7	7	F R
ADD	REG	1000 0SSS	1	4	4	F
ADD	M	86	1	7	7	F R
ADI	DATA	C6 data	2	7	7	F R
ANA	REG	1010 0SSS	1	4	4	F
ANA	M	A6	1	7	7	F R
ANI	DATA	E6 data	2	7	7	F R
CALL	LABEL	CD addr	3	18	17	S R R W W*
CC	LABEL	DC addr	3	9/18	11/17	S R•/S R R W W*
CM	LABEL	FC addr	3	9/18	11/17	S R•/S R R W W*
CMA		2F	1	4	4	F
CMC		3F	1	4	4	F
CMP	REG	1011 1SSS	1	4	4	F
CMP	M	BE	1	7	7	F R
CNC	LABEL	D4 addr	3	9/18	11/17	S R•/S R R W W*
CNZ	LABEL	C4 addr	3	9/18	11/17	S R•/S R R W W*
CP	LABEL	F4 addr	3	9/18	11/17	S R•/S R R W W*
CPE	LABEL	EC addr	3	9/18	11/17	S R•/S R R W W*
CPI	DATA	FE data	2	7	7	F R
CPO	LABEL	E4 addr	3	9/18	11/17	S R•/S R R W W*
CZ	LABEL	CC addr	3	9/18	11/17	S R•/S R R W W*
DAA		27	1	4	4	F
DAD	RP	00RP 1001	1	10	10	F B B
DCR	REG	00SS S101	1	4	5	F*
DCR	M	35	1	10	10	F R W
DCX	RP	00RP 1011	1	6	5	S*
DI		F3	1	4	4	F
EI		FB	1	4	4	F
HLT		76	1	5	7	F B
IN	PORT	DB data	2	10	10	F R I
INR	REG	00SS S100	1	4	5	F*
INR	M	34	1	10	10	F R W
INX	RP	00RP 0011	1	6	5	S*
JC	LABEL	DA addr	3	7/10	10	F R/F R R†
JM	LABEL	FA addr	3	7/10	10	F R/F R R†
JMP	LABEL	C3 addr	3	10	10	F R R
JNC	LABEL	D2 addr	3	7/10	10	F R/F R R†
JNZ	LABEL	C2 addr	3	7/10	10	F R/F R R†
JP	LABEL	F2 addr	3	7/10	10	F R/F R R†
JPE	LABEL	EA addr	3	7/10	10	F R/F R R†
JPO	LABEL	E2 addr	3	7/10	10	F R/F R R†
JZ	LABEL	CA addr	3	7/10	10	F R/F R R†
LDA	ADDR	3A addr	3	13	13	F R R R
LDAX	RP	000X 1010	1	7	7	F R
LHLD	ADDR	2A addr	3	16	16	F R R R R

Instruction		Code	Bytes	T States		Machine Cycles
				8085A	8080A	
LXI	RP,DATA16	00RP 0001 data16	3	10	10	F R R
MOV	REG,REG	01DD DSSS	1	4	5	F*
MOV	M,REG	0111 0SSS	1	7	7	F W
MOV	REG,M	01DD D110	1	7	7	F R
MVI	REG,DATA	00DD D110 data	2	7	7	F R
MVI	M,DATA	36 data	2	10	10	F R W
NOP		00	1	4	4	F
ORA	REG	1011 0SSS	1	4	4	F
ORA	M	B6	1	7	7	F R
ORI	DATA	F6 data	2	7	7	F R
OUT	PORT	D3 data	2	10	10	F R O
PCHL		E9	1	6	5	S*
POP	RP	11RP 0001	1	10	10	F R R
PUSH	RP	11RP 0101	1	12	11	S W W*
RAL		17	1	4	4	F
RAR		1F	1	4	4	F
RC		D8	1	6/12	5/11	S/S R R*
RET		C9	1	10	10	F R R
RIM (8085A only)		20	1	4	–	F
RLC		07	1	4	4	F
RM		F8	1	6/12	5/11	S/S R R*
RNC		D0	1	6/12	5/11	S/S R R*
RNZ		C0	1	6/12	5/11	S/S R R*
RP		F0	1	6/12	5/11	S/S R R*
RPE		E8	1	6/12	5/11	S/S R R*
RPO		E0	1	6/12	5/11	S/S R R*
RRC		0F	1	4	4	F
RST	N	11XX X111	1	12	11	S W W*
RZ		C8	1	6/12	5/11	S/S R R*
SBB	REG	1001 1SSS	1	4	4	F
SBB	M	9E	1	7	7	F R
SBI	DATA	DE data	2	7	7	F R
SHLD	ADDR	22 addr	3	16	16	F R R W W
SIM (8085A only)		30	1	4	–	F
SPHL		F9	1	6	5	S*
STA	ADDR	32 addr	3	13	13	F R R W
STAX	RP	000X 0010	1	7	7	F W
STC		37	1	4	4	F
SUB	REG	1001 0SSS	1	4	4	F
SUB	M	96	1	7	7	F R
SUI	DATA	D6 data	2	7	7	F R
XCHG		EB	1	4	4	F
XRA	REG	1010 1SSS	1	4	4	F
XRA	M	AE	1	7	7	F R
XRI	DATA	EE data	2	7	7	F R
XTHL		E3	1	16	18	F R R W W

Machine cycle types:

F	Four clock period instr fetch	
S	Six clock period instr fetch	
R	Memory read	
I	I/O read	
W	Memory write	
O	I/O write	
B	Bus idle	
X	Variable or optional binary digit	
DDD	Binary digits identifying a destination register	B = 000, C = 001, D = 010 Memory = 110
SSS	Binary digits identifying a source register	E = 011, H = 100, L = 101 A = 111
RP	Register Pair	BC = 00, HL = 10 DE = 01, SP = 11

*Five clock period instruction fetch with 8080A.

†The longer machine cycle sequence applies regardless of condition evaluation with 8080A.

•An extra READ cycle (R) will occur for this condition with 8080A.

F

8051 Instruction Set Summary[1]

[1]Courtesy of Intel Corporation.

MCS®-51 INSTRUCTION SET

8051 Instruction Set Summary

Interrupt Response Time: Refer to Hardware Description Chapter.

Instructions that Affect Flag Settings[1]

Instruction	Flag C	Flag OV	Flag AC	Instruction	Flag C	Flag OV	Flag AC
ADD	X	X	X	CLR C	O		
ADDC	X	X	X	CPL C	X		
SUBB	X	X	X	ANL C,bit	X		
MUL	O	X		ANL C,/bit	X		
DIV	O	X		ORL C,bit	X		
DA	X			ORL C,bit	X		
RRC	X			MOV C,bit	X		
RLC	X			CJNE	X		
SETB C	1						

[1]Note that operations on SFR byte address 208 or bit addresses 209-215 (i.e., the PSW or bits in the PSW) will also affect flag settings.

Note on instruction set and addressing modes:

- **Rn** — Register R7–R0 of the currently selected Register Bank.
- **direct** — 8-bit internal data location's address. This could be an Internal Data RAM location (0–127) or a SFR [i.e., I/O port, control register, status register, etc. (128–255)].
- **@Ri** — 8-bit internal data RAM location (0–255) addressed indirectly through register R1 or R0.
- **#data** — 8-bit constant included in instruction.
- **#data 16** — 16-bit constant included in instruction.
- **addr 16** — 16-bit destination address. Used by LCALL & LJMP. A branch can be anywhere within the 64K-byte Program Memory address space.
- **addr 11** — 11-bit destination address. Used by ACALL & AJMP. The branch will be within the same 2K-byte page of program memory as the first byte of the following instruction.
- **rel** — Signed (two's complement) 8-bit offset byte. Used by SJMP and all conditional jumps. Range is −128 to +127 bytes relative to first byte of the following instruction.
- **bit** — Direct Addressed bit in Internal Data RAM or Special Function Register.
- ***** — New operation not provided by 8048AH/8049AH.

Mnemonic		Description	Byte	Oscillator Period
ARITHMETIC OPERATIONS				
ADD	A,Rn	Add register to Accumulator	1	12
ADD	A,direct	Add direct byte to Accumulator	2	12
ADD	A,@Ri	Add indirect RAM to Accumulator	1	12
ADD	A,#data	Add immediate data to Accumulator	2	12
ADDC	A,Rn	Add register to Accumulator with Carry	1	12
ADDC	A,direct	Add direct byte to Accumulator with Carry	2	12
ADDC	A,@Ri	Add indirect RAM to Accumulator with Carry	1	12
ADDC	A,#data	Add immediate data to Acc with Carry	2	12
SUBB	A,Rn	Subtract Register from Acc with borrow	1	12
SUBB	A,direct	Subtract direct byte from Acc with borrow	2	12
SUBB	A,@Ri	Subtract indirect RAM from ACC with borrow	1	12
SUBB	A,#data	Subtract immediate data from Acc with borrow	2	12
INC	A	Increment Accumulator	1	12
INC	Rn	Increment register	1	12
INC	direct	Increment direct byte	2	12
INC	@Ri	Increment direct RAM	1	12
DEC	A	Decrement Accumulator	1	12
DEC	Rn	Decrement Register	1	12
DEC	direct	Decrement direct byte	2	12
DEC	@Ri	Decrement indirect RAM	1	12

8051 Instruction Set Summary (Continued)

Mnemonic		Description	Byte	Oscillator Period
ARITHMETIC OPERATIONS (Continued)				
INC	DPTR	Increment Data Pointer	1	24
MUL	AB	Multiply A & B	1	48
DIV	AB	Divide A by B	1	48
DA	A	Decimal Adjust Accumulator	1	12
LOGICAL OPERATIONS				
ANL	A,Rn	AND Register to Accumulator	1	12
ANL	A,direct	AND direct byte to Accumulator	2	12
ANL	A,@Ri	AND indirect RAM to Accumulator	1	12
ANL	A,#data	AND immediate data to Accumulator	2	12
ANL	direct,A	AND Accumulator to direct byte	2	12
ANL	direct,#data	AND immediate data to direct byte	3	24
ORL	A,Rn	OR register to Accumulator	1	12
ORL	A,direct	OR direct byte to Accumulator	2	12
ORL	A,@Ri	OR indirect RAM to Accumulator	1	12
ORL	A,#data	OR immediate data to Accumulator	2	12
ORL	direct,A	OR Accumulator to direct byte	2	12
ORL	direct,#data	OR immediate data to direct byte	3	24
XRL	A,Rn	Exclusive-OR register to Accumulator	1	12
XRL	A,direct	Exclusive-OR direct byte to Accumulator	2	12
XRL	A,@Ri	Exclusive-OR indirect RAM to Accumulator	1	12
XRL	A,#data	Exclusive-OR immediate data to Accumulator	2	12
XRL	direct,A	Exclusive-OR Accumulator to direct byte	2	12
XRL	direct,#data	Exclusive-OR immediate data to direct byte	3	24
CLR	A	Clear Accumulator	1	12
CPL	A	Complement Accumulator	1	12

Mnemonic		Description	Byte	Oscillator Period
LOGICAL OPERATIONS (Continued)				
RL	A	Rotate Accumulator Left	1	12
RLC	A	Rotate Accumulator Left through the Carry	1	12
RR	A	Rotate Accumulator Right	1	12
RRC	A	Rotate Accumulator Right through the Carry	1	12
SWAP	A	Swap nibbles within the Accumulator	1	12
DATA TRANSFER				
MOV	A,Rn	Move register to Accumulator	1	12
MOV	A,direct	Move direct byte to Accumulator	2	12
MOV	A,@Ri	Move indirect RAM to Accumulator	1	12
MOV	A,#data	Move immediate data to Accumulator	2	12
MOV	Rn,A	Move Accumulator to register	1	12
MOV	Rn,direct	Move direct byte to register	2	24
MOV	Rn,#data	Move immediate data to register	2	12
MOV	direct,A	Move Accumulator to direct byte	2	12
MOV	direct,Rn	Move register to direct byte	2	24
MOV	direct,direct	Move direct byte to direct	3	24
MOV	direct,@Ri	Move indirect RAM to direct byte	2	24
MOV	direct,#data	Move immediate data to direct byte	3	24
MOV	@Ri,A	Move Accumulator to indirect RAM	1	12

MCS®-51 PROGRAMMER'S GUIDE AND INSTRUCTION SET

8051 Instruction Set Summary (Continued)

Mnemonic		Description	Byte	Oscillator Period
DATA TRANSFER (Continued)				
MOV	@Ri,direct	Move direct byte to indirect RAM	2	24
MOV	@Ri,#data	Move immediate data to indirect RAM	2	12
MOV	DPTR,#data16	Load Data Pointer with a 16-bit constant	3	24
MOVC	A,@A+DPTR	Move Code byte relative to DPTR to Acc	1	24
MOVC	A,@A+PC	Move Code byte relative to PC to Acc	1	24
MOVX	A,@Ri	Move External RAM (8-bit addr) to Acc	1	24
MOVX	A,@DPTR	Move External RAM (16-bit addr) to Acc	1	24
MOVX	@Ri,A	Move Acc to External RAM (8-bit addr)	1	24
MOVX	@DPTR,A	Move Acc to External RAM (16-bit addr)	1	24
PUSH	direct	Push direct byte onto stack	2	24
POP	direct	Pop direct byte from stack	2	24
XCH	A,Rn	Exchange register with Accumulator	1	12
XCH	A,direct	Exchange direct byte with Accumulator	2	12
XCH	A,@Ri	Exchange indirect RAM with Accumulator	1	12
XCHD	A,@Ri	Exchange low-order Digit indirect RAM with Acc	1	12

Mnemonic		Description	Byte	Oscillator Period
BOOLEAN VARIABLE MANIPULATION				
CLR	C	Clear Carry	1	12
CLR	bit	Clear direct bit	2	12
SETB	C	Set Carry	1	12
SETB	bit	Set direct bit	2	12
CPL	C	Complement Carry	1	12
CPL	bit	Complement direct bit	2	12
ANL	C,bit	AND direct bit to CARRY	2	24
ANL	C,/bit	AND complement of direct bit to Carry	2	24
ORL	C,bit	OR direct bit to Carry	2	24
ORL	C,/bit	OR complement of direct bit to Carry	2	24
MOV	C,bit	Move direct bit to Carry	2	12
MOV	bit,C	Move Carry to direct bit	2	24
JC	rel	Jump if Carry is set	2	24
JNC	rel	Jump if Carry not set	2	24
JB	bit,rel	Jump if direct Bit is set	3	24
JNB	bit,rel	Jump if direct Bit is Not set	3	24
JBC	bit,rel	Jump if direct Bit is set & clear bit	3	24
PROGRAM BRANCHING				
ACALL	addr11	Absolute Subroutine Call	2	24
LCALL	addr16	Long Subroutine Call	3	24
RET		Return from Subroutine	1	24
RETI		Return from interrupt	1	24
AJMP	addr11	Absolute Jump	2	24
LJMP	addr16	Long Jump	3	24
SJMP	rel	Short Jump (relative addr)	2	24

8051 Instruction Set Summary (Continued)

Mnemonic		Description	Byte	Oscillator Period
PROGRAM BRANCHING (Continued)				
JMP	@A+DPTR	Jump indirect relative to the DPTR	1	24
JZ	rel	Jump if Accumulator is Zero	2	24
JNZ	rel	Jump if Accumulator is Not Zero	2	24
CJNE	A,direct,rel	Compare direct byte to Acc and Jump if Not Equal	3	24
CJNE	A,#data,rel	Compare immediate to Acc and Jump if Not Equal	3	24

Mnemonic		Description	Byte	Oscillator Period
PROGRAM BRANCHING (Continued)				
CJNE	Rn,#data,rel	Compare immediate to register and Jump if Not Equal	3	24
CJNE	@Ri,#data,rel	Compare immediate to indirect and Jump if Not Equal	3	24
DJNZ	Rn,rel	Decrement register and Jump if Not Zero	2	24
DJNZ	direct,rel	Decrement direct byte and Jump if Not Zero	3	24
NOP		No Operation	1	12

Answers to Selected Problems

CHAPTER 1

1. **(a)** $0110_2 = 6_{10}$
(b) $1011_2 = 11_{10}$
(c) $1001_2 = 9_{10}$
(d) $0111_2 = 7_{10}$
(e) $1100_2 = 12_{10}$
(f) $0100\ 1011_2 = 75_{10}$
(g) $0011\ 0111_2 = 55_{10}$
(h) $1011\ 0101_2 = 181_{10}$
(i) $1010\ 0111_2 = 167_{10}$
(j) $0111\ 0110_2 = 118_{10}$

3. **(a)** $1011\ 1001_2 = B9_{16}$
(b) $1101\ 1100_2 = DC_{16}$
(c) $0111\ 0100_2 = 74_{16}$
(d) $1111\ 1011_2 = FB_{16}$
(e) $1100\ 0110_2 = C6_{16}$

5. **(a)** $86_{16} = 134_{10}$
(b) $F4_{16} = 244_{10}$
(c) $92_{16} = 146_{10}$
(d) $AB_{16} = 171_{10}$
(e) $3C5_{16} = 965_{10}$

7. **(a)** $1001\ 1000_{BCD} = 98_{10}$
(b) $0110\ 1001_{BCD} = 69_{10}$
(c) $0111\ 0100_{BCD} = 74_{10}$
(d) $0011\ 0110_{BCD} = 36_{10}$
(e) $1000\ 0001_{BCD} = 81_{10}$

9. **(a)** $t_p = 1/2$ MHz $= 0.5\ \mu$s
(b) $t_p = 1/500$ KHz $\times\ 2\ \mu$s
(c) $t_p = 1/4.27$ MHz $= 0.234\ \mu$s
(d) $t_p = 1/17$ MHZ $= 58.8$ ns

11. An open switch $= \infty\ \Omega$
A closed switch $= 0\ \Omega$

13. $V_{out_1} \approx 0$ V, $V_{out_2} \approx 5$ V

15. Pin 14 (V_{cc}) and Pin 7 (Gnd)

CHAPTER 2

1.

A	B	C	X
0	0	0	0
0	0	1	0
0	1	0	0
0	1	1	0
1	0	0	0
1	0	1	0
1	1	0	0
1	1	1	1

3. $2^8 = 256$
5. The output is HIGH whenever any input is HIGH; otherwise the output is LOW.
10.

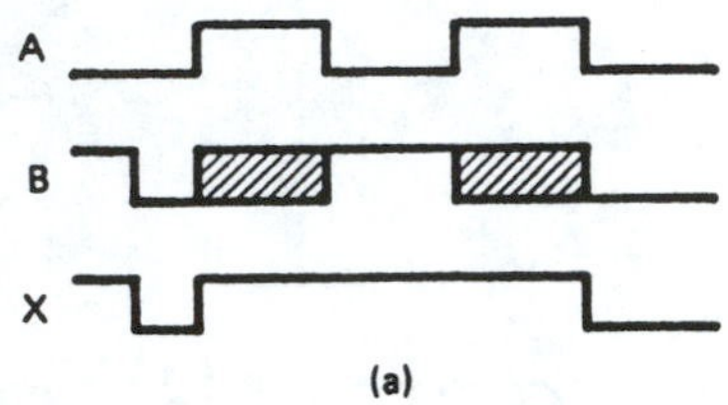

(a)

(cont.)
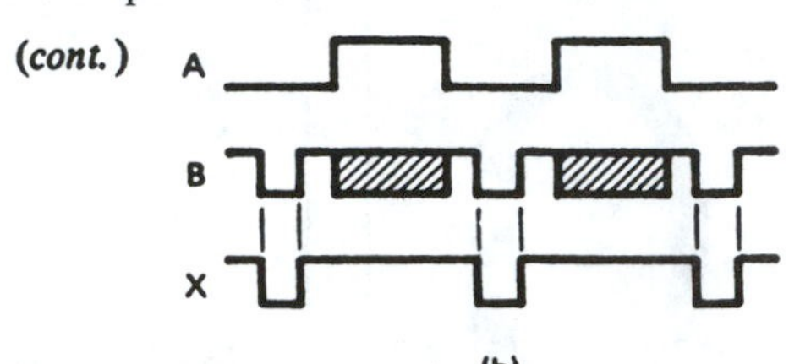

(b)

11. Four
13. Two
15. $X = \overline{A}, Z = A, X = 1, Z = 0$
17. $X = \overline{AB}, Y = \overline{CD}$
21. It disables the other two inputs when it is DOWN for the NAND and UP for the NOR.

23.

A	B	C	X
0	0	0	1
0	0	1	0
0	1	0	0
0	1	1	0
1	0	0	0
1	0	1	0
1	1	0	0
1	1	1	0

25. **(a)** Exclusive-OR produces a HIGH output for one or the other input HIGH, but not both.
(b) Exclusive-NOR produces a HIGH output for both inputs HIGH or both inputs LOW.
31. $U = C_pAB \quad W = BC \quad V = \overline{C}\,\overline{D} \quad X = C_pCD$
35. To provide pulses to a digital circuit for troubleshooting purposes.
37. HIGH, to enable the output to change with pulser (if gate is good).
39. Pin 2 should be ON; the Enable switch is bad, or bad Enable connection.
41. LOW; to see inverted output pulses. (Otherwise output would always be LOW.)
43. Pins 4 and 10 should be HIGH. The inverters connected to those pins are bad.
45. The inverter is not working.
47. Pins 8 and 12 should be LOW. The NORs connected to those pins are bad.
49. Sink current is higher.
51. **(a)** $t_p = 100\ \mu s + 120\ \mu s = 220\ \mu s$ **(b)** $t_r = 2\ \mu s \quad t_f = 3\ \mu s$
$f = 1/t_p = 4.55$ KHz
(c) $t_{PLH} = 8\ \mu s \quad t_{PHL} = 6\ \mu s$
55. **(a)** 7400 HIGH-state (min levels) = 0.4 V
LOW-state (max levels) = 0.4 V
74LS00 HIGH-state (min levels) = 0.7 V
LOW-state (max levels) = 0.3 V
(b) The 74LS00 has a wider margin for the HIGH-state. The 7400 has a wider margin for the LOW-state.

57. The open-collector FLOAT level is made a HIGH-level by using a pull-up resistor.

59. **(a)** TTL uses bipolar transistors. **(b)** CMOS uses Field-Effect transistors.

61. **(b)** Noise margins:

74HCMOS	74ALSTTL
HIGH = 1.4 V	HIGH = 0.7 V
LOW = 0.9 V	LOW = 0.4 V

(c) 74HCMOS-to-74ALSTTL will work, but 74ALSTTL-to-74HCMOS will not work because ALS may output 2.7 V when HC expects at least 3.5 V (HIGH).

63. No trouble with current ratings.

CHAPTER 3

1. **(a)** W = (A + B)(C + D)
X = AB + BC
Y = (AB + B)C
Z = [AB + B + (B + C)]D

3.

A	B	C	D	M	N	Q	R	S
0	0	0	0	0	0	0	0	0
0	0	0	1	1	0	0	1	1
0	0	1	0	1	0	0	0	0
0	0	1	1	1	1	0	1	1
0	1	0	0	0	0	0	0	0
0	1	0	1	1	1	0	1	1
0	1	1	0	1	0	0	1	1
0	1	1	1	1	1	1	1	1
1	0	0	0	0	0	0	0	0
1	0	0	1	1	1	0	1	1
1	0	1	0	1	0	0	0	1
1	0	1	1	1	1	0	1	1
1	1	0	0	1	0	0	0	1
1	1	0	1	1	1	0	1	1
1	1	1	0	1	0	0	1	1
1	1	1	1	1	1	1	1	1

A	B	C	P
0	0	0	0
0	0	1	0
0	1	0	0
0	1	1	1
1	0	0	0
1	0	1	1
1	1	0	0
1	1	1	1

5.

M = 0	S = 0
N = 1	T = A
P = AB	U = 1
Q = C + D	V = A
R = A	W = A

7.

A	C	D	V
0	0	0	0
0	0	1	0
0	1	0	0
0	1	1	1
1	0	0	0
1	0	1	0
1	1	0	1
1	1	1	1

A	B	C	Y
0	0	0	0
0	0	1	0
0	1	0	0
0	1	1	1
1	0	0	0
1	0	1	0
1	1	0	1
1	1	1	1

A	B	C	D	X	Z
0	0	0	0	0	0
0	0	0	1	0	0
0	0	1	0	0	0
0	0	1	1	1	1
0	1	0	0	0	0
0	1	0	1	0	0
0	1	1	0	1	0
0	1	1	1	1	1
1	0	0	0	0	0
1	0	0	1	1	0
1	0	1	0	0	0
1	0	1	1	1	1
1	1	0	0	1	0
1	1	0	1	1	0
1	1	1	0	1	1
1	1	1	1	1	1

C	D	W
0	0	0
0	1	0
1	0	0
1	1	1

11. **(a)** NAND **(b)** NOR

13. $W = \overline{A} + \overline{B}$
$X = \overline{B} + \overline{C}$
$Y = \overline{C}$
$Z = A\overline{B}\,\overline{C}$

19.

A	B	C	W	X
0	0	0	0	1
0	0	1	1	1
0	1	0	1	1
0	1	1	1	0
1	0	0	1	0
1	0	1	1	1
1	1	0	0	1
1	1	1	0	0

A	B	C	D	Y	Z
0	0	0	0	1	1
0	0	0	1	1	1
0	0	1	0	1	1
0	0	1	1	0	1
0	1	0	0	0	0
0	1	0	1	1	0
0	1	1	0	1	1
0	1	1	1	1	1
1	0	0	0	0	1
1	0	0	1	1	1
1	0	1	0	0	1
1	0	1	1	0	0
1	1	0	0	0	0
1	1	0	1	1	0
1	1	1	0	0	1
1	1	1	1	1	0

25. $W = \overline{B}\,\overline{C} + \overline{B}\,\overline{D} + \overline{A}\,\overline{B}$
$X = \overline{C}\,\overline{D} + \overline{B}\,\overline{D} + ABCD$
$Y = A\overline{B} + A\overline{D} + \overline{B}C\overline{D}$
$Z = \overline{C} + B\overline{D} + \overline{A}\,\overline{D}$

27. $X = \overline{A}\,\overline{B} + \overline{A}CD + \overline{B}\,\overline{C} + \overline{A}\,\overline{C}\,\overline{D}$ where
$A = MSB$

CHAPTER 4

5.

2^3	2^2	2^1	2^0	$\overline{0}$	$\overline{1}$	$\overline{2}$	$\overline{3}$	$\overline{4}$	$\overline{5}$	$\overline{6}$	$\overline{7}$	$\overline{8}$	$\overline{9}$
0	0	0	0	0	1	1	1	1	1	1	1	1	1
0	0	0	1	1	0	1	1	1	1	1	1	1	1
0	0	1	0	1	1	0	1	1	1	1	1	1	1
0	0	1	1	1	1	1	0	1	1	1	1	1	1
0	1	0	0	1	1	1	1	0	1	1	1	1	1
0	1	0	1	1	1	1	1	1	0	1	1	1	1
0	1	1	0	1	1	1	1	1	1	0	1	1	1
0	1	1	1	1	1	1	1	1	1	1	0	1	1
1	0	0	0	1	1	1	1	1	1	1	1	0	1
1	0	0	1	1	1	1	1	1	1	1	1	1	0

7. That input is a "Don't Care" and will have no effect on the output for that particular table entry.

9.

Time interval	Low output pulse at:
$t_0 - t_1$	None (E_3 disabled)
$t_1 - t_2$	None (E_3 disabled)
$t_2 - t_3$	$\overline{5}$
$t_3 - t_4$	$\overline{4}$
$t_4 - t_5$	$\overline{3}$

$t_5 - t_6$	$\overline{2}$
$t_6 - t_7$	$\overline{1}$
$t_7 - t_8$	$\overline{0}$
$t_8 - t_9$	$\overline{7}$
$t_9 - t_{10}$	$\overline{6}$
$t_{10} - t_{11}$	$\overline{5}$
$t_{11} - t_{12}$	None (E_3 disabled)
$t_{12} - t_{13}$	None (E_3 disabled)

11. All HIGH

13. The higher number

15. $Y = 1, \overline{Y} = 0$

18.

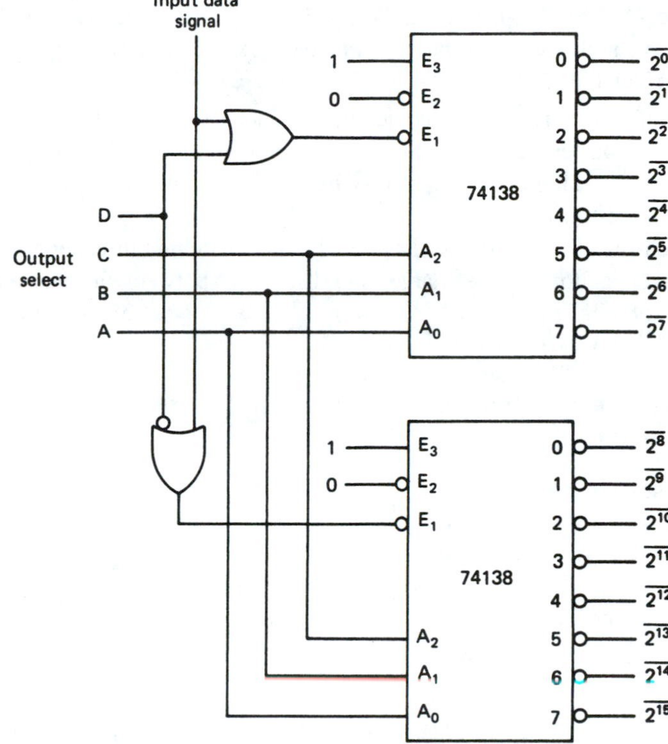

19. **(a)** The 74138 is not working. The outputs should be disabled by the HIGH on pin 5, but the $\overline{3}$ output is LOW.

(b) The 74151 is OK. The data select is set for input I_0, which is 1. Y should equal 1 and $\overline{Y} = 0$, which they do.

(c) The 74139 has two bad decoders. Decoder A is enabled and should output 1011, but does not. Decoder B is disabled and should output 1111, but does not.

(d) The 74154 is OK. The chip is disabled so all outputs should be HIGH, which they are.

CHAPTER 5

5. It is called "transparent" because the *Q*-output follows the level of the *D* input as long as *E* is HIGH. When *E* goes LOW, *Q* "latches," or holds onto the level of *D* before the HIGH-to-LOW edge of *E*.

7. **(a)** $\overline{S_D}, \overline{R_D}$ **(b)** CP, D

8.

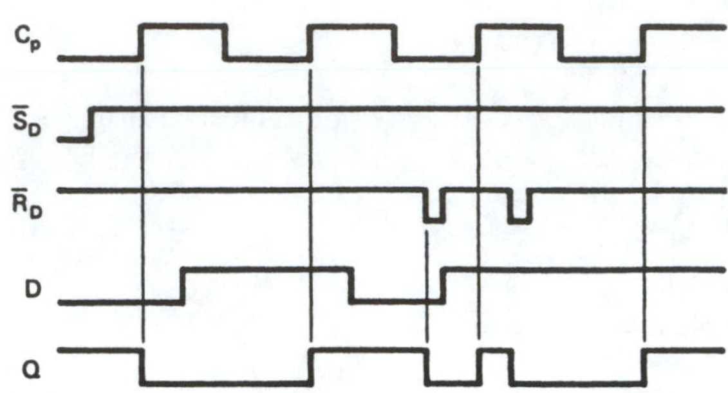

9. The 7474 is edge-triggered, the 7475 is pulse-triggered. The 7474 has asynchronous inputs at $\overline{S_D}$ and $\overline{R_D}$.

11. The triangle indicates that it is an edge-triggered device as opposed to being pulse-triggered.

13. The TOGGLE mode.

23. A buffer is a transparent device that connects two digital circuits; a latch is a storage device that can hold data. A buffer only allows data to flow in one direction; a transceiver is bidirectional.

25. Because a single closure of a switch will look like multiple closures to the digital circuit, causing erroneous results.

27. Because a Schmitt device has two distinct switching thresholds: V_T+ and V_T-, a regular inverter does not. The capacitor voltage charges and discharges between those two levels.

29. To solve for t_{HI}:
$\Delta V = 3.3\text{ V} - 2.0\text{ V} = 1.3\text{ V}$
$E = 6.0 - 2.0 = 4.0\text{ V}$
$t_{HI} = 68\text{K} \cdot 0.0047\mu \cdot \ln[1/(1 - (1.3/4.0))]$
$t_{HI} = 126\ \mu s$

To solve for t_{LO}:
$\Delta V = 3.3\text{ V} - 2.0\text{ V} = 1.3\text{ V}$
$E = 3.3\text{ V} - 0\text{ V} = 3.3\text{ V}$
$t_{LO} = 68\text{K} \cdot .0047\ \mu \cdot \ln[1/(1 - (1.3/3.3))]$
$t_{LO} = 160\ \mu s$

$DC = 126\ \mu s/(126\ \mu s + 160\ \mu s) = 44.1\%$
$f = 1/(126\ \mu s + 160\ \mu s) = 3.5\text{ KHz}$

33. $@0_\Omega\ t_{LO} = 3.26\ \mu s$, $t_{HI} = 7.97\ \mu s$ } from Problem 32
$f = 1/(t_{LO} + t_{HI}) = 89.0\text{ KHz}$
$DC = t_{HI}/(t_{LO} + t_{HI}) = 71.0\%$

$@10\text{K}_\Omega\ t_{LO} = .693 \times 14.7\text{ K} \times 1000\text{ p} = 10.2\ \mu s$
$t = .693 \times (6.8\text{ K} + 14.7\text{ K}) \times 1000\text{ p} = 14.9\ \mu s$
$f = 1/(t_{LO} + t_{HI}) = 39.8\text{ KHz}$
$DC = t_{HI}/(t_{LO} + t_{HI}) = 59.4\%$

35. Using the formula derived in Problem 34:
$0.6 = (R_A + R_B)/(R_A + 2R_B)$
$0.6\,R_A + 1.2R_B = R_A + R_B$
$R_B = 2R_A$
also:
$50\text{ KHz} = 1.44/[(R_A + 2R_B).0022\ \mu F]$
$R_A + 2R_B = 13091$
$R_A + 2(2R_A) = 13091$
$R_A = 2618\ \Omega,\ R_B = 5236\ \Omega$

36. 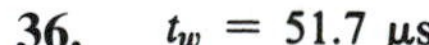
$t_w = 51.7\ \mu s$

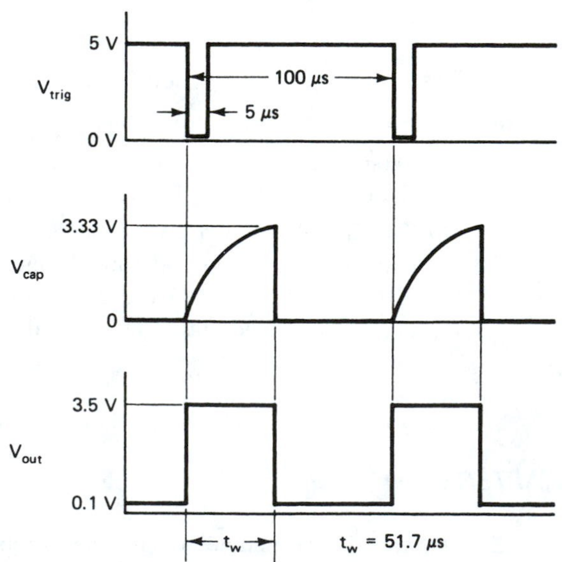

CHAPTER 6

1. Sequential circuits follow a predetermined sequence of digital states triggered by a timing pulse or clock. Combination logic circuits operate almost instantaneously based on the levels placed at its inputs.
3. **(a)** 3 **(b)** 3 **(c)** 1 **(d)** 5 **(e)** 6 **(f)** 4
7. **(a)** 3 **(b)** 15 **(c)** 127 **(d)** 1
9. **(a)** 2 **(b)** 4 **(c)** 4 **(d)** 5
23. Connect a RESET pushbutton across the 0.001 μf capacitor. When momentarily pressed, it will RESET the circuit to its initial condition.
25. With all of the current going through the same resistor, as more segments are turned ON, the voltage that reaches the segments is reduced, making them dimmer. The displayed #8 is much dimmer than the #1.
29.

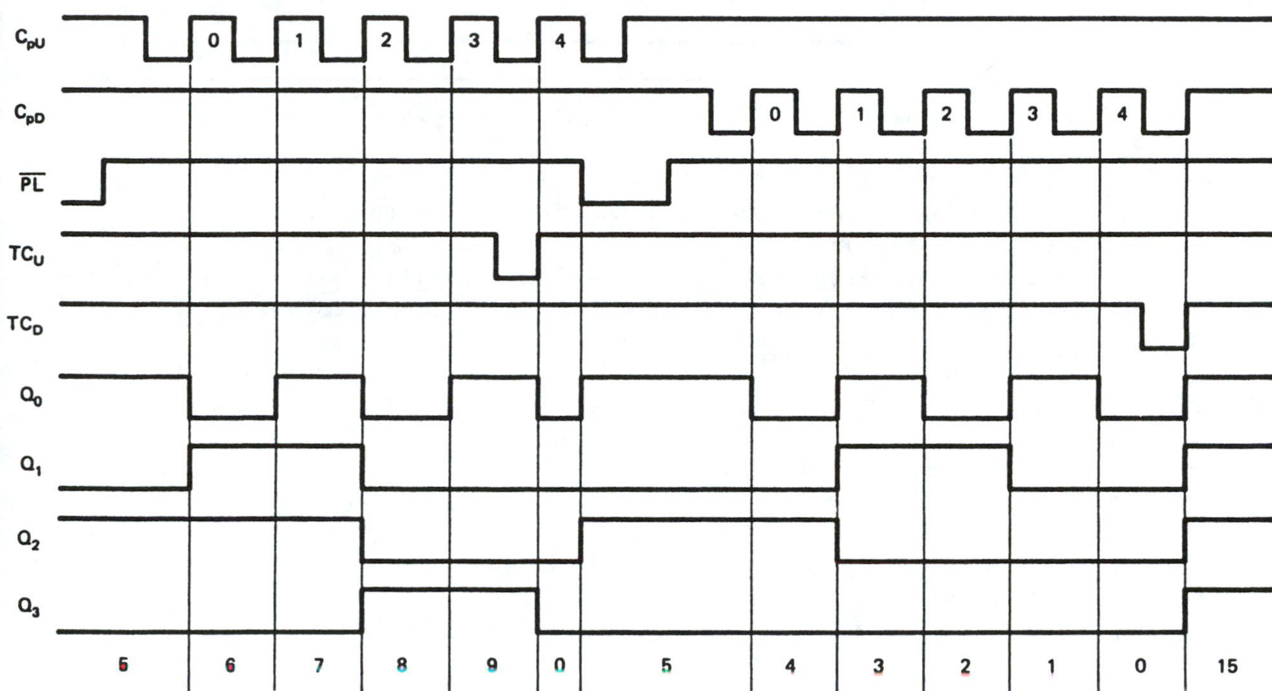

31. Right, Negative
33. 1110, 1111
35. Apply a LOW pulse to $\overline{R_D}$ to RESET all Q-outputs to 0. Next, apply a LOW pulse to the active-LOW D_3, D_1, D_0 inputs.
37. J_3, K_3 are the data input lines. Q_3, Q_2, Q_1, Q_0 are the data output lines. (D_3, D_2, D_1, D_0 are held HIGH.)
39. Five

43.

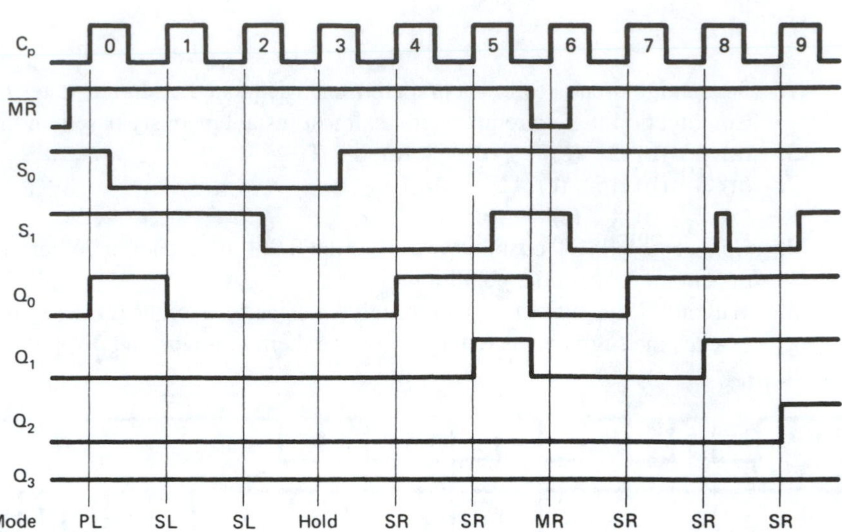

49. The parallel-load and clock inputs are $\overline{PE}$, CP on the 74195 and PE, $\overline{CP}$ on the 74395A. The 74195 provides *J-K* type functions on its first flip-flop via the *J-K* inputs. The outputs of the 74395A are 3-stated, made active by a LOW on $\overline{OE}$.

51. Data is parallel-loaded *asynchronously* with the 74165 by applying a LOW to $\overline{PL}$. Data is parallel-loaded *synchronously* with the 74166 at the positive edge of CP while $\overline{PE}$ is held LOW.

CHAPTER 7

1. Converts physical quantities into electrical quantities.

3. Very high input impedance, very high voltage gain, and very low output impedance.

5. The (−) input is at 0 volts potential.

7. The output voltages (V_{out}) would all double. (V_{out} would be limited, however, by the size of the supply powering the op-amp.)

9. $V_{out} = -2.0$ V

11. *R*/2*R* method

13. 8-bit resolution

15. By making $V_{REF} = 7.5$ V. The range of I_{out} would then be 0 mA to 1.5 mA. The range of V_{out} would be 0 V to 7.47 V.

19. **(a)** 0 volts **(b)** They are equal.

21. $t_{TOT} = 8 \times 1/50$ KHz $= 0.16$ ms

23.

$$\left.\begin{array}{ll} 7.28 & \\ \underline{-5.00} & D_7 \\ 2.28 & \\ \underline{-1.25} & D_5 \\ 1.030 & \\ \underline{-0.625} & D_4 \\ .4050 & \\ \underline{-\ .3125} & D_3 \\ .092500 & \\ \underline{-\ .078125} & D_1 \\ .014375 & \end{array}\right\} \ 1011 \quad 1010$$

$$1011 \quad 1010 = 5 + 1.25 + .625 + .3125 + .078125 = 7.265625 \text{ V}$$

$$\% \text{ ERROR} = \frac{\text{DAC OUT} - \text{ACTUAL } V_{IN}}{\text{ACTUAL } V_{IN}}$$

$$\% \text{ ERROR} = \frac{7.265625 - 7.28}{7.28}$$
$$= -0.197\%$$

25. $\overline{CS}$ (Chip Select) and
$\overline{RD}$ (READ). Active-LOW.

27. **(a)** The 3-state output latches (D_0–D_7) would be in the float condition.
(b) The outputs would float and $\overline{WR}$ (Start Conversion) would be disabled.
(c) It issues a LOW at power-up to start the first conversion.
(d) 1.0 volt

29. @ 20 degrees C, V_{in} = 200 mV, D_7–D_0 = 0001 0100
@ 80 degrees C, V_{in} = 800 mV, D_7–D_0 = 0101 0000

31. The temperature transducer is selected by setting up the appropriate code on the ABC multiplexer select inputs. The voltage level passes to the LF198, which takes a sample at some precise time and holds the level on the hold capacitor. The LH0084 adjusts the voltage to an appropriate level to pass into the ADC. The microprocessor issues $\overline{CS_1}$, $\overline{WR_1}$ then waits for $\overline{INTR}$ to go LOW. It then issues $\overline{CS_1}$, $\overline{RD_1}$ to transfer the converted data to the data bus, then $\overline{CS_2}$, $\overline{WR_2}$ to transfer the data to RAM.

CHAPTER 8

1. Bipolar—faster, MOS—more dense.

3. Static RAMs use flip-flops as basic storage elements, and dynamic RAMs store a charge on an internal capacitor. Dynamic RAMs require refresh circuitry but are more dense and less expensive per bit.

5. **(a)** 2048 **(b)** 2048 **(c)** 8192 **(d)** 1024 **(e)** 4096 **(f)** 16384

7.

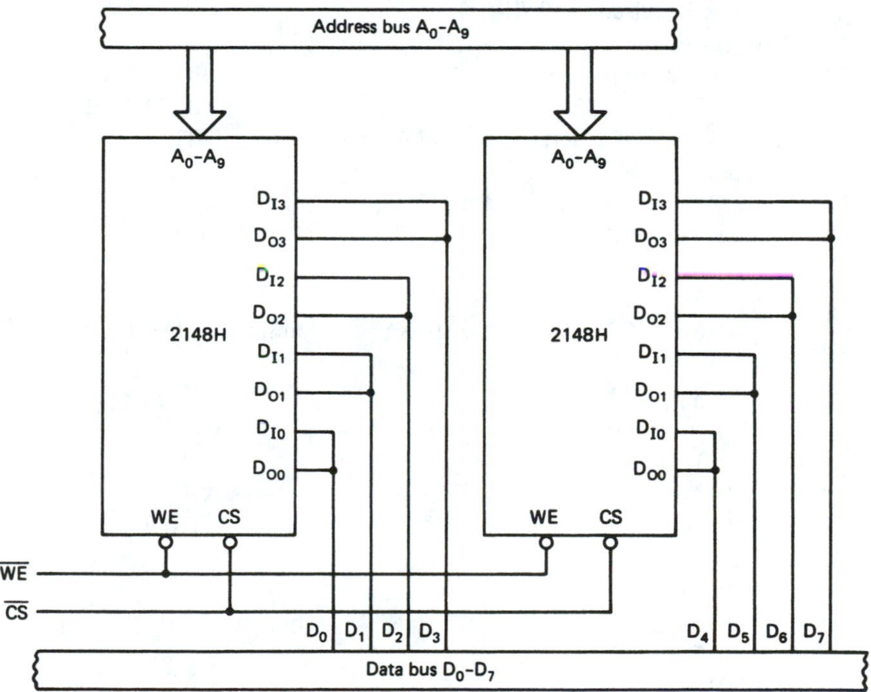

9. The Row Address Strobe ($\overline{RAS}$) signifies that a row address is present on the address lines, then $\overline{CAS}$ is used to signify that the column address is present.

11. The charge on the internal capacitors must be refreshed every 2 ms or sooner to prevent loss of data.

13. **(a)** nonvolatile **(b)** volative **(c)** volatile **(d)** nonvolatile

15. The 74154 1-of-16 decoder.

CHAPTER 9

1. A microprocessor-based system would be used whenever: calculations are to be made; decisions based on inputs are to be made; a memory of events is needed; a modifiable system is needed.

3. The address bus is used to select a particular location or device within the system.

7. The input port has 3-stated outputs so that it can be disabled when it is not being read.

9. 2^{16} (65,536)

11. It stores the contents of the accumulator out to address 6000H.

13. Instruction decoder and register: Register and circuitry inside the μP, which receives the machine language code and produces the internal control signals required to execute the instruction.

15. 1st pulse: Read memory location 2000 (LDA)
2nd & 3rd pulses: Read address bytes @ 2001, 2002 (4000H)
4th pulse: Read data @ address 4000H

CHAPTER 10

1. A high level language (e.g., FORTRAN or BASIC) has the advantage of being easier to write and understand. Its disadvantage is that the programs are not memory efficient.

3. **(a)** LDA *Addr* $IO/\overline{M}$ is LOW **(b)** STA *Addr* $IO/\overline{M}$ is LOW
(c) IN *Port* $IO/\overline{M}$ is HIGH **(d)** OUT *Port* $IO/\overline{M}$ is HIGH

5. U6a and U6b are drawn as inverted-input NAND gates to make the logical flow of the schematic easier to understand.

7. **(a)** IN instruction, $\overline{RD}$ is pulsed LOW. **(b)** OUT instruction, $\overline{WR}$ is pulsed LOW.

9. **(1)** A8-A15 = FEH
(2) $IO/\overline{M}$ = HIGH
(3) $\overline{WR}$ is pulsed LOW/HIGH

11. MVI A,4FH

13. **(a)** MVI D,*data* = 16 **(b)** INR C = 0C
(c) JNZ *Addr* = C2 **(d)** DCR B = 05

15. **(a)**

```
2010   3E   INIT:   MVI A,04H   ;A ⟵ 04H
2011   04
2012   3D   X1:     DCR A       ;A ⟵ A-1
2013   CA           JZ INIT     ;Jump on zero ⟶ 2010
2014   10
2015   20
2016   C3           JMP X1      ;Jump ⟶ 2012
2017   12
2018   20
```

(b)

```
2010   3E   INIT:   MVI A,04H   ;A ⟵ 04
2011   04
2012   3D   X1:     DCR A       ;A ⟵ A-1
2013   C2           JNZ X1      ;Jump on zero (overlined) to 2012
2014   12
2015   20
2016   C3           JMP INIT    ;Jump to 2010
2017   10
2018   20
```

17. CPI 0DH compares the number in the accumulator with 0DH. If A = 0D, the zero flag is set. If A < 0D, the carry flag is set.

19. The first CMA is used because the output LEDs are active LOW. The second CMA is used to restore the accumulator for the count.

21. **(a)** B, C, D, E, H, L **(b)** BC, DE, HL

23. B; destination D; source

25. **(a)** D = 00 E = FF **(b)** D = 40E = 51
(c) D = 41 E = 00 **(d)** D = 40E = 00
(e) D = 9F E = FF **(f)** D = DFE = 00

27. MVI D,38H (56_{10})

29.
```
204E   26  DELAY:   MVI H,04H   ;H <-- 04H
204F   04
2050   16  LOOP3:   MVI D,8CH   ;D <-- 8CH
  :
205C   25           DCR H       ;H <-- H-1
205D   C2           JNZ LOOP3   ;Jump to LOOP3
205E   50                       ;    4 times
205F   20
2060   C9           RET
```

31.
```
FLASH:  MOV B,A
LOOP:   CMA
        OUT FE
        CALL DELAY
        MVI A,FF
        OUT FE
        CALL DELAY
        MOV A,B
        JUMP LOOP
```

33.
```
2000   DB  READCD:  IN FFH      ;Read code
2001   FF
2002   FE           CPI 00H     ;No trouble,
2003   00                       ;  keep reading
2004   CA           JZ READCD   ;  code
2005   00                       ;on zero go to 2000
2006   20
2007   47           MOV B,A     ;etc.
2008
```

CHAPTER 11

1. Serial data input and output
3. **(a)** LOW **(b)** LOW **(c)** HIGH
5. Multiplexed bus: Advantage, reduced pin count. Disadvantage, external demultiplexing is needed.
7. Address
9. Positive edge
13. 74LS138 is enabled:
(1) $\overline{RD}$ or $\overline{WR}$ is LOW
(2) IO/$\overline{M}$ is LOW
(3) A14 + A15 are both LOW
15. 1024; 4 bits
17. 2400-27FF; it exists because the RAMs are still enabled by the address decoder during these addresses.
19. **(a)** 10 T-states $\cdot$ 0.5 μs $=$ 5 μs **(b)** 10 T-states $\cdot$ 0.5 μs $=$ 5 μs
(c) 7 T-states $\cdot$ 0.5 μ $=$ 3.5 μs **(d)** 10 T-states $\cdot$ 0.5 μs $=$ 5 μs
21. F ⟶ Opcode Fetch decoded as LDA
R ⟶ Memory Read, Low order 8 bits of address
R ⟶ Memory Read, High order 8 bits of address
R ⟶ Memory Read, Data at address (byte2 + byte3) is loaded into accumulator.
25. The 8755A should be used during the early stages of system development because it has 2K of EPROM.
27. 0800H–08FFH
29. 20H
31.
```
MVI A,03H                   7 6 5 4 3 2 1 0
OUT 20H      Command Reg    0 0 0 0 0 0 1 1
```

33. 1/10 KHz = 100 μs

100 μs = .5 μs · count length
count length = 200 = C8H

```
MVI A,40H ;(MSB,TM)
OUT 25H
MVI A,C8H ;(LSB)
OUT 24H
MVI A,C0H ;Command Reg
OUT 20H
HLT
```

CHAPTER 12

1.
```
LXI H,20B0H
MOV A,M
INX H
MOV M,A
```

3.
```
LXI H,20C5H
MOV A,M
INX H
MOV M,A
```

5.
```
2000  3E   START:   MVI A,00H   ;Port A ⟶ Inputs
2001  00
2002  D3            OUT 02H     ;Program DDR A
2003  02
2004  3E            MVI A,FFH   ;Port B ⟶ Outputs
2005  FF
2006  DE            OUT 03H     ;Program DDR B
2007  03
2008  26            MVI H,20H   ;H ⟵ 20 (HL=20XX)
2009  20
200A  DB   READEM:  IN 00H      ;Read input switches
200B  00
200C  6F            MOV L,A     ;A(switch settings) ⟶ L
                                  (HL now has 20XX)
200D  7E            MOV A,M     ;M(HL) ⟶ A
200E  D3            OUT 01H     ;M(20XX) ⟶ LEDs(Port B)
200F  01
2010  CD            CALL DELAY;1/4 sec delay
2011  50
2012  20
2013  C3            JMP READEM
2014  0A
2015  20
 .
 .
 .
2050       DELAY                Delay subroutine
```

7. 200E MVI A,E0H

9.
```
START:  LXI B,20A0H   ;BC ⟵ 20A0 (source)
        LXI D,20B0H   ;DE ⟵ 20B0 (destination)
LOOP:   LDAX B        ;A ⟵ M(BC)
        STAX D        ;M(DE) ⟵ A
        INX B         ;BC ⟵ BC+1
        INX D         ;DE ⟵ DE+1
        MOV A,E       ;A ⟵ E
        CPI C0H       ;Does A = C0?
        JNZ LOOP:     ;Jump not zero to loop
        HLT
```

11. **(a)** S, Z, AC, P **(b)** S, Z, AC, P, C **(c)** None **(d)** None **(e)** C **(f)** C **(g)** None **(h)** S, Z, AC, P, C

13. **(a)** A = 7FH
B = 77H
F = 00H
(b) A = 69H
F = 14H
(c) A = 8DH
F = 84H
(d) A = 67H
F = 00H
(e) A = 39H
F = 05H (P,C)

15. ANI F0H after reading the input port.

17. **(a)** CY = 1 A = 83H **(b)** CY = 0 A = 00H **(c)** CY = 1 A = EBH

19.

```
2000 3E  START:   MVI A,FFH     ;DDR B ⟶ outputs
2001 FF
2002 D3           OUT 03H
2003 03
2004 3E           MVI A,80H     ;A ⟵ 1000 0000
2005 80
2006 D3           OUT 01H       ;Turn on left-most LED
2007 01
2008 CD           CALL DELAY    ;Call 1/4 sec delay
2009 50
200A 20
200B 3E  RIGHT:   MVI A,40H     ;A ⟵ 0100 0000
200C 40
200D D3  LOOP1:   OUT 01H       ;Turn LED on
200E 01
200F CD           CALL DELAY    ;Call 1/4 sec delay
2010 50
2011 20
2012 0F           RRC           ;Rotate right
2013 DA           JC LEFT       ;Jump on carry to LEFT
2014 19
2015 20
2016 C3           JMP LOOP1     ;Jump to LOOP1
2017 0D
2018 20
2019 3E  LEFT:    MVI A,02H     ;A ⟵ 0000 0010
201A 02
201B D3  LOOP2:   OUT 01H       ;Turn LED on
201C 01
201D CD           CALL DELAY    ;Call 1/4 sec delay
201E 50
201F 20
2020 07           RLC           ;Rotate left
2021 DA           JC RIGHT      ;Jump on carry to RIGHT
2022 0B
2023 20
2024 C3           JMP LOOP2     ;Jump to LOOP2
2025 1B
2026 20
  .
  .
  .
2050 DELAY:                     ;1/4 sec delay
```

21. end

23. False; Last on, first off

25.
```
DELAY:  PUSH D        ; Save DE rp
        MVI D,8CH
LOOP2:  MVI E,FFH
LOOP1:  DCR E
        JNZ LOOP1
        DCR D
        JNZ LOOP2
        POP D         ; Restore DE
        RET
```

27. It is like a subroutine in that it branches to a specific address, but unlike a subroutine, it can occur at any time during program execution.

29.
```
MVI A,0CH ; A ⟵ 0000 1100
SIM       ; Set interrupt mask
EI        ; Enable interrupts
```

31.
```
003C C3   RST 7.5:  JMP 20CEH      ;Jump to ISR
003D CE
003E 20
    .
    .
    .
2000 31   MAIN:     LXI SP,20A0H   ;Initial SP
2001 A0
2002 20
2003 3E             MVI A,00H      ;DDR A = inputs
2004 00
2005 D3             OUT 02H
2006 02
2007 3E             MVI A,FFH      ;DDR B = outputs
2008 FF
2009 D3             OUT 03H
200A 03
200B 3E             MVI A,0BH      ;RST 7.5 to be enabled
200C 0B
200D 30             SIM            ;Set interrupt mask
200E FB             EI             ;Enable interrupt
200F 00   LOOP:     NOP            ;Do nothing loop
2010 C3             JMP LOOP
2011 0F
2012 20

2050      DELAY:                   ;1/4 sec delay

2080 DB   ISR:      IN 00H         ;Read switches
2081 00
2082 D3             OUT 01H        ;Turn on LEDs
2083 01
2084 CD             CALL DELAY     ;1/4 sec delay
2085 50
2086 20
2087 FB             EI             ;Re-enable interrupts
2088 C9             RET            ;Return
20CE C3             JMP ISR        ;Jump to different RAM area
20CF 80
20D0 20
```

CHAPTER 13

1. **(a)** $I_{out} = 2$ mA (1/256) = 7.81 μA $V_{out} = 7.81\ \mu A \cdot 5\ k\Omega = 39.1$ mV
(b) $I_{out} = 2$ mA (1/2) = 1 mA $V_{out} = 1\ mA \cdot 5\ k\Omega = 5$ V
(c) $I_{out} = 2$ mA (1/2 + 1/4 + etc) = 1.99 mA $V_{out} = 1.99\ mA \cdot 5\ k\Omega = 9.96$ V

3. V_{out}(40H) = (2 mA[1/4] · 5 kΩ) = 2.50 V *T*-states = 7 + 10 = 17
V_{out}(04H) = (2 mA[1/64] · 5 kΩ) = 0.156 V *T*-states = 10 + 7 + 10 = 27
Time (40H) = 17 · .326 μs = 5.53 μs
Time (04H) = 27 · .326 μs = 8.79 μs

5.
```
START:  MVI A,FFH   ; DDR B = output
        OUT 03H     ;
        CMA         ; A <— 00
UP:     OUT 01H     ; Output to DAC
        INR A       ; A <— A + 1
        JNZ UP      ; Count up until zero
        CMA         ; A <— FF
DOWN:   OUT 01H     ; Output to DAC
        DCR A       ; A <— A-1
        JNZ DOWN    ; Count down until zero
        JMP UP      ; Up again
```

7.
```
SC: MVI A,00H
    OUT 01H
    MVI A,01H
    OUT 01H
```

9. Binary-to-BCD

11. Port A is used to drive the segments via transistor current buffers. Port B is used to drive one digit at a time. Port A outputs the segments to be turned on for a digit while it is enabled by Port B. This is done for each of the six digits (one at a time). The μP cycles through each digit fast enough so that they appear to be all on at the same time.

13. Because the segment bus is driven next, and the next digit to be turned on is not known.

15.
```
MAIN:  LXI SP,20C0H    ; Initialize stack pointer
       MVI A,FFH       ; Program Ports A + B
       OUT 02H         ; DDR A = output
       OUT 03H         ; DDR B = output
LOOP:  MVI B,89H       : B <— H segments
       MVI C,F7H       ; C <— 4th digit
       CALL DISP       ; Display H
       MVI B,86H       ; B <— E segments
       MVI C,FBH       ; C <— 3rd digit
       CALL DISP       ; Display E
       MVI B,C7H       ; B <— L segments
       MVI C,FDH       ; C <— 2nd digit
       CALL DISPLAY    ; Display L
       MVI B,8CH       ; B <— P segments
       MVI C,FEH       ; C <— 1st digit
       CALL DISP       ; Display P
       JMP LOOP        ; Repeat
DISP:  Same as Ex: 13–4
```

17. A 0 is output on one row (only). The μP then reads the columns. If a key in the active LOW row is depressed, then the data read in will have a 0 in bit 0, 1, 2, or 3. The program will pinpoint which key. If none of the columns' bits are LOW, then the next row is made LOW.

19.
```
INIT:  MVI A,FFH
       OUT 02H
       CMA
       OUT 03H
```

```
ROW:    MVI A,FDH   ; Row 1 active
        OUT 00H
COL:    IN 01H
        CMP FBH     ; Check for Col 2 Active
        JNZ COL
LED:    MVI A,7FH
        OUT 00H
        JMP LED
```

21. A stepper motor rotates in steps instead of the continuous motion of conventional motors.

23. HIGH

25. **(a)** Because the A register is used in the delay subroutine.
(b) Yes;

```
DELAY: PUSH PSW
       IN 01H
       etc.
LOOP:

       JNZ LOOP
       POP PSW

       RET
```

CHAPTER 14

1. Because it has RAM, ROM, I/O Ports, and a Timer/Counter on it.

3. Extra 4 K ROM, 128 bytes RAM, 16 bit Timer/Counter, 1 interrupt.

5. High

7. 80H–FFH

9. By reading the two bank-select bits (D0H.3, D0H.4) in the PSW.

11. **(a)** A = 40H **(b)** A = 88H **(c)** A = 88H **(d)** A = 40H

13. **(a)** MOV P3,#0C7H **(b)** MOV R7,P1 **(c)** MOV A,#55H **(d)** MOV A,@R0
(e) MOV P0,R1

15.

```
START: MOV A,#20H
LOOP:  MOV P1,A
       INC A
       CJNE A,#91H,LOOP
       JMP START
```

17. **(a)** A = 75H **(b)** A = F5H **(c)** A = EBH

19.

```
MOV A,P0
ADD A,#05H
MOV P1,A
```

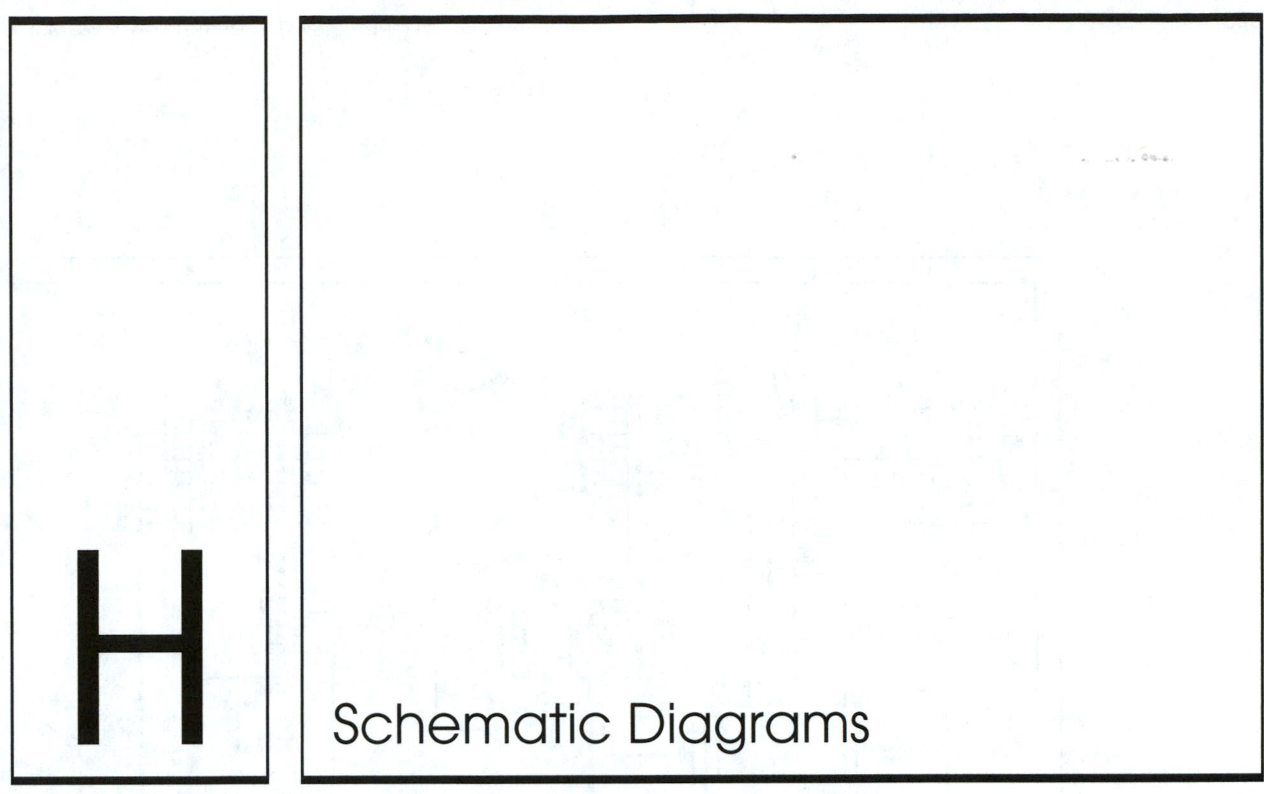

H Schematic Diagrams

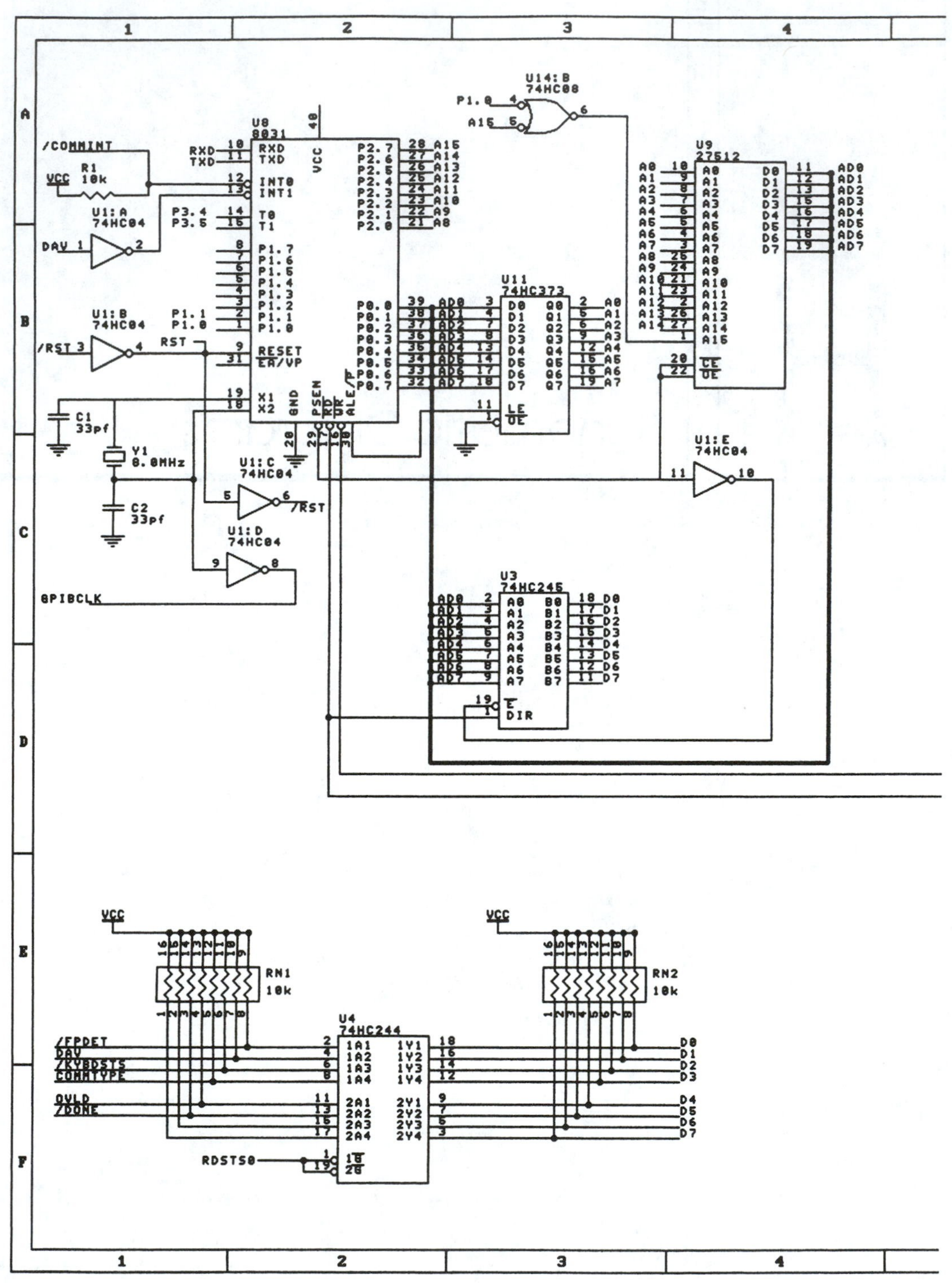
U14:B
74HC08
U8
8031
U9
27512
U11
74HC373
U3
74HC245
U4
74HC244
U1:A
74HC04
U1:B
74HC04
U1:C
74HC04
U1:D
74HC04
U1:E
74HC04
R1
10k
C1
33pf
C2
33pf
Y1
8.0MHz
RN1
10k
RN2
10k
/COMMINT
DAV
/RST
GPIBCLK
/FPDET
/KYBDSTS
COMMTYPE
OVLD
/DONE
RDSTS0
VCC

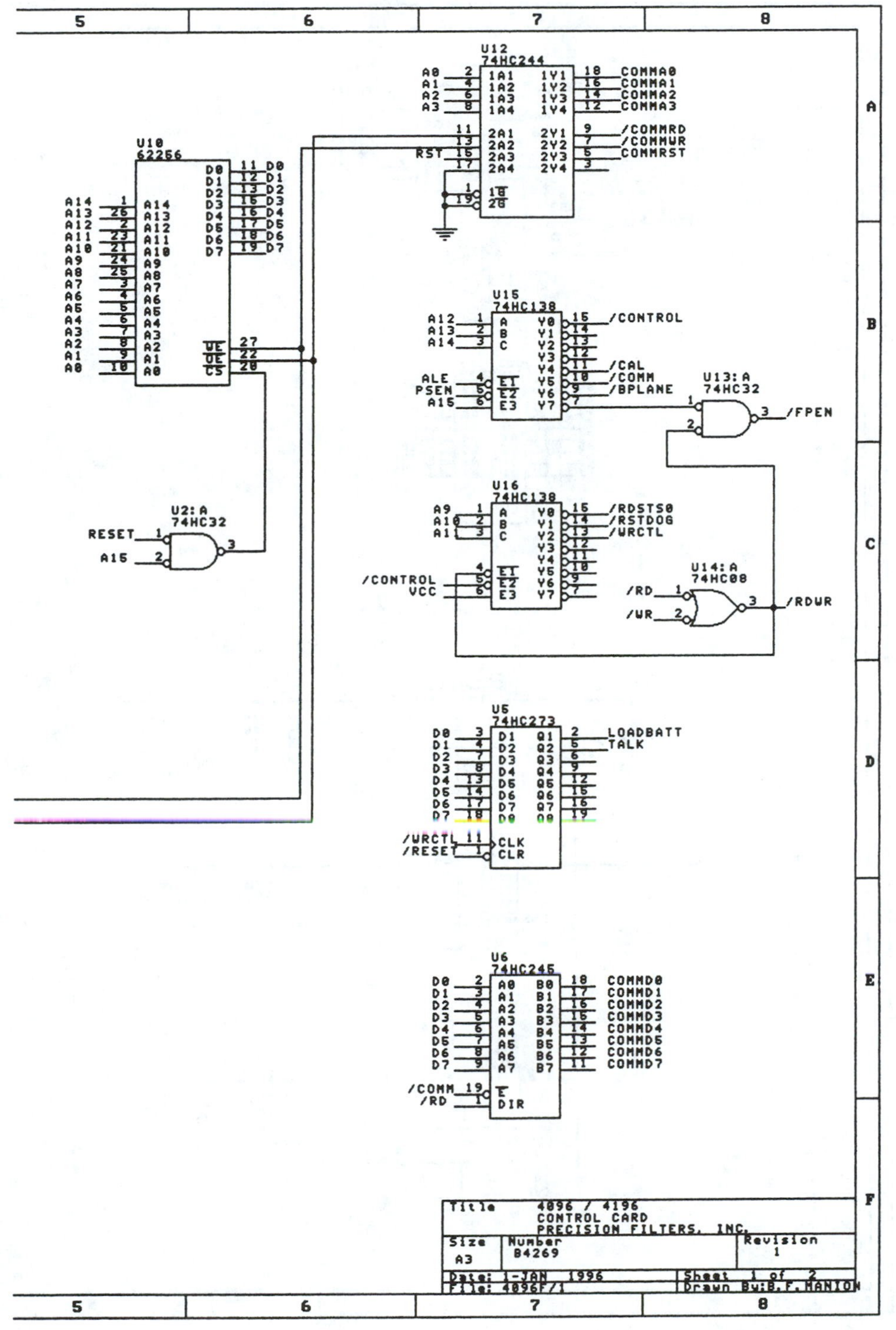
U12
74HC244
COMMA0
COMMA1
COMMA2
COMMA3
/COMMRD
/COMMWR
COMMRST
RST
U10
62256
U15
74HC138
/CONTROL
/CAL
/COMM
/BPLANE
ALE
PSEN
U13:A
74HC32
/FPEN
U2:A
74HC32
RESET
U16
74HC138
/RDSTS0
/RSTDOG
/WRCTL
VCC
U14:A
74HC08
/RD
/WR
/RDWR
U5
74HC273
LOADBATT
TALK
/RESET
U6
74HC245
COMMD0
COMMD1
COMMD2
COMMD3
COMMD4
COMMD5
COMMD6
COMMD7
Title 4096 / 4196
CONTROL CARD
PRECISION FILTERS, INC.
Size A3
Number B4269
Revision 1
Date: 1-JAN 1996
Sheet 1 of 2
File: 4096F/1
Drawn By:B.F.MANION

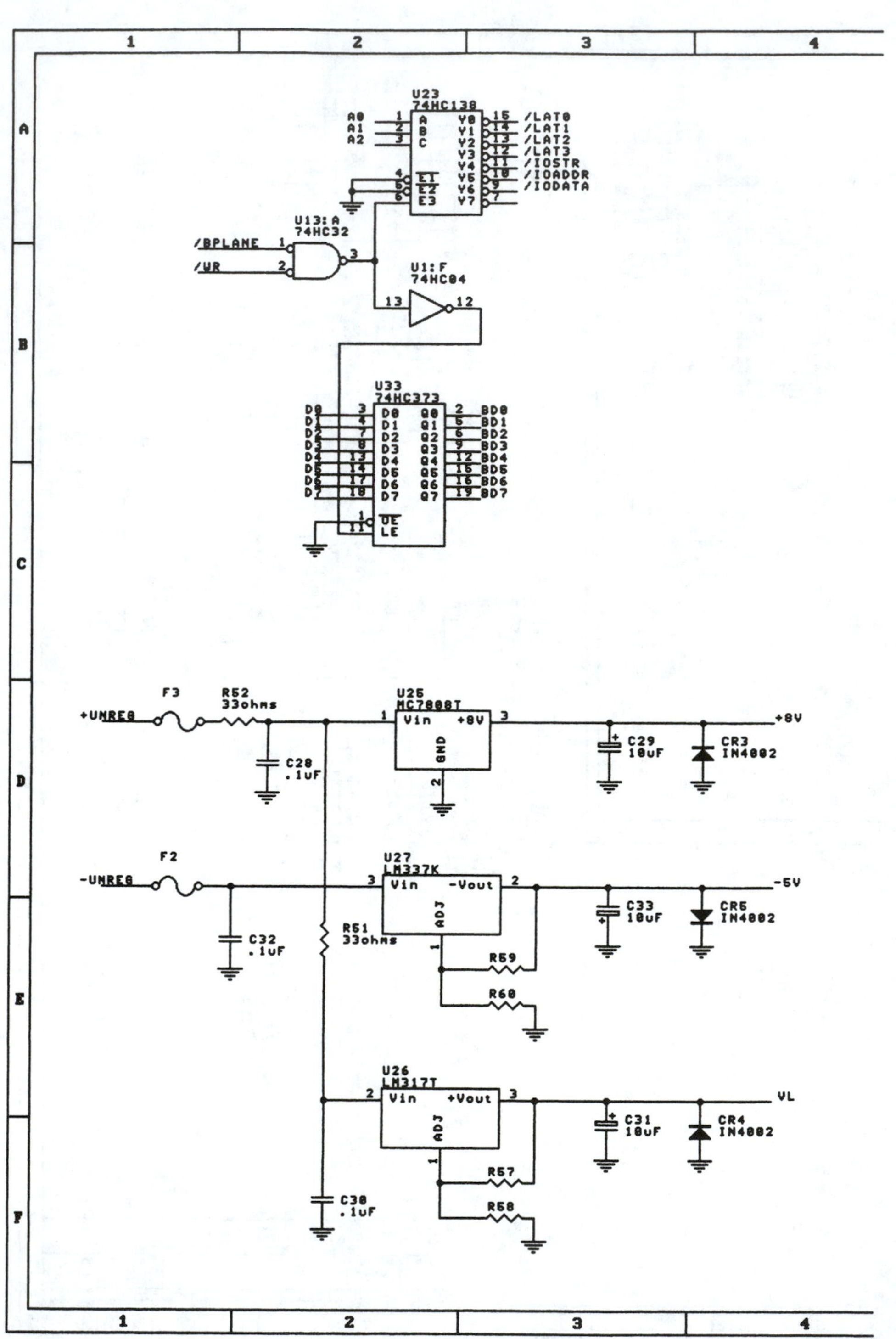
U23
74HC138
A0
A1
A2
/LAT0
/LAT1
/LAT2
/LAT3
/IOSTR
/IOADDR
/IODATA
U13:A
74HC32
/BPLANE
/WR
U1:F
74HC04
U33
74HC373
F3
R52
33ohms
+UNREG
C28
.1uF
U25
MC7808T
+8V
C29
10uF
CR3
IN4002
F2
-UNREG
C32
.1uF
U27
LM337K
-5V
R51
33ohms
C33
10uF
CR5
IN4002
R59
R60
U26
LM317T
VL
C31
10uF
CR4
IN4002
R57
R58
C30
.1uF

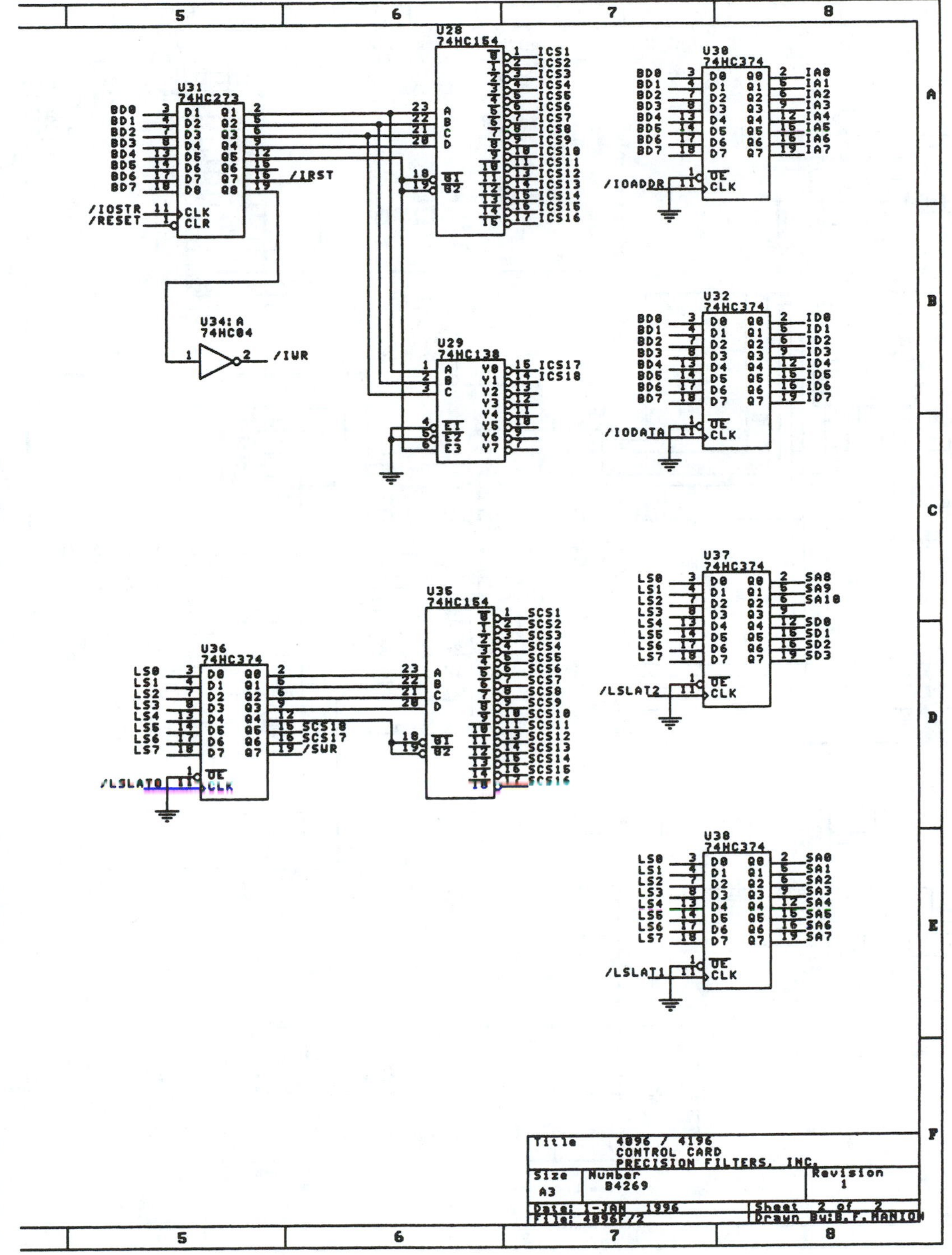
U28
74HC154
ICS1
ICS2
ICS3
ICS4
ICS5
ICS6
ICS7
ICS8
ICS9
ICS10
ICS11
ICS12
ICS13
ICS14
ICS15
ICS16
U31
74HC273
BD0
BD1
BD2
BD3
BD4
BD5
BD6
BD7
/IRST
/IOSTR
/RESET
CLK
CLR
U34:A
74HC04
/IWR
U29
74HC138
ICS17
ICS18
U30
74HC374
IA0
IA1
IA2
IA3
IA4
IA5
IA6
IA7
/IOADDR
U32
74HC374
ID0
ID1
ID2
ID3
ID4
ID5
ID6
ID7
/IODATA
U37
74HC374
LS0
LS1
LS2
LS3
LS4
LS5
LS6
LS7
SA8
SA9
SA10
SD0
SD1
SD2
SD3
/LSLAT2
U35
74HC154
SCS1
SCS2
SCS3
SCS4
SCS5
SCS6
SCS7
SCS8
SCS9
SCS10
SCS11
SCS12
SCS13
SCS14
SCS15
SCS16
U36
74HC374
SCS18
SCS17
/SWR
/LSLAT0
U38
74HC374
SA0
SA1
SA2
SA3
SA4
SA5
SA6
SA7
/LSLAT1
Title 4096 / 4196
CONTROL CARD
PRECISION FILTERS, INC.
Size A3
Number B4269
Revision 1
Date: 1-JAN 1996
Sheet 2 of 2
File: 4096F/2
Drawn By:B.F.MANION

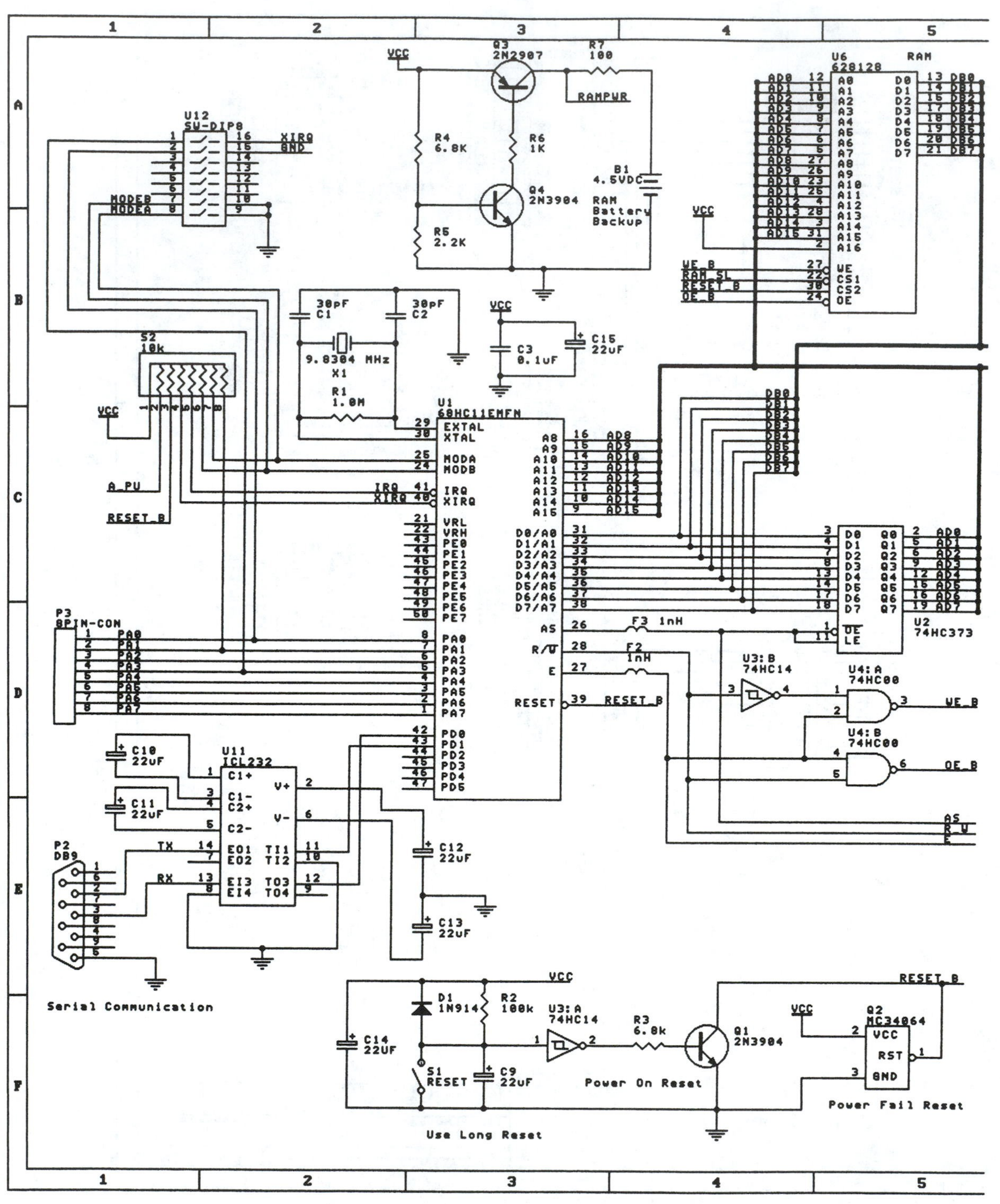

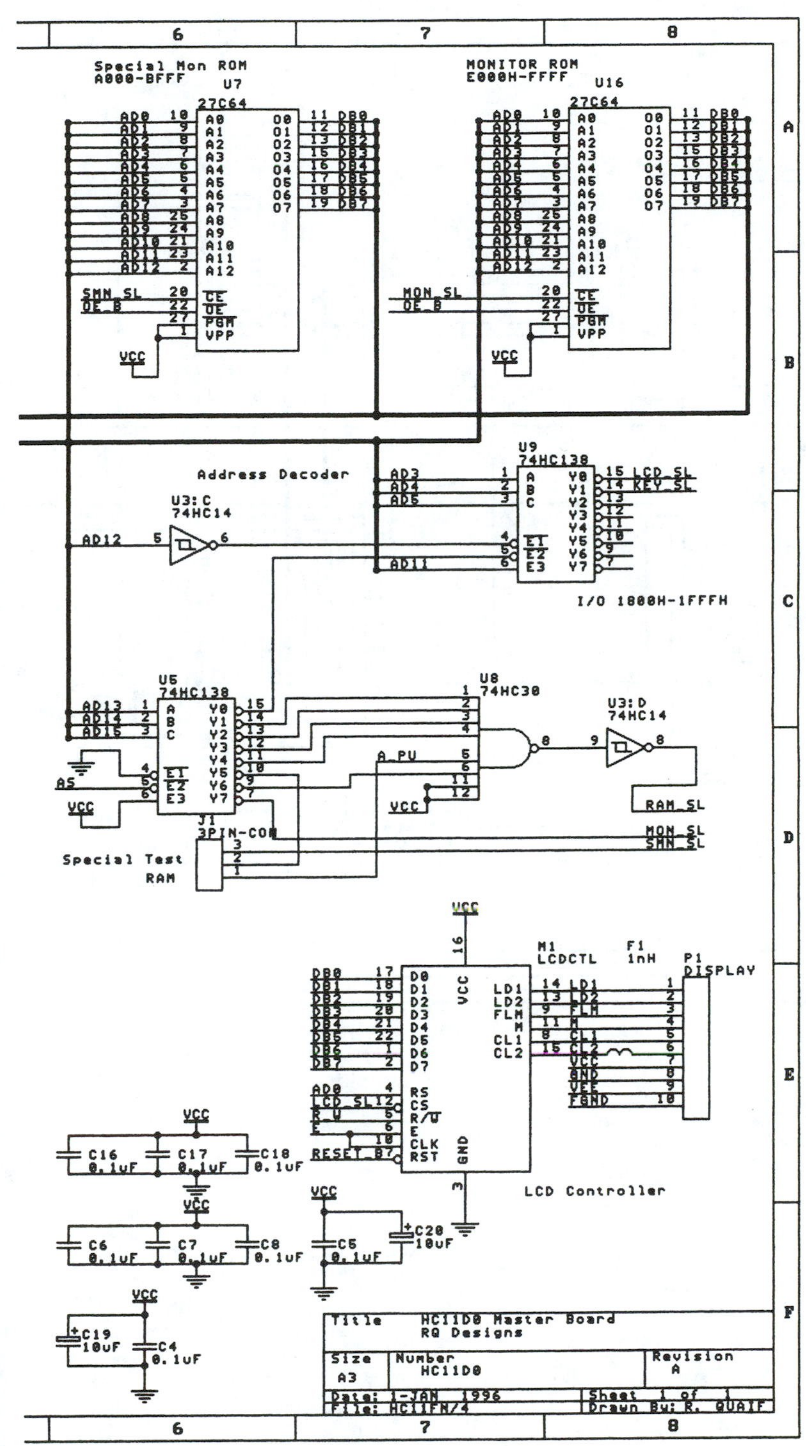
Special Mon ROM
A000-BFFF
U7
27C64
MONITOR ROM
E000H-FFFF
U16
27C64
Address Decoder
U3:C
74HC14
U9
74HC138
I/O 1800H-1FFFH
LCD_SL
KEY_SL
U5
74HC138
U8
74HC30
U3:D
74HC14
RAM_SL
MON_SL
SMN_SL
A_PU
J1
3PIN-CON
Special Test
RAM
M1
LCDCTL
F1
1nH
P1
DISPLAY
LCD Controller
C16 0.1uF
C17 0.1uF
C18 0.1uF
C6 0.1uF
C7 0.1uF
C8 0.1uF
C5 0.1uF
C20 10uF
C19 10uF
C4 0.1uF
Title HC11D0 Master Board
RQ Designs
Size A3
Number HC11D0
Revision A
Date: 1-JAN 1996
Sheet 1 of 1
File: HC11FN/4
Drawn By: R. QUAIF

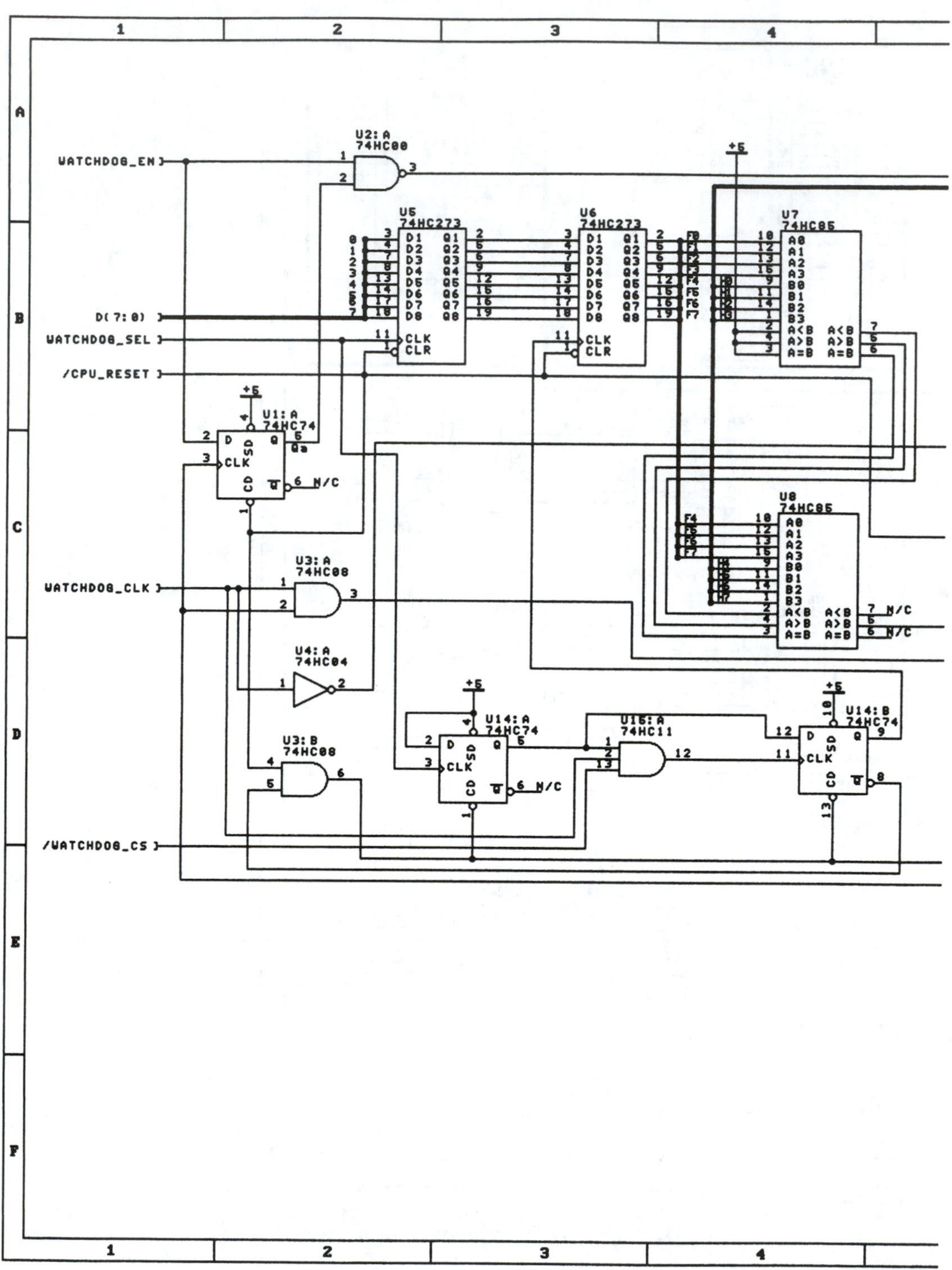
WATCHDOG_EN
U2:A
74HC00
U5
74HC273
U6
74HC273
U7
74HC85
D(7:0)
WATCHDOG_SEL
/CPU_RESET
U1:A
74HC74
N/C
U8
74HC85
U3:A
74HC08
WATCHDOG_CLK
U4:A
74HC04
U14:A
74HC74
U15:A
74HC11
U14:B
74HC74
U3:B
74HC08
/WATCHDOG_CS
+5

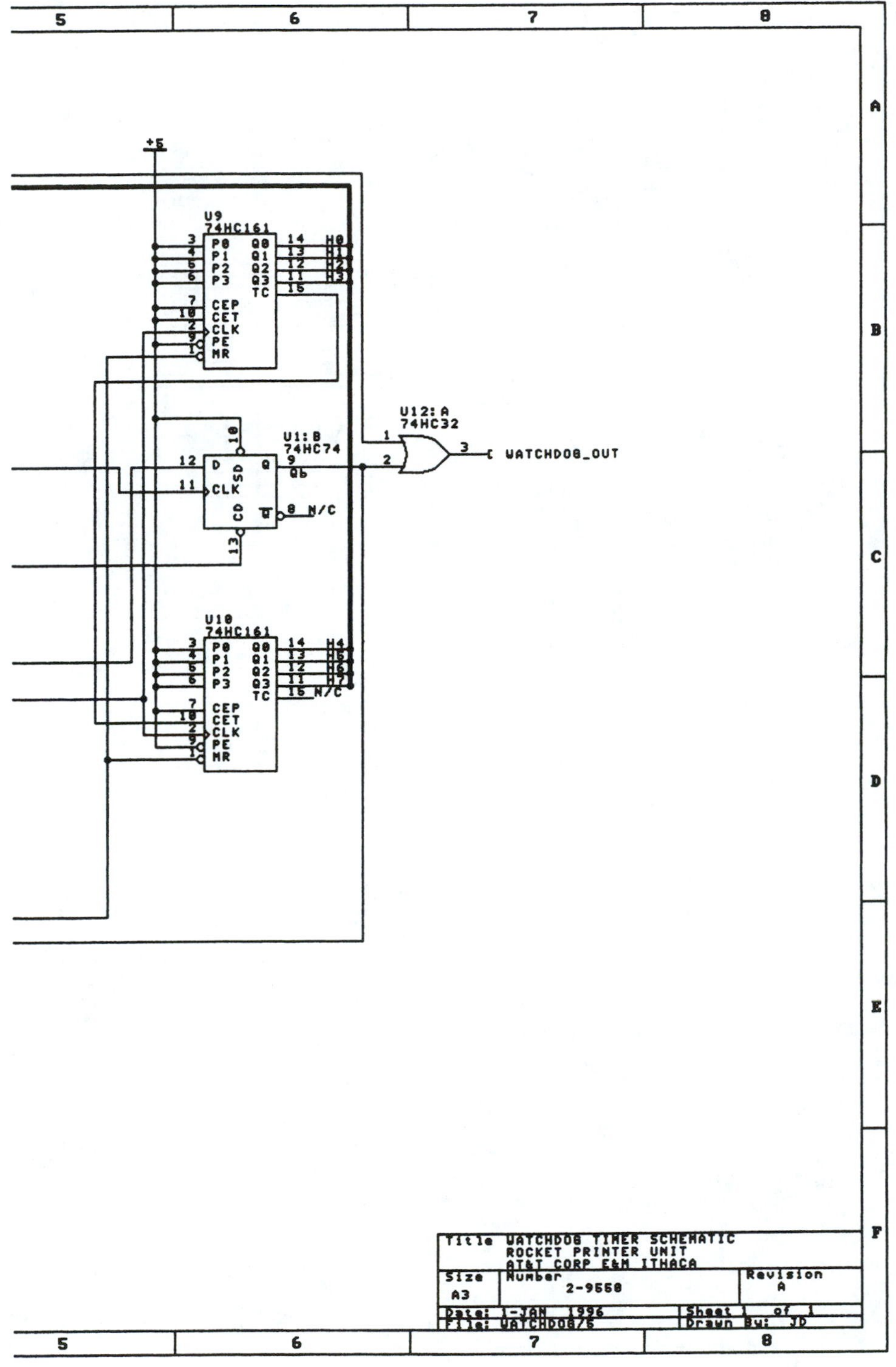
U9
74HC161
U1:B
74HC74
U12:A
74HC32
WATCHDOG_OUT
U10
74HC161
N/C
Title WATCHDOG TIMER SCHEMATIC
ROCKET PRINTER UNIT
AT&T CORP E&M ITHACA
Size A3
Number 2-9550
Revision A
Date: 1-JAN 1996
File: WATCHDOG/6
Sheet 1 of 1
Drawn By: JD

AN 8051-BASED DATA ACQUISITION AND CONTROL APPLICATION

The object of this Appendix is to implement the 8051-based data acquisition and control system shown in Figure I–1. Three modules will be discussed using several external transducers and instruments:

1. ADC module with hex display of the output.
2. DAC module with hex display of the input.
3. 8051 microcontroller module with connection points for (1) and (2).

Assembly-language software will be written and stored on an EPROM connected to the 8051. This software will exercise the ADC/DAC modules and perform other I/O functions.

One of the outcomes of these designs is for their use in colleges for teaching the concepts of data acquisition and control, so careful attention will be paid to the cost, durability, and ease of reproduction of each module.

HARDWARE

The 8051 Microcontroller Module

The 8051 microcontroller module is shown in Figure I–2. The heart of this module is the 8051 microcontroller. Also on the module are input and output buffers, an address latch, and an EPROM interface.

In Figure I–2, the microcontroller used is the 8031, which is a subset of the 8051. The 8031 differs in that it is the ROMless part. Instead of using the 8051, which has an internal mask-programmable ROM, we will use the 8031 and interface an EPROM to it to supply our program statements.

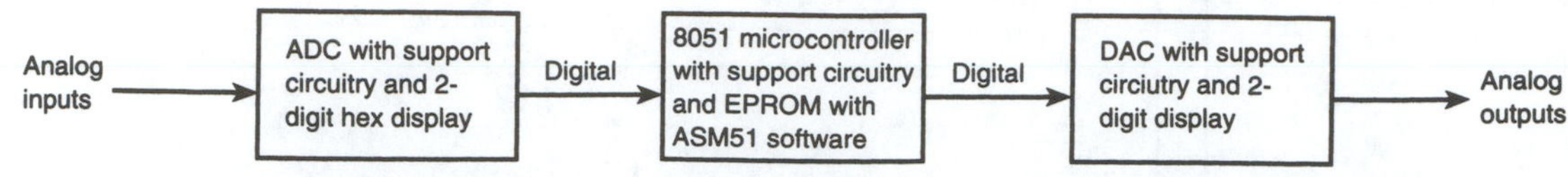

Figure I–1 Modular 8051-based data acquisition and control application

The 8031 will be run with a 12 MHz crystal. This was chosen because most of the instructions take 12 clock periods to execute. This way, accurate timing can be developed based on one microsecond instruction execution time.

The EPROM is interfaced to the 8031 via port 0 and port 2. The 12-bit address required to access the EPROM comes from the 8 bits provided from port 0 and the 4 bits provided from port 2. The data output from the EPROM is sent back into the microcontroller via port 0. Since port 0 has a dual purpose (i.e., address and data), a control signal from the ALE line is used to tell the address latch when port 0 has valid addresses and when it has data. Also, the $\overline{\text{EA}}$ signal is grounded, which signifies that all instructions will be fetched from an external ROM. The PSEN line goes LOW when the microcontroller is ready to read data from the EPROM. This LOW control signal is attached to the $\overline{\text{OE}}$ line of the 2732 EPROM, which changes the outputs at D0 through D7 from their float condition to an active condition. At that point the microcontroller reads the data.

Hardware reset is provided at pin 9 via the R-C circuit of R1–C3. At power-up the capacitor is initially discharged and the large inrush current through R1 provides a HIGH-to-LOW level for PIN 9 to be reset. Resetting can also be accomplished with the push-button, which shorts the 5-volt supply directly to pin 9 any time you want to have a user reset.

Two eight-bit I/Os are provided via port 1 and port 3. In this circuit, port 1 is used as an output port capable of sinking or sourcing up to four LSTTL loads. This is equivalent to a sink current of approximately 1.6 mA, which is normally not enough to drive loads such as LED indicators. Because of this, the 74LS244 (U1) is used as an output buffer. The 74LS244 can sink 24 mA and source 15 mA. The outputs of this buffer are always enabled by grounding pins 1 and 19. This way, whatever appears at port 1 will immediately be available to the LEDs. The LEDs are connected as active-LOW by providing a series 270 ohm resistor to the +5-volt supply. When any of the outputs of U1 go LOW, the corresponding LED will become active. The output socket is also provided on this module so that other devices can be driven from this port at the same time that the LEDs are being activated.

In this circuit, port 3 is used for input. The 8-bit DIP switch is connected to port 3 in conjunction with the 10 k DIP pack. Therefore, each of the bits of port 3 is normally HIGH via a 10 k pullup. They are made LOW any time one of the DIP switches is closed. An input socket is also provided at this port so that external digital inputs can be input to port 3. To use the input socket, all eight of the DIP switches must be in their open position so that the devices driving the input socket can drive the appropriate line LOW or HIGH without a conflict with the LOW from the DIP switch.

The 74LS373 is an octal transparent latch used as an address demultiplexer. Data entered at the D inputs are transferred into the Q outputs whenever the LE input is HIGH. The data at the D inputs are latched when the LE goes LOW. This address latch is used to demultiplex port 0's address/data information. When port 0 has valid address information, the ALE line is pulsed HIGH. This causes the address latch to grab hold of the information on port 0 and pass it through to the Qs and latch onto it. After the addresses from port 0 are latched, the port is then placed in the float condition so that it can be used to receive the data from the EPROM as the PSEN line is pulsed LOW. The LOW pulse on the PSEN line activates the active-LOW $\overline{\text{OE}}$ on the EPROM, which causes its outputs to become active. This data is routed back into port 0 of the microcontroller where it is used as program instructions.

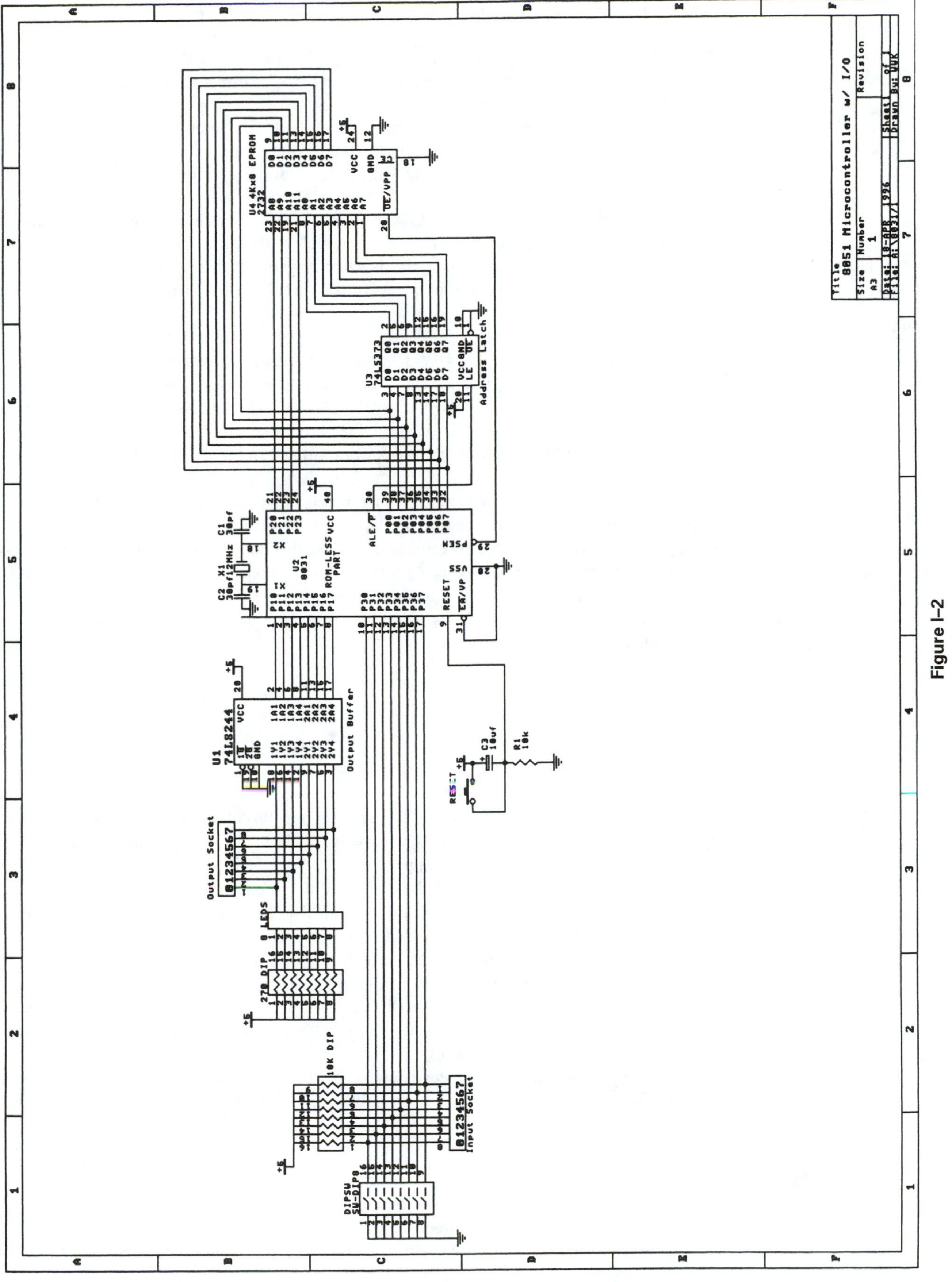
8051 Microcontroller w/ I/O
U4 4Kx8 EPROM 2732
U3 74LS373
Address Latch
U2 8031 ROM-LESS PART
U1 74LS244
Output Buffer
Output Socket
8 LEDS
270 DIP
10K DIP
Input Socket
DIPSW SW-DIP8
RESET
C3 10uf
R1 10k
C1 30pf
C2 30pf
X1 12MHz

Figure I–2

The ADC Interface Module

The heart of the ADC interface module (Figure I–3) is the ADC0804 IC. This analog-to-digital converter provides 8-bit output resolution and can be directly interfaced to a microcontroller. The object of this interface module is to convert analog quantities brought into $V_{in(+)}$ into 8-bit digital outputs. The digital outputs will be routed to an I/O connector, which can be interfaced to other devices or the microcontroller module. It will also be connected to the 2-digit hex display so that you can constantly monitor the 8-bit values that were converted.

The clock oscillator for the ADC is provided via R1–C1. The frequency of oscillation can be calculated by taking the reciprocal of (1.1 $\times$ 10 kΩ $\times$ 150 picofarads), which equals approximately 606 KHz. Since this is a successive approximation ADC we can assume that the conversion will take place in 10 to 12 clock periods, which provides a conversion time of about 20 ms.

The chip select (CS) pin on U1 is connected to ground so that the chip is always activated. The RD pin is an active-LOW output enable. This is connected to ground also, so that the outputs are always active. To start a conversion, the $\overline{\text{WR}}$ line, which is also known as the start-conversion ($\overline{\text{SC}}$), is pulsed LOW-then-HIGH. After the SAR has completed its conversion, it issues a LOW pulse on the INTR line [also known as end-of-conversion ($\overline{\text{EOC}}$)], which will be read by the microcontroller to determine if the conversion is complete.

The $V_{ref/2}$ pin is connected to a reference voltage that overrides the normal reference of 5 volts provided by V_{cc}. By setting the $V_{ref/2}$ at 1.28 volts, the actual V_{ref} will be 2.56 volts. This way, with a V_{ref} of 2.56 volts, and an ADC that has a maximum number of steps of 256, the change in the output will be 1 bit change or 1 binary step for every 10 millivolts of input change. This provides a convenient conversion for the linear temperature sensor (T1) because T1 provides a 10 mV change for every degree C. It also makes the calculations for the linear phototransistor sensor much simpler.

The DAC Interface Module

The heart of the DAC interface module (Figure 1–4) is the MC1408 (U1). This is an 8-bit digital-to-analog converter. The 8 bits to be converted are brought into the A8–A1 digital input side with A1 receiving the MSB. The DAC provides an analog output current that is proportional to that digital input string. The amount of full scale output current is dictated by the R1–R2 reference resistors. With a 15-volt supply and a 10-kΩ resistor, the maximum output current will be 1.5 milliamps.

The analog output current must be converted into a usable voltage. To perform this current-to-voltage conversion, a 741 op amp is used. The current at I_o is drawn through the 10-kΩ feedback resistor of U2. This way, the analog V_{out} will be equal to I_o times 10 kΩ. The analog output voltage can be predicted by using a ratio method. The ratio of the analog V_{out} to the binary input is proportional to the ratio of the maximum analog V_{out} to the maximum binary input (15 V maximum divided by 256 maximum). (Since the supply on the op amp is plus and minus 15 volts, we will never be able to achieve 15 volts at V_{out} but instead will be limited to about 13.5.)

A two-digit hexadecimal display is connected directly to the input to the DAC chip. This provides a way to monitor the binary input at any time.

APPLICATIONS

Three separate applications were used to exercise the three modules. Each of these applications are controlled by the microcontroller and will read analog input values with the ADC and output analog values with the DAC.

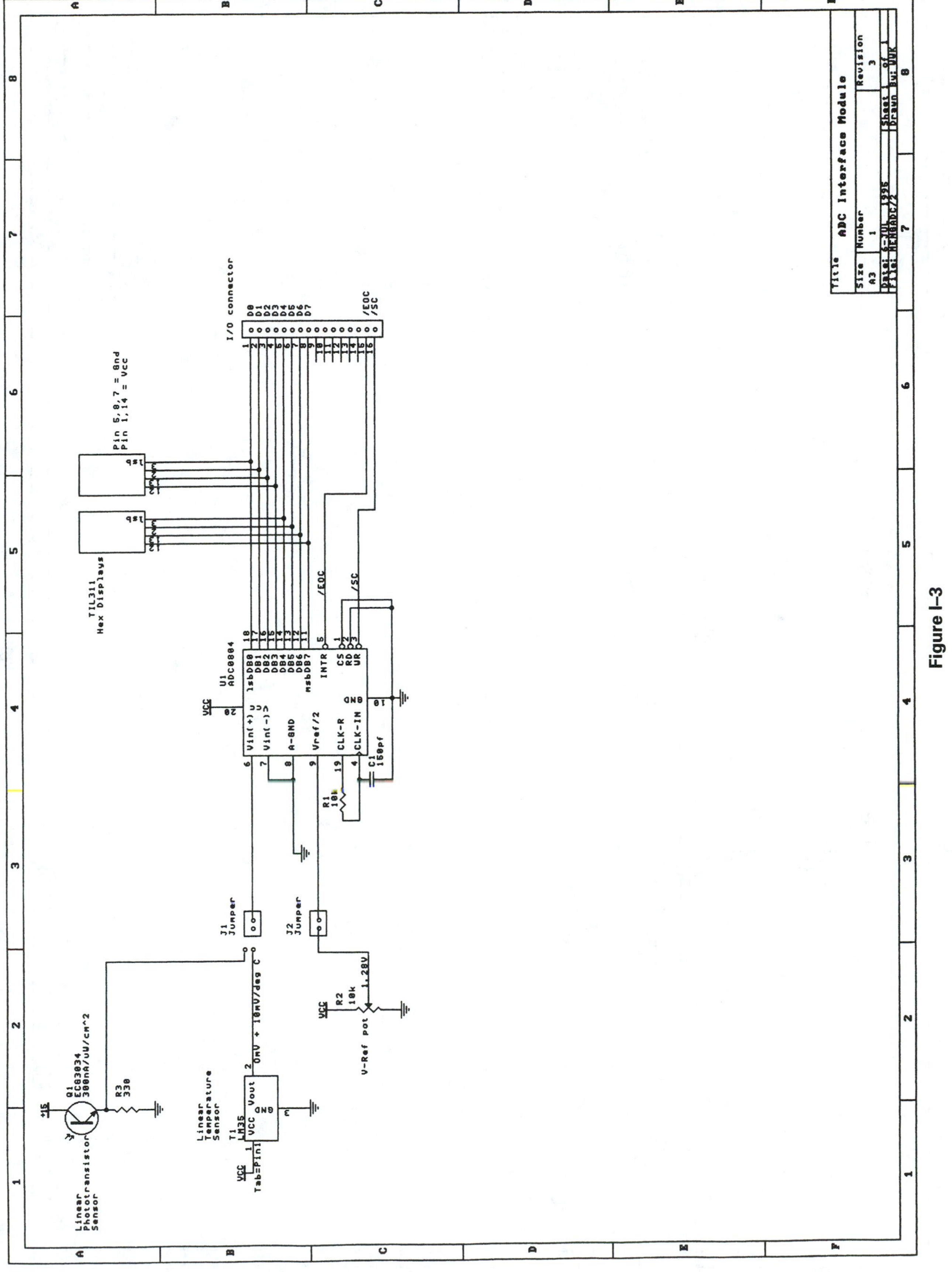

Title
ADC Interface Module
Size A3
Number 1
Revision 3
Date: 6-JUL-1996
Sheet 1 of 1
Drawn By: VVK
I/O connector
D0
D1
D2
D3
D4
D5
D6
D7
/EOC
/SC
TIL311 Hex Displays
U1 ADC0804
Vin(+)
Vin(-)
A-GND
Vref/2
CLK-R
CLK-IN
lsbDB0
msbDB7
INTR
CS
RD
WR
GND
VCC
R1 10k
C1 150pf
J1 Jumper
J2 Jumper
R2 10k
1.28V
V-Ref pot
T1 LM35
Linear Temperature Sensor
0mV + 10mV/deg C
Q1 ECG3034
R3 330
Linear Phototransistor Sensor

Figure I–3

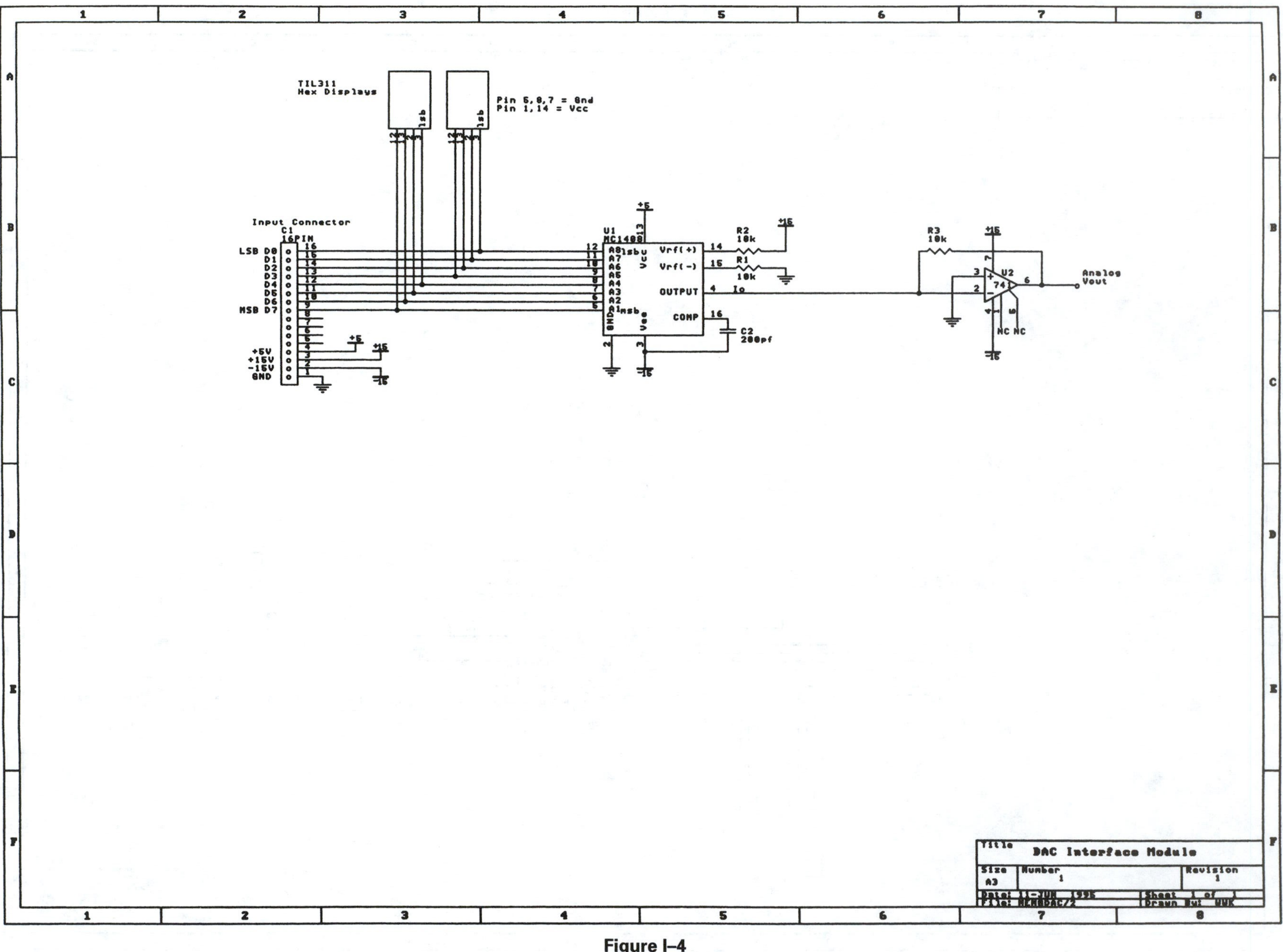
TIL311
Hex Displays
Pin 5,8,7 = Gnd
Pin 1,14 = Vcc
Input Connector
C1
16PIN
LSB D0
D1
D2
D3
D4
D5
D6
MSB D7
+5V
+15V
-15V
GND
U1
MC1408
Vrf(+)
Vrf(-)
OUTPUT
COMP
R2
10k
R1
10k
Io
C2
200pf
R3
10k
U2
741
NC NC
Analog
Vout
Title
DAC Interface Module
Size
A3
Number
1
Revision
1

Figure I–4

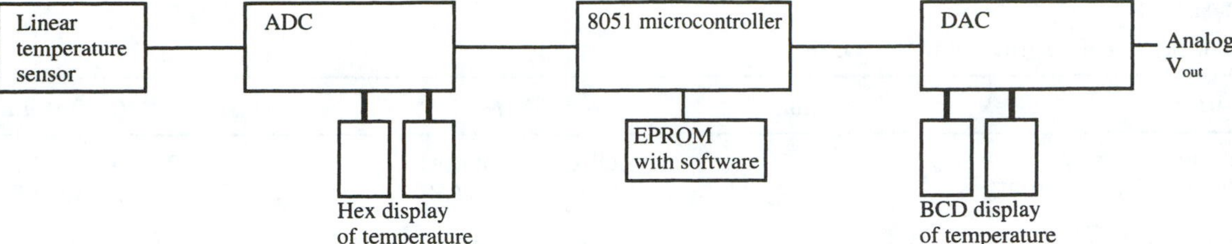

Figure I–5 Centigrade Thermometer

Centrigrade Thermometer

The centigrade thermometer application is shown in Figure I–5. Software is written for the 8051 microcontroller to read the temperature input to the ADC and display a binary coded decimal (BCD) output of the temperature. The LM35 linear temperature sensor is used to measure the temperature. It outputs 10 mV for every degree C. This millivolt value is converted by the ADC into an 8-bit binary string, which is displayed on the hex display and also input to the microcontroller. Since the V_{ref} on the ADC is set at 2.56 volts, then the output of the ADC is an actual hexadecimal display of the degrees C.

The microcontroller is used to start the ADC conversion and then monitor the end of conversion ($\overline{EOC}$) line to wait for the conversion to be complete. Once the conversion is complete, the microcontroller then has to read the digital string and convert the hexadecimal value into a BCD value so that it can be read by the user on the display of the DAC module. The analog output value of the DAC is not used.

The ASM51 software used by this application is given in Table I–1. This listing basically has four columns. The first lists the memory locations 0000–0062 that are programmed in the EPROM. The next shows the actual hex contents of the memory locations, starting with 802E and continuing down through 22. The third shows the assembly language programming starting with the ORG 0008H and ending with the RET statement. In the extreme right-hand column is the comments section for the program.

As you read through the comments section you will see that three subroutines are used to call the three different operations that are going to take place. The first subroutine, ADC, is used to float the buses on port 3 and port 1 and then issue the start conversion ($\overline{SC}$) signal. The "wait" part of that subroutine is necessary to wait until $\overline{EOC}$ goes LOW before reading the data from port 3 into register 0.

After the ADC routine is complete and the digital value is stored in register 0, that value in register 0 must be converted from hex to BCD. To do this, a counter routine is used to count up in BCD until the hex value is reached. At that point, the accumulator will have in it the correct BCD value that was achieved by using the decimal adjust instruction (DA A). After the DAC displays the BCD result, a delay is used for approximately 4 seconds so that there will be a 4 second delay before the temperature is refreshed.

Temperature-Dependent PWM Speed Control

This application is shown in Figure I–6. It uses most of the same hardware as the centigrade thermometer. The object of this application is to monitor the temperature and produce a pulse-width-modulated (PWM) square wave whose duty cycle is proportional to the temperature. This PWM square wave can be used to drive a DC motor. The effective DC value will increase as the duty cycle increases. Since the duty cycle increases with increasing temperature, the speed of the motor will also increase with increasing temperature.

The ASM51 software used to drive this application is given in Table I–2. Basically, this software has two subroutines that will be called. The first one is the ADC subroutine. This subroutine starts the conversion, waits until the conversion is complete, and moves the ADC result into register 0. The next subroutine, the PWM, is used to produce the PWM square wave. To produce a PWM square wave, the output of the DAC will first

TABLE I–1
Centigrade Thermometer Software

Addr	*Hex*	*Label*	*ASM51*		*Comments*
0000			ORG	0000H	; START OF EPROM
0000	802E		SJMP	START	; SKIP OVER 8051 RESTART
					; LINK AREA
0030			ORG	0030H	; START OF PGM
0030	1138	START:	ACALL	ADC	; PERFORM AN A-TO-D CON-
					; VERSION
0032	1159		ACALL	DAC	; DISPLAY RESULT AND PER-
					; FORM D-TO-A
					; CONVERSION
0034	114C		ACALL	DELAY	; DELAY 4 SECOND
0036	80F8		SJMP	START	; REPEAT
				;	
0038	75B0FF	ADC:	MOV	P3,#0FFH	; FLOAT THE BUS
003B	7590FF		MOV	P1,#0FFH	; FLOAT THE BUS
003E	C2B7	SC:	CLR	P3.7	; DROP SC LINE LOW
0040	D2B7		SETB	P3.7	; THEN RAISE HIGH
0042	20B6FD	WAIT:	JB	P3.6, WAIT	; WAIT TILL EOC GOES LOW
0045	C2B6		CLR	P3.6	; CLEAR P3.6 WHICH WAS
					; USED FOR $\overline{\text{EOC}}$
0047	C2B7		CLR	P3.7	; CLEAR P3.7 WHICH WAS
					; USED FOR $\overline{\text{SC}}$
0049	A8B0		MOV	R0,P3	; STORE FINAL ADC RESULT
					; INTO R0
004B	22		RET		
				;	
004C	7F20	DELAY:	MOV	R7,#20H	; OUTERMOST LOOP (32X)
004E	7D00		MOV	R5,#00H	; INNERMOST LOOP (256X)
0050	7EF3	LOOP2:	MOV	R6,#0F3H	; MIDDLE LOOP (243X)
0052	DDFE	LOOP1:	DJNZ	R5,LOOP1	;
0054	DEFC		DJNZ	R6,LOOP1	;
0056	DFF8		DJNZ	R7,LOOP2	;
0058	22		RET		
					; CONVERT HEX TO BCD
					; AND DISPLAY ON DAC
					; MODULE
0059	7400	DAC:	MOV	A,#00H	; ZERO OUT ACCUMULATOR
005B	2401	LOOP3:	ADD	A,#01H	; Acc WILL HAVE VALID BCD,
					; R0=ADC HEX
					; RESULT
005D	D4		DA	A	; DECIMAL ADJUST Acc IF
					; INVALID BCD
005E	D8FB		DJNZ	R0,LOOP3	; DECR R0 WHILE INCR AND
					; ADJUSTING Acc TILL
					; R0=0
0060	F590		MOV	P1,A	; MOV CORRECTED BCD TO
					; DAC DISPLAY
0062	22		RET		;
					;
0000			END		

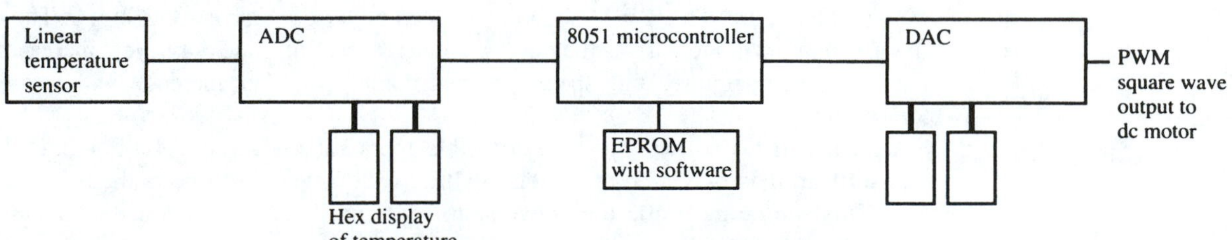

Figure I–6 Temperature-dependent PWM speed control

TABLE I–2
Temperature-Dependent PWM Speed Control Software

Addr	*Hex*	*Label*	*ASM51*		*Comments*
0000			ORG	0000H	; START OF EPROM
0000	8030		SJMP	START	; SKIP OVER 8051 RESTART ; LINK AREA
0030			ORG	0030H	; START OF PGM
0030	7F01	START:	MOV	R7,#01H	; TO ENABLE ADC READ ON ; 1st LOOP
0032	1138	LOOP:	ACALL	ADC	; PERFORM AN A-TO-D CON- ; VERSION
0034	114C		ACALL	PWM	; DRIVE DC MOTOR WITH ; PWM SIGNAL
0036	80FA		SJMP	LOOP	; REPEAT
0038	DF11	ADC:	DJNZ	R7,ERET	; TAKE AN ADC READ EVERY ; 256th TIME FOR ; STABILITY
003A	75B0FF		MOV	P3,#0FFH	; FLOAT THE BUS
003D	C2B7	SC:	CLR	P3.7	; DROP $\overline{\text{SC}}$ LINE LOW
003F	D2B7		SETB	P3.7	; THEN RAISE HIGH (PRO- ; VIDES POSITIVE EDGE)
0041	20B6FD	WAIT:	JB	P3.6,WAIT	; WAIT TILL $\overline{\text{EOC}}$ GOES LOW
0044	E5B0		MOV	A,P3	; STORE ADC RESULT INTO ; Acc
0046	547F		ANL	A,#01111111B	; CLEAR BIT 7 WHICH WAS ; USED FOR $\overline{\text{SC}}$
0048	F8		MOV	R0,A	; MOVE ADC RESULT TO ; REG 0
0049	AF40		MOV	R7,40H	; RESET REG 7 to 64 DECI- ; MAL
004B	22	ERET:	RET		
				;	
004C	759000	PWM:	MOV	P1,#00H	; OUTPUT A LOW
004F	743F		MOV	A,#3FH	; 3FH IS HIGHEST ASSUMED ; ADC OUTPUT VALUE
0051	98		SUBB	A,R0	; SUBTRACT THE ADC ; VALUE FROM 3FH
0052	F9		MOV	R1,A	; R1 (DELAY VALUE) FOR ; LOW OUT (DEC W/ INC ; TEMP)
0053	115D		ACALL	DELAY	; CALL DELAY TO HOLD LOW
0055	7590FF		MOV	P1,#0FFH	; OUTPUT FULL SCALE OUT- ; PUT FOR A HIGH
0058	A900		MOV	R1,R0	; USE TEMPERATURE AS DE- ; LAY MULTIPLIER ; FOR HIGH
005A	115D		ACALL	DELAY	; CALL DELAY TO HOLD ; HIGH
005C	22		RET		; RETURN
					;
005D	7D80	DELAY:	MOV	R5,#80H	; START COUNT OF R5
005F	DDFE	LOOP1:	DJNZ	R5,LOOP1	; R5=INNERLOOP COUNT ; DOWN TO ZERO
0061	D9FA		DJNZ	R1,DELAY	; R1=MULTIPLIER FROM ;PWM ROUTINE
0063	22		RET		
0000			END		

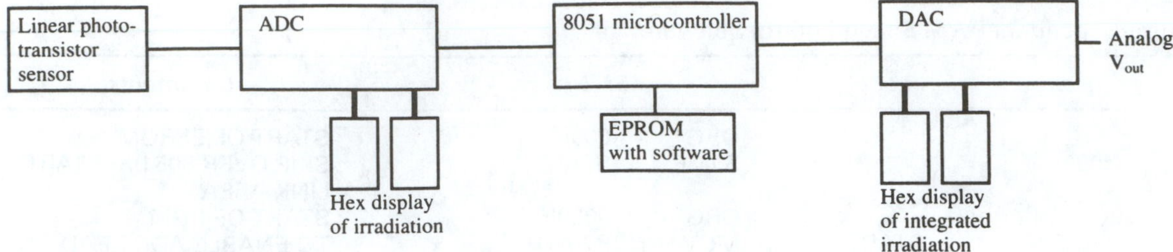

Figure I–7 Integrating solar radiometer

TABLE I–3
Integrating Solar Radiometer Software

Addr	*Hex*	*Label*	*ASM51*		*Comments*
0000		ORG	0000H		; START OF EPROM
0000	8030		SJMP	START	; SKIP OVER 8051 RESTART
					; LINK AREA
0030			ORG	0030H	; START OF PGM
0030	7400	START:	MOV	A,#00H	; RESET ACCUMULATOR
0032	113A	LOOP:	ACALL	ADC	; PERFORM AN A-TO-D CON-
					; VERSION
0034	1158		ACALL	DAC	; DISPLAY RESULT AND PER-
					; FORM D-TO-A
					; CONVERSION
0036	114B		ACALL	DELAY	; DELAY 4 SECOND
0038	80F8		SJMP	LOOP	; REPEAT
003A	75B0FF	ADC:	MOV	P3,#0FFH	; FLOAT THE BUS
003D	C2B7	SC:	CLR	P3.7	; DROP $\overline{SC}$ LINE LOW
003F	D2B7		SETB	P3.7	; THEN RAISE HIGH
0041	20B6FD	WAIT:	JB	P3.6,WAIT	; WAIT TILL $\overline{EOC}$ GOES LOW
0044	C2B6		CLR	P3.6	; CLEAR P3.6 WHICH WAS
					; USED FOR $\overline{EOC}$
0046	C2B7		CLR	P3.7	; CLEAR P3.7 WHICH WAS
					; USED FOR $\overline{SC}$
0048	A8B0		MOV	R0,P3	; STORE FINAL ADC RESULT
					; INTO R0
004A	22		RET		
					; **DELAY 4 SECONDS**
004B	7F20	DELAY:	MOV	R7,#20H	; OUTERMOST LOOP (32X)
004D	7D00		MOV	R5,#00H	; INNERMOST LOOP (256X)
004F	7EF3	LOOP2:	MOV	R6,#0F3H	; MIDDLE LOOP (243X)
0051	DDFE	LOOP1:	DJNZ	R5,LOOP1	;
0053	DEFC		DJNZ	R6,LOOP1	;
0055	DFF8		DJNZ	R7,LOOP2	;
0057	22		RET		
					; INTEGRATE BY ACCUMU-
					; LATING ADC RESULT
0058	8840	DAC:	MOV	40H,R0	; MOVE ADC RESULT TO
					; RAM LOCATION 40H
005A	53403F		ANL	40H,#00111111B	; MASK OFF UNWANTED
					; DATA
005D	2540		ADD	A,40H	; ACCUMULATE NEW ADC
					; VALUE TO PREVIOUS
					; TOTAL
005F	F590		MOV	P1,A	; MOVE TO DAC DISPLAY
0061	22		RET		;
					;
0000			END		

be held LOW (0 volts) for a delay multiple equivalent to 10 hexadecimal. Then the DAC will be sent HIGH, which in this case is one-half the full scale output (80 hex). This HIGH is held for a delay multiple equivalent to the temperature value received from the ADC conversion. This way, as the ADC result increases with increasing temperature, the duty cycle of the PWM will increase, thus increasing the effective DC value sent to the motor.

Integrating Solar Radiometer

The block diagram for the integrating solar radiometer is given in Figure I–7. The object of this application is to measure sunlight intensity and produce a real-time display of the irradiation from the sunlight and then accumulate the irradiation, the same as taking the integral, and display the integrated output as an analog voltage and also on a hex display. To measure the irradiation, an ECG 3034 phototransistor is used. It produces 300 nanoamps per microwatt per centimeter squared. The current is changed to a voltage by the 330 ohm resistor. This analog voltage is converted with the ADC and is then integrated using an ASM51 software program. The integrated output of the ADC will be proportional to the amount of sunlight that struck the measured surface over a certain length of time. This length of time could be one hour, one day, one week, or whatever time is desired.

The software used to execute this application is given in Table I–3. The first subroutine is the ADC subroutine, which is used to convert the irradiation into a digital value. After the microcontroller has this digital value, the DAC routine performs the integration simply by accumulating the ADC result over each period of time. The DAC provides both the hexadecimal equivalent of the integrated irradiation as well as an analog output voltage that is proportional to the integrated irradiation.

CONCLUSION

This three-module method for data acquisition and control proves to be a very inexpensive and simple way to be able to input both analog and digital quantities and to output both digital and analog quantities. Having the hex displays connected to the input and output modules is an effective way of monitoring the real-time activity of the DAC and the ADC. The entire three-module data acquisition set up is very portable in that it only requires a bipolar power supply to operate it. The software used to exercise these modules is very straightforward and is effectively written using the ASM51 assembler on a personal computer and downloaded to an EPROM using a standard EPROM programming software package. Once you have the programmed EPROM, all that is required for a stand alone data acquisition system is a power supply.

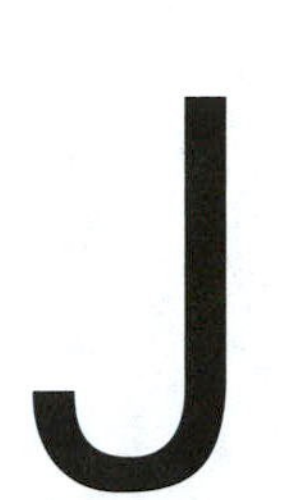

Review of Basic Electricity Principles

Definitions for Figure J–1

$V \equiv$ voltage source that pushes the current (I) through the circuit, like water through a pipe

$I \equiv$ current that flows through the circuit [conventional current flows (+) to (−).]

$R \equiv$ resistance to the flow of current

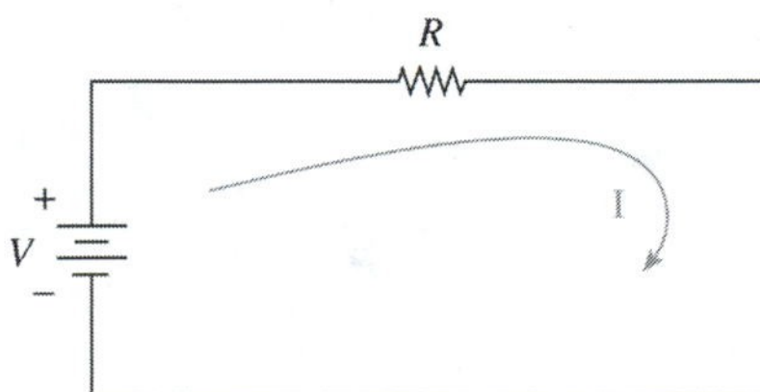

Figure J–1 Series circuit used to illustrate Ohm's Law.

Units

voltage = volts (V), for example, 12 V, 6 mV

current = amperes (A), for example, 2 A, 2.5 mA

resistance = ohms (Ω), for example, 100 Ω, 4.7 kΩ

Common Engineering Prefixes

Prefix	Abbreviation	Value
Mega-	M	1,000,000 or 10^6
kilo-	k	1,000 or 10^3
milli-	m	0.001 or 10^{-3}
micro-	μ	0.000001 or 10^{-6}

Ohm's Law

The current (I) in a complete circuit is proportional to the applied voltage (V) and inversely proportional to the resistance (R) of the circuit (see Figure J–2).

Formulas

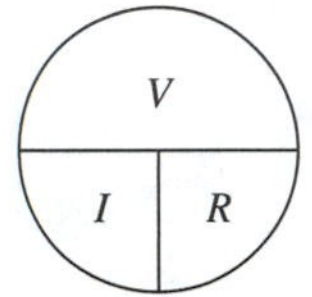

Figure J–2 Ohm's Law circle.

$$I = \frac{V}{R}$$

$$V = I \times R$$

$$R = \frac{V}{I}$$

EXAMPLE J–1

(a) Determine the current (I) in the circuit of Figure J–1 if $V = 10$ V and $R = 2\ \Omega$.

(b) Recalculate the current if $R = 4\ \Omega$.

(c) Describe what happened to the value of the current when the resistance doubled to 4 Ω in (b).

(d) Calculate the voltage required in Figure J–1 to make 2 A flow if $R = 10\ \Omega$.

(e) What voltage would be required in (d) if you only need one-half that current?

(f) If $V = 12$ V in Figure J–1, what resistance is required to limit the current to 2 A?

(g) To limit the current to 1 A in (f), what resistance would you need?

Solution:

(a) $I = \frac{V}{R} = \frac{10\text{ V}}{2\ \Omega} = 5\text{ A}$

(b) $I = \frac{V}{R} = \frac{10\text{ V}}{4\ \Omega} = 2.5\text{ A}$

(c) As the resistance to current flow doubled, the current dropped to one-half.

(d) $V = I \times R = 2\text{ A} \times 10\ \Omega = 20\text{ V}$

(e) $V = I \times R = 1\text{ A} \times 10\ \Omega = 10\text{ V}$

Note we only need one-half the voltage to get one-half the current.

(f) $R = \frac{V}{I} = \frac{12\text{ V}}{2\text{ A}} = 6\ \Omega$

(g) $R = \frac{V}{I} = \frac{12\text{ V}}{1\text{ A}} = 12\ \Omega$

Note that, to reduce the current to 1 A, we needed to increase the circuit's resistance.

EXAMPLE J–2

Apply the values listed below to the circuit of Figure J–1 to determine the unknown quantity.

(a) $I = 2$ mA, $R = 4$ kΩ, $V =$ ________?
(b) $I = 6$ μA, $R = 200$ kΩ, $V =$ ________?
(c) $I = 24$ μA, $V = 12$ V, $R =$ ________?
(d) $I = 100$ mA, $V = 5$ V, $R =$ ________?
(e) $V = 5$ V, $R = 50$ kΩ, $I =$ ________?
(f) $V = 12$ V, $R = 600$ Ω, $I =$ ________?

Solution:

(a) $V = I \times R = 2\text{ mA} \times 4\text{ k}\Omega = (2 \times 10^{-3}\text{A}) \times (4 \times 10^{3}\ \Omega) = 8\text{ V}$

(b) $V = I \times R = 6\ \mu\text{A} \times 200\text{ k}\Omega = (6 \times 10^{-6}\text{A}) \times (200 \times 10^{3}\ \Omega) = 1.2\text{ V}$

(c) $R = \dfrac{V}{I} = \dfrac{12\text{ V}}{24\ \mu\text{A}} = \dfrac{12\text{ V}}{24 \times 10^{-6}\text{ A}} = 500\text{ k}\Omega \text{ (or } 0.5\text{ M}\Omega)$

(d) $R = \dfrac{V}{I} = \dfrac{5\text{ V}}{100\text{ mA}} = 50\ \Omega$

(e) $I = \dfrac{V}{R} = \dfrac{5\text{ V}}{50\text{ k}\Omega} = 100\ \mu\text{A}$

(f) $I = \dfrac{V}{R} = \dfrac{12\text{ V}}{600\ \Omega} = 20\text{ mA}$

EXAMPLE J–3

A *series circuit* has two or more resistors end to end. The total resistance is equal to the sum of the individual resistances ($R_T = R_1 + R_2$). Also, the sum of the voltage drops across all resistors will equal the total supply voltage ($V_S = V_{R1} + V_{R2}$).

Find the current in the circuit (I), the voltage across R_1 (V_{R1}), and the voltage across R_2 (V_{R2}) in Figure J–3.

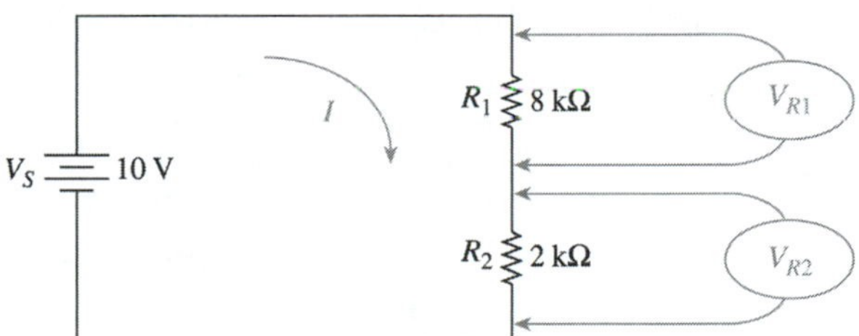

Figure J–3 A series circuit used to derive the voltage divider equation.

Solution:

$$R_T = 8\text{ k}\Omega + 2\text{ k}\Omega = 10\text{ k}\Omega$$
$$I = \frac{10\text{ V}}{10\text{ k}\Omega} = 1\text{mA}$$
$$V_{R1} = 1\text{mA} \times 8\text{ k}\Omega = 8\text{V}$$
$$V_{R2} = 1\text{mA} \times 2\text{ k}\Omega = 2\text{V}$$

Check:

$$V_S = 8\text{ V} + 2\text{ V} = 10\text{ V}$$

Notice that the voltage across any resistor in the series circuit is proportional to the size of the resistor. That fact is used in developing the *voltage-divider equation*:

$$V_{R2} = V_S \times \frac{R_2}{R_1 + R_2}$$
$$= 10\ \text{V} \times \frac{2\ \text{k}\Omega}{2\ \text{k}\Omega + 8\ \text{k}\Omega}$$
$$= 2\ \text{V}$$

EXAMPLE J–4

Use the voltage-divider equation to find V_{out} in Figure J–4. (V_{out} is the voltage from the point labeled V_{out} to the ground symbol.)

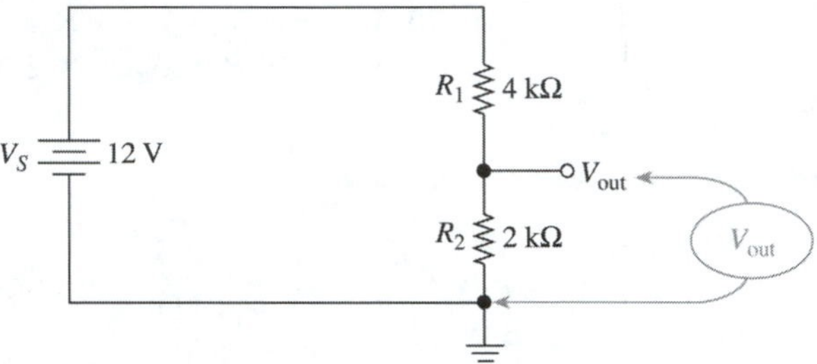

Figure J–4 Circuit used to calculate the output voltage (V_{out}) with respect to ground.

Solution:

$$V_{out} = 12\ \text{V} \times \frac{2\ \text{k}\Omega}{2\ \text{k}\Omega + 4\ \text{k}\Omega}$$
$$= 4\ \text{V}$$

EXAMPLE J–5

A *short circuit* occurs when an electrical conductor is purposely or inadvertently placed across a circuit component. The short causes the current to bypass the shorted component. Calculate V_{out} in Figure J–5.

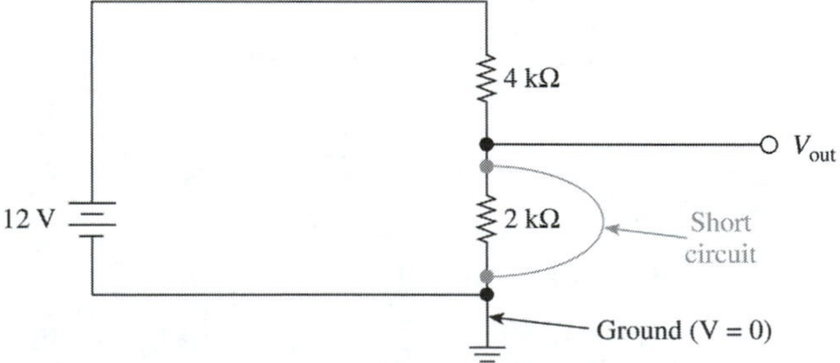

Figure J–5 A short circuit across the 2 kΩ resistor forces the output to zero V.

Solution: V_{out} is connected directly to ground; therefore, $V_{out} = 0$ V. All of the 12-V supply is dropped across the 4-kΩ resistor.

EXAMPLE J–6

Find V_{out} in Figure J–6.

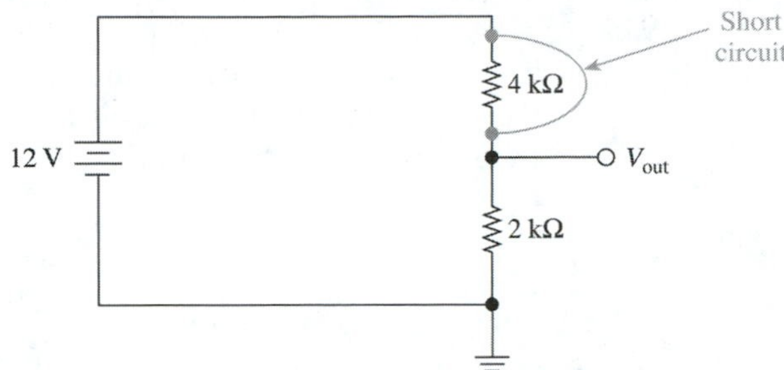

Figure J–6 A short across the top resistor causes the entire 12 volts to reach V_{out}.

Solution: V_{out} is connected directly to the top of the 12-V source battery. Therefore, $V_{out} = 12$ V. The entire 12-V source voltage is now across the 2-kΩ resistor.

EXAMPLE J–7

An *open circuit* is a break in a circuit. This can be done purposely by an electronic switching component, or it could be a circuit fault caused by a bad connector or burnt-out component. This break will cause the current to stop flowing to all components fed from that point, Calculate V_{out} in Figure J–7.

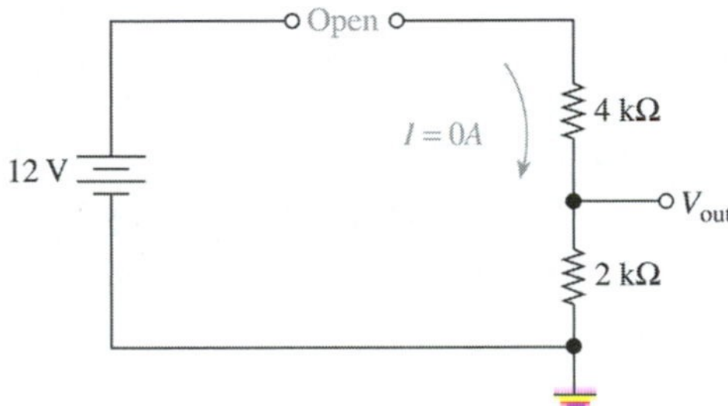

Figure J–7 An open circuit causes current to stop flowing.

Solution: Because $I = 0$ A,

$$V_{2\,k\Omega} = 0\text{ A} \times 2\text{ k}\Omega = 0\text{ V}$$
$$V_{out} = V_{2\,k\Omega} = 0\text{ V}$$

EXAMPLE J–8

Calculate V_{out} in Figure J–8.

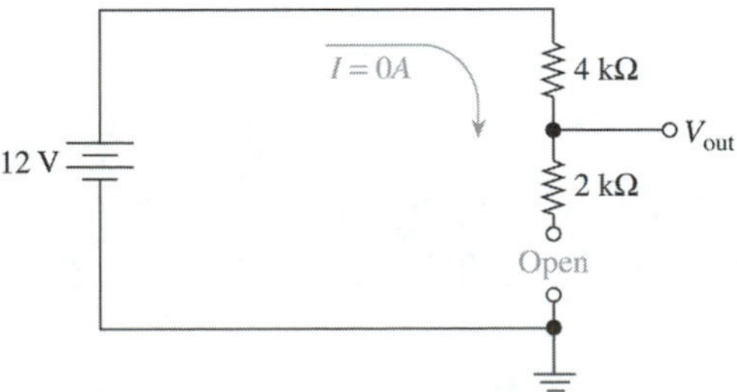

Figure J–8 An open circuit below the measurement point allows the entire supply voltage to reach the output.

Solution: Because $I = 0\text{A}$,

$$V_{\text{drop (4 k}\Omega)} = 0\text{ A} \times 4\text{ k}\Omega = 0\text{ V}$$
$$V_{\text{out}} = 12\text{ V} - V_{\text{drop}} = 12\text{ V}$$

Note:

This is probably the hardest concept to understand. Another way to explain why the entire supply reaches V_{out} is to assume that an open circuit can be modeled by an extremely large resistance, let's say, 10 MΩ. If you then apply the voltage-divider equation to the circuit with 10 MΩ in place of the open, the calculation will be

$$V_{\text{out}} = 12\text{ V} \times \frac{10{,}002{,}000}{10{,}002{,}000 + 4{,}000} = 11.995\text{ V}$$

EXAMPLE J–9

The symbol for a battery is seldom drawn in schematic diagrams. Figure J–9 is an alternative schematic for a series circuit. Solve for V_{out}.

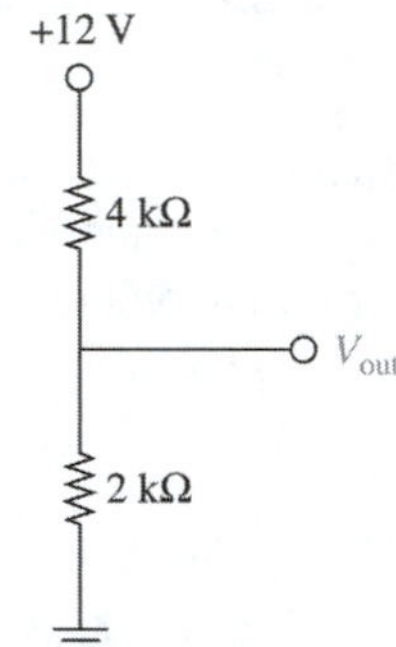

Figure J–9 Series circuit drawn without the battery symbol.

Solution:

$$V_{\text{out}} = 12\text{ V} \times \frac{2\text{ k}\Omega}{2\text{ k}\Omega + 4\text{ k}\Omega}$$
$$= 4\text{ V}$$

EXAMPLE J–10

A relay's contacts or a transistor's collector-emitter can be used to create opens and shorts. Figure J–10(a) uses a relay to short one resistor in a series circuit. Sketch the waveform at V_{out} in Figure J–10(a).

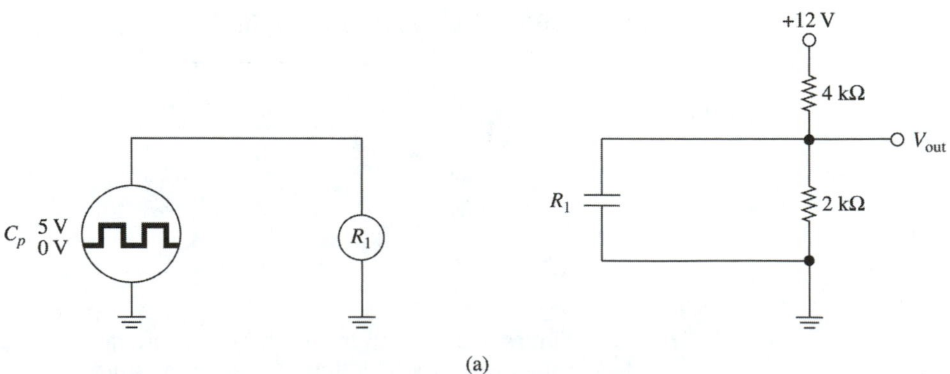

Figure J–10 Using a relay to intermittently short the 2kΩ resistor.

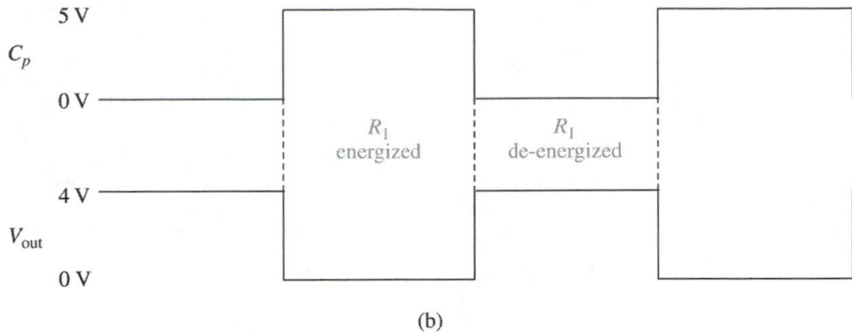

(b)

Figure J–10 *(continued)* Using a relay to intermittently short the 2kΩ resistor.

Solution: When the R_1 coil energizes, the R_1 contacts close, shorting the 2-kΩ resistor and making $V_{out} = 0$ V. When the coil is deenergized, the contacts are open, and V_{out} is found using the voltage-divider equation.

$$V_{out} = 12\text{ V} \times \frac{2\text{ k}\Omega}{2\text{ k}\Omega + 4\text{ k}\Omega}$$
$$= 4\text{ V}$$

The clock oscillator (C_p) and V_{out} waveforms are given in Figure J–10(b).

PROBLEMS

J–1. What value of voltage will cause 6 mA to flow in Figure J–1 if $R = 2$ kΩ (3 V, 0.333 V, or 12 V)?

J–2. To increase the current in Figure J–1, the resistor value should be ________ (increased, decreased).

J–3. In a series voltage-divider circuit like Figure J–3, the larger resistor will have the larger voltage across it. True or false?

J–4. In Figure J–3, if R_1 is changed to 8 MΩ and R_2 is changed to 2 Ω, V_{R2} will be close to ________ (0 V, 10 V).

J–5. If the supply voltage in Figure J–4 is increased to 18 V, V_{out} becomes ________ (6 V, 12 V).

J–6. A short circuit causes current to stop flowing in the part of the circuit not being shorted. True or false?

J–7. The current leaving the battery in Figure J–5 ________ (increases, decreases) if the short circuit is removed.

J–8. The short circuit in Figure J–6 causes V_{out} to become 12 V because the current through the 2-kΩ resistor becomes 0 A. True or false?

J–9. In Figure J–7, the voltage across the 2-kΩ resistor is the same as that across the 4-kΩ resistor. True or false?

J–10. If the 4-kΩ resistor in Figure J–8 is doubled, V_{out} will ________ (increase, decrease, remain the same).

J–11. Refer to Figure J–1 to solve for the unknown quantities in the following table. (*Example*: For part (A), calculate the resistance if $V = 12$ V and $I = 4$ A.)

	Voltage	*Current*	*Resistance*
A	12 V	4 A	
B	8 V	2 A	
C	100 V		20 Ω
D	6 V		2 Ω
E		6 A	2 Ω
F		0.5 A	5 Ω

J–12. Repeat problem 11 for the following table.

	Voltage	*Current*	*Resistance*
A	12 mV	3 μA	
B	6 V	2 mA	
C	100 mV		20 kΩ
D	8 V		2 MΩ
E		6 mA	3 kΩ
F		0.5 A	5 kΩ

J–13. Refer to Figure J–4 to solve for the unknown quantities in the following table.

	V_S	R_1	R_2	V_{out}
A	18 V	6 kΩ	3 kΩ	
B	18 V	3 kΩ	6 kΩ	
C	12 V	20 kΩ	20 kΩ	
D	6 V	1 kΩ	100 kΩ	

J–14. Refer to the original Figure J–4 to solve for V_{out} given the following open- and short-circuit conditions.

		V_{out}
A	Short R_1	
B	Short R_2	
C	Open R_1	
D	Open R_2	

Programmable Logic Devices: Altera and Xilinx CPLDs and FPGAs

OBJECTIVES

Upon completion of this appendix, you should be able to:

- Explain the benefits of using PLDS.
- Describe the PLD design flow.
- Understand the differences between a PAL, PLA, SPLD, CPLD, and FPGA.
- Explain how a graphic editor and a VHDL text editor are used to define logic to a PLD.
- Interpret the output of a simulation file to describe logic operations.
- Interpret VHDL code for the basic logic gates.

INTRODUCTION

As you can imagine, stockpiling hundreds of different logic ICs to meet all the possible requirements of complex digital circuitry became very difficult. Besides having all of the possible logic on hand, another problem was the excessive amount of area on a printed-circuit board that was consumed by requiring a different IC for each different logic function. In many cases, only one or two gates on a quad or hex chip were used.

Then came "programmable logic"—the idea that implementing all logic designs using 7400- or 4000-series ICs is no longer needed. Instead, a company will purchase several user-configurable ICs that will be customized (i.e., programmed) to perform the specific logic operation that is required. These ICs are called programmable logic devices (PLDs).

K–1 PLD DESIGN FLOW

Samples of two PLDs are shown in Figure K–1. They contain thousands of the basic logic gates plus advanced sequential logic functions inside a single package. This internal digital logic, however, is not yet configured to perform any particular function. One way to configure it is for the designer to first use PLD computer software to draw the logic that he or she needs implemented. This is called **CAD** (computer-aided design). The PLD software then performs a process called **schematic capture,** which reads the graphic drawing of the logic and converts it to a binary file that accurately describes the logic to be implemented. This binary file is then used as an input to a programming process that electronically alters the internal PLD connections to make it function specifically as required. Hundreds, or even thousands, of digital logic ICs will be replaced by a single PLD.

(a)

(b)

Figure K–1 Sample PLDs: (a) Altera EPM7128S; (b) Xilinx XC95108.

Another way to define the logic to be programmed into the PLD is to use a high-level language called Hardware Description Language (HDL). A specific form of HDL used by several manufacturers is called **VHDL,** which stands for VHSIC Hardware Description Language (where VHSIC stands for Very High-Speed Integrated Circuit). In this case, the inputs, outputs, and logic processes are defined using a programming language that looks a lot like C++. This method is somewhat more difficult to learn, but depending on the logic, it can be more a powerful—and simpler—tool with which to define complex or repetitive logic.

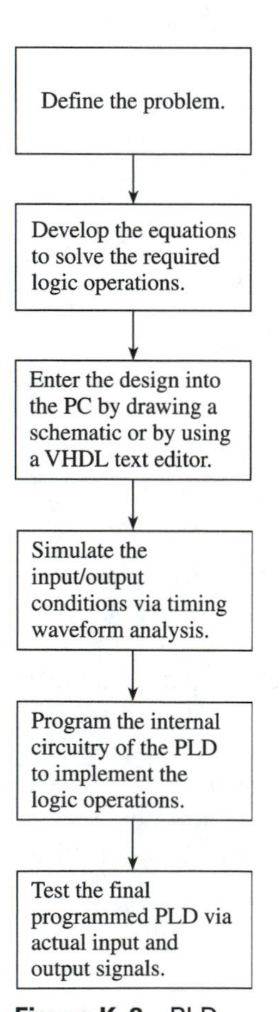

Figure K–2 PLD product design flow.

Figure K–2 illustrates the design flow. First we need to define the digital logic problem that we want to solve. Once we have a good understanding of the problem, we can develop the equations to use in solving the logic operation that we want the circuit to perform.

Once we have completed that work on paper, we will enter the design into a personal computer (PC) by drawing a schematic diagram using the CAD tools provided with the PLD software. In some cases, the design will instead be entered using the VHDL text editor provided. After the PC has analyzed the design, it will allow us to perform a simulation on the actual circuit to be implemented. To do this, we specify the input levels to our circuit, and we observe the resultant output waveforms on the PC screen using the waveform analysis tool provided.

If the computer simulation shows that our circuit works correctly, we can program the logic into a PLD chip that is connected by a cable to the back of our PC. The final step would be to connect actual inputs and outputs to the chip to check its performance in a real circuit.

To illustrate the power of a PLD, let's consider the logic circuit required to implement $X = \overline{A}B + \overline{B + C}$. Figure K–3 shows the logic diagram. As shown, we would need four different ICs to solve this equation. Wires are shown connecting one gate of each IC to one gate of the next IC until the logic requirements are met.

To solve this same logic using a PLD, we would draw the schematic or use HDL to define the logic, then program that into a PLD. One possible PLD that could be used to implement this logic is the Altera EPM7128S (see Figure K–4). After completing the steps listed in Figure K–2, the internal circuitry of the PLD is configured (in this case) to input

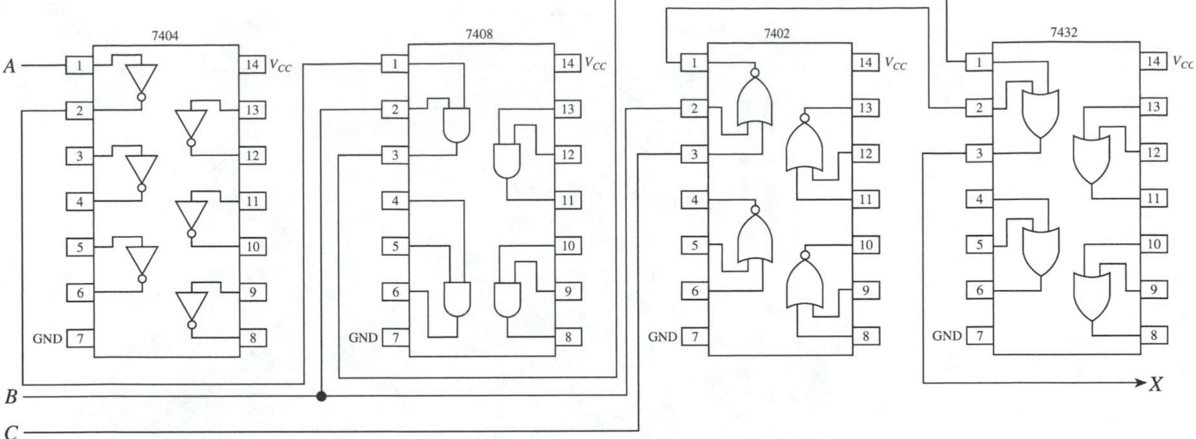

Figure K–3 Implementing the equation $X = \overline{A}B + \overline{B + C}$ using 7400-series logic ICs.

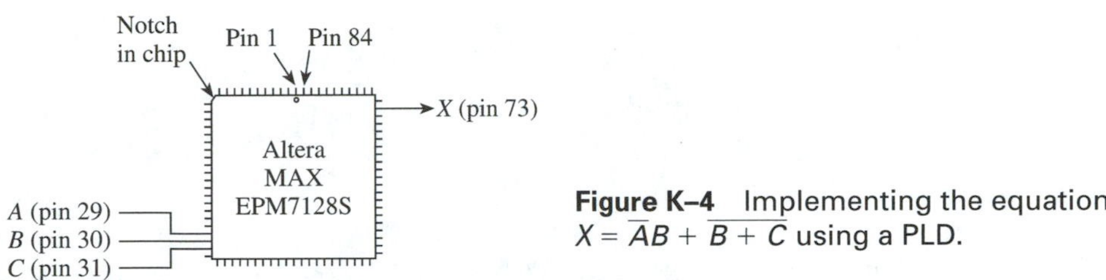

Figure K–4 Implementing the equation $X = \overline{A}B + \overline{B + C}$ using a PLD.

A, B, and *C* at pins 29, 30, and 31 and output to *X* at pin 73. The PLD software selected which pins to use, and as you can see, only a small portion of the PLD is actually used for this circuit.

This particular PLD is an 84-pin IC in a plastic leaded chip carrier (PLCC) package having 21 pins on a side. The notch signifies the upper left corner of the IC. Pin 1 is located in the middle of the upper row adjacent to a small indented circle; subsequent pin numbers are counted off counterclockwise from there. (A photograph of this particular chip is shown in Figure K–1.)

As you may suspect, the price of a PLD is higher than a single 7400-series IC, but we've only used a small fraction of the PLD's capacity. We could enter and program hundreds of additional logic equations into the same PLD. The only practical limitation is the number of input and output pins that are available. Many PLDs are erasable and reprogrammable, allowing us to test many versions of our designs without ever changing ICs or the physical wiring of the gates.

We will learn design entry and waveform simulation in this chapter, and we will continue to explore PLD examples and problems throughout the remainder of this text. The student versions of the software to perform graphic and VHDL entry can be downloaded from the manufacturers' web sites.

Two of the leading PLD manufacturers are Altera Corporation (www.altera.com) and Xilinx Corporation (www.xilinx.com). Throughout the remainder of this appendix, you will have the option to implement and simulate your digital logic using either manufacturers' software package. Also, for a little more than $100, PLD programmer boards can be purchased that physically connect to a PC. This allows you to repeatedly program an actual PLD on the board and then bench test it with your actual inputs and outputs (see Figure K–5).

(a)

(b)

Figure K–5 PLD programmer boards: (a) Xilinx XS95 (by XESS corporation); (b) Altera UP-1.

K–2 PLD ARCHITECTURE

Basically, there are three types of PLDs: simple programmable logic devices (SPLDs), complex programmable logic devices (CPLDs), and field-programmable gate arrays (FPGAs).

The SPLD

The **SPLD** is the most basic and least expensive form of programmable logic. It contains several configurable logic gates, programmable interconnection points, and memory flip-flops. (Flip-flops are covered in Chapter 5.) To keep logic diagrams easy to read, a one-line convention has been adopted, as shown in Figure K–6, which is just a small part of an SPLD, showing two inputs and four outputs. Typical SPLDs like the PAL in Figure K–7 or the PLA in Figure K–8 have more than 12 inputs (plus their complements) and more than 10 outputs. As you can see, the A input is split into two different lines: A, and its complement $\overline{A}$. (The triangle symbol is a special type of inverter having two outputs; a true, and a complement.) The same goes for the B input and any others that are on the SPLD. The W, X, Y, and Z AND gates are programmable to have any of those four lines (A, $\overline{A}$, B, $\overline{B}$) as inputs.

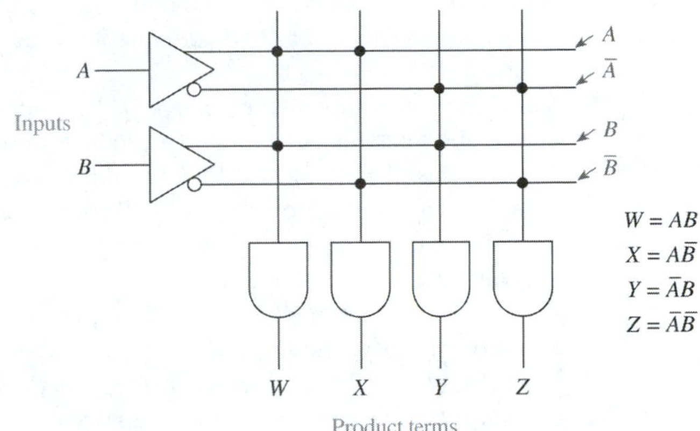

Figure K–6 One-line convention for PLDs.

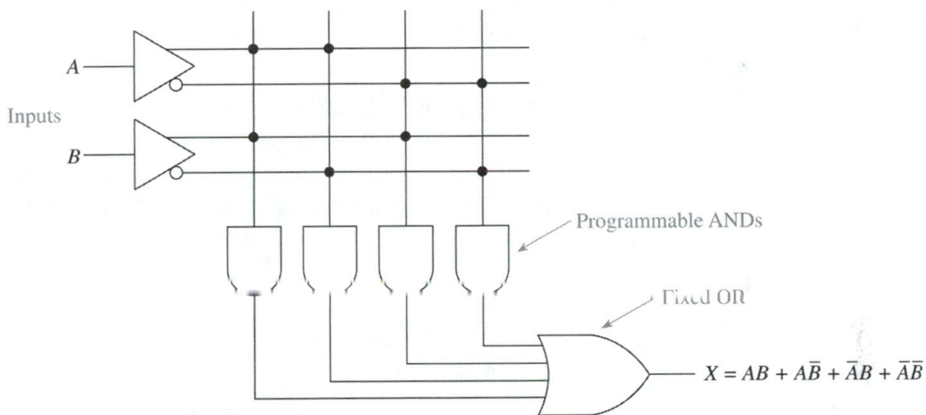

Figure K–7 PAL architecture of an SPLD.

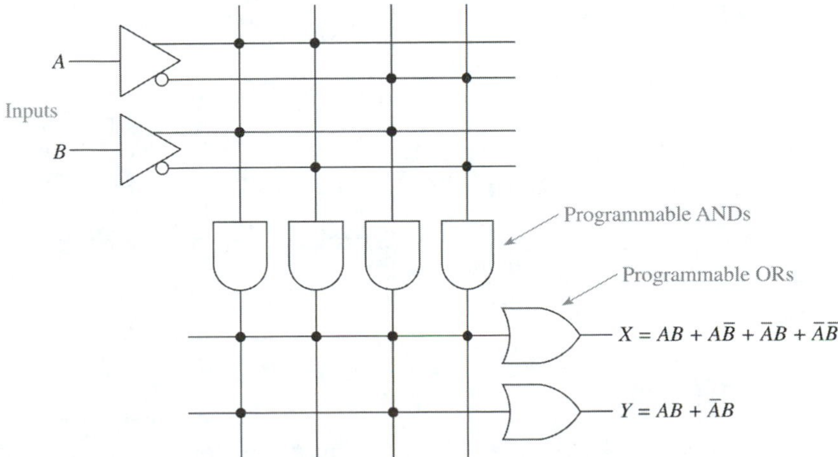

Figure K–8 PLA architecture of an SPLD.

The internal SPLD interconnect points are either made or not made by the PLD programming software. In Figure K–6, the inputs to the W AND gate are connected to A and B. (The connections are shown by a dot.) The inputs to the X AND gate are connected to A and $\overline{B}$, and so on. The outputs of these AND gates are called the **product terms,** because X is the Boolean product of A and B and Y is the Boolean product of A and B.

The product terms in Figure K–6 are not very useful by themselves. The circuit is made more effective by adding an OR gate to the structure, as shown in Figure K–7. This new configuration is the foundation for a programmable logic array (PAL)–type SPLD. As Figure K–7 shows, by OR-ing the four product terms together, we now have the Boolean sum of the four product terms, simply called the **Sum-of-Products** (SOP). The SOP is the most common form of Boolean equation used to represent digital logic. (For more on SOPs, see Section 3–4.)

The programmable logic array (PLA) goes one step further by providing *programmable* OR gates for combining the product terms. Figure K–8 shows a small portion of a PLA. In this illustration, the PLA provides two SOP equations. The inputs to the first OR gate are programmed to connect to all four product terms ($X = AB + A\overline{B} + \overline{A}B + \overline{A}\overline{B}$). The inputs to the second OR gate are programmed to connect to only the first and third product terms ($Y = AB + \overline{A}B$).

The final part of an SPLD is the flip-flop memory section and data-steering circuitry. Flip-flop memory circuitry is used in a type of digital circuitry called *sequential logic.* This type of logic is a form of digital memory that changes states based on previous logic conditions and specific logic control inputs. (Sequential logic is covered in detail in chapter 6.) The data-steering circuitry takes care of input and control signal interconnections and logic output destinations.

The CPLD

The **CPLD** is made by combining several PAL-type SPLDs into a single IC package, as shown in Figure K–9. Each PAL-type structure is called a *macrocell.* Each macrocell has several I/O connection points, which go to the chips' external leads. The macrocells are all connected to control signals and to each other via the programmable interconnect matrix shown in the center of the structure.

Examples of CPLDs are the Altera MAX 7000S series and the Xilinx XC9500 series. (Photographs are shown back in Figure K–1, and data sheets are provided at the

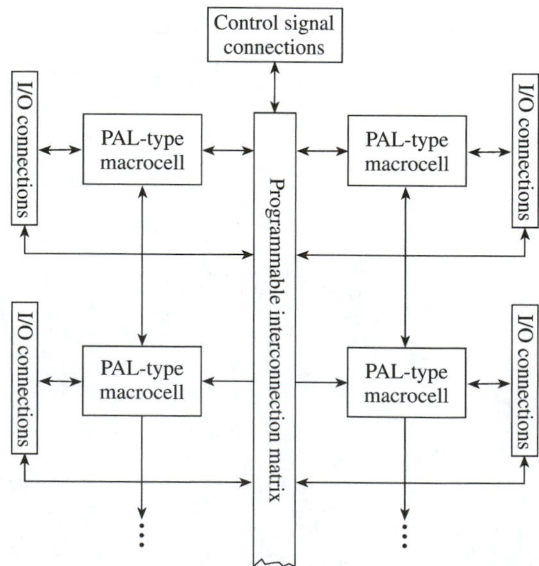

Figure K–9 Internal structure of a CPLD.

manufacturers' websites.) These CPLDs perform the functions of thousands of individual logic gates. Both also feature a **nonvolatile** memory characteristic, meaning that when power is removed from the chip, they will remember their programmed logic and interconnections. (This type of memory is called EEPROM or Flash memory and is covered in chapter 8.) These ICs can be repeatedly programmed to implement new designs or correct faulty ones, thus eliminating the need to rewire circuitry or buy new logic.

The FPGA

The **FPGA** differs from the CPLD in that, instead of solving the logic design by interconnecting logic gates, it uses a **look-up table** method to resolve the particular logic requirement. This technique allows PLD manufacturers to form a more streamlined design, with a much denser concentration of logic functions.

A look-up table is simply a truth table that lists all the possible input combinations with their desired output response. Figure K–10 shows how an FPGA would use a look-up table to output the correct logic level for the equation $X = AB + A\overline{B}$. As you can see in the logic diagram, X will be HIGH if A and B are both HIGH, or if A is HIGH while B is LOW. Instead of forming the logic circuit, an FPGA builds the truth table shown in Figure K–10(b). It then uses the logic levels received at its A and B input pins to pinpoint the location in the truth table to output a logic level at X. For example, Figure K–10(c) shows the result if 1 is input at A and 0 is input at B. The FPGA will point to the result of the third look-up table entry and then output that level, which is a 1. Most look-up tables used in FPGAs utilize four inputs, resulting in 16 possible output results. Modern FPGAs have hundreds of look-up tables, configured with programmable interconnections feeding the I/O pins. This method of simply looking up 1's or 0's allows manufacturers to form a much denser PLD, with hundreds of times more equivalent logic gates than a CPLD contains.

Examples of FPGAs are the Xilinx XC4000 series and the Altera FLEX 10K series. Even though they generally hold a much higher amount of logic than CPLDs, their main drawback is that their memory uses SRAM technology, which is volatile. This means they lose their logic program once power is removed from the chip. (SRAMs are covered in chapter 8.) Thus, their logic program must be loaded into the chip each time the system is powered up.

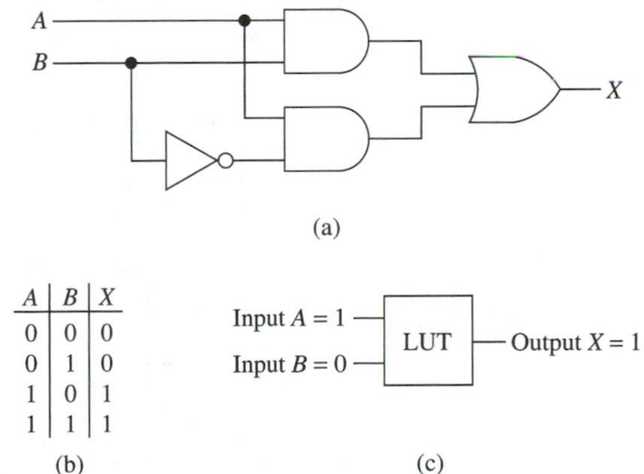

A	B	X
0	0	0
0	1	0
1	0	1
1	1	1

Figure K–10 $X = AB + A\overline{B}$: (a) logic circuit representation; (b) FPGA look-up table method of resolving outputs; (c) look-up table results for $A = 1$; $B = 0$.

K–3 USING PLDs TO SOLVE BASIC LOGIC DESIGNS*

So, the next obvious question is "How do I program a PLD?" We will use the software provided from the manufacturer's web site to design and simulate solutions modeled after Altera and Xilinx CPLDs. Then, if your laboratory has the PLD programmer boards like those shown in Figure K–5, you can test the actual operation of the CPLD with switches and lights. Even without the boards, however, the design and simulation software is a great learning tool for digital logic.

Figure K–11 shows the flow of operations required to design, simulate, and program a CPLD. Several methods are actually available to perform the design entry, but we will address the two most common: schematic, and VHDL. The **schematic editor** enables you to connect predefined logic symbols (AND, NAND, OR, etc.) together with inputs and outputs to define the logic operation that you need to implement. The **VHDL editor** is a text editor that helps you to define the logic in a programming language environment. In a text form, you specify the inputs, outputs, and logic equations that you need to implement.

The next step performed by the software is to compile and implement the design. A **compiler** is a language and symbol translation program that interprets VHDL statements and logic symbols, then translates them into a binary file that can be used later by the simulator and the programmer. The compiler uses several symbol and VHDL library files to obtain the information needed to define the logic entered during the design entry stage. Report files are then generated that describe such things as I/O pin assignments, internal CPLD signal routing, and error messages.

The **waveform simulator** provides a means to check the logic operation of your design. To use it, draw the input waveforms using the CAD tool provided, and the program will show the output response as if these inputs were applied to an actual CPLD. Finally, if

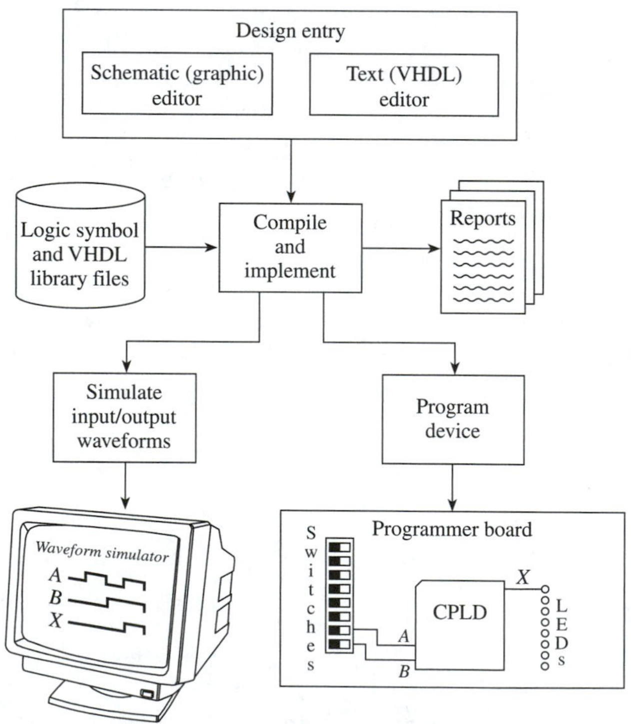

Figure K–11 CPLD design flow.

*Check the following web sites for PLD software and hardware:
www.altera.com (Max + Plus II software and UP-1 programmer board)
www.xilinx.com (Foundation software)
www.xess.com (XS95 and XStend programmer boards)
www.elexp.com [PLDT-1 (Xilinx) and PLDT-2 (Altera) programmer boards]

you have a CPLD programmer board and the waveform simulation was accurate, you can program the CPLD and test it with actual inputs and outputs.

Figure K–12 shows the computer screen displays produced by the CPLD software when implementing the design flow outlined in Figure K–11. These figures are the output screens of two different CPLD design software packages: Altera *MAX + PLUS II,* and Xilinx *Foundation.* Your screens may look slightly different from these depending on which version of the software you are using. These manufacturers usually ship new

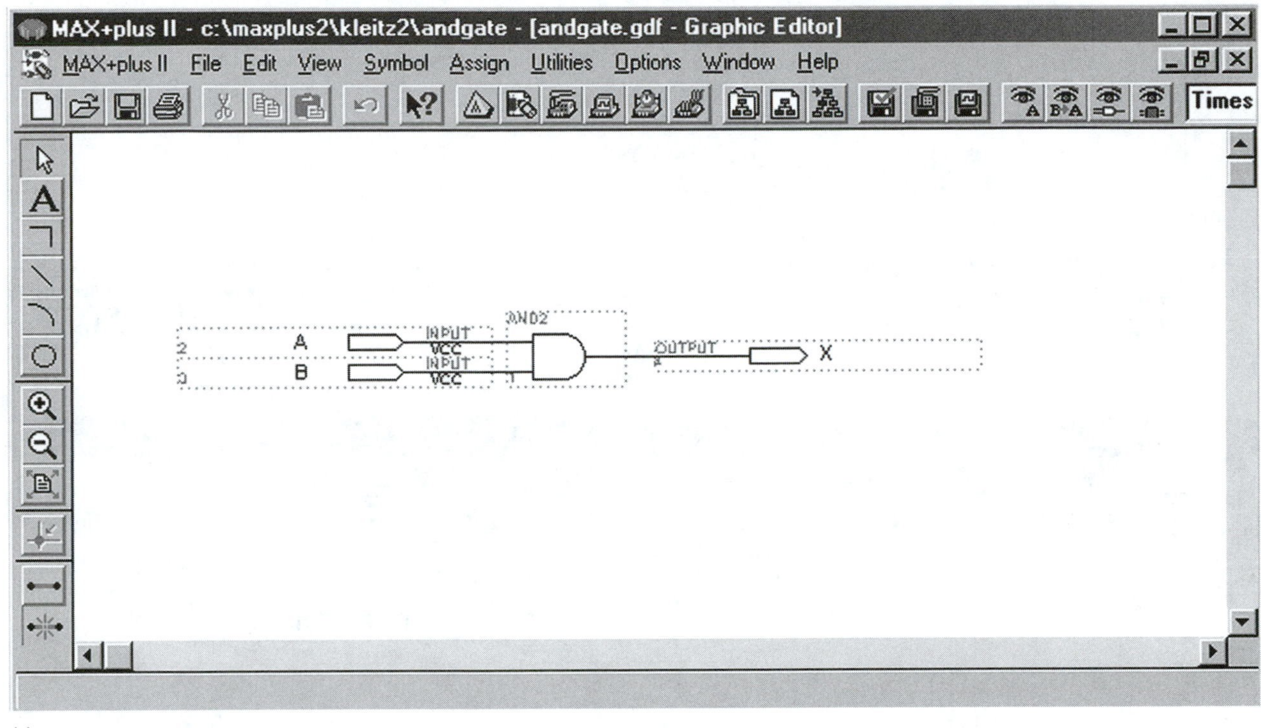

(a)

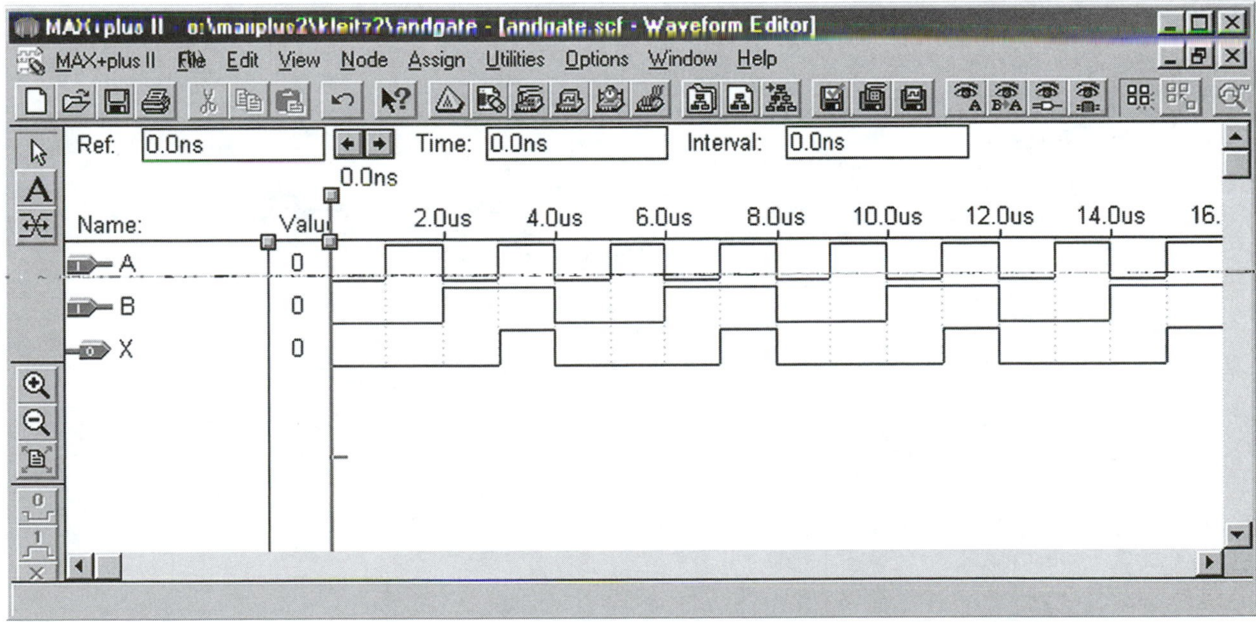

(b)

Figure K–12 Computer screen displays of CPLD software used to design a two-input AND gate: (a) Altera *MAX + PLUS II* graphic editor; (b) Altera *MAX + PLUS II* waveform editor.

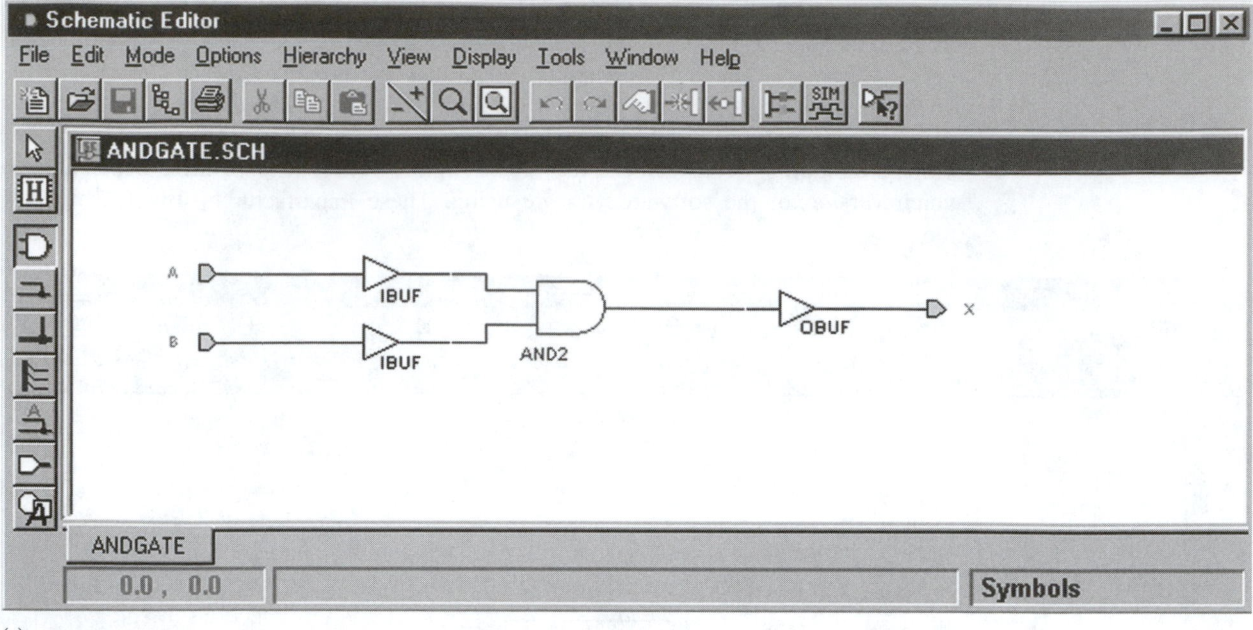

(c)

(d)

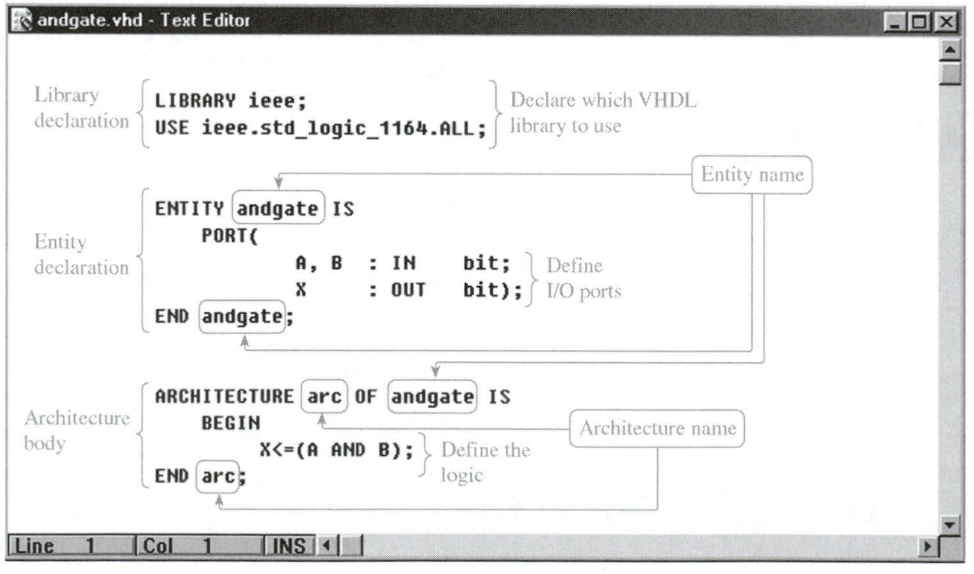


```
LIBRARY ieee;
USE ieee.std_logic_1164.ALL;

ENTITY andgate IS
    PORT(
            A, B  : IN    bit;
            X     : OUT   bit);
END andgate;

ARCHITECTURE arc OF andgate IS
    BEGIN
        X<=(A AND B);
END arc;
```

(e)

Figure K–12 *(Continued)* (c) Xilinx *Foundation* schematic editor; (d) Xilinx *Foundation* waveform viewer; (e) VHDL text editor.

versions several times each year. The updated versions are issued to address the higher-end functions of their newest products and won't affect the basic operations that we deal with in this text.

The design of an AND gate in Figures K–12(a) and K–12(b) was generated by the Altera software. The logic diagram was created using the drawing tools provided by the graphic editor. Figure K–12(b) shows a simulation of the *andgate* logic circuit. To create that output screen, inputs were drawn to test all possible combinations of *A* and *B*. When the simulator was run, those levels were simulated at the inputs to the *andgate* file, and the resultant output at X was drawn.

The Xilinx solution to the same problem is shown in Figures K–12(c) and K–12(d). One difference from the Altera software is that the triangular symbols (called **buffers**) are required to define those pins as inputs and outputs. Whatever signal level is input to the buffer will be output at the same logic level (unlike the inverter, which outputs the complement). The waveform viewer in Figure K–12(d) shows that the output at *X* only goes HIGH when both *A* and *B* are HIGH.

Besides schematic design entry, Figure K–12(e) also shows how to enter the design using a VHDL text editor. VHDL code can be used for either Altera or Xilinx products. Designers can use VHDL, schematic entry, or both to complete their design. VHDL code provides the same capability to perform simulations and device programming as described earlier. The VHDL code in Figure K–12(e) is divided into three sections: **library declaration, entity declaration,** and **architecture body.**

As with most computer languages, the first statements of the program are used to declare the library source for resolving and translating the language codes within the body of the program. The IEEE standard library is used most often by the VHDL compiler to translate references to the inputs, outputs, and equations in the program.

The entity declaration defines the input and output ports to the CPLD. The architecture body defines the internal logic operations to be performed on those ports.

EXAMPLE K–1

Figure K–13 contains several computer screen displays generated by CPLD software. Determine the Boolean equation that is being implemented in each case.

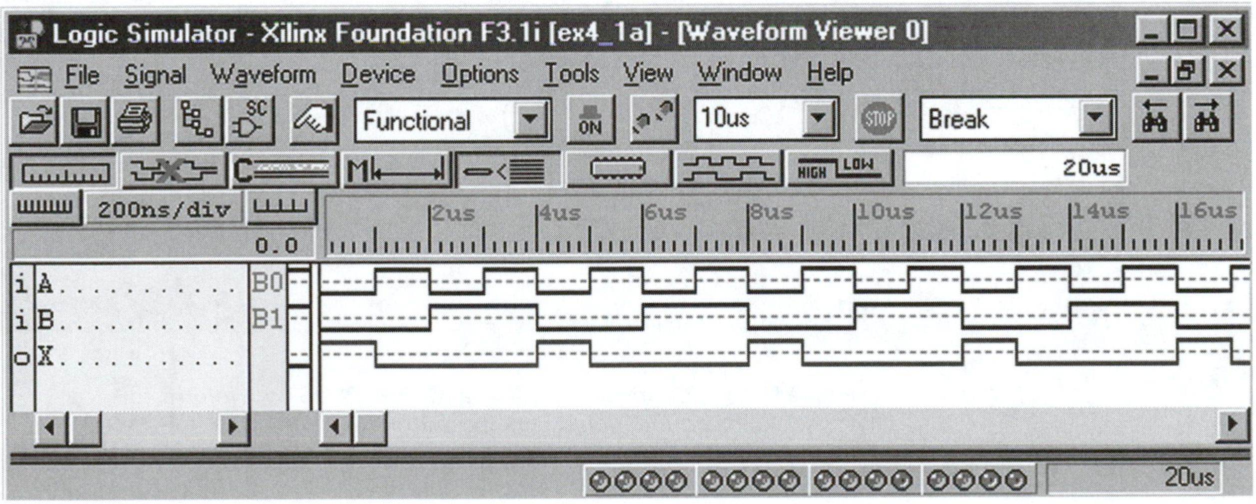

(a)

Figure K–13 Computer screens generated by CPLD software: (a) Xilinx *Foundation* waveform viewer.

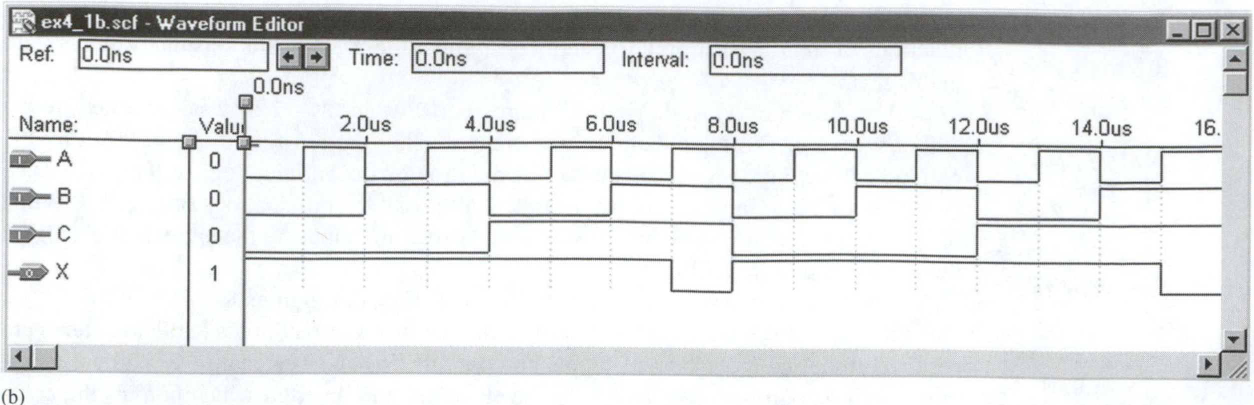

(b)

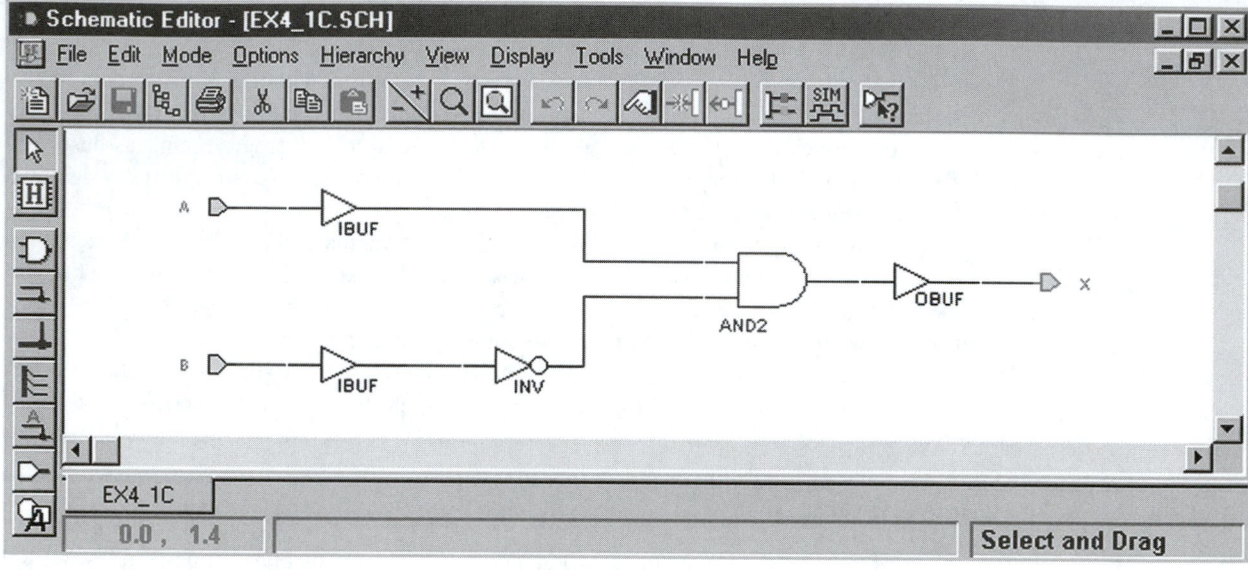

(c)

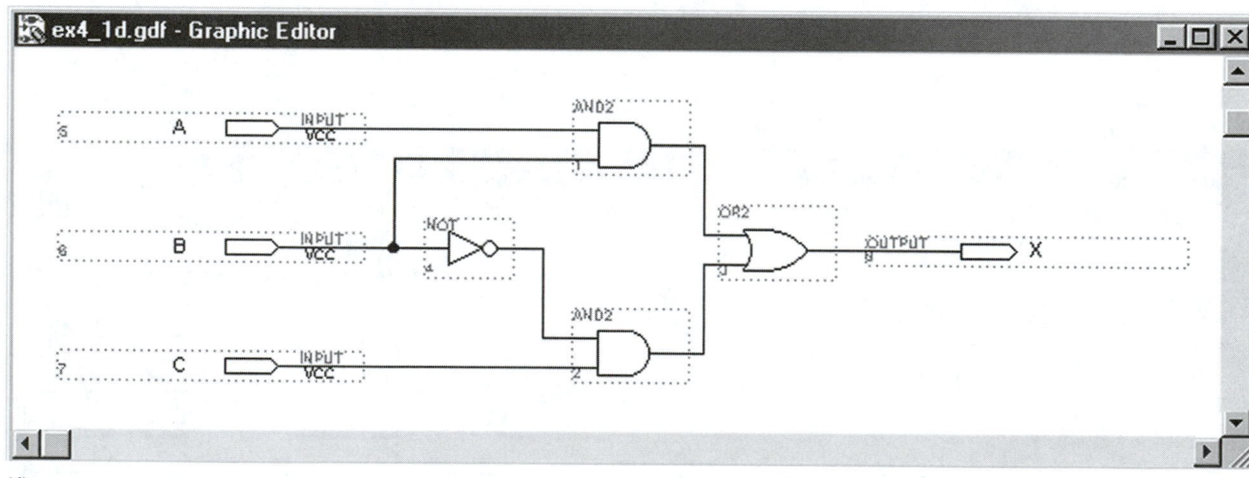

(d)

Figure K–13 *(Continued)* (b) Altera *MAX + PLUS II* waveform editor; (c) Xilinx *Foundation* schematic editor; (d) Altera *MAX + PLUS II* graphic editor.

```
LIBRARY ieee;
USE ieee.std_logic_1164.ALL;

ENTITY ex4_1e IS
    PORT(
            A, B, C    : IN    bit;
            X, Y       : OUT   bit);
END ex4_1e;

ARCHITECTURE arc OF ex4_1e IS
    BEGIN
        X<=(A OR (B AND NOT C));
        Y<= ((A AND B) OR NOT (B OR C));
END arc;
```

[Note: Use underscore in Entity name. (Hyphen is not allowed.)]

(e)

Figure K–13 *(Continued)* (e) VHDL text editor.

Solution:

(a) $X = \overline{A + B}$
(b) $X = \overline{ABC}$
(c) $X = A\overline{B}$
(d) $X = AB + \overline{B}C$
(e) $X = A + (B\overline{C})$
$Y = (AB) + \overline{B + C}$

EXAMPLE K–2

(*Note:* The textbook CD has a **CPLD** tutorial that will walk you through the design and simulation steps required to develop an Altera CPLD. After completing the tutorial, you will be able to duplicate the results of this example.)

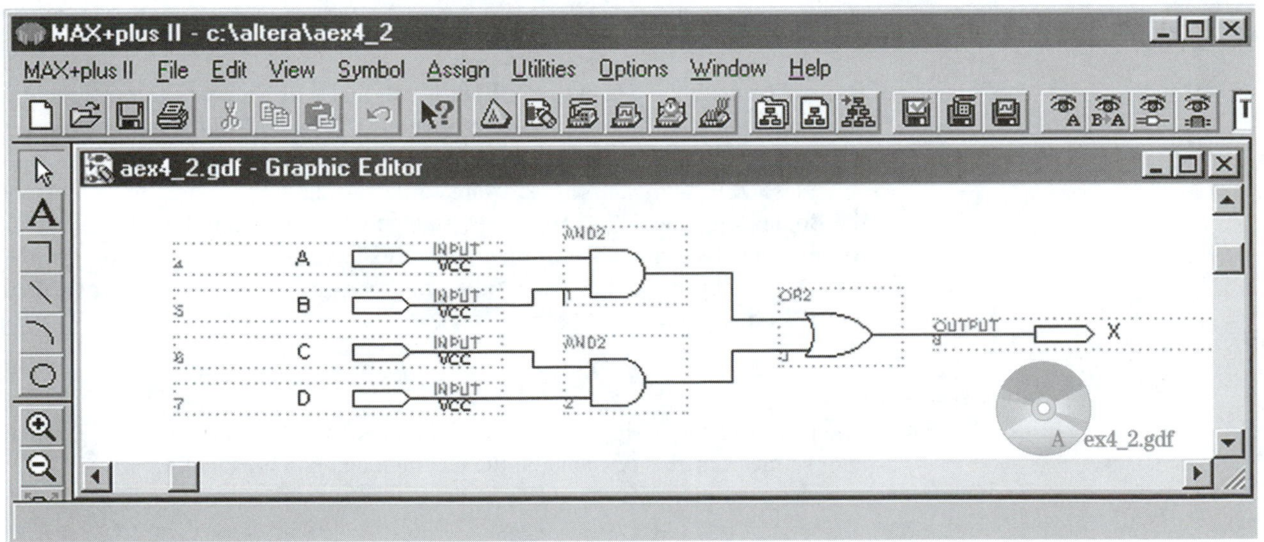

(a)

Figure K–14 Altera solution to the equation $X = AB + CD$: (a) graphic design file.

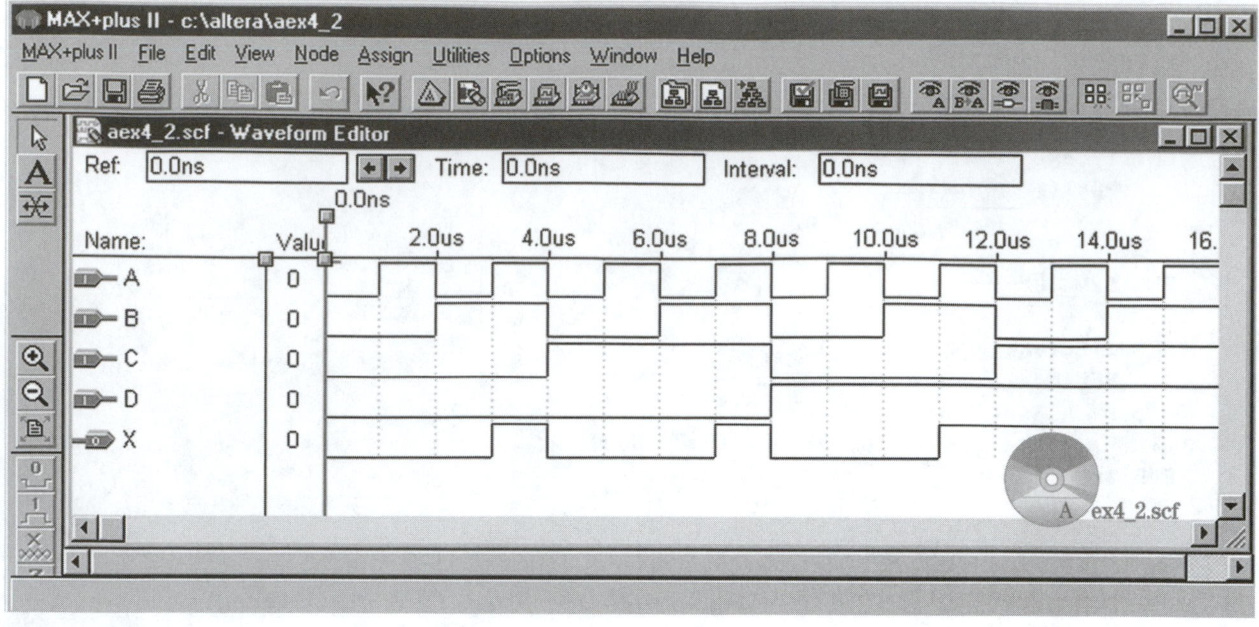

(b)

Figure K–14 *(Continued)* (b) simulator channel file.

Use Altera *MAX + PLUS II* software to design the CPLD logic to implement the Boolean equation $X = AB + CD$. Design the logic using the graphic editor to create a graphic design file (gdf) called **Aex4_2.gdf.** Test the operation of the CPLD logic by using the waveform editor to create a simulator channel file (scf) called **Aex4_2.scf.**

Solution:

The results of the design are shown in Figures K–14(a) and K–14(b). (The gdf and scf files can also be found on the CD included with this text.)

EXAMPLE K–3

(*Note:* The textbook CD has a **CPLD** tutorial that will walk you through the design and simulation steps required to develop a Xilinx CPLD. After completing the tutorial, you will be able to duplicate the results of this example.)

Use Xilinx *Foundation* software to design the CPLD logic to implement the Boolean equation $X = AB + CD$. Design the logic using the schematic editor to create a schematic (sch) file called **Xex4_3.sch.** Test the operation of the CPLD logic by using the logic simulator to create a test vector (tve) called **Xex4_3.tve.**

Solution:

The results of the design are shown in Figures K–15(a) and K–15(b). (The sch and tve files can also be found on the CD included with this text.)

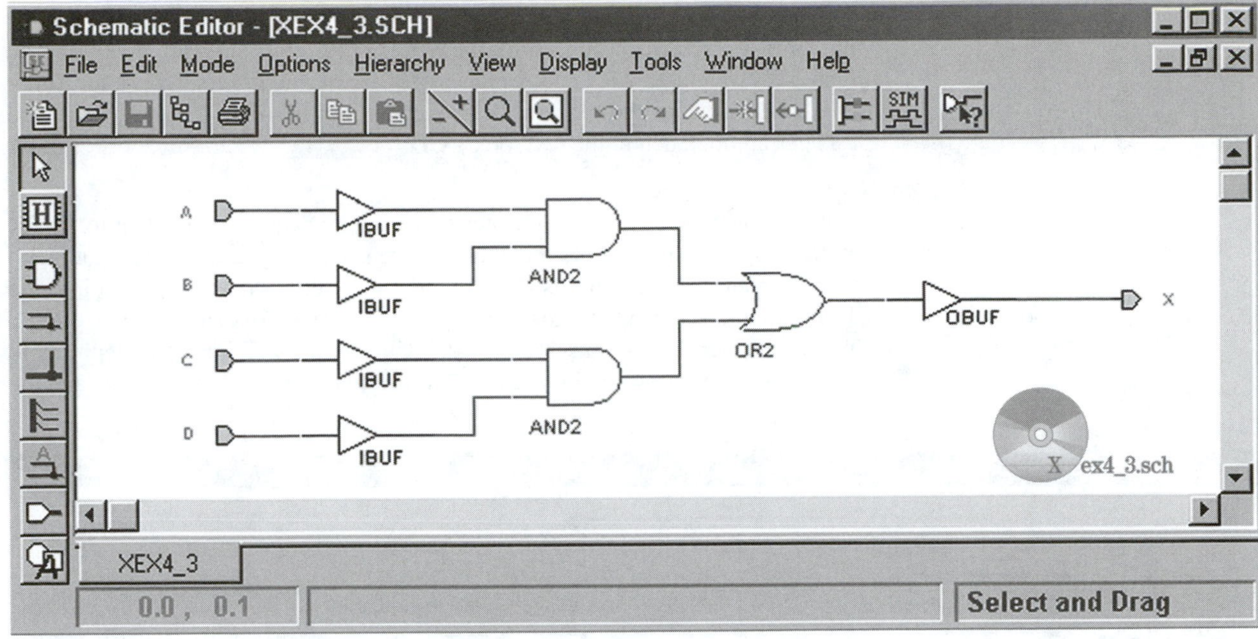

(a)

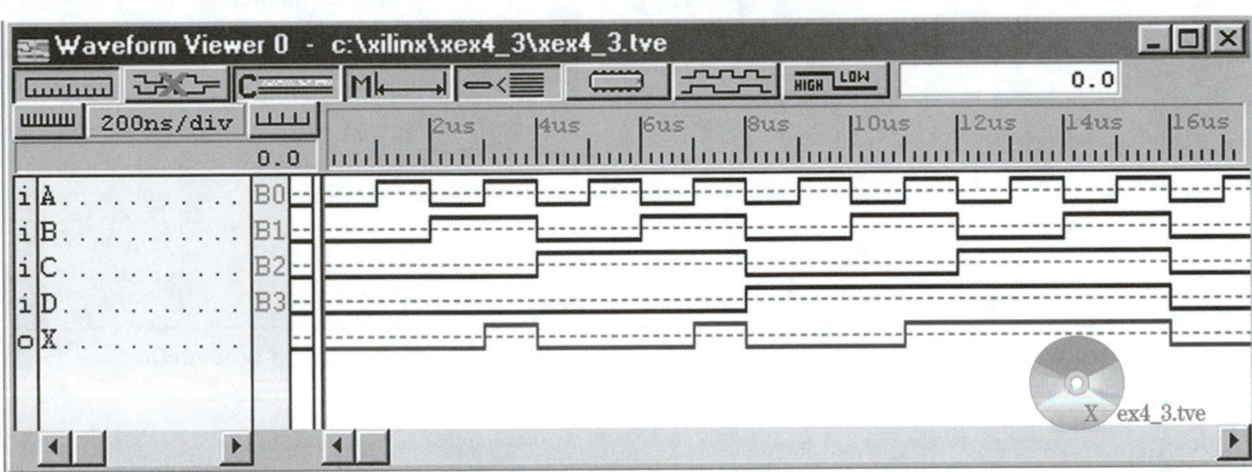

(b)

Figure K–15 Xilinx solution to the equation $X = AB + CD$: (a) schematic file; (b) simulator test vector file.

K–4 CPLD DESIGN APPLICATIONS

Many of the theories and procedures developed in this textbook can be proven by simulating the logic in the CPLD design environment. In the applications that follow, we will use CPLD software to design original circuits to simulate digital logic covered in the textbook. To duplicate these examples yourself, you must first complete the tutorials for Altera or Xilinx contained on the textbook CD.

CPLD APPLICATION K–1

Use the Altera *MAX + PLUS II* or the Xilinx *Foundation* software to prove the validity of the De Morgan's theorem circuits as shown in Figures 3–22 and 3–23. Draw the circuit using the graphic schematic editor, and demonstrate the results by using the waveform logic simulator.

Solution:

The computer screens produced by the *MAX + PLUS II* software are shown in Figures K–16 and K–17. (The results using the *Foundation* software are functionally identical to those shown.) The computer solutions for both software packages are included in the textbook's CD.

Explanation:

Basically, De Morgan's theorem states that a NAND gate is equivalent to an OR gate with inverted inputs, and that a NOR gate is equivalent to an AND gate with inverted inputs. These two sets of equivalent logic circuits were entered into the CPLD graphic schematic editor as shown in Figure K–16. The waveform logic simulator is then used to observe the outputs as all possible inputs are applied to *A, B* and *C, D*. As shown in Figure K–17, the output at *W* is the same as *X,* and the

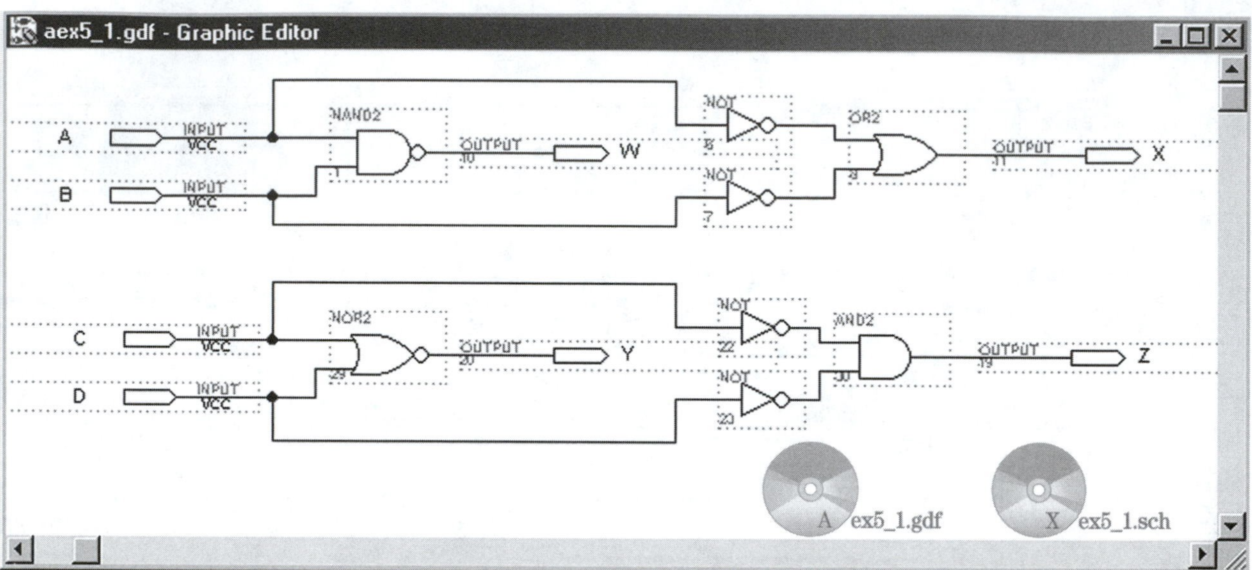

Figure K–16 Graphic schematic editor screen for proving De Morgan's theorem (Altera version).

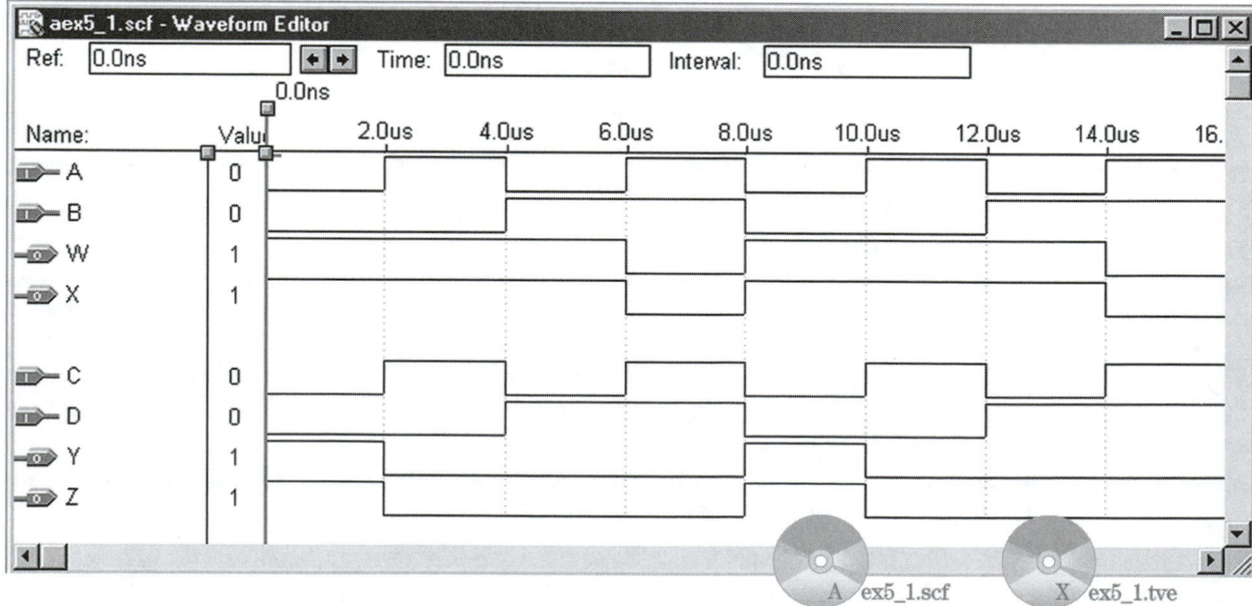

Figure K–17 Waveform logic simulator screen proving De Morgan's theorem (Altera version).

output at Y is the same as Z. This proves that the De Morgan equivalent produces the exact same output as the original.

CPLD APPLICATION K–2

Use CPLD software to simplify the following equation: $X = \overline{AB \cdot AB\,\overline{C + D}}$. Build the circuit using the Xilinx *Foundation* software, and produce a simulation file proving that the reduced function is $X = \overline{A} + \overline{B} + C + D$.

Solution:

The computer screens produced by the *Foundation* software are shown in Figures K–18 and K–19. (The results using the *MAX + PLUS II* software are functionally identical to those shown.) The computer solutions for both software packages are included in the textbook's CD.

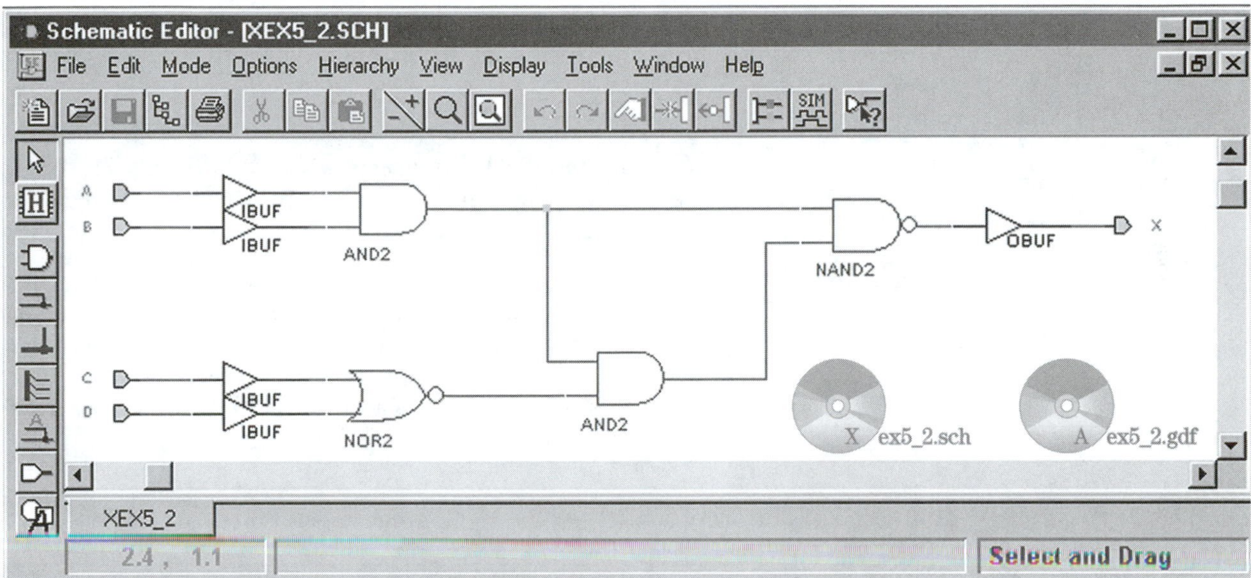

Figure K–18 Graphic schematic editor screen for CPLD Application K–2 (Xilinx version).

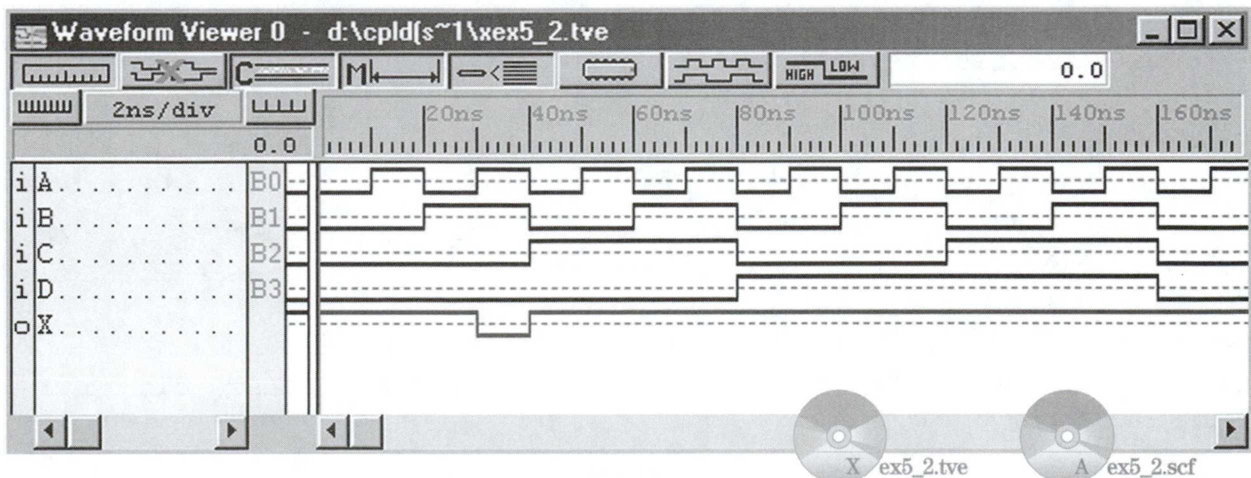

Figure K–19 Waveform logic simulator screen for CPLD Application K–2 (Xilinx version).

Explanation:

Figure K–18 shows the original circuit as it is entered in the graphic schematic editor. According to Application K–2, the circuit reduces to $X = \overline{A} + \overline{B} + C + D$. The simulator output in Figure K–19 should prove that reduction was correct. When you study the X output waveform, you can see that it is HIGH when C is HIGH or D is HIGH. It is also HIGH when A is LOW or B is LOW. This proves that the reduced equation is correct.

CPLD APPLICATION K–3

Instead of using a graphic entry, this time we will use a VHDL text editor to enter our logic design. The VHDL text editor provided with either CPLD development system will produce a file with the extension *.vhd,* which can then be used by the simulator to test the output of the logic design. In Example 3–13, we drew the output waveforms (see Figure 3–34) for the equation $X = AB + B\overline{C} + \overline{A}\,\overline{B}C$. Enter that same equation into either CPLD software package, and perform a simulation using the waveform logic simulator to compare results.

Solution:

The computer screen for the VHDL statements is shown in Figure K–20. This VHDL code is the same for either software package. The simulator output shown in Figure K–21 is from the Xilinx *Foundation* software.

Explanation:

The VHDL statements are similar to the code introduced in Figure K–12(e). The inputs and outputs are defined in the entity declaration, and the logic equation is defined in the architecture body. (Notice the use of an underscore in the entity name instead of a hyphen.) The output at X is identical to that which we determined in Example 3–13.

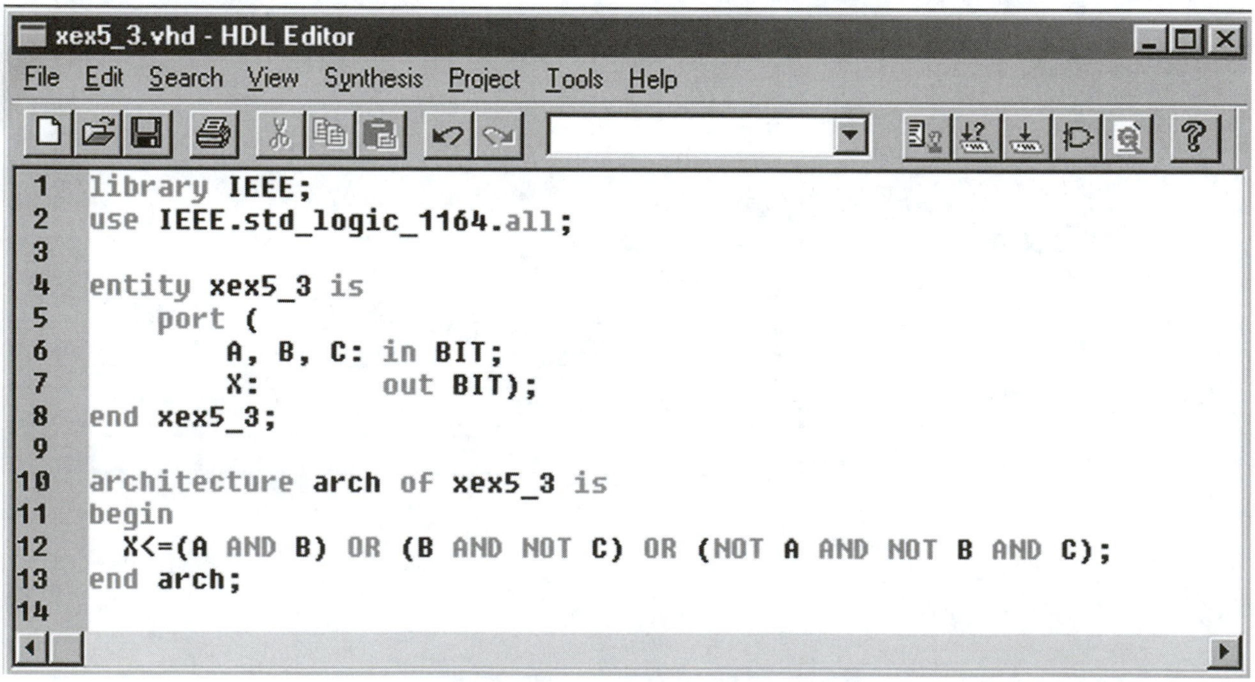

xex5_3.vhd - HDL Editor

File Edit Search View Synthesis Project Tools Help

```
library IEEE;
use IEEE.std_logic_1164.all;

entity xex5_3 is
    port (
        A, B, C: in BIT;
        X:       out BIT);
end xex5_3;

architecture arch of xex5_3 is
begin
  X<=(A AND B) OR (B AND NOT C) OR (NOT A AND NOT B AND C);
end arch;

```

Figure K–20 VHDL code for CPLD Application K–3.

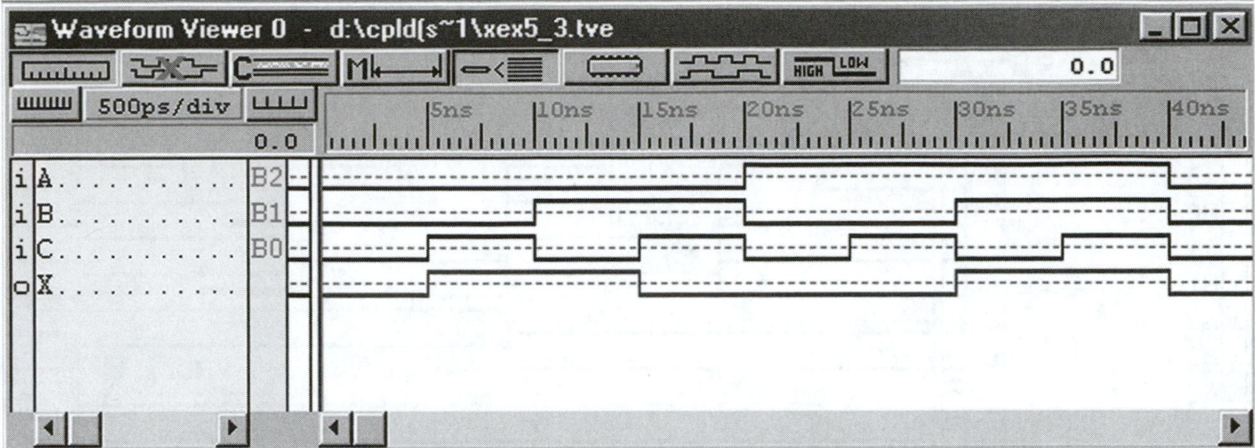

Figure K–21 Waveform logic simulator output for CPLD Application K–3 (Xilinx version).

CPLD APPLICATION K–4

Use the Altera *MAX + PLUS II* or the Xilinx *Foundation* software to prove the operation of the binary comparator circuit presented in Figure 4–1. Draw the circuit using the graphic schematic editor, and demonstrate the results by using the waveform logic simulator.

Solution:

The computer screens produced by the *Foundation* software are shown in Figures K–22 and K–23. (The results using the *MAX + PLUS II* software are functionally identical to those shown.) The computer solutions for both software packages are included in the textbook's CD.

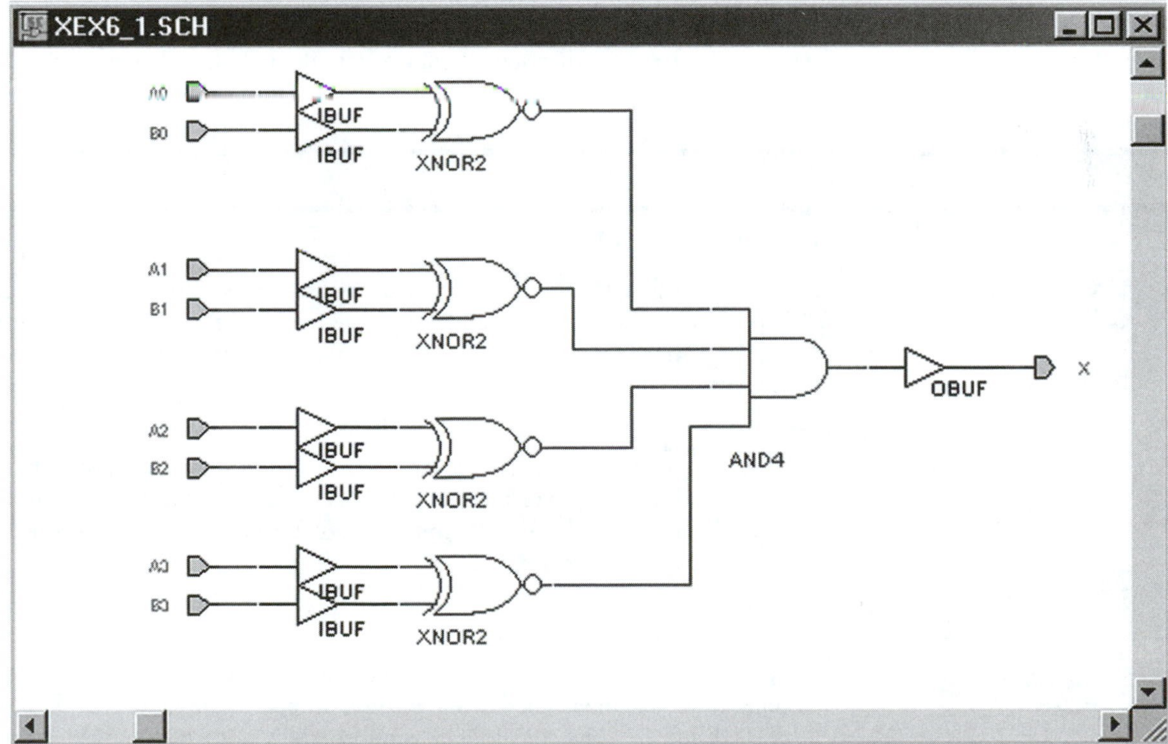

Figure K–22 Graphic schematic editor screen for the binary comparator circuit.

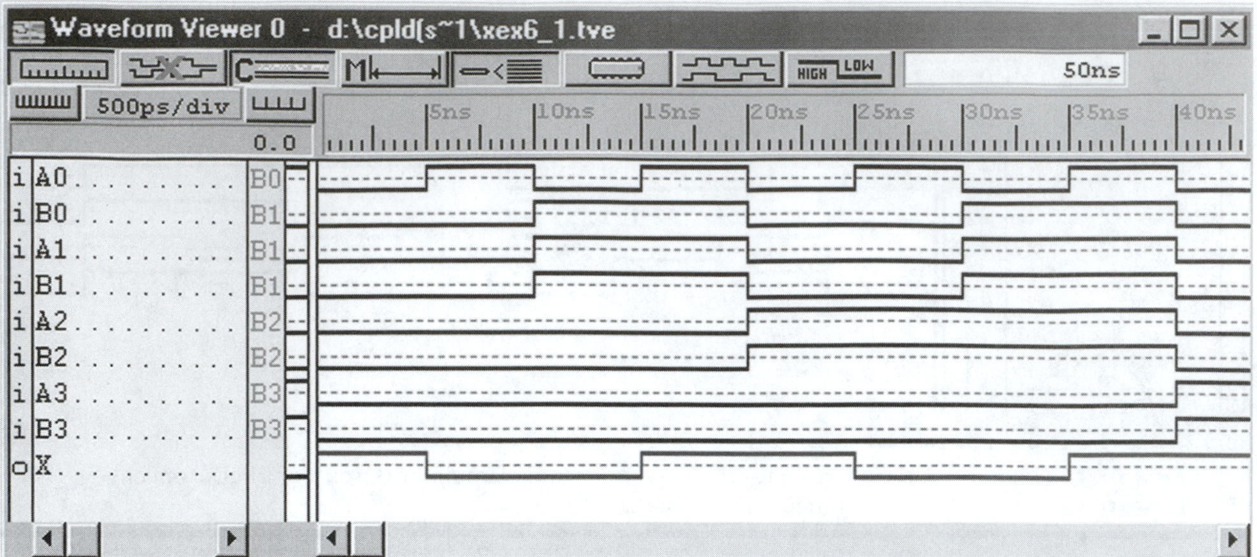

Figure K–23 Waveform logic simulator screen demonstrating the binary comparator.

Explanation:

Figure K–22 is similar to the binary comparator of Figure 4–1. As discussed previously, the output at *X* will be HIGH if all Ex-NORs are outputting 1's. This will occur when bits A_0–A_3 are equal to bits B_0–B_3. Figure K–23 shows the simulator screen for all 8 input bits and the resultant output at *X*. As you can see, the HIGH/LOW levels on waveform A_3 match those of waveform B_3. Also, A_2 matches B_2, and A_1 matches B_1. The A_0 and B_0 waveforms were purposely drawn so that they do not match. This way, we can observe the output for equality as well as inequality. From 0 to 5 ns, A_0 matches B_0, but from 5 to 15 ns, it does not. Also, notice that when they do not match, the output at *X* goes LOW, as we expected. Prove to yourself that the remainder of the waveform (15 to 40 ns) is valid too.

CPLD APPLICATION K–5

Use the Altera *MAX + PLUS II* or the Xilinx *Foundation* software to prove the operation of the full-adder circuit presented in Figure 3–53. Design the circuit using the VHDL text editor, and demonstrate the results by using the waveform logic simulator.

Solution:

The computer screens produced by the *Foundation* software are shown in Figures K–24 and K–25. (The results using the *MAX + PLUS II* software are functionally identical to those shown.) The computer solutions for both software packages are included in the textbook's CD.

Explanation:

Figure K–24 shows the VHDL statements required to define the full-adder. As you can see, three inputs and two outputs are defined in the entity declare. The architecture statements then define the conditions for the inputs that will make the SUM1 and C_{out} bits HIGH. The operators AND, OR, NOT, and XOR have equal precedence in VHDL. Therefore, several sets of parentheses were required to keep the order of operations correct in the C_{out} equation.

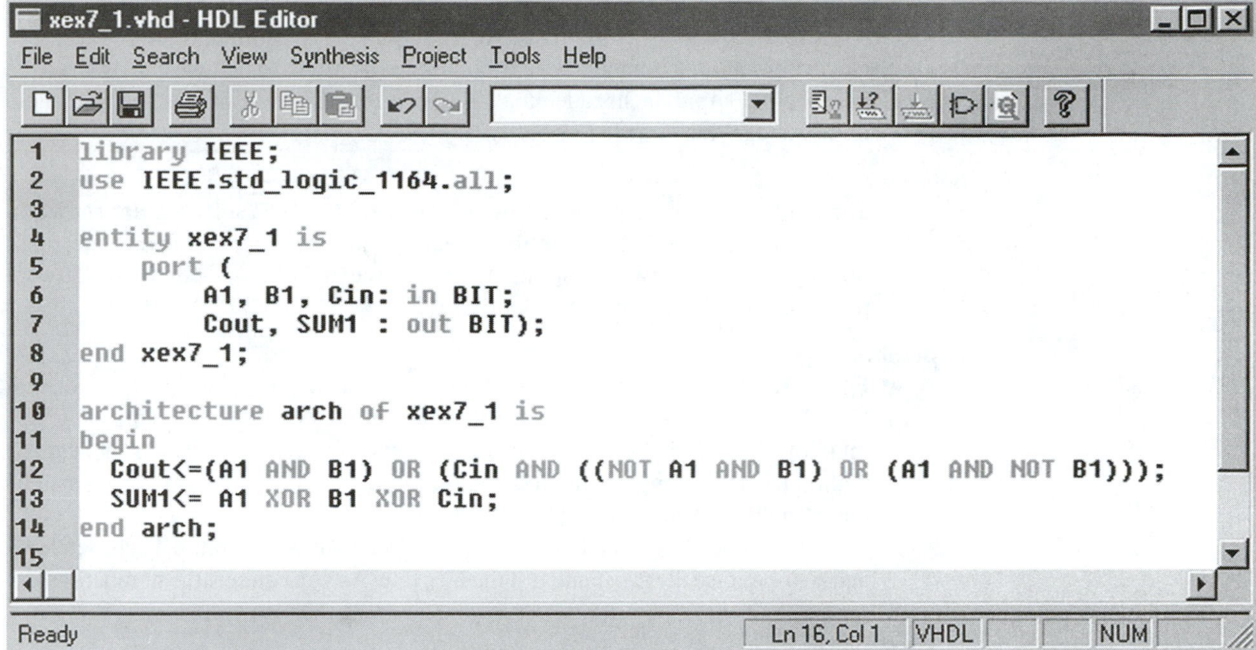

Figure K–24 VHDL text editor screen for the full-adder of Figure 3–53.

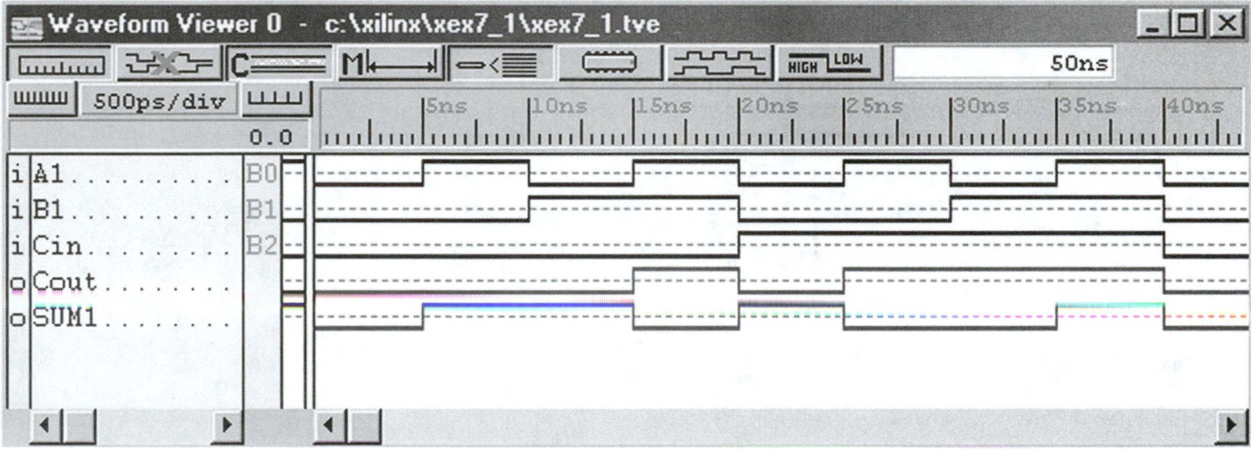

Figure K–25 Waveform logic simulator screen demonstrating the full-adder.

Figure K–25 shows the results of the simulation. As you may recall, the C_{out} bit should only be HIGH if two or more inputs are HIGH. The waveform shown for C_{out} fits these criteria. Also, the SUM1 bit should only be HIGH if there is an odd number of HIGH inputs. This also looks correct in the waveform.

CPLD APPLICATION K–6

Use the Altera *MAX + PLUS II* or the Xilinx *Foundation* software to test the operation of a 4-bit full-adder. These software development systems have a model of a 74283 adder IC in their library. This saves us the task of having to redesign the internal logic of that IC function. *MAX + PLUS II* has all of the most common 7400-series ICs modeled in a macrofunction (mf) library. They are available as a symbol when using the graphic editor. *Foundation* also has the 7400-series IC functions available in its symbols toolbox.

Find the 74283 4-bit full-adder symbol. Connect 4 bits to the *A*-inputs and 4 bits to the *B*-inputs. Observe the SUM outputs on the simulator as you test several different input numbers. Use 4-bit bus connections so that we can observe the inputs and output in hexadecimal. (Help is available from the software help screen under "bus drawing or buses.")

Solution:

The computer screens produced by the *MAX + PLUS II* software are shown in Figures K–26 and K–27. (The results using the *Foundation* software are functionally identical to those shown.) The computer solutions for both software packages are included in the textbook's CD.

Explanation:

Figure K–26 shows the 74283 4-bit adder symbol with its input/output connections. The *A*-number is input on the bus terminal labeled A[3..0]. This is the terminology used to indicate that this terminal has all four lines (A_3, A_2, A_1 and A_0) within it. The *B*-number is input on bus terminal B[3..0]. The four sum bits will be output on bus terminal SUM[3..0].

Figure K–27 shows the variety of numbers that were arbitrarily chosen for *A* and *B* to exercise the adder and test the C_{out} line. As you can see, the numbers are all

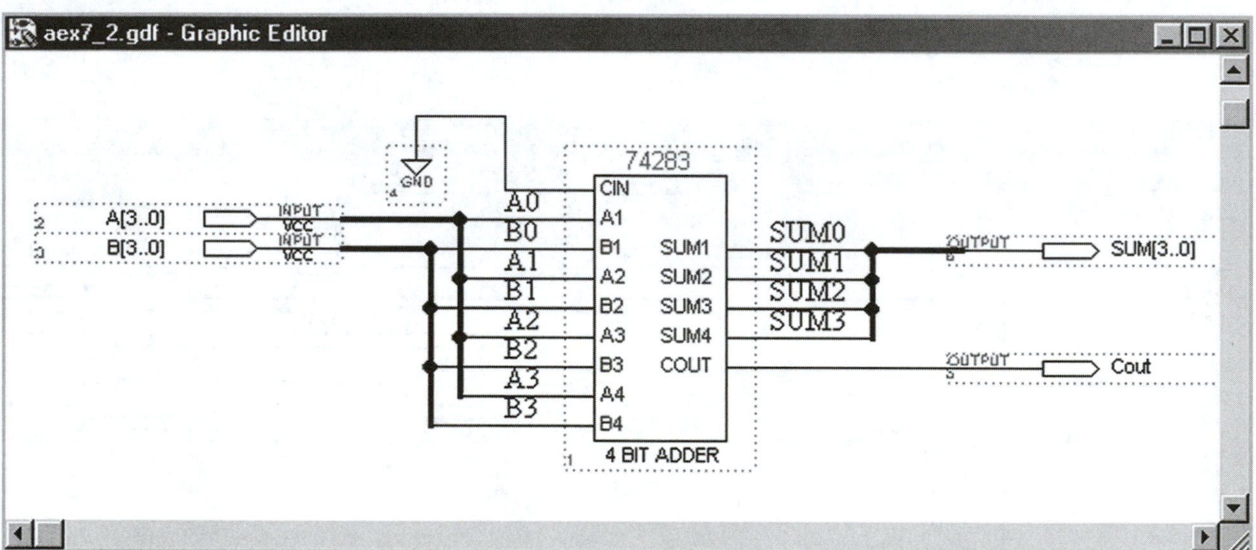

Figure K–26 Graphic schematic editor screen for the 4-bit adder.

Figure K–27 Waveform logic simulator screen demonstrating additions by the full-adder.

listed in hex. For example, in the first addition, instead of adding 0011 + 0110 = 1001, the simulation shows 3 + 6 = 9. The third addition, 6 + 6, would equal 12, but 12 in hex is C. The fifth, sixth, and seventh additions cause a carry out because they are greater than 16.

CPLD APPLICATION K–7

Use the Altera *MAX + PLUS II* or the Xilinx *Foundation* VHDL text editor to design a 3-line-to-8-line octal decoder similar to that shown in Figure 4–5(b) without the enables. Demonstrate the operation of the design by observing the outputs of the decoder on the waveform logic simulator as the inputs vary through all possible combinations.

Solution:

The computer screens produced by the *Foundation* software are shown in Figures K–28 and K–29. (The results using the *MAX + PLUS II* software are functionally identical to those shown.) The computer solutions for both software packages are included in the textbook's CD.

Explanation:

Figure K–28 shows the VHDL statements required to define the octal decoder. As you can see, three inputs and eight outputs are defined in the entity declare. The architecture statements then define the conditions for the inputs that will make

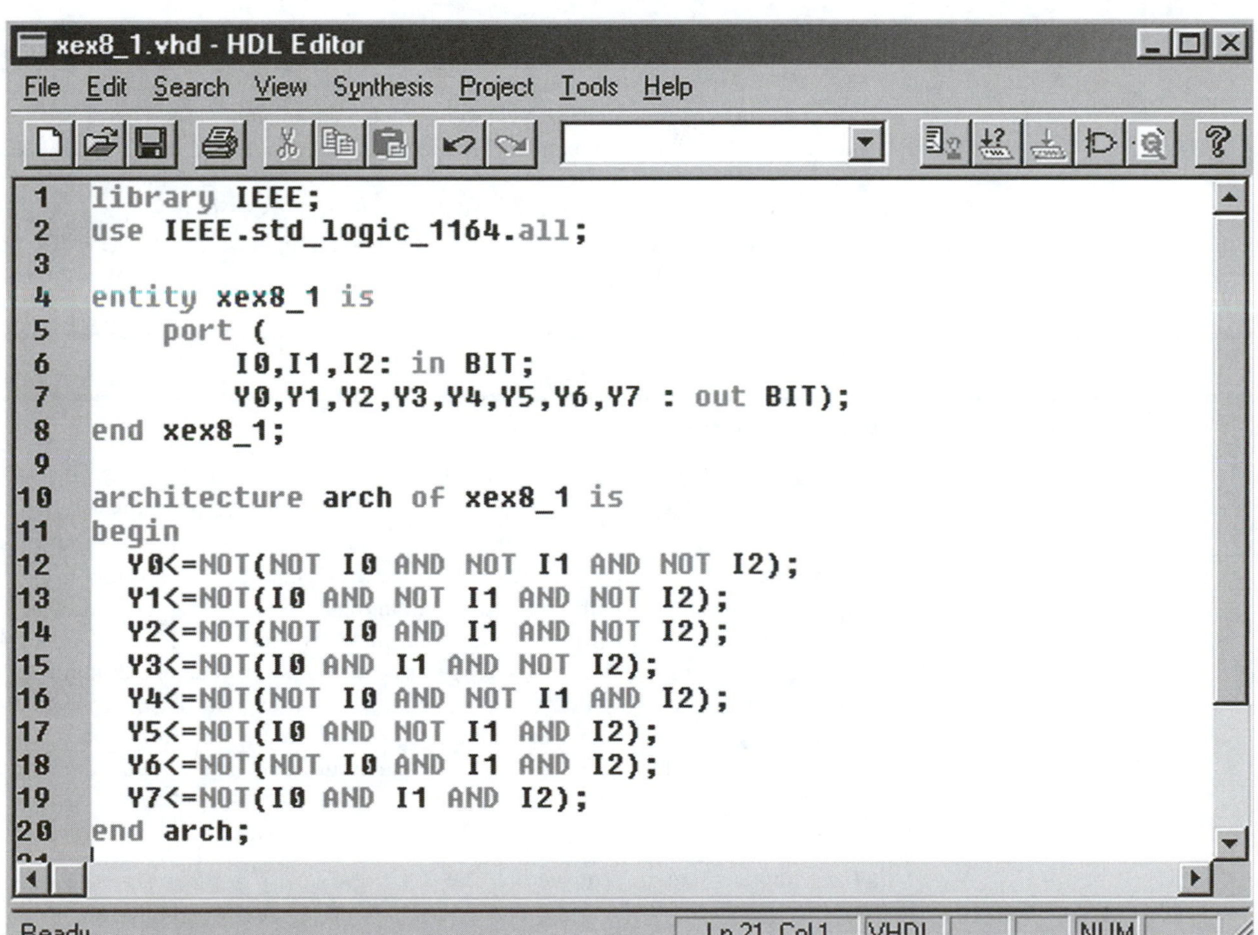


```
library IEEE;
use IEEE.std_logic_1164.all;

entity xex8_1 is
    port (
        I0,I1,I2: in BIT;
        Y0,Y1,Y2,Y3,Y4,Y5,Y6,Y7 : out BIT);
end xex8_1;

architecture arch of xex8_1 is
begin
  Y0<=NOT(NOT I0 AND NOT I1 AND NOT I2);
  Y1<=NOT(I0 AND NOT I1 AND NOT I2);
  Y2<=NOT(NOT I0 AND I1 AND NOT I2);
  Y3<=NOT(I0 AND I1 AND NOT I2);
  Y4<=NOT(NOT I0 AND NOT I1 AND I2);
  Y5<=NOT(I0 AND NOT I1 AND I2);
  Y6<=NOT(NOT I0 AND I1 AND I2);
  Y7<=NOT(I0 AND I1 AND I2);
end arch;
```

Figure K–28 VHDL text editor screen for the 3-line-to-8-line octal decoder.

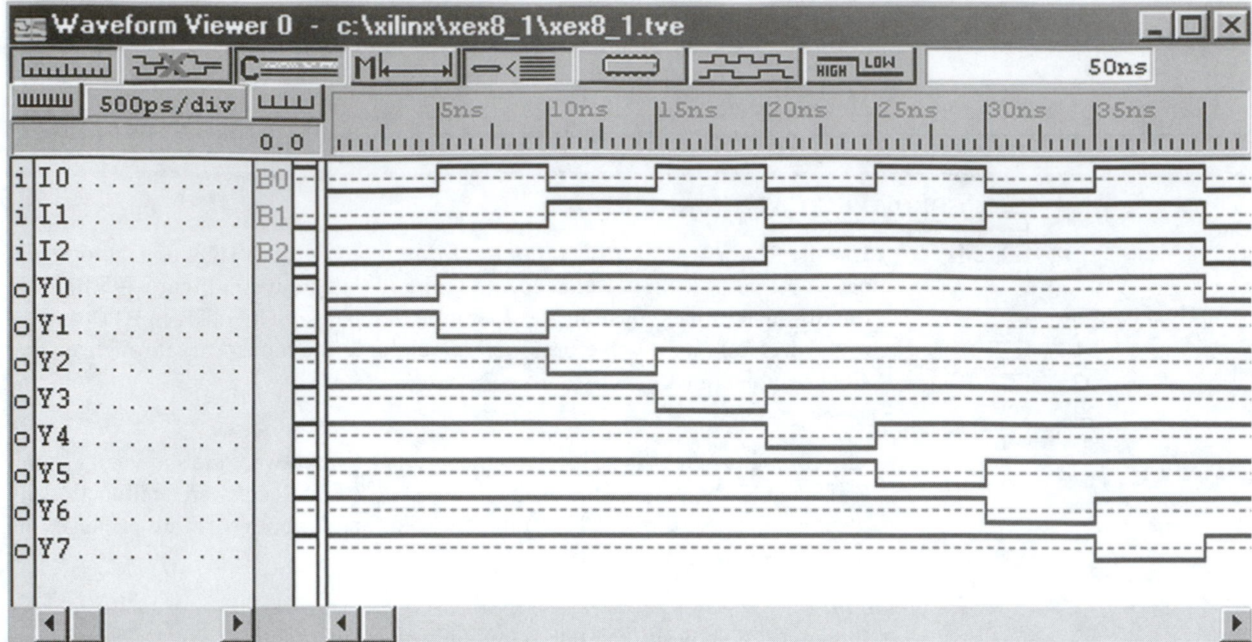

Figure K–29 Waveform logic simulator screen demonstrating the octal decoder.

each *Y* output active. Each equation begins with the NOT command. This has the effect of making the *Y* outputs active-LOW. This method is used because VHDL does not allow you to specify NOT on the left of the equal sign. Therefore, the first architecture statement can be interpreted as "Y_0 goes LOW if I_0 is LOW and I_1 is LOW and I_2 is LOW."

Figure K–29 shows the results of the simulation. As you can see, Y_0 is LOW for I_0, I_1, and I_2 all LOW (000). Y_1 is LOW for I_0 HIGH, I_1 and I_2 LOW (001). Each successive output goes LOW for the corresponding binary value on the inputs.

CPLD APPLICATION K–8

Use the Altera *MAX + PLUS II* or the Xilinx *Foundation* software to test the operation of an 8-line multiplexer. Instead of looking for a specific 7400-series multiplexer, we will use the generic model of a multiplexer that the software provides. Multiple-sized multiplexers, decoders, and encoders are found in the *MAX + PLUS II* macrofunction library and in the *Foundation* symbols toolbar.

Find the generic 8-line-to-1-line multiplexer symbol. (The Altera name is *81mux*. The Xilinx name is *M8_1E*.) Let's work with just four of the inputs by grounding the unused high-order inputs and the high-order data select input. Simulate the multiplexing action by inputting three different clock frequencies, and vary the data select inputs to route each of the input frequencies, one at a time, to the output.

Solution:

The computer screens produced by the *MAX + PLUS II* software are shown in Figures K–30 and K–31. (The results using the *Foundation* software are functionally identical to those shown.) The computer solutions for both software packages are included in the textbook's CD.

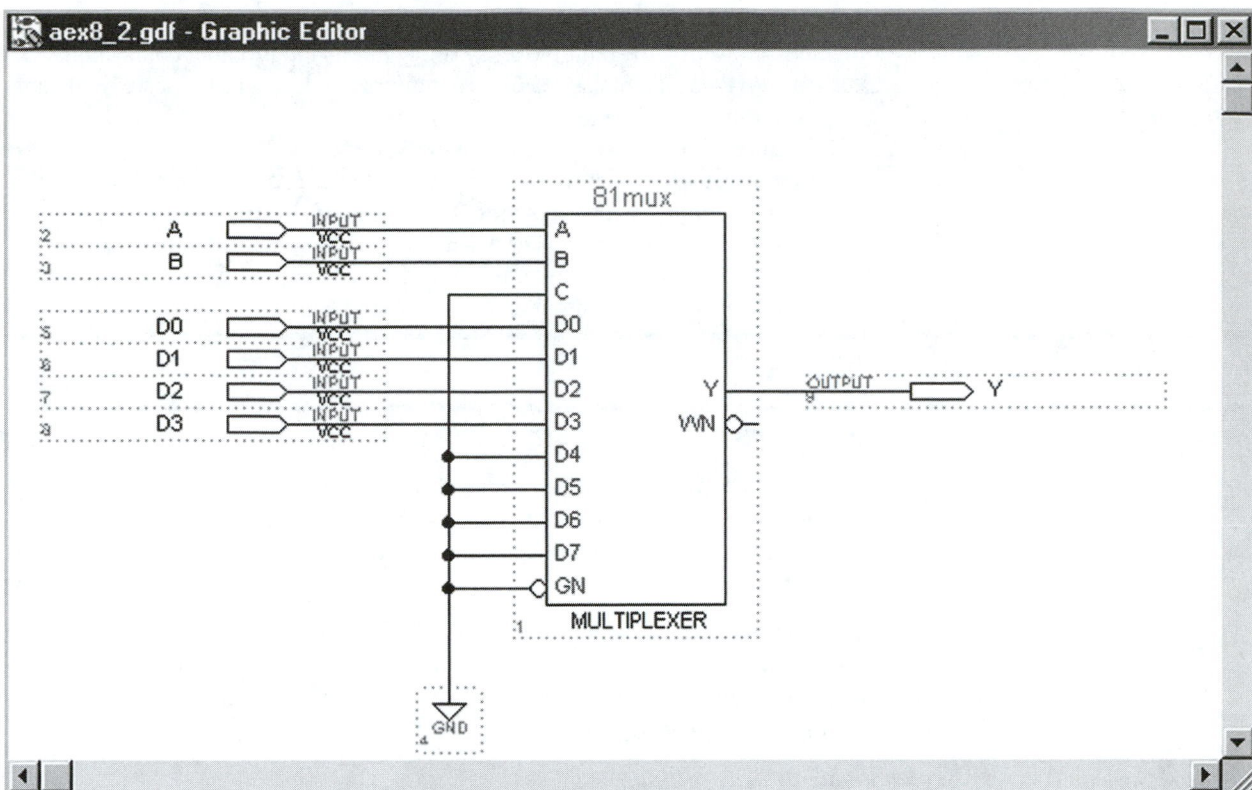

Figure K–30 Graphic schematic editor screen for the 8-line multiplexer.

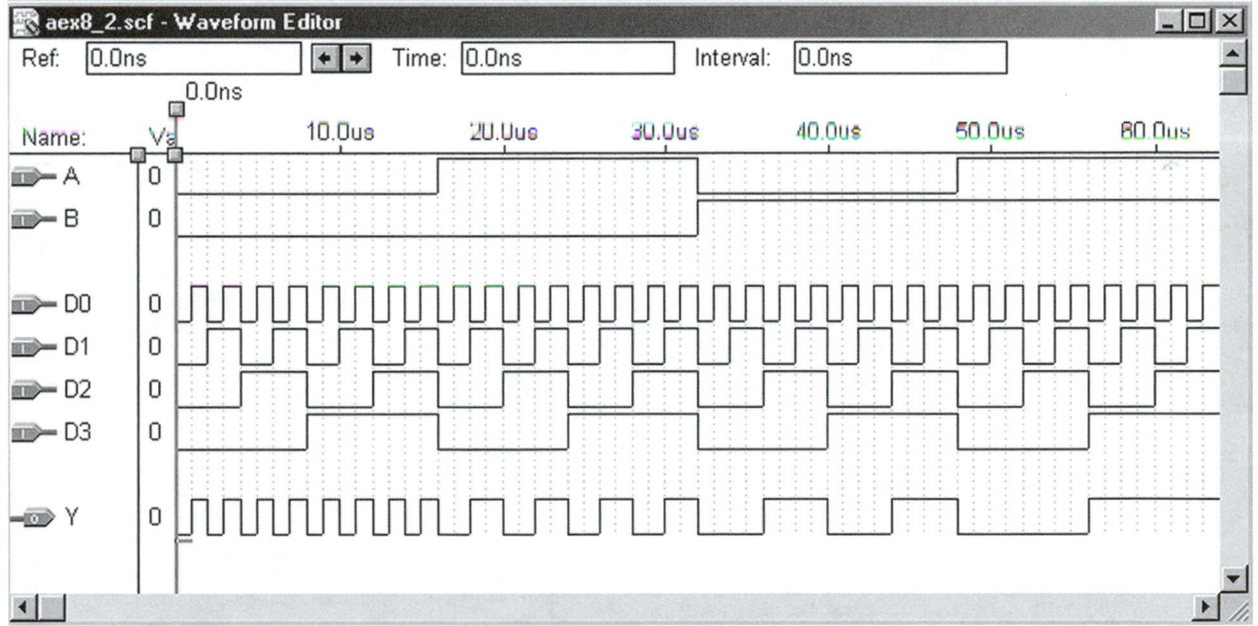

Figure K–31 Waveform logic simulator screen demonstrating the multiplexing action.

Explanation:

Figure K–30 shows the multiplexer symbol with its input/output connections. GN is an active-LOW chip enable. By grounding D_4 to D_7 and C, this circuit functions as a 4-line multiplexer.

Figure K–31 shows the input/output waveforms. From 0 to 16 ns, the data select inputs (A–B) are set to 0–0, which selects the D_0 input. During that period of time, the Y output exactly matches D_0, as it should. Next, from 16 to 32 ns, A–B equals 1–0, which selects D_1. The Y waveform during that time matches D_1, as it should. As you continue in time, the A–B data selectors select the appropriate D input and routes it to the output at Y.

CPLD APPLICATION K–9

Use the Altera *MAX + PLUS II* or the Xilinx *Foundation* software to test the operation of the 74161 connected as a 4-bit counter. Develop a set of simulation waveforms to study the operation of the counter.

Solution:

The computer screens produced by the *MAX + PLUS II* software are shown in Figures K–32 and K–33. (The results using the *Foundation* software are functionally identical to those shown.) The computer solutions for both software packages are included in the textbook's CD.

Explanation:

Figure K–32 shows the input/output connections to the 74161 synchronous counter. This is a 4-bit counter capable of counting 0000 up to 1111. The *A, B, C, D* inputs along with *LDN* allow you to parallel load a number into the counter. The *ENT* and *ENP* inputs are count-enable inputs, which must be HIGH to count. The *CLRN* pin is an active-LOW clear for resetting the outputs to zero. The *RCO* output is an active-HIGH terminal count to signify that you have reached 1111.

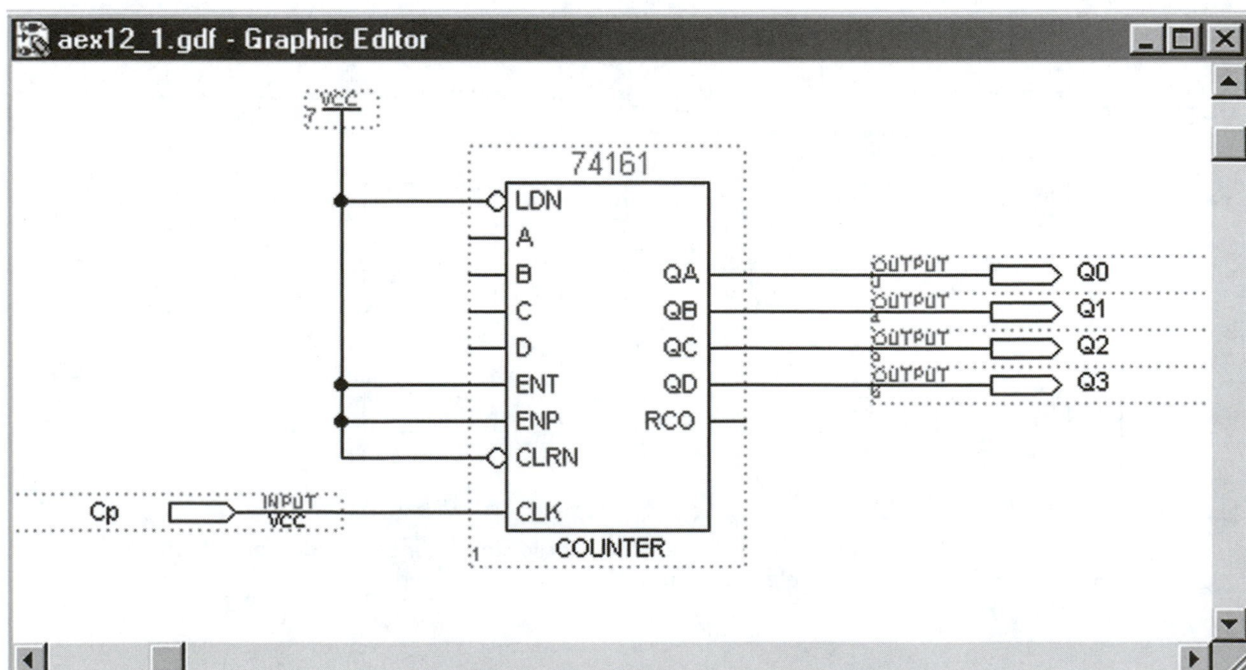

Figure K–32 Graphic schematic editor screen for the 4-bit synchronous counter.

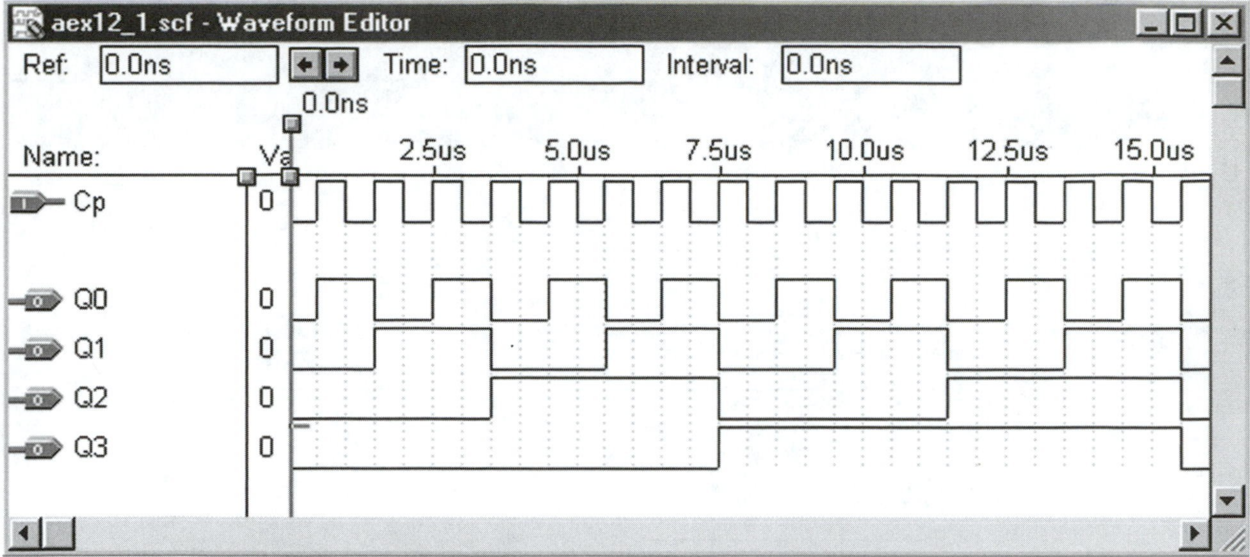

Figure K–33 Waveform logic simulator screen demonstrating the counting action.

The *Cp* clock and the resulting output count are shown in Figure K–33. The count ranges from 0000 up to 1111, as we expected. Also, notice that the *Q* outputs switch state on each positive edge of the clock.

CPLD APPLICATION K–10

Use the Altera *MAX + PLUS II* or the Xilinx *Foundation* software to test the operation of the 74194 connected as a 4-bit shift register as shown in Section 6–10. Exercise its functions by parallel loading 1000, then shift right for four pulses. Then, parallel load 0001, and shift left four pulses. Develop a set of simulation waveforms to study the shifting activity.

Solution:

The computer screens produced by the *Foundation* software are shown in Figures K–34 and K–35. (The results using the *MAX + PLUS II* software are functionally identical to those shown.) The computer solutions for both software packages are included in the textbook's CD.

Explanation:

Figure K–34 shows the input/output connections to the 74194. The data shift left (SLI) and data shift right (SRI) are both grounded so that zeros will be shifted into Q_A during a shift right and zeros will be shifted into Q_D when shifting left. The CLR (clear) terminal is a master reset for all *Q*'s. It is tied HIGH to disable clearing the outputs.

Figure K–35 shows the results of the simulation. Custom waveforms must be drawn for S_0, S_1, *A, B, C,* and *D.* The HIGH/LOW Logical States button on the Xilinx Waveform Viewer produces the Simulator State Selection window, which allows you to customize the waveforms as needed. (Customizing Altera Logic Simulator waveforms is accomplished by drawing the HIGH/LOW states with the mouse as needed.) After parallel loading a 1000, the *A, B, C,* and *D* inputs are set to the Unknown _X (or don't care) state until the next parallel load. The *Q*-output waveforms show the data shifting to the right for four pulses, then parallel loading 0001, then shifting to the left for four pulses.

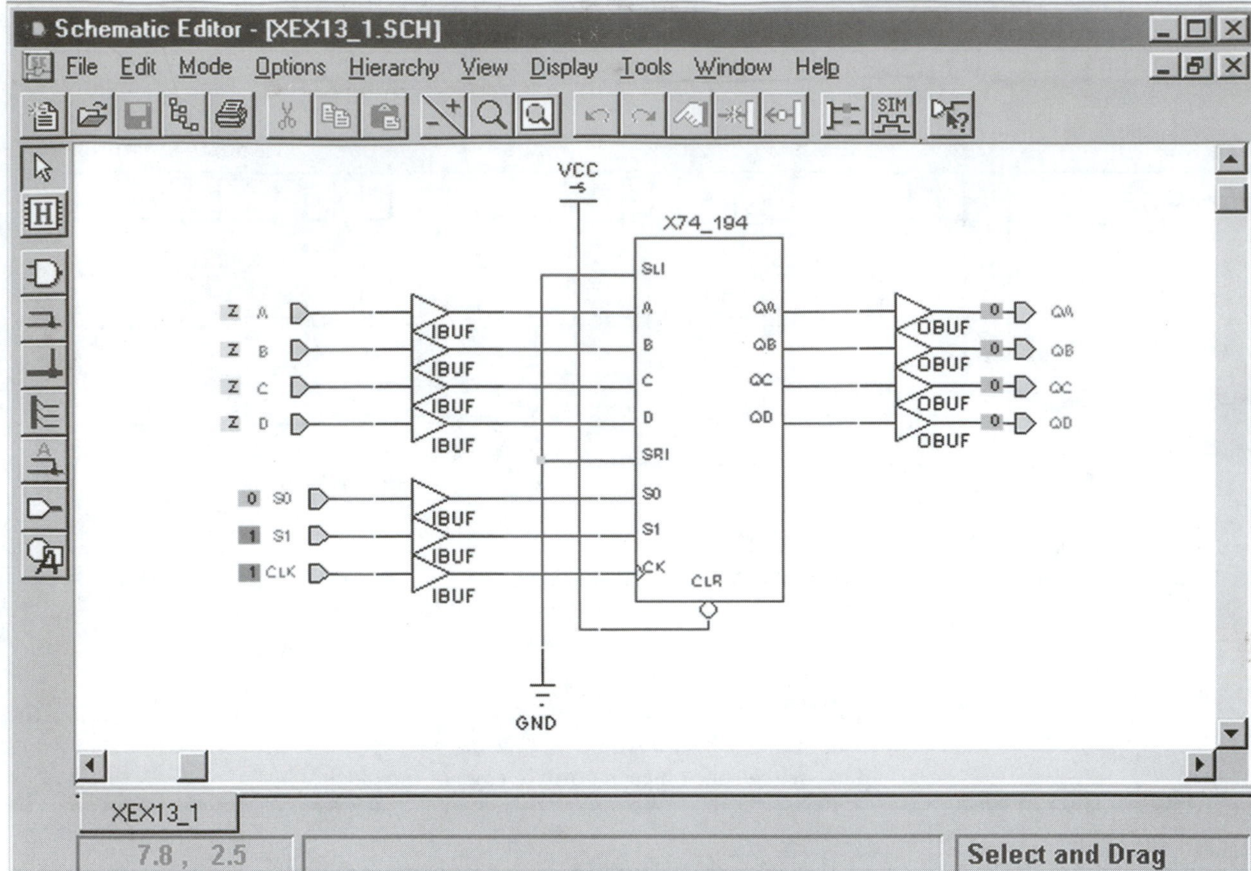

Figure K–34 Graphic schematic editor screen for the 74194 shift register.

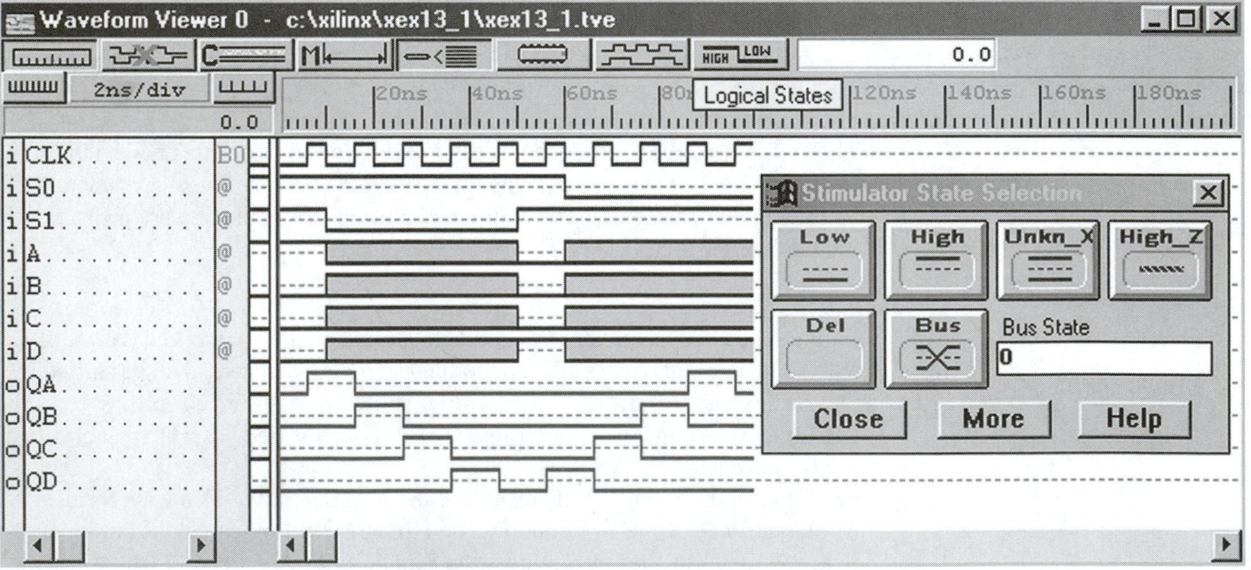

Figure K–35 Waveform logic simulator screen demonstrating the shifting action.

SUMMARY

In this appendix, we have learned that

1. PLDs can be used to replace 7400- and 4000-series ICs. They contain the equivalent of thousands of logic gates. CAD tools are used to configure them to implement the desired logic.
2. The two most common methods of PLD design entry are schematic entry and VHDL entry. To use schematic entry, the designer uses CAD tools to draw the logic to be implemented. To use VHDL entry, the designer uses a text editor to write program descriptions defining the logic to be implemented.
3. PLD design software usually includes a logic simulator. This feature allows the user to simulate levels to be input to the PLD, and it shows the output simulation to those input conditions.
4. Most PLDs are erasable and reprogrammable. This allows users to test many versions of their logic design without ever changing ICs.
5. Basically, there are three types of PLDs: SPLDs, CPLDs, and FPGAs. SPLDs use the PAL or PLA architecture. They consist of several multi-input AND gates whose outputs feed the inputs to OR gates and memory flip-flops. CPLDs consist of several interconnected SPLDs. FPGAs are the most dense form of PLD, solving logic using a look-up table to determine the desired output.

GLOSSARY

Architecture Body: The section in a VHDL program defining the logic functions to be implemented.

Buffer A logic gate used to define an input or output to a logic circuit. The logic level at its input is transferred to its output unchanged.

CAD: Computer-Aided Design. This type of design uses a computer to aid in the drawing and logic development of a logic circuit. It eliminates many of the manual, time-consuming tasks once associated with logic design.

CPLD Complex Programmable Logic Device. A PLD consisting of more than 100 interconnected SPLDs. A single chip can be programmed to implement hundreds of logic equations and operations.

Compiler A language translation software module used by CPLD development systems to convert a schematic or VHDL code into a binary file to represent the digital logic to be implemented.

Entity Declaration The section of a VHDL program defining the input and output ports.

FPGA Field-Programmable Gate Array. The most dense form of PLD. It uses a look-up table to resolve its logic operations. Its main disadvantage is that most FPGAs are volatile, losing their memory when power is removed.

Library Declaration The section of a VHDL program declaring the software libraries to be included in the program. These libraries are used by the compiler to resolve references to the various program commands.

Look-Up Table Used by FPGA logic to determine the output level of a circuit based on the combinations of logic levels at its inputs. It is constructed as a truth table except that its outputs are only HIGH for specific combinations of inputs solving the given logic product terms.

Nonvolatile Internal memory is maintained even when power is removed from the IC.

PAL Programmable Array Logic. Its basic structure contains multiple inputs to several AND gates, the outputs of which are connected to a series of fixed ORs.

PLA Programmable Logic Array. Its basic structure contains multiple inputs to several AND gates, the outputs of which are connected to a series of programmable ORs.

PLD Programmable Logic Device. An IC containing thousands of undefined logic functions. A software development tool is used to specify (i.e., program) the specific logic to be implemented by the IC. PLD is the general term used to represent PLAs, PALs, SPLDs, CPLDs, and FPGAs.

Product Terms Input variables that are ANDed together (e.g., ABC, $AB\overline{C}$:).

Schematic Capture A method used by PLD software to input a design that is defined by a schematic.

Schematic Editor A software tool provided as part of the PLD development package. It provides a way to enter designs by drawing a schematic.

SPLD Simple Programmable Logic Devices. A programmable, digital logic IC containing several PAL or PLA structures with internal interconnections and memory registers.

Sum-of-Products Two or more product terms that are ORed together (e.g., $AB\overline{C} + A\overline{C}\,\overline{D} + BCD$).

VHDL VHSIC (Very High Speed Integrated Circuit) Hardware Description Language. A programming language used by PLD software to define a logic design by specifying a series of I/O definitions and logic equations.

VHDL Editor A software program facilitating entry of text-based instructions comprising the VHDL program.

Waveform Simulator The part of a PLD software development tool that allows users to simulate the input of several signals to a logic circuit and observe its response.

PROBLEMS

K–1. How does programmable logic differ from discrete digital logic like the 7400 series?

K–2. What are two common ways to configure or define logic to PLD programming software?

K–3. What does HDL stand for in the acronym VHDL?

K–4. List the six steps in the PLD design flow.

K–5. How many different ICs would it take to implement the following equations?
- **(a)** $X = AB + \overline{BC}$
- **(b)** $Y = A\overline{B} + B\overline{C} + \overline{C + D}$

K–6. How is pin 1 identified in the PLCC package style used for the PLD in Figure K–4?

K–7. What is the purpose of the PLD programmer boards shown in Figure K–5?

K–8. How many product terms are in the following equations?
- **(a)** $X = A\overline{C} + BC + \overline{A}C$
- **(b)** $Y = A\overline{B}C + B\overline{C}$
- **(c)** $Z = AB\overline{C} + ACD + \overline{B}C\overline{D}$

K–9. How does a PLA differ from a PAL?

K–10. Redraw the PLA circuitry of Figure K–8 to implement the following SOP equations:
- **(a)** $X = \overline{A}B + \overline{A}\,\overline{B} + A\overline{B}$
- **(b)** $Y = AB + \overline{A}B$

K–11. Why is it advantageous to use a CPLD that is nonvolatile?

K–12. Refer to the data sheets on the WWW to determine the number of usable gates and macrocells in each of the following CPLDs:
- **(a)** Altera MAX EPM7128S
- **(b)** Xilinx XC95108

K–13. Instead of interconnecting logic gates, the FPGA solves its logic requirements by using what method?

K–14. Draw a look-up table similar to Figure K–10(b) for the equation $X = \overline{A}\,\overline{B} + A\overline{B}$.

K–15. Because most FPGAs are volatile, what must be done each time they are powered up?

K–16. What are the two most common methods of design entry for CPLD development software?

K–17. What is the function of the compiler in CPLD development software?

K–18. What is the purpose of the buffers in the *andgate* design shown in Figure K–12(c)?

K–19. VHDL allows the user to enter the logic design via a _______ editor.

K–20. Define the purpose of the following three VHDL program segments:
(a) Library
(c) Entity
(d) Architecture

K–21. Write the VHDL entity declare for a three-input AND gate.

K–22. Write the VHDL architecture for a three-input AND gate.

K–23. Draw the logic circuit to be implemented by the following VHDL architecture body:

```
ARCHITECTURE    arc    OF    p4_23    IS
      BEGIN
            X < = (A AND (B OR C));
            Y < = (A OR NOT B) AND NOT (B AND C);
            Z < = NOT (B AND C) OR NOT (A OR C);
      END arc;
```

CPLD PROBLEMS

The following problems can be solved using the Altera *MAX + PLUS II* software or the Xilinx Foundation design software. A tutorial on using this software is provided on the textbook CD.

K–1. Use the Altera graphic editor or Xilinx schematic editor to design the CPLD logic to implement the Boolean equation $X = \overline{AB}$. Test the operation of the CPLD logic by using the Altera waveform editor or Xilinx logic simulator. The solution requires the creation of two of the following files:
(a) Altera graphic design file (gdf) called *AprobK_1.gdf,* and
(b) Altera simulator channel file (scf) called *AprobK_1.scf,* or
(c) Xilinx schematic (sch) file called *XprobK_1.sch,* and
(d) Xilinx test vector file (tve) called *XprobK_1.tve.*

K–2. Use the Altera graphic editor or Xilinx schematic editor to design the CPLD logic to implement the Boolean equation $X = \overline{A + B} + \overline{BC}$. Test the operation of the CPLD logic by using the Altera waveform editor or the Xilinx logic simulator. The solution requires the creation of two of the following files:
(a) Altera graphic design file (gdf) called *AprobK_2.gdf,* and
(b) Altera simulator channel file (scf) called *AprobK_2.scf,* or
(c) Xilinx schematic (sch) file called *XprobK_2.sch,* and
(d) Xilinx test vector file (tve) called *XprobK_2.tve.*

K–3. Use the Altera or Xilinx VHDL text editor to design the CPLD logic to implement the Boolean equation $X = \overline{AB} + BC$. Test the operation of the CPLD logic by using the Altera waveform editor or Xilinx logic simulator. The solution requires the creation of two of the following files:

(a) Altera VHDL design file (vhd) called *AprobK_3.vhd,* and
(b) Altera simulator channel file (scf) called *AprobK_3.scf,* or
(c) Xilinx VHDL design file (vhd) called *XprobK_3.vhd,* and
(d) Xilinx test vector file (tve) called *XprobK_3.tve.*

K–4. Use the Altera graphic editor or the Xilinx schematic editor to design the CPLD logic to implement the Boolean equation $X = \overline{AB} \cdot \overline{B + C}$. Test the operation of the CPLD logic by using the Altera waveform editor or the Xilinx logic simulator. The solution requires the creation of two of the following files:

(a) Altera graphic design file (gdf) called *AprobK_4.gdf,* and
(b) Altera simulator channel file (scf) called *AprobK_4.scf,* or
(c) Xilinx schematic (sch) file called *XprobK_4.sch,* and
(d) Xilinx test vector file (tve) called *XprobK_4.tve.*

K–5. Ten rules for Boolean reduction were given in Table 3–1. The tenth rule states that

1. $A + \overline{A}B = A + B$, and
2. $\overline{A} + AB = \overline{A} + B$

Use the Altera graphic editor or the Xilinx schematic editor to design the CPLD logic to implement both equations in (1) and both equations in (2). Prove that both equations in (1) and both equations in (2) are equivalent by using the Altera waveform editor or the Xilinx logic simulator. The solution requires the creation of two of the following files:

(a) Altera graphic design file (gdf) called *AprobK_5.gdf,* and
(b) Altera simulator channel file (scf) called *AprobK_5.scf,* or
(c) Xilinx schematic (sch) file called *XprobK_5.sch,* and
(d) Xilinx test vector file (tve) called *XprobK_5.tve.*

K–6. Use the Altera or the Xilinx VHDL text editor to design the CPLD logic to implement the Boolean equation $X = A\overline{B}\,\overline{C} + \overline{BC} + \overline{A + C}$. Test the operation of the CPLD logic by using the Altera waveform editor or the Xilinx logic simulator. The solution requires the creation of two of the following files:

(a) Altera VHDL design file (vhd) called *AprobK_6.vhd,* and
(b) Altera simulator channel file (scf) called *AprobK_6.scf,* or
(c) Xilinx VHDL design file (vhd) called *XprobK_6.vhd,* and
(d) Xilinx test vector file (tve) called *XprobK_6.tve.*

K–7. Use the Altera graphic editor or the Xilinx schematic editor to design the CPLD logic to implement the following Boolean equations:

1. $X = \overline{\overline{AB}(A + B)}$
2. $Y = \overline{\overline{A + B} + AB}$
3. $Z = AB + \overline{A + B}$

Determine which two of those equations yield equivalent outputs by studying their waveform logic simulator files. The solution requires the creation of two of the following files:

(a) Altera graphic design file (gdf) called *AprobK_7.gdf* and
(b) Altera simulator channel file (scf) called *AprobK_7.scf* or
(c) Xilinx schematic (sch) file called *XprobK_7.sch,* and
(d) Xilinx test vector file (tve) called *XprobK_7.tve.*

K–8. A chemical processing plant has four HIGH/LOW sensors on each of its chemical tanks [Temperature (*T*), Pressure (*P*), Fluid Level (*L*), and Weight (*W*)]. Several different combinations of sensor levels need to be constantly monitored. Design the CPLD logic to turn on any of the three indicator lights (Emergency, Warning, Check) if the listed conditions are met.

(1) *Emergency:* Shut down and drain the system if any of the following exists:
(a) High *T,* or (high *P* and low *W*) or

(b) High T, or (high P and low L)
(c) High T, or (low P and low W or low L)
(2) *Warning:* Check controls and perform corrections if any of the following exists:
(a) High P, or (high L, and low W) or
(b) High P, or (high W, and low L)
(c) High P, or (low L, and low T)
(3) *Check:* Read gauges and report if any of the following exists:
(a) Any two levels are high
(b) W is high

The solution requires the creation of two of the following files:
(a) Altera graphic design file (gdf) called *AprobK_8.gdf,* and
(b) Altera simulator channel file (scf) called *AprobK_8.scf,* or
(c) Xilinx schematic (sch) file called *XprobK_8.sch,* and
(d) Xilinx test vector file (tve) called *XprobK_8.tve.*

Extra: If your lab is equipped with a CPLD programmer board, demonstrate the final design to your instructor.

K–9. Use the Altera graphic editor or the Xilinx schematic editor to redesign the Binary comparator of Figure 4–1. Bubble push the original design to come up with an equivalent constructed of Ex-ORs instead of Ex-NORs and a NOR instead of the AND. Test the operation of the CPLD logic by using the Altera waveform editor or the Xilinx logic simulator. The solution requires the creation of two of the following files:
(a) Altera graphic design file (gdf) called *AprobK_9.gdf,* and
(b) Altera simulator channel file (scf) called *AprobK_9.scf,* or
(c) Xilinx schematic (sch) file called *XprobK_9.sch,* and
(d) Xilinx test vector file (tve) called *XprobK_9.tve.*

Extra: If your lab is equipped with a CPLD programmer board, then demonstrate the final design to your instructor.

K–10. Use the Altera or the Xilinx VHDL text editor to design the CPLD logic to implement the half-adder shown in Figure 3–49. Test the operation of the CPLD logic by using the Altera waveform editor or the Xilinx logic simulator. Test all possible combinations of A_0 and B_0. The solution requires the creation of two of the following files:
(a) Altera VHDL design file (vhd) called *AprobK_10.vhd,* and
(b) Altera simulator channel file (scf) called *AprobK_10.scf,* or
(c) Xilinx VHDL design file (vhd) called *XprobK_10.vhd,* and
(d) Xilinx test vector file (tve) called *XprobK_10.tve.*

Extra: If your lab is equipped with a CPLD programmer board, then demonstrate the final design to your instructor.

K–11. Use the Altera graphic editor or the Xilinx schematic editor to create the 8-bit adder shown in Figure 3–58. Test the operation of the CPLD logic by using the Altera waveform editor or the Xilinx logic simulator. Demonstrate the hex addition of 22 + 29 = 4B, 29 + 38 = 61, and 9A + 92 = 2C with a HIGH C_{out}. The solution requires the creation of two of the following files:
(a) Altera graphic design file (gdf) called *AprobK_11.gdf,* and
(b) Altera simulator channel file (scf) called *AprobK_11.scf,* or
(c) Xilinx schematic (sch) file called *XprobK_11.sch,* and
(d) Xilinx test vector file (tve) called *XprobK_11.tve.*

Extra: If your lab is equipped with a CPLD programmer board, then demonstrate the final design to your instructor

K–12. Use the Altera or the Xilinx VHDL text editor to design the CPLD logic to implement the function of a 74138 octal decoder. Use the VHDL listing presented in Figure K–28 as a starting point, and then add the three enables that the 74138

uses. Test the operation of the CPLD logic by using the Altera waveform editor or the Xilinx logic simulator. Try all possible combinations of I_0, I_1, and I_2. Then, with the Y_0 output selected, test each of the three enables, one at a time, to prove that they disable Y_0. The solution requires the creation of two of the following files:

(a) Altera VHDL design file (vhd) called *AprobK_12.vhd,* and
(b) Altera simulator channel file (scf) called *AprobK_12.scf,* or
(c) Xilinx VHDL design file (vhd) called *XprobK_12.vhd,* and
(d) Xilinx test vector file (tve) called *XprobK_12.tve.*

Extra: If your lab is equipped with a CPLD prorammer board, then demonstrate the final design to your instructor

K–13. Use the Altera graphic editor or the Xilinx schematic editor to modify the multiplexer *gdf* or *sch* file given in Figure K–30. The modification involves the inclusion of the input enable line for the simulation of the circuit. This line is labeled *E* in the *sch* file and *GN* in the *gdf* file. Add the enable line to the waveform logic simulator screen of Figure K–31 to demonstrate the enable/disable function. The solution requires the creation of two of the following files:

(a) Altera graphic design file (gdf) called *AprobK_13.gdf,* and
(b) Altera simulator channel file (scf) called *AprobK_13.scf,* or
(c) Xilinx schematic (sch) file called *XprobK_13.sch,* and
(d) Xilinx test vector file (tve) called *XprobK_13.tve.*

Extra: If your lab is equipped with a CPLD programmer board, then demonstrate the final design to your instructor

K–14. Use the Altera graphic editor or the Xilinx schematic editor to design a demultiplexing system to route data to one of four different outputs. Your design can be modeled after the 74139 circuit of Figure 4–22. (Altera has a 74139 in its macrofunction library. Xilinx has an X74_139 in its symbols toolbar.) Demonstrate the routing of a data signal to each of the four outputs repeatedly by changing the output destination select lines. The solution requires the creation of two of the following files:

(a) Altera graphic design file (gdf) called *AprobK_14.gdf* and
(b) Altera simulator channel file (scf) called *AprobK_14.scf* or
(c) Xilinx schematic (sch) file called *XprobK_14.sch,* and
(d) Xilinx test vector file (tve) called *XprobK_14.tve.*

Extra: If your lab is equipped with a CPLD programmer board, then demonstrate the final design to your instructor.

K–15. Use the Altera graphic editor or the Xilinx schematic editor to design a MOD-10 counter based on the 74161. Demonstrate the 4-bit counter output on the waveform logic simulator. The solution requires the creation of two of the following files:

(a) Altera graphic design file (gdf) called *AprobK_15.gdf,* and
(b) Altera simulator channel file (scf) called *AprobK_15.scf,* or
(c) Xilinx schematic (sch) file called *XprobK_15.sch,* and
(d) Xilinx test vector file (tve) called *XprobK_15.tve.*

Extra: If your lab is equipped with a CPLD programmer board, then demonstrate the final design to your instructor. (If you are using the on-board push buttons or switches, then they must be debounced.)

K–16. Use the Altera graphic editor or the Xilinx schematic editor to design a MOD-12 counter based on the 74161. Use a bus structure similar to Figure K–26 to represent the outputs so that they will be displayed in hexadecimal (help is available from the software help screen under "Bus Drawing or Buses"). The solution requires the creation of two of the following files:

(a) Altera graphic design file (gdf) called *AprobK_16.gdf,* and
(b) Altera simulator channel file (scf) called *AprobK_16.scf,* or
(c) Xilinx schematic (sch) file called *XprobK_16.sch,* and
(d) Xilinx test vector file (tve) called *XprobK_16.tve.*

Extra: If your lab is equipped with a CPLD programmer board, then demonstrate the final design to your instructor. (If you are using the on-board push buttons or switches, then they must be debounced.)

K–17. Use the Altera graphic editor or the Xilinx schematic editor to design a MOD-10 counter. This counter will be designed using the Altera Library of Parameterized Modules (LPM) or the Xilinx LogiBLOX components. These libraries of components help you to create custom logic functions. You can choose between several generic flip-flops, counters, shift registers, and other common digital logic.

After you choose the counter function for this problem, you will then specify such things as its bit width, its modulus, its up/down direction, and several other optional design criteria, such as asynchronous clear and preset controls. Connect the outputs using a bus structure, and display the count sequence on the waveform logic simulator.

The solution requires the creation of two of the following files:
(a) Altera graphic design file (gdf) called *AprobK_17.gdf,* and
(b) Altera simulator channel file (scf) called *AprobK_17.scf,* or
(c) Xilinx schematic (sch) file called *XprobK_17.sch,* and
(d) Xilinx test vector file (tve) called *XprobK_17.tve.*

Extra: If your lab is equipped with a CPLD programmer board, then demonstrate the final design to your instructor. (If you are using the on-board push buttons or switches, then they must be debounced.)

K–18. Use the Altera graphic editor or the Xilinx schematic editor to design a 4-bit ring shift counter with the 74194 (see Section 6–9). Demonstrate the continuous shift action on the waveform logic simulator. The solution requires the creation of two of the following files:
(a) Altera graphic design file (gdf) called *AprobK_18.gdf,* and
(b) Altera simulator channel file (scf) called *AprobK_18.scf,* or
(c) Xilinx schematic (sch) file called *XprobK_18.sch,* and
(d) Xilinx test vector file (tve) called *XprobK_18.tve.*

Extra: If your lab is equipped with a CPLD programmer board, then demonstrate the final design to your instructor. (If you are using the on-board push buttons or switches, then they must be debounced.)

K–19. Use the Altera graphic editor or the Xilinx schematic editor to design a 4-bit Johnson shift counter with the 74194 (see Section 6–9). Demonstrate the continuous shift action on the waveform logic simulator. The solution requires the creation of two of the following files:
(a) Altera graphic design file (gdf) called *AprobK_19.gdf,* and
(b) Altera simulator channel file (scf) called *AprobK_19.scf,* or
(c) Xilinx schematic (sch) file called *XprobK_19.sch,* and
(d) Xilinx test vector file (tve) called *XprobK_19.tve.*

Extra: If your lab is equipped with a CPLD programmer board, then demonstrate the final design to your instructor. (If you are using the on-board push buttons or switches, then they must be debounced.)

K–20. Use the Altera graphic editor or the Xilinx schematic editor to design a Johnson shift counter. This counter will be designed using the Altera Library of Parameterized Modules (LPM) or the Xilinx LogiBLOX components. These libraries of components help you to create custom logic functions. You can choose between several generic flip-flops, counters, shift registers, and the other common digital logic. After you choose the shift register function for this problem, you will then specify such things as its bit width, parallel load value, and shift direction. Demonstrate the continuous shift action on the waveform logic simulator. The solution requires the creation of two of the following files:

(a) Altera graphic design file (gdf) called *AprobK_20.gdf,* and
(b) Altera simulator channel file (scf) called *AprobK_20.scf,* or
(c) Xilinx schematic (sch) file called *XprobK_20.sch,* and
(d) Xilinx test vector file (tve) called *XprobK_20.tve.*

Extra: If your lab is equipped with a CPLD programmer board, then demonstrate the final design to your instructor. (If you are using the on-board push buttons or switches, then they must be debounced.)

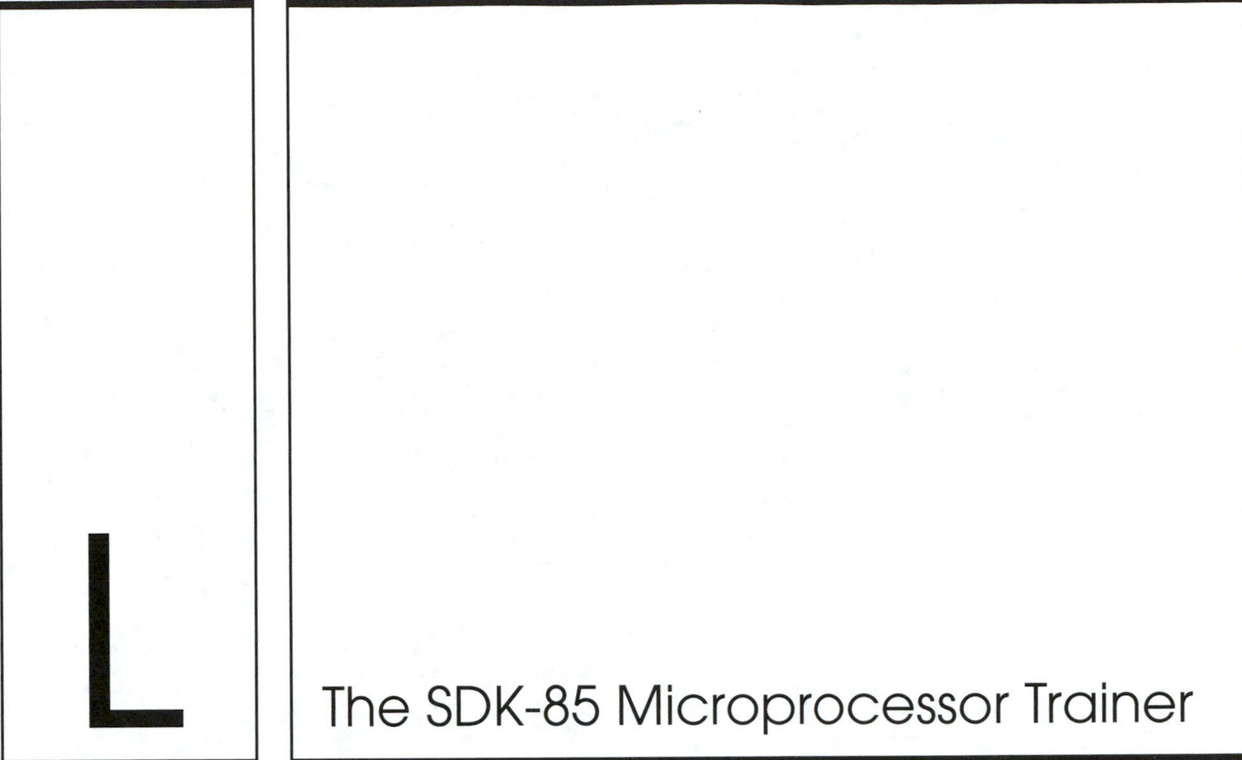

L The SDK-85 Microprocessor Trainer

The SDK-85 is an 8085A-based microprocessor trainer used primarily in colleges for teaching the fundamentals of microprocessor hardware and software. The board layout given in Figure L–1 shows the physical layout of the trainer. As you can see, the 8085A is located near the center of the board and is driven by a 6.144-MHz crystal. User interface to the board is provided via the keypad located in the lower right or a computer terminal connected to the TTY interface expansion circuitry available in the upper right of the board. A 4-digit display for the addresses and a 2-digit display for the 8-bit data are located just above the keypad. Several bus expansion and I/O connectors (J1 through J5) are provided next to the blank prototyping area on the left side of the board. The prototyping area is provided for students to build custom circuitry of their own to interface to the SDK-85 board.

The functional block diagram in Figure L–2 shows the interconnection of the major parts of the trainer. As shown, the 8085 sends and receives signals from all three buses in the system. Chip enabling is controlled by the 8205 address decoder or its replacement, the 74LS138. This IC provides a chip enable for either the 8355, the 8155, or the 8279.

The 8355 (or its replacement, the 8755 EPROM) is a 2K ROM containing the monitor program (operating system) provided by the manufacturer for the SDK-85. It also provides 16 bits of I/O capability.

The 8155 is a 256-byte RAM used to hold the software programs entered by the student. The 8155 also provides hardware counter/timers and 22 bits of I/O.

The 8279 is a keyboard and display controller. The monitor program uses this IC to read the keyboard and drive the 6 LED displays.

Room is also provided on the board for 8212s and 8216s for bus expansion to the J1 to J5 connectors.

The layout of the keypad is given in Figure L–3. The keypad provides the means to enter hexadecimal digits for representing addresses and data values that make up the

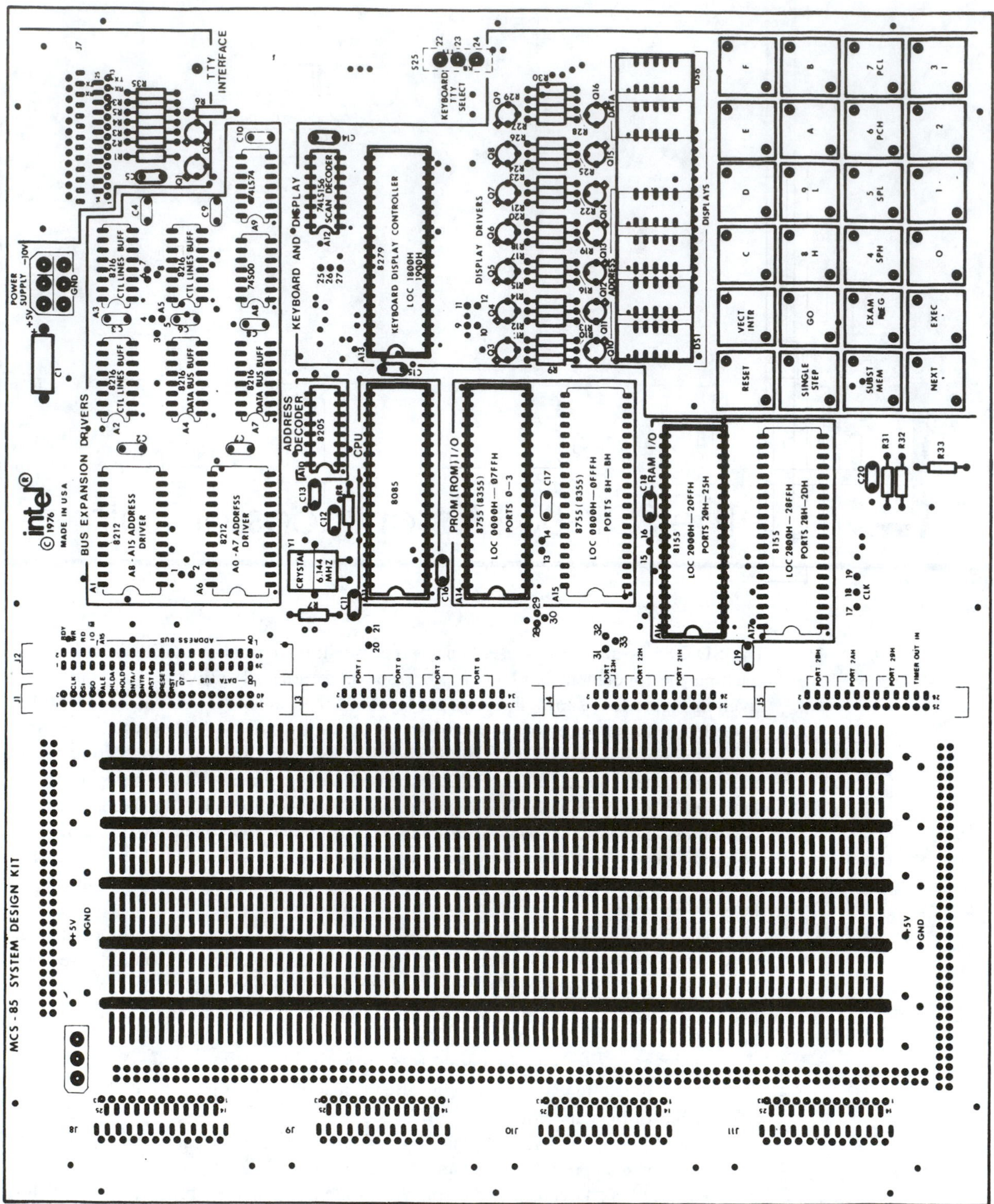

Figure L–1 The SDK-85 board layout. (Courtesy of URDA, Inc.)

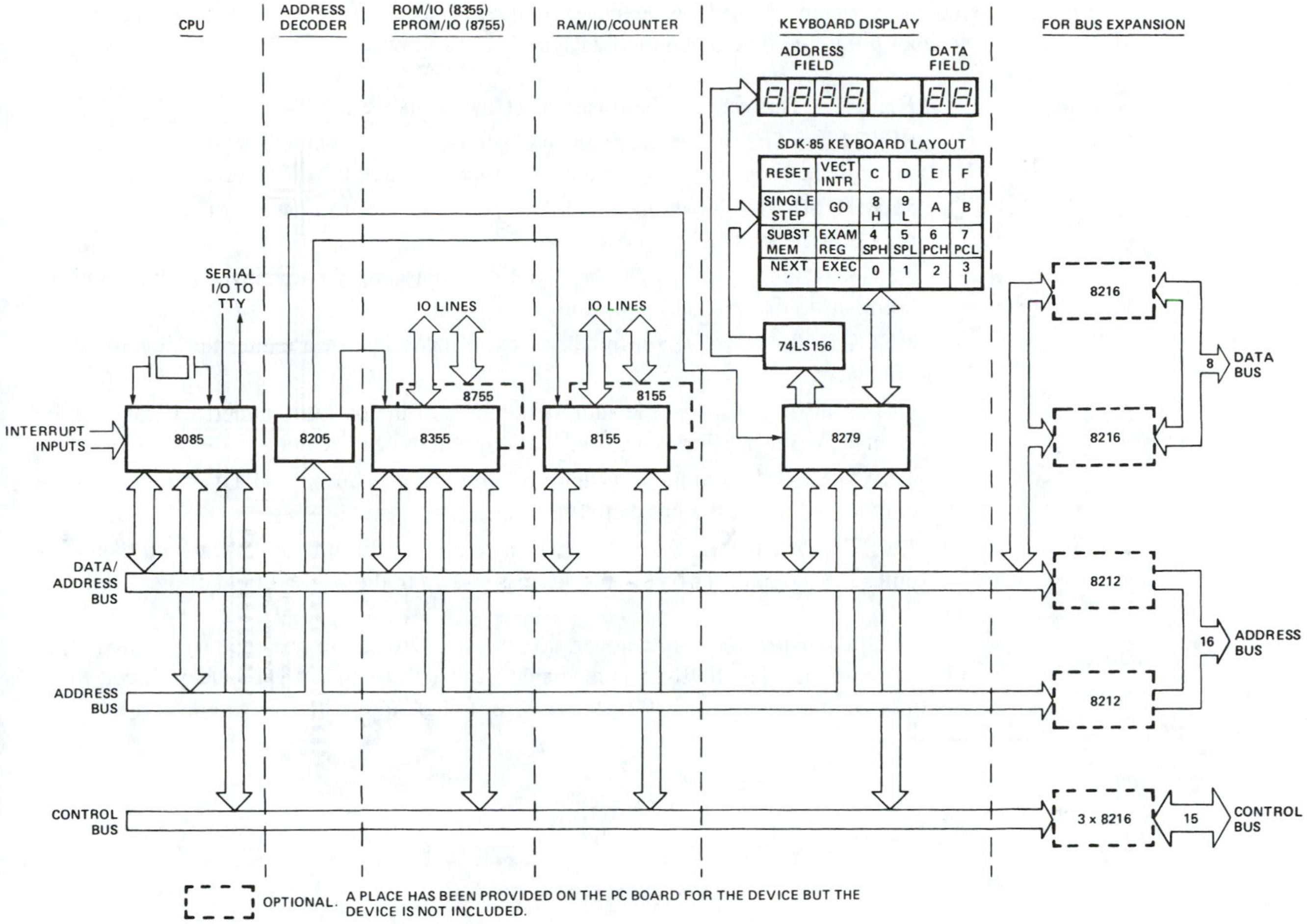
CPU
ADDRESS DECODER
ROM/IO (8355)
EPROM/IO (8755)
RAM/IO/COUNTER
KEYBOARD DISPLAY
FOR BUS EXPANSION
ADDRESS FIELD
DATA FIELD
SDK-85 KEYBOARD LAYOUT
RESET
VECT INTR
C
D
E
F
SINGLE STEP
GO
8 H
9 L
A
B
SUBST MEM
EXAM REG
4 SPH
5 SPL
6 PCH
7 PCL
NEXT ,
EXEC .
0
1
2
3 I
SERIAL I/O TO TTY
IO LINES
IO LINES
8755
8155
74LS156
INTERRUPT INPUTS
8085
8205
8355
8155
8279
8216
8216
8
DATA BUS
DATA/ ADDRESS BUS
8212
16
ADDRESS BUS
ADDRESS BUS
8212
CONTROL BUS
3 x 8216
15
CONTROL BUS
OPTIONAL. A PLACE HAS BEEN PROVIDED ON THE PC BOARD FOR THE DEVICE BUT THE DEVICE IS NOT INCLUDED.

Figure L–2 The SDK-85 functional block diagram. (Courtesy of URDA, Inc.)

RESET
VECT INTR
C
D
E
F
SINGLE STEP
GO
8 H
9 L
A
B
SUBST MEM
EXAM REG
4 SPH
5 SPL
6 PCH
7 PCL
NEXT ,
EXEC .
0
1
2
3 I

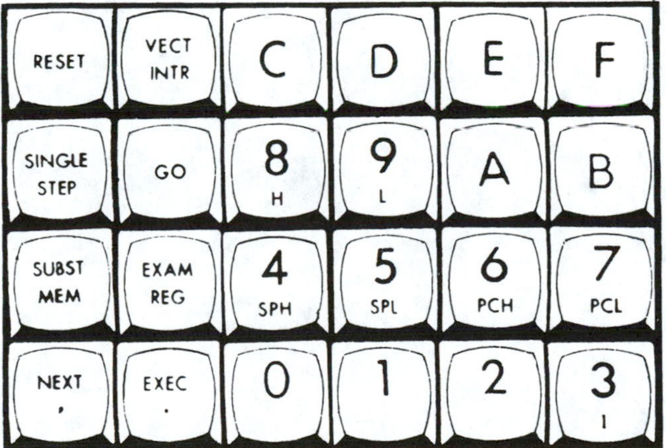

Figure L–3 The SDK-85 keypad. (Courtesy of URDA, Inc.)

student's program. The 8 keys on the left of the keypad are used to enter, execute, and troubleshoot programs. The function of each key is as follows:

RESET—This causes a hardware reset and starts the monitor.

SINGLE STEP—A powerful troubleshooting tool that allows the user to step through the program one statement at a time to watch the effect of each step.

SUBST MEM—Substitute Memory allows the user to read ROM memory and to read or write to RAM memory.

NEXT—This key stores the displayed data byte into the displayed address and increments to the next address location.

VECT INTR—The Vector Interrupt key initiates a hardware interrupt directly to the 8085A.

GO—This key displays the address contents of the program counter. It can then be changed to the starting address of the program to be executed.

EXAM REG—Examine Register is another troubleshooting tool used to look at the contents of any of the internal 8085A registers.

EXEC—Pressing the Execute key transfers control to the 8085A, which then executes the program starting at the address shown in the address field of the display.

Detailed instructions on the operation of the SDK-85 are provided by the manufacturer of the trainer. The SDK-85 is available from URDA Inc., 1811 Jancey Street, Suite 200, Pittsburgh, PA 15206, 1–800–338–0517 or (412) 363-0990. (www.urda.com)

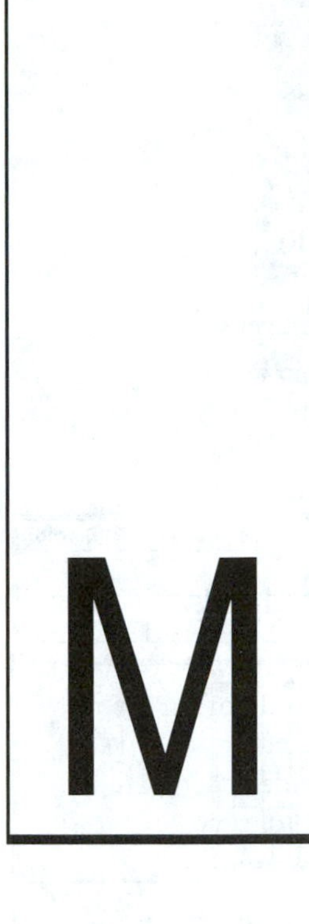

EMAC Primer 8085 Trainer

The Primer shown in Figure M–1 is an 8085 based training tool developed specifically for learning the operation of today's microprocessor based systems. Microwave ovens, stereos, TVs, and almost every other electronic product utilize embedded microprocessor technology. The Primer Trainer demonstrates the principles used by those products, providing you the opportunity to program, interface, and control virtually any device.

The Primer Trainer has the ability to process analog signals (like temperature and voltage) as well as digital signals (like switches and relays). By processing those signals at precisely timed, interrupt driven intervals you can achieve real time embedded control.

The Primer Trainer's 8085 microprocessor is an ideal platform for learning microprocessor theory. The straightforward 8085 architecture is easy to understand and the instruction set is powerful, allowing the use of programming techniques similar to those used for a PC, but much simpler to learn.

HARDWARE FEATURES:

- Based On INTEL IC'S (See Figure M–2).

  Knowledge gained on the Primer applies directly to computers that are widely used in engineering and business applications. ICs used include:
  - 8085 Microprocessor.
  - 8155 Programmable peripheral interface with timer and RAM.
  - 8279 Keypad and display controller.
  - 8251 Optional UART serial controller.
- 20 Key Keypad.
- 6 Digit, 7 Segment, LED Display.
- 8 Position Dipswitch Input Port with I/O Connector.
- 8 Bit Output Port with Status LEDs and I/O Connector.
- Analog to Digital Converter.
- Digital to Analog Converter.

Figure M–1 Primer Photograph (Courtesy EMAC, Inc.)

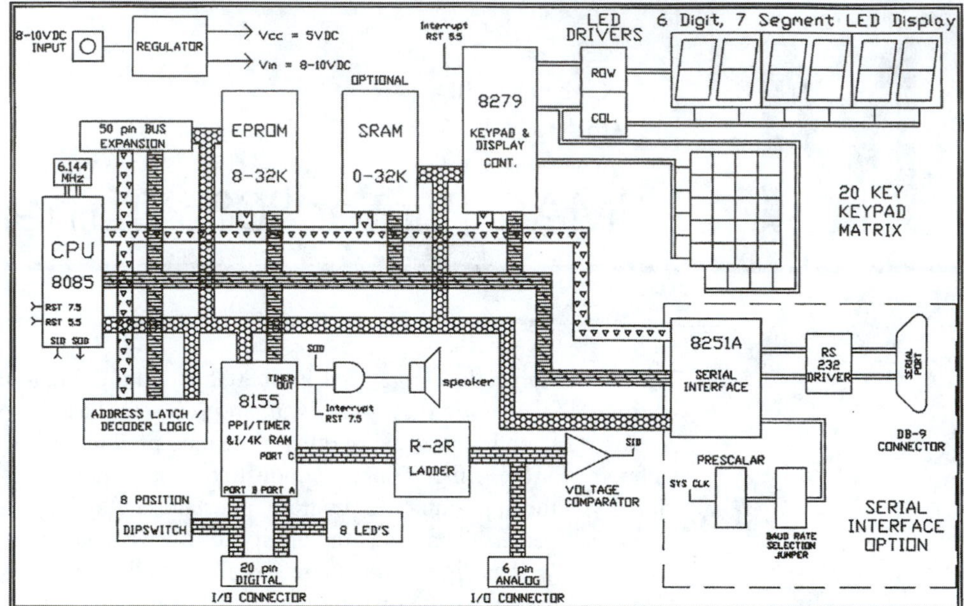

Figure M–2 Primer Block Diagram (Courtesy EMAC, Inc.)

- Timer Output with Speaker.
- 14 Bit Timer with Interrupt Support.
- System Reset Button.
- 50 Pin Bus Expansion Connector Allows Interfacing of Most Devices.
- Available as a kit Kit or Assembled and Tested.
- All ICs are socketed.
- 30 Page Assembly Manual.
- Over 100 Page Self Instruction Manual.
- 70 Page Application Manual.

PROGRAMMING FEATURES:

- 8K EPROM Containing Monitor Operating System (MOS) Allows the User to:
  - Display and edit memory.
  - Display and edit registers.
  - Display and edit top of stack.
  - Single step by instruction.

- Run full speed with breakpoint.
- Run diagnostic to test board.
- Utilize MOS internal subroutines for each I/O device as well as for multiply, divide, getkey, display number, and display ASCII.

- Commented Assembly Language Source Code Listing with Addresses & Opcodes.
- Comprehensive 100-plus Page Self Instruction Manual, Including Numerous 8085 Programming Examples.
- Built In Diagnostic Utility Tests ROM, RAM, Keypad, Display, A/D, D/A, I/O Ports, Speaker & Serial Port

The system may be purchased as a kit, or pre-assembled by contacting

EMAC, Inc.
2390 EMAC Way
Carbondale, Illinois 62901
(618) 529-4525
www.emacinc.com.

Sim8085* Microprocessor Simulator for the PC

The CD enclosed with this textbook contains a software simulation program (SIM8085) that you can install on your Windows 95/98/2000 PC to simulate the operation of the 8085 microprocessor. Figure N–1 shows the screen produced when simulating the mod-10 down counter of Table 10–1.

Operation

The assembly-language program that you want to simulate is entered in the window on the left. Then you press the **Compile** button. This converts the mnemonics into machine code starting at the address specified in the **ORG** statement. The **System RAM** columns show the addresses and contents of the assembled machine code. After compiling, pressing **Run** will run the program full speed until you press the **Pause** button.

Troubleshooting

To troubleshoot a program, you can press **Step** to execute one instruction at a time. Each time that you press **Step** you can check the **CPU Registers**, **Flags**, **SP** (Stack Pointer) and **PC** (Program Counter) to insure that they contain the values that you expect. You can also insert software breakpoints (**BrkPt**) to enable the program to run full speed up to a particular statement, and then pause. To enter a software breakpoint, position the cursor in front of the statement that you want to pause at, and then press **BrkPt**. Then, when you run the program, execution will proceed full speed down to that statement and pause, allowing you to check the register contents up to that point. To continue, press **Run** again.

System RAM Contents

To monitor additional addresses, you can press the **Cols** button to specify the columns that you want to see listed. The **Fill** and **Copy** buttons are also useful for loading the system RAM with any data that you may need for your program execution. **Reset** clears all System RAM.

*SIM8085 software provided by ABCreations

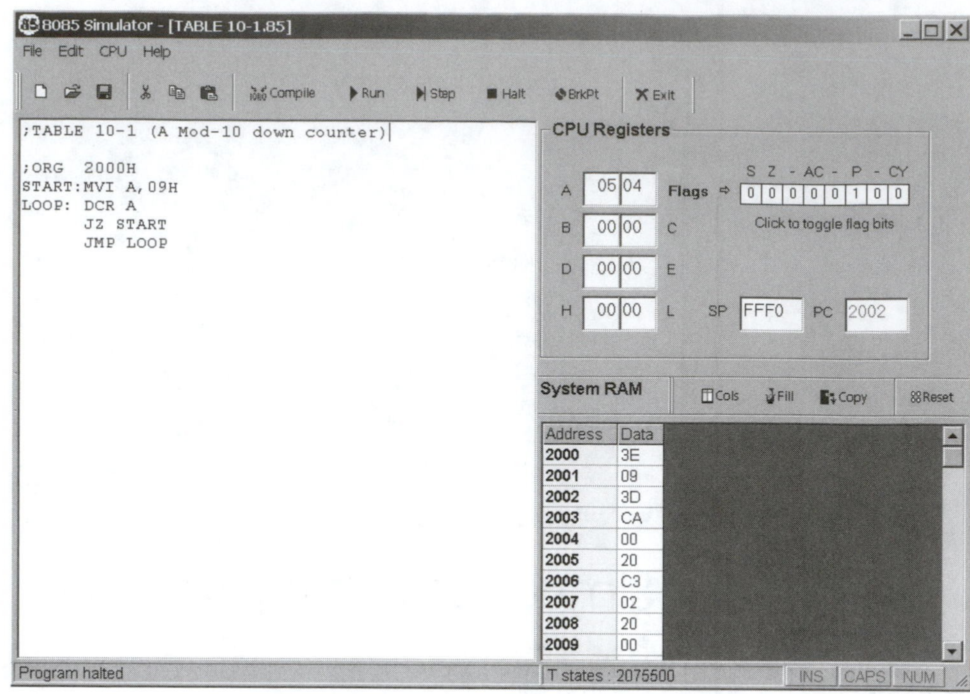

Figure N–1 SIM8085 simulation screen

Textbook Examples

To save time entering programs, several textbook program examples are included on the textbook CD. (See the list in Table N–1.)

TABLE N–1

Textbook Program Examples on the CD

Program	*Description*
Table 10–1	Mod-10 down counter
Table 10–6	Hex up-counter with delay and output
Example 10–1	Mod-10 up-counter
Example 10–9	Flash the output at 6000H the number of times entered at 4000H
Example 12–1	Load memory 20C0+ with even numbers
Example 12–4	Monitor the flag byte
Example 12–5	Arithmetic with the A register
Example 12–7	The DAA instruction
Example 12–8	Using subroutines
Example 12–9	Logic operations
Example 12–12	Rotate commands
Example 12–13	Stack operations

Index

Access time, 283
Accumulator, 309, **314***
Active clock edge, **190**
Active-HIGH, 116, **140**
Active-LOW, 116, **140**
ADC interface to microprocessor, 399–401, 518
Adder circuits, 100–105
Address bus, 307, **314**
Address decoder, 308, 344
Address decoding, 136
Address Latch Enable (ALE), 338–42, **359**
Address latch IC, 341–43
Algorithm, 402, **414**
ALS TTL (*see* TTL)
Analog multiplexer, 272
Analog multiplexer/demultiplexer, 129–30
Analog-to-digital converter (ADC), 252, 260–67
  ADC ICs, 264–67
  Counter-ramp ADC, 261–62
  Microprocessor interface, 399–401, 434
  Parallel-encoded ADC, 260
  Specifications, 258–59
  Successive-approximation ADC, 262–64, **275**
Analog value, 2–3, **20**
AND gate, 24–25
Arithmetic circuits, 100–105
Arithmetic logic unit (ALU), 309, **314**
ASCII code, 10–11, 13–14
Assembler, 318, **334**
Assembly language, 309, **314,** 321
Astable (free-running) multivibrator, 171–73, **191**
Asynchronous, 148, **190**
Asynchronous counter, 202 (*see also* Ripple Counter)
Auxiliary carry flag, 369–71, **390**

Basic electricity principles, 527–34
BASIC language, 320, **333**
BCD-to-seven-segment decoder, 219
BiCMOS, 58
Bidirectional data bus, 307, **314**
Binary-coded-decimal (BCD), 9, **20**
Binary conversion (*see* Conversion)
Binary numbering system, 4, **20**
Binary-to-BCD conversion using software, 402–3
Binary weighted DAC, 253–54
Bipolar transistor (*see* Transistor)
Bistable multivibrator (S-R flip-flop), 171
Bit, **20**
Bit string, **241**
Boolean algebra, 78–84
  associative law, 80
  commutative law, 80
  distributive law, 80
  reduction, 78, **107**
  rules, 80–85
  simplification, 80

* *Note:* Page number in **boldface** type indicates end-of-chapter glossary definition for the term.

Boolean equation defined, 25, **65**
Bubble pushing, 91
Buffer, 61, **65,** 164, **190**
Bus, 165–67, 307
Bus contention, 282, **302**
Byte, 280–82, **303**

Carry flag, 369–71, **390**
CAS (Column Address Strobe for dynamic RAMs), 287–91, **303**
Cascading flip-flops, 201, **241**
Central processing unit (CPU), 309–**314**
Chip, **20**
Clock enable, 235–36
Clock waveform, 15–16
CMOS (Complementary metal-oxide semiconductor), 19, **20,** 55–65
  Comparison table, 59, 62
  Handling MOS devices, 57
  High-speed CMOS, 57–58
  Interfacing, 59–64
  MOSFET, 55
Combinational logic, 77–79, **106**
Command register, 354–55, **359**
Common-anode LED, 218–19
Common-cathode LED, 403–04, **414**
Compact disk, 2, 301
Comparator, 113–15, **140**
Compiler, 318, **333**
Complement, 33, **65**
Contact bounce (*see* Switch debouncing)
Conditional branching, 321–22, **390**
Control bus, 307, **314**
Conversion (number systems)
  BCD-to-decimal, 9
  binary-to-decimal, 4–5
  binary-to-hexadecimal, 8
  decimal-to-BCD, 9
  decimal-to-binary, 5–7
  decimal-to-hexadecimal, 8–9
  hexadecimal-to-binary, 8
  hexadecimal-to-decimal, 8
  successive division, 6–9
Conversion time (of an ADC), 262, **274**
Counter (*see* Ripple counter *and* Synchronous counter)
Counter-ramp ADC, 261–62
CPLD, 535–70
Crystal, 182

DAA, 371–73
DAC interface to microprocessor, 396–99, 518
Data acquisition system, 271–73
Data bus, 307, **314**
Data direction register (DDR), 352–53, **359**
Data distributor (*see* Demultiplexer)
Data selector (*see* Multiplexer)
Data Sheet, 50–53, 441
Data transmission, 238–40
Debouncing a switch (*see* Switch debouncing)
Decimal conversion (*see* Conversion)
Decimal display system (three digit with counter), 221
Decimal numbering system, 3, **20**
Decoder, 115–21, **140**
  BCD decoder IC, 119
  BCD-to-seven segment, 219–21
  Binary-to-octal, 115–16
  Hex decoder IC, 120
  Octal decoder IC, 117–19
Delay line (for DRAMs), 291, **303**
DeMorgan's theorem, 85–91, **106**
Demultiplexer, 127–29
Differential measurement, 264–65, **274**
Digital clock, 14
Digital sequencer, **241**
Digital thermometer, 400–01, 521
Digital-to-analog converter (DAC), 252–58
  Binary weighter DAC, 253–54
  DAC ICs, 255–58
  R/2R DAC, 254–55
  Specifications, 259–60
Digital value, 2–3, **20**
Diode, **20**
DIP (Dual In-Line Package), **20**
Divide-by-*N* counters, 203–07, **242**
Don't care condition, 35, **106**
Down-counter, 223, 321
Duty cycle, 174, 180, **190**
DVD, 301–02
Dynamic RAM, 287–91, **303**
  Controller, 290
  RAS and CAS, 287–91, **303**
  Refresh, 290–91

ECL (Emitter-coupled logic), 59
Edge-triggered flip-flop, 154–60, **190**
EEPROM, 292, **303**
Enable, 29, **65**
Encoder, 121–23, 137, **140**
  Decimal-to-BCD, 121–22
  Octal-to-binary, 121
EPROM, 292–96, **303**
  Address decoding, 294–95, **302**
  Programming steps, 293–94
  Read cycle, 293
EPROM interface to microprocessor, 344–46, 351
Exclusive-NOR gate, 36, **65**
Exclusive-OR gate, 36, **65**

Fall time, 49, **65**
Fan-out, 44–46, **66**
FAST TTL (*see* TTL)
Fault, 30, **66**
555 IC timer, 177–82
  Astable (oscillator) connections, 177–80
  Monostable (one-shot) connections, 180–82
  Pin definitions, 178
Flag byte, 369–71, **390**

Flash converter, ADC, 260
Flip-flop, 146–60, **190**
  Cross-NAND *S-R* flip-flop, 147
  Cross-NOR *S-R* flip-flop, 146
  D flip-flop, 150–54
  J-K flip flop, 154–60
  S-R flip-flop, 146–49
Float, 30, **66,** 164, **190**
Floppy disk, 299–300
Flow chart, 320, **334**
Fold-back area, 346–47, **359**
FPGA, 541
FPLA (Field-programmable logic array) (*see* PLA)
Frequency, 15, **20**
Frequency dividing, 203
Function table, 118, 120, 157, **190**

Gate, 24–29, **66**
Gate loading, 44–46
Glitch, 204, **242**

Hand assembly, 318, **334**
Handshaking, 400, **414**
Hard disk, 299–300
Hertz, 15
Hexadecimal conversion (*see* Conversion)
Hexadecimal numbering system, 7, **20**
High impedance state (*see* Float)
Hold, 149, 155
Hold time, 161, **190**
Hysteresis, 132–33, **140**

I/O ports, 308, 330, 350–58, 367
I/O-mapped I/O, 318, **334,** 349–50
Indirect addressing instructions, 364–66, **391**
Instruction cycle, 347–49, **359**
Instruction decoder, 312–13, **314**
Instruction register, 312–13, **314**
Integrated circuit (IC), 18–**20**
  (*see also Supplementary Index of ICs*)
Interrupt, 383–88
Interrupt mask, 384–85, **391**
Interrupt Service Routine (ISR), 384
Inversion bar, 33, **66**
Inversion bubble, 33
Inverter gate, 19, **20,** 33
IO/$\overline{\text{M}}$, 318–19, **359**

Jitter, 133, **140**
Johnson shift counter, 37–38, **66,** 233

Karnaugh map, 92–100, **106**
Keyboard scanning, 407–08, **414**

Latch, 147–48, **190**
LCD (Liquid crystal display), 298, **303**
Least significant bit (LSB), 6, **21**
LED (Light-emitting diode), 184–85
Linear temperature sensor, 269–70, 400–01
Logic analyzer, 27
Logic probe, 30–32, **66**
Logic pulser, 30–32, **66**
Logical instructions, 374–78, **391**
Look-up table, 295, **303,** 401–02, **414**
Low-voltage CMOS, 58
LS TTL (*see* TTL)

Machine code, 311, **314,** 321
Machine cycle, 347–49, **359**
Magnetic storage, 298–301
Magnitude comparator (*see* Comparator)
Mask ROM, 292, **303**
Masking, 375–77, **391**
Memory, 280–302, **303** (*see also* RAM, DRAM, ROM, EPROM)
Memory address (location), 280, **303**
Memory address pointer, 364–65
Memory cell, **303**
Memory contents, 280, **303**
Memory expansion, 294–95
Memory map, 345–46, **359**
Memory-mapped I/O, 318, **334,** 343–47
Memory timing (*see* Read cycle *and* Write cycle)
Microcontroller (the 8051), 137–39, 417–35, **436**
  Address spaces, 421–23
  Applications, 137–39, 431–35, 515
  Bidirectional I/O, 419, **436**
  Bit addressable memory, 422, **436**
  External Access (EA) to program memory, 420–21, 423–24, **436**
  Instruction set, 424–31, 483–87
  Interfacing to external memory, 423–24
  Pindefinitions, 419–20
  Program Store Enable (PSEN), 420–21, 423–24, **436**
  Register bank, 422, **436**
  Special Function Register (SFR), 421–23, **436**
Microprocessor applications, 396–408
  ADC interface, 399–401
  DAC interface, 396–99
  Digital thermometer, 401
  Driving a multiplexed display, 404
  Driving a stepper motor, 408
  Keyboard scan and decode, 407, 433–34
  Time delay, 326–32
  Waveform generation, 397
Microprocessor architecture, 309–**14**
Microprocessor busses
  Address bus, 307, **314**
  Control bus, 307, **314**
  Data bus, 307, 314
  Multiplexed bus, 338, **359**
Microprocessor defined, 306, **314**
Microprocessor flags
  Auxiliary carry flag, 369–71, **390**
  Carry flag, 369–71, **390**
  Flag byte, 369–71, **390**
  Parity flag, 369–71, **390**

Microprocessor flags *(Contd.)*
  Sign flag, 369–71, **390**
  Zero flag, 321–23, **334,** 369–71
Microprocessor instruction set, 364–88, 461–82
  Arithmetic group, 369–73
  Data transfer group, 364–69
  Interrupts, 383–88
  Logical group, 374–78
  Subroutines and the stack, 378–83
Microprocessor pin definitions for the 8085A, 338–40
Microprocessor simulator, 579–80
Microprocessor support circuitry, 349–50, **359**
Microprocessor timing, 347–49
Mnemonic, 309, **314,** 321
Mode control, 236, **242**
Modulus, 200–07, **242**
Monitor program, 308, **314**
Monostable (one-shot) multivibrator, 173–77, 180–82, **191**
  555 monostable, 180–82
  IC monostable, 173–77
Monotonicity (of a DAC), 258, **274**
MOSFET (Metal-oxide-semi-conductor field-effect transistor), 55–57, **66,** 186
Most significant bit (MSB), 6, **21**
Multiplexed address bus, 338–42, **359**
Multiplexed display, 403–04, **414**
Multiplexer, 123–27
Multivibrator, 170–182, **191**
  Astable (free-running), 171–73
  Bistable (S-R flip-flop), 171
  555 astable, 177–80
  555 monostable, 180–81
  IC monostable, 173–76
  Monostable (one-shot), 173–76, 180–81

NAND gate, 33–34
Negative-edge, 154–60, **191**
Nested loop, 328–29, **334**
Noise, **191**
Noise margin, 47–48, **66**
NOR gate, 35
NOT, 33, **66**

Ohm's law, 528
One-shot (monostable) multivibrator, 173–77, 180–82
Opcode, 321–22, **334**
Opcode fetch, 347
Open-collector output, 53–54, **66**
Operand, **334**
Operating system, 308, **314**
Operational amplifier, 252–53, **274**
Optical storage, 301–02
Optocoupler, 185
OR gate, 25–26
Oscillator (*see* Multivibrator)
Oscilloscope, 14, **21,** 27
Output enable, 164–65, **191**

Parallel enable, **242**
Parallel-encoded ADC, 260
Parallel load, 223–26, 236–37, **242**
Parity flag, 369–71, **391**
Period, 14, **21**
Phototransistor, 185
PLD (Programmable Logic Device), 535–70
Pop, 380–83
Port number, 330–31, **334**
Positive edge, 151, 155, **191**
Power dissipation, 54–55, 59, **66**
Power supply, 183
Primer microprocessor trainer, 575–77
Priority encoder, 122, **140,** 260
Program counter, 312–13, **314**
Program Status Word (PSW), 381, **391**
Programmable-gain amplifier, 273, **274**
Programmable read-only memory (*see* PROM)
Programmable voltage source, 396, **414**
PROM (Fusible-link), 292, 295
  (*see also* EPROM *and* EEPROM)
Propagation delay, 49, 53, **66,** 163, **191**, 202
Pull-down resistor, 205
Pull-up resistor, 54, 61, **66,** 168, **191**
Pulse stretching, 174, 181, **191**
Pulse-width modulation, 521–23
Push, 380–83

Quartz crystal, 182

R/2R DAC, 254–55
Race condition, 160, **192**
RAM (Random-access memory), 282–91, **303**
  Dynamic RAM, 287–91, **303**
  Static RAM, 282–87, **303**
RAM fold-back area, 346–47, 359
RAM interface to microprocessor, 344–46
RAS (Row Address Strobe for dynamic RAMs), 287–91, **303**
Read cycle, 283–85, 342, **359**
Read/write memory (RWM) (*see* RAM)
Recirculating shift register, 230, **242**
Register, 323–26
Register pair, 323–26, **334**
Reset, 146, **191**
Resolution (of a DAC or ADC), 252, 258, **275**
Restart interrupts, 340
Ring shift counter, 232–33
Ripple blanking a display, 220–21, **242**
Ripple counter, 201–13, **242**
  Digital clock, 215–16
  Divide-by-*N* counter, 203–07
  Down-counter, 203
  Egg timer circuit, 216–17
  ICs, 207–17
  System design applications, 213–18
  Three-digit counter, 214–15
Rise time, 49, **66**

ROM (Read-only memory), 291–98
EEPROM, 292, **303**
EPROM, 292–96, **303**
Fusible-link PROM, 292, **303**
Mask ROM, 292, **303**
Rotate commands, 377–78

Sample and hold circuit, 272, **275**
Sawtooth wave, 397, **414**
Schematic, 505–14
Schematic diagrams, 505–13
Schmitt, trigger, 132–36, **140**
Schottky TTL (*see* TTL)
Scientific notation, 15–16
SDK-85 trainer, 571–74
Segment bus, 404, **414**
Semiconductor memory, 279–98, **303**
Applications, 295–98
Memory concepts, 280–82
Memory expansion, 294–95
RAM (Random-access memory), 292–91, **303**
ROM (Read-only memory), 291–98, **303**
Sequential logic circuits, 146, **191**
Set condition, 146, **192**
Setup time, 161–62, **192**
Seven-segment display, 218–21
(*see also* LED and LCD)
7400-series of ICs (*see Supplementary Index of ICs)*
Shift counter, 37–38, **66,** 233
Shift register, 226–40, **242**
Basics, 226–27
ICs, 234–40
Johnson shift counter, 37–38, **66,** 233
Parallel-to-serial conversion, 228–29
Recirculating, 230
Ring shift counter, 232–33
Serial-to-parallel conversion, 231
System design applications, 238–40
Universal shift register, 236–39
Signal conditioning, 267–73, **275**
Sign flag, 369–71, **391**
SIM8085 microprocessor simulator, 579–80
Sink current, 44, **66**
Skewed, 202, **242**
Software, 308, **314**
Solar radiometer, 524
Source current, 44, **66**
Speed-power product, 59
Stack, 378–83, **391**
Stack pointer, 378–83, **391**
Statement label, 321, **334**
Static RAM, 282–87, **303**
Stepper motor, 408–13
Ferromagnetic, **414**
Flux lines, **414**
Pole-pair, **414**
Rotor, **414**
Stator coil, **414**
Step angle, **414**
Storage (*see* Memory)
Strain gage, 270–71, **275**
Structured programming, 373, **391,** 402
Subroutine, 329–32, **334,** 378–83
Successive-approximation ADC, 262–64, **275**
Successive division (*see* Conversion)
Sum-of-products (SOP) expression, 89, 92, **107**
Support chips for microprocessors, 349–50, **359**
Switch debouncing, 168–70, **191**
Synchronous counter, 220–26, **242**
Two-digit decimal counter, 227
Up/down counter IC, 223
Synchronous flip-flop operations, 152–60

Temperature sensor, 268–70, 298
Terminal count, 223, **242**
Thermistor, 268, **275**
Thermometer design, 400–01, 521
Three-state output (tri-state), 164, **192**
Threshold, 132–33
Time delay, 326–32, **334,** 431–33
Timer/counter interface to microprocessor, 356–58
Timer IC (*see* 555 IC timer)
Timing diagram, 15, **21,** 26
Toggle, 154–60, **192**
Totem-pole output, 40, **66**
Transceiver, 167–68, **192**
Transducer, 267–73, **275**
Transfer function, 134, **140**
Transistor, 16–19
As an inverter, 19
Basic switch, 16–17, **21**
Bipolar (*NPN* and *PNP*), 17
MOSFET, 55–57, **66,** 186
Transparent latch, 146, 150, **192**
Trigger, 154–55, **192**
Troubleshooting techniques, 30–32, **66**
Truth table defined, 24–26, **66**
T state, 326–29, **334**
TTL (Transistor-Transistor Logic), **21,** 43–64, **66**
Advanced low-power Schottky TTL (ALS), 55
Basic inverter, 19
Circuit operation, 44
Comparison table, 59, 62
Current ratings, 44–46
FAST TTL, 55
Interfacing, 59–64
Low-power Schottky TTL (LS), 55
Open collector, 53–54, **66**
Power dissipation, 59, **66**
Schottky TTL (S), 54–55
Time parameters, 49
Totem-pole output, 44, **66**
Voltage ratings, 47–48

Undetermined state, 160, 162–63
Up counter program, 322–23
Up/down counter IC, 223

VHDL, 536, 544–45, 547, 552, 555, 557
Virtual ground, 253, **275**
Volatile memory, 291
Voltage-controlled oscillator (VCO), 182, **192**

Waveform generation, 37–38, 397
(*see also* Oscillator)
Weighting factor, 4
Write cycle, 283–85, 342–43, **359**

Zero flag, 321–23, **334,** 369–71
Zero suppression, 220–21

Supplementary Index of ICs

This is an index of the integrated circuits (ICs) discussed in this book. The page numbers indicate where the IC is first discussed. Page numbers in **boldface** type indicate pages containing a data sheet for the device.

7400 Series

7400 Quad 2-input NAND gate, 41, **50–52, 442**
7402 Quad 2-input NOR gate, 41
7404 Hex inverter, 19, 41
7406 Hex open collector inverting buffer, 411
7408 Quad 2-input AND gate, 29, 31
74HC08 Quad 2-input AND gate, 29
74HCT08 Quad 2-input AND gate, 109
7411 Triple 3-input AND gate, 29
74HC11 Triple 3-input AND gate, 29
7414 Hex inverter Schmitt trigger, 133–35
74HC14 Hex inverter Schmitt trigger, 171
7421 Dual 4-input AND gate, 29, 32
74HC21 Dual 4-input AND gate, 29
7427 Triple 3-input NOR gate, 73
7430 Eight-input NAND gate, 73
7432 Quad 2-input OR gate, 29, 32
74HC32 Quad 2-input OR gate, 29
74HCT32 Quad 2-input OR gate, 109
7442 BCD-to-decimal decoder (1-of-10), 119
7447 BCD-to-seven-segment decoder/driver, 219
7474 Dual D-type flip-flop, 151–54
7475 Quad bistable latch, 150–51
7476 Dual J-K flip-flop, 156
74LS76 Dual J-K flip-flop, 156–60
7483 4-bit full adder, 103–04
7485 4-bit magnitude comparator, 114
7486 Quad 2-input exclusive-OR gate, 43
7490 Decade counter, 210
7492 Divide-by-12 counter, 211
7493 4-bit binary ripple counter, 208–09
74H106 Dual J-K negative edge-triggered flip-flop, 163
74109 Dual J-K positive edge-triggered flip-flop, 158
74121 Monostable multivibrator, 174–76
74LS124 Voltage-controlled oscillator, 182
74138 1-of-8 decoder/demultiplexer, 117, 136, 344
74139 Dual 1-of-4 decoder/demultiplexer, 128
74147 10-line-to-4-line priority encoder, 122
74148 8-input priority encoder, 138

74151 8-input multiplexer, 125–27
74154 1-of-16 decoder/demultiplexer, 120, 129, 281
74164 8-bit serial-in parallel-out shift register, 234–35, 238
74165 8-bit serial/parallel-in, serial-out shift register, 236, 240
74192 Presettable BCD decade up/down counter, 223–26
74193 Presettable 4-bit binary up/down counter, 223–26
74194 4-bit bidirectional universal shift register, 236–39
74LS244 Octal buffer (3-state), 166, 342, 344
74LS245 Octal transceiver (3-state), 168
74266 Quad two-input exclusive-NOR, 43
74HCT273 Octal D flip-flop, 188
74LS373 Octal latch with 3-state output, 167, 341
74LS374 Octal *D* flip-flop with 3-state outputs, 166, 281, 343–44
74HCT4543 BCD-to-seven-segment latch-decoder/driver, 219, 298

Linear

555 Timer, 177–82, **457**
741 Operational amplifier, 257, **454**
7805 Voltage regulator, 183

4000 Series

4001 Quad 2-input NOR gate, 41
4011 Quad 2-input NAND, 41
4049 Hex inverting buffer, 41
4050 Hex noninverting buffer, 62
4051 8-channel analog multiplexer/demultiplexer, 130
4069B Hex inverter, 61
4543 BCD-to-seven-segment latch/decoder/driver, 219, 298

Memory

2114 1K × 4 static RAM, 344
2118 16K × 1 dynamic RAM, 288
2147H 4K × 1 static RAM, 283–87
2716 2K × 8 EPROM, 293–94, **344**
2732 4K × 8 EPROM, 294–95, **444**
3242 Dynamic RAM controller, 290

Data Acquisition

LF198 Sample and hold circuit, 272
LH0084 Programmable-gain instrumentation amplifier, 273
LM35 Linear temperature sensor, 187, 269, 401
LM339 Analog comparator, 187
LM185 Precision reference diode, 270
ADC0801 Analog-to-digital converter, 264–67, 399, **448,** 519
DAC0808 Digital-to-analog converter, 257–58, 396
MC1408 Digital-to-analog converter, 257–58, **451,** 520
AM3705 Analog multiplexer switch, 272
NE5034 Analog-to-digital converter, 264–67

PLD

EPM7128S Altera CPLD, 536
XC95108 Xilinx CPLD, 536

8000 Series

8085A 8-bit microprocessor, 309–13, 338
8355 ROM with I/O, 350–54
8755A EPROM with I/O, 350–54
8155 RAM with I/O and timer, 350–51, 354–58
8156 RAM with I/O and timer, 350–51, 354–58
8051 8-bit microcontroller and family, 417–35

Code	Mnemonic	Operand
00	NOP	
01	LXI	B,dble
02	STAX	B
03	INX	B
04	INR	B
05	DCR	B
06	MVI	B,byte
07	RLC	
08	---	
09	DAD	B
0A	LDAX	B
0B	DCX	B
0C	INR	C
0D	DCR	C
0E	MVI	C,byte
0F	RRC	
10	---	
11	LXI	D,dble
12	STAX	D
13	INX	D
14	INR	D
15	DCR	D
16	MVI	D,byte
17	RAL	
18	---	
19	DAD	D
1A	LDAX	D
1B	DCX	D
1C	INR	E
1D	DCR	E
1E	MVI	E,byte
1F	RAR	
20	RIM*	
21	LXI	H,dble
22	SHLD	adr
23	INX	H
24	INR	H
25	DCR	H
26	MVI	H,byte
27	DAA	
28	---	
29	DAD	H
2A	LHLD	adr
2B	DCX	H
2C	INR	L
2D	DCR	L
2E	MVI	L,byte
2F	CMA	
30	SIM*	
31	LXI	SP,dble
32	STA	adr
33	INX	SP
34	INR	M
35	DCR	M
36	MVI	M,byte
37	STC	
38	---	
39	DAD	SP
3A	LDA	adr
3B	DCX	SP
3C	INR	A
3D	DCR	A
3E	MVI	A,byte
3F	CMC	
40	MOV	B,B
41	MOV	B,C
42	MOV	B,D
43	MOV	B,E
44	MOV	B,H
45	MOV	B,L
46	MOV	B,M
47	MOV	B,A
48	MOV	C,B
49	MOV	C,C
4A	MOV	C,D
4B	MOV	C,E
4C	MOV	C,H
4D	MOV	C,L
4E	MOV	C,M
4F	MOV	C,A
50	MOV	D,B
51	MOV	D,C
52	MOV	D,D
53	MOV	D,E
54	MOV	D,H
55	MOV	D,L
56	MOV	D,M
57	MOV	D,A
58	MOV	E,B
59	MOV	E,C
5A	MOV	E,D
5B	MOV	E,E
5C	MOV	E,H
5D	MOV	E,L
5E	MOV	E,M
5F	MOV	E,A
60	MOV	H,B
61	MOV	H,C
62	MOV	H,D
63	MOV	H,E
64	MOV	H,H
65	MOV	H,L
66	MOV	H,M
67	MOV	H,A
68	MOV	L,B
69	MOV	L,C
6A	MOV	L,D
6B	MOV	L,E
6C	MOV	L,H
6D	MOV	L,L
6E	MOV	L,M
6F	MOV	L,A
70	MOV	M,B
71	MOV	M,C
72	MOV	M,D
73	MOV	M,E
74	MOV	M,H
75	MOV	M,L
76	HLT	
77	MOV	M,A
78	MOV	A,B
79	MOV	A,C
7A	MOV	A,D
7B	MOV	A,E
7C	MOV	A,H
7D	MOV	A,L
7E	MOV	A,M
7F	MOV	A,A
80	ADD	B

*8085 Only.

Code	Mnemonic	Operand
81	ADD	C
82	ADD	D
83	ADD	E
84	ADD	H
85	ADD	L
86	ADD	M
87	ADD	A
88	ADC	B
89	ADC	C
8A	ADC	D
8B	ADC	E
8C	ADC	H
8D	ADC	L
8E	ADC	M
8F	ADC	A
90	SUB	B
91	SUB	C
92	SUB	D
93	SUB	E
94	SUB	H
95	SUB	L
96	SUB	M
97	SUB	A
98	SBB	B
99	SBB	C
9A	SBB	D
9B	SBB	E
9C	SBB	H
9D	SBB	L
9E	SBB	M
9F	SBB	A
A0	ANA	B
A1	ANA	C
A2	ANA	D
A3	ANA	E
A4	ANA	H
A5	ANA	L
A6	ANA	M
A7	ANA	A
A8	XRA	B
A9	XRA	C
AA	XRA	D
AB	XRA	E
AC	XRA	H
AD	XRA	L
AE	XRA	M
AF	XRA	A
B0	ORA	B
B1	ORA	C
B2	ORA	D
B3	ORA	E
B4	ORA	H
B5	ORA	L
B6	ORA	M
B7	ORA	A
B8	CMP	B
B9	CMP	C
BA	CMP	D
BB	CMP	E
BC	CMP	H
BD	CMP	L
BE	CMP	M
BF	CMP	A
C0	RNZ	
C1	POP	B
C2	JNZ	adr
C3	JMP	adr
C4	CNZ	adr
C5	PUSH	B
C6	ADI	byte
C7	RST	0
C8	RZ	
C9	RET	
CA	JZ	adr
CB	---	
CC	CZ	adr
CD	CALL	adr
CE	ACI	byte
CF	RST	1
D0	RNC	
D1	POP	D
D2	JNC	adr
D3	OUT	byte
D4	CNC	adr
D5	PUSH	D
D6	SUI	byte
D7	RST	2
D8	RC	
D9	---	
DA	JC	adr
DB	IN	byte
DC	CC	adr
DD	---	
DE	SBI	byte
DF	RST	3
E0	RPO	
E1	POP	H
E2	JPO	adr
E3	XTHL	
E4	CPO	adr
E5	PUSH	H
E6	ANI	byte
E7	RST	4
E8	RPE	
E9	PCHL	
EA	JPE	adr
EB	XCHG	
EC	CPE	adr
ED	---	
EE	XRI	byte
EF	RST	5
F0	RP	
F1	POP	PSW
F2	JP	adr
F3	DI	
F4	CP	adr
F5	PUSH	PSW
F6	ORI	byte
F7	RST	6
F8	RM	
F9	SPHL	
FA	JM	adr
FB	EI	
FC	CM	adr
FD	---	
FE	CPI	byte
FF	RST	7